Encyclopedia of Pseudoscience

ADVISERS AND CONSULTANTS

Jerome Clark
Board of Directors
J. Allen Hynek Center for UFO Studies, Chicago

Dr. J. Gordon Melton
Research Specialist, Department of Religious Studies
University of California, Santa Barbara

Dr. Carl Mitcham
Professor of Philosophy and of Science, Technology, and Society
Pennsylvania State University

Dr. Marcello Truzzi
Director, Center for Scientific Anomalies Research and
Professor of Sociology, Eastern Michigan University

CONTRIBUTORS

Dr. Daniel W. Conway

Dr. Lisle W. Dalton

Dr. R. G. Alex Dolby

Dr. R. Shannon Duval

Honor C. Farrell

Jeff Frazier

Dr. John E. McMillan

Dr. J. Gordon Melton

Terry O'Neill

Kenneth R. Shepherd

Steven Utley

Joyce Williams

Dr. William F. Williams

Encyclopedia of Pseudoscience

Dr. William F. Williams
General Editor
Formerly Visiting Professor
Department of Science, Technology, and Society
Pennsylvania State University
and
Life Fellow, University of Leeds (England)

Facts On File, Inc.

Encyclopedia of Pseudoscience
WRITTEN AND DEVELOPED BY BOOK BUILDERS INCORPORATED

For information contact:

Facts On File, Inc.
11 Penn Plaza
New York NY 10001

Library of Congress Cataloging-in-Publication Data
Encyclopedia of pseudoscience / William F. Williams, general editor.
p. cm.
Includes bibliographical references and index.
ISBN 0-8160-3351-X
1. Science—Miscellanea. 2. Belief and doubt. 3. Parapsychology and science. 4. Skepticism.
I. Williams, William F.
Q157.E57 1999
001.9—dc21 98-47141

Facts On File books are available at special discounts when purchased in bulk quantities for businesses, associations, institutions or sales promotions. Please call our Special Sales Department in New York at (212) 967-8800 or (800) 322-8755.

You can find Facts On File on the World Wide Web at http://www.factsonfile.com

Cover design by David Freedman

Printed in the United States of America

VB Hermitage 10 9 8 7 6 5 4 3 2 1

This book is printed on acid-free paper.

CONTENTS

PREFACE

In producing the material and briefing the writers for this encyclopedia on a controversial subject, certain decisions were made at the outset. A particular viewpoint could have been adopted and all entries written or edited to reflect that particular outlook. Writers could have been selected with the foreknowledge that they would either write to a particular concept or viewpoint on pseudoscience or accept that their entries might be adjusted. The shortcoming of that approach would be that the reader would receive an unrepresentative picture of a subject on which almost no two people agree. Also there would have been a further difficulty: The editor and his associates would have had to agree on the approach to be adopted and would probably have had problems finding suitably adaptable writers.

So, instead, we looked for writers with appropriate interests and skills and accepted that each should be free to take his or her own line. The result more accurately reflects the diversity of views on what constitutes pseudoscience and the opinions people have on the many topics. Wherever possible, references that offer more than one viewpoint on the particular subject are appended to entries so that the reader can explore alternative positions. At the end of the volume, there is an extensive bibliography for those readers who want to advance their knowledge of the whole subject or any part of it beyond the scope of what is possible in an encyclopedia.

The diversity of views on the subject of pseudoscience is well illustrated in the introduction and essays that follow. The introduction attempts to expand on the position of this preface. The general editor's essay "Science or Pseudoscience?" lays out the point of view of an orthodox scientist, a skeptical approach to many of the subjects that this volume contains—one extreme position. The essay by a philosopher, "Ethical Issues in Pseudoscience," takes a very different position: It questions the grounds on which anyone, particularly any orthodox skeptical scientist, describes any genuine position as pseudoscientific. Indeed Professor Mitcham raises the possibility that any orthodox scientific theory is uncertainly founded and, to that extent, by its own criteria could be described as pseudoscientific.

The purpose behind this approach is to demonstrate to the reader that the range of beliefs on this subject is very wide and that there are many more positions that individuals adopt. Our writers are a mixed bag, and their different viewpoints are reflected in the variety of attitudes implicit in the following pages. As with many of this encyclopedia's other topics, the reader can consider the different viewpoints and the supporting evidence. My purpose throughout has been not to dictate a particular set of beliefs but to enhance the reader's ability to consider this highly controversial subject critically.

The General Editor is indebted to the consultants and contributors, especially Joyce Williams, both for her researches and her more than 100 entries. He is grateful also to the staff of Book Builders and, most particularly, to the staff of the Moor Allerton branch of the Public Library—always cheerful and extraordinarily patient—in locating and obtaining the many reference works and sources he required.

William F. Williams, General Editor

INTRODUCTION

What is pseudoscience? What, indeed, is science? What is their relationship to each other? This encyclopedia attempts to answer, or at least address, these very difficult questions through a series of articles and short essays on specific subjects. This introduction attempts to illuminate the area of pseudoscience mainly by illustrating what is perceived as science and, by implication, what will be seen as pseudoscience.

I know of no definition of science that is entirely satisfactory. Any one of those I have met or devised falls short, either because it admits subjects that are generally thought to be inadmissible or because it excludes some that I think should be allowed.

Notwithstanding the above, as I have to start somewhere, I am going to use this definition of science:

> *The practice of generalizing from observations to form a rational explanation—a theory—of a particular part of our world and then testing that theory by experiment, rejecting it if it doesn't match up, accepting it provisionally if it does.*

I say "provisionally" because further experiment or fresh data may disprove the theory. This, then, will be my criterion for deciding what to categorize as science and what as pseudoscience, fraud, or something other than true science. You, the reader, may feel that this approach prejudges the issue, is too dogmatic or too narrow. I don't think so. This way of determining the topics that fall properly into an encyclopedia of pseudoscience discloses a range of topics that are in some measure debatable. That debate is potentially fruitful in helping us each to a better understanding of the areas and the ways in which they might sometimes overlap. This would be so and the topics selected would be much the same no matter what definition I adopted.

Even within the fringe area that is excluded from legitimate science by my statement of what science is, there are many distinct categories:

1. claims for scientific status that do not satisfy the definition by any stretch of the imagination
2. frauds and hoaxes
3. mistaken theories that are sooner or later disproved
4. ideological preemption
5. superstition

The foregoing might be summarized by saying that we have included in this volume, thereby implying that they may be pseudoscientific, matters that are not generally accepted as legitimate science but that are or have been claimed to be so. For good measure we have included topics at the forefront of research that are still controversial.

There is no single definitive position on this subject; how such a project is viewed depends very strongly on the viewpoint of the commentator. A religious fundamentalist will define the borderline between science and pseudoscience in a quite different way from a hard-line, abstract scientist; a creationist has his or her own definition of science and therefore of what is pseudoscientific (see CREATION SCIENCE). Some sociologists regard the whole of science as very suspect, none of it being as objective as most scientists believe. The attitude adopted in this encyclopedia is to analyze the role of science and technology, being prepared to be critical, even skeptical, but concerned that our scrutiny is academically sound—a difficult task but one that we feel is important to attempt. That way lies a positive contribution to how scientists and technologists see their work and understand their role in society; we hope that it will also prove illuminating to those looking at science and technology from the outside.

I should say more about each of the categories I have listed above:

1. **Claims for scientific status:** Many people or groups have seized on an idea or thesis and have claimed scientific status for it without submitting it to the critical inspection that is essential. Take, for example, the

hypothesis of the Flat Earth (see FLAT EARTH SOCIETY). Its proponents claim it to be self-evidently true; any criticism is rejected, often very passionately, using any argument that comes to hand. As the protagonist's position is not rational, rational counterargument is ineffective.

2. **Frauds and hoaxes** hardly need explanation; the history of science has many examples and they continue to occur. The recent obsession with CROP CIRCLES or the several UFO hoaxes are cases in point. The credulous are easy targets.

3. **Mistaken theories** are very much on the edge of our border region. PHLOGISTON or the existence of the ether are examples, both once believed to be the best available explanation of the existing knowledge of their respective fields. Each was examined rigorously for a considerable period, was eventually found wanting, and was abandoned when a better explanation emerged. In their day they could hardly be described as pseudoscience, but a committed advocate of either theory today would definitely be into pseudoscience. The abandonment of the ether theory coincided with Einstein's publication of the special theory of relativity, later extended to the general theory. When a particular theory held at some time in the past gives way to a new explanation, either because improved observations disprove the old view or because some new paradigm or theoretical approach displaces it, the old is not relegated to the status of pseudoscience. This occurrence is merely an example of the provisional nature of scientific knowledge and understanding. Sir Isaac NEWTON's mechanics were subsumed into Einstein's relativity, but nobody in his or her right mind would describe Newton's contributions to our understanding of the physical world as pseudoscience. To delve further back into history and to emphasize the provisional nature of scientific knowledge, consider Aristotle, in third century B.C.E. Greece—the great and very early natural scientist—long before the word "scientist" was conjured. Aristotle believed that the absence of migrating birds, for example swallows, during the winter meant they had been hibernating; he also believed that the winter robin changed in spring into the redstart. Without any knowledge of migration, how else was he to explain the changing bird population over the seasons? His ideas on this subject persisted until well into the last century. Pseudoscience, then? No—his best shot at explaining his observations. Pseudoscience now or in the 19th century? Yes.

4. By **ideological preemption** I mean those cases where a prior commitment to some ideology or religious text is held to overrule any scientific analysis. The two creation stories of Genesis, themselves inconsistent, are held by some to settle the matter once and for all—or should I say twice and for all? The modern theories of geological and biological evolution must somehow be forced into the Genesis mold. It seems more sensible to regard the Genesis stories as the best available theories of their day, giving way in time to better, but still provisional, understanding. Likewise the belief that there is some general sociopolitical theory that takes precedence over scientific enquiry falls into this category. The dogmas of some Marxists or the Aryan racist theories of the Nazi regime are examples (see Karl MARX and NAZI RACISM).

5. **Superstitions** are widespread, and the boundaries of science are not immune to infection. To list and describe all superstitious beliefs would fill a bigger book than this, but we have included some examples that seem both to make the point and to fall within our scope. Lucky numbers with the power to affect the spin of the roulette wheel or the outcome of a lottery lie outside our border area (see LUCKY/UNLUCKY NUMBERS). A number does not have the power to interfere with the physical world.

What can we say about those areas of science that are on the forefront of inquiry and theoretical speculation? Look back for a minute at Alfred Lothar Wegener's theory of CONTINENTAL DRIFT. Some readers, very senior citizens, may remember the reception the theory received from the scientific world early in the century when it was first advanced: It was definitely seen as pseudoscience. Although the continental shelves of Africa and South America make a nice fit, and remarkable correlations of flora, fauna, and geological strata and fossils are present across the adjacent coasts, the mechanism by which these huge landmasses could once have been joined together and then separated by hundreds of miles was not known nor could it be imagined. Only years later with the discovery of plate tectonics could continental drift be adequately explained, and only then was it accepted as legitimate science. In this context a very useful reference is Stephen Jay GOULD's *Time's Arrow, Time's Cycle,* which gives a detailed but thoroughly readable account of how theories of the Earth's formation have changed over the centuries. These ideas were neither pseudoscience nor the delusions of our ancestors but the best that could be done in the light of current knowledge.

To sum up: We are into a very hazy area of knowledge, and the reader may feel that it is all a waste of time, a subject best ignored. Not so. It is an area of knowledge that affects millions of our fellow citizens, often to their disadvantage. It is a field ripe for exploitation by cranks and the purveyors of snake oil. What we should be aiming for is a skeptical and well-informed public that is able to evaluate claims for scientific authenticity, to assess and dismiss what is patently false, and—with some skepticism—to weigh the claims of what seems to be worthy. These principles that apply to the general public must also be valid for the scientific community. Scientists particularly need to have grounds for discriminating between science and pseudoscience. Hoaxers long ago discovered how gullible scientists can be, especially in areas outside their own specialities, and have often called on them to lend credibility and authority to their bogus claims.

From the tone of the above and from many of my own entries in this book, it will be obvious that I approach this subject from the point of view of a scientist. I make no apology for that. Any observer of pseudoscience has a point of view, and science has important things to say on this subject. But I am not claiming a holier-than-thou position. No person is entirely rational—nor entirely irrational: Neither position would be human nor would it allow a sustainable life. This is particularly so of scientists who often appear to live a divided life: on the one hand, in the laboratory, following logic, testing by experiment, weighing the evidence for some theory, and discarding it when the weight of evidence goes against it or when some better theory emerges; on the other hand, outside the laboratory or when drawn into some field outside his or her own, subscribing willingly and uncritically to some idea or doctrine that appeals to him or her, without either using his or her professional analytical skills or adopting the skepticism inherent in the scientific approach. What I am arguing for is that, when weighing up the claims of science and pseudoscience, we should not discard our critical faculties and should approach all claims with a healthy skepticism.

We can all benefit from a well-informed skepticism. To quote Carl SAGAN, an internationally renowned scientist:

> *But the tools of skepticism are generally unavailable to the citizens of our society. They're hardly ever mentioned in the schools, even in the presentation of science—their most ardent practitioner—although skepticism repeatedly sprouts spontaneously out of the disappointments of everyday life. Our politics, economics, advertising, and religions (New Age and Old) are awash in credulity. Those who have something to sell, those who wish to influence public opinion, those in power, a skeptic might suggest, have a vested interest in discouraging skepticism.*

I hope this volume will help to generate such skepticism.

William F. Williams, General Editor

Ethical Issues in Pseudoscience: Ideology, Fraud, and Misconduct

CARL MITCHAM

The word "pseudoscience" is not itself a scientific term. Insofar as it connotes a rigor it does not in reality possess, one might even call it a pseudoscientific term. Moreover, demarcation of the pseudoscientific from the truly scientific takes place not in scientific papers, which report theoretical analysis or experimental results, but in both sophisticated and popular discussions about science and its functions in society.

Indeed, "pseudoscientific" is often used as a rhetorical device to disparage either epistemologically or ethically—but mostly with a moral emphasis—forms of cognition that may or may not make claims to scientific status. The characterization is sometimes more a way to reinforce existing prejudices than to advance knowledge.

In the 1950s science apologist Martin GARDNER could complain that "One curious consequence of the current boom in science is the rise of the promoter of new and strange 'scientific' theories" who "is riding into prominence . . . on the coat-tails of reputable investigators."[1] Gardner sought to combat the spread of what he termed the pseudoscientific ideas of Immanuel VELIKOVSKY, Wilhelm REICH, William Bates, L. Ron Hubbard, and others. He also wanted to revive the "voluntary" or "unwritten code of science ethics developed in the profession of news journalism" that reported only the consensus judgments of the professional science community.[2]

Since the 1950s, however, there has been enough of a decline in the prestige of science that markedly fewer promoters of new and strange ideas attempt to ride the coattails of science. For example, Shirley MacLaine and TV's *X-Files'* agent Mulder increase their appeal by rejecting rather than accepting science. Furthermore, to what extent self-proclaimed "alternatives" to science and "complements" to medicine should be judged "pseudo" rather than simply "other" deserves explicit consideration.

In truth, that which constitutes science (and, by reflection, its alternatives and counterfeits) has itself evolved over time and is to some extent a culturally constructed artifact. As a result, it is crucial to recognize and to subject to critical review the very idea of pseudoscience, with its own (often unreflective) connotations of moralistic criticism. Indeed, this would seem to be part of what may be termed the new ethics of science journalism, which not only enthusiastically reports on achievements in science, from medicine to space exploration, but also airs the dirty linen, from vituperative disagreements among scientists to scientific fraud and misconduct.[3]

Dimensions of Pseudoscience

The anomalous conflation of the Greek *pseudo,* "false," and the Latin-derived "science" originated in the mid-1800s to pronounce "what had been before recognized as a branch of science, to have been . . . composed merely of so-called facts, connected together by misapprehensions."[4] Today it is easily taken to blend, on the one extreme, into what science fiction theorists call "imaginary science"[5] and, on the other, with what in popular parlance is termed "junk science."[6] The former is thought to denote as yet unexplored possibilities, while the latter lacks substance. Indeed, during the 1970s U.S. governmental funding of marginal or useless scientific research would occasionally call forth a "Golden Fleece award" from Senator William Proxmire.

It seems reasonable to propose that in its more clearly delineated forms, the term "pseudoscience" would properly eschew those theories and practices that present themselves as alternatives to, or other than, natural science. At the same time, the term may refer to a variety of human activities and pretensions to demonstrable or systematic knowledge of nature, many of which are not morally culpable. Failed science, methodologically inadequate science, and even superstition glossed with a patina of scientific rationality are no doubt to be avoided. They are also to some degree excusable, given the changing cultural constructions of science and the honorable intentions of their practitioners or proponents. Much failed and methodologically inadequate science, for instance, simply results from ignorance or lack of sophistication and is often a necessary stage in the development of knowledge and more adequate methodologies and/or instrumentation.

What many scientists would regard as superstitions (if not worse) that are supported with allegedly mistaken or deficient scientific explanations are likewise often advanced by persons who have the best intentions, although some ideas that may at first look like superstitions eventually become accepted into science. An example of the former is the so-called scientific investigation (meaning simply systematic investigation) by the Catholic Church into whether someone should be canonized a saint. Arguments from miracles fail to subject evidence to statistical analysis and adopt what is, from the natural scientific perspective, a naive strategy of inductive confirmation, but such arguments in fact make no claim to natural scientific validity. An example of the latter is acupuncture, which—once rejected as "oriental fakery"—is now increasingly (if grudgingly) recognized as engaged with something quite real. Much of what some might denominate as pseudoscience thus lies outside the bounds of moral censure and is subject only to epistemological criticism.

Nevertheless, there are at least two types of pseudoscience in the most general sense that do call for ethical analysis and moral criticism. The first is constituted by those ideologies that make special claims to political appropriation or for the public support of science. These correspond to what might be called foreign ideologies of science—for example, ASTROLOGY, ESP, MENTALISM—and domestic or popular ideologies (among scientists) such as misconceptions or errors within science. The former are given some attention in the present encyclopedia, but the latter are largely overlooked. Yet insofar as scientists ride the coattails of their scientific expertise and achievement to become involved (as they increasingly do) in economic and governmental affairs, surely their actions and pronouncements about such affairs—for example, the proper public policies toward drug research or global climate change—constitute a kind of pseudoscience.

The second type of pseudoscience arises when good or respectable intentions give way by either design or negligence to attempts to mislead or to deceive, resulting in scientific fraud or misconduct. Fraud and misconduct are almost wholly internalist issues, although not for that reason exempt from external scrutiny. Moreover, one aspect of such scrutiny must be an articulation of the ethical principles implicitly present in the rhetorical dismissal of pseudoscience of any kind.

Ideologies of Science

The most virulent ideologies of science are political doctrines such as communism and national socialism that rationalize morally repugnant teachings (in these cases, class and race hatred) with assertions of scientific insight and validity, as explored in the articles on Karl MARX and NAZI RACISM. On such bases these same foreign ideologies affirm their right to be the ultimate arbiters of the internal character and content of all science as well as its social applications. The expulsion of Jews from the community of science in Nazi Germany and the Lysenko affair in the former Soviet Union are simply two of the most egregious results. Racist science and scientific socialism should not be so easily rejected, however, as to contribute to the self-righteous obscuring of those other less obvious ideologies of science with which democratic capitalist cultural formations are entwined. Belief that science is an engine of economic progress and a model for democratic polity constitutes an ideology of its own, one often put forth in pseudoscientific garb.

The term "science," from the Latin *scientia,* originally meant simply demonstrable or systematic knowledge. From one perspective, most knowledge was, is, and will likely remain a haphazard aggregate of isolated bits of information. Most people, for instance, know all sorts of things about the living world—for example, that trees are green and that fleas bite—without being able to explain the physical causes of such phenomena or their interrelationships. One can have a great deal of biologically relevant information (for example, the Aborigines' knowledge of their natural world) without possessing biological science, that is, biological knowledge in a demonstrable and systematic form.

At the same time and from another perspective, very little information is not integrated into some cultural worldview or personal framework. That trees are green and that fleas bite will only be remembered—that is, transferred from short- to long-term memory—if such knowledge is relevant to one's personal life or makes sense by fitting into some larger, more meaningful context. The means for integrating pieces of information can vary from parental authority and personal experience to mythological narratives and epistemological methodologies.

The rise of modern natural science during the early modern period corresponds to the argument that certain specific epistemological methodologies are cognitively superior to all others, at least for certain purposes. Scientism is the view that these methodologies are superior for all purposes *tout court.* A domestic ideology of science is constituted by the less totalitarian but nonetheless powerful view that only certain methodologies are practically beneficial in a way that deserves public, that is, governmental, funding.

It is important to realize that arguments for the public funding of science have a history. The founders of modern natural science only selectively made claim to public funding—although the dream is certainly present from the beginning (see Francis BACON's idea of a tax-funded "Solomon's House" or scientific research institute in his *New Atlantis* of 1624). In the early 17th century, for instance, Galileo may have worked to calculate ballistic trajectories, work that was funded by the military, but his astronomical discoveries were financed out of his own pocket or by private patrons.

It was not until the late 17th and early 18th centuries that Enlightenment intellectuals began to make successful claims for the public funding of science through such institutions as national academies and, ultimately, for the reform of politics along lines emerging within the cosmopolitan "republic of science." Such claims are pseudoscientific, perhaps even philosophical, insofar as they rest not so much on any scientific method or experimentation as on arguments about the worth of science. As philosophical arguments they degenerate into ideology insofar as they defend the self-interests of those scientific intellectuals who advance them by presupposing the superiority of their own values—arguing, in effect, that science is good for society because it creates a society that is good for science.

The mid-20th century witnessed the apotheosis of such arguments. In 1949, shortly after the end of World War II, scientist Michael Polanyi asserted that, "The world needs science today above all as an example of the good life. Spread out over the planet scientists form even today, though submerged by disaster, the body of a great and good society."[7] Scarcely three years earlier the philosopher of science Karl POPPER, in *The Open Society and Its Enemies* (first edition 1946) presented his version of the scientific method, that is, fallibilism, as a model for society. Both views confirmed the sociologist of science Robert Merton's contemporary description of the ethos of science—which involves a rejection of nationalism, a sharing of knowledge, detachment, and organized skepticism—as fundamentally opposed to fascism and communism.[8]

Prior to the articulation of this ideology, science had made only piecemeal and pragmatic advance toward general public support, but in light of such views and of the military achievements of scientific research and engineering development, presidential science adviser Vannevar Bush in *Science, The Endless Frontier* (1948) proposed the governmental funding of science as a major new national priority. Indeed, the geographical frontier that had played so large a role in shaping the democratic ethos in the United States was to be replaced by the cognitive and inventive frontier of technoscientific progress. Precisely some version of this ideology has rationalized the progressive pseudoscientific involvement of scientists and engineers in governmental administration and public policy formation and their unsubstantiated talk about the "spin-off" benefits of even the most pure science.

As an aside, note that in the creation of the National Science Foundation (NSF), the most concrete result of Bush's report, there was considerable debate about the degree to which the public should participate in the control of this new institution. Should all oversight be by scientists alone, or should there be some measure of public oversight? There was, as well, debate about whether the social sciences qualified for inclusion in NSF; after all, to many hard scientists these softer sciences were much more like pseudoscience than true science.

Scientific Fraud and Misconduct

According to Merton, "the virtual absence of fraud in the annals of science, which appears exceptional when compared with the record of other spheres of activity," is not to be attributed to the exceptional virtue of scientists. Instead, it can be credited to "certain distinctive characteristics of science itself";[9] that is, the social institutionalization of the scientific ethos in such activities as publication, the replication of experimental results, and especially peer review are what keep scientists honest. The problem, however, as summarized by science journalists William Broad and Nicholas Wade in *Betrayers of the Truth* (1982), is that fraud and misconduct are not as exceptional as the conventional ideology of science would have it.

By contrast with ideologies of science, both foreign and domestic, scientific fraud and misconduct create what may be termed "internalist forms of pseudoscience." Of these, "scientific fraud" is the easier to define and reject, although "misconduct" is perhaps the more important. Indeed, questions of scientific misconduct are the subject of intense debate within the scientific community and emerged in part over concern about the ways scientists themselves have on occasion cooperated with and even promoted the virulent ideologies of communism and Nazism.

"Fraud" is defined as deliberate deception to obtain unfair or unwarranted gain; it cannot exist without an *intention* to defraud. Simple negligence may yield error and mistake but not fraud. Scientific fraud is thus scientific error intentionally created to further some nonscientific end and, on occasion, constitutes a form of pseudoscience, but what about a situation where technical error is intentionally introduced in order to clarify a point, perhaps by idealizing results?

Cases of outright fraud, because of the bad intentions inherently involved, are relatively easy to dismiss. Much more difficult to deal with are the greyer areas of scientific misconduct. Johann Bernoulli's plagiarizing of his own son's discovery of the Bernoulli equation; the PILTDOWN MAN hoax, apparently perpetrated to give Britain the honor of being the original home of homo sapiens; and William SUMMERLIN's faked results of skin transplants—

these are all fraud. But Galileo, NEWTON, Mendel, PASTEUR, and Nobel physicist Robert Millikan all either exaggerated their results or ignored anomalous data in reporting their work in order to communicate better what they were sure were their essential discoveries.[10] Is this not also a kind of pseudoscience?

Certainly such questions of fraud and misconduct might reasonably be termed "issues of pseudoscience." That they are instead called "fraud and misconduct in science" is revealing of the rhetorical dimension of the whole pseudoscience issue. Under any name, however, attempts to reinforce traditional norms of scientific conduct or to transform those norms have been slow to become explicit concerns within the scientific community; indeed, they have become increasingly prominent only in the last third of the 20th century. Contributing to this emergent discussion of scientific misconduct have been at least four areas of concern.

First, questions about proper conduct have been stimulated by concerns for the way scientists behaved under conditions of national emergency during World War II and their general willingness to allow national security issues to distort the scientific community. Initially this was a problem attributed only to German and then Russian scientists who had worked for or aided Hitler and Stalin. But even in the United States some scientists raised questions about their own involvement in the creation of nuclear weapons.[11] Such questions can be found voiced, for instance, in the *Bulletin of the Atomic Scientists* (founded 1945), the international Pugwash Movement (founded 1957), and the Committee on Scientific Freedom and Responsibility of the American Association for the Advancement of Science (founded 1974). Indeed, the award of the 1995 Nobel Peace Prize to Joseph Rotblat and the Pugwash Movement for their efforts to diminish the part played by nuclear arms in international politics constituted public recognition of the importance of this concern and, in effect, a declaration that such activist attempts by scientists to transform the norms of scientific conduct are not pseudoscience. Yet, attempts by computer hackers to promote anarchistic freedom on the Internet—attempts that no doubt reflect some historical feedback—are often less well received.

Second, a new outline for the proper exercise of responsibility among members of the medical science profession, one that went beyond the traditional and paternalistic Hippocratic Oath by emphasizing free and informed patient consent, was first given explicit articulation by the Nuremburg Code that emerged from the Nazi war-crimes trials. What was originally thought to be a breakdown of scientific integrity through human experimentation among only foreign physicians during the late 1960s began to be recognized as dwarfed in quantity, if not in quality, by syphilis, drug, and radiation experiments in the United States. The progressively more-important principle of free and informed consent in medical research has been further extended into general medical practice by the increasingly problematic choices facing physicians and patients as a result of advances in medical science and technology. The allocation of scarce medical resources and the utilization of therapeutic treatments with unquantified risks has led to increasing involvement of patients in decision making about their own treatment. Indeed, a lay public challenge of medical scientific experts is no longer immediately rejected as based in pseudoscientific or ignorant failure to appreciate scientific knowledge but recognized as natural and appropriate behavior.

Third, the problems of environmental pollution have raised questions about scientific and engineering responsibility, not to mention the allocation of research priorities. Although it was biologist Rachel Carson who, exercising a new sense of public responsibility with her book *Silent Spring* (1962), exposed the widespread negative impacts on nature of many technoscientific advances, she was often labeled as a radical and pseudoscientist, especially by scientists working for the agrochemical industry. It took the general scientific community almost a decade to adopt a more pro-environmental attitude—a transformation brought about as much by external and governmental pressures as by internal scientific reform. For instance, it was not until the 1970s and 1980s that engineering codes of ethics began to include environmental responsibility as guiding principles. Even in the 1990s, disagreements over the implications of, for example, global climate change research are often phrased as charges that, in effect, one side or the other is merely practicing or being misled by pseudoscience.

Fourth, beginning in 1981 governmental hearings chaired by then Representative Albert Gore, Jr., there have been widespread concerns about the scientific misuse of public funds and the scientific abuse of public trust. In part such concerns have been stimulated by a series of journalistic disclosures of scientific misconduct in the forms of overpromising results (as in the "war on cancer," which after 25 years and $20 billion has never been won[12]), cost overruns (as in virtually every big science project, from the space program to the superconducting supercollider), plagiarism,[13] the falsification of research reports, and the breakdown of peer review (including conflicts of interest and breaches of confidentiality).[14] But in part such phenomena are merely the result of government largesse and the resultant proliferation of scientific careerism. The scientific community, having grown to enormous proportions, finds it more and more difficult to police itself. Moreover, welfare can have the same result on the scientific community as among the poor—that of sapping moral strength.

As a result of Gore's hearings and related government investigations, in 1985 the U.S. Congress passed legislation requiring that such major research funders as the National Institutes of Health (NIH) and the National Science Foundation (NSF) formulate explicit guidelines for scientific integrity and explicit mechanisms for the investigation of scientific misconduct. This legislation led eventually to the formulation in 1989 of an official definition of "misconduct" as fabrication, falsification, and plagiarism (FFP) and the creation of an Office of Scientific Integrity (OSI). In 1991 these issues became the feature of a *Time* magazine cover story,[15] and shortly after, both the National Academy of Sciences (NAS) and the American Association for the Advancement of Science (AAAS) undertook efforts to increase awareness within the scientific community of issues of scientific integrity.[16]

Indicating its dissatisfaction with such efforts, however, Congress in 1993 established a Commission on Research Integrity (CRI) to revisit such questions. Following two years of discussion, the CRI proposed a new definition of "scientific misconduct" as misappropriation, interference, and misrepresentation (MIM). Within the U.S. scientific community major disagreements developed between supporters of the more restrictive FFP and the broader MIM definitions—disagreements that, of course, necessarily went beyond strictly scientific issues. It is noteworthy, however, that those who argued for the MIM definitions tended to include a greater proportion of nonworking scientists and were often castigated in language reminiscent of the critics of pseudoscience.[17]

Finally, one may note that technoscientific disasters such as the explosion of the space shuttle *Challenger* in 1986[18] and the questionable treatment of whistleblowers, as well as hyped false discoveries such as COLD FUSION in 1989,[19] have had their own stimulating, if more indirect, impacts on discussions of scientific integrity. The ups and downs of two of the most prominent charges of misconduct—those against AIDS researcher Robert Gallo (that he stole an HIV culture from the lab of French scientist Luc Montagnier) and against Nobel biologist David Baltimore and his assistant Thereza Imanishi-Kari (that they falsified lab records)—have had a chilling effect on scientific support for misconduct investigations.

The Ethical Principles for Criticizing Pseudoscience

One remarkable feature about the traditional criticisms of pseudoscience and more recent discussions of scientific fraud and misconduct has been the degree to which, even among scientists, there has been little effort to consider the general principles of theories involved in such criticisms. Surely there is more involved than simply false advertising—that is, people proclaiming to be scientists when they really are not. Even to those for whom false advertising is the sole issue, one may pose a question about why false advertising is immoral. Is it because of the bad consequences of false advertising, or might it be possible that there is something wrong with false advertising in itself, never mind the consequences?

Somewhat oversimplifying, virtually all moral criticisms of pseudoscience call into play one or both of two basic ethical theories. One is a utilitarian theory that focuses on the results or consequences of actions. The other is a rights-based theory that focuses on the way the actions are performed. To argue that science ought to be practiced in a certain way to insure good results, whether these results are knowledge or economic progress, and that such results are good because they increase human well-being by, in some sense, promoting the greatest happiness of the greatest number is to adopt a utilitarian rationale for scientific integrity. By contrast, to argue that science ought to be practiced in a certain way because only this is in accord with universal moral maxims such as telling the truth and keeping promises and that such practices are moral in a way that is ultimately independent of the results is to adopt what moral theorists call a deontological rationale for scientific integrity.

Take, for example, the issue of fraud. To claim that fraud is wrong because it takes time and money away from science that could achieve results that would really benefit humanity is to take a utilitarian position. To claim that fraud is wrong because of the ill intentions of the person perpetrating it is to adopt a deontological stance. In most cases, as in this example, both theories yield the same results. But there may be subtle differences in emphasis and implication that in some cases will point toward quite different evaluations.

Take, as a further example, the issue of idealizing results which, as mentioned earlier, was practiced by Galileo, Newton, Mendel, Pasteur and Millikan. If the stress were on intentions, one could well argue that noble intentions absolved these scientists from any misconduct. Yet if the stress were on results, one could argue that they set a precedent that has allowed lesser scientists to rationalize and even cynically to promote practices detrimental to science as a whole, thus undermining the fragile contract between science and society on which continued scientific advancement rests.

Consider the still more extreme case of the results of medical experimentation conducted by Nazi scientists. Some of these experiments involved human torture and yielded useful knowledge. Is such knowledge morally tainted because of its deontologically immoral origins, or should we make use of it because the knowledge can have good consequences for other patients?

Finally, return to the issue of pseudoscience. Is pseudoscience to be condemned because of its bad consequences or because of the sometimes immoral character of its intentions and practices? Moreover, is the use of pseudoscientific rhetoric to criticize certain kinds of human conduct and belief acceptable because it results in advancing science, or should such rhetoric be most carefully employed because it fails to recognize and respect the intentions and rights of those it accuses?

Needless to say, such questions are not easily resolved. Indeed, they point toward deep and enduring difficulties in any attempt to lead a consistently moral life. But they are issues that surely ought at least to be raised as part of any comprehensive attempt to assess the scope of pseudoscience.

End Notes

1. Martin Gardner, *Fads and Fallacies in the Name of Science* (New York: Dover, 1957), p. 3. (This is an expanded edition of a book first published in 1952 under the title *In the Name of Science.*)
2. Gardner, *Fads and Fallacies*, pp. 5 and 4, respectively.
3. Although Dorothy Nelkin in *Selling Science: How the Press Covers Science and Technology* (New York: W. H. Freeman, 1987) could still see the press as more sycophantic than critical, such may not have been the majority feeling among scientists even then.
4. See *Oxford English Dictionary*, rev. ed., 1992 (New York: Oxford University Press, 1992).
5. See, e.g., John Clute and Peter Nicholls, eds., *The Encyclopedia of Science Fiction* (New York: St. Martin's Griffin, 1995), entry on "Imaginary Science," pp. 613–15.
6. See, e.g., Peter W. Huber, *Galileo's Revenge: Junk Science in the Courtroom* (New York: Basic Books, 1991).
7. Michael Polanyi, "The Social Message of Pure Science" (first published 1949), in *Scientific Thought and Social Reality: Essays*, ed. Fred Schwartz, *Psychological Issues*, vol. 8, no. 4 (New York: International Universities Press, 1974), pp. 46–47.
8. See Robert K. Merton, "The Normative Structure of Science" (from 1942) and "Science and the Social Order" (from 1937), both included in his *The Sociology of Science: Theoretical and Empirical Investigations*, ed. Norman W. Storer (Chicago: University of Chicago Press, 1973), pp. 267–78 and 254–66.
9. Merton, "Normative Structure of Science," p. 276.
10. For details and references on all these cases, see William Broad and Nicholas Wade, *Betrayers of the Truth* (New York: Simon and Schuster, 1982); and Robert Bell, *Impure Science: Fraud, Compromise, and Political Influence in Scientific Research* (New York: John Wiley, 1992). On Summerlin, see also Joseph R. Hixon, *The Patchwork Mouse* (Garden City, N.Y.: Doubleday, 1976). For a critical response to these negative forecasters, see biologist Alexander Kohn's well-informed *False Prophets: Fraud and Error in Science and Medicine* (New York: Blackwell, 1986).
11. For some update on concerns about the possible military distortion of scientific work, see Carl Mitcham and Philip Siekevitz, eds., *Ethical Issues Associated with Scientific and Technological Research for the Military*, proceedings of a 1989 conference, *Annals of the New York Academy of Sciences*, vol. 577 (1989).
12. For more on this particular boondoggle, see Robert N. Proctor, *Cancer Wars: How Politics Shapes What We Know and Don't Know about Cancer* (New York: Basic Books, 1995).
13. For more on this topic, see Marcel C. LaFollette, *Stealing into Print: Fraud, Plagiarism, and Misconduct in Scientific Publishing* (Berkeley: University of California Press, 1992).
14. For more on the peer-review problem, see Daryl E. Chubin and Edward J. Hackett, *Peerless Science: Peer Review and U.S. Science Policy* (Albany: State University of New York Press, 1990).
15. "Science under Siege," *Time* (August 26, 1991), pp. 44–51.
16. See, e.g., National Academy of Sciences, *On Being a Scientist* (Washington, D.C.: National Academy Press, 1989; 2nd edition, 1995) and Albert H. Teich and Mark S. Frankel, *Good Science and Responsible Scientists: Meeting the Challenge of Fraud and Misconduct in Science* (Washington, D.C.: AAAS, 1992).
17. For a good presentation of the MIM position by one instrumental in formulating it, see C. K. Gunsalus, "Rethinking Unscientific Attitudes about Scientific Misconduct," *Chronicle of Higher Education* (March 28, 1997), pp. B4–B5.
18. For two quite different interpretations of this disaster, see Claus Jensen, *No Downlink* (New York: Farrar, Straus, and Giroux, 1996) and Diane Vaughan, *The Challenger Launch Decision* (Chicago: University of Chicago Press, 1996).
19. See Gary Taubes, *The Short Life and Weird Times of Cold Fusion* (New York: Random House, 1993).

Science or Pseudoscience?

WILLIAM F. WILLIAMS

With the approach of the millennium, there is an increase in the public's attraction to a whole variety of phenomena and activities that lie outside the normal concerns of orthodox science. ASTROLOGY columns and horoscopes in the press have proliferated and can now be found in some of the quality newspapers, which, only a year or two ago, would have strenuously denied them admission. Several pseudoscientific television programs, such as *The X-Files,* are broadcast regularly and are awarded accolades. Political leaders turn to astrologers and spiritualist MEDIUMS for advice about matters of national, and even international, importance. Cults, always present in society, appear to be increasing both in number and in strangeness and are treated with more respect than once was the case.

In these and in other ways, the turn from the second to the third millennium is seen as flagging a change from the rational (appealing to reasoned argument) to the irrational or, perhaps, as precipitating some major event for which this preoccupation with what is seen to be the spiritual will provide some sort of insurance. This is puzzling behavior. For many years, there has been a growing reliance throughout the world on science and its close relative, technology, both representing an essentially rational outlook. Now we appear to be regressing.

All the alternatives—astrology, mediums, and the like—are essentially irrational and form a large part of what in this volume we have labeled pseudoscience. With the end of the second millennium, the world faces several major problems: overpopulation, global warming, energy depletion, refugees, AIDS and other epidemics, the danger of nuclear weapons, and more. That these will disappear on a day when the calendar changes from this century to the next is not just unlikely but impossible. We then have to ask the question: In these circumstances, in what do we put our trust? In science and technology or in some or all of these pseudosciences?

Let us examine the alternatives as sensibly as we can. To examine some particular part of science—either one of the disciplines or one or more discovery, hypothesis, or law—would not be fruitful. What might help is to consider how science proceeds.

There is no one scientific method, but it is fair to say that there is an overriding principle: the commonsense approach of trial and error. Nor are scientists the sole practitioners: We all use essentially this same approach in our daily lives. We try something out, a new product, say, and if we don't like it, we tend to drop it. If we like it, we try it again and perhaps again and again. If it continues to please, we take the view that it can be relied on. Trial and error. The alternative would be to pick products at random and never to rely on our previous experience. Sometimes, of course, we get it wrong: The product changes—or our taste changes.

So it is with science. The scientist accumulates evidence on some aspect of nature, forms a theory to explain that evidence and tries it out. If it doesn't work, he or she abandons it. If it works on the first test, the scientist is encouraged to continue and gathers confidence in the theory if it continues to be satisfactory. But, again, if new evidence or understanding discredits the theory, the scientist must abandon it and look elsewhere. Science is not a miraculous process with certain and unchangeable answers to the problems of nature, but it is a way toward a steadily improving understanding. Science is not a provider of certainties; some things it can predict with remarkable certainty, the time of sunrise tomorrow morning, for example, or the motion of the planets in the solar system. About other things it is less certain. The best it can do is to anticipate intelligently—whether global warming is occurring and what its effects may be, if so, for example. There are other subjects on which its confidence comes and goes; see the entry on the NEBULAR HYPOTHESIS, for example.

In some ways, technology is more difficult to examine. We are extremely reliant on technology, and it is impossible to imagine a world populated by humans without it.

We take for granted the use of tools, printing, food production, and much more, and we assume that they will always be there to sustain us. Even primitive hunter/gatherer societies used a limited range of tools, food production, and clothing to help them survive; they were dependent on technology—and on science, in the sense of knowledge of nature.

The methods of technology are different from those of science, although there is some exchange of knowledge and equipment between the two activities, but the thing that science and technology have in common is the rational approach—reasoned argument, the connection of cause and effect—again essentially that of trial and error. All our technology at any stage has proceeded by trying out a tool, equipment, or process and discarding or improving it step by step. Again, there is no magic route and no perfection.

All technology has both its advantages and its disadvantages. The best that technologists can do is to try to predict both in order to maximize the advantages and reduce the disadvantages as far as possible. Even then, there will be some good and some bad aspects of a new technology that have not been anticipated, and there will be some technologies whose bad sides are so baleful that we would be better to do without them. The obvious example of this last, in my opinion, is nuclear weaponry.

So we have the resources of science and technology that we can use to understand the world we live in and to improve our lives—but no invariable certainty, no perfection, nothing without its downside; still, science and technology rely on the basic process by which we conduct most of our daily lives.

What Alternative Is There?

If the alternative is one or more pseudoscience, then we must ask the same question: What is it that characterizes pseudoscience and how does it proceed? It is not sufficient to pick out one particular pseudoscientific activity or process and, using anecdotal evidence, convince ourselves that it offers an alternative answer to one of our problems. In turning to one or many pseudosciences, we are making a choice between distinct modes of thinking about the world and how we go about living. The immediate and obvious answers to our questions are these:

1. all pseudoscience is irrational; that is, it disregards or contradicts rational principles.
2. pseudoscience does not proceed by trial and error but by revelation.

The first of these answers will be hotly contested by advocates of some particular doctrine or belief, their general contention being that, if we can imagine a possibility, then we must keep an open mind about it until it is disproved. That, they claim, is a rational approach. I beg to differ. On that basis we could demand an open mind on "pigs might fly." That is not rational. Pigs have no wings, and their body weight far exceeds the maximum with which an animal can fly, even when an animal does have wings. The only way in which a pig might fly is in a Boeing 747.

Over the years a whole slew of scientists and philosophers have adopted an open-minded, allegedly rational attitude to extrasensory perception (ESP). But the history of that quasi-academic venture exposes them either to having been dupes or accomplices to trickery or to having abandoned their earlier position. It is not for observers to assume the credibility of a pseudoscientific claim until disproved; the burden of proof for any pseudoscience, say ESP, which requires some extraordinary process, must rest with its believers. It is notable that, faced with increasing skepticism as time passes, the advocates' claims tend to become steadily more modest.

The basic process of pseudoscience is not that of trial and error. The typical pattern is of somebody having an inspiration or of receiving a revealed truth—that dreams tell of the past and/or the future, even of the dreamer's past lives, for example. Fine: There is a place for inspiration and for dreams. Both, as writers on creativity have explained, may help us to unravel some problems with which we have been struggling. Archimedes in his bath, Sir Isaac NEWTON under the apple tree, Friedrich Kekulé dreaming of the structure of the benzene molecule, are all examples. But the inspiration or the dream has to be tested. Sir Karl POPPER argued that any proposition, to be meaningful, to lay claim to truth, must be such that there is a test by which, if failed, the proposition would be disproved. Some pseudoscientific theories do not satisfy this criterion; they are such that there is no test that would dismiss them once and for all. Others defy the trial and error process, but any number of failed tests fail to discourage their adherents.

Again the example of ESP is informative. United States intelligence agencies and the military were attracted to the idea that messages could be passed from mind to mind, eliminating the problems of interception and jamming that ordinary communication systems experience. REMOTE VIEWING, the mental imaging of the disposition of enemy forces miles or thousands of miles away, would avoid the risk and expense of spy planes or reconnaissance. Huge, exhaustive testing costing millions of dollars produced no worthwhile corroboration of the reality of either of these processes, but believers are not deterred; their beliefs are not subject to trial and error. At a more mundane level, failed tests are explained away by the experimenter's tiredness or by the presence of skeptical observers blocking some supernatural mechanism.

In fairness it must be stated that the history of science also records episodes when cherished theories were defended fiercely despite negative test results—giving up some theory in which there is a large intellectual investment is hard—but in science, the trial-and-error process does eventually triumph. In pseudoscience it does not; the believer in pseudoscience will cling to his or her beliefs despite any evidence or opinion to the contrary.

Professor Frank Close, while attending the meeting of the British Association for the Advancement of Science (BAAS) in September 1996, was asked to do a live interview for a mid-European radio station. He prepared by reading up on the papers that had actually been presented at the meeting. He described the experience in a column in the United Kingdom publication *The Guardian* on September 19 as follows:

> *There had been some good stuff. A computer model of how we walk had been tried out on apes, on humans and on Lucy, our oldest known ancestor. . . . Earthquakes, collisions with comets: choose your disaster. Vesuvius is overdue for eruption; many major volcanoes are near to large centres of population; some eruptions will make Mount St. Helens appear puny by comparison; hundreds of thousands of people are at risk, millions may have to be moved from their homes. . . .*
>
> *Somewhere in Transylvania a man with a perfect English accent was presenting me, live: "At the British Association for the Advancement of Science this week they have been debating the paranormal and here with us to discuss it is Professor Frank Close." Vesuvius, Lucy, even death by cosmic catastrophe cut no ice: this was paranormal or nothing, with an interviewer whose persistence put Jeremy Paxman [considered to be England's most forceful TV interviewer] in the minor leagues. Pressed on spoon-bending I offered to demonstrate how it is done without recourse to psychic forces, but as the possibility of rational explanation was not on his agenda, he moved on to ESP. I began to feel sorry for the man whose researchers had prepared him for an interview with a committed New-Age Aquarian Millenniumist, only to find himself with, horror of horrors, a sceptical scientist. "If ESP really worked, you would have known that the BAAS was not about the paranormal," I ventured, at which point he cut the interview.*

Another live interview with a local radio station followed, at which Frank Close was asked to explain why so many disasters happen on Friday the 13th—the day of the broadcast!

Some advocates of various forms of pseudoscience maintain that science and pseudoscience are complementary ways of understanding the world; the two can exist side by side, enriching our understanding of reality, without one ruling out the other. In my opinion, this is not so. Speaking as a self-appointed spokesperson for science and technology, I assert most emphatically that the two world views are totally incompatible; they are in head-on collision. MAGIC, ESP, the paranormal, SPIRITUALISM—the whole panoply of pseudoscience—cannot be squared with the approach of science and technology. We have to make a choice between the two. Confronted with the problems with which we will have to deal in the 21st century, I have no doubt about which I choose.

Living with Anomalies

JEROME CLARK

Human nature abhors an explanatory vacuum, especially when it comes to the profoundly unsettling. Here's an example:

> *Sailing in the Canary Islands on November 30, 1861, the crew of the French gunboat* Alecton *encountered an enormous sea monster. The crew opened fire with rifle and cannon, to no effect. After a long pursuit the ship got close enough so that a harpoon could be hurled into the creature's flesh. A noose was put around its body, but it slid down to the dorsal fin, rendering futile the attempt to carry the monster aboard. All that the would-be captors were able to claim was a small portion of the tail.*

When a report on the encounter was read at the French Academy of Sciences a month later, member Arthur Mangin expressed the consensus view of those present: No "wise" person, he said, "especially the man of science," would "admit into the catalogue those stories which mention extraordinary creatures like the sea-serpent and the giant squid, the existence of which would be . . . a contradiction of the great laws of harmony and equilibrium which have sovereign rule over living nature." In other words, the captain and his crew were a pack of liars.

They were, as it turned out, no such thing. Barely a decade later, a series of strandings of identical creatures on the shores of Newfoundland and Labrador produced incontrovertible proof of the existence of giant squids (the KRAKEN of legend).

We can take pleasure from this little episode's thoroughly satisfying resolution, but it will be a rare pleasure because many reports of extraordinary anomalies do little more than generate endless disputes and fanciful theories. Some of them describe phenomena that no reasonable person could believe exist. Unhappily, reasonable persons—in all times and places—report them.

In 1987, for example, a reasonable person, an experienced outdoorsman, spoke of an encounter with a 9-feet-tall hairy, apelike biped in a wooded area near Salisbury, New Hampshire. This "man of sound body and sober spirit," as a local newspaper characterized him, observed the creature in daylight, under good viewing conditions. That, however, did not stop a local game warden from opining that the witness was so bereft of common sense or functioning perceptual apparatus that he confused the thing he thought he saw with the thing he really saw: a moose.

In fact, or at least in many bizarre allegations, persons all over the North American continent—not just in the American Northwest or the Canadian Far West, lair of the fabled BIGFOOT/Sasquatch—have come forward with comparable tales of humanlike apes or apelike humans and, on rare occasion, been blessed with a scrap of ambiguous physical evidence (usually a hair that, on analysis, may defy ready identification) or odd-looking tracks. For their efforts, the witnesses are called liars or fools as the explanatory vacuum quickly fills and closes.

There is no end to it, either. One finds at least two centuries of UFO reports and centuries or decades of sea serpents, LAKE MONSTERS, relic sauropods, Tasmanian tigers, giant ground sloths, and more. For things that *may* or *may not* exist—reports for which satisfactory explanations are not much in view even as proof remains elusive—the evidence is typically more or less solely testimonial. These are things that, whether real or not, at least *could* be real. They are unlike, say, FAIRIES and "merfolk" (MERMAIDS and mermen), which equally reasonable persons have unreasonably reported.

One set of anomalous claims, in short, speaks merely to the fantastic, another to the utterly incredible. The first merely expands the world; the second reinvents it. A population of Sasquatch deep in the Oregon wilderness is in the former category. Hairy bipeds in New Hampshire—or Texas or South Dakota or Pennsylvania or Indiana or Colorado, or just about any state or province you can name, as the reports are disturbingly ubiquitous—are, on the other hand, barely more probable than the merman that members of a Danish Royal Commission swore they observed for some minutes near the Faeroe Islands one memorable day in 1723. Thirty years later, Bishop Erik Pontoppidan wrote in the classic *A Natural History of Norway,* "Here, in the diocese of Bergen, as well as in the manor of Nordland, there are several hundreds of persons of credit and reputation who affirm, with the strongest assurance, that they have seen this kind of creature."

Well, let us start with the assumption that merfolk do not "exist" in the ordinarily understood sense of the verb. Not only are they manifestly a zoological impossibility, but if they were dwelling in northern oceans, their bodies would have washed ashore, and we would have no shortage of specimens. Confronted with a significant body of testimony from sane and sober observers, some folklorists and marine biologists have proposed an explanation, the sanity and sobriety of which are open to question. The witnesses really saw sea cows and dugongs, we are asked to believe, and somehow (as one scientist writes) they "became 'transformed' into a mermaid by the expectant attention of the superstitious mariners who saw it." Of course, a sea cow is to a half-fish/half-human what a moose is to a 9-feet-tall primate, though at least a moose is not out of place in the New Hampshire woods. According to a survey by British folklorists Gwen Benwell and Arthur Waugh, nearly three-quarters of first-person reports of merfolk occurred far from areas where sea cows and dugongs are known to exist.

In other words, *no* explanation, rational or extraordinary, makes sense. So what are we to do?

We would do well—all of us, be we believers or debunkers—to resist the compulsion to fill the explanatory vacuum. Occult "theories" proposing forces and causes that themselves are (at the very least) debatable offer obfuscation rather than clarification. Explanations that draw on complacent assumptions about witness confusion, stupidity, or dishonesty are not invariably helpful either and conceivably, because they are often prefaced by the adjective "scientific," may have the unfortunate consequence of throwing the authority of "scientific explanation" into question even when, in other circumstances, it is being applied quite properly—none of which is to say, obviously, that misidentifications, hoaxes, and mental disorders play no role in anomaly reporting. Conventional explanations often work. Our concern here, however, is with those that don't.

What we are trying to do is to refrain from admitting our ignorance. Better to throw an explanation or an insult at a witness than to concede that we have not a clue about what happened to him or her. But we do have every right to live our lives *as if* these things don't exist or, in any event, until we have a lot better evidence to go on than testimony, however considerable or interesting, concerning things that always seem to promise more than they deliver.

There are anomalies, and there are anomalies. If Sasquatch or LOCH NESS MONSTERS are there, they will be found one day. They will create a sensation for a few weeks or months, and then they will fade into the twilight zone of zoological journals and nature documentaries. We will think about them as often as we think of mountain gorillas and giant squids, which once claimed comparable legendary status. The same fate, we may safely predict, is not in store for the truly extreme anomalies we have been discussing.

Yet, experiences of the extremes will continue, and that is probably how we should think of them: as experiences, as opposed to events. Events occur in the world we recognize and understand, the world in which we live, work, eat, sleep, fall in and out of love, and attend to the business of being born and dying, all the while surrounded by natural forces and phenomena whose secrets science and reason are slowly unraveling. Events are measurable, documentable. Experiences simply are, well, experienced.

Let us concede the obvious: (1) There are things in this world, including some pretty weird things, that we don't understand, things that are out there somewhere beyond current knowledge and even useful vocabulary; (2) the range of experiences to which mentally sound humans are susceptible is far broader than conventional opinion allows.

So what do we do in the meantime? Let us hear out the witnesses. After all, they were there, and we weren't. If we can give them a plausible, nonextraordinary explanation for what they say happened to them, fine. If not, we can spare them the insult of reinventing their experiences so as to trivialize them into crude misperceptions of ordinary phenomena. There is no reason to ridicule them or call them, without good evidence to that effect, liars. All the while we can keep in mind that, in the absence of other evidence, the most sincere testimony to the most outlandish occurrence or encounter is insufficient to remake the world on its own. All we can say about such things is that they comprise some people's experience of the world—maybe even the experience of some who are reading these pages.

In other words, this world isn't simple. "Weird stuff happens," the folklorist Bill Ellis says. We would all do well to learn to live with it.

The Perspective of Anomalistics

MARCELLO TRUZZI

Center for Scientific Anomalies Research

What Is Anomalistics?

The term "anomalistics" was coined by anthropologist Roger W. Wescott and refers to the emerging interdisciplinary study of scientific anomalies (alleged extraordinary events unexplained by currently accepted scientific theory). The approach to anomalies is today loosely represented by a number of independent organizations and publications, most notably: the Society for Scientific Exploration and its journal, founded by astrophysicist Peter Sturrock; science writer William CORLISS's multivolume *THE SOURCEBOOK PROJECT;* sociologist Marcello Truzzi's CENTER FOR SCIENTIFIC ANOMALIES RESEARCH and its journal *Zetetic Scholar;* science writers Patrick Huyghe's and Dennis Stacy's journal *The Anomalist;* and editor Steve Moore's *Forkan Studies*. Those who take this approach are called anomalists.

Anomalistics has two central features. First, its concerns are purely *scientific*. It deals only with empirical claims of the extraordinary and is not concerned with alleged metaphysical, theological, or supernatural phenomena. As such, it insists on the testability of claims (including both verifiability and falsifiability), seeks parsimonious explanations, places the burden of proof on the claimant, and expects evidence of a claim to be commensurate with its degree of extraordinariness (anomalousness). Though it recognizes that unexplained phenomena exist, it does not presume these are *inexplicable* but seeks to discover old or develop new appropriate scientific explanations.

As a scientific enterprise, anomalistics is normatively skeptical and demands inquiry prior to judgment, but skepticism means *doubt* rather than *denial* (which is itself a claim, a negative one, for which science also demands proof). Although claims without adequate evidence are usually *unproved,* this is not confused with evidence of *disproof.* As methodologists have noted, an absence of evidence does not constitute evidence of absence. Because science must remain an open system capable of modification with new evidence, anomalistics seeks to keep the door ajar even for most radical claimants who are willing to engage in scientific discourse. This approach recognizes the need to avoid both the Type I error—thinking something special is happening when it really is not—and the Type II error—thinking nothing special is happening when something special, perhaps rare, actually occurs.[1] While recognizing that a legitimate anomaly may constitute a crisis for conventional theories in science, anomalistics also sees them as an opportunity for progressive change in science. Thus, anomalies are viewed not as nuisances but as welcome discoveries that may lead to the expansion of our scientific understanding.[2]

The second key feature of anomalistics is that it is *interdisciplinary.* It is so in two ways. (1) A reported anomaly is not presumed to have its ultimate explanation in a particular branch of science. Once all conventional scientific explanations have been rejected, the eventual explanation for an anomaly may turn out to be something new but in an unexpected field; for example, data that reports experiments suggesting telepathy may eventually be best explained by a revision in our assumptions about statistics, or some reports of UFOs might eventually be best explained in terms of neurophysiology rather than astronomy or meteorology. (2) Anomalistics is also interdisciplinary in that it seeks an understanding of scientific adjudication across disciplines. This often involves not only the physical and social sciences but also the philosophy of science. Anomalists search for patterns in the acceptance and rejection of new scientific ideas, and this may involve the history, sociology, and psychology of science, as well as the scientific fields themselves.

What Anomalistics Is Not

Anomalistics may best be understood by comparing it with some of the alternative approaches to anomalies. These would include three major organized groups; proponents, mysterymongers, and scoffers.

Proponents of anomaly claims range from those who are involved with the occult and mystical to those who seek scientific legitimacy and are what I have termed "protoscientific."[3] Anomalistics is primarily concerned with the claims of protoscientists, for they seek entry into the scientific community and agree to play by the rules of scientific method. Perhaps the most advanced of the protosciences is PARAPSYCHOLOGY because it, unlike CRYPTOZOOLOGY or "UFOLOGY," has obtained an affiliation with the American Association for the Advancement of Science. Protoscientific proponents are concerned with specific areas of anomalies and usually champion their significance for a single science (for example, parapsychologists see themselves as revisionists of psychology), whereas anomalists may be interested in the same specific anomaly but frame it in a broader context and recognize that the anomaly examined may ultimately best be explained by another branch of science (for example, the data of parapsychology may turn out to be understood best in terms of quantum physics rather than psychology). Anomalistics attempts an integrative overview of all the protosciences and their relations with the accepted sciences.

Going further than the proponents, some who claim anomalies can correctly be described as *mysterymongers.* Many students of anomalies, such as those associated with Fortean groups (followers of the writer Charles FORT, who catalogued what he called the "damned facts" that science dogmatically ignores) fit into this category. These claimants enjoy calling attention to what seem to be unexplained phenomena that show the limitations of science. Their writings give the distinct impression that, even if a radically new scientific explanation could be found for these phenomena, it would produce disappointment rather than celebration. At their most extreme, their attitude is fundamentally antiscientific, for such mystery-mongers want to embarrass rather than advance science. Unlike anomalists, who see anomalies as a wake-up call that tells us of a need for new, improved, and more-comprehensive scientific theories, mysterymongers seek the extraordinary for its own sake. The mysterymonger mainly wants to be entertained by nature's "freak show," which stands outside the main entrance to science's central "circus."

In diametric opposition to the mysterymongers, who love things unexplained, there are the *scoffers,* who seem to loathe a mystery. Although many in this category who dismiss and ridicule anomaly claims call themselves skeptics, often they are really pseudoskeptics because they *deny* rather than *doubt* anomaly claims.[4] While taking a skeptical view toward anomaly claims, they seem less inclined to take the same critical stance toward orthodox theories; for example, they may attack alternative methods in medicine (such as a lack of double-blind studies) while ignoring the fact that similar criticisms can be leveled against much conventional medicine (these scoffers rarely complain about the absence of double-blind tests for the results of surgery).

Many claims of anomalies are bunk and deserve proper debunking, so anomalists may engage in *legitimate debunking.* However, those I term "scoffers" *often* make judgments without full inquiry, and they may be more interested in discrediting an anomaly claim than in dispassionately investigating it.[5] Because scoffers sometimes manage to discredit anomaly claims (through such methods as ridicule or ad hominem attacks) without presenting any solid disproof, such activities really constitute *pseudodebunking.*

A characteristic of many scoffers is their pejorative characterization of proponents as "promoters," and sometimes even the most protoscientific anomaly claimants are labeled as practitioners of "pathological science." In their most extreme form scoffers represent a form of quasi-religious scientism that treats minority or deviant viewpoints in science as heresies.[6]

What Anomalists Do: The Four Functions of Anomalistics

(1) *Anomalistics seeks to aid in the evaluation of a wide variety of anomaly claims proposed by both scientists and protoscientists.* It seeks to bring historical and sociological perspective to the issues, calling attention to nonrational factors and sources of bias often present among both proponents and their critics. It acts as a watchdog for violations of scientific rigor by all parties involved in the "litigations" over anomaly claims. Anomalistics recognizes that most anomaly claims are probably mistaken, and it stresses distinguishing between anomalies merely *alleged* and those *validated.* It recognizes, too, that evidence always varies in quality and degree, and seeks to assess the weight of all evidence without the complete dismissal of weak evidence (such as anecdotal or experiential reports) that many scientists too often simply reject as totally inadmissible.

(2) *Anomalistics seeks to understand better the process of scientific adjudication and to make that process both more just and rational.* A valid anomaly is just a fact in search of theory to explain it and is extraordinary only relative to what we view as ordinary. Anomalistics therefore recognizes that a claim can only be considered anomalous in the context of a specific scientific theory.

An anomaly for which we can specify a theory that should be able to house or accommodate it but does not is termed a *nested anomaly.*[7] Nested anomalies seem to contradict some accepted theories' expectations and so may be denied by those theories. For example, a valid case of clairvoyance would be a nested anomaly because it violates currently accepted perception theory in psychology. It is important to recognize that a nested anomaly in the context of one area of theory in science may be considered less extraordinary in the framework of some other area of scientific theory; for example, clairvoyance, viewed as a nonlocal information transfer, may seem more tenable (less anomalous) to a scientist working within quantum physics.

There are also *unnested anomalies*, those that do not contradict any accepted scientific theory but only appear bizarre and unexpected; for example, the discovery of a unicorn (here meaning merely an otherwise normal horse with a single horn) may be highly improbable, but such an animal would violate no accepted laws in zoology (as might a centaur). Because unnested anomalies seem strange or weird merely in terms of our psychological expectations, their degree of scientific anomalousness (extraordinariness) has been exaggerated by both the mystery-mongers and the scoffers who dispute them.

(3) *Anomalistics attempts to build a rational conceptual framework for both categorizing and assessing anomaly claims.* It examines the various approaches to extraordinary claims and differentiates those that stem from scientific, nonscientific and antiscientific perspectives.[8] It gives much attention to developing a typology of anomalies and "unpacking" many of the concepts routinely used in discussing them. Anomalistics distinguishes *extraordinary events* from *extraordinary theories about events*. In looking at the former, it separates issues over the *credibility* of the narrator, the *plausibility* of the narrative and the *probability/extraordinariness* of the event. It clarifies commonly confused terms, such as the "supernatural," the "natural," the "preternatural," the "abnormal," and the "paranormal."[10] Perhaps most significantly, anomalistics distinguishes between *cryptoscientific* and *parascientific* anomalies.[11] Cryptoscientific claims refer to extraordinary things or objects (for example, a yeti or a UFO), whereas parascientific claims refer to extraordinary processes or relationships between what may be quite ordinary things (for example, a claim of mental telepathy or of a planetary influence on human personality).

Such categorizations have important implications for our understanding of the assessment of anomaly claims. For example, a cryptoscientific claim is at least theoretically easy to validate (for one need capture and produce only a single giant sea serpent to establish its existence), but it may be difficult to falsify (for the thing may be avoiding detection or be elsewhere in the world); whereas a parascientific claim may theoretically be easy to falsify (for example, an hypothesized relationship may not appear in an experiment), but it may be difficult to validate (for alternative explanations must be rejected and replication is usually demanded).

(4) *Anomalistics seeks to act in the role of amicus curiae ("friend of the court") to the scientific community in its process of adjudication.* Because anomalistics has no vested interest in either the existence or nonexistence of any claimed anomaly, it is possible for it to concentrate on inquiry and the search for empirical truth rather than upon advocacy. Whereas other groups concerned with anomalies should properly act as attorneys for or against the claims, they sometimes improperly try to take the roles of judge and jury, too. Anomalists more modestly try to stand somewhat outside the disputes and examine the adjudication process itself. Their position is comparable to the role of an amicus curiae in the legal system. In effect, anomalists file independent briefs to help the "court" (in this case the scientific community at large) arrive at better judgments. For example, anomalists may shed light on such issues as how much and what sort of evidence should be necessary to prove what sort of anomaly and whether the burden of proof science demands should merely be for a *preponderance of evidence* or (as is too often and perhaps even unfalsifiably demanded) *proof beyond any reasonable doubt*. It also can help us ascertain when evidence for or against an anomaly is either merely *suggestive* (interesting), *compelling* (appears significant and likely), or *convincing* (appears to be valid).

End Notes

1. Marcello Truzzi, "Editorial [re Type I and Type II errors and the paranormal]," *Zetetic Scholar 3/4* (1979), p. 2; Truzzi, "Editorial [Extra-scientific factors and Type II error]," *Zetetic Scholar 8* (1981), pp. 3–4.
2. Truzzi, "Discussion: On the Reception of Unconventional Scientific Claims," in Seymour M. Mauskopf, ed., *The Reception of Unconventional Science* [AAAS Selected Symposium 25], Boulder, Colo.: Westview Press, 1979.
3. Truzzi, "Definitions and Dimensions of the Occult: Towards a Sociological Perspective," *Journal of Popular Culture 5* (1972), pp. 635–46.
4. Truzzi, "Editorial: On Pseudo-Skepticism," *Zetetic Scholar 12/13* (1987), pp. 3–4.
5. Ray Hyman, "Pathological Science: Towards a Proper Diagnosis and Remedy," *Zetetic Scholar 6* (1980), pp. 31–43.
6. Truzzi, "Pseudoscience," in Gordon Stein, ed., *Encyclopedia of the Paranormal* (Amherst, N.Y.: Prometheus Books, 1996), pp. 560–75.
7. Ron Westrum and Marcello Truzzi, "Anomalies: a Bibliographic Introduction with Some Cautionary Remarks," *Zetetic Scholar 2* (1978), pp. 69–78.
8. Truzzi, "Definitions and Dimensions of the Occult: Towards a Sociological Perspective," *Journal of Popular Culture 5* (1972), pp. 635–46; Truzzi, "Pseudoscience," in Gordon Stein, ed., *Encyclopedia of the Paranormal* (Buffalo, N.Y.: Prometheus Books, 1996), pp. 560–75.
9. Truzzi, "On the Extraordinary: An Attempt at Clarification," *Zetetic Scholar 1* (1978), pp. 11–19.
10. Truzzi, "From the Editor: Parameters of the Paranormal," *The Zetetic* [now *The Skeptical Inquirer*] 1 (1977), pp. 2, 4–8; Truzzi, "Editorial: A Word on Terminology," *Zetetic Scholar 2* (1978), pp. 64–65.
11. Truzzi, "Zetetic Ruminations on Skepticism and Anomalies in Science," *Zetetic Scholar 12/13* (1987), pp. 7–20.

References:

Hyman, Ray. "Pathological Science: Towards a Proper Diagnosis and Remedy." *Zetetic Scholar* 6 (1980), pp. 31–43.

Truzzi, Marcello. "Definitions and Dimensions of the Occult: Towards a Sociological Perspective." *Journal of Popular Culture* 5 (1972), pp. 635–46.

———. "From the Editor: Parameters of the Paranormal." *The Zetetic* [now *The Skeptical Inquirer*] 1 (1977), pp. 2, 4–8.

———. "On the Extraordinary: An Attempt at Clarification." *Zetetic Scholar* 1 (1978), pp. 11–19.

———. "Editorial: A Word on Terminology." *Zetetic Scholar* 2 (1978), pp. 64–65.

———. "Editorial [re Type I and Type II errors and the paranormal]." *Zetetic Scholar* 3/4 (1979), p. 2.

———. "Discussion: On the Reception of Unconventional Scientific Claims." In Seymour M. Mauskopf, ed. *The Reception of Unconventional Science* [AAAS Selected Symposium 25]. (Boulder, Colo.: Westview Press, 1979).

———. "Editorial [Extra-scientific factors and Type II error]." *Zetetic Scholar* 8 (1981), pp. 3–4.

———. "Zetetic Ruminations on Skepticism and Anomalies in Science." *Zetetic Scholar* 12/13 (1987), pp. 7–20.

———. "Editorial: On Pseudo-Skepticism." *Zetetic Scholar,* 12/13 (1987) pp. 3–4.

———. "Pseudoscience." In Gordon Stein, ed. *Encyclopedia of the Paranormal.* (Amherst, N.Y.: 1996), pp. 560–75.

Westcott, Roger W. "Anomalistics: The Outline of an Emerging Area of Investigation." Paper prepared for Interface Learning Systems, 1973.

———. "Introducing Anomalistics: A New Field of Interdisciplinary Study." *Kronos* 5 (1980), pp. 36–50.

Westrum, Ron and Marcello Truzzi. "Anomalies: A Bibliographic Introduction with Some Cautionary Remarks." *Zetetic Scholar* 2 (1978), pp. 69–78.

Encyclopedia of Pseudoscience

ABDUCTIONS, ALIEN Abductions of humans by extraterrestrials, usually to a spaceship and often for the apparent purpose of physical examination of the humans, after which they are returned to Earth, usually, though not always, to the place they were taken from. One of the earliest and best-known claims of alien abduction was that of Betty and Barney Hill, a New Hampshire couple who in 1961 experienced strange events and a time "anomaly" while traveling home on a lonely rural road. At first, the Hills were only aware of losing two hours on the journey home. Later, with the help of HYPNOSIS, they recalled seeing a hovering UFO and being taken aboard a spaceship. John G. Fuller's 1967 book *The Interrupted Journey* publicized their adventure.

Many other claims of abduction have been made in many countries around the world. Many of these reports include claims of gynecological examinations of female victims, along with (sometimes) implantation of alien embryos and/or removal of human or alien embryos and encouragement of the human female to hold an apparent alien–human hybrid infant. Others report sexual examination of male victims, often including sexual intercourse with an alien, and implantation of tiny devices of unknown purpose in the victims's nasal passages or elsewhere. Many people who claim to have been abducted have scars of unknown origin, have experienced multiple abductions, or are related to others who have had similar experiences. Most reports suggest that the abductions are frightening experiences, although a few victims consider them benign. Almost invariably the abductions are recalled only when the victim's memory is jogged by something—a televised account of an abduction or a mysterious scar noted for the first time, for example—and almost invariably, the details are brought out with the help of hypnosis.

A few cases have been highly publicized. These include those of novelist Whitley Streiber, who has discussed his abduction in several books; Betty Andreasson Lucas, a multiple abductee whose mother and daughter also claim to have been abducted and whose story has been recounted by artist and abduction specialist Budd Hopkins; and Travis Walton, a young man who disappeared from a group of friends one night in 1975 and reappeared, dazed and weakened, five days later several miles from where he disappeared. The accounts abductees give bear many similarities, suggesting several possibilities, these among them: All have been exposed to similar material through the media, or most of these people have indeed had the experiences they claim.

In 1995 MUTUAL UFO NETWORK (MUFON), a prominent organization devoted to the study of UFOs, began the Alien Abduction Transcription Project, attempting to collect, catalog, and, with the aid of computers, index various aspects of the phenomenon. Among other things, project directors hope that, through careful and detailed analysis of the data, they will be able to ascertain the purpose of the abductions.

Reports of abductions appear to be on the increase. A few mainstream scientists accept the likelihood of alien abductions happening as the victims recount them. Notable among these is Harvard psychiatry professor John Mack, who in 1995 found himself under fire and risk of censure from his university because he publicly gave credence to abduction accounts. Few scientists speak out as frankly in favor of the phenomenon. Most, if they consider it at all, assert that there are more mundane explanations for what the alleged victims believe they have undergone. Among the most popular explanations are these: (a) People are being influenced by what they read and see in the media; (b) The hypnotists who work with many alleged victims are influencing them; (c) The abduction experiences are a form of hallucination or hypnagogic (waking) dream; (d) The accounts are outright hoaxes or are the product of mental illness. Interestingly, several recent psychological studies have shown that people who claim abduction and other anomalous experiences are no more fantasy-prone than those who do not.

See also FLYING SAUCERS, CRASHED; UFOLOGY.

For further reading: C. D. B. Bryan, *Close Encounters of the Fourth Kind: Alien Abductions, UFOs and the Conference at MIT* (Alfred A. Knopf, 1995); Philip J. Klass, *UFO Abductions: A Dangerous Game* (Prometheus Books, 1989); Andrea Pritchard, et al., eds., *Alien Discussions: Proceedings of the Abduction Study Conference Held at MIT* (North Cambridge Press, 1994); and The Roper Organization, *Unusual Personal Experiences: An Analysis of the Data from Three National Surveys* (Roper, 1992).

ABIOGENESIS The creation of life from inorganic forms of matter. One current theory as to how life originated on Earth is that it derived from inorganic matter that was already in place. Some chemists suggest that "self-replicating" matter—in other words, living matter—could have been created out of existing nonliving molecules. They suggest that life first originated as a series of chemical compounds floating in a "primordial soup." There is some evidence to support this theory: Scientists have been able to create amino acids and other basic proteins in the laboratory from inorganic compounds; however, they have not been able to produce anything as complex as a self-replicating living organism.

Two current problems with the theory of abiogenesis are, first, that there is no evidence of how inorganic molecules—which cannot reproduce themselves—became self-replicating organic cells; second, there are also no detectable historical clues as to what form the simplest type of self-replicating compound took. The earliest fossils found are of living organisms that are themselves already very complex.

ABOMINABLE SNOWMAN Popular name for the YETI, a large, hairy hominid believed by many to live in the Himalaya Mountains. The term comes from a mistranslation of the Tibetan name for the creature (*Metoh-kangi*, "manlike creature that is not a man," misinterpreted as "wild man of the snow") and from the creature's alleged "abominable" odor.

See also ALMAS; BIGFOOT.

ABRAMS, ALBERT (1863–1924) Californian physician who devised the alternative healing system later called RADIONICS. Abrams studied medicine at Heidelberg University, in Germany and following graduation became a professor of pathology at Cooper Medical College in San Francisco, where he studied cancerous tissue. He reportedly discovered that the tissue gave off a distinct radiation and that the radiation changed in the transformation from a healthy to a diseased condition. He went on to develop an overall theory that all living things radiated an electromagnetic field. Abrams claimed to be able to measure these variances with an instrument he called an oscillocast, more popularly referred to as a BLACK BOX; blood samples of the patient were placed in the black box, which measured their radiation against those from the blood of a healthy person.

During the last decade of his life, Abrams shared his findings with his medical colleagues in several books, including *Human Energy* (1914) and *New Concepts in Diagnosis and Treatment* (1922). He also developed his diagnostic work into a treatment system called "electronics" and assembled instrumentation that he claimed could not only detect radiation but also emit "healthful" radiation to assist diseased tissue to return to a healthy state.

After World War I, Abrams's theories and procedures were examined by the British Royal Society of Medicine, which in 1922 issued a negative report. The following year the American Medical Association also tested Abrams's work and denounced it. As a result, his practice was closed. He died the next year, but his ideas were picked up in the 1930s by Ruth DROWN in the United States and George De La Warr in Great Britain. His work has continued as a fringe medical practice ever since. Researchers who have put Drown and other radionics advocates to the test have been unable to verify their diagnostic claims.

ACUPRESSURE A form of bodywork similar to ACUPUNCTURE. Based upon the same philosophical principles and view of the human body but differing in that no needles are used. Acupressure also draws upon the same Taoist principles of yin and yang. The universe is seen as divided between the yin (feminine/negative) and yang (masculine/positive), everything in the universe being a relative mixture of the two. Given the nature of these polarities, the acquisition of balance is a prized goal. Chinese physicians have mapped the human body in relation to these principles. In addition, the universe is animated by a life force, *chi* (also spelled *ki* or *qi*), which flows through the body along pathways (termed "meridians"). Located along the meridians are certain points that connect them to the organs. It is assumed that the free flow of *chi* is essential to a healthy life and that illness is a sign that the flow has been blocked or become otherwise unbalanced.

As with many traditional Chinese medical practices, the passage of acupressure to the West came through Japan. Acupressure had been practiced in Japan as one of a variety of massage techniques, all of which fell from official favor in the 19th century. Laws were passed restricting its practice after government officials concluded that massage was being used more for simple pleasure than any health benefits. That law was not repealed until 1955. The spread of acupressure was soon revived in Japan and then quickly spread to the West, where it drew upon the wave of interest in Chinese medicine following U.S. President Richard Nixon's visit to China in the early 1970s.

Claims for the benefits of acupressure have drawn on those made for acupuncture, but have tended to be metaphysical in nature and merge with those ascribed to massage in general. It is a popular form of bodywork among those who follow holistic health practices and, although not accepted by mainline physicians as having any medical value, it has also been seen as, at worst, harmless since its practitioners neither prescribe medicines nor invade the body. A number of variations on acupressure such as ACU-YOGA have also emerged in the West.

ACUPUNCTURE One of the oldest systems of therapy in the ALTERNATIVE MEDICINE complex. Patients are treated by having needles stuck into their skin at designated acupuncture points and then twirled manually or agitated with the aid of a low frequency electrical current. The acupuncture points, called "meridians," are claimed to lie along invisible energy channels that are said to be connected to internal organs, some of which are, in effect, nonexistent. The treatment is believed to manipulate and balance the body's flow of "life energy" (called *chi* or *qi*) flowing through the channels. Many of these ancient systems share the conviction that all treatments have to adjust the imbalance of the body's "vital force."

Acupuncture is said to have been recognized in China more than 3,500 years ago when it was considered possible that soldiers who survived arrow wounds in battle were recovering from other long-standing ailments. The word "acupuncture" is a European term meaning to prick with a needle. It was coined by a Dutch physician who introduced the practice into Europe after a stay in Nagasaki, Japan, in 1683. In the early 19th century British doctors used the treatment for pain relief and fever. The first edition of the British medical journal *The*

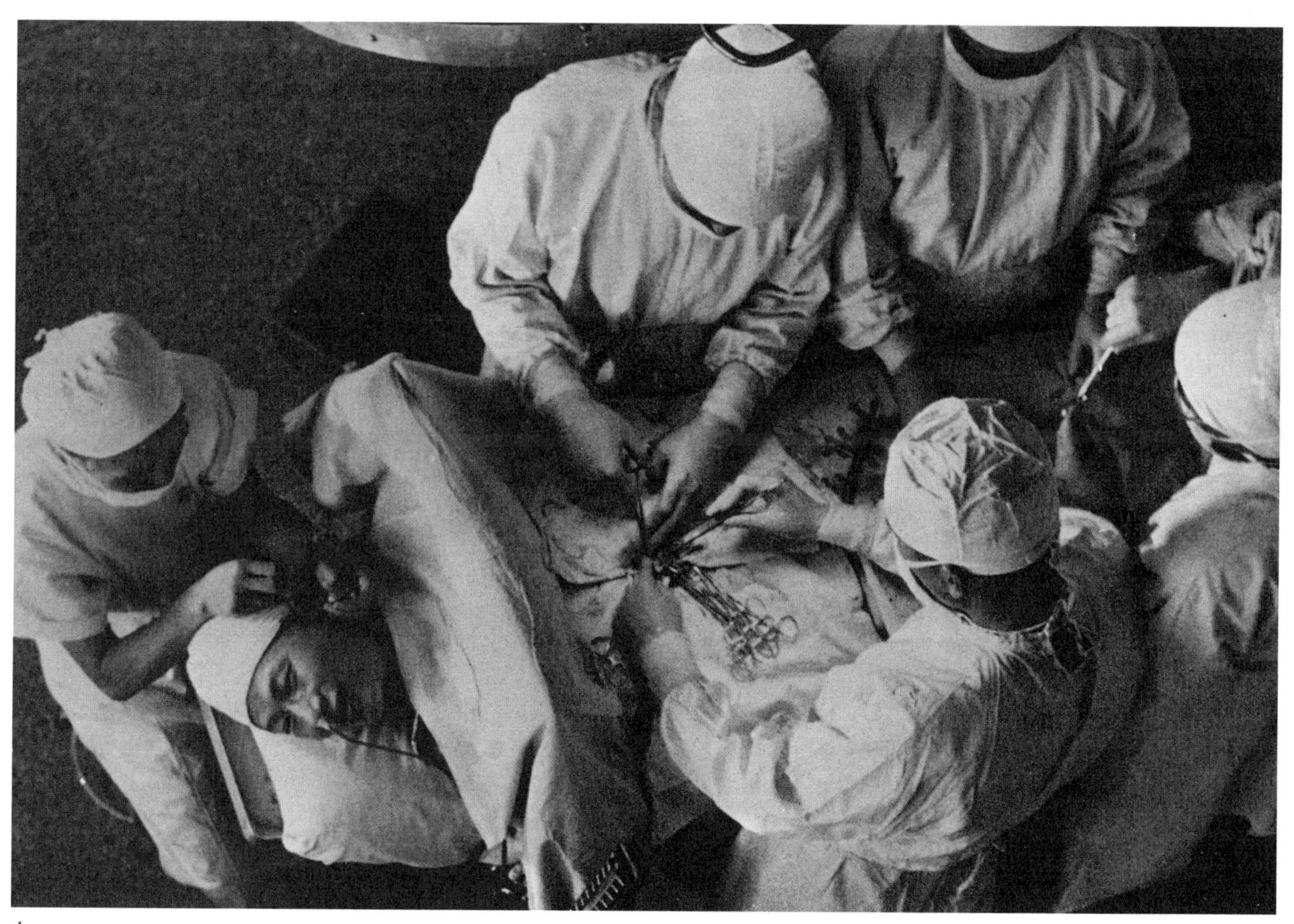

Acupuncture

Lancet (1823) reported that it was successful in the treatment of rheumatism.

Traditional Chinese medicine sees the human body as a balance between two opposing but complementary natural forces, the female force called yin and the male force called yang. The yin force is described as passive and tranquil and is said to represent coldness, darkness, wetness and swelling. The yang force is described as aggressive and stimulating and is said to represent, heat, light, dryness, and contraction. An imbalance of yin and yang is thought to be the cause of all body ailments and diseases; for example, too much yin may cause dull aches and pains, chilliness, fluid retention and tiredness; too much yang may cause sudden sharp pain, inflammation, spasms, headaches and high blood pressure.

Some practitioners of acupuncture reject the pseudoscientific trappings of yin and yang and the notion of an energy flow or life force, but they believe that the needles do relieve pain. Orthodox researchers posit that the practice generates endorphins, chemicals similar to narcotics, but they add that, although pain is reduced, there is no evidence that the application of the needles can influence the course of any organic disease. They also add that cutting off pain may cut off the body's warnings that all is not well. The point is not always to stop pain itself but to explore the cause of that pain and deal with it.

See also ACUPRESSURE; VITALISM.

ACU-YOGA An alternative form of bodywork that combines the practice of ACUPRESSURE with the breathing and postures (asanas) of hatha YOGA. Acu-yoga is generally self-administered after a person has learned the basic yoga postures and the location of the acupressure points. The floor is used to apply pressure to those parts of the body that cannot be reached by the hand. Acu-yoga was developed by Michael Reed Gach, a leading acupressure practitioner who also trains physicians and other medical professionals in various bodywork techniques. He heads the Acupressure Institute in Berkeley, California. As with many holistic health practices, the metaphysical and scientific claims mix in a complex fashion, but the practice of acu-yoga relies upon the claims for general overall body invigoration by the practice of both hatha yoga and acupressure.

ADAMS, SAMUEL HOPKINS (1871–1958) Muckraking journalist remembered for his attacks upon NOSTRUMS. Adams attended Hamilton College and, following his graduation in 1891, began his journalistic career with the *New York Sun*. His writing caught the attention of S. S. McClure, who hired him away from the *Sun* to his own magazine. At *McClure's* he was introduced to the crusading journalism for which it was known. After only a short time, however, Adams moved on to *Collier's*. During the summer of 1905, Adams became deeply involved in the investigation of the patent medicine industry. Legislation in the form of the first Pure Food and Drug Act was sitting in Congress, but languished due to the lobbying effort of the patent medicine manufacturers and the meat industry.

Adams called the patent medicine business the "Great American Fraud." Americans were spending more than $75 million annually on what were in his understanding little more than alcohol and opiates. Many of what today are illegal narcotics—opium and morphine, for example—were perfectly legal and widely used in the 19th century. Adams suggested that more alcohol was distributed annually in medicines than in liquor. He placed the blame on the con men running the business, but he reserved a substantial part of his criticism for advertising people, who had no ethics in writing copy for their advertisements, and the newspapers, which accepted the ads and the revenue they brought. His simple solution was for his colleagues to stop accepting the patent medicine ads. He centered his attack upon the more popular products such as PARUNA, CELERY compounds, and Lydia PINKHAM's popular remedy for female problems (which in the end turned out actually to contain some helpful ingredients). As he noted the alcohol content of the average patent medicine, Adams also aimed his journalistic pen at "hypocrite" church and temperance leaders who vocally endorsed these products.

Adams called upon the government to intervene. First, he suggested that these products should be fairly labeled and sold only as alcoholic beverages (a suggestion later written into law). He also pushed for passage of the Pure Food and Drug bill. Adams's series of ten articles combined with the efforts of the American Medical Association and the scandal of the meat industry that was so vividly portrayed in Upton Sinclair's novel *The Jungle*, persuaded President Theodore Roosevelt to back the new legislation, which, when it finally got to the floor of the House of Representatives, passed 240–17. Adams went on to a lengthy career in journalism and never lost interest in the nostrum question. He lived to see the broad federal legislation that took the more harmful products off the market.

AGASSIZ, LOUIS RODOLPHE (1807–1873) Swiss-born naturalist, teacher and scientist. Louis Agassiz is perhaps best remembered today as the founder of Harvard's Museum of Comparative Zoology in 1859. He also promoted the study of science in the United States and helped found the National Academy of Sciences in 1863. Famous for his teaching skills, he trained a whole generation of U.S. scientists. Agassiz was a supporter of both CREATIONISM and CATASTROPHISM. He was one of the major American scientists who opposed Charles DARWIN's theory of EVOLUTION, and his own theory of ice ages pointed out flaws in Sir Charles LYELL's theory of gradualism.

Agassiz received his medical degree from the University of Munich in 1830. He then moved to Paris and studied with Baron Georges CUVIER, the founder of the science of comparative anatomy. Using Cuvier's principles, Agassiz published a five-volume work, *Recherches sur les poissons fossiles* (1833–43), that provided descriptions of most of the fossil fish then known to science. In 1846, Agassiz accepted a position at Harvard University, where he taught until his death in 1873 at age 66.

Throughout his lifetime, Agassiz supported the creationist and catastrophic theories he learned from Cuvier. In 1830s Europe, prevailing views held that the Earth had arrived at its present geologic state only through uniform and gradual changes. Through his observations, Agassiz showed that Europe and America had been covered by large sheets of ice (in the form of glaciers). These ice ages, Agassiz showed, were catastrophic events that disproved the prevailing theory of uniformity and gradualism. He also argued less successfully against the evolutionary theories of Charles Darwin, maintaining that the history of life had been established by God at the beginning of time and that it progresses according to a divine plan. Agassiz died unhappy, isolated, and frustrated, having lived to see most of his students and colleagues accept Darwin's theories.

AGE OF THE EARTH The number of years Earth has existed, resulting in a subject of controversy. Creationists and many fundamental religious sects believe that Earth's age is only several thousand years, while orthodox science puts the age at about 5 billion years. Establishing the age of Earth has always been marked by tension between the scientific and religious communities. Whether or not a divine force lies behind the historical event of Earth's creation fuels the debate and helps determine whether study of Earth's age falls within the realm of science or pseudoscience.

The central figure who believed in a youthful Earth was Archbishop Ussher. In the 17th century, he carefully worked out from the Old Testament (with computations where the text was uncertain) that the Earth was formed in 4004 B.C.E., establishing a precise date and time for the event. Sir Charles LYELL, the geologist, changed this date. He could not see how the fossil record of life could possibly fit into such a timespan. In the 1890s, Lord Kelvin (1824–1907), the eminent physicist, applied the best science of his day to the problem. He reasoned that the Earth, initially a body of molten material, would cool according to the laws of thermodynamics, and calculated that it would take between 20 and 400 million years to reach its present temperature—not a precise estimate, but more acceptable to uniformitarians like Lyell and much longer than several thousand years. In reasoning thus, Kelvin assumed that no internal heat was being added to the cooling mass. Soon afterwards, it was discovered that radioactive material had been decaying throughout the Earth's existence, generating heat and thereby invalidating Kelvin's calculations, making his estimate too low.

The radioactive process opened up another possible dating method. Radioactive materials decay exponentially. In a time characteristic of each substance (known as its half-life), half the substance decays, forming another substance, its decay product. In that same time half of what remains decays, and that same time later, half of the remainder, and so on. So the time of formation of a rock containing a radioactive substance and its decay product can be dated by measuring amounts of the radioactive substance and decay product—assuming that no dispersion of either has taken place. The Uranium238-Lead206 combination with a half-life of 4.5 billion years has proven ideal for this purpose. Measurements made of decay of ancient rocks, supported by similar measurements on Uranium238-Lead207 and Thorium232-Lead208, give the date of their formation as about 4.6 billion years. On this basis, the Earth must be at least that old.

Establishing the age of the Earth has been marked by tensions between the scientific and religious communities. Whether or not a divine force is behind the historical events of the Earth's creation fuels the debate as to whether the study of the Earth's age is within the realm of science or pseudoscience.

See also CREATION SCIENCE; CATASTROPHISM.

ALCHEMY Derived from the Arabic *al-kimia,* the Egyptian art that strove to change substances from the known and commonplace to something other in the attempt to uncover universal secrets. It began in ancient China and was practiced in Asia, the Middle East, and Europe for millennia until it gave way to modern science about 400 years ago. It was based originally on the idea that everything was made up of four elements—earth, air, fire, and water—mixed in different proportions in different substances; changing their proportions would change the substance, also known as transmutation. Medieval alchemists searched to uncover the following universal secrets: (1) the *elixir* of life, which would confer immortality; (2) the *panacea,* which would cure all ills; (3) the *philosopher's stone,* which would turn base metals into gold; and (4) the *alkahest,* which would melt anything and be very useful in experiments and also in war.

Although regarded with some disdain today, the work of the experimental alchemists should not be dismissed lightly. In their searches they heated, pounded, mixed and tested everything they could find and in doing so discovered much about many different materials; they established many of the chemical processes—distillation, fusion, calcination, solution, sublimation, putrefaction, fermentation—that we take for granted today. They accumulated expertise and knowledge that formed the basis for much

Alchemy

of today's chemistry. They paved the way for much of today's science, including the transmutation of elements.

Because many lacked a sound knowledge of the constituents of matter, they became frustrated by their lack of success by experimental methods—or at best very limited success. As a result, many alchemists turned to MAGIC, SPELLS, and INCANTATIONS. Because they were concerned that such potentially powerful knowledge could be dangerous in the wrong hands, they published information about their discoveries in ambiguous, allegorical form that only the initiated can understand. Consequently, later alchemists were never quite sure what their precursors had actually achieved. This secrecy also aroused suspicions and accusations of Satanic practices. This grew and led to condemnation by the Catholic Church on the one hand—in 1317 Pope John XXII issued a proclamation against alchemy—and fraudulent practices by unscrupulous tricksters on the other.

Thus in the Middle Ages, magic and alchemy had become associated with demonic arts, necromancy and the like. It was only with the Renaissance and a new evaluation of human significance in the world that more respectable forms of magic could be reinstated as worthy of humankind and to be practiced without shame. Magic in the Renaissance became an intellectual achievement, praised by many important intellectuals, and it retained this status until the time of Johannes Kepler, Francis BACON, Pierre Gassendi, and René Descartes.

Sir Robert Boyle in *The Skeptical Chymist* (1661) exposed the unreality of the alchemists' dreams, distinguishing clearly what were elements, what were compounds, and what were mixtures, and laying to rest the old four elements, at least by genuine researchers, for good.

At about the same time the tricksters and their claims for access to untold riches were denounced by the Jesuit Anathasius Kircher, an alchemist for many years, as "a congregation of knaves and impostors." The claims of the knaves and impostors were not limited to transmutation of metals but covered the whole range of alchemical practices: eternal youth, immortality, cures for diseases and disabilities—in short, anything magical that could be sold to the gullible.

Nevertheless, despite denouncements, trickery continued for many years—it could be highly profitable, but it could also be highly dangerous. There are recorded cases of alchemists being imprisoned, tortured and even committing suicide when they were unable to substantiate their claims. Nevertheless the claims of alchemists, both genuine and fraudulent, were taken sufficiently seriously as late as the 18th century for both the French Royal Academy of Science and the Royal Society in Britain to investigate them, finding none supportable.

During the Renaissance, scholars made a distinction between supernatural forms of magic and natural magic, practitioners of which were seeking the hidden but natural properties of substances to produce spectacular and unexpected effects. In the same way true and false alchemy were distinguished. False alchemy made all sorts of impossible claims, such as curing disease instantaneously and bestowing eternal youth. True alchemy only sought what were then seen to be natural properties.

In the 16th and 17th centuries the mechanical philosophers, who were to become the core of the new natural philosophy of the scientific revolution, were not usually mystically inclined, but they were concerned with the ultimate nature of matter, with the kind of transformations that could be achieved by the experimental manipulation of matter. They took the practical side of alchemy seriously and tried to represent alchemical ideas in their own atomistic terminology in which all was reduced to matter and motion. Many accepted, for example, the possibility of transmutation and showed how it might occur within their purview. This phase of the relation between alchemy and natural philosophy lasted until late in the seventeenth century. Such leading natural philosophers as Robert Boyle and Sir Isaac NEWTON took alchemy seriously in this way. Those investigators of the 17th and 18th centuries who were most interested in reproducing the practical claims of the alchemical tradition were also pioneers of the new science of chemistry. They concluded that the more grandiose claims of the alchemists were in error; for example, they eliminated approach after approach to the problem of transmutation.

However, there was one striking difference between the new mechanical philosophers and the older alchemists: The mechanical philosophers were opposed in principle to secrecy and increasingly adopted a conception of knowledge as something for the public benefit. They believed that everything should be published and open to public scrutiny. In this way obscurities and mistakes could be exposed and eliminated in critical public discussion. One figure in this story is especially interesting—Isaac Newton, perhaps the most famous natural philosopher of the 17th century or any century. His achievements are taken as paradigms of scientific research. His published work was disciplined in method—nothing was claimed that had not been argued for by a combination of induction and deduction, the latter modeled on the deductive method of geometry. In his private life, however, Newton was deeply interested in alchemy. He took copious notes on alchemical books and manuscripts and he carried out prolonged and detailed experimental studies, believing that his own experiments would be most productive if they proceeded in conjunction with the study of records of the ancient past.

As part of his natural philosophy, Newton was interested in the interactions of very small particles, knowledge of which he believed to be hidden in allegorical form in alchemical writings. It was the *most* mythical alchemical writings that he thought were the most important to study—as these were thought to represent the oldest part of alchemy.

Newton's studies and experiments provided important insights into what was and was not possible by alchemical manipulations. Nevertheless he never published anything directly concerned with alchemy.

Today we tend to regard alchemists as "knaves and impostors," happily now a thing of the past. However, like Newton and the contemporary mechanical philosophers, we should draw a distinction. There were two distinct groups: the genuine seekers after knowledge to whom today's scientists are indebted for their contributions to scientific understanding, and the rogues and villains who exploited claims to secret knowledge for their own ends, the pseudoscientists of their day.

See also TRANSMUTATION OF ELEMENTS.

For further reading: J. Read, *Prelude to Chemistry: An Outline of Alchemy* (MIT Press, 1966); R. G. A. Dolby, *Uncertain Knowledge* (Cambridge University Press, 1996); and Stuart Gordon, *The Book of Hoaxes* (Headline, 1996).

ALEXANDER TECHNIQUE Medical/psychological theory that body posture affects the health of the whole body. Frederick Matthias Alexander (1869–1955) was a popular Shakespearean actor from the Australian province of Tasmania. During a long series of performances, Alexander lost his voice. His doctors could find no physical cause for his problem, so Alexander decided that his illness must be caused by something he was doing to himself while he was acting and reciting. He started on a program of self-examination, using mirrors, and discovered that his posture—especially the way he held his head and neck—drastically affected the way he spoke. By changing the position of his head in relation to his neck and the rest of his body, Alexander recovered his voice. He also discovered that patterns of tension throughout his body affected his voice.

Alexander developed his technique based on the theory that a person's regular body posture in movement or stillness can become part of an unconscious pattern. If the tension becomes habitual, the person's body begins to understand it as "correct" and allows it to continue without the person becoming aware of it. Alexander labeled this "faulty sensory perception." His technique aims to correct it by removing tension from the neck and lower back; by breaking the tension pattern, the Alexander technique allows a person to develop new habits that are healthier and put less stress on the body.

Some physicians believe that the Alexander Technique can improve a person's health and well-being. Others reject it, believing that any good effects are psychological and do not arise directly from using the technique. However, actors, dancers, and singers have found the Alexander Technique useful in correcting posture problems that can affect their performances. Adherents claim that it can increase their energy and change their mental state. Because it reduces stress, they claim that it can cut down on the frequency and severity of illnesses as well.

ALMAS A term for a hairy hominid allegedly found in the mountains of Mongolia and Russia and similar in appearance to the Himalayan creature called the YETI or ABOMINABLE SNOWMAN. The word "Almas" comes from the Mongolian language and means "wildman." The first recorded sighting was described in the journal of a 15th-century Bavarian nobleman, Johann (Hans) Schiltberger, while he was a Mongolian captive. Schiltberger wrote that a wild people, entirely covered with hair, except for face and hands, lived in the mountains. Many other sightings have been reported in the centuries since, and the Almas was formally studied by Russian scientists in the early part of the 20th century. Reported sightings dramatically decreased in the latter part of the century, although as late as 1974 a Mongolian shepherd reported seeing in the Asgat Mountains a half-human, half-beastly creature covered with reddish black hair. Unfortunately, no strong physical evidence—photographs, a living or dead specimen, or casts of tracks—exists to prove the Almas's existence.

Some cryptozoologists and anthropologists consider the Almas an element of Mongolian folklore; others suggest that the creature might be a leftover from the Neanderthals; still others believe it is a relative of the many similar creatures reported in Asia and North America. However, most scientists do not believe sufficient evidence exists to prove the existence of these creatures.

See also BIGFOOT; CRYPTOZOOLOGY.

ALTERED STATES OF CONSCIOUSNESS An umbrella term used to describe an extraordinary physical or mental state, not usually experienced in daily life. Such states comprise hypnagogic hallucinations (as in DAYDREAMING), fantasizing, hallucinations, HYPNOSIS, trances, DEATHBED VISIONS, spirit POSSESSION, and OUT-OF-BODY EXPERIENCE. Such states occur among all peoples, and are usually brought about by changes in body chemistry, which can occur naturally—through disease, by taking hallucinatory drugs, through practices like fasting, drumming, twirling the body, shaking the head, or through psychological suggestion.

There are two ways of thinking about such altered states. The traditional ancient explanation proposes that the states are temporary change of location for the individual who believes he or she is actually entering into and participating in a spirit world. Modern psychological explanations recognize these states as being localized within and generated by the brain's cognitive system.

Some anthropologists who have studied shamanism believe the purpose of using altered states of consciousness within the indigenous metaphysics of societies is to control and influence the weak and powerless, thus dominating the social policy of a tribe or people.

ALTERNATIVE MEDICINE All therapies that are an equivalent substitute to modern scientific medical practice. Equivalent when applied to medical practice means having equal value for a particular curative purpose.

In 1983 the British Medical Association set up a scientific committee to "consider the feasibility and possible methods of assessing the value of alternative therapies, whether used alone or to complement other treatments." The major difficulty with this task was that the term covers such a wide range of practices, some used purely as complementary medicine and others used totally as an alternative treatment. The committee made four main points:

1. Alternative practitioners generally offered patients more time and were more prepared to listen than regular doctors.
2. Patients who entered the alternative system had the feeling that therapists were more compassionate and more concerned and provided care for the whole person rather than just addressing an illness.
3. Patients liked touching as a way of healing, as in massage, acupressure, and osteopathy, rather than letting modern technology come between themselves and their doctors.
4. Many patients felt comforted and reassured by the "laying on of hands" and other magical practices. They liked the strange words and the unfamiliar qualities inherent in a paranormal approach and felt that they were being linked into a powerful unknown healing force.

The committee concluded that conventional medicine was failing patients. Busy doctors did not spend enough time with their patients to look at their problems in detail and perhaps consider them as whole persons. However, on the other side, the committee thought that not everything the patients wanted was always in their best interest, and it was concerned that many alternative systems had poor training programs. Furthermore, none have strict ethical guidelines, similar to those which regulate conventional doctors.

The committee was also concerned about many alternative diagnostic techniques, such as IRIDOLOGY which

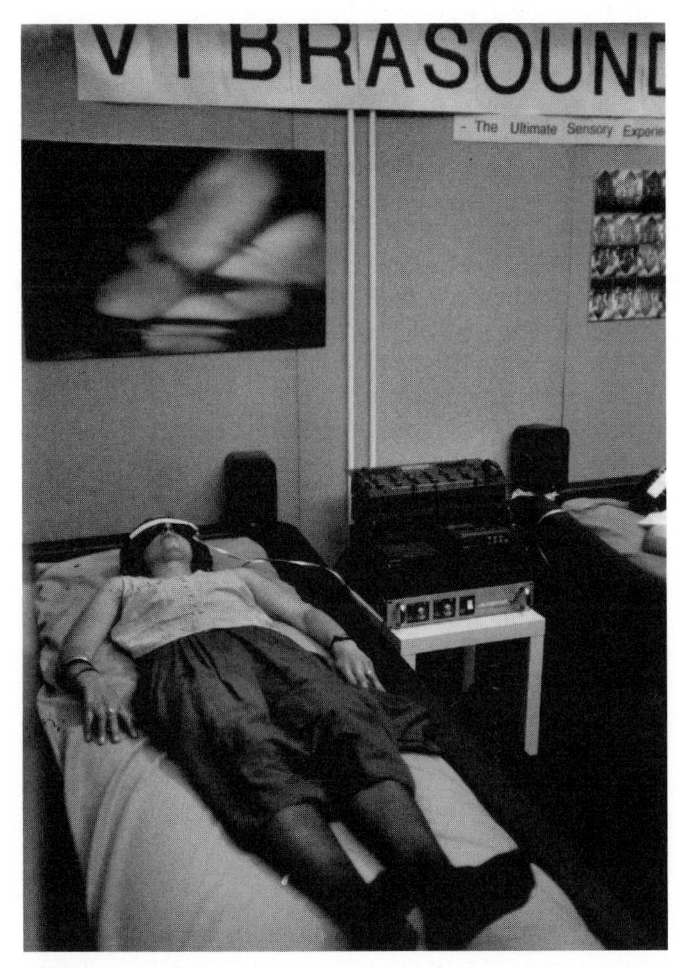

Alternative Medicine

diagnoses by identifying the location of certain flecks in the eyes, linking them to specific diseases. Especially worrying to scientific doctors was "intuitive diagnosis," which does not even need the patient to be present but claims to call on powers unknown to science. Diagnosis is a very important part of primary medical care and should be in the hands of men and women who have a wide knowledge of many sides of medicine and access to a network of colleagues with whom they can confer. Although many patients reported good effects from using certain alternative systems, the effectiveness of the therapies themselves are impossible to assess scientifically, and successes could be put down to the placebo effect. A placebo is a substance, known to have no pharmacological effect, given to the patient as an illusory treatment, to which he or she often responds positively for a short time.

Many alternative approaches are based on VITALISM, an ancient notion that the functions of the body operate well when the "life force" flows freely. The life force is a substance distinct from the laws of physics and chemistry. A good example of this force is found in ACUPUNCTURE, in which needles are said to balance the *chi*.

Yet another pseudoscientific theory of health is put forward by people of the NEW AGE MOVEMENT. They claim that all illness is caused by a separation from the Divine. Healers know how to channel psychic energy from a past healer or spirit, and give out the magical powers to the "AURA," an energy field alleged to surround each human body, of any patient who needs it. Sick people are then, by using meditation and visualization techniques, able to activate the special power enabling them to heal themselves.

Many alternative treatments have proved to be very helpful; for example, BIOFEEDBACK in the hands of a responsible patient under good physician care. But others are highly suspect; one—"Colonic Irrigation"—has caused death. Quackery also finds its way into the alternative health system, that is, the promotion and sale of unproven health products or services, such as "Cellular Therapy."

Herbalist remedies range from the totally useless to the "perhaps may do you some good." Many herbs contain chemicals that have not been completely cataloged. Some may turn out to be useful therapeutic agents, but others could prove to be toxic. For simple ailments like sleeplessness and constipation, the herbalists may well be worth a visit, but many who practice lack adequate training in the diagnosis of disease and consequently do not know what they are treating.

For further reading: K. A. Butler, *A Consumer's Guide to "Alternative" Medicine* (Prometheus, 1992); J. Raso, *"Alternative" Healthcare: A Comprehensive Guide* (Prometheus, 1994); V. E. Tyler, *The Honest Herbal* (Haworth, 1993); and S. Barrett and W. T. Jarvis, *The Health Robbers: A Close Look at Quackery in America* (Prometheus, 1993).

AMERICAN SOCIETY FOR PSYCHICAL RESEARCH (ASPR) Oldest of the U.S. organizations engaged in psychical research and PARAPSYCHOLOGY. The ASPR exists to advance the understanding of phenomena alleged to be paranormal: TELEPATHY, CLAIRVOYANCE, PRECOGNITION, PSYCHOKINESIS, and related occurrences that are not at present thought to be explicable in terms of physical, psychological, and biological theories. The ASPR was founded in 1885, but support was difficult to maintain in this controversial field; for financial considerations, in 1889, the society affiliated with the SOCIETY FOR PSYCHICAL RESEARCH in London. During its first generation, the research work was led by Richard Hodgson. Following his death in 1905, the society, never strong, was dissolved and continued as a branch of James Hervey Hyslop's American Institute for Scientific Research at Columbia University.

When Hyslop died in 1920, the ASPR regained its independent status, and Dr. Walter Franklin Prince became the Society's director of research and editor of its publications. He carried on a variety of investigations prior to his observations of Mina S. Crandon, better known as Margery, who claimed to be a medium. The ASPR board was strongly behind Margery, but Prince believed she was a fraud. The issue came to a head in 1925 when J. Malcolm Bird, who had written several items favorable to Margery, was appointed as co-research officer with Prince. Infuriated, Prince resigned in 1925 and with other disaffected members founded the rival BOSTON SOCIETY FOR PSYCHICAL RESEARCH (BSPR).

Bird continued as a research officer for the ASPR but suddenly in 1930 resigned from his position. Many years later it was revealed that he had had second thoughts on Margery. He had submitted a confidential report to the board suggesting that Margery had approached him to become a confidant in producing some of her phenomena. Subsequently, Bird and his last manuscript on Margery disappeared.

Following the merger of the BSPR with the ASPR in 1941, George Hyslop, the son of J. H. Hyslop, became president. He demanded the full exposure of Margery's fraudulent activity and worked to reestablish the standards demanded during his father's years of leadership. Hyslop would be succeeded by Gardner Murphy, who spent 20 years as president of the ASPR and became its dominating figure, bringing new prestige to the organization and recruiting talented researchers. During this period laboratory parapsychology emerged as the cutting edge of psychical investigations and the Parapsychology Association was established (1957) as the major professional association for scholars engaged in psychical research.

The ASPR is one of the stable organizations in U.S. psychical research and the home of leading lights such as Gertrude Schmeidler and Karlis Osis. Publication of the society's *Journal* and *Proceedings* commenced in 1907 and has continued uninterruptedly to the present. The society's headquarters building in New York City houses a specialized research library for the society's members. The ASPR is a nonprofit, open-membership institution. Members are not required to accept the reality of any particular paranormal phenomena.

See also MARGERY CONTROVERSY.

For further reading: R. Laurence Moore, *In Search of White Crows* (Oxford University Press, 1977); and D. Scott Rago, *Parapsychology: A Century of Inquiry* (Dell, 1975).

AMITYVILLE HORROR Title of book and movie about an alleged haunting in the 1970s. On November 13, 1974, a young man named Ronald DeFeo ran from his house in Amityville, New York, screaming that someone had murdered his family. DeFeo himself was later convicted of shooting his parents, two brothers, and two sisters. The next owners of the DeFeo home, a large, pleasant, Colonial-style house, were George Lutz, his wife Kathy, and their three young children. They moved in on December 18, 1975. Twenty-eight days later, they fled the home, saying it was haunted.

Very quickly, the Lutz's attorney, William Weber, and a nationally known author, Jay Anson, got involved, helping the Lutzes tell—and sell—their story. Filled with sensational details, the successful book became a popular movie. Some of the incidents the Lutzes claimed occurred included cloven-hoofed tracks in the snow outside the house; malevolent entities that levitated Kathy and tried to possess her; an encounter between one of the children and a demonic pig; green slime that appeared in various places in the home; a hideous stench; the sounds of an invisible marching band tramping through the house; and personality changes in the Lutz family. In addition, Anson's book said, a local priest who had blessed the house in an effort to drive the demons out was himself driven out by a shouting demon, and subsequently came down with a mysterious, enervating illness.

The Amityville events are one case where both skeptics and most parapsychologists agree: The case was a fraud from beginning to end. One of the Lutzes' apparent motives was money; another was said to be George Lutz's psychological problems. Several independent investigators went to Amityville and interviewed people quoted in the book. The police claimed they had never investigated the case as the book stated; the priest claimed he had never had a demonic encounter at the house; damages the house had supposedly sustained were not in evidence; the Lutzes' account changed significantly from its first telling to the written version; and numerous other details were shown to be false.

The book sold millions of copies and had a sequel, the movie sold millions of tickets and had five sequels, and the Lutzes moved to another home, which they claimed was also haunted. This time their story was not as well received. Other owners of the former DeFeo house say that nothing unusual occurred there.

ANGEL HAIR One of several types of mysterious "falls," anomalous items that fall from the sky, often associated with UFOs. Angel hair has been described variously as glass-like filaments that dissolve when touched; spiderweblike fluffy masses that fall on bushes, streets, and trees; cottony, radioactive fibers; silken threads as long as 50 feet; metallic, tinsel-like strands; and short, weak cottonwool-like fibers. Some falls have been fleeting, but others have allegedly occurred over a period of two or more hours. Because of its fragility and transience, the substance often disappears before scientists have had a chance to analyze it.

Many reports of angel-hair falls were made in the 1950s. Typical was the account of Mrs. W. J. Daily of Puente, California, as reported in a 1957 issue of *FATE MAGAZINE*. Mrs. Daily said that she observed a "huge disk or ball-shape object through binoculars. She saw it turn reddish and then expel a fine downy substance that drifted toward earth. When she tried to touch it with her hands, it disappeared, leaving no trace." She was, however, able to pick up a sample of the stuff with a stick. She placed the substance in glass jars and handed it over to a U.S. Air Force investigator, but no analysis results were revealed.

Chemists analyzed angel hair that fell in northeastern Japan in 1957. They found it to be an organic substance of indeterminate makeup, but it definitely was not spiderwebs, which it resembled. Dr. Francis A. Richmond, professor emeritus of chemistry at Elmira College in New York, analyzed fibers that blanketed several square blocks in Horseheads, New York, in February 1957. Richmond found that the fibers were radioactive. Other investigators suggested that the fibers may have been cotton fibers transmuted by an explosion or that they may have come from a nearby milk processing plant, whose airborne effluvia had precipitated in this unusual weblike manner because of odd atmospheric conditions.

Various other explanations have been proffered for angel hair. Some of it resembles the metallic chaff military planes used to use to frustrate radar detection devices. Some authorities have suggested that it is merely some form of pollution or perhaps debris that has fallen from aircraft. This latter explanation would also account for the "UFOs" that are sometimes seen at the same time as the fall or immediately preceding it. Various atmospheric conditions could cause an ordinary plane to look unusual.

Other kinds of anomalous falls have included ice, frogs, rocks, fish, jelly-like blobs, and many other kinds of unusual objects. Although many of these have never been satisfactorily explained, most authorities think there is a natural explanation—or several—for them and that the falls are not paranormal.

See also UFOLOGY.

ANIMAL MAGNETISM According to physician Franz Anton MESMER, the force permeating all animal life, including human. It is the mechanism by which HYPNOTISM acts. The idea that there is some substance or field of force that runs through all nature has been (and still is) very widespread. Although in this form it is a 19th-century Western idea, its essence is found in many other cultures across time and place; for example, the Chinese yin and yang, the constituents of *qi*, are very similar. The idea was taken up by Mary Baker Eddy and, in an adapted form called MALICIOUS ANIMAL MAGNETISM, was incorporated into Christian Science.

See also MACROBIOTICS.

ANIMAL PSI An old belief that animals have some form of psychic understanding with human beings and that they use some method, other than normal animal to human communication, that involves perhaps a sixth sense. Stories have been told to suggest that this rapport exists not only between domesticated pets and their owners but also between humans and wild animals, some of which are otherwise naturally very fierce and powerful. Animals often feature in tales of the supernatural, and psychic researchers believe that dogs in particular have the ability to recognize the presence of departed spirits. Animals have also been credited with extraordinary powers of precognition, giving warnings to each other of such disasters as earthquakes or forest fires.

A group of psychical researchers is seriously considering the possibility that animals do have unusual and psychic powers. Recently, parapsychologist William Roll has claimed to have used animals as detectors in experiments with astral travel.

Zoologists, too, have come to the conclusion that animals have extraordinary powers. Yet there is still a vast amount that science does not know about how animals feel and how they think. Even quite lowly species have a built-in "intelligence" sufficient to make them adapt to environments, defend themselves against predators, obtain food, find a mate, and raise their young. But it is only quite recently that we have had the scientific knowledge to investigate specific abilities like the sonar system of the bat, which enables them to locate themselves in the dark by transmitting high frequency sound waves that bounce off objects and back to the bat. DOLPHINS and whales use a similar sonar system that enables them to find their way in water. For example, the female gray whale has a natural ability to make an 11,000-mile journey from the Chukchi Sea, north of Alaska, to a warm lagoon off the coast of Mexico to give birth, and six weeks later, with a young calf, to make the return journey to her arctic feeding grounds. Fragile monarch butterflies make a relay journey, staged over three generations, from Central Mexico to Canada and back. We all know about the amazing feats of migration that have been a natural feature of birds' lives since they developed from archaeopteryx in the Jurassic period, many millions of years before human existence.

Today, the study of the behavior of humanized animals is a very specialized area. A whole new science of veterinary psychology is opening up that studies how pets reconcile their ancestral behavioral history with the pressures of living in an artificially restricted environment. In addition, dogs in particular are being controlled and having their behavior modified, not by their pack leader, another dog, but by their often irrational owners. Many humans observe their pets in a very anthropomorphic way and some even put animal behavior down to phenomena like ESP or animal psi. If the pet owners themselves believe in such pseudoscientific ideas, it is easy for them to project onto animals their own beliefs and preconceptions. Pet clinics have recently opened where a pet-psychic can be engaged to read the minds of disturbed and unhappy animals by letting a pendulum swing over them. The therapists also claim to be able to diagnose and cure any animal by post; all that is needed is a photograph and a few strands of the animal's hair, and they will be able to dowse the problem and suggest a cure.

In the past there has been an occasional claim by pet owners of a very clever animal showing a specific psychic power, but as in the case of Clever Hans, all have been shown to be either the animal responding to unconscious cuing on the part of the owner, frauds, or hoaxes.

See also COUNTING HORSES, RADIESTHESIA.

For further reading: Georgina Ferry, ed., *The Understanding of Animals* (Basic Blackwell Limited, 1984); and William G. Roll, *Theory and Experiment in Psychical Research* (Ayer, 1975).

ANTIGRAVITY MACHINES Devices that their inventors claim will overcome gravity in some unspecified way outside our normal understanding of machine design. Rockets, gliders, balloons, and heavier-than-air planes are ways by which we overcome gravity to fly or, in the case of space rockets, escape from the Earth's gravitational field (or, in general relativity theory terms, escape from the local timewarp). But this way of overcoming or escaping gravity is not what is meant in this context by an antigravity machine. It is a machine that will cancel out gravity in its locality and thus, for example, be able to escape the earth's attraction without the huge expenditure of fuel required by a space rocket. Roger BABSON and his Gravity Foundation sought such a machine without success.

Babson and his collaborators in the foundation were not the only ones to seek such a machine. There have been many hopeful inventors, none successful so far. "Hopeful inventors" brush aside those who claim that their task is hopeless and point to other dreams that were once seen to be unrealizable: the TRANSMUTATION OF ELEMENTS, voyages to the Moon and Mars, or nuclear energy. But this fails to distinguish between the impracticable and the impossible—projects that pose great practical difficulties and projects that are unrealizable on fundamental grounds. PERPETUAL MOTION MACHINES and antigravity machines fall into this second category as ideas that will tempt many but will remain dreams.

ANTIMATTER A theory proposed in 1928 by the Cambridge University physicist Paul Dirac that predicted a partner for the electron that would have a positive charge to counter the electron's negative charge. Dirac was the first to speculate about the existence and behavior of subatomic particles consistent with both the special theory of relativity and with quantum mechanics. Experiments conducted in 1932 confirmed Dirac's theory with the discovery of the positron, and as a result, Dirac won the 1933 Nobel Prize.

Theoretical physicists believe that for every type of particle there may exist an antiparticle with all its properties reversed—each a mirror image of the other—but only a very few antiparticles have been produced experimentally, and the existence of a few others can only be inferred from observations of high energy nuclear reactions. When a particle meets its antiparticle (for example, an electron e^- and a positive e^+ collide), then annihilation energy is produced (in the e^-/e^+ case, a burst of gamma rays) and the particles are destroyed in the process. The amount of energy produced E_{ann} is, as predicted by the Einstein relation between mass and energy,

$$E_{ann} = 2M_pc^2$$

where M_p is the mass of each of the particles involved and c is the velocity of light in a vacuum.

Antimatter is used by science fiction writers as a fuel for space ships, including the USS *Enterprise* on *Star Trek*. The annihilation caused by the collision of matter and antimatter releases all the potential energy stored in the particles. Science fiction writers fantasize that the energy from such an explosion could be used to propel a spaceship at speeds approaching a significant fraction of the speed of light (300,000 kilometers [186,000 miles] per second). Speeds of that order would be needed to make space travel outside the solar system even remotely feasible. If such a spacecraft were to travel through HYPERSPACE, it would need to use antimatter as a source of power. However, given current means of producing antiparticles, the energy consumed in making them would be far greater than the energy produced through matter–antimatter reactions. Fermilab, near Chicago, Illinois, can produce 450,000 billion antiprotons annually at an average of 50 billion antiprotons per hour. The amount of energy released by this yearlong production of antimatter, however, would not even light a single lightbulb.

Scientists can and have been creating antimatter on the subatomic level for decades at centers such as Fermilab. These centers are using matter–antimatter collisions to investigate subatomic physics and to identify new subatomic particles. In December 1995 and January 1996, scientists led by the German physicist Walter Oelert produced the first nine atoms (rather than particles) of antimatter—an isotope of hydrogen consisting of an antiproton and a positron—at the European Laboratory for Particle Physics (CERN) in Geneva, Switzerland. The antiatoms only existed for about 40 billionths of a second before destroying themselves by colliding with particles of ordinary matter.

For further reading: Stephen J. Hawking, *A Brief History of Time* (Bantam Books, 1988); Lawrence M. Krauss, *The Physics of Star Trek* (Basic Books, 1995).

APHRODISIACS Derived from the name of the Ancient Greek goddess of love, Aphrodite, a drug, a potion, or any other agent that is said to induce sexual potency in human beings, especially men. Sexual performance in men reaches a peak in late adolescence and from then on undergoes an inevitable steady decline throughout life. Consequently there has been an agelong search for substances to restore men's youthful teenage vigor, but so far there is no evidence at all that any of the remedies are effective.

Although many drugs and foods are supposed to have an aphrodisiac effect, scientists have been unable to confirm that they work for everyone all the time. Because of this, they have dismissed most aphrodisiac stories as folklore. In fact, most materials popularly believed to enhance sexual desire actually depress it if taken in large amounts.

According to Alfred Kinsey, of the Institute for Sex Research at Indiana University, in his book *Sexual Behavior in the Human Male* (1948), sexual performance in men, in contrast to women, peaks at sixteen or seventeen years of age, and from then on, undergoes an inevitable steady decline throughout life. The Kinsey report did debunk the then prevalent myth that the male libido abruptly ceases in midlife. The downward descent of the male libido tends to be gradual and never touches zero; there is always the rare male who claims to still be sexually alert and potent long after his peers.

Kinsey's report on the human female (1953) tells a very different story, revealing her sexual desires to remain unchanged throughout life, up to the menopausal age and beyond.

The consequence of men's very early decline from their sexual peak has been an ongoing search for substances to restore their vigor, but so far there is no evidence at all that any of the remedies are effective. PLINY THE ELDER, the first-century Roman scholar who described many of the nostrums of his time, advised men to eat animal testicles to overcome impotence, and many tales have been gleaned from classical literature of fertility rites with goats, where Roman youths induced satyriasis (excessive male sexual desire) by eating GOAT GLANDS and even wolf glands. During the Middle Ages people believed that, because the world was created providentially for human beings, every animal and plant carried in its shape or color an indication of its usefulness to humans. Sixteenth-century Swiss physician and alchemist PARACELSUS, following this ancient doctrine of *signatures*, established the role of chemistry in natural medicine by looking at substances that had similar appearances to the ailing part of the human body. Thus a likeness between the male genitalia and certain plants or animal parts led to the assumption that, for example, the mandrake (first noted in Genesis), ginseng, and asparagus, or powdered rhinoceros horn or elephant tusks, would rejuvenate the male parts.

Aphrodisiacal properties have been attributed quite erroneously to many foods such as oysters, mushrooms, and some varieties of fish, especially shellfish. The suggestive power of MAGIC has also been tried, but change in potency after any treatment of this kind must be put down to the power of suggestion, as in the placebo effect.

The use of cannabis is said to increase the pleasure experienced by both sexes, but again, it has no effect on potency. An excess of alcohol is well known to act more as an inhibitor than a promoter of sexual drive. Treatment with androgens, the male sex hormones, is ineffective and has the disadvantage that it may cause cancer of the prostate. Cantharidine extracted from cantharis, a brilliant green beetle, the so-called Spanish fly, is a dangerous vesicant that causes blistering of the skin and, if taken internally, inflammation of the urinary tract resulting in painful uncontrollable erections and the passage of bloody urine. Large overdoses can cause gastrointestinal bleeding, and even death. Yohimbine is unproved in trials when compared with placebos and in excess is toxic.

Today there is a growing and extensive use of mechanical devices such as implants of permanent flexible splints, inflatable devices, and vacuum systems applied to the outside of the penis to promote blood flow to help an erection. Penile injections are also used to relax muscles and encourage blood flow, but these are considered to be a temporary expedient to break the cycle of anxiety that often follows inability to achieve an erection.

Even though it is not considered a disease, the gradual loss of potency in men has always been looked upon as an ailment needing treatment. The latest literature on the subject suggests that the condition not be called "impotence," but instead "erectile dysfunction" or "sexual arousal disorder." The newest cure is a pill containing a drug called sildenafil, marketed under the name Viagra, to be taken an hour before intercourse. The makers claim that it is not actually an aphrodisiac, because desire has to be there. Its function is to improve and prolong erections, giving the same sustaining effect claimed by Casanova for his cup of chocolate. Its success has been widely documented in medical trials and in practice. Notwithstanding this attempt at medicinal restorative, the belief in a magic potion still is, as it always has been, one of the most enduring pseudosciences in medicine.

See also HOMUNCULUS.

For further reading, A. C. Kinsey, et al., *Sexual Behavior in the Human Female* (W.B. Saunders, 1953); A. C. Kinsey, et al., *Sexual Behavior in the Human Male* (W.B. Saunders, 1948); Richard Alan Miller, *The Magical and Ritual Use of Aphrodisiacs* (Destiny Books, 1985); Raymond Stark, *The Book of Aphrodisiacs* (Stein & Day, 1981); and Peter V. Taberner, *Aphrodisiacs: The Science and the Myth* (Croom Helm, 1985).

APPARITIONS A supernatural manifestation of a dead person or animal or even of a live person or animal living too far away to be within normal range of the observer. Contrary to common belief, most apparitions are of the living (not the dead) and few are visual. Most involve the sensing of a presence, sometimes using touch or smell or the hearing of strange thumps, raps, or moans.

Throughout history every civilization has had, or still has, its beliefs about apparitions, GHOSTS, or spirits of some kind. These beliefs are intertwined with the religion, myth, and folklore of the country of origin. Many reasons are given to account for apparitions. A favorite belief is that they are unhappy restless spirits of the dead, trapped here on earth with nowhere else to go and in need of appeasing or placating. Another belief is that these spirits are evil and will haunt the same location time and time again until they are exorcised by adjuration. This does not mean that they are to be cast out but rather that they are put on oath by invoking a higher authority to oblige them to behave. Yet another view holds that important apparitions are uncommon and are seen only once or twice. It is thought that they manifest themselves to us to tell us something. Examples of these special kinds of manifestations are the Marian apparitions at Lourdes in France and at Fatima in Portugal. People who believe strongly in survival after death tend also to believe in the OUT-OF-BODY EXPERIENCE and the NEAR-DEATH EXPERIENCE. In folklore there are many archetypal apparitions, such as the phantom ship sailing with no living souls aboard and headless hunters roaming the countryside following dangerous phantom hounds.

Since the late 19th century, apparitions have been studied extensively by psychical researchers, parapsychol-

Apparitions

ogists, and others, yet none have come up with any significant positive conclusion. Two 19th-century researchers, Edmund Gurney and Frederic William MYERS of the SOCIETY FOR PSYCHICAL RESEARCH (SPR), did, however, agree that apparitions had no physical reality. By this they meant that they could walk through solid walls; some researchers claimed they were able to put their hands through walls. But there the conclusions of the two men diverged: Gurney believed that they were the product of telepathy from the dead to the living, projected out of the percipient's (person who sees the apparition) mind in the form of an apparition; Myers believed that apparitions consist of a "phantasmogenic center," a locus of energies clairvoyantly expended by the agent (the person whose apparition is seen) and sufficiently strong to modify the space of the percipient. Andrew MacKenzie, a modern psychical researcher and author of *Apparitions and Ghosts* (1971), found when examining hallucinatory cases that about one-third of the experiences occurred just before or after sleep or when the mind was in a state of relaxation and the guard was down, opening the way for impressions to rise from the subconscious—occasionally taking the form of apparitions.

Dr. Mike Dash, publisher of the *FORTEAN TIMES*, once a believer but now a skeptic, says in his book *Borderlands* that phenomena like apparitions and ghosts "are the product of human imagination . . . are internally generated fantasies rather than paranormal events," and that believers take to the scene with them "expectant attention . . . seeing what they want or expect to see." He thinks that the core of the whole problem is "the complete fragility of human testimony."

See also DEATHBED VISIONS; FORT, CHARLES HOY.

For further reading: Mike Dash, *Borderlands* (Heinemann, 1997); Paul Kurtz, ed., *A Skeptic's Handbook of Parapsychology* (Prometheus 1985).

APPLIED KINESIOLOGY A form of bodywork that combines elements of CHIROPRACTIC and ACUPRESSURE in kinesiology, the art/science of muscle testing, as a means of diagnosing and treating physiological conditions and anatomical problems. Applied kinesiology was developed by Dr. George Goodheart, a chiropractor who had come to believe that every human disease has a structural manifestation in a specific muscle weakness. As early as 1964, he first suggested that a misaligned spine was caused by a weak muscle and that, in response, the same muscle on the opposite side of the spine tightens. He also suggested that, rather than attempt to loosen the tight muscle, the weakened muscle should be strengthened. A weak muscle would tend to affect a particular organ.

Goodheart later combined his theories with insights from Chinese medicine, which saw health as the free flow of universal energy, or *chi,* along an invisible set of pathways called meridians. The free flow of *chi* is attained in Chinese forms of treatment by stimulating points along the meridians with needles or pressure. Goodheart felt that the poor muscle tone he had observed corresponded to the diagnosis of unbalanced and blocked movement of *chi* along the meridians. Applied kinesiology thus came to stimulate specific parts of the body, thereby strengthening weakened muscles and opening blocked meridians.

In 1973 Goodheart introduced the practice termed "therapy localization." He would ask patients to place their hand over a suspected acupuncture point of suspected blockage or imbalance. He then tested the suspected muscle that was either abnormally loose or tight. When the patient's hand was over the meridian point where energy was blocked, reportedly, the corresponding muscle would immediately change; for example, a loose muscle would tighten and a tight muscle would loosen. Once the problem was located, treatment followed traditional chiropractic and acupressure procedures. Applied kinesiology has given birth to a total health care program called Touch for Health.

Goodheart founded an International College of Applied Kinesiology in Detroit, Michigan, and the practice spread primarily among chiropractors. As the spread of applied kinesiology has been helped by its combining the strengths of two widely practiced forms of alternative healing, it has been equally hindered by having to continue using the essentially questionable assumptions upon which both are based.

See also ALTERNATIVE MEDICINE.

APPORTS Objects materialized out of thin air, usually by a spiritualist MEDIUM. Apports became an important part of the SPIRITUALISM craze that swept much of the Western world in the late 19th century. The conflict between science and still strongly held mystical and religious beliefs led to an unprecedented interest in the mystical in all strata of society. When a few spiritualist mediums began to hold séances that seemed to confirm the possibility of communication with "the other side," people started to flock to them. It seemed that some mediums were able not only to talk with spirits or to be the medium of the spirits' communication with living persons but also to provide physical evidence of the spirits' presence. Sometimes this only included such things as mysterious rappings or movements of the table around which the guests were seated, but in the presence of some mediums, the spirits seemed to want to be more concrete in proving their existence: They rained coins onto the table; they materialized as a mysterious substance called ectoplasm that oozed out of a medium's mouth or nose; they presented fruit and flowers, jewels, and sometimes even live animals to the people present.

Usually these MATERIALIZATIONS took place in dimly lit circumstances, and careful investigation proved in several instances that the materializations were frauds. Fraudulent mediums, often with the aid of hidden assistants, used stage magicians' tricks—mirrors, objects hidden in sleeves, skillful use of various draperies, optical illusions, hidden gramophones, and distraction, for example—to fool their clients.

Sometimes, apports seem to have no fraudulent explanation. Many POLTERGEIST cases, in which apports are an element of the phenomena, have not been explained. Certain Indian mystics as well, including Sai Baba, a Hindu mystic living today, have materialized apports and brought about other "miracles" without evidence of fraud. Nevertheless, scientists do not believe in apports. Some of the more liberal may speculate that these mysterious objects can be explained by some aspect of quantum physics, which has many elements that are still not fully understood; others continue to believe that they are hoaxes not yet uncovered.

See also TELEPORTATION.

AQUATIC APE THEORY OF EVOLUTION See MORGAN, ELAINE.

ARCHEUS Renaissance term for a primordial life force. Archeus (pronounced "ARK-ee-us") was first coined by 16th-century German philosopher PARACELSUS. He used it to refer to his own concept of a life force that gave life to all other organisms and individuals. Like Paracelsus's concept of medicine, which he saw as a fight against the invading organism rather than against an imbalance within the body itself, the Archeus idea was based on religious and mystical ideas rather than scientific principles.

ARKEOLOGY The name given to the search for the ark upon which the biblical Noah reportedly rode out the universal flood. Many conservative Christians believe the story of Noah and the Flood in the Book of Genesis (Chapters 6–8) to be a literal report of historical events. In the story, Noah and his family and a number of mating pairs of animals survived the Flood in the ark, which came to rest on one of the mountains of Ararat. Most liberal Christians consider the story mythical. If the ark had survived to the present and was located, it would serve to verify a more literal interpretation of the Genesis text.

The initial problem for arkeology is the location of Ararat. In biblical times, Ararat was the whole of what today is Armenia. Thus Genesis literally says that the ark came to rest on one of the mountains of Armenia. The Caucasus mountains run through the country, offering numerous possible locations. One tradition, however,

Arkeology

identifies the mountain as present-day Ararat. It is located in Turkey very near the borders of Armenia and Iran. The mountain has a two peaks, the higher one reaching more than 5,000 meters (16,000 feet) with the top 900 meters (3,000 feet) always covered with snow. During the centuries, there have been stories of the existence of the ark near its peak. A new era of searching began in 1876 after James Bryce discovered a piece of wood near the summit and declared his belief that it was from the ark. Then, in 1892, Chaldean clergyman John Joseph Nouri claimed that he had not only found the ark but had entered it and measured it. The Bible was correct, he said: It was 300 cubits (450 feet) long.

Through the early 20th century, a number of expeditions searched for the ark, but not until 1952 did an expedition led by French industrialist Fernand Navarra return with pieces of wood that were at first carbon dated to 3,000 B.C.E.; later they were more thoroughly scrutinized and dated at 800 C.E. Since his return in 1955, several other expeditions have been launched, though work on the mountain has frequently been hindered by political unrest. When expeditions could not be launched, other forms of data were assembled, such as aerial photographs. Several movies have been made based upon the speculative evidence.

The latest wave of interest in the survival of Noah's ark on Mount Ararat has generated much speculation but no hard data or convincing evidence. Modern skeptics have argued that the story of the ark is impossible: There is no geologic or historic evidence of a universal flood, or even of a local flood that could have covered Ararat. Shipbuilders also question a boat of 300 cubits (450 feet) made of wood. Over 300 feet, a wooden ship becomes most unstable and leak prone. Biologists have added the observation that the ark could, however tightly packed, hold only a small minority of all of the different animal species now known to exist. Given the mountain of arguments against the literal truth of the Noah account, only the production of hard evidence of a 5,000- to 6,000-year-old wooden boat on Mount Ararat (or one of its neighbors) would provide material for reconsideration of current opinion concerning the historical truth of the Noah story.

AROMATHERAPY The use of aromatic substances made from plant parts to calm or stimulate the nerves and to heal, aromatherapy has been practiced in one form or another since ancient times. Aromas, like all smells, are an effect of a chemical reaction between particles of vapor and a liquid on the olfactory organs in the nasal cavity. The particles dissolve, and the resulting chemical solution stimulates the organs, which lead to the olfactory nerves on the bulblike ends of the brain's olfactory lobes. Aromatherapy holds that what smells good to one must be good for one—that aromatic and therefore beneficial molecules entering the body via the lungs are taken into the bloodstream and distributed throughout the system to do their healing work.

The substances used in aromatherapy are generally essential oils (also called volatile oils) derived from flowers, fruits, herbs, spices, resins, and wood. Health-food stores and "metaphysical shops" stock oils to suit every taste and budget. The oils are used in scented candles, potpourris, hydrosols, diffusers, bath and massage oils, cosmetics, and perfumes. They are diluted in almond, canola, or other carrier oils for direct application to the skin. As many as 300 different oils are used by professional aromatherapists. Clove, geranium, peppermint, rosemary, and thyme are used as stimulants, chamomile and lavender as relaxants, eucalyptus and lemon as antiseptics. Combinations of oils are suggested for treating a variety of ailments.

Claims made for aromatherapy's effectiveness embrace both the unglamorous but usefully mundane and the somewhat astonishing: Sassafras kills lice; clary sage is recommended for premenstrual syndrome, menstrual cramps, and postpartum depression; pure rose relieves depression, too, and is said also to open the heart to love; not only is ylang-ylang a strong sedative, but it also increases sexual potency; fear of flying can be overcome by inhaling from a plastic bag into which one drop each of geranium and lavender has been placed; afterward, a dash of rosemary in a bath can relieve jet lag.

Still, problematic though some of the properties ascribed to aromas may be, aromatherapy can be said to reduce psychological stress and, thus, promote some real measure of physical, mental, and spiritual health, by inducing relaxation and aesthetic pleasure.

See also HERBAL MEDICINE.

ASHKIR-JOBSON TRIANION GUILD A small British psychical research organization founded specifically to develop a mechanical apparatus for communication with the spirits of the dead. It was named for the founders A. J. Ashdown, B. K. Kirby, and George Jobson, who were somewhat obsessed with scientifically proving survival of bodily death and made a pact that whoever passed away first would attempt to communicate with the other two in the form of a brief message. Jobson died in 1930, and three months later the confidential message was received through a MEDIUM who was not formerly acquainted with any of the three. After Ashdown and Kirby established a relationship with the medium, Mrs. L. E. Singleton, Jobson, supposedly speaking through Singleton, gave the instructions for constructing various instruments to facilitate communication.

The Ashkir-Jobson Trianion Guild was founded to construct and market the several instruments described by Jobson. Among these were the Ashkir-Jobson vibrator, which produced a continuous musical tone for the purpose of creating a harmonious atmosphere at séances. It operated by clockwork and could send out a steady sound for up to three hours. The device activated an A-note tuning fork. Other instruments for actual communication were termed the COMMUNIGRAPH and the REFLECTOGRAPH.

The guild continued for several decades; eventually, Singleton became the guild's warden. The several devices sold by the guild enjoyed a brief popularity but never provided the source of regular evidential communication for which the originators had hoped.

ASIANS IN AMERICA BEFORE COLUMBUS Speculation regarding the origins of Native Americans. As early as the 1590s, a Spanish Jesuit proposed an Asian origin for the Native American tribes. In the early 17th century, Englishman Edward Brerewood drew parallels between the Indians and the inhabitants of Northeast Asia. Other early writers speculated that the Native Americans were descended from the 12 LOST TRIBES OF ISRAEL; still others believed that they were of Grecian, Anatolian, Phoenician, Roman, or Egyptian origin. A few even suggested that they had been transported to the Western hemisphere by the devil.

Later research showed that Brerewood was basically correct: Some Native Americans, especially the Aleut and Inuit who inhabit the polar regions, show marked similarities to Asian tribespeople. Aleutian languages are also very similar to Asian languages spoken in Siberia and other regions of Northeast Asia. One of the physical characteristics shared by Asians and Native Americans is the "shovel incisor"—the front teeth are concave on the inside, as though they had been hollowed out. However, Native Americans also have some distinctive physical traits: Many have a prominent, arched nose as opposed to the relatively flat nose of most East Asians, and they lack the eyelid folds that give East Asian eyes their characteristic appearance. Anthropologists have concluded that the ancestors of the Native Americans crossed a land bridge between Siberia and Alaska several times before the end of the ice ages caused the bridge to flood.

When these ancestors crossed the land bridge is still a subject of much speculation and argument. At the beginning of the 20th century, most anthropologists, led by the Smithsonian Institution's Ales Hrdlicka, believed that *Homo sapiens* was a relative newcomer to the American continent—probably within a couple of thousand years. In 1926, however, near the town of Folsom, a New Mexican cowboy named George McJunkin discovered shaped flint arrowheads in conjunction with the bones of extinct bison. A 1932 discovery of more stone points at Clovis, New Mexico, helped push human

Asians in America

occupancy in the New World back to about 12,000 B.C.E. Discoveries since that time have suggested dates as long ago as 36,000 to 40,000 B.C.E. for the arrival of humans in America.

Some scholars suggest that Asians had contact with the New World after the arrival of the Native Americans and before the Columbian voyages of the late 15th century. Dr. Barry Fell, in his book *Saga America,* declares that he has found evidence of Chinese and Japanese influences dating from the medieval period. At a place called Atlatl Rρck, Nevada, Fell has discovered pictographs that he believes are ancient Chinese ideograms. In the Smoke River valley in Idaho, Fell claims to have identified Japanese *hiragana* characters. If these interpretations are correct, contact between the Asian and American continents has a considerable history. Most anthropologists, however, do not accept Fell's views; they believe that contact between Asia and America was fitful at best between the time of the original migrations and the 19th century.

For further reading: Robert Claiborne, *The First Americans* (Time-Life Books, 1973); Barry Fell, *Saga America* (Times Books, 1980).

ASTRAL/ETHERIC BODY Invisible energy duplicate; a double of the physical body, made up of dense, vibrating matter that can be compared to atoms and molecules. Some believe that the physical and etheric bodies are inextricably connected. The etheric (or ætheric) body, often called the astral body, contains chakras (loosely, the sense organs of the etheric body) that supply a person with the psychic energy that some believe underlies and is essential to all life. While the physical body eats, runs, breathes, and performs the other functions of mundane life, the etheric body supplies the very essence of life itself. When the etheric body is weakened or harmed, the physical being of the individual is severely affected. Some believe the etheric body allows people to think, dream, and attain spiritual ecstasy.

Belief in the etheric body has been held since ancient times. The writings and art of early Egypt, India, Greece, and Rome, for example, contain many references to this mystical and invisible body. Occultists sometimes distinguish between the etheric and astral bodies, asserting that both are part of the aura that surrounds all living bodies. They say that the etheric body affects an individual's physical health, while the astral body affects the individual's emotional and spiritual health.

The idea of the etheric body was popularized by the Theosophists of the early 20th century. They adapted ideas from early Hindu belief to create this mystical concept. Today, many New Agers include the etheric body in their belief system.

The etheric body, it is said, can be harmed by negative energy absorbed from other individuals (negative thinkers, emotional drainers) and from the environment (pollutants, food additives), as well as by straying from a spiritual path (allowing oneself to be distracted by stress, materialism, and other non-spiritual attitudes), among other things. If this happens, the etheric body must be healed through activities such as meditation, cleansing the body of unwholesome influences (alcohol, red meat, and so forth), and using spirituality-enhancing tools such as CRYSTALS. Use of these things will bring the etheric body back into balance, restoring its strength to protect and enhance the life it surrounds.

The etheric body remains a metaphysical or occult belief. Although there is little evidence to lend it scientific support, some relate it to the mild electromagnetic field that has been shown to surround living things.

See also KIRLIAN PHOTOGRAPHY; OUT-OF-BODY EXPERIENCE.

ASTROLOGY The forecasting of earthly and human events by means of observing and interpreting the fixed stars, the Sun, the Moon, and the planets. As a science, astrology has been utilized to predict or affect the destinies of individuals, groups, or nations by means of what is believed to be a correct understanding of the influence of the planets and stars on earthly affairs. As a pseudoscience, astrology is considered to be diametrically opposed to the findings and theories of modern Western science.

Astrology has a long history and has been—and still is—influential in every society, exerting a sometimes extensive and peripheral influence on many civilizations, both ancient and modern. Although hotly contested by modern science, it is nevertheless subscribed to by many people—probably more than half the world's population, albeit without any great commitment.

Astrology is not a single body of doctrine or beliefs. For some it defines an inescapably predetermined future; it foretells what is going to happen. For others it foreshadows what might happen if a person, group, or nation does not take appropriate action. For yet others it is a source of advice, access to which helps forward planning. For many it goes no further than casual scanning of a horoscope published in a daily or weekly newspaper, more as entertainment than a guide to action. Others will not act unless the portents are favorable. Those consulting the message of the stars range from the humble man or woman in the street to heads of state.

Astrology can be categorized into four groups according to use:

1. individual, based on the birth chart of an individual
2. collective, using the birth chart of a company, city or country
3. divination, using the chart for a particular moment
4. natural, using astrological data to plan agriculture or predict weather

The use of astrology differs from country to country; for example, in India where astrology still retains a position in

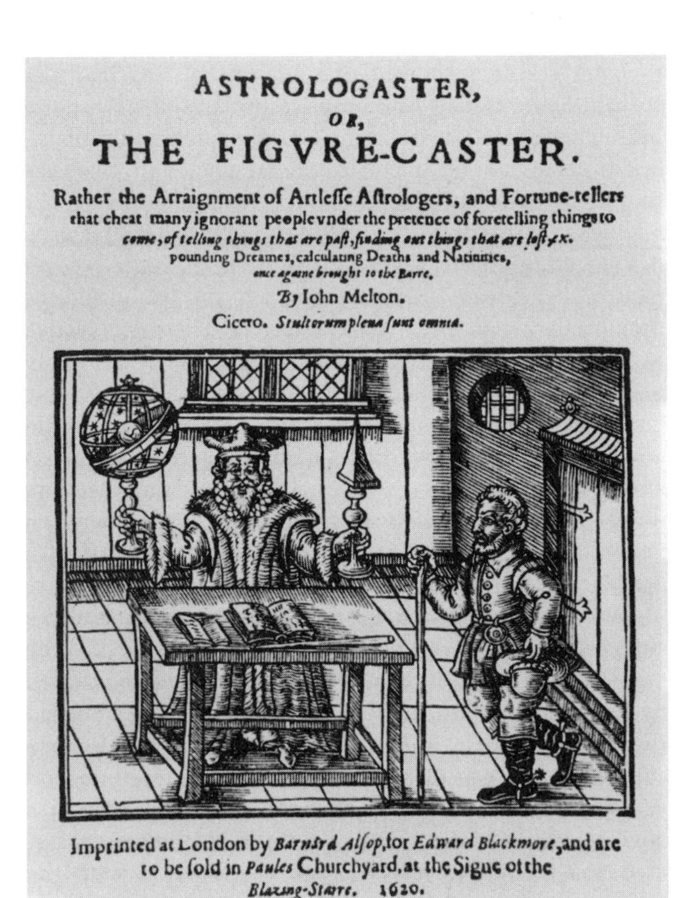

ASTROLOGASTER,
OR,
THE FIGVRE-CASTER.
Rather the Arraignment of Artleſſe Aſtrologers, and Fortune-tellers that cheat many ignorant people vnder the pretence of foretelling things to come, of telling things that are paſt, finding out things that are loſt, &c.
pounding Dreames, calculating Deaths and Natiuities,
once againe brought to the Barre.
By Iohn Melton.
Cicero. Stultorum plena ſunt omnia.

Imprinted at London by Barnard Alſop, for Edward Blackmore, and are to be ſold in Paules Churchyard, at the Signe of the Blazing-Starre. 1620.

Astrology

the sciences and some universities offer advanced degrees in the subject, most seek astrologers' advice on any matter of importance: marriage, health, relationships, work, elections, finance, farming, affairs of state, and so forth. In the United States the main concerns are self-understanding and inner growth, jobs and relationships, and predicting a favorable (or unfavorable) time to take some action.

What might be called the mechanics of present-day astrology derive from the system developed by the Greeks in the Hellenistic period (300 B.C.E. to 300 C.E.) The Sun's circle around the Earth (astronomers then placed the Earth as the center of the universe) was divided into 12 equal parts or houses, each of 30 degrees, each assigned a sign of the zodiac: Aries, the Ram (March 21–April 19), Taurus, the Bull (April 20–May 20), Gemini, the Twins (May 21–June 21), Cancer, the Crab (June 22–July 22), Leo, the Lion (July 23–August 22), Virgo, the Virgin (August 23–September 22), Libra, the Balance (September 23–October 23), Scorpio, the Scorpion (October 24–November 21), Sagittarius, the Archer (November 22–December 21), Capricorn, the Goat (December 22–January 19), Aquarius, the Water Carrier (January 20–February 18), and Pisces, the Fish (February 19–March 20). The all-important birth dates or event dates for each house are shown in brackets. Each zodiacal sector is dominated by a planet, and each has a connection with a part of the human body. Each sector is subdivided into three of 10 degrees and each of these into four subsectors of $2\frac{1}{2}$ degrees. Each planet has a point on the circle where its influence is high, opposite to which is the position where that planet's influence is low.

Other elements are added to the basic chart: opposites, substances, human characteristics, the position of a zodiac sector relative to the eastern horizon at a particular time, the position of a planet with respect to the Sun, and so on. Casting a horoscope depends on the interpretation an astrologer places on the interrelationships between all these factors and can produce different predictions for any person or event.

Today horoscopes are printed daily in the press and are available on the Internet. There are astrology programs for personal computers and CD-ROMs with interactive tutorials. In a world in which modern science and technology so dominate our lives, it may seem surprising that astrology still commands such widespread belief.

Astronomer Carl SAGAN pointed out that astrology must satisfy a need that science does not and will remain while that need remains.

See also NEOASTROLOGY.

For further reading: Julia and Derek Parker, *The Compleat Astrologer* (Beazley, 1975); Percy Seymour, *Astrology: The Evidence of Science* (Lennard Publishing, 1988); and John Anthony West, *The Case for Astrology* (Viking Arkana, 1991).

ASTRONAUTS, ANCIENT A term coined by some writers in the 1950s, as the concept of flying saucers spread, to suggest that extraterrestrials had been visiting Earth since prehistoric times. It was claimed that visitors from outer space had left their mark in some of the massive constructions of the ancient world, were represented in ancient artwork, and were the origin of myths of encounters with supernatural beings. While advocated by popular flying-saucer writers such as George Hunt Williamson, M. K. Jessup, Brinsley le Poer Trench, and Desmond Leslie, it was not until the 1969 publication of *Chariots of the Gods?: Unsolved Mysteries of the Past* (English edition, 1970) by Swiss writer Erich VON DÄNIKEN that a significant international following for the idea developed. Von Däniken followed his initial effort with a series of additional books including *The Gold of the Gods* (1973), *In Search of Ancient Gods: My Pictorial Evidence for the Impossible* (1974), and *Miracles of the Gods: A Hard Look at the Supernatural* (1975). These books in turn spawned a number of books by other writers.

In 1973, Illinois attorney Gene Phillips founded the Ancient Astronaut Society to focus speculation on Von Däniken's ideas. It regularly holds an annual conference and sponsors trips around the world to sites mentioned in the ancient astronaut books so that members can personally observe the artifacts believed to be evidence of ancient astronauts.

Critics of the ancient astronaut theory emerged through the mid-1970s and largely discredited Von Däniken's theories. They pointed out that there were perfectly mundane and more satisfying explanations for the monuments and the artifacts he claimed were the work of ancient astronauts. Ufologists also moved to separate the idea of ancient astronauts from the UFO controversy. By the early 1980s, few advocates of the theory could be found, though Von Däniken and Zecharia Sitchin, the most scholarly of the ancient astronaut writers, have continued to publish.

See also ABDUCTIONS, ALIEN; FLYING SAUCERS, CRASHED; UFOLOGY.

For further reading: Ronald Story, *The Space Gods Revealed: A Close Look at the Theories of Erich Von Däniken* (G. P. Putnam's Sons, 1969).

ATLANTIS AND LEMURIA Legendary lands purported to be inhabited by races of people with extraordinary paranormal powers. Atlantis was thought to be a lost continent in the Atlantic Ocean and its people to have abilities far beyond our own, to be the originators of all our technological skills. Lemuria was believed to be a large island in the Indian Ocean on which the atmosphere was very dense and the water more fluid than elsewhere. Its people had no speech but communicated through telepathy. Belief in the past reality of these imaginary places is now part of the NEW AGE MOVEMENT. A measure of how seriously the New Agers treat the matter occurred recently in the Austrian courts when Judy Knight, a New Age MEDIUM, was awarded sole access to "Ramtha," a psychic channel to Atlantis, over the claims of her rival Judith Ravell.

New Agers acquired the notion of lost continents through many 19th-century texts, the principal being Madame BLAVATSKY's six-volume theosophical treatise *The Secret Doctrine,* published between 1888 and 1938. She claimed that her work was taken from ancient Indian mystical texts and was actually written in the almost unintelligible ancient language of the Atlantis people, which she deciphered. According to Blavatsky, Atlantis was the home of the "Fourth Root Race" and Lemuria the home of the "Third Root Race." The root race theory has no basis in science.

Most scholars credit the idea of a place called Atlantis to the Greek philosopher Plato (427–347 B.C.E.). In two of the Socratic dialogues recorded by Plato, *Timaeus and Critias,* Socrates tells his students to imagine a story that demonstrates the exemplary behavior of a perfect society when under duress. One of his students, Critias, sketched out a hypothetical barbarian society located in a place called Atlantis on an island in the Atlantic Ocean, beyond the Pillars of Hercules. These "pillars," today's Mount Acho (south) and the Rock of Gibraltar (north), marked the boundary of the classical world. The Critias dialogue showed how Athens, the perfect society, would conduct itself with dignity and fairness when faced with a decadent and depraved aggressive enemy such as Atlantis. In Critias's argument Athens was preserved because her people abided by all the rules of good conduct, and the people of Atlantis were defeated because, sapped by luxury and pomposity, they had become confused and were unable to do so. There was no such locality as Atlantis; it was just a Socratic postulate in order to make clear a moral point.

Throughout the Middle Ages and the Renaissance, Atlantis remained in the minds of classical scholars, but in the 19th century, writers began to mention it as a geographical location. In 1881, Ignatius Donnelly wrote a book called *Atlantis: The Antediluvian World.* Donnelly's thesis was that the continent of Atlantis, which was at one time in the Atlantic Ocean, was where humankind arose from barbarism to develop the world's first civilization. Its culture was said to be outstanding, with a religion that worshiped the Sun and with highly advanced scientific and technological knowledge. The story goes on to explain how colonizers spread all over the world and, because of their superior knowledge, became models for our gods. About 13,000 years ago there was a volcanic eruption and Atlantis was submerged. To support his theory, Donnelly cobbled together a great mass of material, showing similarities in the cultures of the ancient Egyptians and the Mexican Indians, for example: knowledge of embalming, the building of pyramids, and the use of a 365–day calendar.

According to the Latin poet Ovid (43 B.C.E.–18 C.E.), Lemuria was an ancient Roman religious festival to placate the malevolent ghosts of a group of people who, for one reason or another, were trapped on Earth. Some had died without a surviving family member to bury them; others had died prematurely. Added to these were executed criminals, victims of murder and other forms of violent death, including drowning. The spirits were appeased each year at the three-day festival of Lemuria, when black beans were tossed over shoulders and brass cymbals were banged to frighten the spirits away. According to legend the festival was inaugurated by Romulus after he murdered his brother Remus.

Notwithstanding that Lemuria is a religious festival, Madame Blavatsky firmly believed that it was a geographical location and that the Third Root Race flourished there. She claimed Lemurians were apelike giants who lived in caves, were unable to reason and calculate, and lived by instinct. They had no speech but communicated by telepathy and had the ability to lift enormous weights through the power of their will. Col. James Churchward, a British man who served in India, wrote about Lemuria in the early 20th century. He called the lost continent Mu and claimed that his information was based on a collection of tablets written in the ancient Mu script. He gained access to the tablets through a friendship with a temple priest who knew of them in a monastery in India. With much difficulty and the aid of his friend, he finally deciphered the tablets and at the age of seventy began to write his first book *The Continent of Mu* (1926). Others followed.

Geologists agree that no great continental sinkings such as those involved in the Atlantis and Lemuria myths occurred during human existence.

See also ROOT RACES.

For further reading: Martin Gardner, *Atlantis and Lemuria, Fads and Fallacies in the Name of Science* (Dover, 1957); and L. Sprague de Camp, *Lost Continents: The Atlantis Theme in History, Science and Literature* (Dover, 1970).

ATOMISM The theory that the universe consists of tiny, invisible particles that cannot be destroyed. In Classical Athens (461–429 B.C.E.), the idea of atomism was part of the great philosophical debate over the nature of the matter that makes up the universe. The followers of Pythagoras, a philosopher of the previous century, declared that the essence of matter lies in the abstract concept of numbers, not in its substance. The Pythagoreans believed that, by understanding numbers, they could understand the nature of the universe. Opposed to the Pythagoreans were the materialists, who believed that the universe was essentially solid and material. Among the most radical materialists were the atomists.

The two principal exponents of atomism were Leucippus and Democritus. Both of them believed that the universe consisted of tiny particles that could not be seen or felt but whose existence could be deduced through the exercise of reason. The atomists declared that these tiny atomic particles are the ultimate basis of matter. There are an infinite number of these atoms, according to the atomists, and they are identical, indestructible, and indivisible. The only difference between a man and a rock in the atomist viewpoint is in the number and arrangement of their atoms.

AURAS Colorful but invisible fluctuating biomagnetic or psychic energy fields said to emanate from and surround all living things. Auras are perceptible only to those with the gift of CLAIRVOYANCE or with special equipment.

Through their colors and intensity, the auras supposedly indicate the mental and physical state of the being emitting them. The colors range through the rainbow's spectrum, from rosy red to orange, yellow, green, blue, and purple; as emotions change, so do the colors. Different colors seem to be related to different kinds of emotions; for example, red is associated with strong passions, whether anger or lust, while green is associated with calm. In general, the more intense the color, the stronger the emotions being experienced. As one's health changes, so do the intensities and sizes of the auras; generally, the further the auras extend from the physical body, the stronger the subject's health. If colors and sizes are weak, small, and dark, the subject is usually considered to be in failing health, with the possibility of death being imminent. Some psychics claim to be able to diagnose specific kinds of physical or mental illness through their analysis of the auras.

Some believers distinguish between several different types of auras. Early in this century, Walter J. Kilner, a British physician, divided auras into three classes: the etheric double (a narrow, dark band of energy immediately next to the body from which it emanates), the inner aura (next to the etheric double), and the outer aura. Others divide auras into five categories, each of which reflects a different part of the being emitting them: the etheric or health aura, the astral or emotional aura, the Karmic or intellectual aura, the character aura, and the spiritual aura. Different clairvoyants claim to be able to see some or all of these.

Auras have been described for many centuries—some suggest that the halos artists depict circling the heads of saints represent auras—but there is no scientific evidence for their existence. Some scientists or metaphysicians have attempted to prove their existence through experiments with different kinds of viewing media and with psychic experiments. Kirlian photographs are perhaps the most convincing evidence to date; yet most scientists suggest that these photographs and other kinds of evidence are merely reflections or depictions of the electromagnetic energy emitted from all living things. In other words, they view these colorful emanations as purely physical energy, with no mystical or psychic aspects attached.

See also ASTRAL/ETHERIC BODY; KIRLIAN PHOTOGRAPHY.

AURORA ISLANDS Nonexistent archipelago of islands in the seas surrounding Antarctica, reported and charted by early explorers. According to several historians of exploration, 18th-century explorers discovered several islands to the southeast of the Falkland Islands in the South Atlantic Ocean. The islands were first sighted in 1762. In 1774, they were named the Aurora Islands after the ship that found them again. Two Spanish ships, the *San Miguel* and the *Atrevida*, saw them in the 1790s, described them, and charted their location, complete with latitude and longitude.

By the early 19th century, however, the islands had disappeared. In 1820, an expedition led by Captain James Weddell searched for them without success. The Aurora Islands remained on many maps until 1870, but no ship ever sighted them again.

AUTOMATIC WRITING Writing accomplished without the writer's effort or consciousness, sometimes with the aid of an instrument such as a PLANCHETTE. Often the writer is in a trance state; sometimes he or she has simply "emptied" his or her mind in order to be receptive to messages that allegedly come from the spirit world, with the writer acting as a tabula rasa MEDIUM or channel.

Automatic writing was especially popular during the Spiritualist movement of the 19th century. It featured heavily in many séances and other meetings between client and medium. Generally the medium would be the automatic writer, although in some instances the medium encouraged the client to allow himself or herself to receive "automatic" messages.

According to their authors, entire books have been received through automatic writing. An example is a mid-19th-century book "written" by Samuel Adams through the mediumship of Joseph D. Siles. The entire text flowed from Siles's hand in the shaky handwriting of Adams's late years. The famous Seth books, by Jane Roberts, written in the 1960s and 1970s, started out as automatic writing and then became automatic speaking—Seth "spoke" through Roberts, whose husband recorded the words as they came from her mouth. Roberts claimed that while she was writing a poem in 1963, she suddenly felt as though she lost consciousness and had been taken over by another entity;

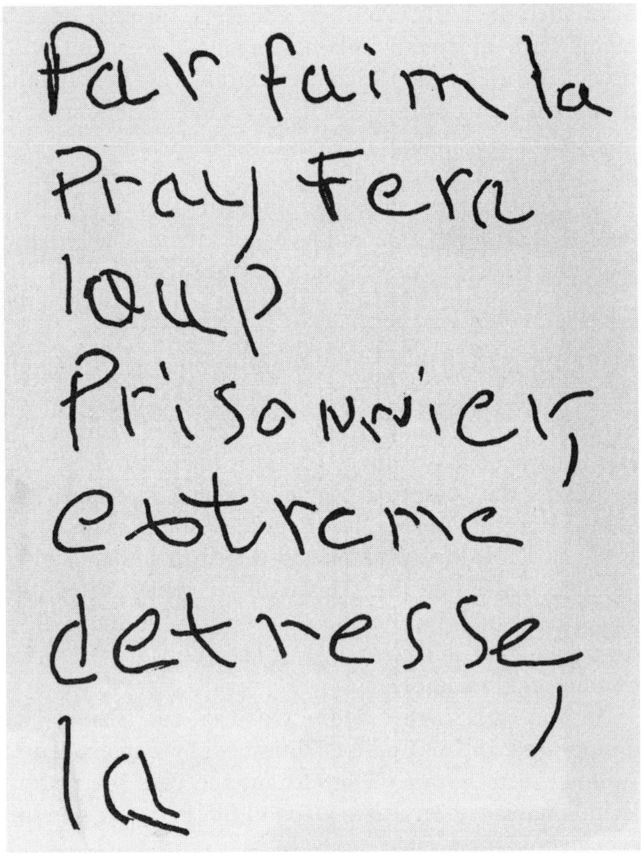

Automatic Writing

she was no longer writing her own words but the words of a being she later called Seth. Soon after, Roberts made direct efforts to contact Seth through a OUIJA BOARD and to capture his messages. (Use of the Quija board is itself a form of automatic writing: The participants hope to receive messages from the spirit world through words spelled out on the board by the planchette which, as it moves around the board, is touched by one or more of the participants.) Eventually, Seth would enter Roberts's body and speak directly from her mouth, in a deep, accented male voice.

Remarkable material has materialized through automatic writing; so has a lot of nonsense and fraud. Possible explanations for nonfraudulent automatic writing include TELEPATHY and the tapping of the subject's unconscious; in other words, the messages come from living minds, not from the spirit world. Many psychologists would opt for the unconscious-mind explanation, suggesting that all kinds of things, whether learned or imagined, can be dredged up by blanking out conscious thought and letting the words flow onto paper. In fact, some therapists use this technique today to aid reluctant or inhibited patients in opening up.

See also CHANNELING; SPIRITUALISM.

AUTO-SUGGESTION See COUÉ, EMILE.

AVEBURY A small village in Wiltshire, southwest of England; the site of the Avebury Circle, the largest ancient stone circle in the world. The circle was built around 2000 B.C.E. at about the same time or perhaps a few years earlier than STONEHENGE, a smaller but better-preserved monument of a different type that is located 17 miles to the south.

The monument originally had several components. The main circle consisted of 98 very large stones surrounding two smaller circles of about 30 smaller stones each. Unlike Stonehenge, the stones retained their natural shape. There are some clues that there might have been a third small 30-stone circle. All stones lie within a huge circular ditch and bank, 330 meters (1,400 feet) in diameter; the bank is about 6 meters (20 feet) high and the ditch 9 meters (30 feet) deep, with a 5 meter (15 foot) ledge between bank and ditch. The bank was originally higher and steeper, and the ditch was about twice as deep. The construction of bank and ditch would have taken more than a million human-years of labor. There are four entrances through the bank and ditch, roughly southeast, southwest, northwest, and northeast. The large circle is just inside the ditch. Most of its stones (weighing up to 50 tons) and almost all the stones of the smaller circles have long since been broken up and used in the buildings of the village, which lies inside the circle. Concrete posts now mark where stones once stood. The inner circles, each about 90 meters (300 feet) in diameter and 15 meters (50 feet) apart, contained smaller stone structures. Two lines of stones 24 meters (80 feet) apart, link Avebury with another structure, the Sanctuary, a mile away. The Sanctuary was demolished in the 18th century, and the only traces we now have of it are the holes that once held the stones.

Much of what is known about Avebury is owed to the antiquarian John Aubrey, who in 1684 recognized its existence and made a first rough record and William STUKELEY, who in the 18th century made more detailed and very careful measurements and drawings before too much demolition had taken place. For Aubrey and Stukeley and for us today, the site poses some interesting questions.

There has been much speculation during the years about how such a massive undertaking could have been mounted 4,000 years ago and what was the motive. In the medieval period, before Aubrey realized that it was a human-made structure, local residents believed it was the work of the devil—every 25 years they held a ceremony in which one of the great standing stones was thrown down and smashed. Both Aubrey and Stukeley favored a religious purpose and associated it with the Druids. That seems unlikely because the dates do not match; there is no record of the Druids before the third century B.C.E.

Archaeologists today accept that we do not know the answers to these questions and probably never will.

AYURVEDIC MEDICINE In Sanskrit the "science of life," an ancient Indian medical system, holistic and preventative, of supposedly divine origin. Ayurveda is based on the collected works of various authors, most of which date back to 1500 B.C.E. While some works are relatively more recent and their authors known, many of the older scripts' authors are either mythical or anonymous.

These classic ayurvedic volumes, some of which are still used as textbooks at Indian universities today, contain extensive case histories, detailed surgical records and illustrations, philosophical wisdom, and information on herbalism, anatomy, hygiene, causation, pharmacology, toxicology, and disease therapy. The works of ayurveda represent humankind's first and perhaps most comprehensive set of medical sourcebooks.

Because of its deep connection with Hinduism (in legend, ayurveda was given by Indra, the king of the gods, to the vedic sage Atreya) ayurveda has a highly philosophical nature. Ayurveda is often unconcerned with the clearly manifested illness, seeking rather to restore what it perceives as a body's balance of both spiritual and physical elements. Ayurveda dictates that, to prevent future illness while eliminating the present malady, one must create and preserve a sort of philosophical and somatic homeostasis.

Ayurveda defines 25 basic elements, including psyche, ego, intelligence and even ur-matter and nonmatter, but most predominant are the five elements of fire, earth, air, water, and ether. These elements (which are not to be confused with the ancient Greek elements) in their varying permutations and combinations, make up the triform body-typing system of *Tridosha.* Tridosha states that human beings can be categorized by their dominant elements, or *doshas. Vata* types, those with an affinity for the elements ether and air, are wispy and elfin. *Pitta* types, those associated with fire, are balanced and energetic, though prone to overexertion and its physical symptoms. *Kapha* types are of water and earth and are often thick-bodied, determined, and reliable.

Avebury

Much of ayurveda addresses elemental imbalances among dosha-types with culinary solutions: Foods that correspond to certain elements may inhibit or stimulate the functions of tissues, organs, and systems that are also elementally related. Massage is encouraged as a weekly activity, as is oil bathing, sweat treatment, and regulated sexual and athletic activity. Various emetic and purgative treatments are also prescribed from time to time to "flush out" systems. In combination with the other aforementioned ayurvedic practices, they purportedly aid elemental balance and flow.

While critics question ayurveda's validity, they cannot doubt its popularity. Ayurveda has gained popularity in the West because of its veneer of "ancient wisdom." To some the practice has been sufficiently proven after several millennia of allegedly successful healing in India. Ayurveda is connected with Hinduism and New Age spiritualism. In this way, its disdain of modern medicine (and subsequent celebration of intuition and "self-awareness") is in harmony with popular mistrust of established health care, which has been mounting for more than a half-century.

For further reading: Hans H. Rhyner, *Ayurveda: The Gentle Health System* (Sterling Publishing Co., 1994); and Bernard Seeman, *Man Against Pain* (Chilton Books, 1962).

B

BABSON, ROGER (c. 1880–1955) A successful businessman who founded the Gravity Research Foundation in 1948. The foundation's principal interest is to find some way of nullifying gravity, either by developing some material that cuts off gravity (in the way that a closed metal container will screen the contents from electric fields) or by creating a machine that is not affected by gravity. It also acts as a center for scientists working on gravity. Through a series of Babson Reports, the foundation advanced several hypotheses about gravity and its effects: methods of harnessing the power of gravity; chairs and pills to aid blood circulation; effects on behavior and thinking caused by small variations in gravity resulting from the relative positions of Sun, Moon, and Earth; the effect of a person's weight on his or her temperament; inclining the floors of buildings to enable gravity to clear bad air. In 1940 Babson was the Prohibition Party's candidate for president.

See also ANTIGRAVITY MACHINES.

BACH FLOWER REMEDIES Homeopathic technique developed in the 1930s by Dr. Edward Bach (pronounced *batch*) to treat specific emotional ailments or "imbalances." Bach, a British bacteriologist and physician, became disillusioned with the side effects of drug therapy and turned to HOMEOPATHY, a type of healing using natural substances and minuscule amounts of illness-causing substances. Bach sought a remedy for emotional imbalance, which he believed caused physical illness. A combination of serependipitous observation of animals licking dew from flowers and subsequent research led him to develop 38 flower remedies—very mild flower infusions created by floating flowers in water and exposing them to sunlight—which, he asserted, used plants' energies to bring balance to human ones. He believed that it is not so much the biochemical makeup of the flowers that effects cures—it is their psychic energy.

Bach's remedies are used alone or in combination to alleviate certain kinds of fear, anxiety, loneliness, and other kinds of emotions that lead to stress, illness, and pain. They are used to treat both humans and animals. Since Bach, several others have developed additional flower essences that are used in emotional and physical healing; these became particularly popular in the 1970s and continue to be used today by those who believe in homeopathic remedies.

A 1979 study of the Bach remedies found a distinct pattern of well-being among the flower users that was not shared by the placebo users. Nevertheless, most medical doctors and other scientists doubt the efficacy of Bach remedies.

For further reading: Edward Bach, *Heal Thyself* (Fowler, 1931).

BACKSTER, CLEVE (1924–) Polygraph expert who advocated the idea that plants were sensitive to human thoughts. Backster moved from his work with the Central Intelligence Agency to become director of a Polygraph Institute and then founder of the Cleve Backster School of Lie Detection in New York City. In the late 1960s he conducted a series of experiments to determine whether plants reacted to the destruction of living cells. Backster was building upon research conducted by Indian scientist Sir Jagadis Chandra Bose early in this century and more recent research that suggested that plants react to music and light. Backster raised the issue of the paranormal nature of plant sensitivity. In a paper published in the *International Journal of Parapsychology* in 1968, he concluded that he had demonstrated the existence of a primary perception (ESP) in plant life.

Based upon his early positive results, Backster received a large grant to found the Backster Research Foundation to continue to investigate plant sensitivity. During the early 1970s, there was some hope among Backster's supporters that such research would provide a basis for affirming a "form of instantaneous communication among all living things that transcends the [presently known] physical laws."

While Backster's initial data were impressive, they were criticized as methodologically flawed. His results could not be duplicated, and various mundane explanations were offered to account for his original findings. By the early 1980s, research had proved a dead-end and was discontinued.

BACON, FRANCIS (1561–1626) English statesman and philosopher of science. At a time when the religious fanaticism of the Counter Reformation had driven modern science from its birthplace of Italy, Bacon became its founding father in England, while at the same time achieving the highest political office under King James I. Not strictly a scientist himself, Bacon instead provided the method and inspiration of modern science, sponsoring in particular the inductive method.

Bacon's father, Sir Nicholas Bacon, was lord keeper of the great seal of England under Elizabeth I. At the age of 12, Francis entered Trinity College, Cambridge, and later attracted the attention of the queen because of his intellectual precocity. Sir Nicholas died in 1579, and the eighteen-year-old Francis, being the youngest son, found himself penniless. He turned to the study of law, a career offering much to a poor youth with royal connections, and by the age of twenty-three he had already found a seat in the House of Commons. Near to being appointed attorney general to the queen, Bacon criticized her taxation policy in Parliament, and was denied the appointment. From this he concluded that sincerity in politics was unprofitable.

With the death of Elizabeth and the succession of James I, Bacon was made solicitor general, then attorney general, and soon afterward, lord keeper of the great seal. At the age of 57 he was named lord chancellor, ennobled as a baron, then advanced to a viscountcy. At the age of 60, however, he was indicted for taking a bribe from a litigant, found guilty, and was ruined, having all his offices under the court stripped away in disgrace. He spent his last years writing books and working out schemes for the advancement of science.

Bacon published books at intervals throughout his life. His celebrated *Essays*, largely covering moral themes, were first published in 1597 when Bacon was 36. He published the final edition in 1625. His philosophical works were published after he had achieved high office under James I. They include *The Advancement of Learning* (1605), *Cogitata et Visa* (1607), *De Sapienta Veterum* (1609), the *Novum Organum* (1620), *De Augmentis Scientiarum* (1623), and *The New Atlantis* (1627).

The new science envisioned by Bacon was not just an academic and intellectual tool to help humanity understand nature. Its purpose was to lend humanity mastery over nature. To Bacon, the promotion of science rested upon its being organized on a large scale and lavishly financed. Bacon's statesmanship and political theory were in accord with this approach; he believed in the necessity of a large, modernized, and centralized nation-state and a powerful, dominant monarchy. Bacon hoped that by proving himself a wise and successful statesman, he might persuade the king to subsidize the scientific institutions he proposed to establish. He did not, however, succeed in winning the sympathies of James I, a reactionary monarch uninspired by progressive dreams. Bacon died leaving behind him colossal debt. He had lived extravagantly, excusing this as well as his ambition and corruption, with the claim that he needed splendor in order to have power, and that he needed power in order to do good in the world.

See also BACONIANISM.

BACONIANISM A philosophy of science, based on the writings of Francis BACON, rejecting natural philosophy and sensory knowledge in favor of an empirical and inductive approach to understanding the universe and nature. Fighting traditional medieval philosophy and the Renaissance cult of rediscovered classics, Bacon argued that the metaphysical philosophers of his time had made no progress since antiquity, and in fact knew less than their Greek predecessors. A new philosophy for the new world demanded that men should amass and judiciously interpret data and conduct experiments, with the goal of mastering nature by learning its secrets through planned and organized observation.

This new philosophy, which Bacon called "the Great Instauration," was presented in six parts: a thorough classification of the existing sciences; a new inductive logic as the paramount means of interpreting nature; collection of empirical data and conducting of experiments; a series of examples illustrating the successful working of the new method; a list of generalizations that could be derived inductively from the study of nature; and the new philosophy presented as a complete science of nature. Thus Bacon, having no real scientific background, laid out for the world not so much an account of what he himself had done, as an outline of what others ought to do.

Bacon identified certain tendencies of thought that must be corrected if knowledge were to be advanced. He divided these false notions or "Idols" into four distinct types: "Idols of the Tribe," including sensory perception and emotion; "Idols of the Cave," referring to the individual's idiosyncratic experience and means of interpretation; "Idols of the Market Place," which speaks to the errors arising from conversation and the fallibility of words; and "Idols of the Theatre"—those false impressions left by the various dogmas of old philosophical systems.

Bacon is perhaps most famous as the founding father of the modern "scientific method." His method was laid out in what he called the three "Tables of Investigation." The first, called the rule of presence or the "Table of Affirmation," consisted of assembling for study all known instances of a phenomenon. People must further study the absence of phenomena, captured in the "Table of Negation." The third table, the "Table of Comparison," entails study of variations in different phenomena to see if there were any correlation in observed changes. With this rudimentary system, Bacon introduced the notion of science as a systematic study, as well as a way of separating pseudoscientific beliefs from the realm of science.

BALL LIGHTNING A mysterious natural electrical phenomenon that reportedly occurs in the form of a burst of electrical energy resembling a fireball. The first modern study of ball lightning was written by Russian scientist G. W. Richman, who himself was killed while studying the phenomenon in 1754. Ball lightning is described as a sphere of light that varies in size from a few inches to several feet. It hovers or moves about in the air, free from any physical conductor (a characteristic that distinguishes it from St. Elmo's Fire). Ball lightning ranges widely in color and length of occurrence. Frequently an occurrence ends in an explosion.

Explanations of ball lightning vary widely. Some doubt the existence of the phenomenon, reducing the reports to hallucination or misidentified mundane phenomena. Others have moved to more esoteric explanations such as antimatter, a hypothesis suggested by E. T. F. Ashley and C. Whitehead in a 1971 article in *Nature*. It is difficult to agree on an explanation for the phenomenon for two reasons: First, an occurrence of ball lightning is a short-lived and unpredictable phenomenon; second, it is difficult to assign a single explanation to a phenomenon whose manifestations show such radical differences in color, length, and method of demise (exploding or gradually fading away).

Because ball lightning is an enigmatic aerial phenomenon, ufologists have called attention to the similarity of some elements of the discussions of both phenomena. Arguments about the existence or nonexistence of ball lightning are much the same as those used to argue the existence or nonexistence of UFOs. The data suggesting the existence and describing the nature of both consists largely of anecdotal accounts of sightings, presenting varying and inconsistent evidence. Both have been confused with more familiar phenomena.

See also UFOLOGY.

BAQUET A large circular tub used in the early days of MESMERISM. It was developed around 1780 by a friend of Franz Anton MESMER. Inside the baquet were some bottles partially submerged in water. The tub was covered with a lid through which holes had been punched. Iron rods ran from the interior of the tub through the holes to the perimeter; here patients who wished to receive a dose of the "magnetic fluid" generated by the baquet could grasp them. When in operation, a group of patients would sit around the tub while a piano played. Patients sitting adjacent to the baquet frequently experienced symptoms that included convulsions, uncontrolled laughter, and vomiting. These were attributed to the initial action of the fluid on the body and considered a sign that the healing process had begun.

In 1784 the Royal Society of Medicine determined to examine the phenomenon of mesmerism and the occurrences around the baquet; the commission, headed by visiting American Benjamin Franklin, denied the existence of any magic fluid.

The commission's findings prompted Michael A. Thouret to develop the idea of ANIMAL MAGNETISM. Thouret argued that mesmeric phenomena were not caused by any ferro-magnetism (hence denying any relation to the common magnet or to the iron rods of the baquet) but to power inherent in living beings. Thus Mesmer's student, the Marquis de Puységur, who began his work in 1785, replaced the baquet with a tree that he had magnetized. Patients were connected to the tree with ropes rather than iron rods. Mesmerism was brought to England in 1785 when Dr. Bell set up a baquet in his healing center.

BARING-GOULD, SABINE (1834–1924) One of the greatest of the Victorian eccentrics and polymaths. He was a devout clergyman, a far-flung traveler, a collector of folk songs, an architect, an artist, an archaeologist, and a renowned novelist and poet. He is best remembered for his famous hymn "Onward, Christian Soldiers," but he also wrote more than 100 books, including a 16-volume work entitled *Lives of the Saints,* 30 novels, several collections of travel writing, and an opera libretto. Baring-Gould was always intrigued by ancient and lost things, ranging from Norse sagas to wilderness preservation. During the 1890s, he led a group of amateur archaeologists in an excavation of local MEGALITHS and proposed that they had been erected during the Bronze Age, 5,000 years before.

However, Baring-Gould was also a great collector of myths, legends, and fairy tales. Some of his contempo-

Baring-Gould

raries believed that he was a practicing magician with an interest in the occult. In one of his works, he reported that he had himself seen FAIRIES when he was only four years old, laughing and climbing all over his parents' carriage. Another work, *The Book of Were-Wolves*, reports legends of LYCANTHROPY that he discovered during his frequent trips to the continent. His writings, especially his autobiographies *Early Reminiscences* and *Further Reminiscences* contain much digressive information on local folklore and MAGIC, including uses for the mandrake root and HERBAL MEDICINE.

BARRETT, SIR WILLIAM FLETCHER (1844–1925) The first person to make a significant scientific examination of DOWSING. A professor of physics in the Royal College of Science, Dublin, Barrett published his theories in *The Divining Rod* (1926). He suggested that clairvoyance by the dowser stimulated an involuntary muscular reaction that caused the dowsing rod to turn. Barrett was involved in PSYCHIC RESEARCH and his explanation hovered between the physical and the psychic. On the one hand, he thought that the dowser's involuntary reaction that flipped the rod might be the result of auto-suggestion that arises from cues picked up from the surroundings. On the other hand, he suggested that some kind of clairvoyance might be involved, the dowser receiving signals from the hidden water.

Although Barrett's basic theory of the movement of the dowsing rod is today generally accepted, his own ambivalence on the subject is mirrored in the two prevailing views. The skeptics hold that the explanation lies in the dowser picking up clues that trigger the involuntary muscular action; the other party will not accept that, maintaining that the dowser receives some paranormal information that triggers the muscular action.

For further reading: William F. Barrett, *The Divining Rod* (University Books, 1967); and Evon Vogt and Ray Hyman, *Water Witching U.S.A.*, 2d ed. (University of Chicago Press, 1979).

Barrett

BATES METHOD See SIGHT WITHOUT GLASSES.

BEAUTY TREATMENTS Activities and processes based on chemistry and hucksterism that form the basis of one of the most profitable businesses in the Western industrialized world.

The ideal market, briefly defined, is one that is never saturated and that will bear inflated prices. (One is tempted to add that the product must also be unnecessary in any practical sense.) Many consumables—paper, photographic film and paper, seeds, and so forth—satisfy the first of those conditions but not the second (or third). Supplies for beauty treatments satisfy all.

One of the ways in which this situation has been achieved and maintained is by exploiting the social authority and some outward appearances of science—a classic case of SCIENTISM. The area inside the main entrance of most major department stores is filled with cosmetics and other beauty products. The assistants at the counters often wear white coats of clinical appearance and/or are equipped with computers or other scientific-looking gadgets to calculate the "right" materials for the customer's skin type, hair, lips, eyes, and so on. All this supposedly backs up the implicit assertion that the substances on sale will produce the desired effect: remove wrinkles, moisturize, promote healthy hair—in short, render the user more beautiful and younger looking.

The jargon employed is seductive and generally impressive to the customer: vitamins, provitamins, lipids, antioxidants, balanced pH, time-release formulas, moisturizers, and the like. To underline the "scientific" character of their business, cosmetics firms have sites on the Internet. Few of their claims, if any, stand up to serious examination; few of their products have any real effect. Most cost less than their elegant containers, but year after year they sell for higher and higher prices. The vendors rely on the assumption by the customer that the costlier the product, the better and more effective it must be.

Fortunately for us, the human skin is a very good barrier to invasion from outside; it is our first line of defense. That being so, the chances of any applied substance breaching that barrier are small, in most cases zero. There is a risk, though—by no means negligible—that substances applied will interfere with the skin's natural function, blocking the pores, destroying surface layers—doing more harm than good.

Shampoos are an interesting case in point. Human hair consists of filaments of keratin, effectively dead material once extruded from the scalp. One poll found that hair's appearance is the most important beauty concern of men and is second only to wrinkles for women. Satisfying those concerns, largely generated by advertising, generates an extremely lucrative market for the cosmetic and chemical industry. Internationally—and most of the firms are international—that translates into profits of thousands of millions of dollars per year. The principal ingredients of the shampoos that produce this huge income are the same as the basic ingredients of washing-up liquid and are very cheap. The head, like any other part of the body, needs to be kept clean, both the hair and the scalp. Occasional washing in soap and hot water, followed by thorough rinsing, is sufficient. Washing removes the natural sebaceous secretions that give the hair a slightly shiny appearance and leaves it looking dull for a day or two. Pomades, brilliantines, and oils were used for many years to correct this dullness, but for several decades now manufacturers have come up with another solution: conditioners. "Conditioners give body and life to hair." What they actually do is coat the hair with a very thin layer of silicone, which advertisements have persuaded us looks "healthy." But that alone, despite the high prices, would not be sufficiently profitable, so advertisements further persuade us that the process of shampooing plus conditioning is so good for the hair that we should do it more often—ideally every day, applying shampoo twice each time. This actually damages the scalp's natural oil production and leans more heavily on conditioning. Then some clever chemist found a way of incorporating the conditioner with the shampoo but delaying its action until the rinse: "Wash and Go." Add to all this dyeing, bleaching, perming—all creating problems which require costly solutions.

Beauty treatments for both men and women have a long history, going back to the ancient Egyptians at least. What is special about the present is the huge size both of the industry, the extent to which it has coopted science and technology, and the manner in which it is producing specialized ranges of cosmetics and hair treatments for different ethnic groups.

For further reading: Rita Freedman, *Beauty Bound* (Lexington/Columbus, 1988).

BERGSON, HENRI-LOUIS (1859–1941) French philosopher and metaphysician. Henri Bergson's concepts of how the human mind works influenced later philosophers and metaphysicians such as Pierre TEILHARD DE CHARDIN, Jacques Maritain, and William JAMES. Bergson began his teaching career at a school in Clermont-Ferrand in 1883 and in 1900 moved to the Collège de France, where he continued to teach until he retired in 1921. He served as a diplomat to the United States during World War I and worked as head of the League of Nations Committee for Intellectual Cooperation (the precursor to UNESCO) during the 1920s. He was awarded the Nobel Prize in literature in 1927.

Bergson's theories stressed the importance of "experienced duration" in human history and affairs. *Time and Free Will,* his 1889 doctoral thesis, set forth the concept that history consists of individual experiences through time that cannot be measured accurately. This idea was in opposition to another major theory of the day that humanity and history were subject to understandable mathematical or mechanical rules. Bergson's *Matter and Memory* (1896) and *An Introduction to Metaphysics* (1903) both drew on his concept of "experienced duration" to suggest ways that human minds and bodies interacted and to stress the importance of intuition—understanding how "experienced duration" works. Bergson's essentially dualistic philosophy held that man perceives matter through the intellect as divisible and measurable; however, through intuition and memory, time *(la durée réelle)* is apprehended as indivisible.

In 1907 Bergson made his greatest impact on the scientific world with the publication of *Creative Evolution.* The book rejected Darwinian natural selection, concluding that biological life is too dynamic and creative for natural selection to have been the primary deciding influence on EVOLUTION. Modern scientists have rejected Bergson's theories in favor of natural selection.

BERINGER, JOHANN (1677–1738) A science professor who was the victim of a famous hoax. Beringer believed that the Earth was only a few thousand years old and that the biblical account of the Flood was a real event. In the early 18th century, fossils were considered to be creatures made extinct by the Flood or the castoffs of God's experiments. Beringer became the victim of an elaborate hoax that had been developed to destroy his reputation: His students and two of his colleagues—Ignatz Roderich, a mathematician, and Georg von Elkhart, a librarian—fashioned strange objects with unlikely animal forms (a bird with a fish's head, for example) and presented them to him as fossil finds. When he accepted them as genuine, his tormentors produced even more unlikely forms—stones with emblems and Hebrew letters, and even the name "Jehovah" engraved on them. In 1726 Beringer published a scholarly book in Latin, *Lithographia Wirceburgensis,* painstakingly listing, illustrating, and explaining all these finds. Soon after the book was published, he found the most peculiar stone of all—one with his own name on it.

An inquiry was held and one of Beringer's assistants confessed. The peculiar stones had been carved by Roderich and Elkhart who persuaded Beringer's student fossil hunters to bury them in local quarries and then rediscover them. Initially, Beringer refused to believe his colleagues' confession, but when he realized that he had been fooled, he sued the perpetrators. In an effort to save his reputation, Beringer went bankrupt buying up copies of his book and burning them. Nevertheless, some copies survived; the book became a classic of the discovered-hoax genre, with a new edition published in Germany in 1767. In 1963, the University of California Press issued a translation: *The Lying Stones of Dr. Johann Bartholomew Jim Beringer.*

BERLITZ, CHARLES (1914–) Vice-president of the famous Berlitz Schools of Language (founded by his grandfather) and popular advocate of a variety of controversial claims concerning the existence of extraordinary phenomena including the BERMUDA TRIANGLE and the PHILADELPHIA EXPERIMENT.

Born November 23, 1914, Berlitz graduated from Yale in 1936 and became an accomplished linguist, speaking more than 30 languages. He served as a captain in counterintelligence during World War II.

His initial foray into the paranormal came with the publication of *The Mystery of Atlantis* (1969), a popular survey of the "lost-continent" topic then in the news because of Edgar CAYCE's predictions that Atlantis would rise.

In 1974 he released one of his most controversial titles, *The Bermuda Triangle,* in which he argued for the existence of a genuine mystery area located in the ocean between Puerto Rico, the tip of Florida, and the island of Bermuda. Berlitz's book climaxed two decades of growing fascination with the Bermuda Triangle. Interest dissipated the following year after publication of Lawrence Kusche's *The Bermuda Triangle Mystery—Solved,* which offered a comprehensive and satisfying solution to the "mystery." Berlitz attempted to answer Kusche and other critics in *Without a Trace* (1977) but found little support.

Berlitz then delved into UFOs and apocalypticism. In 1978 he returned to a story he had mentioned briefly in *The Bermuda Triangle. The Philadelphia Experiment* (1979) tells of a supposed World War II experiment testing Einstein's unified-field theory. In this experiment a ship disappeared from a dock in Philadelphia and reappeared a few minutes later in Norfolk, Virginia. It then disappeared from Virginia and reappeared in Philadelphia. Berlitz's book became a best-seller and was made into a movie. Quite apart from the several excellent reasons for not believing that the experiment was ever attempted, much less had succeeded, the story now appears to have been a complete hoax.

BERMUDA TRIANGLE Name coined by writer Vincent H. GADDIS in 1964 for an area of the Caribbean Sea and Atlantic Ocean where hundreds of unexplained disappearances of ships and planes allegedly have occurred. The area extends roughly between the three points of Florida, Bermuda, and Puerto Rico. Different accounts extend the mysterious area as far east as the Azores Islands, as far north as New York, as far west as Mexico's Yucatán Peninsula, and as far south as the coast of Venezuela. Disappearances here have been reported at least since the late 19th century through the present day, although the heyday of Bermuda Triangle vanishings was undoubtedly the 1940s through the 1960s.

Perhaps one of the most famous incidents in the Triangle was the December 1945 disappearance of Flight 19. Five U.S. Navy bombers and their 14 crew members took off from their Ft. Lauderdale, Florida, base for routine target practice. About the time they were to return to base, the flight commander radioed that his instruments were not reading properly and that the bombers could see no sign of land. In fact, they reported that nothing, including the sea, looked as it should. The flight was never heard from again. Despite extensive searches, no sign was found of the planes or the men aboard. A second aircraft, sent out to search for the missing planes, also disappeared.

This disappearance was typical of many that have been reported. Commonly, instruments that are known to operate properly become erratic; skilled navigators, sailors, and pilots lose their bearings and—if they manage to get radio messages to bases—report unusual conditions such as bizarre cloud formations and strangely roiled water. If the flights do not return, no signs of survivors or debris are found. It is probably this last factor that has lent so much credence to the legend of the Bermuda Triangle. Most often, if a ship or a plane is lost at sea, some trace is found, even if it is only an oil slick or a few timbers. But in this region, it is not uncommon for no sign whatsoever of the lost vessel or its crew to turn up.

A wide range of explanations has been offered for the alleged mysteries of the Bermuda Triangle. These range from the bizarre (giant sea monsters, extraterrestrials, and time/space warps) to the vagaries of nature (sudden storms, fireballs, deepwater eddies, underwater natural-gas explosions, and windshear) to simple human error, equipment breakdown, and incompetence. In 1975, a representative of Lloyd's of London wrote to *FATE MAGAZINE* that their statistics did not support the idea that more unexplained disappearances occur here than in any other area of the world with similar conditions. The U.S. Coast Guard, which has thoroughly investigated many of the disappearances, asserts that environmental factors can account for the majority of them.

There are other areas in the world's oceans that also bear notorious reputations. Probably the next most famous is the DEVIL'S SEA, an area off the east coast of Japan. Like the Bermuda Triangle, an unusual number of air and sea vessels is said to disappear there. Again, as with the Bermuda Triangle, authorities prefer the natural explanations of weather, human error, and equipment malfunction—if indeed an unusual number of disappearances does occur there.

For further reading: Charles Berlitz, *The Bermuda Triangle* (Doubleday, 1974); Lawrence Kusche, *The Bermuda Triangle Mystery—Solved* (Prometheus Books, 1986).

BERNARD, RAYMOND (1890–?) French metaphysician and mystic. Raymond Bernard was one of the modern leaders of the Ancient and Mystical Order Rosae Crucis (AMORC), or the Rosicrucian Order. Bernard was the author of a series of four books based on Rosicrucian philosophy: *Strange Encounters, The Secret Houses of the Rose and Croix, A Secret Meeting in Rome,* and *The Invisible Empire.* The first two volumes and the fourth were translated into English in 1981. Bernard had been a member of AMORC's inner circle in France but resigned and created a new organization called the *Cercle International de Recherche Culturelles et Spirituelles* (CIRCES). While AMORC was aimed at increasing the spiritual knowledge and abilities of individuals, CIRCES was intended to provide a means for spiritual researchers to work as a group. The original group of CIRCES researchers were members of AMORC—a number of them were members who had risen in the AMORC hierarchy. CIRCES later broke with AMORC over questions of precedence. The two organizations now work separately on their different goals.

BESANT, ANNIE (1847–1933) The supporter of politically radical causes—atheism, the National Secular Society, BIRTH CONTROL, the Fabian Society—who later espoused mystical religious cults. In the 1890s Annie Besant came under the influence of MADAME BLAVATSKY and in 1897 joined the Theosophical Society, in which she became a very active leader. She was elected president of the society in 1907, a post she held until her death. Besant spent much of her time at the society's international base in India and became a supporter of Indian independence.

THEOSOPHY is essentially a mixture of SPIRITUALISM and REINCARNATION, maintaining that humanity has evolved through a series of ROOT RACES to its present, fifth, stage. Besant had access, via a male medium, to guidance by spirits, "Hidden Masters and Brothers"—in particular, Masters Hoomi, Morya, Rishi Agastya, the Lord of the World, and Lord Maitreya—who steered her through the tortuous and often bitter politics of the international society.

With C. W. Leadbetter, Besant coauthored books on chemistry and the history and geology of the world, in which their guiding spirits dictated the texts. *Occult Chemistry* explains the chemical behavior and structure of substances without regard to the conventional disciplines or to experimentation. Besant and Leadbetter discovered the structure of atoms and molecules while sitting on a bench in Finchley Road, London, England. When they needed to refer to actual materials, they would make astral visits to exhibits in museums.

For further reading: Peter Washington, *Madame Blavatsky's Baboon* (Secker and Warburg, 1993); Ann Taylor, *Annie Besant* (Oxford University Press, 1993).

BIBLE SCIENCE ASSOCIATION One of the most conservative organizations promoting CREATIONISM. The Bible Science Association (BSA) was founded in 1963 by Walter Lang, a pastor in the Lutheran Church–Missouri Synod. The organization still draws much of its support from that denomination. Basing its arguments on a literal interpretation of Genesis, the organization supports the idea of a young Earth (less than 10,000 years) and literal flood. In the mid 1960s, BSA started publishing articles arguing for GEOCENTRISM, the idea that the Earth is stationary and the solar system revolves around it. The organization has proved one of the most effective at mass distribution of creationist literature and the mobilizing of creationists to the cause. In the early 1990s, 27,000 people received the *Bible Science Newsletter,* the publication of the group.

The BSA publishes creationist literature for mass distribution as well as a set of reading books for children and youth. It sponsors a variety of seminars, tours, and meetings in addition to its biennial National Creation Conference.

BICAMERAL MIND The two chambers of the brain. The actual relationship between the brain and the mind has been for many centuries one of the most difficult problems of both philosophy and science, but recent research into how the two sides of the brain work in unison, or—after injury—separately, is beginning to throw some light on the subject.

The brain is divided into two hemispheres—the left and the right—and is connected to the nervous system in a crossover fashion: The left hemisphere controls the right side of the body, and the right hemisphere the left side. Because of the crossing over of the pathways, any damage or a stroke in the left hemisphere will affect the right side of the body and vice versa. In the mid-19th century, doctors learned from observation of brain injuries that visuospatial thought was mainly located in the right hemisphere and language-related capabilities mainly in the left. Because language and speech are so closely linked with thinking, reasoning, and the higher mental functions, the general view developed that the right half of the brain was the less-advanced part, with a lower level of capability than the other half. Thus the left half came to be called the major hemisphere and the right half the minor hemisphere.

It is only quite recently that a different, more complex view has started to emerge. This developed when neuroscientists began to study the function of the corpus callosum, a thick nerve cable composed of millions of fibers that cross-connect the two hemispheres. Through a series of animal studies in the 1950s, it was established that the corpus callosum provides a communication pathway between the two hemispheres. Nevertheless if the corpus callosum pathway was severed, the two brain halves were found to be able to function independently without any significant observable effect. During the 1960s a series of extended studies caused many scientists to revise their view of the relative capability of the two halves of the brain. Scientists now think that each hemisphere perceives reality in its own quite different way, neither side being superior to the other. Both hemispheres are involved in higher cognitive functioning, with each half of the brain specializing in different modes of thinking, both highly complex. The left half is convergent, rational, and analytic, and the right half divergent, intuitive, and imaginative. Subsequent findings over the last 30 years have substantiated this view. Our brains are two consciousnesses mediated and integrated by the connecting cable of nerve fibers between the hemispheres.

Early on, human beings must have felt that the body's right side or hand possessed different characteristics from the left. In Western languages, the right hand has been connected with what is good, moral and *right*—the Latin word is *dexter,* meaning "skill," and the French, *droit* meaning "good," "just," or "proper." The left hand has always been associated with otherness and feelings that are out of control—the Latin word for the left is *sinister,* the Anglo-Saxon, *lyft,* meaning "weak" or "worthless," and the French *gauche,* meaning "awkward." These pejoratives probably reflect the prejudices of a right-handed majority against a left-handed minority. At the beginnings of human thought, concepts of duality tended to bisect the world, but the primal division was between good and evil, right and wrong. The big step forward that has occurred over the last half-century is not to view the two sides of the bicameral mind evaluatively, but as two different approaches to the thinking process, each path adding a necessary dimension to the other.

Recent studies show it is important that educators help students use and develop both sides of their brains, especially during maturation when, it is thought, neurons are prone to regression and childrens' potential could remain unrealized. While teachers are aware of the importance of intuitive and creative thought, however, school systems still favor left-hemispheric modes of teaching. Emphasis in examinations is always on the literary and numerate skills of reading, writing, and arithmetic, with imaginative, perceptual, and spatial skills trailing behind.

A simplistic, unscientific version of how the hemispheres function has resulted in the fallacious notion that people can be simply designated as right-brain or left-brain types. Right-brain types are believed to be capable of entering into other or altered states of consciousness by way of mysticism, CLAIRVOYANCE, and SPIRITUALISM. Such people, the self-styled intuitives, believe themselves capable of making connections with minds from inhabitants of other planets. They "know" themselves to possess the truth because they believe they have a direct insight into immediate knowledge without recourse to any information or intellectual weighing of evidence.

See also DUALISMS.

For further reading: Richard L. Gregory, ed., *The Oxford Companion to the Mind* (Oxford University Press, 1987); Michael C. Corballis, *The Lopsided Ape: Evolution of the Generative Mind* (Oxford University Press, 1991); Tony Buzan, *The Mindmap Book* (BBC Books, 1993).

BIG BANG Theory of the creation of the universe. The big bang theory, as proposed by the late Belgian astrophysicist Georges Lemaitre (1894–1966), states that the universe came into being in a huge explosion sometime between 10 and 20 billion years ago. In the beginning, all the matter that currently exists or has ever existed was contained in a state called a *singularity,* a spot of tremendous density and tiny volume. When the singularity exploded, it threw out all the matter that makes up stars, galaxies, and planets—and their inhabitants. Scientists theorize that the new universe was so hot that the first simple atoms of hydrogen did not form until one million years after the initial event. Stars and galaxies followed about one billion years later, and the solar system in which we live did not form until the universe was about 10 billion years old.

Albert Einstein's theory of general relativity, published in 1915, provided the impetus for the big bang theory. After Lemaitre proposed it, the big bang theory was itself rejected as pseudoscience by some researchers. Scientists in the 19th and early 20th centuries believed that the universe was both infinitely large and infinitely old. They believed that, although individual stars and planets moved, the universe as a whole was motionless. Einstein's theory overturned this model, suggesting instead that the universe is expanding constantly, presumably from some initial event. Astronomical observations of distant galaxies have since proven him correct. The American astrophysicists Arno Penzias and Robert Wilson began measuring universal background radiation in 1965, looking for evidence of the intensely hot period of the initial explosion. Penzias and Wilson found that this radiation was almost completely uniform across the entire sky, and realized that it was the remnant of the initial big bang. They were awarded the Nobel Prize in 1978 for their work.

Although scientists are fairly certain that the universe was created by the explosion of some sort of singularity, they are unable to determine what might have caused it. If all the matter that ever existed were compressed into one singularity, the very nature of space and time itself would have been warped. Events predating the big bang—if any were possible without time or space to happen in—would not have had an effect on the universe as it exists today, and therefore there is no way of measuring them. Most scientists acknowledge this problem by saying that time itself began at the big bang. A number of theologians and religious leaders have declared the big bang theory consistent with religious belief about the origins of the universe, including some of those presented in the Bible.

BIGFOOT Huge, hairy hominid reportedly sighted throughout the North American continent. Some believe it to be a relative of the YETI, the ALMAS, and other wild, furcoated, humanlike creatures found in folklore and sightings around the world. Others think it is a creature of "monster envy," either consciously or unconsciously created by those who want America to have its own ABOMINABLE SNOWMAN. Although Bigfoot is most often thought of as inhabiting the Pacific Northwest, reports have come from all over the United States, Canada, and even Mexico. Some Pacific Northwest Native American tribes held a traditional belief in a *Sasquatch,* or "hairy man," and Bigfoot is often connected to this traditional being.

Those who have evaluated the several hundred Bigfoot reports that are made virtually every year say that the species clearly has two genders (at least one witness claims to have been abducted and held captive by a Big-

Bigfoot

foot family, with two parents and two children); that the males may be more than 8 feet tall and weigh up to 600 pounds; and that their feet are proportionally larger than humans' feet. The creatures also are said to have medium-to-long hair, usually described as reddish-brown and covering most of the body, to make unique sounds, and to have an unpleasant odor. They are considered shy and generally herbivorous and nonviolent, although a few attacks by Bigfoots have been reported.

Some sightings have been accompanied by physical evidence. This includes hair and feces, but the most common evidence is footprints, many of which have been captured in plaster by Bigfoot investigators. The prints vary widely, not only in the quality of their reproduction, but in their very details. Most footprints show five toes, but a few show six; some are structured very much like human feet, but some are more like those of an ape or even a bear; some have faint but distinct dermal ridges (like the whorls of fingerprints), but most do not; some clearly show the depth and textural variations that would be typical of a large creature walking, but some do not. To date, no mainstream scientist has fully accepted any of the existing Bigfoot prints as credible evidence of the creature's existence.

In 1967 a brief motion picture appeared of what was alleged to be a female Bigfoot walking into the woods near Bluff Creek, California. As the Bigfoot walks away, she turns and looks at the photographer, Bill Patterson, an actor, Bigfoot hunter, and amateur cameraman. Cryptozoologists (see CRYPTOZOOLOGY) and Bigfoot hunters were immensely excited about the film, but as is often the case, mainstream scientists tended to dismiss it. Even after three decades, the film remains controversial, but most scientists and even many cryptozoologists dismiss it. Science is not yet convinced that Bigfoot roams America's forests.

For further reading: Grover S. Krantz, *Big Footprints: A Scientific Inquiry into the Reality of Bigfoot* (Johnson Books, 1992); John Napier, *Bigfoot* (E.P. Dutton, 1973).

BILIOUS PILLS A medical compound that was first patented by Samuel Lee, Jr., of Windham, Connecticut. Lee received his patent for "Bilious Pills" in 1796. A mixture of gamboge, aloes, soap, and potassium nitrate, the pills were touted for their curative value for bilious (liver) complaints and yellow fevers, jaundice, dysentery, dropsy (edema), worms, and "female complaints," but found major use as a cure for dyspepsia (i.e., common indigestion). Three years later, another Samuel H. P. Lee, he of New London, Connecticut, received a patent for another formula, which he also marketed as Bilious Pills.

The first-mentioned Samuel Lee eventually found himself with even more competition. Bilious Pills were among the most popular of 19th-century NOSTRUMS. The pills seem to have enjoyed their success due to widespread overeating among the urban public. There was a popular idea that all food contained a universal element that undergirded the life process. As a result, quantity rather than quality was stressed. Only in the 20th century would knowledge of the different nutritive values of different foods and the need for a balanced diet be widely understood.

BINET, ALFRED (1857–1911) A French psychologist often held responsible for the frequent misuse of intelligence measurements and to that extent a pseudoscientist, but that is unjust. Binet originally intended to become a lawyer, but switched to medicine. Put off by the horrors of the operating theater, he became a psychologist. By the end of the 19th century, he had become France's leading experimental psychologist. His experience had convinced him that:

1. No aspect of human ability was immutable; while each person had certain innate abilities, he or she was still capable of learning.
2. The study of the individual was important; each person had many varied and complex aspects, which were ignored in statements about groups: men, women, whites, blacks, and so on.
3. A person's intellectual ability was multifaceted and could not be measured on a simple linear scale, by a number.

He experimented at home with his two young daughters, giving them a variety of tests and observing their responses—and, in particular, their differences. That reinforced his view of the necessity of "individual psychology," the name he gave to his subsequent activities. "Individual psychology," he wrote, "studies those properties of psychological processes which vary from one individual to another. It must determine those variable processes, and then study to what degree and how they vary across individuals."

In the late 1880s and early 1890s Binet identified several aspects of intellectual ability and how they developed as the subject grew from infant to child to adult. With his assistant Victor Henri, he set out 10 possible "faculties" that might comprise an individual's mental equipment, ability in which might vary appreciably both within the individual and between persons: memory, imagery, imagination, attention, comprehension, suggestibility, aesthetic sentiment and moral sentiment, muscular strength and willpower, motor ability, and hand-eye coordination. Binet and Henri hoped to formulate tests that would measure these various aspects of a person's intelligence in a reasonably short time, but they failed. Success in this area has eluded psychologists to this day.

In 1904 Binet turned his attention to a related problem: how to identify those children who are learning–

disabled and needed special education. He became a member of a government commission set up to deal with the problem. With his collaborator Theodore Simon, Binet produced a series of tests ranging from the very simple to the fairly complex, gradually modifying them in the light of experience with both retarded and normal children, yielding the first Binet-Simon Scale in 1905. The tests were psychological, depending as little as possible on learned knowledge but assuming some familiarity with commonplace French life and culture. In today's jargon they were culture specific, only suitable for French children. They discovered that, with both normal and subnormal children, performance depended on age, that there were overlaps and inconsistencies. However, they now had a scale that showed what the average child of a given age could do, and with which they could determine that a certain child was of average "intelligence" or was so many years behind (or ahead) of his or her age group.

Even at this stage Binet did not pretend to know what constituted "intelligence." He knew only that he and Simon had a method by which retarded French children could be classified more reliably and uniformly without the vagaries of an individual physician's judgment. In 1908 and in 1911 Binet and Simon published improved versions of the scale. What they set out to test were a range of those faculties listed above plus *judgment,* which they thought of prime importance. Binet knew that the tests and scale were not precise, that children performed differently at different times and under different circumstances, and that the assumptions were not always valid. Above all, there was considerable variability in different mental abilities and it was impossible to represent intelligence by a single number. Binet preferred to use mental level rather than mental age which, he felt, implied greater precision than was justified. It was not until a year after his death that psychologists in the United States divided a child's mental age by his actual age to give an intelligence quotient. They then replaced some of Binet's tests by others more suited to their purpose and maintained that the IQ thus obtained was fixed for life—all of which would have horrified Binet. Binet believed that individuals might always improve their intelligence, and in his last years devised exercises, "mental orthopedics," designed to improve intelligence. Some of these were incorporated into primary school curricula and into games for Cub Scouts and Brownies. Kipling used one of them as a training exercise for his young hero—Kim's Game.

See also INTELLIGENCE TESTS.

For further reading: Raymond E. Fancher, *The Intelligence Men: Makers of the IQ Controversy* (Norton, 1987); and Steven Jay Gould, *The Mismeasure of Man* (Norton, 1981).

BIOFEEDBACK The operation of a feedback loop in a biological system. In general, the term "feedback" defines a method of controlling any system by reinserting into it the results of its past performance—a feedback loop. This sort of feedback is common in biological systems, and no creature could survive without it. Through feedback humans, like animals, learn how to control their limbs and coordinate their hands and eyes, as well as solve problems.

The term is also used to describe a series of techniques aimed at providing us with information on the connection between psychological and physical processes that are not usually under our voluntary control. The lie detector is just one of many biofeedback machines developed to monitor changes in physical states. Many types of instruments can be used, but the simplest system includes a hand-held thermometer, an electric skin resistance gauge, and a display box. The technology does not harm the body; it just provides information.

Therapists also use these machines to allow patients to observe their bodily changes. The patient is then taught to control the body by calming the mind, using methods such as visualization, relaxation, deep breathing, or MEDITATION. Biofeedback equipment is also available for home use; users can teach themselves to relax and watch their body responding favorably. This process is called biofeedback training. It concentrates on turning the mind away from the tensions and strains of life, and controlling the body's physical responses—elevated blood pressure, increased heartbeat, muscular tension, and so forth. Doctors have found that in the cases of some patients with hypertension, biofeedback is so successful that they could be taken off drugs.

Medical studies indicate that biofeedback may become the preferred alternative to more invasive types of treatment. It is one of the few alternative therapies that is based on scientific principles. Most of the instrumentation was originally designed for use in scientific and medical research, and the effects it measures are accepted by doctors as accurate information on bodily processes. Biofeedback therapy has been shown to work. Patients suffering from high blood pressure have been able to bring it down to normal using the technique, and migraine sufferers have found welcome relief. Furthermore, biofeedback therapy has the added bonus of emphasizing the patient's role in determining the course of treatment and in taking an active part in it.

See also BIOMETER.

For further reading: *Readers Digest Family Guide to Alternative Medicine* (Readers Digest, 1991); and Elmer and Alyce Green, *Beyond Biofeedback* (Delacorte, 1977).

BIOMETER An instrument developed by 19th-century French psychical researcher Hyppolite Baraduc to measure paranormal psychokinetic forces coming from the human body. Baraduc's biometer consisted of a needle suspended by a thread. When a subject's hand was brought near the apparatus, the needle's movement supposedly indicated a variety of personal conditions—physical, mental, and even moral. The biometer was criticized for not dealing with possible movement due to air currents or the heat of the human body. Later varieties of similar instruments have tended to be enclosed in glass for that very purpose.

BIONS The energy vesicle that is supposedly the basic unit of life. It is smaller than a cell, a liquid-filled membrane that pulsates continuously with ORGONE ENERGY. The bion and orgone energy are both discoveries or inventions of Dr. Wilhelm REICH, a 20th-century Austrian physician and psychiatrist. He believed that bions propagate like bacteria, and he produced photomicrographs showing aggregates of bions in the process of transforming into protozoa, small single-celled organisms. Conventional biologists have not been able to reproduce these experiments and believe that Reich's material might have been contaminated with the organisms he claimed to see formed from bions.

BIORHYTHMS Part of a theory that suggests that human life is strongly affected by three biological cycles. The idea of such cycles developed in stages. Their existence was first proposed by the German Wilhelm FLIESS, a contemporary of Sigmund FREUD. He proposed two such cycles, and an Austrian engineer named Alfred Telscher suggested a third. Herman Swoboda then added to their idea the observation that the cycles began at birth. By tying them to the birth date, any person's present relationship to his or her cycles could be quickly and easily calculated. Although the idea of biorhythms existed since the turn of the century, it only became popular in the 1970s. By that time several instruments had been developed to calculate quickly and interpret individual biorhythm charts.

The biorhythm chart tracks the course of a 23-day cycle of physical strength; a 28-day cycle of emotional sensibility, intuition, and creative ability; and a 33-day cycle of mental activity, reasoning, and ambition. The chart indicated when people are at their best in any of the cycles. Practitioners believe that individuals are especially strong when two or three cycles reach their peak at approximately the same time.

In the 1970s, several attempts were made to verify the existence of the three cycles. No evidence was found, and their use was largely discarded by the end of the 1980s.

BIRTH CONTROL Methods adopted to prevent conception before, during, or after engaging in sexual intercourse. Today, the method by which birth control is achieved or attempted ranges from consumption of pills to utilization of chemical and physical barriers, including vasectomy and sterilization, to the method still advocated by the Roman Catholic Church of limiting intercourse to days of the month when the woman is not fertile. In the whole history of contraceptive methods, some smack of pseudoscience.

A principal source of information on the subject is the surviving work of writers who described the practices of their day. Literature on the subject goes back to the ancient Egyptians, Greeks, and Romans. Soranus of Ephesus, a Greek gynecologist practicing in Rome, advised coughing, jumping, and sneezing after intercourse to expel semen, and holding the breath—whether during or after intercourse is not clear. A method of douching with cold water after coitus is also ancient. The Romans advocated physical obstruction of sperm, using oil and soft wool; the Greeks favored vinegar. Samuel Pepys in his diaries refers to his use of crude condoms, less perhaps for contraceptive purposes than to prevent infection by venereal disease. Other writers refer to the use of condoms made of leather or animal gut, surely representing a triumph of hope over experience. On July 26, 1664 Pepys solicited and received as many as 10 strange precepts from a company of ladies with whom he was making merry, including avoiding hot suppers; drinking juice of sage and spiced ale and sugar; not hugging his wife too tightly; avoiding straitlacing; keeping the stomach warm and the back cool; wearing cool drawers, and making his bed higher at the foot than at the head.

With the passage of time, and improved understanding of human physiology, it might be expected better advice would be forthcoming, but in the 19th century, Frank Harris (1854–1931) in his *My Life and Loves* (1925) was still giving his many conquests just as strange advice, recommending douches and the rhythm method—which he had all wrong. Other 19th century writers recommended barriers, using various substances including alcohol, alum, baking soda, carbolic acid, green tea, hemlock, iodine, opium, prussic acid, quinine, sugar of lead, tannin, and zinc. None of these was affective although some claim might be made to the assay being scientific, since it was a form of testing, albeit with a very high error rate.

People continued to cleave to superstitious practices, such as tying knots at bridal ceremonies, eating dead bees, and wearing amulets. It would be comforting to think that such methods are a thing of the past, but they persist in both industrial and less-developed parts of the globe. Nowadays, as the processes of conception and birth are sufficiently understood, there is no excuse for resorting to pseudoscientific methods of birth control, physical or mystical.

See also COITUS RESERVATUS.

BLACK BOX Popular name for a diagnostic machine originally created by Albert ABRAMS. In the late 19th century, as a professor of pathology at Cooper Medical College in San Francisco, California, Abrams turned his attention to cancer. Out of his study, he conceived the notion that each disease had a special vibration, and he built a machine to measure that vibration. During the first decade of the new century, he began to use the "dynamizer," which allowed diagnosis from a sample of a patient's blood. This blood sample was placed inside the machine, and a healthy patient was then attached to the machine by a wire. Abrams would tap on the stomach of the person next to the machine, which would reportedly measure the vibrations along the spine. By reading the gauges on the dynamizer, he could render a diagnosis. Abrams termed his system "electronics;" it has more recently been called RADIONICS.

Building on his hypothesis, Abrams built the ERA (for "Electronic Reaction of Abrams") machine or "oscillocast." He maintained that it could not only diagnose but treat; it sent healthy vibrations back to the diseased patient. He leased oscillocasts to his colleagues. Toward the end of his career, Abrams proposed a diagnostic technique based on handwriting. From his examination of autographs, he concluded that a number of famous 19th-century figures had been suffering from syphilis.

Both the British Royal Society of Medicine and the American Medical Association (AMA) moved against Abrams. The AMA sent him a blood sample from a healthy guinea pig that he diagnosed as coming from a patient with cancer, a streptococcus infection, and sinus problems. The AMA also opened one of the black boxes, in which they found an ohmmeter, a rheostat, a condenser, and a magnetic interrupter but could discover no way that the machine could produce its claimed results.

The AMA's action effectively stopped Abrams's operation in the United States, and little was heard of the black box for some years. However, in the 1930s, the black box was picked up by Dr. Ruth B. DROWN in the United States and George De La Warr in England. De La Warr improved on Abrams's black box by eliminating the need for the person sitting next to the device. He substituted a rubber pad that he stroked with his fingers. He claimed that as the pad was stroked, the needle on the ohmmeter would tend to stick at various points indicative of diseased conditions. De La Warr found himself denounced by the medical community in spite of the enthusiastic support of people who felt they had been helped by his diagnostic and healing work.

Drown, a chiropractor, developed Abrams's techniques in the United States, and De La Warr even wrote two books supportive of Drown's techniques. The British government granted her a patent for her version of the black box, but in the 1940s the U.S. Food and Drug Administration declared it fraudulent. In 1950, following a request by Drown's supporters, the device was tested at the University of Chicago. The device failed, and the American Medical Association widely publicized the results. From this point on, anyone using Drown's techniques in the United States was liable to legal action for medical malpractice. Support for Drown continued, but actual diagnosis and treatment based on her ideas moved to the Caribbean, outside of the jurisdiction of the U.S. government.

Modern support for the black box acknowledges a "paranormal" element as integral to radionics, noting that the radiations being measured are similar to those felt by a dowser; hence the person operating the machine must have some paranormal powers. Unlike dowsing, however, which has often shown interesting successes in locating water in unexpected places, radionics has yet to document a record of accomplishment.

BLACK HOLES Stars so massive that gravity attracts and holds everything in the vicinity, including light; if no light can escape, they appear black. Their possible existence is suggested by relativity theory applied to cosmology. The argument runs: If a massive star, with its high gravity, continues to mop up surrounding matter, it will eventually become so large that its gravity will prevent anything from escaping. Relativity predicts that light is also affected by gravity. There has been experimental confirmation of this: Light from a star passing close to a planet is deflected slightly by gravitational attraction. Light too will be unable to escape from so large a star; we would be unable to see it either by emitted or reflected light. It would be a perfect absorber, a perfect "black body," totally invisible and only detectable if at all by its tendency to absorb any matter approaching. Such an object is known as a black hole. It is a theoretical construct, a hypothesis; there is no direct evidence of one's existence.

Such ideas that violate common sense, particularly when accepted by the scientific community, inevitably attract even stranger associations. For example, the theoretical physicist John TAYLOR in his book on the subject *Black Holes* (1973), after explaining the theory in some detail, goes on into flights of the imagination. He imagines black-hole energy driving spaceships. He sees beings, sucked into black holes, exiting into an alternative universe. Perhaps souls could travel via black holes into another existence—a sort of REINCARNATION. Several other writers have joined in these speculations and added more: white holes as the complement of black holes; the two joined by a wormhole—matter going in the black, through the wormhole, and out of the white (as antimatter)—and so on.

There is nothing wrong with indulging in speculation, providing it is seen as speculation, not given such authority as to mislead the nonscientific public. At best, one should base speculation on sound foundations. To quote Martin GARDNER from *Science, Good, Bad, and Bogus* (1989): "Today black holes are the fashionable playthings of clever astrophysicists. Tomorrow their models may collapse to take their place alongside Phlogiston and the epicycles of Ptolemy."

See also PHLOGISTON.

For further reading: Isaac Asimov, *The Collapsing Universe* (Walker 1977); Nigel Calder, *The Key to the Universe* (Viking 1977); John G. Taylor, *Black Holes: The End of the Universe?* (Random House, 1973); and Martin Gardner, *Science: Good, Bad, and Bogus* (Prometheus, 1989).

BLACKSTRAP MOLASSES One of Gayelord Hauser's wonder foods. Hauser (1895–1984) was born in Germany, emigrated to the United States in 1911, but returned to Europe soon afterwards, apparently with incurable tuberculosis of the hip. Acting on advice that he must eat only natural foods, he turned to a diet of fresh fruit and vegetables. His hip healed. In the early 1920s he returned to the United States, changing his name from Helmut Eugene Benjamin Gellert Hauser to Gayelord Hauser, and became an advocate of NATUROPATHY and NAPRAPATHY. Naprapathy is a system of dietary measures and treatment by manipulation of the connective tissues and adjoining structures. Convinced of the healing powers of natural foods and counseled by Benjamin Lust, sometimes described as the father of American naturopathy, Hauser advocated five wonder foods: skim milk, brewer's yeast, wheat germ, yogurt and blackstrap molasses.

Hauser has definitely influenced eating habits, and with the exception of blackstrap molasses, his wonder foods are still high on the list of health food devotees. However, the U.S. Food and Drug Administration has tried to curb unsubstantiated claims for these foods.

Hauser's inordinate claims for blackstrap molasses—the residue from sugar refining—are expressed in *Look Younger, Live Longer,* where he asserted it would help cure a number of afflictions, including baldness, insomnia, and low blood pressure, and would aid in digestion, restoration of hair, the postponement of aging, and much more.

Gayelord Hauser, who lived 89 years, promoted his ideas very successfully in books, on radio and television, and in the magazine *Gayelord Hauser's Diet Digest.* He was also partner in a firm producing his special foods and medicines.

See also FOOD FADS.

For further reading: Martin Gardner, *Fads and Fallacies* (Dover, 1957).

BLAVATSKY, MADAME (HELENA PETROVNA) (1831–1891) Cofounder of the Theosophical Society (1875). Born in Russia, Helena Petrovna Blavatsky led a colorful and controversial life. She has been called both a mystic saint and a charlatan. After a very short-lived marriage at age 17 to Nickifor Blavatsky, the adventurous Madame Blavatsky traveled the world for several years, eventually settling in New York City in 1873. By that time she had spent time in Egypt, where she had cofounded the occult-focused Société Spirite, which was closed down for fraud within two years.

Blavatsky

While living in New York, she met and became friends with journalist Colonel Henry Steel Olcott. Formerly skeptical, Olcott was completely converted by Madame Blavatsky's séances. The two remained together as colleagues and friends for the rest of her life. In 1875 they helped found the Theosophical Society, an organization dedicated to promoting Eastern studies, universal brotherhood, and occult wisdom. Although Blavatsky, Olcott, and many of their followers had had a strong interest in SPIRITUALISM, their new organization distanced itself from that movement, as well as from conventional scientific and religious philosophies. They moved the organization's headquarters to India for several years, where they became great celebrities and where they helped promote Indian cultural pride and the preservation and accessibility of Hindu works.

Madame Blavatsky seemed to possess remarkable psychic powers, materializing flowers, lost brooches, letters, and other things out of thin air and CHANNELING spirits of the dead. Her demonstrations were severely attacked by critics. Their criticisms were given fuel when letters allegedly written by her were sold by her housekeeper to a London newspaper. The letters detailed ways to make it appear that spirits and objects were being evoked. However, after an investigation the SOCIETY FOR PSYCHICAL RESEARCH (SPR) labeled her an "accomplished and interesting imposter," an accusation Madame Blavatsky, Colonel Olcott, and the Theosophists adamantly denied. Nevertheless, Madame Blavatsky and Colonel Olcott fled to Europe to put the disturbing controversy behind them. (Much later, the SPR's damning report was itself challenged and today remains controversial.)

Madame Blavatsky wrote several widely influential and allegedly spirit-inspired books, most notably *Isis Unveiled* (1877) and *The Secret Doctrine* (1888), both of which promoted her—and the Theosophical Society's—beliefs. These included a wide-ranging mixture of spiritualism, eastern religious beliefs (REINCARNATION, the evolution of the soul, and karma, for example), and science. While many mainstream religionists consider Madame Helena Blavatsky to have been a fake, thousands of Theosophists still regard her as an inspiring figure and her writings as guidelines for their faith. The Theosophical Society's headquarters today are located in Colombo, Sri Lanka. The organization has branches all over the world.

See also THEOSOPHY.

For further reading: Sylvia Cranston, *H.P.B.: The Extraordinary Life and Influence of Helena Blavatsky* (Putnam's, 1993); Paul K. Johnson, *The Masters Revealed* (State University Press of New York, 1994); and Marion Meade, *Madame Blavatsky: The Woman Behind the Myth* (Putnam's, 1980).

BLEEDING STATUES See WEEPING STATUES.

BOND, FREDERICK BLIGH (1864–1945) An architect and authority on medieval church construction and ancient woodwork. He claimed extrasensory perception while pursuing research on the ruined Abbey at Glastonbury in England. The abbey was thought to be the earliest Christian church in England with connections by legend to King Arthur and the Knights of the Round Table.

Bond was employed by the Church of England in the early years of the 20th century to stabilize the standing walls of the abbey and to help produce a guidebook showing the layout of the visible foundations. During this period he made a detailed study of the ancient historical records and gathered a great deal of data to understand better the construction and history of the church. Unknown to his employers, he began to write a book describing how he obtained site information by calling on the collective memory of spirits of dead monks and stonemasons who had worked on the abbey during their stay on earth. He also contacted historical figures, all with the aid of his close friend, retired navy captain, MEDIUM, and automatist, Captain John Allen Bartlett. The book *Gate to Remembrance: A Psychological Study* was published in 1918, much to the horror of the Church of England, which throughout the previous century had fought against what it perceived to be the evil influence of SPIRITUALISM.

In 1922, the trustees of the Glastonbury site severed their connection with Bond. Bond was a maverick in archaeology. He refused to follow the standard archeological procedures, causing objects to be misplaced. He was thought to have moved articles because they were not found where he said they should have been. To escape the controversies surrounding him, Bond moved to America, and did not return home until 1835.

Archaeologists today give Bond high praise for his work on the abbey at Glastonbury, but they ascribe his undoubted successes to his hard work in acquiring knowledge of medieval church construction, not to voices from the past or to spirit writing. Reading transcripts of Bond and Bartlett's AUTOMATIC WRITING shows that their medieval monks spoke and wrote in a mock ancient English tongue that would not have been known at the time.

See also PSYCHIC ARCHAEOLOGY.

For further reading: William W. Kenawell, *The Quest at Glastonbury: A Biographical Study of Frederick Bligh Bond* (Helix, 1965).

BOOK OF DYZAN The spirit source of Madame BLAVATSKY's book *The Secret Doctrine,* (1888). Acting as a MEDIUM she would consult her astral guides, the Masters, and translate sections of *The Book of Dyzan,* which she claimed was written in Senzar (a hitherto and still unknown language). The second volume of *The Secret Doctrine* is, in effect, the spirits' version of a theory of EVOLUTION, a sort of reply to Charles DARWIN (1809–82). According to the book, humankind is descended from spiritual beings on another planet—the moon—gradually taking physical form through a succession of ROOT RACES. The present phase in this development, human history, is just one step in the spirits' evolution through a series of rebirths moving from planet to planet.

See also BESANT, ANNIE; THEOSOPHY.

BORLEY RECTORY A "haunted" house in Essex, England, that became the focus of several PSYCHIC RESEARCH studies of GHOSTS. The house was constructed in 1863 by the Reverend Henry Bull and remained in the family until 1927. By that time the building had attained a reputation as haunted. In 1928, the Reverend G. E. Smith, frustrated at the unwillingness of church members to attend meetings at the rectory, decided to invite psychical researchers to study the site and dispel the rumors. The investigation was carried out by psychical researcher Harry Price.

Price attempted to gather accounts of unusual occurrences in the house. Members of the Bull family had reported seeing the apparition of a woman in a flowing black robe such as nuns would wear. There were numerous confirming reports of people seeing the nun over the years. Harry Bull had also reported seeing the apparition of a headless man—also confirmed by later reports. The reporter who wrote the original story on the house had himself heard footsteps in the empty rooms of the large rambling structure. Among the most spectacular occurrences were the numerous POLTERGEISTS.

Price spent 20 years studying the rectory case: He lived in the rectory for years, interviewed as many of the people who had lived there as he could find, and published two books on the subject. During his residency, a MEDIUM came into the house, telling a story of a French nun who had lived in a convent near Borley who had been murdered and interred in the grounds where subsequently the rectory was constructed. Then in February 1939 the house was gutted by fire. Four years after the fire, Price excavated the cellar and found a skull and jawbones, which he believed could have been parts of the nun's body.

Price died in 1948. In the early 1950s, the SOCIETY FOR PSYCHICAL RESEARCH asked Eric J. Dingwall, K. M. Gildney, and Trevor Hall to reappraise the Borley case and Price's work. Their report, published in 1956 concluded that Price's books contained many uncorroborated accounts, that he had exaggerated and distorted the testimony of the witnesses, and, most important, had manufactured evidence. The accusations against Price remained unanswered for more than a decade. In 1969, R. J. Hastings looked again at the Price documents and publications. He concluded that while Price may have drawn unwarranted conclusions, he had shown no dishonesty in his research and writings. He had left all of his documents and artifacts at the University of London, where they would be readily accessible for anyone to examine, hardly the act of a fraud.

The case of Borley Rectory thus remains important in the history of psychical research and will probably remain a topic of debate for parapsychologists for many years. It will continue on the list of unresolved cases as the rectory is no longer available for study and the records are decidedly inconclusive.

For further reading: Harry Price, *The End of Borley Rectory* (Harrap, 1946); Harry Price, *The Most Haunted House in England* (Longman Green, 1940).

BOSTON SOCIETY FOR PSYCHICAL RESEARCH (BSPR) A prominent psychical research organization of the 1920s and 1930s that grew out of a critical response to the MARGERY CONTROVERSY.

In opposition to claims made by the AMERICAN SOCIETY FOR PSYCHICAL RESEARCH (ASPR) that authenticated the psychic powers of Mina Crandon, who was referred to as "Margery," Walter Franklin Prince and his supporters withdrew from the ASPR and founded the Boston Society for Psychical Research. Among other founders were psychologist William McDougall, Lydia W. Allison, and Elwood Worcester. Worcester, a prominent Boston clergyman, had largely been responsible for reviving spiritual healing within the Episcopal Church. During the 15 years of its existence, the BSPR would create an impressive record of research and publication through its *Bulletin* series. Its most famous publications include Prince's volumes on Pearl Curran, the MEDIUM for the spirit entity "Patience Worth," and a collection of testimonies entitled *Noted Witnesses for Psychic Occurrences.* The Society also published Joseph Banks RHINE's initial volume of parapsychological research, *Extrasensory Perception,* in 1934.

The program of the society suffered greatly following the death of Prince in 1934. The BSPR and the ASPR were formally reunited in 1941.

BOUVET ISLAND One of several islands that appeared and reappeared in the navigation charts through the 19th century, Bouvet Island is named for Pierre des Loziers Bouvet, the pioneer explorer of the Antarctic, who reported seeing the island in 1738 and noted its location at approximately 1,500 miles southwest of the Cape of Good Hope. He described the isolated bit of land as being approximately five miles across and covered by a glacier. However, in both 1772 and 1775, when Captain James Cook returned to the area, he was unable to locate the island. It was not until 1808 that two vessels hunting seals came across what seemed to be Bouvet. In 1822, an American, Captain Benjamin Morrell, reported that he had landed on the island and killed a number of seals. Then, in 1825, not only did other ships land at Bouvet, but one of them, commanded by Captain George Norris, reported sighting another island northeast of Bouvet, which he named Thompson Island, and four smaller islands around it.

The area of the sea where Bouvet was supposedly located was outside the more popular sea lanes, and thus few ever visited Bouvet. In 1843, a British ship sent to the region with Bouvet on its schedule of stops failed to find any land. Two years later, to settle the matter finally, a second ship was sent, but again no land was sighted. It was agreed that Bouvet and the other islands must be phantoms. They were promptly removed from the sea charts. Thus the situation remained until Bouvet reappeared to the crews of the *Golden West* (1878), the *Delia Church* (1882), the *Francis Allen* (1893), and the *Valdivvia* (1898). The Captain of the *Francis Allen* even made several sketches of Thompson Island.

By World War I, Bouvet Island had returned to the charts and appears on maps of the area to this day. But of Thompson and its satellites, no sign has been found. The depth of the water suggests that they are not hidden below sea level. It is still unknown what the captain of the *Francis Allen* saw and sketched. The most common explanation has been a mirage or some other atmospheric anomaly.

For further reading: Vincent Gaddis, *Invisible Horizons: True Mysteries of the Sea* (Chilton Books, 1965).

BRAIN WASHING See MIND CONTROL.

BRENDAN, SAINT (484–578) An Irish monk who attempted to find Paradise by sailing west. He founded a community at Clontarf in western Ireland, served as its abbot, and worked as a missionary in Scotland and Wales. Some of his other exploits are recorded in a medieval tale, the *Navigatio Sancti Brendani Abbatis (Voyage of Saint Brendan the Abbot),* which tells of the monk's attempt to find Paradise on Earth. Together with 17 comrades, Brendan sailed in a leather boat called a curragh to a variety of fantastic places, including the Island of Sheep, the Paradise of Birds, the Island of Delights, and the edge of Hell, before finally working his way to the outskirts of Paradise itself. Brendan then returned to Ireland, where he rejoined his community and died among his brothers. The story of Brendan's holy island sparked the medieval imagination; there are more than 120 surviving manuscripts of the *Navigatio,* indicating its popularity, and cartographers continued to put Saint Brendan's Island on their maps until 1759.

The *Navigatio* is an example of Paradise literature, a subclass of religious and fantasy writing. Many works of Paradise literature feature pilgrims having amazing adventures with giants, talking birds, and dangerous animals. Brendan's voyage, however, may have more basis in truth than most examples of the genre. In 1976 Timothy Severin, a writer focusing on exploration, decided to test the seaworthiness of the type of boat Brendan is supposed to have used. Through the summers of 1976 and 1977, Severin and his crew sailed their curragh *Brendan* from the saint's traditional point of departure to the Faroe Islands north of Scotland, to Iceland, and along the Greenland coast to the island of Newfoundland off the Canadian coast. Severin's trip proved that medieval monks could have made a voyage such as the one described in the *Navigatio*. Severin further suggested that Irish monks may have settled Iceland, Greenland, and on the eastern shores of North America during the medieval period, but for most historians that idea remains speculation.

BRIDEY MURPHY A reputed person who lived in Belfast, Ireland, in the early 19th century and later reincarnated as Ruth Simmons (a pseudonym of Virginia Tighe). The story of Bridey Murphy became the subject of a best-selling book, *The Search for Bridey Murphy* (1956), which set off an immediate debate about REINCARNATION and raised the issue of the viability of HYPNOSIS and PAST LIFE REGRESSION as a means of obtaining otherwise inaccessible information.

The book chronicled Simmons's sessions with hypnotist Morey Bernstein who reportedly "regressed" her to a previous life where she assumed the persona of Bridey Murphy and described her life in a small Irish village. Simmons had never been to Ireland. Many interesting details were quickly verifiable, such as her passing reference to the tuppence, an obscure Irish coin in circulation between 1797 and 1850. The account set off a search for the information that could either verify or falsify the information given by the hypnotized girl.

In support of Bridey Murphy, researchers found that she gave a reasonably accurate picture of early 19th-century Ireland. More important, she offered some obscure facts—for example, she named two grocers and a rope company that had operated in Belfast—that were only verified by detailed research.

In the summer of 1956, *Chicago American* ran a series of articles dealing with Simmons's early life in Chicago that discounted her claim. It was revealed that she had lived across the street from a woman whose maiden name was Bridie Murphy and that places and people in her neighborhood had been woven into her hypnotic reports. The accounts presented in the newspaper articles provided strong evidence that the Bridey Murphy story was a product of cryptomnesia, forgotten memory. It has been the majority opinion that Simmons responded to Bernstein's commands and leading questions to create a narrative that drew together many fragments from her life and created a coherent story. That opinion is by no means unanimous, however, and some parapsychologists, especially those who still give consideration to survival research, feel that the Bridey Murphy case is still open.

For further reading: Morey Bernstein, *The Search for Bridey Murphy* (Doubleday 1956); M. V. Kline, ed., *The Scientific Report on "The Search for Bridey Murphy,"* (Viking Press, 1956).

BRINKLEY, JOHN ROMULUS (1885–1942) Surgeon known as "GOAT GLAND" Brinkley for his sex rejuvenation remedy. As a young man, Brinkley worked in the slaughterhouse, where he developed the idea that animal's sex organs could aid impotence. He obtained a doctor's degree from the Eclectic Medical University and later acquired a diploma in medicine and surgery from the Royal University in Pavia, Italy. He set up practice in Milford, Kansas, offering men whose sexual prowess was flagging a new life with transplanted sex glands from a goat. His advertisements ran under a slogan, "Just let me get your goat, and you'll be a Mr. Ram-What-Am for every lamb."

In 1930 Brinkley began to advertise his goat-gland remedy over the radio. His program brought him unwanted attention from Kansas Medical Association, which filed a complaint against Brinkley charging him with alcoholism, malpractice, and unprofessional conduct. The Kansas Medical Board revoked Brinkley's license to practice on the grounds that the operation he offered was medically useless, and the Federal Communications Commission refused to renew the license of his radio station. Brinkley responded by hiring licensed physicians to work with him, and the hospital he had established continued to perform the operation without Brinkley's personal involvement in the surgical procedure. He then moved his entire operation to Mexico, where he established a clinic and broadcast over one of Mexico's unregulated stations. He had made a number of political allies with gifts of free radio time, and called upon them in hours of crisis. Then, he countered all of his detractors by announcing a campaign for governor of Kansas. Unable to get on the ballot, he received enough write-in votes for the state political powers to recognize his widespread popularity.

During the succeeding years the American Medical Association's Morris FISHBEIN branded Brinkley a quack. Brinkley sued, but lost. The court ruled that he *was* in fact a quack. He was able to continue selling his surgical process, however, until 1941 when a U.S. agreement with Mexico led to the closing of the radio station.

BROWN MOUNTAIN LIGHTS Multicolored lights that exhibit small movements and have been sighted since 1913 near Morganton, North Carolina. They are a long-established example of the phenomenon of GHOST LIGHTS. The U.S. Geological Survey determined that the lights could be from distant car or locomotive headlights and/or brushfires. Another investigation in the 1970s by the Oak Ridge Isochronous Observation Network concluded that some—not all—of the lights might be refractions from locations beyond Brown Mountain.

BRUNONIANISM A cosmological system of thought, based on the magical and animistic ideas of 16th-century Italian philosopher Giordano Bruno. At a young age, Bruno entered the Dominican order in Naples. He was eventually accused of heresy and fled in 1576, traveling throughout Europe for the rest of his life.

Heavily influenced by Agrippa's textbook of Renaissance magic, *De Occulta Philosophia,* Bruno's teachings are full of magical images, incantations, and other occult procedures. Bruno envisaged a universe of infinite space and innumerable inhabited worlds, overseen not by a single god but by the "God in things," with a "profound magic" or psychic life of nature as the principle of movement. Although "against" mathematics, Bruno defended Copernican theory against popular sentiment on animistic and magical grounds. Describing Copernicus as "only a mathematician" who had little understanding of his own scientific discoveries, Bruno saw Copernican heliocentricity as confirmation of a return to an "Egyptian" philosophy of universal animation.

For Bruno, the imagination reflects the divine mind of the universe and is the sole cognitive power, thus sweeping away Aristotelian concepts of the divided psyche. Cultivation of imaginative powers is the technique for achieving the personality of a magus. This philosophy ultimately led to Bruno's demise. In 1591, Bruno went to Venice to teach the arts of memory but was arrested and put on trial. Transported to Rome for a second trial, he was imprisoned there for eight years, sentenced as a heretic and "Egyptian," and burned at the stake.

BUEREN, GODFRIED (n.d.) A 20th-century West German patent attorney who took the idea of the HOLLOW EARTH DOCTRINE (HOLTWELTLEHRE) and applied it, not to the Earth but to the Sun. According to him the very hot exterior of the Sun, the part we see, encloses a cool interior containing another sphere supporting vegetation and, presumably, other forms of life. It is, of course, hidden from us by the outer incandescent (6,000°C) shell; only occasionally do we get glimpses of the cool interior sphere through sunspots, short-lived holes in the shell. So convinced was he of the truth of his theory—it is not clear what evidence, if any, he had for it—that he offered a prize of 25,000 marks (about $6,000 then, the equivalent of about $100,000 today) to anyone who could disprove it. The German Astronomical Society took his offer seriously and demolished the theory very convincingly. Bueren, not surprisingly, did not accept the society's disproof and would not hand over the prize. The society took him to court, won their case, and obtained judgment for the prize plus interest and legal costs. The case evoked wide interest and was reported in an issue of *Time* magazine (February 23, 1953).

BURT, SIR CYRIL (1883–1971) One of the foremost educational psychologists of his generation. He served the London County Council for many years as their official psychiatrist, and he occupied the chair of psychology at University College, London, from 1932 until 1950. Burt was responsible for testing and interpreting the results of mental tests for thousands of schoolchildren in the greater London area. After his retirement from University College, he published several papers in which he tried to establish that intelligence was hereditary and not environmental. He did this by comparing the results of IQ TESTS taken by close relatives who were raised in different backgrounds.

Burt's most famous study compared the results of tests given to identical twins who had been raised apart. He claimed to have found 53 pairs of identical twins who were raised separately, and in each case the twins showed similar IQ scores. One's intelligence, according to Burt, was based almost entirely on who one's parents were. In 1969, psychologist Arthur Jensen used Burt's figures to try to prove that differences in intelligence between blacks and whites in America were the result of inherited differences, not of inequities in schooling.

After Burt's death, psychologists began to look more closely at his information. First, a Princeton psychologist named Leon Kamin discovered that Burt's figures were impossibly close to his desired conclusion. In 1976, a reporter for the *London Sunday Times* discovered that the two women who collected and processed Burt's data either never existed or could not have collaborated with Burt when he wrote the articles. In addition, Burt apparently falsified some of his figures and claimed to be the originator of "factor analysis," actually the work of another psychologist named Charles Spearman. These revelations, summed up in L. S. Hearnshaw's biography *Cyril Burt Psychologist,* destroyed Burt's reputation and dealt a severe blow to the hereditarian theory of intelligence.

For further reading: L. S. Hearnshaw, *Cyril Burt Psychologist* (Hodder & Stoughton, 1979); Stephen Jay Gould, *The Mismeasure of Man* (W.W. Norton, 1981).

C

CANALS OF MARS See MARS CANALS.

CARDIFF GIANT Famous American anthropological hoax. The Cardiff Giant was developed in 1869 by a cigar manufacturer named George Hull. A confirmed charlatan, Hull successfully presented a 10-foot, crudely carved gypsum statue—purportedly uncovered by well diggers in Cardiff, New York—as a man's petrified fossil. For several months in 1869 and 1870, the statue fascinated the public. P. T. Barnum, the circus impresario, went so far as to have a duplicate of Cardiff Man made.

Hull left a clear, easily traceable trail. He purchased a block of gypsum in Fort Dodge, Iowa, and shipped it to Chicago, where two artists shaped it into the likeness of a man. Hull also tried to age it artificially by roughening the surface with darning needles and sulphuric acid. He then moved it to Cardiff, where it was "discovered" in October 1869 in a stone quarry outside of town. In November, Hull and the respectable businessmen and politicians he had persuaded to become his partners carried the Cardiff Giant to Syracuse and placed it on public display. Hull maintained the fraud for several more months. Early in 1870, however, the sculptors who had carved the statue revealed the truth and ended the hoax.

The Cardiff Giant differed from other scientific frauds because it was quickly exposed as a counterfeit. No paleontologists or geologists ever accepted that an entire human body, including its soft flesh, could ossify into a solid gypsum statue. Although some scientists believed that the Cardiff Giant might have been a statue carved by an ancient race, most regarded it as a poorly constructed fake. The eminent paleontologist O. C. Marsh, for example, quickly labeled it "of very recent origin and a decided humbug." The Cardiff Giant is currently on display in the Farmers Museum in Cooperstown, New York.

CARNIVOROUS TREES Trees or other large plants capable of consuming animal tissues. There have been several accounts of cannibalistic trees. Dr. Carl Liche reported in the September 26, 1920, issue of the *American Weekly* that he had witnessed a large floral plant consume a young woman in 1878 in Madagascar. The young woman, whose fate was pictured in imaginative accompanying artwork, was supposed to have been a member of the Mkodos, a little-known but cruel tribe. In 1925 the same paper offered a second story of a carnivorous plant, this time a tree species on Mindanao, in the Philippines. In another article, W. C. Bryant claimed that he had wandered into a taboo area on the island only to have one of the cannibalistic trees reach out, its leaves making a hissing sound. His guide, who knew of the trees, knocked Bryant to the ground and out of reach of the potentially fatal leaves. Bryant's story was accompanied by a drawing of the tree. He would later note that the large number of bones shown around the tree were not really present and that he had moved toward the tree originally as a welcome shelter from the Sun.

Later surveys of the isolated sections of Madagascar and the Philippines disproved this account. The idea of a plant large enough to consume a human ultimately derives from the several plants, such as the Venus's–flytrap, which consume insects or even small pieces of meat placed on a leaf. However, the accounts of actual encounters with such plants are little more than tall tales related for their entertainment value.

CASTANEDA, CARLOS (1931?–) Author of a series of books about a Yaqui Native American sorcerer named Don Juan. Carlos Castaneda is supposed to have met Don Juan while doing research for his master's degree in cultural anthropology at the University of Southern California. The first volume in the series, *The Teachings of Don*

Carnivorous Trees

Juan (1968), was based on his master's thesis. Other volumes include *A Separate Reality: Further Conversations with Don Juan* (1971), and *Journey to Ixtlan* (1972). In them, Castaneda tells about his long apprenticeship to Don Juan. Through the use of hallucinogenic drugs and special training, Don Juan initiated Castaneda into different levels of consciousness in an effort to introduce him to an unseen reality. His training included introductions to practicing witches and sorcerers, who worked in an intangible world. Castaneda routinely portrays his own occasional blunders in his books, and reports Don Juan's blunt condemnations of his work.

Critics differ on whether Castaneda's works are primarily anthropological (chronicling an aspect of Native American spirituality), philosophical (suggesting new ways of viewing reality), or fictional. They raise questions about the status of anthropology as a science or pseudoscience. Some critics suggests that Don Juan is simply a creation of Castaneda's imagination. Others have detected elements of Oriental mysticism and European philosophy amid the Native American elements of Don Juan's teachings. *Horizon* reviewer Richard de Mille points out that Castaneda's notes sometimes contradict themselves. Other critics suggest that if Castaneda's work had in fact been fictional, he would have done a much more coherent job of presenting and organizing his experiences. Despite such doubts, most critics recognize and celebrate Castaneda's literary talent.

See also ALTERED STATES OF CONSCIOUSNESS.

CATASTROPHISM A geological theory that explained the creation of the physical features of Earth as the result of sudden, violent acts. Its major advocate during the 18th century was the French naturalist Baron Georges CUVIER. Cuvier was best known as a comparative anatomist, who reconstructed fossil species from their partial remains. Cuvier was able to prove, from his reconstructions, that some animals that had once existed had become extinct. At the same time, his examination of rock formations around Paris showed that there were distinct breaks in the rocks, representing a great physical change in Earth's history. Cuvier found that each of these breaks was associated with the extinction of some fossil species. He concluded that each of these extinctions was the result of a great catastrophe. New species were divinely created or migrated from unaffected areas after each catastrophe.

Cuvier's theory won support from religious thinkers because it seemed to support the biblical view of history. However, as the 19th century progressed, Cuvier's views were replaced by those of the Uniformitarianists—including the English theorists James Hutton and Sir Charles LYELL—who believed the Earth was so old that there was enough time for great changes to have happened gradually. The uniformitarians recognized that natural disasters, such as earthquakes and volcanic eruptions, occurred, but they believed that such catastrophes only affected small areas. By the late 19th century, catastrophism had largely been discarded in favor of the theory of gradual processes.

In the 20th century, however, a modern form of catastrophism became popular. During the 1950s the works of the Russian-born American Immanuel VELIKOVSKY suggested that the shape of the solar system had changed during a relatively recent period in the world history and that this change had had a catastrophic effect on life on Earth. In *Worlds in Collision* (1950), Velikovsky suggested that the planet Venus was originally a comet that had had several encounters with the Earth and Mars before settling into its present orbit, causing worldwide catastrophes—including the extinction of the dinosaurs.

Scientists quickly rejected Velikovsky's theories, but modern scientific thinking accepts the idea that catastrophic events may have played a significant role in the Earth's history. The U.S. geologist A. W. Grabau traced sudden changes in the oceans throughout history and prehistory. More recently, the Alvarezes, father and son, developed a theory that the mass extinction at the end of the Cretaceous era that killed off the last of the dinosaurs was the result of the collision of an asteroid or a comet with the Earth.

CATTLE MUTILATIONS Dead animals with missing organs found in mysterious circumstances. Beginning in 1973, in separate locations in Kansas and Minnesota, cattle ranchers reported the discovery of dead animals that apparently had had various body parts surgically removed. The cattle had commonly lost eyes, ears, lips, sex organs, rectum, and/or tail; often there was a seeming blood loss. Rumors circulated of Satanic cults and bizarre rituals. Some initial autopsies performed on the Kansas animals determined that the animals had died of natural causes. In spite of such prosaic findings, reports of similarly mutilated cattle bodies spread across the Western United States and Canada through the remainder of the decade.

In 1980 a documentary film by Linda Howe, *Strange Harvest,* brought together information about a number of the dead animals. It largely ignored the autopsy results and tied the phenomena to UFOs. Howe continued to cover reports of mutilations, and by the early 1990s proponents of the UFO theory suggested that some extraterrestrials had given the U.S. government advanced technology in return for permission to conduct biological experiments that included both cattle mutilations and the abduction of human beings.

The theory flew in the face of the available evidence. In 1980, the same year that *Strange Harvest* was released, a study of reported cattle mutilations in New Mexico found no evidence of mutilations. Animals had died of common range causes—disease, poisonous plants, and rattlesnakes. Parts removed from the carcass

Cattle Mutilations

were consistent with predator damage (i.e., the softest part of the visible flesh was removed), and under the microscope, teeth marks were clearly visible. Several years later, two journalists, Daniel Kagan and Ian Summers, conducted a thorough survey of the phenomenon and concluded that the picture of the West as being plagued by an alarming number of cattle mutilations was a myth. In spite of the lack of evidence of any widespread cattle mutilations and the loss of interest in reports by the authorities, the myth has continued to find advocates of the likes of Howe, who as late as 1989 wrote a vividly illustrated book championing the idea, but presenting no new evidence to contradict the earlier findings.

For further reading: Daniel Kagan and Ian Summers, *Mute Evidence* (Bantam Books, 1984).

CAYCE, EDGAR (1877–1945) Clairvoyant, psychic diagnostician, who was also known as The Sleeping Prophet. For a period of 43 years, Cayce performed some 15,000 diagnoses and prescribed treatments while in a sleeplike trance. His diagnoses were said to have been accurate in more than 90 percent of his cases. He was also able to see the human AURA, describe past lives for some clients, and predict future events. He is said to have foreseen the stock market crash of 1929 and the beginning and end of World War II. Cayce spoke of the resurgence of a "lost continent," ATLANTIS; forecast numerous storms, earthquakes, and volcanic eruptions; he also predicted a "free and God-fearing Russia," and the eventual democratization of China.

A man with no medical training and little formal education, Cayce began his "psychic diagnosis" at the age of 16. Working as a book salesman, he lost his voice and consequently his job. No doctor seemed able to help him so finally, with the aid of a hypnotist, Cayce prescribed his own treatment while in a hypnotic trance (see HYPNOSIS). He followed the treatment and his voice was restored. Successful diagnoses and treatment of his wife and son reportedly convinced Cayce further of his own mystical powers. He began to diagnose and recommend treatment for neighbors who requested it, charging little or no fee. Cayce apparently found he could visualize not only the health of subjects hundreds of miles away but also their surroundings. He was soon performing "readings" for people around the nation, often free of charge. Cayce was arrested twice: once for practicing medicine without a license and once for fortune-telling.

Cayce's readings on disease and curative measures have received more attention in recent times than they did during his lifetime. His work is under ongoing scrutiny at the Association for Research and Enlightenment at Virginia Beach. Numerous books detail Cayce's work, most notably Jess Stearn's *The Sleeping Prophet* (1967).

CELERY Leading ingredient in a host of 19th century patent medicines designed to treat what were vaguely described as nervous disorders. Numerous SARSAPARILLAS were already on the market when the first celery-based product, Paine's Celery Compound, was introduced in the 1870s. It was marketed by Wells & Richardson, a company formed in 1872 by a group of Civil War veterans, including General William Wells, who made his reputation at Gettysburg. The formula for the celery brew was purchased from a Burlington, Vermont, widow and soon began to sell far beyond expectations. Besides celery, the compound contained coca (cocaine), hops, and 21 percent alcohol. Bolstered by an 1883 pamphlet—"Great Things, What They Are, Where They Are"—Wells & Richardson soon promoted the compound as a natural and permanent cure for nervous diseases.

Though remaining the market leader, Paine's Celery Compound had competition by 1890. Sears & Roebuck had their own version called Celery Malt Compound, which was advertised as a "Recognized Nerve and Brain Medicine."

Muckraker Samuel Hopkins ADAMS attacked celery compounds in his famous 1905 series in *Colliers* on NOSTRUMS. Nevertheless, they were able to survive the regulations of the 1906 Pure Food and Drug Act. As with many medicinal compounds with high alcohol content, they did well during Prohibition (1919–33), but declined markedly through the 1930s and disappeared soon thereafter.

CELESTIAL BED A couch that had the magical property of improving the potency and fertility of couples who slept in it. Scottish doctor John Graham invented the celestial bed in London around 1780. To make his claim plausible, he added strange coils to the bed, constructed to provide small electric shocks. His clients used the bed in an atmosphere of erotic perfumes, music, and colored lights. Graham charged as much as 100 pounds per occasion (equivalent to many thousands of dollars today) and became quite successful. His accomplice in this dubious venture was Emma Lyon, later known as Lady Hamilton, the mistress of Lord Nelson. Skepticism grew, interest diminished, and his business collapsed. The bed was eventually sold at auction.

CELESTIAL SPHERES Ancient Greek astronomers' conception that the heavens are a series of crystal spheres and circles, centered on Earth, supporting the stars and planets. In the early fourth century B.C.E., Eudoxus of Cnidus, a pupil of Plato, suggested that each planet was attached to a separate sphere. Other philosophers believed that the planets were in fact small crystal buttons and that light shone through them, reflecting the eternal fire that existed outside the human world. These ideas were adopted and refined by Aristotle in the late fourth century B.C.E. Aristotle's ideas, in turn, formed the basis of many ancient and medieval cosmologies, including those of Apollonius of Perga (c. 262–190 B.C.E.), Hipparchus (190–120 B.C.E.), and Ptolemy (100–165 B.C.E.)

Aristotle's geocentric concept of the universe reigned unchallenged until the 16th century, although astronomers had noticed curious motions in the heavens that required refinements of his theory. At certain times, planets appeared to move backwards in their orbits. The movements of the Sun and the Moon were especially erratic. In 1547, a Polish monk named Nicolaus Copernicus proposed a heliocentric universe, centering on the Sun rather than on Earth. In 1607, the Austrian astronomer Johannes Kepler showed that planetary orbits were not true circles but were ellipses instead.

The concept of celestial spheres was discarded as Kepler's views spread. However, the term and some of the old measurements were adopted into a standardized coordinate system in modern astronomy. Like the ancient concept, the modern celestial sphere is centered on Earth. It rotates on the same axis as Earth, with celestial north and south located above the respective poles. The Sun's path along this celestial sphere is known as the ecliptic, and the points at which it crosses the celestial equator mark the spring and autumn equinoxes.

See also GEOCENTRISM; HELIOCENTRISM.

CENSUS OF HALLUCINATIONS An 1889 social survey inquiring into the frequency and nature of reported contacts with the dead. The census grew out of an early interest of the SOCIETY FOR PSYCHICAL RESEARCH (SPR) in GHOSTS and APPARITIONS. Among the first important works of the SPR was *Phantasms of the Living* (1886) by Edmond Gurney, Frederick William Henry MYERS, and Frank Podmore. It was built from the accounts of 702 people drawn from a random sample of 5,705 individuals. One major conclusion of the volume concerned the coincidence of apparitions and the death of the person seen in the apparition. Some responses to the book complained that the sample was too small. Thus, the SPR leadership decided to conduct a much larger survey.

The Census of Hallucinations canvased 17,000 individuals from whom 1,684 accounts of apparitions were received. The survey was compiled for publication by Eleanor Sedgwick, whose husband Henry Sedgwick chaired the committee in charge of the census. The results, which filled Volume 10 of the *Proceedings* of the SPR (1894), further confirmed the conclusions of *Phantasms of the Living*, and remain as one of the important accomplishments of PSYCHICAL RESEARCH. In 1899 French researcher Charles Richet conducted a similar survey with like results. The census did not address the objective nature of apparitions, an issue that continues to be debated to the present.

CENTER FOR SCIENTIFIC ANOMALIES RESEARCH A center to facilitate communication between scholars and researchers concerned with scientific inquiry into the anomalous and the paranormal. The center was founded in 1981 by Marcello Truzzi and Ron Westrum, both sociologists at Eastern Michigan University at Ypsilanti, Michigan. In the early 1970s Truzzi had published a newsletter, *The Zetetic*, which opened a discussion on the scientific appraisal of paranormal and anomalous phenomena. Six years later, he became one of the founding members of the COMMITTEE FOR THE SCIENTIFIC INVESTIGATION OF THE CLAIMS OF THE PARANORMAL, and *The Zetetic* began to function as the committee's periodical. However, in 1977, Truzzi left the committee, which he saw as biased against anomalies rather than centered upon an open-minded skepticism based on the best scholarship.

In 1978 Truzzi published a new periodical, the *Zetetic Scholar*, to continue the scientific study of the anomalous and paranormal. Three years later, he established the Center for Scientific Anomalies Research as an arena for the debate being nurtured by the new journal. Truzzi has suggested that the burden of proof for various claims for the paranormal and for anomalous phenomena rests with those supporting such claims and that the demand for proof must be equal to the extraordinary character of any phenomena advocated. While it was assumed that most claims would be denied, it was also assumed that such denial came after a period of discussion and that, on occasion, such phenomena might meet the standards of proof.

The *Zetetic Scholar* was issued through 1987, and while the center has formally remained in existence, it has become moribund. Truzzi and those associated with the center have continued to be active participants in the discussion on scientific anomalies.

CHAMBERLAIN, HOUSTON STEWART (1855–1927) Pioneering theorist for what became nazi racial beliefs in the 1930s. Though born in Britain and educated in an English boarding school, Chamberlain was influenced by a German tutor whom he met in 1870. The tutor communicated a love of the fatherland, and Chamberlain eventually married Eva Wagner, daughter of German composer Richard Wagner, and settled in Switzerland to pursue his university studies.

In 1899 he published *Foundations of the Nineteenth Century*. In it he presented a racial theory of history based upon two basic notions: first, that humanity is divided into several distinct races, each having its own physical structure and mental and moral capacity; and second, that history is best understood as a struggle between these different races. Chamberlain marked historical epochs by the coming to prominence of a dominant racial type. He saw European history in the late 19th century as the product of six important influences from previous epochs: Greek philosophy and art; Roman law and organizational skill; the revelation of Christ; racial chaos following the fall of the Roman Empire; the negative influence of the Jews; and the creative and regenerative mission of the Teutonic (or Aryan) race. Modern European civilization was built on foundations laid by the Germanic people. Chamberlain's ideas also drew upon traditional anti-Semitic ideas which he pushed to new extremes, rejecting the notion that Jesus was Jewish.

To some extent, Chamberlain's notions reflected the general racism of the era in which he lived, ideas which permeated European colonialism and anti-black and anti-Asian legislation in the United States. However, he promoted these ideas into a universal vision of northern European superiority that provided some of the ideological underpinnings upon which Hitler ultimately built his program of racial annihilation.

See also NAZI RACISM.

For further reading: Geoffrey G. Field, *Evangelist of Race: The Germanic Vision of Houston Stewart Chamberlain* (Columbia University Press, 1981).

CHAMP LAKE MONSTER allegedly living in Lake Champlain. The Iroquois Indians called the monster *chaoussarou*. The Iroquois believed it had mystical qualities, including the ability to put observers into a hypnotic trance. Their stories, formed long before white settlers came to the region, told of herds of *chaoussarous* frolicking in the lake. In modern times only a single beast—or, rarely, two—has been seen at one time. According to some sources, the first European to sight Champ was the explorer Samuel de Champlain. Other sources, however, say that Champlain's report was of nothing more than a large garfish.

Champ's description has changed over time, but it is generally described as a long-necked, thick-bodied, serpentine creature. Many sightings have been reported in the 20th century, including an impressive group sighting in 1945 by the passengers and crew of the S.S. *Ticonderoga*.

In 1977 Sandra Mansi took a photo of the monster, which has been a source of controversy ever since. Fearing ridicule, Mansi kept the photo to herself for three years; then she sent it to Joseph W. Zarzynski, a teacher and cryptozoologist. Mansi's photo appeared to show a snake-like neck and head and humped back moving away from the camera. Zarzynski championed the photo, believing it to be one of the best existing photos of Champ or any other lake monster. Not everyone agrees, however. Some skeptics continue to believe that the photo is a hoax or an illusion. If the photo is genuine, it shows a creature similar in many respects to the popular perception of the LOCH NESS MONSTER, which also is the subject of controversial photographs. The creatures bear a strong resemblance to the supposedly extinct plesiosaur or possibly a zeuglodon. As with many elusive "crypto-animals," no concrete evidence, beyond the photos, exists. No one has ever killed or found the carcass of a Champ–like animal, and no spoor has been found.

For further reading: Peter Costello, *In Search of Lake Monsters* (Coward, McCann & Geoghegan, 1974); Joseph W. Zarzynski, *Champ: Beyond the Legend* (Bannister Publications, 1984).

CHANNELING The transmitting of information and advice through a human being (referred to as the channel) to someone in this world from a spirit guide in another world. Channeling is claimed to be distinct from SPIRITUALISM, communication from spirits of a deceased person through a MEDIUM, in that the channel contacts a wise unembodied spirit entity, who may be a deceased person but is usually an evolved entity such as an ascended master, to be given teaching of a philosophical kind. Each channel only has contact with one, or perhaps two, spirit entities to whom he or she turns for guidance.

It is claimed that anyone can learn to become a channel, and there are now books on do-it-yourself channeling. All that is needed is the power to meditate, and in this way it is possible to enter the collective unconscious and so draw on the pool of wisdom to which everyone has access.

Some believe that communications from "the other side" can be found on radio and television, while others are convinced that messages are left for them on their personal computers and fax machines.

See also MONTGOMERY, RUTH and PLEIADES.

For further reading: Sanaya Roman and Duane Packer, *Opening to Channel—How to Connect with Your Guide* (Kramer, 1987).

CHASTITY BELTS Supposedly spurious devices used to prevent infidelity and for BIRTH CONTROL. The first chastity belts were developed and used in Italy in the 14th century, and the custom spread across Europe. Attempts to use such devices made them the object of satirical stories and obscene jokes throughout the 15th and 16th centuries. In 1889, however, the skeleton of a female still wearing a chastity belt was discovered in a 15th-century Austrian grave. In the early 20th century, chastity belts became the subject of museum displays and private erotica collections. The most famous public collection was that displayed at the Cluny Museum in Paris, while the most notorious assemblage was in the possession of Egyptian

King Farouk I—though the latter collection was rivaled by that of eccentric American millionaire Ned Green, who owned several diamond-encrusted chastity belts.

By the 19th century, the demand for chastity belts was much greater than the supply, and a small industry developed to fulfill collectors' needs. Given the cruelties and crudities of the Middle Ages, people were prepared to believe husbands widely employed these devices. However, by the middle of the 20th century, it was found that the overwhelming majority of those in existence were forgeries. Not only would the wearing of chastity belts have been terribly uncomfortable, but they would have made simple acts like sitting down impossible. More importantly, they would have been impossible to keep clean, in spite of the hole through which feces, urine, and other bodily fluids could theoretically pass.

In the years following World War II, curators of British and French museums were bombarded with American tourists who wanted to see chastity belts. The British Museum responded straightforwardly that almost all existing examples were fakes; the Cluny Museum, concluding that it had assembled mostly modern "artifacts," removed its collection from public view in 1950.

CHIROPRACTIC An alternative medical practice, deriving from osteopathy, that rejects conventional medicine, particularly the use of drugs, and the germ causes of disease; it hypothesizes instead that illnesses result from disorders of the joints, muscles, and—particularly—the spine. The originator of this theory was David D. Palmer from Davenport, Iowa. From 1895 to 1905, he practiced healing by ANIMAL MAGNETISM. When he thought that he had cured deafness by realigning the spine, he began to treat symptoms by dealing with spinal misalignment.

His son, B. J. Palmer, set up a school for chiropractors, wrote the textbook *The Science, Art and Philosophy of Chiropractic,* published a magazine, and produced a chart that showed which cervical vertebrae should be manipulated to treat which disease: for diphtheria, the third, fifth, and seventh vertebrae; for scarlet fever, the sixth and twelfth dorsals, and so on. Palmer invented an electrical device—a neurocalometer—that he claimed diagnosed the spinal locations that required treatment.

Contemporary practitioners of chiropractics use a variety of spinal charts and publish their own textbooks and journals. Some manipulate joints and muscles particular to their treatment; others work with different joints and muscles; still others make use of X-rays and other conventional diagnostic techniques.

Established chiropractic colleges confer diplomas on qualified practitioners when these students successfully complete the program. Mainstream physicians who accept the validity of chiropractic treatment will refer patients who need relief from lower back pain.

CHRISTIAN SCIENCE Among the first of several 19th–century religious teachings that utilized the emerging authority of science to define its place in popular culture. Christian Science was the name given by Mary Baker Eddy (1821–1910) to her discovery of God, the mind, and the mind's superiority over matter. Eddy had been a sickly person most of her life, but in 1862 had discovered one Phineas Parkhurst Quimby (1802–66), a former mesmerist turned mental healer who resided in Portland, Maine. She became one of Quimby's students, and from him received relief of her symptoms, but found they returned soon after she left his center. Shortly after Quimby's death in 1866, Eddy fell on the ice and seriously injured herself. She was in great pain and many thought she might die. Instead, her days of recovery were spent in reflection on all she had learned, which culminated in her revelation of truth. As a result of that revelation, she was instantly healed. She set about the task of teaching students and putting her ideas on paper, first as a booklet, the *Science of Man* (1870), and then her textbook, *Science and Health* (1875), later renamed *Science and Health with Key to the Scriptures.*

Eddy's use of the term science raised basic questions of definition. While calling her system "science," she used the word in a manner at some variance from that of orthodox scientists, who understand science to be a body of systematized knowledge based upon observation of the world. Eddy defined Christian Science as the unfolding of true metaphysics, of explaining God, or Mind, and God's attributes. Science proves its truth by demonstration. The demonstration of Christian Science was in the healings experienced by those who accepted Eddy's teachings. Christian Science begins with God and attempts to explicate the laws that flow from a realization of God's omnipotence. As such, it stands against human science, as commonly understood, based as that is on what Eddy considered to be a false principle: the reality of matter and the fallible observation of the universe.

Taking the basic understanding of "science" as in Christian Science, though differing on various important points from Eddy's teachings, came other "science" religions, most prominently divine science (see DIVINATION) and RELIGIOUS SCIENCE (or Science of Mind). Both "divine science" and "science of mind" were terms used by Eddy as synonyms for Christian Science.

CHRISTOPHER, MILBOURNE (n.d) A professional conjuror of the 20th century who, following in the footsteps of Houdini, put considerable effort into exposing frauds, especially claims of the occult. His investigations embraced a wide range of frauds and deceits, from psychic animals to clairvoyants, MEDIUMS, dowsers, POLTERGEISTS, astral projections, and firewalkers. His exposures are very detailed, giving much information, without disclosing too many professional secrets about how the tricks are performed. He is the author of three books on the subject: *ESP, Seers and Psychics* (Crowell, 1970), *Mediums, Mystics and the Occult* (Crowell, 1975), and *Search for the Soul* (Crowell, 1979).

See also CLAIRVOYANCE; DOWSING.

CHROMOTHERAPY Often termed Color Therapy, a type of ALTERNATIVE MEDICINE that uses color as a treatment for various ailments. Usually part of a general category of healing called REFLEXOLOGY, color is used sometimes purely by itself and sometimes in conjunction with such enhancers as selected musical sounds or the eating of certain naturally colored foods. Reflexologists who advocate color therapy tell us that to remain vibrantly healthy the body's energy requires replenishment from the sun's energy in its pure form. Imbalances occur in specific parts of our bodies when one or more colors are not vibrating to their full capacity. Reflexology opens the body to receive the full spectrum of color for balance to be reestablished.

Practitioners believe that Color Therapy is a very ancient method of healing, one used by the inhabitants of the vanished continent of ATLANTIS and also practiced by the Egyptians in their specially constructed "healing rooms." In the early 19th century, following GOETHE'S COLOR THEORY, many in Germany and Russia thought that color had mystical powers. Both Wassily Kandinsky and Rudolf STEINER thought that color had both a physical and psychological effect that was highly therapeutic. With the rise of scientific medicine, these ideas fell out of favor, and chromotherapy was rarely used.

With the rise of various alternative medical treatments, however, color therapy is again becoming popular. Today there are many methods of color therapy. One procedure uses a special Reflexology Torch, which is the size of a pencil with a distinctive head into which is fixed a quartz crystal. The practitioner then inserts a disk of the appropriate color and switches it on, flooding the crystal with color. The treatment usually takes place in a darkened room with the color applied for 20 to 30 seconds, followed for a similar period of time by its exact complementary color (the color that lies directly opposite it on the color circle). The complementary is thought to act as a fixative. Specific colors, along with their exact complementaries, are used for treatment of each ailment. For example, a certain kind of red is used for the cure of infertility and also for piles, and its fixative is turquoise; blue is for goiter and stiff necks, with orange the fixative; yellow is used for diabetes and indigestion, with violet its fixative.

Despite its recent resurgence, chromotherapy is not accepted as a valid treatment by conventional medicine.

For further reading: Chris Stormer, *Reflexology—The Definitive Guide,* (Hodder and Stoughton, 1995); and Pauline Wills, *The Reflexology Manual (Headline, 1995).*

CHURCH OF SCIENTOLOGY Church to promote spiritual growth founded by the late science fiction writer L. Ron Hubbard in 1954. The Church of Scientology is currently headquartered in Los Angeles and has a combined membership of about 8 million in 70 countries around the world. The basic tenets of the faith, as set forth in the Church's book *What Is Scientology?,* state that a person is primarily a spiritual rather than a material being, and that people are basically good, not evil. "Man does not *have* a spirit," states *What Is Scientology?* "he *is* a spirit." Scientology, according to the church's statements, is intended to better mankind spiritually by increasing individual self–awareness. It does this through healing psychological and psychosomatic wounds, thus making people healthier and less likely to suffer from debilitating diseases.

Scientology's primary aims are similar to those of Hubbard's DIANETICS, a form of psychological counseling he first summarized in *Dianetics: The Modern Science of Mental Health* (1950). However, scientology is not the same thing as Dianetics. According to *What Is Scientology?* Dianetics tries only to adjust the effects the spirit has on a person's body. It can solve problems arising from the unconscious mind, and it can supposedly help prevent accidents, injuries, and some illnesses. Scientology rests on a more spiritual basis. It states that people consist of three parts: body, mind, and *thetan,* the immortal spirit of a person. The Church of Scientology claims to address mankind directly as a spirit, intending to bring people into a realization of their basic immortal spiritual nature.

The Church of Scientology tries to release the thetan from its deeply-rooted painful memories, known as *engrams.* The unconscious mind, in Hubbard's thinking, continuously records an individual's life experiences. These experiences are recorded in a chronological sequence, as a movie camera records actions and sounds in a series of connected, but individual, pictures on a single strip of film. An engram occurs when something either physically or mentally damaging happens to an individual. Engrams, according to Hubbard, can occur at any time in an individual's life—including before someone is born; some engrams are believed to have been formed even before conception.

Through a church ritual called *auditing,* believers bring their engrams out of their unconscious, or *reactive,* minds into their conscious minds, where they become less damaging. A trained church official known as an auditor prompts the believer to move back through the record of his or her unconscious memory and to utter these engram-making experiences out loud. After these engrammatic experiences are recounted, they begin to lose their power, until eventually they are totally erased. In theory, once the thetan escapes from the negative power of these engrams, the individual can realize his or her full spiritual potential. The church also uses a special device called an E-meter, a sort of lie detector, used by auditors to guide believers in their spiritual quest.

Many people have questioned L. Ron Hubbard purposes in founding the Church of Scientology. The science fiction critic Sam Moscowitz claimed that Hubbard once told him that the easiest way to get rich was to found your own church. Hubbard's son, L. Ron Hubbard, Jr., claimed in *L. Ron Hubbard: Messiah or Madman?* that his father was extremely possessive of both dianetics and scientology and was often angry when he thought that others were trying to profit from his discoveries.

Hubbard exhibited this possessiveness in the early 1950s, when he lost control of Dianetics. He founded the Church of Scientology after he lost control over the Dianetics Foundation of Wichita, Kansas. The Dianetics Foundation went bankrupt in 1952 and was bought by a businessman who removed it from Hubbard's control. The science fiction writer founded scientology two years later and spent a great deal of money trying to discredit the Wichita group. The church has published many of Hubbard's later writings on scientology and distributes tapes and teaches courses on scientology.

Other people question whether scientology actually works on everybody, as it claims. The process of auditing and erasing engrams is long and complicated. A person who has erased his or her major engrams is known as a *release.* However, it takes months or years of auditing before an individual escapes all the consequences of his or her engrams. Such a person is known as a *clear,* and it is only in the clear state that the thetan realizes its full potential. Clears are very rare. Dr. J. A. Winter, an early convert to dianetics, could not find a single one—and people he examined who were supposed to be clear had relapsed into psychoses or exhibited behaviors that made him suspect that they were not really healed.

Despite its title, the Church of Scientology is not a scientific organization. It states fundamentals of belief rather than suggesting and testing hypotheses, and it neither authorizes nor performs experiments. It claims to be a fortress of discovered truth rather than an organization seeking after truth. The results obtained by church auditors using the E-meter are difficult to understand and uncorroborated by other methods. The church has received testimonials from many famous people it has helped, such as the actor John Travolta, but without a way of testing its theories and practices, scientology cannot be classified as a science.

For further reading: Bent Corydon and L. Ron Hubbard, Jr., *L. Ron Hubbard: Messiah or Madman?* (Lyle Stuart, Inc., 1987); Martin Gardner, *Fads and Fallacies in the Name of Science* (Dover Publications, 1958); L. Ron Hubbard, *Scientology: The Fundamentals of Thought* (Bridge Publications, 1988); L. Ron Hubbard, *What Is Scientology?* (Bridge Publications, 1992); Harriet Whitehead, *Renunciation and Reformulation* (Cornell University Press, 1987); and J. A. Winter, *Dianetics: A Doctor's Report* (New Julian Press, 1987).

CLAIRVOYANCE A type of paranormal or psychic phenomenon that manifests itself in keenness of mental perception—instant insight. It is derived directly from an external object or event rather than from the mind of another who is observing it, as in mind reading. Clairvoyance views the happening at the time of the event and not before the occurrence, as in PRECOGNITION, or after the event, as in retrocognition.

Clairvoyants believe themselves to have EXTRASENSORY PERCEPTION (ESP), which enables them to obtain, either empirically or by rational inference, information that is not available to the senses or to others. Some psychics claim to be able to see, and even scan over, distant places while sitting quietly and closing their eyes. This ability is called REMOTE VIEWING and has been subjected to close scrutiny by the military because of its potential use in mapping enemy territory; the idea was dropped when nothing useful was found. Psychic research that used clairvoyants to identify cards in sealed envelopes or that presented them with special ESP cards bearing five different symbols have all failed to establish conclusively the existence of psychics' special gifts.

See also TELEPATHY.

For further reading: C. E. M. Hansel, *The Search for Psychic Power: ESP and Parapsychology Revisited* (Prometheus, 1989); Jenny Randles, *Sixth Sense: Psychic Powers and Your Five Senses* (Robert Hale, 1987); and James E. Alcock, *Science and Supernature* (Prometheus, 1990)

CLARKE, ARTHUR C. (1917–) Science fiction author and investigator of strange phenomena. Arthur C. Clarke is probably best known for his science fiction epic *2001: A Space Odyssey* (1968) and its sequel *2010: Odyssey Two* (1982). However, he has also earned a reputation as a science writer, explaining ideas to a popular audience. Clarke published an article called "Extraterrestrial Relays" in *Wireless World* of October 1945, suggesting that an artificial satellite orbiting Earth's equator at an altitude of about 35,800 kilometers could be used to transmit radio and television signals around the globe. Such a satellite, traveling at nearly 11,000 kilometers per hour, would orbit Earth once every 24 hours, and appear to stay in one place over Earth's surface. Almost 20 years later, on February 14, 1963, the United States launched its first communication satellite, *Syncom 1,* into geostationary orbit over the equator, bringing Clarke's idea to life.

With the publication of *The Exploration of Space* in 1952, Clarke began his career as an independent writer and as a proponent of science. Other volumes—including *The Challenge of the Spaceship: Previews of Tomorrow's World* (1959), *Profiles of the Future: An Inquiry into the Limits of the Possible* (1962), and *The Promise of Space* (1968)—also established him as a predictor of advancements in science. In the 1980s, Clarke began to host television series that explored the paranormal: *Arthur C. Clarke's Mysterious World* (1980) and *Arthur C. Clarke's World of Strange Powers* (1984).

CLEVER HANS PHENOMENON See COUNTING HORSES.

CLONING From the Greek *klon* meaning "twig" or "graft," clones are duplicates of organisms derived from a single individual. Until recently the cloning of a whole animal from a cell of an adult animal was considered to be a pseudoscientific fiction; the only scientific cloning that was possible was gene cloning, meaning the production of identical copies of a specific gene, or piece of DNA, for use in genetic engineering. The method involves inserting the piece of DNA in the form of a plasmid, a small loop of DNA, into a bacterium. When the bacteria multiply, the plasmid duplicates itself many times and so produces a large number of gene clones.

The procedure of gene cloning has been commonly used in plant breeding, with some experimentation in animals. In an experiment with frogs in the 1970s, a team led by John Gurdon at the University of Cambridge, England, transplanted nuclei from the skin cells of adult frogs into frogs' eggs that were deprived of their own nuclei. In this experiment some embryos grew into tadpoles, but none reached adulthood.

Following the Cambridge experiment, another team of scientists, this time led by Dr. Ian Wilmut at the Roslin Institute near Edinburgh, Scotland, started to look into the possibility of cloning sheep's embryos. In this work, successfully carried out in 1996, an egg was taken from an ewe, and the nucleus containing the genetic material was removed. Then the nucleus was taken from a cell of an immature embryo of a different ewe. The genetically engineered egg was then implanted in the womb of a surrogate mother ewe and allowed to develop. In 250 attempts, two of the embryos developed into lambs—Morag and Megan.

Early embryo cells are much easier to clone than cells from adult sheep, but it was the necessary first step in developing the Wilmut technique. In 1997 he and his team published the results of the first cloning of an adult animal by taking a single cell from a ewe's udder and from it growing a lamb. Mammary cells were used because they are rich in special proteins and enzymes that are more adaptable for this purpose than cells from other tissue; it is not known whether any type of adult cell could be used. The single cells were then soaked in a chemical bath to make their nuclei assume a quiescent state. The division of each nucleus's genetic material was slowed down to a rate that was acceptable to the egg cells into which they were to be placed without losing any of the genetic material. They were then implanted into these eggs, which were from another ewe and from which the nuclei had been removed. In this way each mammary-gland cell's nucleus was turned into a visible embryo that was then implanted into a surrogate mother. That ewe then gave birth to a lamb, Dolly, which is genetically identical to the ewe from which the mammary gland cell was taken, bearing its entire gene complement. The process is as yet very hit-and-miss; it took 277 fusions to create Dolly.

The announcement about Dolly took everyone by surprise, and people immediately began to speculate about the significance of her creation. Britain's Human Genetic Advisory Commission immediately put it on the agenda for its next meeting, and the U.S. National Bioethics Advisory Panel was told by the White House to review the implications of the research. The atomic scientist and Nobel Peace Prize winner Joseph Rotblat went further and called for an international ethics committee to be set up to look into the wider issues of scientific research.

A sheep is almost as complex an animal as a human being. So, in principle, this same process could be carried out with humans, producing identical twins vertically, instead of horizontally as now happens from one womb. In the minds of many, this raises disturbing ethical problems that have often been explored in science fiction, such as Huxley's *Brave New World* and the film *The Boys from Brazil* (the latter mooted the horrifying prospect of cloning Hitler.) But cloning is a very arduous and expensive process, and we must ask why would anyone want to produce human beings in this way. If the belief is that absolute duplicates of either Hitler or Einstein could be produced, then this ignores the fact that the clone's development would be as much the result of its social environment as of its genes. It must not be forgotten that nurture as well as nature has its place in this process. To offset the understandable worries over the consequences of the technique, scientists point out that producing carbon-copy humans was never the intention of the research and indicate its many advantages—for example, to study aging, or to engineer medicines or spare parts for use in surgery.

There is, however, a further problem that involves both long- and short-term risks: the problem of maintaining societal control over genetic engineering in general. Uncertain technologies are today everywhere mixed up with traditional processes in the biology of agriculture and farming and, of course, in the biological control of our own reproduction. We might be able to ban the cloning of human beings, but given the multiple implications of genetic engineering, there is no defined target to ban. If we continue sidestepping natural selection to any great extent, the biodiversity of all species, animal and vegetable, could be put at risk.

CLOUD BUSTING Any device or process that will cause clouds to empty—to rain. The hope that humans might control weather has a long history. Rain, or the absence of rain, can have very serious consequences. As rain comes and goes of its own accord, beyond human control, it can be frustrating. Societies in many parts of the world, particularly those in regions subject to drought or to torrential rain, have sought to control it by various sorts of MAGIC and, latterly, through science and technology. Rain dances by Native North Americans are an obvious example. Medicine men in many places have also used different types of magic to bring rain. In more recent times, charlatans have developed devices to bust clouds. In the early 1950s Wilhelm REICH invented a device that consisted of nothing more than a battery of hollow pipes and tubes. He claimed that it sucked ORGONE ENERGY out of clouds, causing them to break up and produce rain. He installed five such machines in the United States—two in North Carolina and three in Rangeley, Maine; their alleged success established him as a modern rainmaker.

There have also been research programs aimed at persuading clouds to let down their water content at selected times and places. When an area has been desperately short of rain for a long time, threatening its agricultural economy, it is frustrating to see clouds form, build up, and then either move off or melt away without rain resulting. The likely explanation for this weather pattern is that the clouds lack nuclei—dust or similar particles—about which moisture could condense until it forms large enough droplets to overcome the upward convection currents within the cloud and descend as rain. So attempts were made, starting in 1946, to seed promising clouds with very small silver iodide or solid carbon dioxide crystals in sufficient quantity to produce rain. These precipitation nuclei were injected into the clouds from small aircraft, rockets, cannon, or ground generators. The results were mixed. In one particularly successful series of seeding experiments in Florida, three times as much rainfall was obtained from seeded clouds as from unseeded. There is the problem that, if and when one area gains from increased rainfall, a neighboring area may lose. At the other end of the scale, seeding has been used to reduce the effects of too much rain and accompanying wind and to minimize hail.

COELACANTH "Living fossil" fish, member of an ancient suborder called crossopterygians ("fringe-finned" fish). Coelacanths were long thought to have become extinct during the same period in which the dinosaurs

Coelacanth

died out. Then in 1938 fishermen landed a living coelacanth off the coast of South Africa. The scientific world was stunned. The discovery of this fish (*Latimeria chalumnae*) raised the possibility that other creatures long thought extinct could still exist. The coelacanth has fins that somewhat resemble legs in both appearance and function. This led some scientists to conjecture that this fish (a forerunner of the amphibians) is ancestor to land animals, including humans; most, however, believe that the lungfish is more likely the water–land link.

Since 1938, about 200 coelacanths have been caught, most of them in the Indian Ocean off Grand Comoro Island. None of the 5-foot-long fish has survived more than a few hours, generally because of injuries inflicted accidentally or purposely (coelacanths have nasty teeth) by fishermen. Ironically, in fact, today the creature once thought to be extinct is in danger of becoming extinct in reality. In 1991, Dr. Hans Fricke, of the Max Planck Institute for Behavioral Physiology in Seewiesen, Germany, reported in the journal *Nature* that coelacanths, which live in deepwater lava caves near shore and which hunt for prey at depths up to about 700 meters (about 2,300 feet), are steadily decreasing in population, largely because of their encounters with fishermen. See also EVOLUTION.

COITUS RESERVATUS A method in which intercourse is maintained for long periods during which the woman may have orgasm several times while the man succeeds in holding back orgasm. It has long been practiced in the East, both as a method of BIRTH CONTROL and for the increased satisfaction it gives the woman. It was adopted by the Oneida community in the United States as part of its moral code and EUGENICS policy.

In 1843 John Humphrey Noyes founded the Putney Corporation of Perfectionists in the village of that name in Vermont. He believed that Christ expected perfection from his followers and set down a strict moral code for members of the corporation. However, its behavior antagonized the Putney villagers, and in 1848 it moved to Oneida Creek, Madison County, New York—becoming the Oneida Community. Noyes described his code as Bible Communism: All property was communally owned, and children were raised by the community. In these respects the community's practices resembled those adopted by Israeli kibbutzim a century later. The community practiced a system of complex marriage. The members' sex life was, in effect, communal. Single-partner sex was frowned on as selfish; everyone should love everyone of the opposite sex. Every man in the community was theoretically the partner of every woman, though the women appeared to have preferences. Intercourse could take place between any couple, but procreation was only allowed according to eugenic principles to improve the stock of the community. If propagation is to be scientific, there must be no confusion between the two acts—intercourse and procreation—and procreation must never be involuntary. To separate the two acts required that the male partner not attain orgasm while allowing his partner to do so—coitus reservatus.

Noyes believed that this practice was actually beneficial to the male. Both boys and girls were taught the technique in puberty by the older members of the opposite sex. Although the community broke up after 30 years, coitus reservatus was adopted by a number of other organizations, believing that it was beneficial. In particular a U.S. doctor, Mrs. Alice Bunker Stockham, calling the practice *karezza*, advocated it in a book by that same name, which was published in 1896. It is not a practice in much favor today, nor is there any scientific evidence that it is beneficial to the male; some authorities claim that its prolonged practice is harmful. The claimed scientific basis for eugenics, the essences of Noyes's code, is at best dubious and again is not accepted by the scientific community today.

COLD FUSION The 1989 discovery by Martin Fleischmann of the University of Southampton and Stanley Pons of the University of Utah of a low temperature method of fusing atomic nuclei, with the consequent release of huge amounts of energy. The process of nuclear fusion, much sought after because of its promise of virtually endless amounts of energy free of the radioactive waste problems of traditional nuclear power plants, was hitherto believed only to be possible at extremely high temperatures, around 100 million degrees Celsius. Working at such high temperatures was very difficult and expensive, and no practical solution to the problems of doing so had been found—nor were they likely to be found in the foreseeable future. If Pons and Fleischmann were right, the high-temperature approach would be bypassed and the world's energy problems might be over.

The process of cold fusion was a simple one. It consisted of an electrolytic cell containing water that had more than the usual proportion of heavy water—water in which the usual hydrogen atoms are replaced by deuterium, the isotope of hydrogen, which is twice the usual weight, (D_2O instead of H_2O). The cell has palladium electrodes. On current being passed, both hydrogen and deuterium were released at the cathode (the negative electrode). The experimenters believed that the palladium absorbed the deuterium atoms, forcing them sufficiently close together to cause them to fuse. In support of their claim, they said that they had detected neutrons being emitted—to be expected if fusion occurs—and that surprisingly large amounts of heat were produced.

Other physicists tried to reproduce the results, but many dismissed the claim on a priori grounds: The barriers obstructing the fusion of two atomic nuclei are so great that overcoming them at low temperatures, whatever the process, was ruled out. If cold fusion were to

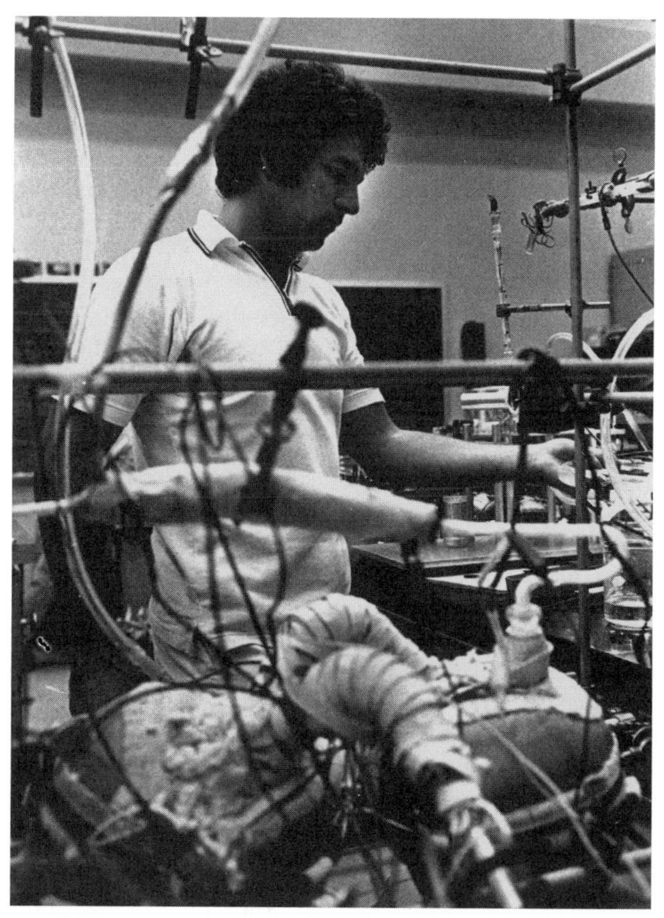

Cold Fusion

happen in this way, it would require abandoning the knowledge of nuclear physics that we have accumulated over 90 years.

In 1991, the *New Scientist* journal concluded that Fleischmann and Pons had short-circuited normal scientific procedure and failed to consult nuclear physicists before making their claims, treating the nuclear evidence for fusion with extraordinary carelessness.

Today there are still a few scientists who are investigating the possibility of cold fusion, but the scientific community has effectively abandoned the idea.

COLOR, THEORY OF See GOETHE'S COLOR THEORY.

COLORADO MAN Spurious prehistoric man. Following his more famous CARDIFF GIANT hoax, P. T. Barnum ordered the construction of another primitive human specimen that he could exhibit. He had learned from the criticism of the earlier effort and directed that the new man be constructed out of more convincing materials, including actual bones, and that the arms be longer than the legs. The finished product was shipped to Colorado, where a Barnum confidant, a "geologist" who just happened to be in the state, discovered it. Coincidentally, Barnum was visiting in Colorado to deliver temperance lectures. After receiving word that the giant man had been found, Barnum immediately announced to the press and the public that he would pay $20,000 to acquire the giant.

Barnum was immediately attacked by many of the same people who had uncovered the Cardiff hoax. On seeing the new discovery, paleontologist O. C. Marsh of Yale University immediately pronounced it a fake. He pointed out that the body showed none of the characteristics of a fossilized specimen. In particular, he pointed out that the abdominal region, unsupported by bone, would have collapsed and would not manifest the normal outline of a living person, as the Colorado Man did. Marsh's conclusions ended Barnum's plan to exhibit Colorado Man. After showing his "giant" around in the West, he discarded it.

COMBE, GEORGE (1788–1858) Scottish lawyer and author of popular books on PHRENOLOGY. One of 13 children, Combe grew up in a poor family in a partially industrialized section of Edinburgh, Scotland. His father owned and operated a small brewery in the building where the family lived. George was educated in local schools, and apprenticed in a law office before starting his own practice. In 1816, Johann SPURZHEIM went to Scotland to lecture on phrenology, and in spite of initial doubts, Combe was converted. Upon studying the new "science of the mind" in greater detail, he began to lecture and write essays espousing its virtues.

In 1828 Combe produced the first edition of his best-known work, *The Constitution of Man in Relation to External Objects,* which would eventually go through dozens of editions and become one of the best-selling English language books of the mid-19th century. In it, Combe expounded an elaborate, often proscriptive, scheme that detailed the relationship of the human brain to morality, intellect, and behavior. Combe did not have the medical background that phrenologists such as Franz Joseph GALL and Spurzheim did, and though he would champion their anatomical and physiological claims, his work emphasized phrenology as a system of practical moral philosophy. Like Spurzheim, Combe felt that phrenology could be used for social reform and aggressively promoted its use in education, penology, and mental institutions. In 1836 he made an unsuccessful bid for the Chair of Logic at Edinburgh University, and a year later the newly founded University of Michigan offered him a position, which he politely declined.

Following his marriage at age 45 to a wealthy woman, Combe devoted himself fully to phrenologist pursuits. These included lecture tours of the continent and the United States, the latter recounted in his *Notes of the United States of North America, 1838–9–40.* In spite of harsh critical reaction to phrenology led by anatomists and the clergy, Combe enjoyed a considerable popular reputation through the mid-19th century.

For further reading: Roger Cooter, *The Cultural Meaning of Popular Science: Phrenology and the Organization of Consent in Nineteenth-Century Britain* (Cambridge University Press, 1984); Charles Gibbon, *The Life of George Combe* (Macmillan and Co., 1878).

COMMITTEE FOR THE SCIENTIFIC INVESTIGATION OF THE CLAIMS OF THE PARANORMAL (CSICOP) An international organization that "encourages the critical examination of paranormal and fringe-science claims from a responsible, scientific point of view and disseminates factual information about the results of such inquiries to the scientific community and the public." It numbers among its Fellows and Consultants scientists, educators, magicians, health professionals, engineers, philosophers, journalists, linguists, and psychiatrists. It has subcommittees on ASTROLOGY, electronic communication, health claims, PARAPSYCHOLOGY, and UFOs. Its official journal is the *Skeptical Inquirer.* Paul Kurtz, emeritus professor of philosophy, State University of New York at Buffalo, is its chairman; Barry Karr is the executive director, and Lee Nisbet is the special projects director. In its journal it emphasizes that it is a nonprofit scientific and educational organization and that it does not reject claims on a priori grounds, antecedent to inquiry, but examines them objectively and carefully. It is a powerful and influential organization, providing judgments on many scientific and pseudoscientific issues.

COMMUNIGRAPH An instrument created in the early 1930s to facilitate the reception of scientifically evidential communications with the spirits of the dead. It was a product of the ASHKIR-JOBSON TRIANION GUILD, which grew out of the work of three British researchers, A. J. Ashdown, B. K. Kirby, and George Jobson. Jobson died in 1930 and purportedly gave instructions for the construction of a communigraph through a MEDIUM.

The instrument consisted of a pendulum hanging from underneath a table. The swinging pendulum could make contact with a set of small metal plates representing the alphabet. Upon making contact, a circuit was closed and the corresponding letter was illuminated upon the face of the table. According to claims, even in the absence of a medium, people could sit in a circle around the communigraph and the pendulum would begin to swing on its own (through the spirits') volition and messages would be spelled out.

In the early 1930s several efforts were made to use the Communigraph the most noteworthy being those of Lady Zoe Caillard. Lady Caillard believed that she had received a number of messages from her late husband, which became the substance of two books, *Vincent Caillard Speaks from the Spirit World* (1932) and *A New Conception of Love* (1934). The Communigraph was used throughout the 1930s but did not supply the regular opening to the spirit world for which its creators had hoped and was largely abandoned during World War II. The guild also built the REFLECTOGRAPH, a more complicated device.

CONDON REPORT (1968–1969) 1,485-page report evaluating the U.S. Air Force's two-decade accumulation of material on UFOs (unidentified flying objects). In 1947, the U.S. Air Force began secretly to compile information on UFO reports. During the next two decades, the air force's UFO project underwent several reorganizations and name changes, the final name being Project Blue Book. The information gathered was evaluated and discussed not only by the air force but by the Department of Defense. By 1966, many reports of UFO sightings and even abductions had generated a great deal of public excitement. Anxious to put these reports to rest, Congress appointed a committee, headed by University of Colorado physicist and UFO skeptic Dr. Edward Condon, to study the air force's research with the overt aim of recommending whether any more effort or money should be put into the UFO research project. The committee evaluated 87 of the 25,000 reports the air force had gathered and, even though more than 20 of the 87 cases were listed as unsolved, recommended that "further extensive study of UFOs probably cannot be justified on the expectation that science will be advanced thereby." As a result of the Condon Report, Project Blue Book, an official air force collection and evaluation of UFO reports was ended in 1969. It was soon revealed that the committee's covert purpose had been to end UFO research, and it did succeed in that purpose, at least officially. Other organizations continued the research privately and in a less public way than before, so did the air force and other government organizations.

See also ABDUCTIONS, ALIEN; UFOLOGY.

CONTINENTAL DRIFT The idea that continents move horizontally. Once considered pseudoscientific, this tenet is now part of mainstream science. Before the modern revolution of plate tectonics of the 1950s and 1960s, the idea that the continents might move horizontally around the globe seemed ludicrous; yet *prima facie* evidence had long been known, in the approximate match between the adjacent coasts of Africa and South America. A very full case for continental drift was made in the early 20th century by the German glacial meteorologist and geologist Alfred Lothar WEGENER. He assembled much of the modern evidence, in particular the same fossils and rock strata on the matching coasts of Africa and South America, which was so important for later scientific acceptance. His arguments were discussed most fully in the mid-1920s. Most who considered them rejected them as the work of a crank, in part because he did not have a satisfactory theory of the mechanism of the movement of continents. Wegener also totally ignored counterarguments and counterevidence and appeared too willing to interpret ambiguous evidence favorably.

In the 1950s, improved techniques for measuring Earth's magnetism gave some disturbing results: In some periods, Earth's magnetic poles appeared to be in different places simultaneously. That only made sense if the continents had been located differently and had subsequently drifted. In 1929, M. Matuyama showed that the North and South Poles had changed places from time to time. In the 1960s F. J. Vine and D. H. Matthews showed that there were magnetically reversed stripes on the ocean floor, suggesting that rock was being laid down at a split in the ocean floor, forcing the continents apart. Further evidence proved that Earth's surface consisted of several plates, some drifting apart, other crashing into each other to form mountain ranges, or to produce earthquakes and volcanoes. In the 1980s very sophisticated measurements of the distance between fixed points on adjacent continents, using lasers, satellites and distant galaxies, confirmed that plates do move relative to each other at about 2 centimeters (roughly 1 inch) a year.

For further reading: A. Hallam, *A Revolution in the Earth Sciences: From Continental Drift to Plate Tectonics* (Clarendon Press, 1973); and R. G. A. Dolby, *Uncertain Knowledge* (Cambridge University Press, 1996).

CONVICTION SCIENCE A term used to describe a common characteristic of both scientists and pseudoscientists—the unshakeable conviction of rightness, whatever the evidence. Sometimes this conviction is so strong that it defies disproof. The holder *knows* his idea is right, and any denial is dismissed with varying levels of contempt. The denier has not understood, or is blinded by prejudice, or the contrary evidence is unsound, and so on. In some cases, believers have been proven correct, as in the case of Alfred Lothar WEGENER, who was convinced of the reality of CONTINENTAL DRIFT long before there was sufficient experimental evidence to substantiate it.

See also POLYWATER, N-RAY.

COOK, FREDERICK ALBERT (1865–1940) U.S. explorer best remembered for making false claims about his achievements. Cook was surgeon to both the Peary Arctic expedition (1891–92) and the Belgian Antarctic expedition (1897–99). From 1903 to 1906, he led an expedition to Mount McKinley (6187 meters [20,300 feet]) in Alaska, the highest mountain in North America. Despite his lack of surveying skills, he mapped and explored a very inaccessible region. Not satisfied with this achievement, Cook claimed to have reached the peak of Mount McKinley in 1906 and produced a photograph as supporting evidence. That photograph clearly did not substantiate his claim. Three years later he made another claim—that he had reached the North Pole in company with two Inuit.

Cook later spent time in prison for oil-land fraud. When he died in 1940, he was still maintaining that his claims were genuine and was pursuing law cases against those whom he saw as maligning him.

See also EXPLORATION HOAXES.

CORLISS, WILLIAM R. (1926–) Leading contemporary spokesperson for anomalistics and head of the SOURCEBOOK PROJECT. Corliss attended Rensselaer Polytechnic Institute and received his M.S. degree in physics in 1953. In 1963 he became a freelance writer and began to gather the material for what would become the Sourcebooks, covering events on the margins of science. Corliss had become interested in what he has termed "outlaw science" in 1951 after reading George McCready PRICE's *Evolutionary Geology and the New Catastrophism.* Price, a biblical creationist, had, in the midst of making his case for a literal interpretation of the Genesis creation account, presented a number of anomalous geological phenomena that, he argued, undermined conventional paleontology. Although not accepting Price's theology, Corliss nevertheless became fascinated with the idea of anomalous occurrences and the unconventional theories that had been presented to explain them. Price's book led him to the writings of Charles Hoy FORT and the Forteans.

Corliss's approach to gathering and cataloguing anomalous events differed significantly from that of the Forteans, who had tended to emphasize some of the more extraordinary and bizarre stories—accounts of UFOs, monsters, and so on. Corliss emphasized odd events and artifacts that have been considered by but remain puzzling to mainstream science. For example, in his *Unknown Earth: A Handbook of Geological Enigmas* (1980), he covers such topics as dinosaur tracks on the roofs of coal mines, naturally occurring pyramids, tektites, and notable mass extinctions of fossil species.

The Sourcebook Project was formally initiated in 1974 with the publication of *Strange Phenomena,* the first of more than 20 volumes that appeared over the ensuing years. While several Forteans had developed catalogues of anomalous events, Corliss has been the first to catalog them systematically in a manner that relates meaningfully to more orthodox science. Corliss has explained that his publications cover only one-fifth of the material he has uncovered and that he has been able to deal systematically only with the American and British journal literature. He has added approximately 1,200 new reports to his files annually. Some of these are noted in the Sourcebook newsletter, *Science Frontiers.*

For further reading: William R. Corliss, ed. *Ancient Man* (The Sourcebook Project, 1978); William R. Corliss,

Unknown Earth: A Handbook of Geological Enigmas (The Sourcebook Project, 1980); William R. Corliss, *The Moon and the Planets,* (The Sourcebook Project, 1985); and William R. Corliss, *Neglected Geological Anomalies,* (The Sourcebook Project, 1990).

CORRESPONDENCE, THEORIES OF One of the basic theories of "truth," where the goal of enquiry stands in contrast to "falsity" and not in contrast to "opinion." The correspondence theory is the most commonly used theory of the nature of truth because it says quite simply that a statement is true if it corresponds with reality, with the facts of how things are. For example the statement: "The cat is on the mat" is true only if indeed the cat **is** on the mat and untrue if the cat is **not** on the mat. Thus a statement is true if it corresponds with the situation or state of affairs that verifies it.

Correspondence theories appear in other disciplines. In mathematics the one-to-one (1:1) correspondence is an important principle. At its simplest it is a 1:1 correspondence between two series, for example in the following, the first series is an addition of 3 at each step, and the second series is a multiplication of 3 at each step:

3 6 9 12 15 . . . . . .

and

3 9 27 81 243 . . . .

Thus addition in the first series *corresponds* to multiplication in the second. In secret codes, the production and deciphering of which is a concern of mathematicians, there is a 1:1 correspondence between a cipher and whatever letter or number or thing it represents. In sophisticated mathematics the test of whether there is a 1:1 correspondence between A and B is very important.

In physics the idea of a correspondence principle was invoked by Niels Bohr in 1918 as a way of resolving the apparent conflict between classical and quantum physics. In classical physics energy, matter, and motion are continuous and changes are smooth. In quantum physics energy states are discrete, and packets of energy as radiation are emitted when there is a change or jump from one state to another. Classical physics had been very successful in explaining a great range of phenomena but failed to explain others. It was to solve these remaining problems that Planck, Einstein, and Bohr turned to a quantum approach, which implied a discontinuous model of matter. Bohr proposed a correspondence principle to resolve the apparent conflict. It said, in effect, that classical physics ruled in the macroscopic world and quantum physics in the atomic and sub-atomic world and, most importantly, that where the two overlapped—or joined—they should both come to the same conclusions. In the 1920s when Shrödinger and Heisenberg successfully applied the principle, Einstein—who in 1920 had written to Max Born: "that one has to solve the quanta by giving up the continuum, I do not believe"—was pleased to see the matter resolved.

The correspondence principle has wide relevance in science generally. Newtonian mechanics has been successful in unifying terrestrial and celestial mechanics, in explaining the behavior of automobiles, baseballs, and the solar system—with one exception, the planet Mercury. Relativity deals well with systems involving velocities near that of light in a VACUUM and in addition explains Mercury's orbit. When choices arise in developing relativity theory, an appeal to the correspondence principle ensures that two theories will agree where they overlap and that no successful part of Newtonian mechanics is overturned.

The importance of correspondence in science can be summarized thus: When a new scientific paradigm is adopted it must satisfactorily explain, to essentially the same conclusion, those problems which the old paradigm had solved.

For further reading: Gerald Holton, *Thematic Origins of Scientific Thought: Kepler to Einstein* (Harvard University Press, 1974).

COSMOLOGIES, VARIANT Derived from two ancient Greek words, *kosmos* meaning universe and *logos* meaning reasoning, the study of any scientific, mythological, or religious theory of the universe, particularly those involving heavenly bodies.

Until about 400 years ago, thinking about the origin and working of the universe was a matter of speculation and religious faith. Creation myths are found in most cultures and are usually centered on the Earth, with the specific creation spot, the umbilicus, located in that part of the world where the myth-creators live. Myths vary in composition and complexity. Nevertheless, there are some basic elements that are similar: Most include the origin of a god or gods and elaborate on his or their role in the creation of Earth and humankind; sometimes the creation comes out of nothing, sometimes from another form of preexisting matter. In many stories, the creation of Earth, sea, plants, animals, and humans came out of a primeval chaos that was not neutral but was believed to have its own constitution that forever opposed being put in order, hence its propensity to break out into violent activities such as earthquakes, volcanoes, and violent storms—hence also the destructive practices of humans, as shown in their continuous struggle between good and evil. On the divine level and in the various dualistic systems of religion, the activity of chaos is represented by devils.

Probably the best-known metaphysical account of the beginning of the world is in Genesis, in which God in six days created the world from nothing. In this account, on the sixth day God formed man "of the dust of the ground" and placed him, Adam, in a beautiful garden called Eden. Eve was later created out of one of Adam's ribs and was "called woman because she was taken out of man."

A few ancient scholars did speculate about the heavens, in what we might today call a scientific manner, looking for rational theories, but they tended to place intellectual reasoning above the investigation and analysis of actual phenomena (rationalism as opposed to empiricism), and so gave descriptions of phenomena and their causes, which were almost as fanciful as mythological accounts. The Babylonians (c. 3000 B.C.E.), for example, held that in the center of Earth was a hollow mountain supported by an ocean. Above the ocean curved the solid vault of heaven, resting on the ocean and dividing it from the upper waters. Sometimes the vault of the sky leaked and the waters fell through onto Earth as rain. The ancient Greeks from the philosopher Thales of Miletus (c. 600 B.C.E.) onwards to Ptolemy of Alexandria (c. 150 C.E.) followed and elaborated on this idea. Although Philolaus put forward the view that Earth was a sphere revolving daily round a central fire (not the Sun), and Heracletus thought that Earth did actually move around the Sun, most Greek scholars believed that Earth was the immobile center of the universe.

What we now see as *scientific* cosmology began 400 years ago with Copernicus (1473–1543) and Galileo (1564–1642), who deposed Earth from its central position in the universe. Copernicus made only 27 observations himself, a very small number for an astronomer. But, relying mainly on existing information about the planets, he became convinced that their behavior could be better explained by assuming that they pursued paths around the Sun, not around Earth. Knowing full well that he risked condemnation by the Church in Rome, he was very careful to say that this concept was only a mathematical convenience and was not meant to conflict with the church's unqualified support of the Ptolemaic geocentric system. Even then he delayed publication of *De Revolutionibus Orbium Celestium* until he was on his deathbed. Nevertheless the Church put the book on the censored list and forbade anyone to subscribe to the views it advanced. Forty years later Galileo, from his own observations with the newly developed telescope, realized that the geocentric theory, incorporating the Christian vision of the circular perfection of God's handiwork, was not sustainable and that the heliocentric (Sun-centered) model was more credible. This heretical declaration immediately put him in conflict with the Church.

The 17th-century German astronomer Johannes Kepler (1571–1639) added to the scientific understanding of the solar system, working from data collected by Danish astronomer Tycho Brahe (1546–1601) years earlier. Kepler discovered that the planets followed elliptical orbits and not the so-called perfect circular orbits. Then, and only then, did order become apparent. The planetary orbits obeyed three laws: (1) each planet moves in an elliptical orbit with the Sun at one focus; (2) the imaginary line connecting the planet to the Sun sweeps out the same area in a given time; and (3) the squares of the periods of the different planets are proportional to the cubes of their average distances from the Sun.

Sir Isaac NEWTON helped explain both planetary motion and terrestrial physics with his laws of motion and theory of gravity. The planets orbit in the Sun's gravitational field and the behavior of falling bodies under Earth's gravity obeyed the same scientific laws. This theory, that the science of Earth was the same as the science of the heavens, overturned the previously held Aristotelian view that one set of laws pertained for Earth and another for the heavens. The 200 years from Copernicus to Newton saw astronomy and cosmology transformed from mythology and religion into scientific disciplines based on observation, classification, theory, and testing.

See also HOYLE, Sir Fred.

For further reading: Arthur Koestler, *The Sleepwalkers* (Penguin, 1988).

COTTINGLEY FAIRIES One of the most unusual hoaxes in the history of photography, perpetrated by two preteen English girls in the summer of 1917. During World War I, Frances Griffiths, age 10, returned home thoroughly wet one afternoon, explaining to her unsympathetic mother that she had fallen into a brook while playing with the local FAIRIES. Her 13-year-old cousin Elsie Wright, feeling sorry for Frances, resolved to borrow her father's camera and a single photographic plate and photograph the fairies as proof of Frances's story. When Arthur Wright developed the girl's plate, he saw a picture of Frances surrounded by four tiny dancing women. A month later Elsie supplied him with another picture, featuring herself encouraging a tiny figure to jump into her lap. Arthur Wright suspected the girls of playing tricks with the camera and refused to allow them to take any more pictures.

Three years later the Cottingley Fairies came to public notice when Elsie's mother Polly Wright attended a lecture on folklore and fairy beliefs. Mrs. Wright supplied copies of the photos to Edward L. Gardner, a London investigator into the occult. Gardner had the photos reproduced by a friend, H. Snelling, and gave a public lecture on them. He also brought them to the attention of Sir Arthur Conan DOYLE, the author of the Sherlock Holmes stories and an enthusiastic supporter of SPIRITUALISM. During the summer of 1920, Gardner met with Elsie and gave her a modern camera. He received in return three more pictures of fairies.

Both Doyle and Gardner believed in the authenticity of the pictures. Doyle published defenses of the photographs in the English magazine *The Strand* in December 1920 and March 1921. The following year he published an entire book on the subject, *The Coming of the Fairies* (1922), and Gardner followed with one of his own, *Fairies: The Cottingley Photographs and Their Sequel* (1945), 23 years later.

It was not until 1983 that Frances and Elsie confessed that all five photographs were fakes. Elsie had made the figures, and the girls had photographed them together. The two girls had originally intended to deceive only their parents, but when the photos became public, the two women resolved to conceal the truth until after the three major supporters—Doyle, Gardner, and Gardner's son Leslie—had died.

COUÉ, EMILE (1857–1926) A French pharmacist who at his clinic at Nancy in 1920 introduced a method of psychotherapy based on autosuggestion (self-induced suggestion). His method, which ran counter to his two great contemporaries, Sigmund FREUD and Carl Gustav JUNG, was to encourage each patient to set his or her personal goal and then to repeat frequently: "Every day, and in every way, I am becoming better and better."

Coué always stressed that he was not primarily a healer but a teacher who taught others to heal themselves. His lectures in England and the United States attracted a large following, but his ideas were eventually overshadowed by psychoanalysis, developed by Sigmund Freud. Later, psychoanalysis was found to encourage dependency because it relied on the psychoanalyst to interpret or analyze what the patient was saying. Coué's patients, on the other hand, were self-reliant.

Coué claimed that his method of autosuggestion could actually effect organic changes in the patient. Many doctors and scientists at the time scoffed at the idea, but modern studies have shown that the mind is important in the healing process. If the brain can accept as true what our inner talk tells it, the body will react accordingly. One example brings results in seconds: Subjects are told to close their eyes and imagine they're standing on top of a skyscraper; most people experience a sharp increase in pulse rate, some get queasy and a few actually fall over or become very sick. Similarly when watching high precipice or roof-top shots on television some people report going weak at the knees, although they are sitting safely at home.

Emerging studies show that there are links between what we say to ourselves and what we accomplish. Two psychologists—Shad Helmstetter, who wrote *What to Say When You Talk to Yourself* (1994), and Pamela Butler, author of *Talking to Yourself*—believe that people can get achieve more if they believe that they can do so. In his book *Learned Optimism,* another advocate of positive thinking, Martin E. P. Seligman, director of clinical training in psychology at the University of Pennsylvania, shows that pessimism and optimism are not fixed inborn traits but learned "explanatory styles" that can be unlearned. While a pessimist will construe a defeat as permanent, catastrophic, and evidence of personal ineptitude, an optimist will interpret the identical misfortune as temporary, controllable, and rooted in circumstances, or plain bad luck. Seligman's message is that learned behavior like pessimism can be unlearned.

There are many forms of mental illness that cannot be identified as coming from one dominant single cause. Some seem to be triggered by social events in the patient's life, some, like "baby blues," seem to arise from chemical changes in the body, and yet another explanation, "genetic predisposition" is the favorite disease concept of our times. Consequently, it would be wise to regard mental illnesses as caused by a complex interaction of many factors. But whatever the causes, a positive attitude should prove helpful, especially for those who are not severely depressed but just merely discouraged or overcome by gloom for a period; the relearning of one's attitude must be part of any plan to beat stress. Coué's phrase from the past: "Every day, in every way, I am becoming better and better" seems today to be coming through loud and clear.

For further reading: Hugh McNaughten, *Emile Coué: The Man and His Works* (Society of Metaphysicians, 1995), and Norman Vincent Peale, *The Power of Positive Thinking* (Mandarin, 1990).

COUNTING HORSES An interesting occurrence in the field of animal cognition that received plenty of observational evidence but no scientific approval. A famous case was that of Clever Hans, a horse that demonstrated to audiences many times that he could add, subtract, and even solve complicated mathematical problems by tapping out numbers with his hoof and stopping when he had reached the correct answer. The horse's trainer, retired German schoolteacher Wilhelm Von Osten, was not a trickster and sincerely believed that his horse was doing this unaided. But when the case was investigated by psychologist Oskar Pfungst, it was revealed that Von Osten and others who were brought in to ask the animal questions, would involuntarily jerk their heads when Clever Hans had tapped out the right answer. In this way the horse was cued. Further investigation discovered that Clever Hans was able to detect human head movements of one-fifth of a millimeter.

Information about the investigation of Clever Hans was published in 1911 and, according to the science writer Martin GARDNER, by 1937 there were more than 70 thinking animals making a living on the stage or circus

Counting Horses

for their owners. As well as horses there were cats, dogs, pigs, a dolphin, and two learned London geese.

For further reading: Martin Gardner, *Science, Good, Bad and Bogus* (Prometheus Books, 1981); and W. Broad and N. Wade, *Betrayers of the Truth: Fraud and Deceit in Science* (Oxford University Press, 1982).

COVENS A gathering of witches or practitioners of the Wicca religion. Covens usually have a fairly constant membership—often, according to some traditions, of 13 members. This number supposedly relates to Christ and his 12 apostles, but this idea probably arose during the Inquisition, when witches were persecuted as heretics; the idea of witches perverting Christ's "coven" or "convention" of disciples to their own satanic purposes would have appealed to the witch-hunters of that day. Witch covens during that time were accused of performing "black" (satanic) masses, which included wild orgies and the sacrificing of animal and human victims, casting evil spells on innocent Christians, and performing other nefarious deeds. Today, historians believe these accusations lacked factual basis, but were instead the work of overzealous church and court officials, who exaggerated facts, tortured accused witches until the kind of information they sought was proffered, and made up "evidence" to serve their cause of ridding society of heretics, nonconformists, and misfits.

Some historians say the whole idea of covens as a genuine part of WITCHCRAFT only arose in modern times, through the work of people such as Margaret Murray, an English anthropologist who popularized the idea of benign witchcraft based on her interpretation of pre-Christian pagan practices. Murray's books (including *The Witch-Cult in Western Europe,* 1921, and *The God of the Witches,* 1931) emphasized a link with nature, worship of a Goddess and a Horned God, and Celtic traditions and adaptations of them. Murray's ideas were further adapted and popularized by Gerald Gardner in the 1950s and following decades.

Followers of this revived "Wiccan" religion might practice alone or gather in covens to combine their "powers," derived from nature, to worship, cast protective spells, refresh members' psychic energy, and heal physical and other illnesses. Although Wiccans claim a strong historical tradition significantly predating Christianity based largely on anthropologist Murray's work, historians and anthropologists question the accuracy of this claim. Christina Hole (*Witchcraft in England,* 1977) is one of many scholars who challenge Murray's scholarship. This detail, however, does not deter dedicated Wiccans, who believe in the precepts of their religion regardless of its historical authenticity.

For further reading: Margot Adler, *Drawing Down the Moon* (Beacon Press, 1986); Laurie Cabot, *Power of the Witch* (Bantam, 1989); and Gerald Gardner, *Witchcraft Today* (Magickal Childe, 1991).

CRANIAL OSTEOPATHY A therapy promoted by Dr. William Garner Sutherland in the 1940s. The skull is formed of several plates with slight separations between them. They are displaced during birth and may not return to their proper positions subsequently, sometimes causing a misalignment of the facial bones, especially the jawbones. They also may be displaced later in life by blows, whiplash effects in automobile accidents, or even by pressure exerted during dental surgery. Sutherland believed that such displacements affect various parts of the body through pressure on nerves originating in the brain. Cranial osteopathy aimed to correct these displacements and their consequent pressure. The treatment consists of very gentle manipulation of the skull and the jawbones to restore them to a proper alignment. Osteopaths who practice cranial osteopathy claim some success in treating neuralgia and headaches, including migraine. There is also some evidence of success in treating high blood pressure and stomach ulcers.

Orthodox medical practitioners are skeptical about both the diagnosis and the treatment but tend to regard the treatment as harmless. Diagnosis is viewed differently: Unless the osteopath is also a qualified medical practitioner, there is a danger that some other more serious disorder may be overlooked.

CRANIOSACRAL THERAPY A comprehensive therapy that purports to treat the whole person by carefully balancing the membranes, muscles, bones and fluids, which together make up the craniosacral system.

The therapy was developed from cranial osteopathy, which was promoted by Dr. William Sutherland in the 1940s. Practitioners explain that the craniosacral system contains the vital dural membranes surrounding the brain, the spinal cord including the spinal fluid, and the fascia enveloping every organ, muscle, nerve, and blood vessel throughout the body. They believe that these are joined together in a symmetrical rhythmic movement associated with the rhythmic secretion and absorption of the cerebrospinal fluid. Consequently imbalances or restrictions anywhere within their vital system cause stress, strain, and dysfunction in associated areas throughout the body.

Treatment is exceptionally gentle. The practitioner simply places his or her hands very lightly on the patient's head or on any other appropriate area and applies pressure with a gently twisting movement. This basic movement both diagnoses the patient's condition and also corrects any malfunctions by stabilizing the imbalances at the very core of the trouble. This basic action will then reflect outward to remove any other restrictions, asymmetries, and imbalances and so restore proper functional balance to all parts of the body.

Craniosacral therapy claims to help anything from simple back pain to complex functional and emotional disturbances, persistent aftereffects of operations or physical injuries from accidents, and any other symptoms of "obscure origin." Therapists claim an especially important role in the treatment of birth trauma in babies, which they say could lead to cerebral palsy, autism, epilepsy, squinting, hyperactivity, reduced mental ability, learning difficulties, dyslexia, poor memory, susceptibility to allergies or just general constitutional weakness.

Because the therapy treats the patient as a whole individual and with sympathy and understanding, it is very effective as a helpful rather than a curative treatment. Patients are said to feel better, generally, and are less stressed and worried about their condition after a session, a very necessary state of mind to dispel impediments to the self-healing process. But like all alternative treatment, it should be complementary to conventional medicine. Anyone intending to resort to this therapy would be wise to consult his or her general practitioner about the condition beforehand. In this way, symptoms of "obscure origins" can be investigated, and patients can get the best of both worlds.

CREATION SCIENCE Or scientific creationism, a movement among fundamentalist Christians, especially in the United States since the 1970s, that maintains that the world was created by God, either exactly as described in the book of Genesis in the Bible or in some variation of that account allowing for a different time scale. Today's creationists go one step further—they assert that there is convincing scientific evidence in support of this view and press for scientific creationism to be taught in public schools alongside or in place of EVOLUTION. For creationists, evolution by means of natural selection is either false or mere speculation—that is, unsupported by the scientific evidence.

The idea that the world was created by a divine being by fiat has a long history in the intellectual community. Many thinkers of the 17th and 18th centuries accepted the idea that the world had been created in the fairly recent past. In his scholarly work, *Annales veteris testament, a prima mundi origine deducti* (Annals of the Old Testament, deduced from the first origin of the world), in 1653, James Ussher the archbishop of Armagh, Ireland, dated the creation as taking place on October 23, 4004 B.C.E., at 12.00 noon. His dating was done by very carefully and painstakingly plodding through the Old Testament record and, when that failed, by cross-referencing to other historical records of antiquity. Ussher's choice of the date 4004 B.C.E. derived from comparison of calendars; finally, he assumed that the creation of light must have occurred at midday. Ussher's whole work was meticulously researched, using the best information then available and commanded the respect of his contemporaries.

In the following century, as puzzling evidence that did not fit easily into the Ussher theory began to accumulate, theorists split into two groups: evolutionists and creationists. The great French anatomist, Baron Georges CUVIER, explained the fossil record of extinct species by a succession of divine creations and extinctions. The evolutionists challenged such views, holding that the biblical story was just that, a myth. It may well have been Darwin's growing realization, as his researches continued, that his theory of evolution by natural selection was going to put the cat among the pigeons (more appropriately perhaps, the humans among the apes), that caused him to delay publication so long, until he had abundant evidence to support his theory.

As the evidence accumulated during the 18th, 19th, and 20th centuries and the vast majority of biologists accepted the fact of evolution, creationism maintained its adherents primarily among certain fundamentalist religious sects.

Creationism came to public prominence in the United States in 1925 when the Tennessee legislature passed the Butler Act, which made the teaching of any theory denying divine creation illegal. A small group of citizens in

Dayton then charged a local teacher, John Thomas Scopes—with his agreement—with violating this law. What might have been a small local affair became a national matter when the American Civil Liberties Union seized the opportunity to challenge the state's law and hired Clarence Darrow, a famous trial lawyer and agnostic, to defend Scopes. The great Christian politician and orator, William Jennings Bryan, spoke for the prosecution. The ensuing courtroom battle, which became known as the Scopes Monkey Trial, has passed into folk history, particularly in Hollywood's version in the play and the two film versions of *Inherit the Wind.* Scopes lost, was convicted, and fined $100; the conviction and fine were later overturned on a technicality. Bryan died of a heart attack five days after the case ended. A fundamentalist Bryan College, where creationism is taught and evolution is not, was established in Dayton in his honor and thrives to this day. The Butler Act remained in force until it was repealed in 1967.

There is a superficial plausibility in creationist arguments (a) for a shortened timescale for life on earth, (b) for no species to have emerged from any other by "descent with modification," (c) for the human species to have coexisted with the other species of the fossil record, and (d) for a global catastrophic flood. Fundamentalists now present these arguments as counter to the evolutionists' Darwinian theory, as *scientific* creationism. In 1963 the Creation Research Society was founded to promote the cause of creation science, and by the late 1960s creationists were claiming that evolutionists in the public schools were teaching a rival religion—secular humanism—forbidden by the First Amendment. They demanded that schools give equal time to scientific creationism, essentially the biblical version of creation.

In 1973 the Tennessee legislature passed a bill requiring public schools to give evolution and creationism equal treatment. Textbooks used in these schools had to print a declaration emphasizing that statements made about the origins of humankind were not to be taken as facts. (The Bible was declared a reference work and therefore exempt.) The law was later judged unconstitutional. In the late 1970s legislatures in 14 states introduced bills that required the teaching of creationism together with evolution. In Arkansas and Louisiana these bills were passed into law but were overturned by federal courts in 1982. In 1984 the Texas board of education removed its own rule restricting the teaching of evolution in state-supported schools.

For further reading: Langdon Gilkey, *Maker of Heaven and Earth: Christian Doctrine of Creation in the Light of Modern Knowledge* (University Press of America, 1986); Langdon Gilkey, *Nature, Reality and the Sacred: The Nexus of Science and Religion* (Augsburg Fortress, 1993); D. T. Gish, *Evolution? The Fossils Say No!* (Creation-Life Pubs., 1979); L. R. Godfrey, *Scientists Confront Creationism* (Norton, 1983); Stephen Jay Gould, *Hen's Teeth and Horse's Toes: Further Reflections in Natural History* (Norton, 1983); Dorothy Nelkin, *Science Textbook Controversies and the Politics of Equal Time* (MIT Press, 1977); Ronald L. Numbers, *The Creationists* (University of California Press, 1992); and Gordon Stein, ed., *The Encyclopedia of the Paranormal* (Prometheus, 1996).

CREATION SCIENCE RESEARCH CENTER A facility to promote the teachings of CREATION SCIENCE. The Creation Science Research Center was founded in 1970 by Nell Seagraves, her son Kelly Seagraves, and Henry Morris as a division of Christian Heritage College, the independent fundamentalist school associated with Scott Memorial Baptist Church, San Diego, California. Shortly after the founding of the center, differences developed between the Seagraves and Morris: Morris, the center's director, wanted it to focus on education and scholarship; the Seagraves pushed for more promotional activity. The issue came to a head when the Seagraves approved the publication of some material that Morris viewed as sloppy and too expensive. The Seagraves and a majority of the center's board withdrew, severed their connection with the college, and reorganized in other facilities. Morris reorganized those who supported him as the Institute for Creation Science.

The center's subsequent direction proved too marginal even for the many supporters of creation science, and during the next few years support dropped. Kelly Seagraves wrote a book, *Sons of God Return* (1975), which linked UFOLOGY, Flood geology, and demonism. The center has developed the argument that EVOLUTION is essentially equivalent to a religious worldview and that it is illegal to teach it exclusively in the public schools. To do so amounts to having a public establishment of religion.

Creation Science

The center continues to exist but has slowly ceased to be a force within the larger national creationist movement.

For further reading: Kelly Seagraves, *Sons of God Return* (Pyramid Books, 1975); and Ronald L. Numbers, *The Creationists: The Evolution of Scientific Creationism* (University of California Press, 1993).

CRIMINAL GENES The idea that there is a gene that predisposes its possessors to criminal or aberrant behavior. In the 19th century, Italian physician Cesare LOMBROSO claimed to have identified features that characterized criminals: "the enormous jaws, high cheek bones, prominent superciliary arches, solitary lines in the palms, extreme size of the orbits, handle-shaped ears," characteristics shared by criminals, savages, and apes, which led them to exhibit "insensibility to pain, extremely acute sight, tattooing, excessive idleness, love of orgies, and the irresponsible craving of evil for its own sake, the desire not only to extinguish life in the victim, but to mutilate the corpse, tear its flesh and drink its blood."

Modern proponents of the idea that criminality is inherited now focus on genes. Quite recently, researchers testing inmates of a U.S. prison found that there was a higher-than-usual incidence of men carrying an extra Y on the sex chromosome (the female sex chromosome has XX; the normal male sex chromosome, XY). These abnormal individuals had XYY. The claim was then made that the extra Y—extra maleness—led to criminal tendencies. The motivation of such studies is to identify those individuals with these criminal traits and to persuade or induce them not to reproduce; in this way, the criminal behavior in society will diminish and perhaps eventually disappear. It is one stand of EUGENICS, the belief that social and medical problems can be bred out. All such claims have eventually been discredited or disproved.

For further reading: Stephen Jay Gould, *The Mismeasure of Man* (Norton, 1981); Steven Rose, Richard C. Lewontin, and Leon J. Kamin, *Not in Our Genes* (Penguin, 1984); and Richard C. Lewontin, *Biology and Ideology: The Doctrine of DNA* (HarperCollins, 1993).

CROOKES, SIR WILLIAM (1832–1919) One of England's outstanding scientists who during the early years of his career became involved in PSYCHICAL RESEARCH and with one of the most controversial MEDIUMS of modern SPIRITUALISM, Florence Cook. After graduating from the Royal College of Chemistry, he went to work in the Meteorological Department at Radcliffe Observatory. His first major contribution was the discovery of the element thallium in 1861. He was elected to the Royal Society two years later. Crookes invented the spinthariscope, the radiometer, and the Crookes tube (which allowed the study of cathode rays). He founded the *Chemical News* and the *Quarterly Journal of Science*. In acknowledgment of his

Crookes

scientific accomplishments, he was awarded the Royal Gold Medal (1875), the Davy Medal (1888), and the Sir John Copley Medal (1904). He was knighted in 1897, received the Order of Merit (1910), and served a term as president of the Royal Society.

Crookes became interested in spiritualism following the death of his brother in 1867. He had some preliminary sittings with various mediums prior to his announcement in 1870 that he would begin a study of spiritualist phenomena. Crookes first investigated Daniel Douglas Home and became convinced that his LEVITATION and related psychokinetic phenomena were caused by a psychic force. Beginning in December 1873, Crookes worked with Florence Cook, a MATERIALIZATION medium. At one point he had himself photographed with "Katie King," the entity who materialized in the séances. He wrote glowingly of what he had experienced at these séances and defended Cook against charges of fraud. Crookes also conducted a series of studies with the American medium Annie Eva Fay. He then dropped further psychical research and turned his attention to the work that later brought him so much fame and honor.

Crookes continued to affirm his results in studying the mediums, and for three years (1896–99) served as president of the SOCIETY FOR PSYCHICAL RESEARCH (SPR). Quietly, he continued to attend séances throughout his life. His support was important to the SPR, and his refusal to renounce psychical research became a problem to many of his colleagues. While the controversy of his championing of Cook largely died after he withdrew from active research, it revived for a time in the 1880s when Mary Showers, who had assisted in the Cook séances, was on two occasions caught in trickery.

Crookes died in 1919. Three years later it was revealed that Cook's mediumship was fraudulent. She had had an affair with Crookes and had used the séances as a cover for their meetings. About this same time, stage magician Harry HOUDINI published his account of an interview with Annie Eva Fay. He claimed Fay had revealed the methods by which she had fooled Crookes.

In *The Spiritualists* (1962), Trevor H. Hall charged that Crookes was involved in a fraud to hide his affair with Cook. Whether Crookes was duped by Cook's tricks or was himself part of a fraud has not yet been determined.

CROP CIRCLES Strange circular depressions in fields of grain crops. In England during 1976, patterns of unknown origin were observed, impressed on fields of grain. At first the patterns were simple circles but, as years passed, the number, locations, and complexity of these circles increased. The idea that some human agency might be responsible was dismissed: There were just too many and they were too widespread and sophisticated for them to be the product of someone playing a prank, some sort of a joke. They appeared in short time during the night, and there were no footprints of pranksters leading to or from the pictograms. Besides, who would want to do such a thing and why?

Having ruled out a human origin, people, some of whom had scientific training, began to examine the sites in detail and postulate possible causes. Some of the pat-

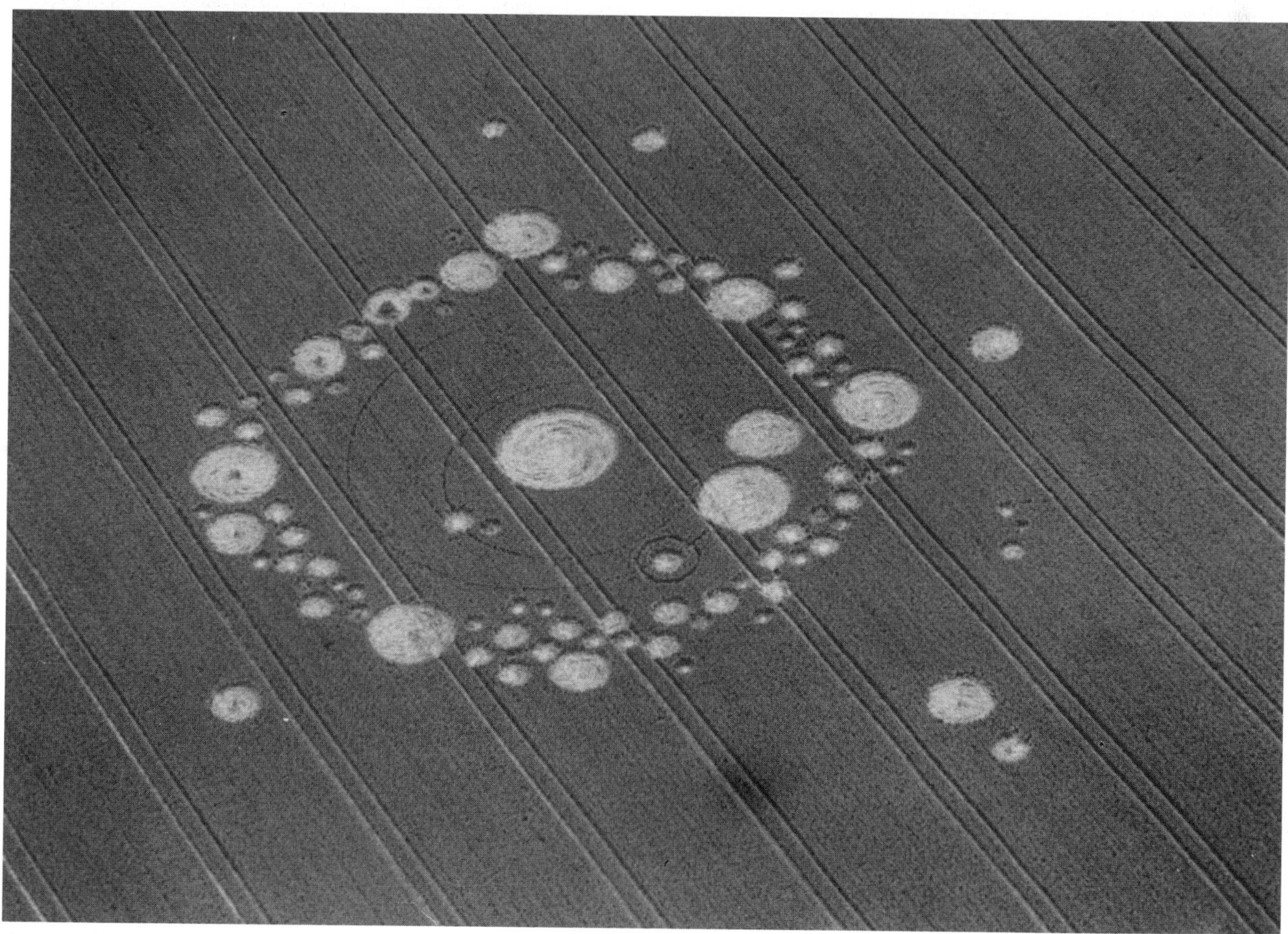

Crop Circles

terns showed a central circle surrounded by four symmetrically placed smaller circles—clearly, it was concluded by some, caused by a FLYING SAUCER and its four landing pods. Another explanation was unusual wind patterns; Dr. Meaden of the Circles Effect Research Group (CERES) suggested that the circles were caused by an electrically charged whirlwind, or plasma vortex, which when it collapses does so with a powerful downward gust of wind. That might be a possible explanation for simple circle, but not for the more complex patterns. As these speculations and the interest they aroused spread, a new discipline was formed—cerealogy—with its magazines, societies, and books.

The balance of argument swung in favor of those positing UFOs and aliens. The patterns were too large, too varied, and too complex for them to have a natural cause; clearly some higher intelligence was—had to be—responsible. The patterns might be a sort of geometrical language by which visiting aliens were telling us something. The phenomenon excited the interest of members of the NEW AGE MOVEMENT, who came in their hundreds to inspect these crop circles, and especially to witness some new occurrence. Enthusiasts waited all night, equipped with the latest gadgetry, in the hope of seeing a new pattern being produced by aliens in their space ships. The media made the most of the story. Nobody actually saw a circle being formed, nor any UFO or alien before, during or after the event. Despite that, dowsers authenticated their alien origin and channelers made contact with the aliens responsible. ORGONE ENERGY was detected within the circles.

The phenomenon grew. At first a few circles, then many more, all within 30 or 40 miles of London, then further afield into the hundreds, with 112 reported in 1988, 305 in 1989, and almost 1000 in 1990. The phenomenon eventually spread to the United States, Canada, Bulgaria, Hungary, Japan, and the Netherlands. The pictograms, especially the more complex of them, began to be cited increasingly in arguments for alien visitation. Ability in extremely sophisticated mathematics was detected in these figures, which could only be the work of a superior intelligence. One matter on which almost all of the contending cerealogists agreed was that the later crop figures were much too complex and elegant to be due to mere human intervention, much less to hoaxers. Extraterrestrial intelligence was apparent at a glance.

In 1991, Doug Bowers and Dave Chorley from Southampton, England, announced that they had been making crop figures for 15 years. They had been amused by UFO reports and thought it might be fun to spoof the UFO believers. At first, they flattened the wheat with the heavy steel bar that Bowers used as a security device on the back door of his picture–framing shop. Later, they used planks and ropes.

Bowers and Chorley were delighted with the response, especially when scientists and others began to announce their considered judgment that no merely human intelligence could be responsible. Carefully, they planned more elaborate nocturnal excursions, sometimes following meticulous diagrams they prepared in watercolors. They closely tracked their interpreters. When a local meteorologist deduced a kind of whirlwind because all of the crops were deflected downward in a clockwise circle, they confounded him by making a new figure with an exterior ring flattened counterclockwise.

Eventually, Bowers and Chorley tired of the increasingly elaborate prank. So they confessed. They demonstrated before reporters how they made even the most elaborate and insectoid patterns.

Crop circles are a classic case of pseudoscience in the CONVICTION SCIENCE category. Even today, while the craze has died down, there are many who are convinced that, though *some* circles may have been the work of hoaxers, there were far too many for all to have been.

For further reading: Carl Sagan, *The Demon-Haunted World: Science as a Candle in the Dark* (Headline, 1996); Jim Schnabel, *Round in Circles* (Prometheus Books, 1996); and Pat Delgado, *Crop Circles: Conclusive Evidence?* (Bloomsbury, 1992).

CROSS-CORRESPONDENCES A form of communication from the dead through MEDIUMS; these are believed by many psychical researchers to be among the best evidence of survival after death. The correspondences are established by bringing together the meaningless utterances of two or more mediums working independently of each other, utterances that only become meaningful when combined. Classical cross-correspondence cases are among the most famous researched by the SOCIETY FOR PSYCHICAL RESEARCH.

The incidents of cross-correspondences began soon after the deaths of Henry Sidgwick (1900), Edmund Gurney (1888), and F. H. W. Myers (1901), the leaders of the society's first generation. They involved several automist–mediums including Helen de G. Verrall, Alice Fleming (Mrs. Holland in the literature), Margaret Verrall, and Winifred Coombe-Tenant (Mrs. Willett in the literature). Alice Johnson, the society's research officer, seemed to have first posed the hypothesis that Myers might be the agent behind such a scheme when she discovered fragments in the messages from several mediums that only made sense when considered together.

Illustrative of cross-correspondences is a reference in a communication reportedly from the deceased Myers to Margaret Verrall referring to St. Paul on the road to Damascus and mentioning Ernest Renan's *Chemin de Damas*. Two days later Coombe-Tenant received a message to write to Verrall a letter with the words *Eikon Renam*,

Eikon Renam (a misspelling of *Renan*). It was later discovered that in his *Chemin de Damas,* Renan had discussed Paul's trip to Damascus on which he saw Jesus the Eikon. Fitted together, the two essentially meaningless fragments made an entire message.

An American case of cross-correspondence was seemingly initiated in 1928 by Walter, the spirit control of Mina Crandon, publicly known as "Margery." Simultaneous messages were received by Margery in Boston and by three other mediums in Maine, New York City, and Niagara Falls. While some of the cross-correspondence manuscripts have been examined, it is laborious work that requires some training in classical literature, and the great majority of the manuscripts remain unstudied.

See also MARGERY CONTROVERSY.

CROWLEY, ALEISTER (1875–1947) Modern exponent of ceremonial MAGIC. Crowley was a product of the late-19th-century magic revival which assumed that there was an invisible power, analogous to the "magnetic" power discussed by the mesmerists, that was the agent of magical operations. He later developed his own system of magical theory and practice, which he termed *thelema* (from the Greek word for will). He defined magic (which he usually spelled *magick*) as the art of producing change in the world using the universe's underlying cosmic power through an intentional act of the will.

Two events largely gave structure to Crowley's life after he left the HERMETIC ORDER OF THE GOLDEN DAWN, the organization that introduced ceremonial magic to him. In 1909, in Cairo, Egypt, he received a document called *The Book of the Law* through AUTOMATIC WRITING from an entity called Aiwass. *The Book of the Law* included the basic principles of thelemic magic, including its definitive statement, "Do what thou wilt shalt be the whole of the law." This statement, often misunderstood as a license to do whatever one wanted to, is interpreted by thelemites as a call to discover one's destiny (or true will) in life and then to subordinate everything else in life to the fulfillment of that destiny.

In 1912, during some magical operations, Crowley also came to understand the basic structure of modern Western sexual magic, the use of sexual excitement to raise and focus magical energy. During succeeding years, he experimented widely with the practice and became the leader of the Ordo Templi Orientis whose inner teachings detailed the operations of sex magic.

Today, Crowley has become a cult hero in various counterculture movements, though the number of people who actually attempt to follow his magical teachings is quite small—the largest ceremonial magic group having membership counted in the hundreds. His system did, however, strongly influence the development of modern neo-pagan WITCHCRAFT. Crowley's magic utilized the psychological insights available in the early 20th century. He understood that an essential element of magical development was exploration and mastery of one's own inner psychological dimension. In that endeavor, he used a variety of mind-altering drugs. He largely jettisoned the supernatural element in the teachings of medieval magic and, in particular, cautioned his students against considering "spirit entities" evoked and invoked in magical operations as having any objective reality.

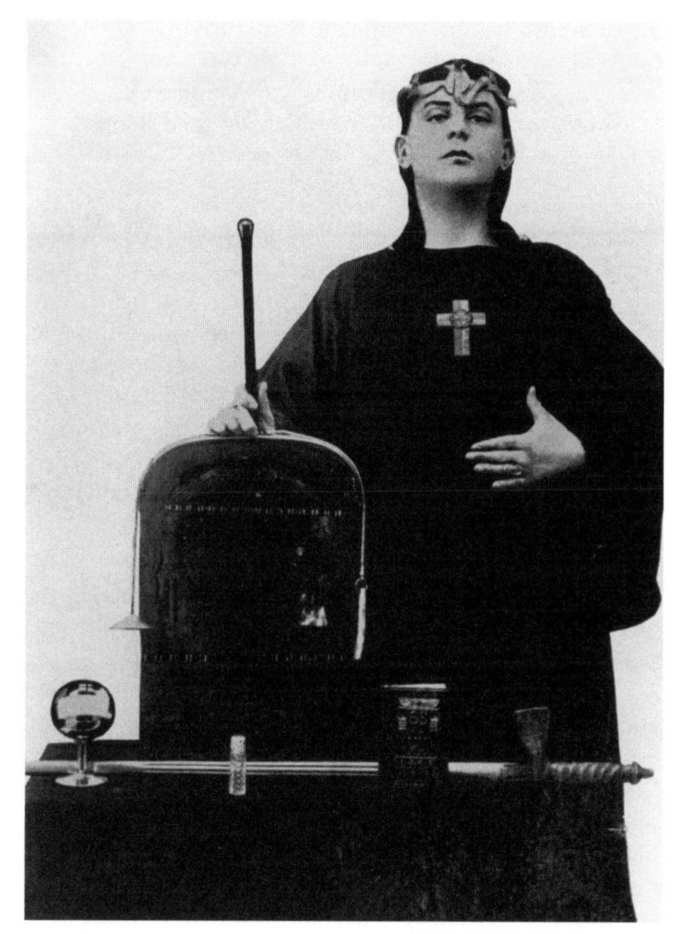
Crowley

CRYOGENICS The production and use of low temperature—temperatures in the range of -150°C to -273°C (-238°F to -459°F), which is roughly a range of 120°C (220°F) above absolute zero. It has many applications, and new ones are constantly being added:

1. the use of liquid oxygen to fuel rockets
2. liquid nitrogen to store human embryos in fertility clinics
3. the transport of liquefied natural gas by tanker and ship and through insulated pipelines
4. liquid natural gas as automobile fuel

5. food preservation for storage and transport
6. cryogenic surgery
7. superconducting materials in many applications: power generation and distribution, transport, and so on
8. the storage of sperm and body parts

Long-term storage of body parts is very much in the realm of pseudoscience: It has been suggested as a way of enabling astronauts to survive a many-year space voyage en route to another star system. Long-term storage is currently used to freeze corpses until medical technology can cure the condition from which they died. Mainstream science believes that the complex and varied nature of the human body makes impossible the freezing of humans and their subsequent return to normal.

For further reading: Randall F. Barron, *Cryogenic Systems*, 2nd ed. (Oxford University Press, 1985).

CRYONICS The storing of dead bodies through the process of CRYOGENICS for the purpose of later reviving individuals.

Some patients elect to have only their heads frozen and stored, believing that when revival occurs, this will be sufficient for reconstructing the whole person. Two firms in the United States, Alcor and the Cryonics Institute, have put the procedure into commercial operation. Mainstream scientists, however, question the validity of the procedure: They point out that ice crystals damage cells beyond repair. To be effective, scientists must have not only solved this problem but also found a cure for the disease that killed the patient.

But cryonics has many enthusiastic advocates and will doubtless continue. Enthusiasts dismiss the doubters as having no understanding of the speed and extent with which human ingenuity overcomes such obstacles.

CRYPTOZOOLOGY A controversial branch of zoology focusing on animal species that have been reported to exist but are as yet unknown to science. The most famous hidden or unknown species include LAKE MONSTERS (such as the LOCH NESS MONSTER), obscure hominids (such as BIGFOOT), and SEA SERPENTS.

The term *cryptozoology* was coined by French zoologist Bernard HEUVELMANS. Fascinated by such creatures since his youth, he pursued his interest over the years after receiving his Ph.D. in zoology. His research culminated in an initial study, *On the Track of Unknown Animals* (1955). The volume sold more than a million copies. In the wake of the popular response, he was finally able to pursue his interest full-time and to develop a network of other cryptozoologists. His research led to a second volume, *In the Wake of Sea Serpents* (1968), and eventually to the formation of the INTERNATIONAL SOCIETY OF CRYPTOZOOLOGY in 1982.

The society defined its mission as the search for the possible existence of known animals in areas in which they were currently believed not to exist and the discovery of the continued existence into modern times of animals presumed to have become extinct in the distant past. The unexpected presence of an animal is the characteristic that makes it a focus of cryptozoological concern.

The existence of cryptozoology is to some extent an artifact of the emergence of zoology as a separate area of scholarly inquiry in the 18th century. There was a legitimate attempt to separate known species of animals from the fabulous creations of myth and legends. Biologists often spoke with contempt of the idea that large as yet unknown animals could exist, and they held up to ridicule colleagues who chose to examine the possibility of their existence. But over the decades a few such significant animals were discovered. The most impressive new entries into the biologist's textbook included the gorilla, the giant squid, and the giant panda. Admittedly the number of new species, especially as we move closer to the present, has been few but nonetheless startling. They include the mountain gorilla (discovered in 1903), the okapi (1900), and the COELACANTH (1938). The coelacanth was particularly significant for the future development of cryptozoology as a clear example of a species known from the fossil record and believed to have become extinct some 60 million years ago.

Modern cryptozoology is frequently dated from the 1892 publication of Antoon Cornelius OUDEMANS's study of *The Great Sea-Serpent* in which he attempted critically to evaluate the numerous reports from mariners of a large serpentlike creature residing in the sea. Oudemans would conclude, and many of his contemporaries agreed, that many of the reports were accurate accounts and that a single animal lay behind them. Most did not accept his conclusion that the animal was probably a peculiarly large, long-necked seal. Cryptozoological concerns would be passed along to 20th-century researchers by Charles FORT as one of a number of unexplained anomalous natural phenomena, along with Fort's particular disregard of what he saw as mainstream science's refusal to consider phenomena that did not fit its current worldview.

In the years immediately after Fort's death, the Loch Ness monster received worldwide attention after the accumulation of some 20 reports over a six-month period in 1933, so did the discovery of the coelacanth in 1938. In the 1940s Forteans such as biologist Ivan Sanderson pointed to these and other reports, for example those of a sauropodlike creature from Africa, the MOKELE-MBEMBE, to suggest the possible continued existence of dinosaurs.

In the wake of Heuvelman's work, numerous books discussing cryptozoological issues appeared. Most dealt with the possible existence of a specific hypothesized

Cryptozoology

species, such as the Loch Ness monster or the YETI, although a few attempted to survey the field. The cryptozoologists associated with the International Society have sought to verify reports of unusual species. Roy MACKAL's volume on the Loch Ness monster is a classic statement of this more conservative cryptozoological position. Others have offered elaborate paranormal explanations both for the reports of unusual animals (from werewolves to dragons and satyrs) and for the difficulty that scientists have had in locating them.

For further reading: Roy Mackal, *The Monsters of Loch Ness* (Swallow Press, 1976); and Bernard Heuvelmans, *On the Track of Unknown Animals* (Hill and Wang, 1958).

CRYSTAL GAZING A specific technique in DIVINATION that uses a crystal ball to focus the attention of the "scryer" (the name given to this kind of fortune-teller). A real crystal ball is a polished globe of rock-crystal or clear quartz, but these are costly and most scryers now use a less expensive, commercially available ball of molded glass. Other objects such as polished black mirrors and bowls of water have been used.

The history of scrying is full of recommendations for a great variety of helpful objects. According to John Aubrey, the 17th-century writer, the very best stone to help achieve a psychic state is the "shrew stone," a smaller and much more expensive version of the crystal ball made of round, cut semi-precious stone beryl, and, he adds, one that is slightly tinged with red.

Modern crystal gazers believe that there is nothing supernatural or occult about the ball itself; they see it as only a means to an end, and they believe that it is their own visionary experience that is important.

By gazing fixedly at the ball and trying to detach their minds from the everyday world of sense experience, scryers often achieve a state of self-hypnosis. This is most easily accomplished in a darkened room with one source of light, preferably from a fire. Perfumes are often used, or varieties of incense are burned. While focusing on the crystal, most scryers do not see moving visions but are merely aware of opaque clouds drifting around the interior of the ball; sometimes the clouds are white, sometimes a single color. Some practitioners attach their own meanings to these colors; others use conventional color interpretations; white and silver as degrees of good fortune; gold, prosperity; gray and black, degrees of ill fortune; green, emotional success; blue, career success; red, danger, or sexual passion, and so on. Less frequently, scryers believe that they can enter strange worlds of consciousness and wander freely through time and space. Those who think themselves to be specially gifted not only see whole moving pictures, but feel they are entering into and participating in the scene.

Mainstream science sees no validity in the practice.

For further reading: Francis X. King, *The Encyclopedia of Fortune Telling* (Hamlyn, 1988); and Bob Couttie, *Forbidden Knowledge: The Paranormal Paradox* (Lutterworth Press, 1988).

CUFORHEDAKE BRANE-FUDE A NOSTRUM that became the subject of the first court case under the U.S. Pure Food and Drug Act of 1906. Cuforhedake Brane-Fude was developed by Robert N. Harper around 1888 while a student at the Philadelphia College of Pharmacy. Earlier in the decade, a coal-tar derivative, acetanilid, had been discovered to have some effect in bringing down fever and deadening pain. Harper developed a nostrum that included it as a major ingredient. Adding doses of alcohol, caffeine, potassium bromide, and other substances, he marketed it under the name Cephalgine and sold it as "brain food."

After graduation, Harper moved to Washington, D.C., where he became a successful businessman and president of the retail drug association. Following the passage of the Pure Food and Drug Act, the Bureau of Chemistry, which is responsible for enforcing the new law, charged that Harper had fraudulently marketed his product as "brain food." It also maintained that the product included a large percentage of acetanilid, which had been shown to be harmful when taken in large doses. In one of his articles for *Collier's*, journalist Samuel Hopkins ADAMS had named some 22 victims of acetanilid poisoning.

In 1908, Harper was convicted on the key issue of describing his product as brain food. He was fined $700, the maximum fine under the law, but escaped a jail sentence. Government officials were happy that the new law was effective. Harper, however, merely adjusted the labeling of his product and continued to market it as before.

CUVIER, BARON GEORGES (1769–1832) The French father of paleontology, the science of past plants and animals based on fossil evidence. Considered to be one of the greatest scientists of his day, Cuvier was adamantly opposed to the orthodox Christian view of creation and was committed to empirical science, trying to understand the incomplete fossil record. A catastrophist he believed that the fossil record, with its many gaps, indicated a cyclic history and also that there had not been one creation, as described in Genesis, but several periods, each laying down its geologic record, each destroyed by some catastrophe and buried, requiring that a fresh start be made. He concluded that the creation stories from many religions told of events (in the Bible, it was Noah's flood) from the most recent of these catastrophes.

Cuvier was the first to search the fossil record for evidence of the extinction of species. To do that he developed a classification theory that laid the foundation of comparative anatomy. Although he thought that Earth was millions of years old, Cuvier did not believe in a slow process of evolution. In 1812, he said "There is little hope of discovering new species of large quadrupeds." That implied an Earth that was millions of years old, not the few thousand years of the theologians. He then had to explain why and how. The huge gaps in the fossil record persuaded him that the catastrophe theory best fitted the available evidence, with least recourse to speculation, least dependence on missing links, steps in an evolutionary chain for which there was no fossil record. Within any one period between catastrophes, he saw species as unchanging, not evolving; changes occurred when a new set of species was created after a catastrophe.

Cuvier produced a huge volume of painstaking, sound and reliable research; he established paleontology and comparative anatomy on firm bases. In addition to his scientific work, he played a role in French educational reforms, and from 1819 to his death presided over the Interior Department of the Council of State. He deserves to be remembered and honored as one of the great 19th century scientists, on a par with Charles DARWIN and Charles LYELL.

For further reading: Stephen Jay Gould, *Hen's Teeth and Horse's Toes* (Penguin/Norton, 1984).

CYCLES THEORIES Cycles theories have emerged in a number of scientific disciplines. In biology, Wilhelm FLIESS theorized that two cycles governed living cells. In men the male cycle of 23 days was dominant; in women the dominant female cycle was 28 days. These cycles affected everything from birth to death, physical and mental health, career prospects, sex, life, and so on. In his major work *The Rhythm of Life: Foundations of an Exact Biology* (1923), Fliess advances a formula to explain how these two cycles combine to impose various compound cycles: $(23x + 28y)$, in which x and y can be positive or negative integers. So the familiar 28-day menstrual cycle has $x = 0$ and $y = 1$. If x is 4 and y is 11 the sum is 365 days, another familiar cycle. What had escaped his attention is that x and y can be selected to give every number from 1 to as high as you have the patience to calculate. A theory which contains everything explains nothing.

In the late 20th century, BIORHYTHMS were held to influence or control human lives. A third cycle was added to the original pair: the intellectual cycle of 33 days. The male cycle governs physical strength and endurance and what are held to be masculine attitudes; the female cycle governs the feminine virtues of nurturing, sensitivity, and mental alertness; and the intellectual cycle governs mental capabilities. People are advised to adjust their actions to these cycles.

Cyclic theories also loomed large in geology for many years. Eighteenth-century Scottish geologist James Hutton thought that the geologic record was best accounted for by a cyclic theory. Early 19th-century French scientist Baron Georges CUVIER believed that a reported cycle of creation, catastrophic destruction, and regeneration was the most supportable explanation of the fossil record with its many apparently unconnected, disconnected stages. These catastrophists were opposed by those favoring the alternative, a gradual change from one primitive beginning, unfolding slowly to the present—a uniformitarian view, of which there were several versions. It was Charles DARWIN's theory of EVOLUTION by a process of natural selection that swayed the balance in favor of the uniformitarians and spelled an end to cyclic geological theories.

For further reading: Stephen Jay Gould, *Time's Arrow, Time's Cycle* (Harvard University Press, 1987/Penguin, 1990); and Martin Gardner, *Science, Good, Bad and Bogus* (Prometheus Books, 1989).

D

DARSEE, DR. JOHN (1948–) The most notable medical faker of the 20th century. Only 31 years old when he joined the faculty at the Medical School of Harvard University in 1979, Darsee brought with him a reputation of hard-working persistence, intelligence, and integrity. Within the first two years at Harvard, Darsee produced almost 100 papers and abstracts.

The discovery of Darsee's fraud began in 1981. Colleagues noted that he had disposed of a dog carcass without first removing the heart, unusual because Darsee was in cardiology research. They retrieved the dog and stored it. They now monitored Darsee's activities and discovered him faking an electrocardiogram tracing in another animal. When his activities were reported to his superiors, Darsee confessed but argued that it was an isolated incident; he was the victim of stress and overwork. The authorities at Harvard gave him a slap on the wrist and overlooked his problem for the moment. More significantly, researchers at the National Institutes of Health found that Darsee's results were not only unconfirmed by research elsewhere, but flatly contradicted what others were finding. This discovery not only tarnished Harvard's reputation, but also sent a seismic wave through the world of medical research. Darsee stated that he had no memory of faking any data.

Not only were all of Darsee's papers removed from research files, but he was fired from Harvard and barred from receiving any federal grants from the National Institutes of Health. He took a nonresearch position at a hospital in Schenectady, New York. Darsee's case, given the crucial nature of his research program, received widespread publicity and pointed to a significant problem that has emerged in the highly competitive field of scientific research (in every discipline) where position and money are dependent upon both a record of publication and the discovery of positive results.

DARWIN, CHARLES ROBERT (1809–1882) British naturalist and biologist whose theory of organic EVOLUTION revolutionized science, theology, and philosophy. Born in Shrewsbury, he attended the universities of Edinburgh and Cambridge, there studying medicine and theology, respectively. Near the end of his undergraduate studies, he was befriended by J. T. Henslow, professor of botany at Cambridge, who later secured for Darwin the post of naturalist "without pay" aboard the HMS *Beagle*. Between 1831 and 1836, the ship journeyed through the Southern Hemisphere, during which time Darwin was able to make extensive observations of the flora, fauna, and geological formations of widely separated points around the globe. This experience laid the course for Darwin's work thereafter and was the foundation for many of his theories. Upon his return, Darwin resided in London for six years, where he became acquainted with many leading scientists of his day: Sir Charles LYELL, Sir Joseph Hooker, and T. H. Huxley were among his closest friends. In 1842 Darwin moved to the secluded village of Down, in the county of Kent, where for the remaining 40 years of his life, he conducted research and wrote the monumental works that would make him famous.

One of the greatest biologists of the 19th century, Darwin produced numerous works throughout his life on a diverse range of subjects, including volcanic islands, coral reefs, orchids, the movement of plants, variation of domesticated animals and plants, the action of earthworms in the soil, evolution, and many more. His most highly acclaimed works, *On the Origin of Species by Means of Natural Selection* (1859) and *The Descent of Man* (1871), began a revolution in thought as profound and far-reaching as that initiated by Copernicus. In these two works, he established the theory that all living things, including human beings, developed from a few very simple forms, or perhaps from one form, by a gradual process of descent with modification. Additionally, his theory of natural selection, supported by a large body of evidence, accounted for this process and explained the "transmutation of species" and the origin of adaptation. Popular reactions to his theories focused on their religious and ideological implications, often with such vehemence that Darwin shrank from addressing the controversy himself, which he found to be utterly distasteful. Darwin never claimed to have originated the concepts of organic evolution, mutability of species, or natural selection. His work instead supplied the first scientific proof that such theories were applicable to the living world.

DAVIS, ANDREW JACKSON (1826–1910) Spiritualist, born in Blooming Grove, New York. Poor and uneducated, Davis was apprenticed to a shoemaker in Poughkeepsie, N.Y., at the age of 15. Under the tutorship of a Mr. Livingston in that town, Davis developed supposed clairvoyant powers and, in March, 1844, claimed under prolonged trance to have conversed with spiritual beings and mentors. While in trance, he spoke in depth on many abstruse subjects and diagnosed and prescribed for the sick, who came from around the country to consult him. In 1845, at the age of 19, he dictated to a Rev. William Fishbough, while asleep, *The Principles of Nature, Her Divine Revelations, and a Voice of Mankind,* which was published in a 500-page volume and widely read, especially in New England and New York. The book contained many original ideas about life here and in the hereafter, some beautifully expressed, some contradicting the Bible, and some simply incomprehensible. Davis tried lecturing, unsuccessfully, and therefore devoted himself to his writing, based on *The Principles of Nature . . . ,* which remained his most notable work. His books, all of which he claimed to have written under the influence of spirits, include *The Great Harmonia* (6 volumes, 1850–61); *The Philosophy of Spiritual Intercourse* (1851); *The Present Age of Inner Life* (1854; 2d ed. 1870); *The Approaching Crisis, a Review of Dr. Bushnell on Spiritualism (1852); The Penetralia* (1856); *The Magic Staff—An Autobiography* (1857); *The Harbinger of Health* (1862); *Appetites and Passions* (1863); *The World's True Redeemer* (1863); *Morning Lectures* (1865); *Tales of a Physician* (1867); and *Stellar Key to the Summer Land* and *Arabula, the Divine Guest* (1867).

DAYDREAMING A state of reverie indulged in while awake. Often called fantasizing, it is definable as a comparatively well-organized, but often illogical, process of sensory thinking where the daydreamer is to all intents and purposes awake, but nevertheless loses partial contact with his or her surroundings. It is often characterized by a free-flowing internal debate that frequently ends up with talking out loud to oneself. Fantasy is extremely egocentric, dramatic, and often pleasurable. It is thought that what we are doing when daydreaming is trying to actualize our self-image in a positive, self-enhancing way.

Parapsychologists have claimed that during the reverie, supernatural forces move in and orchestrate the various scenarios of the daydream, that something "out there" is taking over. Another viewpoint, in keeping with current research on cognition, is that there is nothing paranormal about this particular state of mind; other more naturalistic explanations are offered.

Most, but not all, psychologists distinguish between "unconscious fantasy," the type of daydreaming described in the introduction, which is an imaginative activity that supports and enlivens all thought and feeling, and "neurotic fantasy," which is pathological and could be very dangerous. It is, however, disputed where the line between the two is to be drawn, or even whether a line can or should be drawn. The answers to these questions rely to a great extent on how the subject is defined within the different schools of psychology. Some schools believe fantasizing to be a cathartic activity through which explicit sexual desires and aggressive impulses can be satiated. This is the view that leads to the practice of drawing no lines between different forms of daydreams, because only when all are allowed into the cathartic safety net will all the undesirable antisocial urges of humankind be held back. To this school of thinking, repressing fantasies is dangerous. But many psychologists think differently, recognize the risk of treating all daydreaming in the same way, and believe that lines should be drawn between different types of fantasy because the safety net could, and often does, break down. They argue that, for example, the case of a man who listens to demanding voices from within that tell him to rape and then murder a young girl is too awful to be left for the cathartic cure. These thoughts might be, as Sigmund FREUD suggested, wishful thinking on the part of the fantasizer. Cases like this, where someone seeks help, present enormous difficulties for doctors and priests because the person concerned has not, as yet, done any wrong and cannot be punished or put away for what he *might* do.

A more neutral school of thought sees fantasizing as a Thurberlike way by which urban dwellers deal with a complicated technological world that they can make little sense of, so they escape into the more pleasurable, often humorous world of Walter Mitty. The character of Mitty is essentially adolescent in that he reveals the hopes and fears of a youth who, perhaps because of ineptness or shyness, uses daydreaming as a temporary substitute for facing up to real life. Jung believed that daydreams, like night dreams, occur in childhood as a way of dealing with conflicts in the family that are too difficult and painful to deal with directly.

There are many other beliefs that surround the act of daydreaming. For example, some say that the exercise of acting and arguing through the unresolved difficulties in life and work is a problem-solving activity that tests possibilities, and functions like a memory-searching filing system. Analogous to this is the idea that the act of fantasizing can help the process of creativity by first making chaos and then rearranging a new order out of the chaos. A modern refutation of these positive views believes that many fantasies, by creating either wildly optimistic or, conversely, catastrophic scenarios, actually compromise the future of the daydreamer. When dreamers eventually come down to Earth, they may be disappointed with what they find.

See also DREAMS.

For further reading: J. Allan Hobson, *The Chemistry of Conscious States* (Little Brown and Company, 1994).

DEAN SPACE DRIVE One of a number of ANTIGRAVITY MACHINES. This device was invented by Norman L. Dean, a mortgage appraiser for the Federal Housing Association in Washington, D.C., in the late 1950s. It produced a lift by spinning weights. Its name comes from the belief that such a device could provide the motive power for a spacecraft capable of interplanetary travel. However, such capability for such a device is not only unlikely but impossible.

See also PERPETUAL MOTION MACHINES.

DEATH RAYS Emissions, visible or invisible, that cause death and destruction. In H. G. Wells's groundbreaking science-fiction novel *The War of the Worlds* (1898), the invading Martians repel an attack by British forces with an invisible ray that causes whatever it touches to catch fire. In the early 20th-century science-fiction comic strip *Buck Rogers in the 25th Century,* the hero faces destruction from the disintegrator rays of the Tiger Men of Mars. In the television program *Star Trek,* the crew of the starship *Enterprise* wards off attack with its powerful phasers. All three of these examples show how science-fiction writers have popularized the concept of a death ray—a powerful, if intangible, beam that causes instantaneous destruction and death.

The concept of a "death ray" probably started with the discovery and exploration of radiation in the 19th century. The French scientists Marie and Pierre Curie discovered that the element radium gave off a glow in the dark, a form of visible radiation that was later shown to be a secondary effect of radioactive emissions. In 1896, the German physicist Wilhelm Röntgen detected an invisible type of radiation that penetrated the opaque wrapping and blackened photographic plates. He called these rays X rays. To the popular mind, this radiation—which can damage a person's health—suggested that there might be a form of radiation that could kill a person outright.

Scientists have produced many different kinds of death rays, none of which is capable of quickly killing people (individually or collectively) at the present time. Like Wells's Martians, ultrasonic radiation—sound waves beyond the range of human hearing—can start fires. Phasers resemble modern lasers, beams of concentrated light produced by stimulating radiation. Buck Rogers's disintegrators mimic the natural decay of radioactive elements. In each case, however, the time element is greatly exaggerated. Doctors agree that long-term exposure to many radiations can damage or kill people, but in the short term, no current technology can produce a comparable death ray.

DEATHBED VISIONS OUT-OF-BODY EXPERIENCES that are associated with near-death events. People who have been resuscitated from cardiac arrest or have had life-threatening accidents have later made graphic claims that they have floated upward into a second body and were able to observe what was happening to their physical body on the bed below. Because of recent advances in modern medical technology, more seriously ill people are being, to use a metaphor, pulled back from the brink. Consequently, deathbed visions are beginning to be more of a common experience; doctors, psychiatrists, and others who look after sick and dying patients now have an opportunity of researching the phenomenon.

Whether or not one accepts that something mysterious is happening at this time, because of the subjective nature of the experience it is difficult, if not impossible, to find out what it is. Nevertheless, these claims are made and so must be taken seriously. The point at issue is not whether the informant is sincere, but whether the substance of the second, floating person is actually a separate nonphysical mind or soul that has escaped from the dying body below. If this could be proved, then the Platonic-Cartesian dualistic view of the nature of humankind would also be proved. If the phenomenon is purely a psychological one where the patient is involved in some form of death-imagining similar to, but not the same as, dreaming, then the dualist view must remain unproved.

However, we must believe the patients who declare that what they experience is vivid, realistic, and memorable, yet quite unlike the dreams they generally have at night. The narratives follow no standard pattern except the capability of their relaters to have observed their own bodies from outside. The ability to see is always mentioned. Some say that the second body was a static observer who was attached to the first by a silver cord, while others say that they felt truly separated, capable of turning over and around in space, and even of floating through the ceiling or walls. Some remembered entering a dark tunnel and hearing an earsplitting roaring sound, while others were certain that everywhere was suffused with a bright white light and all around was eerily silent, giving them a feeling of timelessness. It is interesting that some patients have reported similar out-of-body experiences during rest or meditation, when taking drugs, or even during normal everyday activity. A few have claimed to be able to induce the experience at will and to have it under their control. But one patient is reported to have said that the deathbed vision was unlike any other in that the reentry was very painful.

Psychologists studying out-of-body experiences in general are sure that they do not take place during normal REM sleep, the rapid-eye-movement type of sleep that indicates dreaming. Researchers know this because they have monitored the start of dreaming by attaching electrodes to volunteers in sleep laboratories and waking them up to question the content of their dreams. If these deathbed visions are analogous to any kind of dream, they resemble most the fantasizing in DAYDREAMING, where the dreamer knows at the time that it is a dream. If some sort of psychological state of mind could be confirmed, there would be no need to suppose a second spirit body. Then the theory that deathbed visions indicate survival after death would be hard to support. Some researchers trying to establish survival after death have tried to determine that there is an objective double that leaves the body and is capable of independent thought and action. But even if the physical body is, at the time of the vision, under great stress, it is still alive, and there is no reason to think that it is not capable of organizing its thinking sufficiently well to fantasize.

The evidence so far suggests that deathbed visions are a psychological response to unusual, and perhaps frightening, conditions. Rather than searching for the elusive second body, we would do better to concentrate on trying to understand subjective experiences, including the chemical, and nonchemical, changes that could occur in all altered states of consciousness within which deathbed visions occur.

For further reading: Susan J. Blackmore, *Beyond the Body* (Heinemann, 1982); William Barrett, *Death-Bed Visions* (Methuen, 1926); and K. Osis and E. Haraldsson, *At the Hour of Death* (Avon, 1977).

DÉJÀ VU Literally "already seen," the vague but nevertheless definite feeling that a place or event is familiar and has been experienced before.

Déjà vu has been effectively explained as a way in which many people experience cryptomnesia—forgotten memories. Having been impressed by an experience that was lost to conscious memory over time, a person may feel the charge of a similar event or location that brings up a feeling of familiarity without calling up the full content of the prior experience.

Déjà vu experiences would be of only passing interest were they not utilized as an argument for REINCARNATION, a view which has gained a significant and growing following through the 20th century. It has been suggested that déjà vu experiences are an indication of hidden memories of a previous life, lived in the place associated with the memories. Were the déjà vu experience to be explored through HYPNOSIS, for example, then the full account of the previous life might come forth as in the BRIDEY MURPHY case. Through hypnosis, Ruth Simmons, a housewife, recounted a prior life as a woman in Belfast, Ireland in the early 19th century. It was later discovered that much of what she recounted was forgotten memories of people, places, and events from her childhood in Chicago. Once released, such memories can have such a strength and such a convincing value to the person experiencing them that the person can easily come to believe in their literal truth rather than more mundane explanations.

DE LOYS'S APE A fictitious, large, New World ape. The discovery of the previously unknown mountain gorilla in 1903 set the stage for the acceptance of reports by oil geologist François de Loys in 1920 that he had discovered an ape in the hinterlands of Venezuela. The importance of the discovery was underscored by the fact

de Loys's Ape

that no apes (as opposed to monkeys) were known to exist in the Americas.

According to de Loys's story, he and his associates encountered the apes while exploring for oil. The apes were angry at the human intrusion into their territory and began to show their hostility by making loud noises and throwing tree branches and feces at them. When two apes appeared to approach the geological party in an attack mode, de Loys shot one of them and drove the other away. The dead ape was a female, almost 5 feet tall, and covered with reddish hair. The native member of the expedition claimed no knowledge of the creature. Because he had no means of preserving the body, de Loys propped it up and took a photograph. He said he had attempted to preserve the hide and the head but had lost them before returning to civilization. Only the photograph survived. The photo was presented to anthropologist Georges Montandon, who dubbed it a new species, *Ameranthropides loysi*. De Loys wrote an article for the *Illustrated London News* hailing his discovery as filling in a gap in the evolutionary record.

The European scientific community was not as generous as Montandon and very quickly responded with doubts over de Loys's discovery. The Paris Academy of Science was among the first to label it a pure hoax. Members identified the creature in the photo as a spider monkey whose tail had been hidden from the camera (apes do not have tails). While most simply dismissed the discovery, a few continued to believe that possibly de Loys's ape existed.

In 1951 the seemingly long-dead issue of de Loys's ape was raised again by Roger de Courtville, a mining engineer who had traveled extensively through Venezuela. He claimed that on at least two occasions he had encountered de Loys's ape, once in 1938 and again in 1947. To back up his claim he produced several sketches he had made and a new photograph. The photograph turned out to be a doctored version of the original de Loys photo. As no subsequent reports have surfaced in the last four decades, de Loys's apes have been added to the list of scientific hoaxes.

DELUGE MAN A number of bones uncovered in the summer of 1856 by quarry workers removing limestone from a pit in the Neander Valley near Düsseldorf, Germany. The bones were discovered in a small cave about 18 meters (about 60 feet) above the river, and some of them were saved for a local scientist named J. K. Fuhlrott. Fuhlrott knew enough anatomy to recognize the bones as those of a man, but a man very different from anyone living in the 19th century. The bones of the limbs were extraordinarily thick as compared to those of modern humans, and the skullcap had a very thick brow ridge and a slanted forehead—features that seemed to Fuhlrott to indicate that the remains were very old. He concluded that the bones were those of a person (now known as *homo sapiens neanderthalis*) who had died in the biblical Flood—a man of the Deluge, or "Deluge Man."

The idea of Deluge Man was connected to the 18th-century concept of CATASTROPHISM. The French naturalist Baron Georges CUVIER promoted the idea that great natural catastrophes preceded periods of extinction. New species were divinely created or migrated from unaffected areas after each catastrophe. Cuvier's theory won support from religious thinkers because it seemed to support the biblical view of history. However, as the 19th century progressed, Cuvier's views fell into disfavor. After the publication of Charles DARWIN's *On the Origin of Species by Means of Natural Selection* (1859), Deluge Man was discarded in favor of a different concept: humankind as a species that had evolved from other, related species over a long period.

For further reading: George Constable, *The Neanderthals* (Time-Life Books, 1973); Willy Ley, *The Lungfish, the Dodo, and the Unicorn: An Excursion into Romantic Zoology* (Viking Press, 1948); Willy Ley, *Salamanders and Other Wonders: Still More Adventures of a Romantic Naturalist* (Viking Press, 1955); and Tom Prideaux, *Cro-Magnon Man* (Time-Life Books, 1973).

DEMONS Spirits beneath the status of gods and subject to them. Belief in such spirits has existed in all religions throughout history, but not all imagined them to be evil. In ancient Greece, for example, the word *daimon* meant a divine power, usually an individual protector who intervened between the gods and mortals. During the Hellenistic period, there was also a tradition of individual demons, but they were thought to be controlled through a hierarchy.

It was the monotheists of early Judaism who first characterized all gods and spirits, other than their own true God, as evil and postulated the existence of a single chief adversary, wholly evil in the way that God was all good. A common view in other monotheistic religions, like Zoroastrianism, was that there was permanent battle going on within the universe between the powers of Light and Dark. In Christianity, the chief demon who represented absolute evil was a fallen angel from within the religion itself. He came to be known by various names: the Devil, Satan, Lucifer, Beelzebul (sometimes corrupted to Beelzebub), the Demiurge, the Prince of Darkness, and many others. In Christian thought, the chief task of this fallen angel was to tempt men and women away from the path of righteousness and so to obtain increased power over his main adversary, the Archangel Michael, the leader of God's heavenly host.

In predominantly Christian countries in the past, devil worship, or Satanism, was practiced, but at this time there

Demons

was a vast difference between worshiping the devil and believing in evil. Initially the cult was a reaction among pagans against Christianity, and most of its practices were mirror images of Christian practices. Believers used the distinctive rite of the "Black Mass," a parody of the Eucharist, which was celebrated by an unfrocked priest chanting the mass backwards before black candles and an inverted crucifix.

Jesus expelled demon by a word and stated that this act was the sign of the coming of God's kingdom, perhaps performing the first EXORCISM. Today people who are said to be possessed by demons, and certain houses where evil spirits are thought to dwell, can be exorcised by most Christian churches with a special ceremony wherein a priest drives the demons out "in His name." Pope John Paul II himself is reported to have exorcised a demon from a young woman on March 27, 1982. It is usually thought that if we believe in God and angels, then we must believe in the devil and demons. In recent times many churches have discouraged speculative elaboration on the subject, yet despite all the progress in understanding the human condition and developments in psychology, the practice of externalizing evil still remains with us. The French poet Baudelaire summed up the matter by presenting it as an eternal anxiety: "The devil's deepest wile is to persuade us that he does not exist."

Although demons are thought to be solely a religious phenomenon, they do have a parallel in the supernatural world of SPIRITUALISM, where the spirits of the departed are thought to retain an interest in the mortal world, sometimes for good and sometimes for evil. The main difference is that there is no system of spirit hierarchy in spiritualism in the way there is in religion. Spirit invocation, also known as conjuring, is still practiced worldwide through the rites of black magic. The method widely used for calling up the spirits of dead and demons from hell is the "magic circle." Two or more concentric circles are drawn on the ground, and around the inside ring are written various magical words and titles for names that cannot be spoken or written. The purpose of calling up the devil is to enlist his powers of countermagic against other supernatural evils, usually to free a possessed person or a MEDIUM from the clutches of another unspecified evil. This practice of fighting fire with fire must be seen to have its dangers, but apparently the risk is warranted when set against a channel remaining possessed.

Science takes a different intellectual path from religion or myth, which recognize powers in the universe that have character, identity, and the ability to change universal rules to make natural phenomena act in unnatural ways. The scientist treats nature as nonarbitrary and impersonal; the governing principles of science are a number of unchanging objective and discoverable laws that cannot be controlled by the whims of, or battles between, gods and demons.

See also NEOPAGANISM.

For further reading: Sagan, Carl, *The Demon-Haunted World: Science as a Candle in the Dark* (Random House, 1996).

DERMO-OPTICAL PERCEPTION Often shortened to DOP, a claim for EYELESS SIGHT, whereby blind or blindfolded people are said to have the mysterious ability to read or recognize and describe colors without using their eyes. Sometimes the ability is said to be CLAIRVOYANCE, but usually the claim is that parts of some people's skin, especially fingertips and toes, possess light-sensitive organs that act as retinas.

The idea that blind people have the ability to see color by touch was discussed as early as 1772 and written about in Boswell's *Life of Samuel Johnson.* At that time, it was believed that different colors produced different kinds of surfaces, but the differences were said to be so fine that it was doubted whether they could be distinguished. In the 19th century, it was thought that each color has its own degree of heat and that it was this that could be detected by a blind person.

That blindfolded people could see caught the popular imagination in the age of vaudeville and the music hall, and the MAGIC trade has, for over a hundred years, marketed dozens of ingenious blindfolds with simple devices to allow peeking down the nose when the head is tilted back, in what has been called the "sniff" position. The skills used to devise all forms of deception for the magic trade is termed MENTALISM, and magazines devoted to conjuring, and organizations like The Magic Circle, quite rightly lauded the ingeniousness of this art. Mentalists have contrived many ways to perform the trick of eyeless sight, all of them allowing the blindfolded person to see sufficiently well to carry out his or her act. Dermo-optical perception dissociates itself from mentalism and stage magic and claims that, although deception has been and is still used, there actually are people who can genuinely see this way. Skeptics however disagree.

In the 1960s, a succession of Russian women came forward claiming eyeless sight. On January 16, 1964, a Leningrad newspaper reported that experiments conducted on Ninel Sergyeyevna Kulagina at the Psychoneurological Department of the Leningrad Kirovsk District had demonstrated that optical vision can be achieved by the fingertips. *Life* magazine sent a journalist to the Soviet Union, and hired science writer Albert Rosenfeld to report on the experiments. Rosenfeld accepted the findings. No magicians or mentalists were asked to advise or give an opinion. The very idea that it might be a Magic Circle type of trick was never considered. Following Kulagina, dozens of other Russian DOP claimants came forward. One, a nine-year-old Kharkov girl, Lena Bliznova, convinced scientists that she could read with her fingers a few inches off the page, then with her toes and finally with her shoulders!

Afterward, many scientific papers were written trying to explain DOP. Skeptics pointed out that blindness is a loose concept and that many technically, legally blind people have a degree of awareness of light and dark. Science writer and amateur magician Martin GARDNER suggested that the blindfolded subjects had simply peeked. He recommended that all those being tested for DOP should wear a ventilated aluminum box that he designed to prevent "nose peeks."

Before he became a psychic, the well-known Israeli metal bender Uri GELLER used to perform what magicians call a blindfold drive. This is a standard magic act of driving a car through crowded streets while blindfolded. In an interview in *Meta Science Quarterly* (Autumn 1980), Geller remarked that he had always relied on his psychic powers and that he had never resorted to magician's tricks to allow any form of peeking. When asked about his blindfold drives, he replied "It was a form of telepathy . . . I stopped doing this because it can be so easily duplicated by magic."

For further reading: Martin Gardner, "DOP—A Peek Down the Nose," *Science, Good, Bad, and Bogus* (Prometheus, 1981).

DE ROUGEMONT, LOUIS (n.d.) A European who claimed to have traveled to Australia, where he became the king of a cannibal tribe. Louis de Rougemont suddenly burst upon the public scene in 1898 with the story of his previous three decades in Australia. He claimed to have been shipwrecked in the 1860s. He survived on an otherwise uninhabited island for some weeks until a subsequent storm landed four aborigines: a man, a woman, and their two children. Together they constructed a boat that got them to the mainland. When he arrived he was not only greeted ceremoniously, but also invited to become the tribe's ruler. He held that position for many years, and according to his account, feasted at times on the bodies of the tribe's enemies. After some years, he became homesick and, leaving the tribe behind, began the trek back to civilization.

On his way to an area populated by Europeans, he became sick with a fever. Unable to get warm, he killed a buffalo, cut it open, and climbed in. The otherwise nauseating experience reportedly cured him. With stories like this one, de Rougemont quickly attained celebrity status in England and was invited to lecture to both popular and learned audiences such as the British Association for the Advancement of Science. His story was initially accepted at face value; it fitted into the then current understanding of tribal cultures.

The incident with the buffalo was but the most unbelievable of the several elements of de Rougemont's account of his Australian venture, but it led to the unraveling of his tale. As a result of its skepticism about this single incident, a newspaper began to double check all of de Rougemont's story. It soon unraveled the truth: De Rougemont had, it appeared, been to Australia once, but little of the rest of the account of his life among the aborigines coincided with the facts. As it turned out, the would-be king of the cannibals had learned most of what he knew of the land from books in the British Museum. After the hoax was exposed, de Rougemont disappeared from public sight.

DESCENT OF WOMAN, THE See MORGAN, ELAINE.

DE TROMELIN CYLINDER A simple device used to measure psychokinetic power. It was invented by the Count de Tromelin at the beginning of the 20th century. It is similar in operation to the BIOMETER developed by Hyppolite Baraduc just a few years earlier. It consisted of a paper cylinder with crosspiece of straw that revolved on a fine point. The movement of the human hand close to the piece of straw or the mental will of the subject was supposed to be capable of making the straw move. As with Baraduc's device, de Tromelin's ran into criticism from colleagues who argued that he had not eliminated movement caused merely by air currents or the heat of the human body. It was later superseded by similar instruments, such as the STHENOMETER, that were enclosed in glass, thus eliminating those criticisms.

DEVIL'S SEA Located off the southeastern coast of Japan, it is considered to be a Japanese equivalent of the BERMUDA TRIANGLE. Like its parallel region in the Atlantic Ocean, it is supposed to be a section of the ocean that periodically swallows planes and ships without a trace—usually before they have a chance to put out a distress call. The Devil's Sea became popular during the early 1970s at the same time as the Bermuda Triangle concept. Credulous people, such as the writer Ivan T. SANDERSON, suggested that craft in the Devil's Sea went missing because of some wrinkle in the space-time continuum, through magnetic or gravitational aberrations, or through the activity of extraterrestrials or a mysterious underwater race.

During the 1970s, however, a librarian named Larry Kusche began to trace the Devil's Sea stories back to their original sources. He discovered that the first use of the term "devil's sea" appeared in *The New York Times* in 1955, and was only a poetic description of the loss of several fishing boats during the early 1950s. The only "mysterious force" involved in the sinking of the boats was a tidal wave caused by the eruption of an underwater volcano in 1952. Kusche also communicated with Japanese officials and found that none of them knew of the term "devil's sea." In his article "The Bermuda Triangle and Other Hoaxes," published in the October 1975 issue of *Fate* magazine, Kusche declared that the Devil's Sea was nothing more than an exaggeration based on the loss of several fishing boats over a period of five years.

For more information Elizabeth Nichols, *The Devil's Sea* (Award Books, 1975); and Larry Kusche, *The Bermuda Triangle Mystery—Solved* (Prometheus Books, 1986).

DIANETICS A system for understanding the human mind developed by well-known 20th-century science fiction writer L. Ron Hubbard (1911–1986). Hubbard speculated that the true human being (which he termed the Thetan, analogous to the soul of Greek philosophers) inhabits the body. Hubbard taught that the natural ability of the Thetan to express itself positively in the world through the body had been disturbed by the operation of the mind, specifically a part of the mind called the reactive mind. The reactive mind is a stimulus-response mechanism created by the Thetan that records impressions at the unconscious level. It then provides programmed responses to specific repeated stimuli. Along with other data, the reactive mind holds images of pain, injuries, and destructive movements of one's life. Individuals, acting out of the responses of the reactive mind, are often led to destructive acts that run counter to their own best interests, blocking the potential of their Thetans to operate in the world.

Dianetics seeks to unblock the potential of the Thetan through auditing. This is a method of counseling in which the individual works with various techniques that force a confrontation with the counter-survival information housed by the reactive mind (termed engrams), usually in one-on-one sessions with a counselor (called an auditor). It is reminiscent of Sigmund FREUD's early method of abreactive theory of identifying and bringing to consciousness traumatic experiences around which painful associations had accumulated. In some of these sessions, the auditor is assisted by an E-meter, a device that includes a Wheatstone bridge and can measure the body's contrasting resistance to a flow of electricity, thereby functioning as a lie detector. Progress through the auditing program leads to the state of "clear," in which the effect of engrams is largely erased from individual behavior. From that point on, the Thetan can begin to operate more positively in the world.

Dianetics spread following the publication in 1950 of Hubbard's book, *Dianetics: The Modern Science of Mental Health,* primarily as an alternative form of psychotherapy. The book became a best-seller, and Hubbard quickly published a series of books expanding upon his basic perspective. Dianetics attracted a popular audience but was largely rejected by the medical and psychiatric community.

During the early 1950s, Hubbard continued to develop his thoughts about the operation of the Thetan, which led him into metaphysical and religious realms. He created methods by which the Thetan could remove the last of the inhibiting engrams and learn to operate as a completely free entity, including the ability to exteriorize from the body. The expanded thought and program for the Operating Thetan was called scientology, and in 1954 Hubbard founded the Church of Scientology as the proper vehicle for presenting these teachings to the world. Dianetics continues as the introductory program for people becoming associated with the church.

Controversy over Dianetics led in 1962 to the Food and Drug Administration (FDA) seizing the E-meters used in auditing. The FDA claimed that they were unauthorized devices in the diagnosis and treatment of disease. Litigation took a decade, but the church finally won and the court ordered the E-meters returned. The FDA controversy was one phase of an ongoing controversy between the Church and various governments. A conflict with the U.S. Internal Revenue Service, the longest in the body's history, was finally settled in the church's favor in the early 1990s.

For further reading: Martin Gardner, *Fads and Fallacies in the Name of Science* (Dover Publications, 1958); and L. Ron Hubbard, *Dianetics: The Modern Science of Mental Healing* (The American Saint Hill Organization, 1950).

DIETS See FOOD FADS.

DILUVIALISM A theory pertaining to Noah's Flood that attributes certain geological features of the earth to this specific Deluge.

Stories that describe a universal deluge that obliterated mankind and changed the earth's surface are found in many religions. The best known are the ancient Mesopotamian Epic of Gilgamesh, and the story of Noah in Genesis.

At the end of the 17th century a few Christian theologians were anxious to make the sacred biblical account compatible with new discoveries in science. To this end, an Anglican clergyman, Thomas Burnet, published *The Sacred Theory of the Earth* (1684–90) in which he stated that the Bible was unerring and must be taken literally. Believing God to be all-good and all-powerful, Burnet reasoned that he had created a perfect world and did not need to use miraculous intervention to fix it. Earth's history could only be adequately explained when natural physical causes were found for the events related in the Bible.

Burnet thought that the Noachian flood had to be global; it could not have been local because there would have been no need for Noah to build an ark if he could have retreated to another part of the world. With this in mind, Burnet attempted to assess the amount of water in the oceans by the classical method of sounding and concluded that the seas could never have buried the continents. Rejecting the notion that God temporarily created new water, as in the category of miracles, Burnet posited that Earth's crust must have cracked, letting underlying water rise from the abyss. He started to look upon Earth as the shattered and distorted remains of what was once a perfect place. The original perfect Earth could not have been like the jagged and imperfect place it is now. It must have had a surface that was smooth, with rivers flowing, from the assumed slightly higher ground of the north and south poles, toward the warmer central parts. This, he surmised, was the site of the paradise of Eden. A planet of such perfect radial symmetry, with no irregularities to tilt its axis, must have rotated upright, and this is why the climate of Eden enjoyed perpetual spring. The Deluge indeed signaled paradise lost.

Ingenious as this rationalization was, Burnet's *Sacred Theory* later ran into many problems—not least was the problem of fossils, which he had never considered. Naturalists studying strata noticed that fossils of simple forms are found in older, deeper, strata, and the more complex fossil forms are found in younger strata lying nearer the surface. After Charles DARWIN, geologists expected this to be so, as it conformed with evolutionary theory.

George McCready PRICE, the modern opponent of Darwinism, in *The New Geology* (1923) explained that creation happened a few thousand years before Christ in six days, as described in Genesis. When the great Flood came, all life other than what went into the ark with Noah was drowned and deposited in the earth, the fossils being just a record of preflood flora and fauna. In Price's argument, any stacking up of fossil beds into evolutionary strata, with the fossils dating each stratum, is a circular argument because the same theory looks to the strata to date the fossils. Price makes allowance for marine life to be buried first, followed later and a little higher by land life, and, last, by bird life. He cited the "upside-down" areas as places where traditional geologists have invented imaginary faults and folds to explain the reverse order of the fossils. But there is plenty of rock evidence to corroborate a fault or fold without resorting to fossils. Flood geology runs into many difficulties by trying to impose on science axioms and dates derived from the early books of the Bible. Science sees the physical evidence as primary, and does not impose any belief system on its the empirical findings.

Today many in the NEW AGE MOVEMENT believe in a variety of equally pseudoscientific theories that have been advanced to account for the waters of the Deluge. Members of the FORTEAN SOCIETY support Isaac Newton Vail's theory. In *The Waters Above the Firmament* (1886), Vail explains how every planet has to pass through a phase in which it has, like Saturn, rings around it, and in Earth's case this ring was the source of the floodwaters.

The debate entered a new phase when U.S. geologists Walter Pitman and Bill Ryan tried to prove scientifically that there was indeed a real flood but that it was local rather than universal. They thought there could have been a regional catastrophe that was so momentous that it entered the mythology of several Mesopotamian cultures. Pitman and Ryan concentrated their research on the Black Sea, which at one time had become virtually dried out before being reflooded through the Bosporus strait. This was confirmed when oceanographers doing seismic profiling of the Black Sea bed discovered evidence of ancient beaches and shorelines well below present sea level. Carbon dating of seeds and roots that were found in the mud of the ancient beaches put their date around 7,550 years ago. This does not tie in with the estimated date of the Noachian or Gilgamesh universal floods, but does provide a possible origin for the persistent myths of a catastrophic flood in early Mesopotamia.

See also ARKEOLOGY.

For further reading: Stephen Jay Gould, *Time's Arrow, Time's Cycle: Myth and Metaphor in the Discovery of Geo-*

logical Time (Penguin Books, 1987); and Tim M. Berra, *Evolution and the Myth of Creationism* (Stanford University Press, 1990).

DINOSAURS, CONTEMPORARY The belief that there are still dinosaurs roaming the earth today. Dinosaurs became extinct by some catastrophic event or major change in the environment 65 million years ago. There are, of course, many saurians still extant that have existed more or less in their present form since before the dinosaur extinction: crocodiles, alligators, lizards. They are not what is meant by contemporary dinosaurs, generally held to be in various inaccessible places that allow them to evade detection. The belief may well have been reinforced by the discovery in 1938 of a living COELACANTH in the Indian Ocean; zoologists thought this fish had been extinct for 60 million years. If it could happen to a fish, why not to a dinosaur?

The most widely publicized and most believed of these creatures is the LOCH NESS MONSTER. The first recorded sighting of this creature, possibly a living plesiosaur, was by St. Columba in 565 C.E., and thereafter there were few sightings. Reports increased in the 20th century. "Nessie" has since become an international celebrity, photographed and reported in newspapers and on radio and television stations across the world. Eminent scientists have lent their support to Nessie's "incontrovertible" existence; well-financed and equipped expeditions have pursued it. Hoaxers have admitted to their pranks, but the search continues. The most recent photograph was printed in the *Nottingham Evening Post* in September 1995.

That year, the search intensified and three sightings/soundings were reported in the December issue of the *FORTEAN TIMES*. In April 1995, sonar on the boat the *Royal Scot* had reported a strange object in Lock Dochfour, a shallow lake at the north end of Loch Ness, the first time that Nessie or some relation had been reported in Loch Dochfour. In July 1995, a submarine exploring Loch Ness recorded strange "grunts" 140 meters (450 feet) down, close to the west bank, and member of the scientific team hazarded the opinion that "Perhaps the sounds are the monster's mating call."

In September 1995 the *Morgawr* (*sea giant* in the Cornish language) was sighted close to the coast 5 kilometers (3 miles) south of Falmouth. Gertrude Stephens, a local resident, related that the monster had a small head that moved up and down, and a long neck. It was at least 6 meters (20 feet) long, had a dark yellowish green conical body, and a broad, flat, pear-shaped tail. It sank down, tail first, very quickly. There had been occasional sightings of this monster since 1976.

Such stories are infectious, and sightings of similar creatures are reported from hundreds of other lakes in many countries. The May 1996 issue of the *Fortean Times* published an article on sightings of dinosaurs over several recent years in Central Africa in lakes Tele, Dakatua, Bangweulu, and Tebeki, and on the Ubangi and Likouala-aux-Herbes rivers, and in Lake Bala, North Wales. As the author of the article, Mike Dash—at first a believer, now a skeptic—concludes: "Despite an intriguing body of evidence gathered at some personal risk, dinosaur survival in the Congo cannot honestly be reckoned proven. On the one hand, equatorial Africa does appear to have survived largely unchanged since the end of the dinosaur era; there have been no major climatic changes, no mountains have risen and no new seas have inundated the land. On the other, 70 million years is a long time for what appears to be a limited population to survive while adapting to the arrival of new rivals in the food chain and without suffering a severe deterioration of its gene pool."

There is also a legendary animal, the Great Lake Monster, in Lake Storsjön at Östersund in central Sweden. Legend was that the monster would be released when a certain inscription on a stone tablet was deciphered. That happened in 1820; since then there have been occasional sightings of the monster, said to be 20 meters (69 feet) long, up to 1996. In 1895 a company was formed to trap it but was unsuccessful. Now the monster is protected by local legislation.

No one has yet produced a carcass of Nessie or a similar animal, nor a reliable and incontrovertible photograph or sonar record. There are also doubts about how breeding colonies of such large animals could exist undetected in so many places—or even in one—for so long.

See also MACKAL, ROY.

For further reading: Henry B. Bauer, *The Enigma of Loch Ness: Making Sense of a Mystery* (Johnston, 1991); Steuart Campbell, *The Loch Ness Monster: The Evidence* (Birlinn, 1996); Tim Dinsdale, *Loch Ness Monster*, 4th ed. (Routledge, 1982); Stuart Gordon, *The Book of Hoaxes: An A-Z of Famous Fakes, Frauds and Cons* (Headline, 1995); and Nicholas Witchell, *The Loch Ness Story* (Corgi, 1989).

DIVINATION The art and practice of fortune-telling. The ascertaining of hidden knowledge, inaccessible to the rest of humanity through ordinary sensory investigation, by the diviner using supernatural or magical means. For example, the foretelling of future events, the finding of lost objects, or the revelation of hidden character traits. Divination is found in all civilizations, ancient and modern, and in all places. Sometimes it exists with the endorsement of religion, sometimes not. Christianity, Islam, and Buddhism are generally opposed to it: They say that the diviner's mediation between this world and the spiritual world assumes the mantle of a priest without the responsibility, the training, and the backing of a church and that this could be harmful.

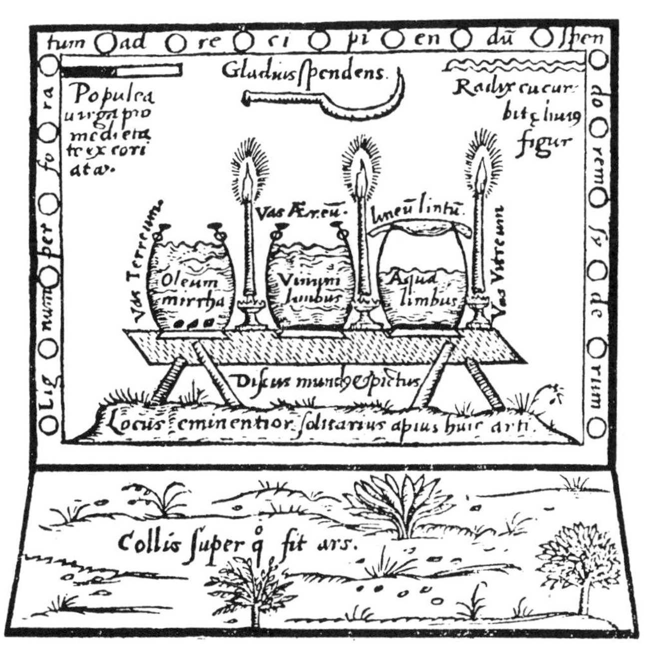

Divination

There are two main systems of divination: the *interpretative,* in which various accumulated data are assessed, as in ASTROLOGY or the TAROT CARDS, and the *oracular,* in which the fortune-teller enters into a special psychic state, such as a trance, and transmits information to the client. But the differences are not clear-cut because diviners in both groups believe themselves to have some psychic powers.

Some critics say that diviners possess nothing more than a shrewd knowledge of human nature. They have learnt to "cold read" their clients by initially using generalized statements that anyone would be able to apply to themselves, then following up with more specific statements, according to how the client replies. Some critics say there is no such thing as divination; fortune-tellers—like the rest of us—get caught out in the rain, have avoidable accidents, and misjudge the characters of others. At best they are party performers and a bit of a lark; at worst they are frauds extracting money from the gullible.

See also FUTUROLOGY.

For further reading: Robert A. Steiner, *Don't Get Taken! Bunco and Buncum Exposed—How to Protect Yourself* (Wide-Awake Books, 1989).

DIVINE SCIENCE One of several 19th-century "science" religions. Divine Science was founded in 1888 by Melinda Cramer, a former Quaker residing in San Francisco in the 1880s. She seems to have absorbed Christian Science from two of Mary Baker Eddy's students, Miranda Rice, who opened an office as a Christian Science practitioner in San Francisco in 1883, and Emma Curtis Hopkins, a student of Mary Baker Eddy, who taught a basic class in Christian Science healing in San Francisco in 1887. Cramer developed her own synthesis, which was distinct from the levels of both Eddy and Hopkins, and took as its main theme the omnipresence of God. Divine Science developed a second main center in Denver and, following Cramer's death in 1906 (the result of an accident she experienced during the San Francisco earthquake), the center of the movement moved to Colorado, where it has since remained.

Divine Scientists have seen in their teachings a union of reason and faith. To them "divine" indicates that the truth proceeds directly from God and hence is inspired and excellent in the highest degree. "Science" indicates that it contains comprehensive information. Science investigates truth for truth's sake. It rejects disjointed and unsupported facts, and embodies only those which are proven. According to Fannie James, the original Divine Science leader in Denver, Divine Science is scientific because its teachings are proved in our experience—for example, the experience of physical healing.

Divine Science thus makes no pretense of competing with secular experimental science, but does hark back to an older definition of science as the systematic organization of fact in one area of knowledge.

For further reading: Charles S. Braden, *Spirits in Rebellion* (Southern Methodist University Press, 1963); J. Gordon Melton, *New Thought: A Reader* (Institute for the Study of American Religion, 1990).

DOLPHINS Mammals whose history has been speculated upon, with two very different stories resulting. First, the Darwinian scientific explanation tells how their ancestors originally came out of the sea when other forms of life started to inhabit the land, and then later as mammals they returned to the sea to make a living. Then there is the much older mythological pseudoscientific explanation, favored by people of the NEW AGE MOVEMENT, that claims that dolphins came from a part of the planetary system of Sirius, and were sent here to help humankind to flourish and be happy.

The Darwinian evolutionary explanation understands dolphins to be small whales belonging to the same aquatic order as the great whales, *Cetacea.* Although living their entire life in the sea, dolphins are not fish but warm-blooded mammals who returned to the sea 65 million years ago in search of food, probably when the forests dried up during the heat and drought of the Pliocene era. During the following millions of years, as their lifestyle altered, their entire bodily structure became adapted to their new environment. Hair was lost, nostrils became blowholes, and because they had to swim in the sea and no longer walk on the land, their front limbs developed into flippers while the hind limbs fused into tail flukes. But two very important aspects of their land existence did not change: they still breathed air and they remained viviparous, capable of giving birth to, and suckling, live young.

As they took to the sea, the ancestors of today's great whales fed on plankton while those of today's little whales, the dolphins, fed on fish. The difference in diet, coupled with the abundance of living space, eliminated any competition between them, and the dolphins were able to adapt peacefully into their new environment, with the consequence that they developed a gentle and mild disposition. This outlook shows in their historic association with human beings, where it is on record that dolphins have always sought an affinity with us, often adopting the role of friends and protectors. Their sociable and affable nature asks nothing from us but rather shows their wish to live pleasurably in harmony with all.

The currently favored mythology surrounding the dolphin is a mixture of many ancient traditions and cultures, usually connecting Earth with the binary star Sirius. Sirius is a very hot, blue-white star, with a surface temperature of 10,000°C (18,000°F) and yet legend has it that "beings" from that land visited Earth many thousands of years ago bringing with them wisdom and knowledge far in advance of our own. The building of the pyramids of Egypt and much more early advanced technology is said to come from these neophytes (literally "newly planted ones") who were made into priests by the pharaohs.

New Age psychics have channeled the essential energy of both whales and dolphins and now believe that they have come from one of the water planets in the region of Sirius. When it was predicted that the human race had only an 80 percent chance of surviving, it was thought that dolphins with their high intelligence and advanced form of communication were sent here to help human beings in their struggle through life. There is a connection here to Christian mythology, in which the dolphin represents resurrection and salvation, but the New Agers have added the very modern idea that dolphins were also sent to teach us how to bring joy and spontaneity into our lives and to encourage us to exist alongside other species without hostility and domination. These stories are indeed very charming; but the evidence shows that cetaceans share their remote ancestry with land-based mammals of this planet.

Through scientific studies, we are learning more and more about the dolphin, its intelligence, its *joie de vivre,* its relationship with humankind, and its wonderful sonar system of seeing and hearing. These studies are beginning to shed light on this uniquely sentient and highly intelligent animal evolved from a more ancient grouping than our own. Pseudoscience has offered paranormal explanations of dolphin behavior, all from an anthropocentric point of view. But now through the eyes of science we are realizing that observations of dolphin behavior can be explained much better by referring to the dolphin's own cognitive structure rather than viewing the species as extra-terrestrials put on Earth to benefit humans.

For further reading: Whitlow Au, *The Sonar of Dolphins* (Springer-Verlag, 1993); and Ronald M. Lockley, *Whales, Dolphins and Porpoises* (David and Charles, 1979).

DOWSING Method of finding things, often hidden or underground, with a simple device such as forked stick or a pendulum. The dowser holds the tool and watches or feels for movement that indicates the sought material or object. Perhaps most commonly known as a way to find underground water sources, dowsing has also been used to find petroleum, gas, and other minerals; lost objects; locations of secret treasures or mines; and even missing people—although most dowsing is performed for nonliving things. The common image of a dowser is that of a man holding out before him, by its forks with stem pointing ahead, a Y-shaped stick or "divining rod," and walking around until the stick quivers and points downward. The place pointed to is where the water or other object of the search may be found. Good dowsers are said to be able to tell how deep the water is and to gain a general impression of quantity from the way in which their stick moves.

Not all dowsers use forked sticks. Some use a pendulum suspended from a thread or chain. The pendulum is allowed to hang straight down and is often held over a map. When the pendulum begins to move, it leads the dowser to his or her object. Other things may also be used as dowsing tools; it is thought that the power of the technique rests not so much in the tool as in the user. Some dowsers use no tool at all.

Dowsing has been done for centuries. Recorded descriptions of it go back at least as far as ancient Egypt, and it was commonly used as a way to find water and other things until the 20th century, when scientific skepticism dismissed it. Most scientifically minded people consider it more of a superstition than a true detecting technique. Even so, some petroleum companies, governments, and water-well drillers still use professional dowsers today to help them in their work. It is extremely expensive to dig oil and water wells, so employing a dowser may be considered adding to the probability of the accuracy of the geological reports. Some claim that dowsers are more accurate than geological reports.

One explanation for dowsing is that the dowser, more sensitive or more perceptive than others to subtle energies emanating from different elements, is able somehow to tune in to the energies from the sought object; the dowsing tool may enhance and focus this energy for the dowser. One problem is the apparent subjectivity of the craft: Different dowsers use radically different methods and often arrive at different conclusions. This inconsis-

Dowsing

tency, along with the inability of science to verify the supposed subtle energies, calls dowsing into question. Skeptics suggest that dowsers who do find water, oil, or treasure have merely been lucky.

Another term for a dowser is "water witch," reflecting the historical view that dowsing is a paranormal ability akin to witchcraft and occult divining.

For further reading: Christopher Bird, *The Divining Hand* (E. P. Dutton, 1979).

DOYLE, SIR ARTHUR CONAN (1859–1930) Remembered best for his fictional character, Sherlock Holmes, and for his involvement in SPIRITUALISM. He received a medical degree from Edinburgh's Royal Infirmary and practiced medicine in Hampshire until 1891 and then left medicine to write his detective novels, short stories and historical romances. Doyle had an interest in psychic phenomena throughout his life, but he took up the cause energetically during World War I. He spoke and wrote as an ardent advocate of spiritualism, publishing many articles and books right up to the time of his death, including a two-volume *History of Spiritualism* in 1926. He also subscribed wholeheartedly to the reality of the COTTINGLEY FAIRIES, pendulum DIVINATION, DOWSING, ectoplasm, spirit writing, GHOSTS, and many other psychical phenomena. In this he was entirely uncritical and impossible to disillusion.

He was on friendly terms with the magician Harry HOUDINI and convinced that Houdini made his escapes by dematerializing his body; oozing out of his chains, straitjacket, casket or whatever imprisoned him; and then becoming material again. Despite Houdini's explaining that his escapes were made by trickery, Doyle did not believe him.

After his son died in World War I, Doyle's growing preoccupation with the paranormal may have been influenced by his hope of making contact with his son through a medium. There is no record of his ever doing so, but his commitment to the spread of psychic phenomena shows the depth and sincerity of his conviction. He was not just seeking solace for his bereavement, but firmly believed in the existence of a whole world outside our everyday physical existence. This belief is all the more strange coming from the creator of Holmes and Watson, characters wedded to the rational and scientific method. Holmes gathered evidence, analyzed it inductively and deductively, eliminated the impossible, arrived at the most likely explanation, however improbable, and then proceeded to test it. Dr. Watson, a medical man, observed and recorded and approved of Holmes's methods. This was quite at odds with Doyle's own worldview. The fiction-writing side of Doyle is usually emphasized in biographies and the activities of the several Holmes societies, while the real-life non-rational side is either forgotten or downplayed. When both are brought out, we are left with a contradictory impression.

For further reading: John Dickson Carr, *Life of Sir Arthur Conan Doyle* (Carroll & Graf, 1987); and Michael Coren, *The Life of Sir Arthur Conan Doyle* (Bloomsbury, 1996).

DREAMS Any train of thoughts, images, or fancies that pass through the mind during sleep. There are many diverse beliefs about the nature of dreams, some dating from antiquity. Others go back to the 19th century and are associated with the rise of psychology as a separate subject within medicine. It was not until the 1950s that a laboratory technique was devised to study dreams.

Many ancient people believed that during sleep the soul left the body to live for a time in a special dream world. Because humans were thought to be without souls at this time, sleepers were never awakened lest their souls be lost on their hurried return home to the body. In classical Greece, dreams became associated with healing, and many sick people went to a special temple to sleep through the night to take the dream cure.

In many societies dreams were thought to be messages from the gods, a source of divination predicting the future. This idea is illustrated on Sumerian and Egyptian monuments; recently a Babylonian dream guide was found in the ruins of the city of Nineveh. Many ancient texts such as the Old Testament are full of prophetic dreams. Among pre-Islamic peoples, dream divination influenced daily life so much that the practice was formally forbidden by Muhammad, father of Islam.

Presumptions that dreams had supernatural attributes persisted until the mid-1800s. There were, however, a few notable early skeptics such as Aristotle, who put forward the idea that dreams were an extension of the waking state and arose from external stimuli. The Roman statesman Cicero courageously attacked dream divination in *De divinatione*. Nevertheless it was not until the 19th century that the French physician Alfred Maury (1817–92) after carefully studying thousands of reported recollections of dreams, went back to Aristotle and concluded that dreams arose from external stimuli.

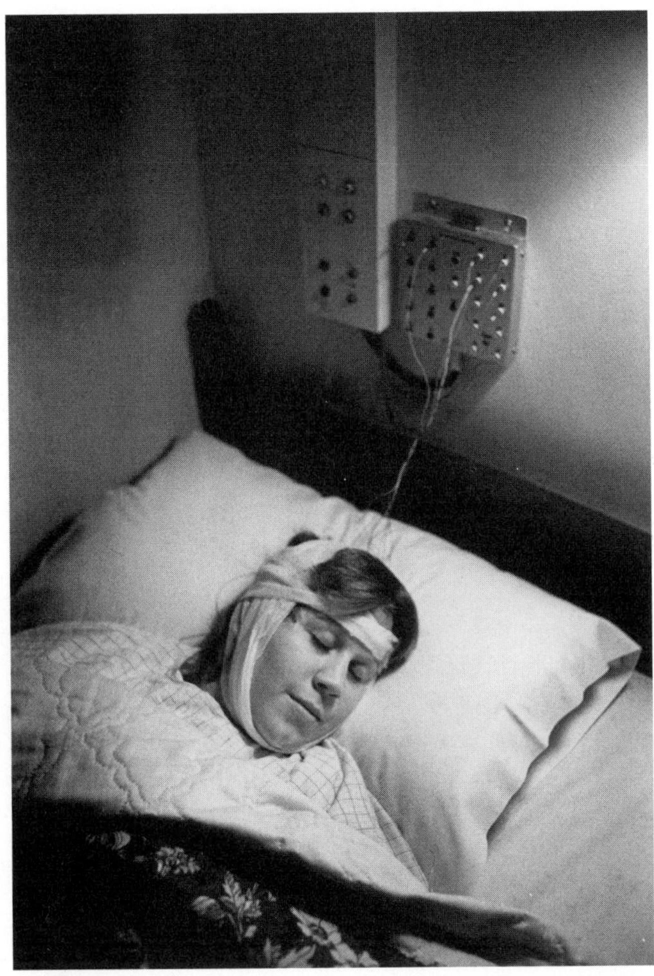

Dreams

In the 19th century, following the example of Maury, dreams were regarded in a less mystical way, and were subsequently located within the discipline of psychology, which dealt with behavior in mankind and animals. Both Sigmund FREUD and Carl Gustav JUNG, in the course of their clinical work on the treatment of mentally disturbed people, used a new approach called psychoanalysis, and the analysis of their parents' dreams formed a major part of their therapy. The split between these two men resulted in two very different psychoanalytic interpretation of dreams.

Freud believed dreams were expressions of wish fulfillment and quoted the old proverb: "Pigs dream of acorns and geese dream of maize." His hypothesis was that dreams partially reflected waking experience, *but* the thinking we do during sleep was primitive and regressive, and had the effect of causing us to remember what we wished to forget, or repress, in waking life. For him, dreams had two contents: the *manifest content*—the straightforward, obvious thing that was dreamt—and the *latent content,* the underlying thoughts, wishes, recollections and fantasies, often of a sexual nature, remembered from early infancy. He called the activity which transforms the latent into the manifest the *dream work.* In short, for Freud the dream was a heavily disguised form of infantile wish fulfillment expressed as a hallucinatory experience in the course of sleep.

For Jung, dreams were not signs but symbols (a sign *represents*—stands in for the object; a symbol *resembles*—stands for, represents, or denotes more than its obvious immediate meaning). Jung believed that many of the symbols in dreams were not individual constructs but were an involuntary part of the shared collective consciousness that had united all humankind throughout time. Dreams had always been the main source for Jung's lifelong study of historical symbols, so it was not surprising that he found that dreams of quite unsophisticated patients exhibited symbolic parallels with myths and legends of the past, and that these dreams also depicted ancient archetypes from traditions and cultures about which his patients could not have possibly known.

Two books written by John William DUNNE, a British inventor and philosopher—*An Experiment with Time* (1927) and *The Serial Universe* (1934)—inspired many readers to record their dreams to check on their predictive value. Dunne saw time as an unfolding of preordained events, and put forward the idea that the universe was one long continuum stretching out in time with no barriers—except those we in our ignorance impose, between past, present and future. Dunne's views on predictive dreaming are no longer taken seriously.

A new field in dream research using laboratory techniques began in 1953 with the discovery that rapid eye movements (REMs) during sleep signaled dreaming. Researchers in Chicago found that two hours per eight-hour adult sleeping period were usually devoted to dreaming (in newborns, it was half the total sleeping time). By repeatedly waking volunteers during REM time, sleep researchers found that their subjects all made up these dreaming time later in the night. From this the scientists concluded that dreaming was a necessary function in healthy individuals, that we all have dreams, but that they are rapidly forgotten on waking. Often, dreams seemed to be vastly influenced by arbitrary factors. Anxiety-provoking movies, seen prior to sleep, were found to lead to dreams having related themes. Heavy meals, consumption of alcohol, and heated arguments also provoked more intense dreams, but, as expected, the dream content varied from individual to individual.

When asked to explain dreams, the researchers say, somewhat tentatively, that according to the present state of knowledge dreams are by-products of the processing of information that has come in during the day and needs to be incorporated into the cognitive system. Modern researchers say that, despite ideas posited by Freud and Jung, they have not found any evidence that a more useful understanding of personality can be gained from dreams than can be more easily determined from waking behavior.

With the coming of New Age thinking, many people are returning to ancient beliefs about dreams, using ideas from Artemidorus, Nostradamus, Thomas Tryon's 1695 *Book of Dreams* and from *Old Moore's Dream Book,* along with many recently published self-help books on how to understand "the secrets of dream interpretation."

For further reading: J. Allan Hobson M.D., *The Dreaming Brain* (Basic Books, 1988); Bert O. States, *Seeing in the Dark: Reflections on Dreaming* (Yale, 1996); and Montague Ullman and Stanley Krippner, *Dream Telepathy: Experiments in ESP* (McFarland, 1989).

DROWN, DR. RUTH B. (1891–1965) A member of the American Naturopathic Association and a licensed osteopath, Drown treated patients from her radio room in Los Angeles by a process that she called Drown Radio Therapy. She used an electronic machine of her own devising to diagnose patients from samples of their blood kept on blotting paper. The sample was then placed in another machine and tuned to the patient, then healing radiation was transmitted to him or her wherever he or she was, over thousands of miles, if necessary. A detailed explanation and justification of Drown's system is given in her books, *The Science and Philosophy of the Drown Radio Therapy* (1938), *The Theory and Technique of the Drown Radio Therapy and Radio Vision Instruments* (1939), and *Wisdom from Atlantis* (1946).

Dr. Drown made great claims for the success of her diagnosis and treatment and also claimed to be able, using equipment that she had invented, to obtain radio photographs of internal organs and to transmit radiation that stopped bleeding. Not only did she have many patients convinced of the efficacy of her treatment but she also sold her machines to other practitioners.

In tests conducted at the University of Chicago, Drown proved unable to substantiate her claims. The trial was reported in the February 18, 1950, issue of the *Journal of the American Medical Association.* Nevertheless, Dr. Drown continued to practice and to sell her machines.

DUALISMS From the Latin *dualis* meaning "containing two"; the belief that the universe has a double nature, and that therefore "the real" is either of two kinds of, or controlled by, two powers. The opposite of this, monism, from the Greek *monos,* "sole," is the doctrine that everything is of the same fundamental kind. The term "dualism" has been applied to oppositions in many subjects, therefore the use here of the plural "dualisms." In fact dualism can be said to be the dominant trope of Western culture.

Two ancient Greek philosophers, Parmenides and Heraclitus, were monists. Parmenides advanced a monism based on Being, describing motion of change as unreal, while Heraclitus advanced a monism based on Becoming, seeing nothing as static—everything in flux—and change as the eternal law of things. The teachings of these two philosophers are important because Plato (427–347 B.C.E.), in seeking to join the two together, produced his metaphysical dualism. He did this in such a way that Parmenides' Being became the "truth" or "essence" of the Heraclitean flux. Plato viewed the whole universe as having this double nature, notably making distinctions between:

1. appearance and reality
2. soul and body
3. ideas and material objects
4. reason and the evidence of the senses

All these opposites imply that behind the world as we perceive it, there lies an "ideal" world that Plato saw as more real, and therefore superior to, that of the world of appearances. Any dualism that is hierarchal, like Platonism or Christianity, always makes one world better than, or superior to, the other. Religions such as Zoroastrianism, Manichaeism, or Gnosticism that believe the universe to be ruled by good and evil principles (in effect that there is a good deity and a bad one) are also termed dualistic.

Most adherents to monism see the main problem in dualism to lie in the question of how and where the supposed two distinct and separate properties of the universe interact; for example, where does the mind really mesh with the physical mechanics of the brain and body? Where does the ghost lie in the machine? Descartes, a 17th-century (dualist) philosopher, decided that mind and body must interact somewhere; he selected the pineal gland, in the middle of the head, as the interface with the soul. But this did not explain how two totally different aspects could possibly influence each other. Today some philosophers and psychologists see dualistic views as posing more problems than they solve.

Scientists mostly reject the idealism inherent in dualism and in their profession work on the principle of a closed system, on there being physical causes for physical events, allowing no place for the incursion of influences from outside the physical world. An extension of this idea is that all wishes, emotions, feelings, thoughts, decisions, and so on are processes that take place in, or are propensities of physical creatures. Not only are these processes always accompanied by changes in the brain and the central nervous system, but they *are* those changes. The person who is sensing and feeling and thinking is a living body and not something else living inside the body called a mind or soul. To introduce an entity between the activity of the body and the thoughts or feelings it generates is to introduce a HOMUNCULUS into the system.

Dualist clashes occur between physics and metaphysics, between the natural and the supernatural, and between science and pseudoscience.

For further reading: C. D. Broad, *The Mind and Its Place in Nature* (Paul, Trench, Trubner, 1925); and P. Churchland, *Matter and Consciousness* (MIT Press, 1988).

DUNNE, JOHN WILLIAM (1875–1949) Author of *An Experiment With Time* (1927) who thought dreams foretold future events. Dunne served for a short time in the British army and was an inventor and designer of some of the early airplanes.

Beginning in 1898, Dunne became convinced that his dreams sometimes foretold future events as well as including information from the past. He became convinced, first, that this foresight—PRECOGNITION was his word—was common. To test his theory, he recorded his dreams as soon as he woke and invited family and friends to join him. He laid down rules to minimize the effects of coincidence and rules about what was an admissible foresight—rules which could be bent occasionally. And he found that there were sufficient successes to warrant extending the project. He also found that by concentrating on some subject, for example a book he was about to read, he could find evidence of precognition while awake—to dream was not essential.

Over the next twenty years, Dunne satisfied himself of the reality and normality of precognition, he then began devising a theory to explain it. As he put it: "Was it possible that these phenomena were not abnormal, but *normal?* That dreams—dreams in general, all dreams, everybody's dreams—were composed of *images of past experience and images of future experience blended together in approximately equal proportions?* That the universe was, after all, really stretched out in Time, and that the lop-sided view we had of it—a view with the "future" part unaccountably missing, cut off from the growing "past" part by a traveling "present moment"—was due to a purely mentally imposed barrier which existed only when we were awake? So that, in reality, the associational network stretched, not merely this way and that way in Space, but also backward and forward in Time."

Drawing on Einstein's theory of relativity, he concluded "that the nature of time allowed the observer a four-dimensional outlook on the universe." He then produced a very laboriously worked out theory of Time using diagrams vaguely related to Minkowski's space-time diagrams (a graphical way of representing Einstein's special relativity theory). And finally he recorded a set of experiments on dreams with seven dreamers (including himself), the results of which he saw as convincing evidence in favor of precognition.

Dunne was not the only individual to promote the idea of precognition. His importance lies in the popularity his theories enjoyed, perhaps because of their apparently sophisticated scientific nature. *An Experiment With Time* sold many thousands of copies, went into several editions, was a constant topic of conversation, and was commended by many scientists, among them Sir Arthur EDDINGTON.

E

EARTH-CENTERED ASTRONOMY The theory, developed by Aristotle in the fourth century B.C.E., that the universe consisted of five elements and eight spheres. The elements were earth, air, fire, water, and ether, the substance of which the spheres themselves were composed. Earth itself (known to be spherical because its shadow could be seen on the Moon) was stationary and held an immobile position at the center of the eight spheres. The eight etherial spheres revolved around the immovable Earth, carrying with them the Sun, the Moon, each of the known planets (five at the time), and all the stars.

Aristotle's system of celestial mechanics was based on commonsense naked-eye observations of how the stars appear to move through the sky. Later sky-watchers, however, noted some discrepancies between Aristotle's system and their own observations. Sometimes the planets seemed to stop in the sky or even to move backward. In the second century C.E., for example, the Alexandrine astronomer Claudius Ptolemy explained these motions through the use of *epicycles*—small spheres within the great etherial spheres in which the planets rotated separately. Ptolemy's star charts, which used as many as 90 epicycles to account for variations in Aristotle's celestial spheres, were used by mariners for hundreds of years as navigational aids.

The Earth-centered astronomy devised by Aristotle and Ptolemy met the needs of the astronomers of late antiquity and the medieval period. During the Renaissance, however, new theories were advanced to explain the observed motion of the stars and planets. Chief among these were the observations and theories of the Polish monk Nicolas Copernicus, who proposed in 1543 a theory that placed the Sun, not Earth, at the center of the universe. The title of his work, however, paid homage to Aristotle's original concept of the etherial spheres—it was called *On the Revolution of the Celestial Spheres.* Copernicus's ideas slowly gained acceptance, and by the 17th century, Aristotle's Earth-centered theory had been discarded by the scientific community in Europe.

EARTHLIGHTS Anomalous lights that appear close to the earth, usually in rural or remote areas, and that show movement and reappear in the same location night after night, often for years at a time. Earthlights have been reported around the world and over many centuries. Also called GHOST LIGHTS or spook lights, they are sometimes associated with GHOSTS or spirits of the dead. Described by some as "dancing," and "ball-like," they frequently appear near power lines, but just as often near hilltops, isolated buildings, and other sites remote from electrical power sources. Reports of their color vary. The lights seem to react to other light sources and to observers trying to approach them. They often can be seen only from particular angles.

Among the best-known earthlights are those near Marfa, in western Texas, appropriately called the Marfa lights. They were first reported in 1883, and they have been reappearing regularly ever since. They dance and chase, and their origin has never been satisfactorily determined. Earthlights are reported in many states and in countries around the world.

Many earthlights have been found to have natural explanations. These include automobile, home, and city lights reflecting off the air in certain atmospheric conditions, and swamp gas. A somewhat more sophisticated explanation is that they are caused by ionized gas formed by the high voltages created by seismic stress; this view is supported by the fact that earthlights have been found near various fault lines (for example, the Brown Mountain lights in North Carolina and the lights over several of the Scottish lochs that sit on fault lines). Electromagnetic energy escaping from transformers and high-power lines explain some earthlights.

Another theory is that they are some kind of anomalous intelligent energy, perhaps extraterrestrial, perhaps from our own world. This idea is based in part on the apparent playfulness of some of the lights—they have been reported as playing hide-and-seek with observers, receding and approaching, and even chasing each other at times. Some have been reported as actually approaching an individual and staying for some moments before disappearing or whisking away.

Researcher Paul Devereux suggests that the lights are "energy forms produced by processes at the very limits of our current scientific comprehension." He asserts that they are the same mysterious force known to ancient cultures (as *chi* in China, as *kurunba* to Australian aborigines, for example) as underlying everything in the material world. Belief in this force, he and others suggest, has been repressed or ignored by our science-dominated age. Devereux believes that the energy that causes earthlights may well be the key to future human evolution. New Age advocates, who believe humans are on the verge of the next step in evolution, have adopted some of Devereux's ideas.

Skeptics say that the seismic stress theory is flawed because many earthlights are not near fault or other high-stress areas. The intelligent-energy theories also have no scientific backing; they are more a matter of spiritual belief. Most scientists who have looked at the phenomenon remain certain that prosaic explanations do exist even though they may not yet be known.

For further reading: Paul Devereux. *Earth Lights Revelation* (Blandford Press, 1990).

EASTER ISLAND Isolated island in the southeastern Pacific Ocean covered by hundreds of enigmatic statues. In 1722, a Dutch ship touched on the shores of a tiny island in the South Pacific Ocean, some 3,700 kilometers (2,300 miles) west of the Chilean coast. The sailors discovered that the inhabitants of the island had erected hundreds of man-shaped statues, ranging in height from one to more than 21 meters (3–70 feet). The statues, called *moai* by the natives, faced variously toward the coast or toward local villages. By the time Captain James Cook and his British crew touched on the island in 1774, most of the statues had been thrown down from their platforms. The statues continue to mystify people today: What upheaval caused their collapse? Why were they carved in the first place? And most mysteriously, what brought the people who carved them to this isolated spot, thousands of miles from the nearest land?

Most anthropologists believe that, between the Dutch and the British visits to Easter Island, a bitter civil war divided the islanders against one another. They base this conclusion partly on stories handed down by island families from generation to generation and partly on the pattern of destruction seen among the statues. The function of the statues is a matter of much speculation. Some anthropologists suggest that the statues were linked to a form of native religion or ancestor worship. Others have suggested that they were family totems, erected to impress others with the family's power. Still others have suggested that the statues were erected out of boredom because there was nothing else to do to pass the time. A more fanciful explanation suggests that the statues were erected by or in honor of extraterrestrial visitors to the island.

Easter Island

Where the Easter Islanders came from has also been a topic of much discussion. Most anthropologists believe that the original islanders came from Polynesian islands far to the north and west, arriving in traditional Polynesian outrigger canoes sometime in the 12th century C.E. These were the "Long-ears," islanders with elongated earlobes, who carved and erected the first statues. Some time later another group of settlers—known as the "Short-ears"—arrived and were enslaved by the Long-ears. Anthropologists suggests that a Short-ears revolt helped destroy the Long-ears culture sometime in the 18th century.

The Norwegian anthropologist Thor HEYERDAHL visited Easter Island in 1955, experimented with erecting some of the fallen statues, and formed his own theory about the origins of the Long-ears. Heyerdahl suggested that the Long-ears came, not from Polynesia, but from South America—specifically, Peru—around 300 C.E. He based this conclusion on three facts: (1) The South American cultures had a history of stone carving, and the Polynesians did not. Furthermore, the Easter Island statues resembled ones found in Peru. (2) The Easter Islanders had a written language, recorded on wooden *rongo-rongo* boards. The Polynesians had no written language; the Peruvians did. (3) Heyerdahl himself demonstrated that such a trip was possible in his famous KON-TIKI trip in 1947.

Many anthropologists have never accepted Heyerdahl's conclusions. They point out that the native language of Easter Island is related to that of Polynesia, not to any known South American indigenous tongue. The fact that Heyerdahl proved that a trip from South America to Polynesia was possible does not prove that such a trip took place. Most anthropologists prefer the explanation that the island's immigrants came from Polynesia. But whatever the islanders' origins, scientists agree that the statues were erected by applied manpower and not through any extraterrestrial agency.

For further reading: Thor Heyerdahl, *Easter Island: The Mystery Solved* (Random House, 1989); Catherine Orliac, *Easter Island: Mystery of the Stone Giants* (Abrams, 1995); and Kathrine Pease Routledge, *The Mystery of Easter Island: The Story of an Expedition* (AMS Press, 1978).

ECLECTICISM A school of natural medicine popular in the 19th century United States. Eclecticism was developed, in part from THOMPSONISM by Wooster Beach. In 1825 Beach emerged out of obscurity as he began to write against the practices of physician colleagues who used bloodletting and the strong remedies of the day and started to advocate the use of natural remedies, primarily roots and herbs. In 1827 he opened an infirmary in New York City and in 1937 established the New York Medical Academy (later the Reformed Medical College of New York), the parent institution of eclecticism.

A second school, established in Worthington, Ohio, in 1833, relocated to Cincinnati in 1843. By 1900 there were nine eclectic colleges. Eclecticism prospered because of the public's reaction to the harsh procedures and remedies advocated by the majority of physicians and because several natural medicines it prescribed had obvious physiological effects. However, by the turn of the century, various medical improvements, including the development of anesthesia and antiseptics, the demand for improved medical education, and the political power of the American Medical Association began to have their effect. By the beginning of World War I, only four eclectic schools remained. Three of these closed by the end of the decade.

During the 1920s, eclecticism was merged into a new movement, NATUROPATHY, developed in the 1890s from the German WATER CURES. Naturopathy gradually incorporated into itself a variety of practices that required neither the use of chemically created medicines nor surgery. The Naturopathic Association of America (later the American Naturopathic Association) was formed soon after the turn of the century.

For further reading: Morris Fishbein, *Fads and Quackery in Healing* (Blue Ribbon Books, 1932).

EHRET, ARNOLD (1866–1922) German doctor of NATUROPATHY who developed the Mucusless Diet Healing System, still popularly advocated in health-food circles. Ehret was born in Freiburg, Baden, Germany. Shortly after graduating from the University of Freiburg, he was drafted into the army but was soon released because of his heart condition. As his health continued to degenerate, he turned to various natural healing centers, including the famous WATER CURE center headed by Fr. Sebastian Kneipp. Out of his search, he began to create his own understanding of the means to health. Healing was based on three key elements: fasting, a proper diet, and climate. He advocated fasting as a means of cleansing the body of impurities. He advocated a diet centered on raw fruit and natural foods that contained no albumin, which he believed produced mucus. As naturopathy emerged he was especially critical of colleagues who allowed substances containing albumin in their diets. He also traveled the world observing the fruits and vegetables grown in different areas as he sought to improve upon the basic diet he advocated.

Ehret developed a specific diet about which he lectured widely. He also built a sanitarium in Switzerland where diet was used to heal people. Beginning in 1914 he publicized his diet throughout the United States. He also prepared a manuscript outlining the basic aspects of his mucusless diet. After the war, he remained in California, where he had been treating a growing clientele. Unfortunately, he died in 1922 as the result of an accident; he was otherwise in perfect health. Shortly after his death, some of his supporters published his small book, *Arnold Ehret's Mucusless Diet Healing System*. It is still in print, though it serves only a small group of devoted vegetarians.

ELECTRIC-SHOCK TREATMENT Also referred to as electroconvulsive therapy (ECT); in use since the 1930s as a treatment for depression that consists of the application of weak electric currents to the head through electrodes. The belief is that the frontal lobe of the brain is the part most affected in depression and that, by damaging it temporarily, the depression may be eased and, with repeated treatments, cured. The patient has to be anaesthetized or have a muscle relaxant and even restrained because the shock produces convulsions. Sometimes the patient goes into a coma. There is some criticism of its widespread use because ECT can cause permanent brain damage—it is unquestionably a blunt instrument—especially loss of memory (it may break the link between long-term and short-term memory) and reduction of intelligence. The treatment is still used widely today, not so much as before, and now primarily for severely depressed patients and as a last resort.

Scientists do not understand how ECT works; it is therefore on the fringe of science. Brain research is still in its early stages. When more is understood, ECT may be explicable and be seen to be legitimately scientific. Alternatively, it may be discredited and dismissed as pseudoscience.

ELECTROMAGNETIC FIELDS In the last half of the 20th century, concern grew that being in the vicinity of electric power installations caused a variety of illnesses. The occupants of houses close to high voltage overhead distribution lines complained of persistent inexplicable headaches, depression, and other health problems. Some areas around the lines appear to have produced unexplained clusters of leukemia. These problems have been attributed to the unusually high electromagnetic (e/m) low frequency (l/f) fields around such equipment. Evidence to support such a claim is anecdotal. Even if a statistical correlation were to be shown, the connection may just be accidental and may not demonstrate a causal relationship. However, public concern has been sufficient for the matter to be taken seriously.

In Sweden, epidemiological studies showed a possible link of high electric voltage with cancer and in 1992 the Board for Industrial and Technical Development thereupon stated that it would *assume* a connection to be established. Scientists in the United States and Great Britain who studied the problem failed to establish a link. Nevertheless, calls continue for further study.

The matter remains unresolved. On the one hand, worries now extend to a greater range of symptoms and sources of possible deleterious effects: radio and television transmitter aerials, cellular phones, domestic equipment (kettles, blankets etc.), power substations, electricity meters, clocks—anything using l/f mains or r/f electricity, around which there will be electric and magnetic fields, however small. On the other hand, investigations in several countries have failed to establish that there is a case to answer, let alone a causal link. The Institute of Electrical Engineers in Britain, for example, after exhaustive examination by a working party drawn from a broad range of expertise and without axes to grind, concluded "there is nothing in the currently available evidence to prove the existence of the effects claimed. If they do exist, then their incidence within the population, taken as a whole, must be very small." Nevertheless, the media continues to explore the issue and there is a very active campaign to take prudent protective measures now. In January 1994 a BBC television program on the subject "featured the health effects of electromagnetic fields, including the Studholme family where a young boy died of leukemia, apparently because of a nearby electricity sub-station and an electric meter behind a wall near his bed." The family are suing the utility company for their son's death. There have been cases taken to court in the United States, not one of which has been upheld. In 1994, a United Kingdom High Court Judge ruled that there should be a full Judicial Review of the National Grid Company's plans to lay additional high voltage cable through part of north London.

Three physicists in the Technology and Science Division of the National Grid Company—John Swanson, David Renew, and Nigel Wilkinson—conducted a survey of the evidence for this effect and published a report, *Power Lines and Health.* A review of the report in *Physics World* (November 1996) summarized it thus: (It) reviews the evidence to date and discusses possible causative mechanisms for e/m fields to affect health and in particular initiate cancer. The authors conclude:

> *Clearly it would be both arrogant and rash of physicists to argue that because we have not been able to think of a possible physical mechanism, it is impossible for there to be an effect. However, the absence of both a mechanism and reproducible laboratory results inevitably means that the epidemiological results, already somewhat weak, are viewed with much greater skepticism. Several national or international bodies have surveyed this question and have concluded that the evidence linking environmental electric and magnetic fields with cancer is not convincing. However, such bodies often call for further research. That research needs the skills of physicists alongside those of biologists, epidemiologists and engineers.*

ELECTRONIC VOICE PHENOMENON Also known as Raudive voices; the phenomenon of mysterious voices, apparently from dead individuals, impressed on tape recordings made on unmodified standard recording apparatus. The phenomenon was discovered by Friedrich Jürgenson in 1959 but was extensively pursued and popularized by Konstantine Raudive, a Latvian psychologist and parapsychologist. Jürgenson, a painter and film producer, discovered the phenomenon one day after recording the song of a bird on his recorder. Playing it back, he heard a human voice. He repeated the process, after initially concluding that the voice was due to a fault in his recorder. However, when the phenomenon reappeared, and he listened to what was being said, he concluded that he had received a message from his deceased mother.

After having learned of Jürgenson's experience, Raudive contacted him and convinced him to cooperate in five years of further research in attempts to receive paranormal voices on tape. The two men disagreed and separated in 1969. Raudive published the results of their research, which became widely known, after the appearance of an English edition, as *Breakthrough: An Amazing Experiment in Electronic Communication with the Dead* (1971). The phenomenon was of some interest to parapsychologists, who saw the possibility of conducting survival research without the need of a medium.

A variety of attempts were made through the 1970s to replicate Raudive's seemingly spectacular results, but the results failed to prove compelling to all but a few. Most voices were faint, brief messages, speaking at higher than normal speed. Even some of those who joined heartily in the initial experiments concluded that the results were ambiguous, at best. Critics have suggested that the source of the voices is quite mundane, not the least being the wishful thinking of the experimenter.

For further reading: Peter Bander, *Carry on Talking: How Dead Are the Voices?* (Colin Smythe, 1972); and Konstantin Raudive, *Breakthrough: An Amazing Experiment in Electronic Communication with the Dead* (Taplinger, 1971).

ELECTROPHYSIOLOGY The use of electricity to aid the better functioning of the body. Using electrotherapy to relieve pain is said to have arisen in Ancient Rome when a physician put a patient's painful gouty foot on a live electric eel. At the end of the 19th century when direct electric-current supply was becoming available in people's homes, a treatment called GALVANISM became fashionable. It was found that pain could be reduced by attaching electrodes to the skin and applying a direct-current voltage. Later, when an alternating current supply was introduced, a similar treatment followed. The treatment was especially used in sports injuries to reduce inflammation in strains, sprains, sprains, and contusions by improving circulation.

Electroconvulsive therapy (ECT)—applying a voltage with surface electrodes on the head across the brain—was used extensively in the late 1960s and 1970s to treat depression. At the time there was no theoretical basis to justify it, and recently the treatment has come under considerable criticism because it is believed to have caused permanent brain damage to some patients, especially loss of memory and intelligence.

Today the electrotherapy most widely used is transcutaneous nerve stimulation (TNS), which activates the nerves that block pain. The treatment is very safe; a small battery-powered machine is used, and although it is usually carried out by a physiotherapist, machines can be bought or borrowed from clinics for self-administered treatment. Weak electrical impulses are sent through the skin via two rubber pads coated in jelly and taped to the painful area. The therapy is mainly used to treat lumbago, sciatica, and sports injuries, but it can also be used instead of drugs to relieve pain in childbirth. The use of TNS is now standard treatment in orthodox as well as ALTERNATIVE MEDICINE, and most people can avail themselves of it, provided the cause of the pain has been investigated.

Pulsed high-frequency (PHF) electromagnetic waves are sometimes used as a treatment for acne, eczema, psoriasis, and other skin problems. The practitioner directs the pulsed waves from a machine to the troubled areas for a half-hour and then treats the liver and the adrenal area. The electromagnetic waves can enter the body to a depth of 15 centimeters (6 inches). Some clinics offer a similar treatment to aid in weight loss. Electrical pulses are directed into the skin over the muscles to make them twitch with the hope that this will burn off calories. Electro-ACUPUNCTURE is also offered. This is a variant on orthodox acupuncture, but it does not appear to have any advantage over the original.

ELLIS, HENRY HAVELOCK (1859–1939) An eminent British psychologist of the early 20th century. He published the landmark work *Studies in the Psychology of Sex* in seven volumes between 1897 and 1928—the final, complete edition was not published until 1936—as well as many other books and papers. His promotion of a more open attitude to sex and to sex education as well as support of women's rights provoked fierce opposition. In the United States his major works on the psychology of sex could only be sold legally to members of the medical profession in 1935. In many university libraries, until recently, Ellis's major works were kept under lock and key and were only available to medical faculty. In 1898 George Bedborough, a bookseller, was prosecuted at the Old Bailey for the crime of having a copy of Ellis's *Sexual Inversion* on his shelves. *Sexual Inversion* is a study of homosexuality with an emphasis on the need for society to be less condemnatory and more understanding.

The abuse that Ellis encountered on publication of his major work was not so much leveled at the probity of his science as at his effense against the puritanical sexual mores of his time, especially in the United States. The legal constraint was effectively a taboo, removing his work on sexual behavior and sexual education from the sight of the people for whom it was intended. To that extent, it placed him and his work on the fringes of legitimate science.

In addition to his books on sexual behavior, Ellis wrote about genius, criminality, DREAMS, Spain, EUGENICS, conflict, obscenity, and literature. In *A Study of British Genius* (1904), he claimed that brilliant men shared certain characteristics: They often were shy, eldest children; had delicate constitutions, high-pitched voices, and problems writing legibly, and disliked social contact. Their mothers were markedly pious and their fathers were comparatively undistinguished. His study was criticized on two fronts: first that his knowledge of statistics was inadequate, and second that he saw himself as a model.

In *The Task of Social Hygiene* (1912), Ellis advocated a voluntary eugenic register to make a program of selective breeding possible. The idea was to procreate a scientifically reformed elite in which the snobbery of birth, money and rank would give place to health, intelligence, and ability. Although very critical of Cesare LOMBROSO, the 19th-century Italian physician who believed that the predisposition to crime was genetic, and although Ellis was persuaded that criminality had social causes, he nonetheless was attracted to the idea of a criminal type and thought that sometimes there was a connection between criminality and genius.

In 1897, Karl Pearson, then professor of biometrics at University College London, attacked Ellis for his theory that because of greater birth difficulties of male babies (their heads are larger than those of females), men would show greater physical variations than women. Pearson said that he was going to "lay the axe" to this "pseudo-scientific superstition."

Far and away Ellis's most important and abiding contribution to our culture was his very thorough and scientific analysis of sexual behavior and the impetus that it gave society to move toward a more open acceptance of sexuality.

For further reading: Phyllis Grosskurth, *Havelock Ellis: A Biography* (Allen Lane/Alfred A. Knopf, 1980); and Paul Robinson, *Modernization of Sex: Havelock Ellis, Arthur Kinsey, William Masters and Virginia Johnson* (Cornell University Press, 1988).

ESP See EXTRASENSORY PERCEPTION.

ETHER See EARTH-CENTERED ASTRONOMY.

ETHERIC BODY See ASTRAL/ETHERIC BODY.

EUGENICS Efforts to improve the human population by means of selective breeding or other genetic manipulation. In a general sense, the idea has been considered in many cultures, usually by elites, privileged classes, or social reformers. In the Western cultural tradition, the notion can be traced to Plato, who envisioned carefully controlled breeding as a means of building an ideal society.

The term "eugenics" itself, however, was only coined in 1883 by English intellectual Francis GALTON and is generally used to designate such efforts from the late 19th century to the present. This period saw eugenics flourish in many countries, including the United States, Great Britain, France, and, most notoriously, Germany. In most cases it was promoted as an important new science, and proponents availed themselves of the latest developments in statistics, biology, psychology, and other academic disciplines to bolster their claims. Many key advocates had solid scientific credentials and pushed to have their ideas integrated into governmental policy.

Practically from the start, however, eugenics had detractors who kept up a sustained criticism of the movement on moral, social, and political, as well as scientific grounds. After World War II, the links between eugenic ideas and wartime atrocities led to a strong public reaction against eugenics programs. This severely stunted the growth of eugenics as a social and scientific movement, although proponents are still active in many countries.

Historians generally regard Galton as the most important early figure in the modern eugenics movement. As early as 1869, in his book *Hereditary Descent,* he had begun to sketch out his basic theories, and he continued publishing and promoting the idea throughout the late 19th century. With the help of admirers like Karl Pearson, he also pioneered the use of statistics and mathematical models to prove the soundness of eugenics. Galton and other eugenicists came to recommend a twofold scheme for introducing their ideas into public policy: (1) a "positive" program of tax breaks and other monetary incentives to encourage the educated professional classes to have more children; and (2) a "negative" program that advocated that reproductive restrictions, including sterilization, be forced upon groups deemed socially undesirable, such as criminals and persons with poor mental ability. In subsequent debates over the social merits of eugenics, advocates and critics tended to focus on this negative program.

Resistance to eugenic policies on ethical grounds, particularly from religious groups, slowed acceptance of the new science throughout the late 1900s. Around the turn of the century, however, the efforts of Galton and his admirers began to garner support. Often eugenics was combined with an interest in proving the scientific superiority of particular races. During the next two decades, enthusiasm for eugenics became widespread as organizations aimed at providing funding for eugenics research formed in Great Britain, Germany, the United States, and other countries. These groups tended to be small, but they generally enjoyed the support of wealthy and distinguished patrons who included scientists, intellectuals,

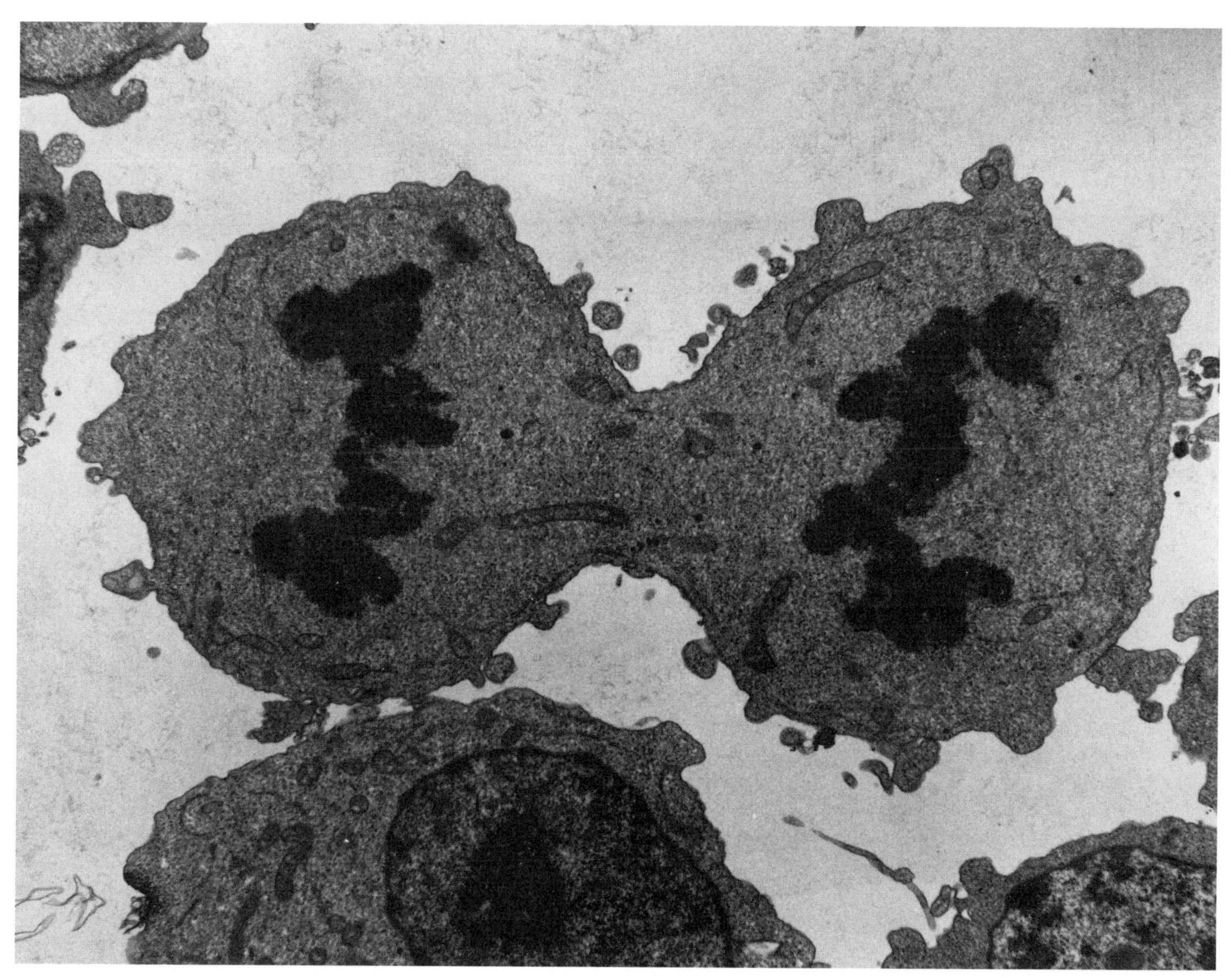

Eugenics

professionals, educators, and public lecturers. The popular press further disseminated eugenic theories.

Supporters of the movement are often stereotyped as anxious elites who saw this "science" as an important means of ideological and social retrenchment. This was often true, but it should not obscure the fact that actual interest was much more broad and varied. Advocates included socialists who supported eugenics as part of a larger effort to bring scientific management to social problems, libertarians who used eugenics to argue against rigid marriage conventions (which too often bound together bad genetic matches), and supporters of women's rights who demanded the freedom to choose eugenically superior spouses. Obstetricians, particularly in France, often promoted eugenics in combination with movements for hygienic reform and improved prenatal care. Similar concerns for the quality of offspring and public health drew many women into the eugenics movement as active participants in both national organizations and local advocacy groups.

Though eugenicists shared the belief that the overall genetic stock of humanity could and should be improved, advocates often debated the relevance of particular scientific theories to this end; for example, factions in the eugenics movement differed sharply over the precise mechanisms of heredity and organic EVOLUTION. Among their ranks could be found Darwinians, Lamarckians, orthogenicists, and Mendelians. Eugenicists also constantly refined and updated their ideas with developments in a variety of disciplines, including psychology, medicine, statistics, and demography. This was especially true in regard to the science of genetics, which had many links to the eugenics movement. American Charles Davenport, for example, produced many important early studies of trait inheritance and, with other scientists, worked to integrate basic eugenic thought with Mendelian genetics.

By the 1930s many of the eugenicists' key scientific concepts were regularly being challenged. Such criticism highlighted the shortcomings of three tendencies of eugenic thought. The first was the way eugenics advocates tended to discount or ignore the influence of environment on the development of behavior and intelligence. Most eugenicists were willing to admit that environment had something to do with the development of talent and intelligence, and some even promoted a parallel program *euthenics* to explore how to improve social institutions. For the most part, however, they tended to believe, with Galton, that inheritance predetermined most important human qualities. This often invoked the ire of liberals and reformers who had long argued that changes in the social environment were necessary to improve the human condition.

The second criticism addressed the way eugenicists tended to oversimplify complex, multicausal conditions such as mental disease by regarding them as readily identifiable, heritable (usually recessive) "traits." As genetics became more sophisticated, scientists came to believe that many important traits were in fact influenced by many different genetic factors, as well as environmental causes; they could not be reduced to simple Mendelian dominant/recessive explanations.

The third critique challenged the categories that eugenicists used to classify people and the motives behind them. Often these complaints targeted the strategies used by eugenicists to quantify intelligence as well as the propriety of policies designed eventually to eliminate the less intelligent; for example, the concept of "feeblemindedness," popularized by eugenicist Henry H. Goddard in the 1910s and 1920s, later came to be regarded as highly problematic. One particular damning report issued by the American Neurological Association in 1935 challenged eugenic notions of the heritability of feeblemindedness, as well as criminality and mental disease. Its authors further charged that the motives of eugenicists were often more "social" than scientific, and they called into question the forced sterilization policies then on the books in more than 25 U.S. states and many European countries.

Though thorough and often damning, the scientific critique of eugenics was long in coming and often limited by the reluctance of many key scientists to denounce publicly the movement or its policies. In many ways scientific might not have been as important as historical and social factors in turning public opinion against the movement. Moral criticism, which variously condemned eugenics for violating notions of natural law, restricting basic freedoms, racism, and cruelty to the less fortunate, was often more compelling to the antieugenic forces. For many observers, the rampant deployment of sterilization policies in Germany during the interwar period represented the inherent dangers of the eugenics movement. Likewise, part of the Nazi ideology that rationalized mass genocide during World War II stemmed from notions of racial superiority and purity promoted both explicitly and implicitly by the eugenics movement. Since the war, supporters of even mild programs of eugenics have been greeted with suspicion, if not hostility, in most Western countries.

In addition to the shadow of nazism, the ascendancy of liberalism after World War II, and subsequent movements championing the rights of historically disadvantaged groups, served to further cast eugenics into disrepute. This occurred because the social and political perspectives of these movements generally favored environment over heredity—a corollary to the notion that social iniquities were due to historical factors more than to biological ones. In this climate, assertions about the heritability of various human qualities, particularly intelligence and criminality, have generally found it tough going in the public sphere. These issues, however, are periodically debated by social scientists, politicians, government officials, and other interested parties, and it seems likely this will continue in the future.

Despite the defensive posture taken by the eugenics movement since the 1940s (and its loss of most scientific support over a decade earlier), it is still influential in some enclaves and enjoys some, mostly private, financial support for research; for example, the Pioneer Fund based in New York City still grants money for eugenic and related projects. As with earlier efforts, however, critics often charge that the motives behind these eugenic projects are racially biased, socially elitist, and scientifically unsound.

Much more scientifically respectable are the considerable advances in knowledge about human genetics achieved by the ongoing Human Genome Project. Rapid progress in this and other areas of biotechnology has led to much discussion about the potential uses of such knowledge in the future. Some scientists and ethicists have speculated that the availability of genetic information might spark a revival of eugenic activity. This time around, however, it is thought eugenic influence will be more subtle—parents who carefully monitor and control the attributes of their children before they are born, or insurance companies that increase premiums on persons with genetic conditions. Thus while the explicit promotion of eugenics rarely goes unchallenged on social, moral, and scientific grounds, the basic idea that some genetic attributes should be cultivated and others discouraged continues to exercise considerable social influence.

See also NAZI RACISM, RACIAL THEORIES, HENRY HAVELOCK ELLIS.

For further reading: Mark Haller, *Eugenics: Hereditarian Attitudes in American Thought* (Rutgers University Press, 1963); Daniel Kevles, *In the Name of Eugenics: Genetics and the Uses of Human Heredity* (Knopf, 1985); and Daniel Kevles and Leroy Hood, *The Code of Codes: Scientific and Social Issues in the Human Genome Project* (Harvard University Press, 1992).

EVOLUTION, PROGRESSIVE The widespread popular scientific idea, current since the late 18th century, that a progressive force might occur in nature. Both Erasmus Darwin (grandfather of Charles DARWIN 1809–82) and France's Count George de Buffon (1707–78) favored such versions of evolution, including the idea that plant and animal species have changed during the existence of life on Earth. Its application to the progressive evolution of species was developed most fully by Jean-Baptiste LAMARCK (1744–1829) around the beginning of the 19th century. Lamarck accepted the belief that species are continually evolving from the simple to the complex within the (separate) hierarchies of plant and animal species. The main mechanism of change, discussed in his *Philosophie Zoologique* (1809), was that the intensive use of an organ, especially if accompanied by effort, "the will to change," motivated organisms to acquire desirable characteristics, which would then be inherited in the more developed form. According to this theory, giraffes, stretching to reach the leaves at the tops of trees, became long necked. It is in that sense—the inheritance by offspring of characteristics acquired and developed during the parents' lifetime—that this version of evolution is described as "progressive."

In the early 19th century, mainstream scientists, especially men such as Baron Georges CUVIER (1769–1830) in

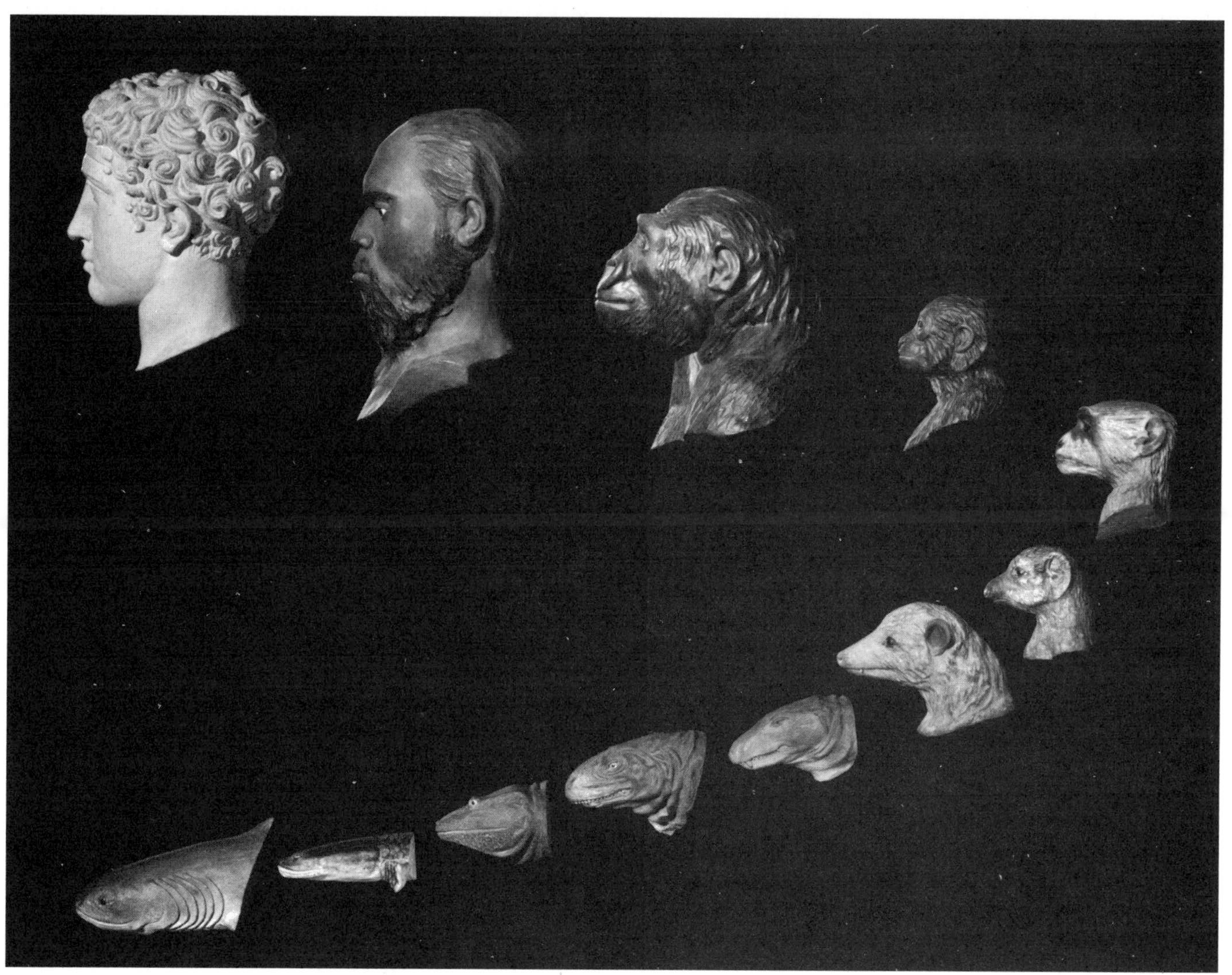

Evolution

France and Sir Charles LYELL (1797–1875) in Britain, judged the idea of evolution to be contrary to evidence. Nevertheless, some writers continued to speculate on evolution, most famously in the anonymous *VESTIGES OF THE NATURAL HISTORY OF CREATION* (1844). This work was labeled "pseudoscience" by those reviewers who defended the then-scientific orthodoxy.

With his *On the Origin of Species by Means of Natural Selection* (1859), the product of 20 years of meticulous work, Charles Darwin and his early supporters initially had to distance the new emerging scientific orthodoxy of *evolution by natural selection* from the older view of progressive evolution. Instead of the inheritance of lifetime acquired characteristics, Darwin reasoned as follows: (1) variation occurs naturally and spontaneously among the individuals of a species; (2) resources are inadequate to support the indefinite expansion of populations; (3) populations remain fairly stable, fluctuating by relatively small amounts about an average; (4) offspring generally outnumber their parents; (5) in successive generations the excess offspring must fail to survive for the population to remain stable; (6) the best-adapted variants survive and transmit their characteristics to succeeding generations; and (7), moreover, this process of natural selection favors the emergence of different variants—and ultimately of new species—as environmental conditions change. Although Darwin carefully avoided extending his theory to human evolution in the *Origin of Species,* he did so later in *The Descent of Man and Selection in Relation to Sex* (1871).

Origin of Species raised much opposition from clergymen, politicians, and scientists who supported the catastrophist and creationist views of natural history. The biggest problem faced by Darwin and his adherents was in explaining step 6 of the theory: What is the mechanism by which characteristics are transmitted from parents to offspring? That it happens was not in dispute—blue-eyed parents have blue-eyed children; red-legged birds have red-legged offspring—and Darwin had accumulated a mass of experimental evidence establishing the fact of inheritance. But how does it happen? That divides into two questions: How are characteristics handed down through generations? What is the mechanism that effects that transmission?

The answer to the first of these two questions was already available. Gregor Johann Mendel (1822–84), an Augustinian monk, had conducted a lengthy series of experiments with peas in his garden at Brünn, Austro-Hungary (now Brno in the Czech Republic), and discovered the principles that govern heredity. However, he had trouble getting his work published. He first sent it to a leading biologist who was unhappy with the mathematics, and returned the paper to Mendel with his criticisms, in effect rejecting it. His work was eventually published in 1865 and 1869, and then only by the local natural history society. There, unfortunately, Mendel's findings languished in obscurity until 1900, when three biologists in different countries—Hugo de Vries in the Netherlands, Karl Correns in Germany, and Erich Tschermak von Seysenegg in Austria—each independently rediscovered Mendel's laws of inheritance. Each searched the literature, found Mendel's papers, and gave him credit for his original discovery. So biologists now had an answer to the first question.

In 1901 Hugo de Vries (1848–1935) published volume 1 of *Die Mutationstheorie (The Mutation Theory)* (he published volume 2 in 1903), in which he offered an alternative to Darwin's gradual adaptation, positing that changes took place in leaps—mutations. "The new species thus originates suddenly; it is produced by the existing one without any visible preparation and without transition."

Many naturalists, particularly the so-called biometricians—those engaged in quantitative biological research—opposed mutationism. During the 1920s and 1930s, theoretical geneticists such as R. A. Fisher, J. B. S. Haldane, and Sewell Wright, used mathematical arguments to provide a framework for the integration of genetics into Darwin's theory of natural selection. This effort, which spelled the end for mutationism, culminated with Theodosius Dobzhansky's (1900–75) profoundly influential *Genetics and the Origin of Species* (1937). There, he promoted a new understanding of the evolutionary process as one of genetic change in populations.

By 1950 Darwin's theory had been universally accepted by biologists, although an answer to the second question was still lacking. In 1953 James Watson and Francis Crick worked out the structure of DNA (deoxyribonucleic acid), the hereditary material contained in the chromosomes of every cell's nucleus. Here was the basis of an answer, on the details of which biologists have been working ever since that time. There is an incalculable wealth of information about evolutionary history stored in the DNA and proteins of living things. Since the 1960s, when such techniques as electrophoresis and selective assay of enzymes became available for the rapid and inexpensive study of differences among proteins, knowledge of biological evolution has been tremendously advanced by the application of molecular biology.

Despite the emerging consensus, controversies still occur as scientists seek to refine their understanding of the mechanism of evolution. Those who favor gradualism take the view that macroevolution (change above the species level, i.e., the emergence of new genera, families, orders, etc.) is merely the cumulative effect of microevolution that results in the production of new species, which in turn diverge so much from the original that new genera separate out, and so on. But in the 1970s a new theory was proposed by Stephen Jay GOULD and Niles Eldredge called punctuated equilibria. They argued against evolution by gradual change and maintained that most species are static for most of their existence, having established a rough equilibrium with their environment. Evolutionary changes sufficient to establish a new species occur suddenly, often in small, isolated communities in response to a major environmental change. The species either dies out or some variant best suited to the new environment survives and very quickly—quickly on a geological timescale that is—gives rise to a new species. The scarcity of graded sequences in fossil groups, particularly of so-called missing links, is often cited in support of punctuated equilibrium.

So, while in the later 19th century the two views, progressive evolution and evolution by natural selection, temporarily blurred into one another—both progressive and natural selection evolution were believed to operate, with some interaction—for many years biologists, as the evidence has built up, have almost unanimously favored natural selection, dismissing LAMARCKIANISM as pseudoscientific, especially in its later version, LYSENKOISM.

See also MALTHUS, THOMAS ROBERT; WALLACE, ALFRED RUSSEL.

For further reading: R. G. A. Dolby, *Uncertain Knowledge* (Cambridge University Press, 1996); C. C. Gillispie, *Genesis and Geology* (Harper, 1951); and Stephen Jay Gould, *The Panda's Thumb: More Reflections in Natural History* (Norton, 1980).

Exorcism

EXORCISM The action of expelling evil spirits (see DEMONS) by adjuring them to abandon the person, place, or object that has come under their control. Technically the word *exorcism* describes a ceremony that is *only* practiced by the Christian Church to expel demons. But the idea of expelling evil spirits goes back to primitive times when the magical rites and practices of WITCHCRAFT were used; for example, the Bushmen of the Kalahari basin in southern Africa used HERBAL MEDICINES and communal dances to send away evil spirits in the form of invisible magical slivers of wood that they thought had become attached to members of their tribe, causing them to exhibit abnormal behavior. For such ceremonies to be successful it was necessary for the whole tribe to agree to participate. Anthropologists believe that the custom had the social function of keeping the tribe together and its members on good terms with each other.

In Europe and the Middle East in the early classical period, the practice of expelling evil spirits by means of various set prayers and formulas was common. A few ancient Greek, Roman, and Arabic scholars such as Hippocrates and Plato interpreted mental deviations as natural phenomena, but their ideas did not prevail. With the advent of Christianity, the belief in demons persisted and with that belief came the necessity for exorcism. Jesus exorcised demons with one word and said that this act was a sign of the coming of God's kingdom; so the Christian tradition grew that exorcisms were always carried out "in His name." In the first two centuries of the Christian era, the power to exorcise was considered to be a special gift not necessarily bestowed on a cleric, and consequently alchemists and witches vied with those from religious communities for the honor of filling the position. In about 250 C.E., the Roman church took exorcism under its control, appointing a special class of the lower clergy called exorcists to do the job and at the same time carefully regulating it by canon law. The duties of the exorcist included the expelling of GHOSTS or troubled spirits from haunted houses or haunted places. But more important, they came to include the laying on of hands to purge those possessed by abnormal mental or physical states, especially insanity. To combat what was seen as the devil's work, specific types of behavior were labeled as sinful, and those possessed were tortured in an effort to exorcise the demons from their bodies.

In the middle of the 1800s, reform movements were pushed to abandon the concept that the insane were possessed and to put forward instead the idea of insanity as a mental illness and to have it accepted as such in law. Nevertheless exorcism, using cleansing water that has been blessed by priests, still remains part of Christian baptismal services, and in many churches a mother after childbirth must be "churched" before she is allowed back into the congregation. But, there has been a very important shift of ideas behind these ceremonies: the observances do not now presuppose a state of possession, but are prayers asking for the restraint of evil.

Science deals with natural not supernatural phenomena, and does not recognize the existence of metaphysical demonic powers in the universe. Therefore, the ritual of expelling them would seem a futile exercise.

For further reading: Carl Sagan, *The Demon-Haunted World: Science as a Candle in the Dark* (Headline, 1996).

EXPLORATION HOAXES Fraudulent claims of discoveries of some hitherto unknown part of Earth and its inhabitants. Just as the astronomer has the task of discovering the structure of the universe and the microbiologist that of microorganisms, the explorer's task is finding out about and mapping the structure, topology, and demography of a specific locality on Earth's surface. It is unfortunate that quite a few explorations have been found, on later examination, in one way or another, to be deceptions.

Frederick COOK is a classic case of a man who claimed more than he had achieved. He started his career as a physician and joined the Peary Arctic expedition of 1891–92 as a surgeon. Robert Edwin Peary was a U.S. explorer who, in 1909, said that he had reached the geographic North Pole when in fact he knew that he was hundreds of miles away. Maybe Cook learned to be dishonest from his association with Peary, or maybe both men let their standards slip in later life. Some say that exploring got into Cook's blood, but others say it was more likely to have been the fame and the glory that attracted him. Whatever the explanation for his behavior, the claims made by Cook for the two explorations he led—to have climbed Mount McKinley in Alaska in 1906 and to have reached the North Pole with two Inuit in 1908—were both found to be flawed. The expedition to Alaska had been a scientific success: He had penetrated and mapped previously unknown territory, but he never actually reached the summit of Mount McKinley. Cook had proclaimed he would climb it and, not having done so, the avoidance of loss of face might have been more important to him than the good work he had in fact accomplished. Otherwise, why did he make the false claim to have reached the summit of McKinley? Was he ashamed of climbing the wrong mountain (it is said of him that he did not know how to use a theodolite)? Was he so inept that he did not realize that his "summit" photographs would eventually be identified as taken from an adjacent minor peak?

Richard E. Byrd was another man who claimed more than he had achieved, when he had already achieved much during his lifetime in discovering unknown regions in the Antarctic. In 1926 Byrd, then a retired admiral from the U.S. Navy, decided to fly over the North Pole and back while Roald Amundsen was attempting to cover the same ground. Byrd told reporters before he set off that he expected the round-trip to take about 20 hours. But less than 16 hours later, Byrd returned to claim the prize. Answering his doubters, Byrd explained that his unprecedented speed was due to a helpful tailwind on the way north which, when they reached the pole, had immediately changed to help them once more on the journey south. Much later Byrd's co-pilot admitted that he and Byrd had been nowhere near the Pole in 1926; he also offered the information that on previous flights Byrd, who had supposedly been the navigator, had never taken a single sextant reading. These revelations emerged after Byrd's death in a book of memoirs by Balchen, *Come North With Me*. Lawyers for the Byrd family sued—the defamatory parts were withdrawn—but later reinstated, and the controversy continued.

A more contemporary fraud involved the reported discovery in 1971 of the world's only surviving Stone Age tribe in the rain forests of Mindanao in the southern Philippines. The claim was made by Manda Elizalde, then presidential assistant for tribal minorities to Philippine President Ferdinand Marcos. The tribe, numbering only 26, was called the TASADAY TRIBE. We were told that the people had no art or crafts, like pottery or cloth, but that they were a very gentle people with no words for weapons, hostility, or war. Books and learned articles were written about them, and NBC made a television documentary called *Gentle Tasaday*. The Philippine government was persuaded to protect them; many acres of the mahogany forest where they lived and allegedly swung from the trees were declared a reserve. The limited access allowed a certain amount of logging to go on unobserved, with the result that Elizalde made his fortune. The scam came to an end in 1988, the year Marcos fell from power. The tribe were then revealed to be just poor members of the local community whom Elizalde had promised handsome rewards if they collaborated with him. "The tribe" received very little, but Elizalde escaped with $55 million.

Hoaxes are an example of how many false scientific facts and theories have their roots in deception and fraud. Hoaxers mislead the public and sometimes the scientific community into mistaken beliefs about the world and so start a chain of erroneous knowledge—pseudoscience—which is often very hard to eradicate once it is enshrined in the literature.

For further reading: W. Broad and N. Wade, *Betrayers of the Truth* (Oxford University Press, 1982); Stuart Gordon, *The Book of Hoaxes* (Headline, 1995); Gordon Stein, *Encyclopedia of Hoaxes* (Gale, 1993); and Nick Yapp, *Hoaxers and their Victims* (Robson Books, 1992).

EXTRASENSORY PERCEPTION (ESP) The belief that certain people have an awareness of an external state, condition or event without recourse to any of the bodily senses. ESP is the cognitive side of psychic phenomena in contrast to the physical aspect, which is called PSYCHOKINESIS (PK). The term includes CLAIRVOYANCE, having knowledge without recourse to empirical means; TELEPATHY, or thought transference, the awareness of another person's thoughts and emotions and the ability of persons to exchange ideas and feelings when not in sound or sight of each other; and PRECOGNITION, the knowledge of a future event that cannot be logically inferred from the current situation.

All these and similar phenomena have been scientifically investigated since the late 19th century; the term by which these investigations were once known was PSYCHIC RESEARCH, but a more modern name is PARAPSYCHOLOGY.

Most of the evidence of ESP that has been collected is from experiments involving card guessing. A special deck of cards is used, called ZENER CARDS after Karl E. Zener, who adapted the idea from the cards devised by Sir Oliver Lodge (1851–1940) in 1884 to put psychic research on a statistical, scientifically testable basis. Each Zener pack has 5 "suits" of cards, 25 cards in all, and each card has on its face one symbol: a cross, a square, a circle, a star, or a wavy line. The subject to be tested, shielded from the cards, attempts to name the card being transmitted. A better than chance percentage of correct calls on a large number of attempts would be considered evidence for ESP.

In any experiment where pure chance operates—where there is no ESP operating—the results of successive runs are distributed around an average, with the bulk near the average, and the better- or worse-than-average results thinning out on either side. When many hundreds of these runs are plotted on a graph, they show this distribution, the graph resembling the shape of a bell—the Bell Curve, or Normal Distribution. If the outcomes are better than chance, as expected in ESP experiments when paranormal agents are said to be involved, the bell would be greatly distorted. According to C. E. M. Hansel in *ESP: A Scientific Evaluation* (1966), some facts have been gleaned from years of conducting these experiments:

1. Subjects, when attempting to guess card symbols, have obtained scores that cannot be attributed to chance.
2. Some of those taking part in ESP experiments have indulged in trickery.
3. Subjects who obtain high scores cannot do so on all occasions.
4. Subjects tend to lose their ability to obtain high scores. This loss often coincides with the termination of an experiment.
5. A successful subject is sometimes unable to obtain high scores when tested by a critical investigator.
6. Some investigators often observe high scores in the subjects they test; others invariably fail to observe such scores.
7. A subject may obtain high scores under one set of experimental conditions and fail to do so under other experimental conditions.
8. No subject has ever demonstrated his/her ability to obtain high scores when the test procedure is completely mechanized.

It is most revealing to see in this evaluation the extent to which the investigator is unconsciously or otherwise interfering with the experiment. Scientists have observed this effect in all types of experiments; even in strictly run scientific experiments where data are being collected on inanimate objects, the experimenter can become biased. Just wishing for one kind of result over another because it confirms what is desired to be proven is sufficient. ESP investigators have been known to be selective and to ignore those results that are chance. Worse, they will explain that the subject is having a bad day or that there is some disturbing effect coming between the transmitter and the receiver. On one occasion when results were going badly for the ESP subject, the investigator thought he recognized PRECOGNITION—that is, the subject was guessing a card ahead of the one that was being transmitted—and wished these attempts to be recorded as positive scores. Whatever this proved, critics said that it certainly should not be allowed to be counted within the confines of this particular experiment. Careful and protracted examination of claimed positive results of ESP have found them all to be open to criticism. As well as the more obvious—cheating and inadequate controls of experimental conditions—there is also the overriding objection: the illogicality of it all.

Despite this skeptical view, shared by most orthodox scientists, there are several well-established parapsychology programs in U.S. universities and a few in European universities. The American Association for the Advancement of Science accepts parapsychology as a legitimate science, and parapsychologists regularly contribute papers to the association's annual conferences. Several well-respected scientists and engineers believe that they can demonstrate real, but small, psi effects—John B. HASTED, Robert JAHN, Sir Alister HARDY, and John G. TAYLOR. Richard Broughton, a parapsychologist, has reviewed all the recent evidence, is very critical of the skeptics, and concludes that parapsychology today, including ESP—whatever may have been its shortcomings in the past—can claim to show a body of respectable research, supported by sound arguments. In his considered opinion parapsychology/ESP is not a pseudoscience.

See also PSIONICS.

For further reading: Richard S. Broughton, *Parapsychology: The Controversial Science* (Ballantine, 1991);

Antony Flew, *The Logic of Mortality* (Blackwell, 1987); and C.E.M. Hansel, *The Search for Psychic Power: ESP and Parapsychology Revisited* (Prometheus Books, 1989).

EXTRATERRESTRIAL INTELLIGENCE, COMMUNICATION WITH (CETI) The search by scientists of several nations for some sign of intelligent life on other worlds.

Some believe that intelligent life exists on other planets. With an estimated 400 billion stars in our own galaxy and a guess (based on present theories of star formation) that a tenth of them may have planets—one per system of which may prove suitable for life analogous to that on Earth—there is a very high possibility that there is life elsewhere in our galaxy alone. If we speculate that one in 10,000 of such planets might go on to develop intelligent life that is capable of communicating between the stars, and should they last long enough for a reasonable proportion to be communicating, then it is worth looking for interstellar communications by electromagnetic radiation—by radio transmission. This calculation is controversial because it compounds a number of plausible speculations into an extreme extrapolation of current thinking. Nevertheless, several scientists, including Carl SAGAN, thought that there was a sufficient possibility to be worth exploring. Sagan drew up a petition, which was signed by 72 prominent scientists from 14 countries, calling for a search to be funded. Congress in 1982 allocated $1.5 million to NASA to begin a long-term search for signals from extraterrestrial beings.

The CETI program found no evidence that intelligent life exists on other planets, yet the mainstream science community views the idea as plausible.

For further reading: A. Asimov, *Extraterrestrial Civilizations* (Pan, 1981); Carl Sagan and I. S. Shklovski, *Intelligent Life in the Universe* (Dell, 1967); Carl Sagan, *Communication with Extraterrestrial Intelligence* (MIT Press, 1973).

EYELESS SIGHT The ability to see without the use of the eyes. The idea has been championed by a variety of scientists and writers under a variety of names including DERMO-OPTICAL PERCEPTION, paroptic, hyperesthesia, synesthesia, cutaneous vision (skin vision), extraretinal vision, and biointroscopy. The idea was introduced into the modern world by 20th-century French author Jules Romains (pen name of Louis Farigoule), though scattered references had already appeared in literature concerning blind people who supposedly had the ability. Romains came to believe that people had a little-known faculty of seeing in a manner usually associated with psychics. He published his findings in 1920 in a book, *Vision Extra Rétinienne* (later translated and published in London as *Eyeless Sight: A Study of Extra-Retinal Vision and Paroptic Sense,* 1924).

The response to the book from French scholars was almost totally negative. Not only was Farigoule denied any funds to continue his research, but he was denied access to further blind subjects. He changed his name and went on to become a world famous poet and novelist. A decade later a Brazilian, Manuel Shaves of São Paulo, claimed that he had found about a dozen individuals, among a total of 400 blind subjects he had tested, who had some degree of eyeless sight, at least to the level of distinguishing colors.

In the early 1960s, Russian psychologist I. M. Goldberg began work with psychic Rosa Kuleshove. Stories of her reputed abilities found their way to Moscow from her rural home in the Ural Mountains. She was invited to the Biophysics Institute of the Soviet Academy of Sciences in Moscow. Kuleshove demonstrated a seeming ability to read ordinary print with the fingers of her right hand and determine the colors of various objects. She was not blind, but rigid controls were taken to ensure her inability to use her eyes in the test.

After an initial round of tests, she returned to her rural home but wanted to return to the city and make new claims about her abilities. On her return to Moscow, she was unable to fulfill her extraordinary claims and was on one occasion caught attempting to succeed by fraud. However, she was retested on the original abilities, which confirmed her basic talents. From these initial tests, the concept of dermo-optical perception entered the literature. Other people were discovered who demonstrated a similar ability, and a program for testing them emerged that continued through the 1970s.

In 1965, Dr. Vichit Sukhakarn, who had opened an institution for blind children in Bangkok, Thailand, claimed that he had found subjects who could see through their cheeks and whom he had trained with hypnosis. His work was followed by that of Dr. Yvonne DuPlessis at the *Centre d'Eclairagisme*. She trained blind volunteers to perceive objects at a distance (paroptic perceptions) and by touch. Before his death, Jules Romains was able to visit the center and see the continuation of his work from the 1920s.

Eyeless sight is one of the better-documented phenomena of PARAPSYCHOLOGY, though there is some complaint that the data are marred by methodological problems and some cheating.

For further reading: Jules Romains, *Eyeless Sight* (Citadel Press, 1924); and Sheila Ostrander and Lynn Schroeder, *Psychic Discoveries Behind the Iron Curtain* (Prentice Hall, 1970).

F

FAD DIETS "Get-thin-quick" diets that promise substantial weight loss in a relatively short period of time. Many fad diets are starvation or semistarvation diets, requiring 600–800 calories per day or fewer. Fad diets usually stress one food or food type to the exclusion of others and require drastic changes in the dieter's eating patterns. They are typically "one-size-fits-all" diets, promising success for any person following the prescribed diet regime. Because fad diets emphasize such a limited range of foods, they tend to be both poor in nutrition and difficult to sustain for long periods.

Fad diets are often successful in the short term. Nutritionists agree, however, that they are an ineffective way to achieve and maintain weight loss. They usually involve drastically cutting calories. Typically, they are deficient in essential vitamins, minerals, and other nutrients, and the deprivation they demand may result in reactive overeating. Most significantly, fad diets rarely teach dieters what to eat after the diet is over. Therefore, when the dieter returns to normal eating patterns, lost weight is usually quickly regained.

Some fad diets can be dangerous. Attempting to sustain a diet that is not well balanced nutritionally for an extended period of time can result in illness and permanent damage to the dieter's health. For example, the dieter may experience isolated nutrient deficiencies, toxicities, breakdown in muscle and vital organ tissue, ketosis, electrolyte imbalance, heartbeat irregularities, and a rise or fall in blood pressure.

Fad diets can be particularly dangerous for teenagers, as the adolescent years are ones of rapid growth and development. Failure to provide the body with the necessary calories, protein, iron, and calcium can jeopardize a teenager's ability to achieve genetically determined height, and may cause gallstones, hair loss, weakness, and diarrhea. Fad diets have even been known to prove deadly. According to the American Heart Association, a fad liquid protein diet was blamed for at least 60 deaths in 1977.

Despite the danger, fad diets are big business. For example, according to a 1995 report from the Institute of Medicine, tens of millions of dieters are spending a total of more than $33 billion yearly on weight-reduction products, including dieting liquids, powders, and pills, in the United States alone.

Most of what we now call fad diets originated in the 1930s, coming in many varieties. Some of the more common ones are:

Liquid Diets. One of the first may have been "Dr. Stol's Diet-Aid, the Natural Reducing Food," sold in Chicago in the 1930s. Liquid diets reemerged in the 1970s and have enjoyed heightened popularity in the 1990s. Currently there are several commercial brands of liquid meals on the market. Most suggest that dieters replace two meals a day with a diet shake. Assuming that the third meal of the day is kept to 600 calories, total daily calorie intake is around 1200. Although today's diet shakes are improving in nutritional quality and are therefore less dangerous than many other fad diets, they are not substantially more effective. Dieters usually regain lost weight when they return to solid foods.

Single-Food Diets. One of the most famous single-food fad diets is the grapefruit diet. Dating back to the 1930s, the grapefruit diet trades on the belief that grapefruit contain a special fat-burning enzyme. There have been many variations of the grapefruit diet since the original Hollywood version of the 1930s: Typically this diet is followed for three weeks; the total calories allowed do not exceed 800 per day. More recently, the Cabbage Soup Diet has appeared on the single-food fad diet scene. Dieters make the soup at home and may eat unlimited quantities, in addition to a specified food each day. Nutritionists advise that neither cabbage nor grapefruit possesses any special far-burning qualities, and they do not endorse single-food diets. However, both grapefruit and cabbage soup may usefully be incorporated as a part of a balanced diet that will encourage permanent weight control.

Combined Foods. The combined-foods school of dieting also originated in the 1930s with a diet book authored by William H. Hay. Combined-foods diets reached the height of their popularity in the 1980s. These diets claimed that dieters could realize dramatic weight loss by eating, or not eating, certain foods in particular combinations. The Beverly Hills Diet, authored by Judy Mazel, is the most famous example of the combined-foods diet genre. However, there is no conclusive evidence that the body processes foods in special combinations any differently than it processes the same foods randomly or in other combinations.

Fasting and Modified Fasting. Fasting not only has an ancient history as a symbolic religious act but also as a form of dieting. Most of the weight a dieter loses when fasting, however, is, in fact, water. Fasting can be a dangerous form of dieting. Some dieters who fast develop severe electrolyte abnormalities, which can lead to irregular heartbeat, loss of consciousness, and, in some rare instances, death. Fasting is not conducive to exercise, which is necessary to maintain weight loss. A modified fast allows liquid intake and is usually maintained from one to seven days. The dieter usually partakes of fruit juice, water, clear broth, and tea. Modified fasting may cause dizziness, however, and a drop in metabolism. In most cases, weight loss is not permanent.

High Protein/Low Carbohydrate Diets. Most popular in the 1970s, these fad diets promoted the theory that a high carbohydrate intake interfered with the body's ability to burn fat. The Complete Scarsdale Medical Diet is the best known of the high protein/low carbohydrate diets. Whereas most fad diets are relatively short in duration, advocates of this fad diet encourage dieters to follow the high protein/low carbohydrate regime indefinitely. Medical experts claim that these diets are usually overburdened with saturated fat and cholesterol and may tax the heart and kidneys.

Cancer Treatment Diets. The two best known of these are:

1. The Gerson Diet devised by Max Gerson in the United States in the 1920s. Thirteen fresh juices are taken every hour, one from oranges, four from carrots and apples, four from greens, and four from calves' liver. All have to be organic. Caffeine enemas and castor oil are taken to cleanse the system; food is limited to salads, oats, and baked potatoes. The idea is to flush out poisons, improve the body's balance, and enable it to fight the invading organism. This diet can have severe side effects.
2. The Bristol Diet in Britain, devised by Dr. Alec Forbes, offered by the Bristol Cancer Help Centre, is basically vegetarian: fresh fruit, vegetables, whole grains, pulses, goat's milk yogurt, plus a little fish, poultry, and eggs, all organically produced. Drinks are limited to herb teas, fresh juices (especially carrot juice), and mineral water. There are serious doubts about the efficacy of diets for cancer treatment. According to the *Family Guide to Alternative Medicine*:

> *"Alternative therapies are controversial and cannot replace normal treatment. Many patients say that anticancer diets have helped by giving them some positive action they can take. However, there is no scientific evidence that they work. One recent study has even found that some patients who followed the Bristol Centre's programme in addition to receiving orthodox treatment had a poorer survival rate than those with similar cancers who only had conventional therapy. The reason is not known, but some doctors have suggested that a strictly vegan or vegetarian diet may be too extreme."*

Successful dieting for slimming means losing weight and keeping it off. The common denominator in all fad diets is their failure to bring about permanent weight loss. However, it must be noted that even weight-loss diets endorsed by nutritionists and physicians meet with limited long-term success. The same culture of immediate gratification that makes fad diets wildly popular may also inhibit proper weight control. Nonetheless, nutritionists advise that losing weight slowly and gradually while incorporating changes in eating habits has been proven to be the most successful form of dieting for weight reduction. Once the desired weight has been achieved, the dieter should follow a doctor's or nutritionist's recommendations for weight maintenance so that lost weight is not regained. The most successful method involves a combination of decreasing total food intake with an appropriate exercise program. Sound diets should include a variety of nutritionally balanced foods. Dieters are encouraged to keep weight-loss expectations realistic. Most safe diets allow weight loss of 1 to 2 pounds a week, and do not require fewer than 800 calories a day.

For further reading: Patrick C. Pietroni, ed., *Reader's Digest Family Guide to Alternative Medicine* (Reader's Digest, 1991).

FAIRIES Inhuman, but humanlike creatures, sometimes with paranormal powers. Encounters between humans and fairies appear in folklore around the world. These fairies differ in form and appearance: Some are tiny, others are tall, some are ugly, others are beautiful. It was not until Victorian times that fairies were sentimentalized as tiny benevolent creatures with butterflylike wings, like those that appear in the COTTINGLEY FAIRIES photographs. In traditional tales they are always regarded as anthropomorphic, unpredictable, and dangerous. These fairies commonly haunted lonely roads, woods, fields, and other wild places. Superstitious people credited them with supernatural powers, including the ability to drive humans mad. They believed that the fairies were best avoided and went out of their way to keep from offending them.

There is no solid evidence for the existence of fairies, although there have been sightings for centuries. The Reverend Robert Kirk, a 17th-century Presbyterian minister, published a very comprehensive account of fairy lore, *The Secret Common-Wealth,* in 1692. Although he never saw fairies himself, he was convinced that they were real beings, between men and angels but usually invisible to human eyes. The Victorian writer, churchman, and eccentric Sabine BARING-GOULD declared that he had seen a troop of "dwarfs about two feet high" when he was four years old, who ran laughing and shouting alongside his family's carriage. (Baring-Gould's parents, he reported, saw nothing.) As recently as 1919, a 13-year-old Wisconsin boy reported that a troop of 20 little bald men crossed his path one summer night. In Iceland, belief in the existence of fairies is still so strong that psychics are employed to negotiate with the "hidden folk" whenever a new construction project is launched. As recently as 1959, sightings of fairies were reported in Ireland. Modern sightings have also taken place in Oregon, England, Scotland, and Wales.

The origin of the belief in fairies is hotly debated. Some modern theorists claim that fairies are (or were) in fact extraterrestrial visitors or UFOs. The late Victorian writer W. Y. Evans-Wentz, as well as the modern occultist Leslie Shepard, suggested that fairies exist in a parallel universe, a kind of separate reality that occasionally merges with our own, or that can be perceived at certain times by certain types of people. Others suggest that all encounters with fairies are hallucinatory and exist only within the mind.

Another theory, first proposed by Robert Kirk, is that fairies are somehow related to angels—perhaps fallen angels not evil enough to fall all the way to hell. Most folklorists, however, believe one of three separate theories. Fairies may originally have been members of conquered or displaced tribes, driven from their ancestral homes into wild terrain. Their descendants, glimpsed occasionally by the children of their conquerors, may have been mistaken for magical beings. Another theory suggests that fairies were the remnants of ancient folk gods or spirits. Still another suggests that they were connected with an ancient religion of the dead. All scientific efforts to prove the existence of fairies, such as the Cottingley photographs, have failed.

For further reading: Katharine Briggs, *An Encyclopedia of Fairies: Hobgoblins, Brownies, Bogies, and Other Supernatural Creatures* (Pantheon Books, 1976); and Stuart Sanderson, ed., *The Secret Common-Wealth & A Short Treatise of Charms and Spells by Robert Kirk* (Rowman & Littlefield, 1976).

FAITH HEALING The act of healing the sick or crippled by appealing to the faith of the sufferer or to divine intervention. It is an activity with a long history, going back to biblical times at least. In the New Testament, Christ heals miraculously on several occasions, curing the blind, the sick, and the lame. Through the centuries many faith healers, sometimes claiming personal powers, sometimes claiming to act for Jesus, have practiced the art. Faith healing takes various forms—the laying on of hands, casting out of devils (particularly in mental illness), and so forth—and is supposed to be effective against almost any disease or affliction, physical or mental: epilepsy, tuberculosis, cancer, blindness, deafness, psychosis, phobias, and so on. Sometimes, the healer claims to be the one and only true practitioner, sometimes he or she accepts that many may have the same power, providing only that the basis of faith is the same.

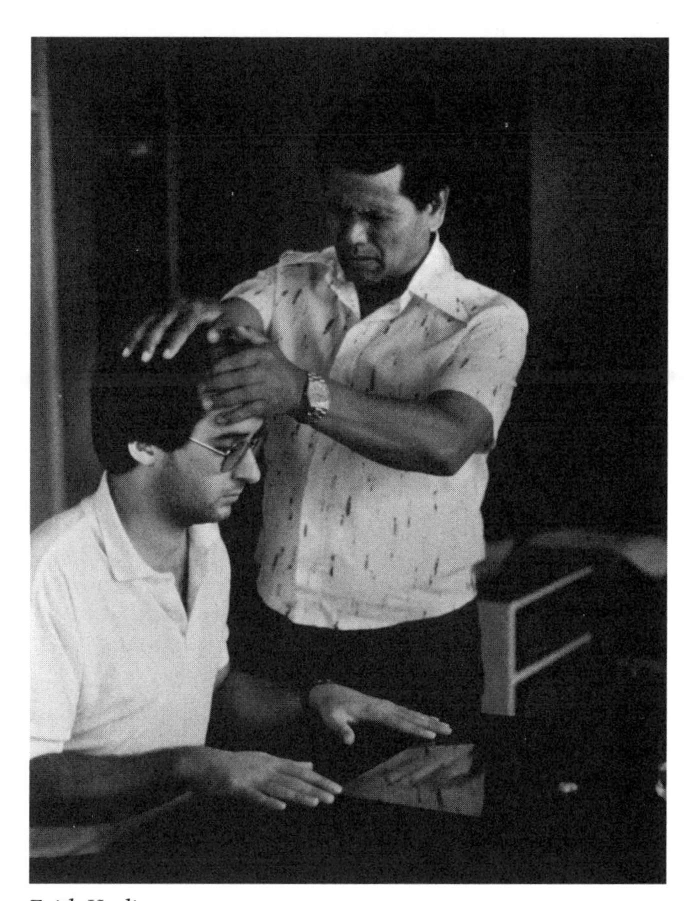

Faith Healing

Sometimes, the claims made are virtually limitless; on other occasions healers accept that there are illnesses or infirmities they cannot cure and which must be referred to a conventional doctor.

In recent years, the number of faith healers has grown, attracting many believers; television and radio have been exploited to expand their celebrity and influence. Currently in the United States, names that come to mind are Kathryn Kuhlman, Billy Graham, Oral Roberts, Jimmy Swaggart, Jim Bakker, Herbert Armstrong, Billy James Hargis, Ruth Carter Stapleton (President Jimmy Carter's sister), Pat Robertson, and many more. Nor, of course, is the practice confined to the United States. Similar activities occur in other countries, though not so extensively.

The practice raises several questions. Is this process outside those of conventional medicine one that effects genuine cures of real medical conditions? Are the illnesses which are claimed to be cured imagined or psychosomatic? Are the cures examples of a sort of placebo effect, or, alternatively, do they take credit for remission or recovery which would have occurred without any intervention? Which of the practitioners are genuine, and which are frauds? Critical observers are convinced that many faith healers sincerely believe in their powers. Billy Graham, for example, is generally accepted as believing that what he does brings his converts to a born-again Christian faith and effects real healing. A particular example of a long-standing healing by faith is that of the Grotto at LOURDES, France. The many thousands of sick people who annually make the pilgrimage to Lourdes in the hope of a cure are undoubtedly genuine in their faith that there is no question of trickery or fraud. But there are some faith healers who have seen the opportunity to cash in, both literally and metaphorically, on a good thing.

Carl Sagan in *The Demon-Haunted World* (1996) has an interesting viewpoint: "The spontaneous remission rate of all cancers, lumped together, is estimated to be something between one in ten thousand and one in a hundred thousand. If no more than five per cent of those who come to Lourdes were there to treat their cancers, there should have been between fifty and 500 'miraculous' cures of cancer alone. Since only three of the attested sixty-five cures are of cancer, the rate of spontaneous remission at Lourdes seems to be lower than if the victims had just stayed at home. Of course, if you're one of the sixty-five, it's going to be very hard to convince you that your trip to Lourdes wasn't the cause of the remission of your disease. . . . *Post hoc, ergo propter hoc* (after this, therefore because of this). Something similar seems true of individual faith-healers."

For further reading: Larry Dossey, *Healing Words* (Harper, 1995).

FALSE MEMORY SYNDROME (FMS) A term that came into use following cases in the late 1980s where memories of abuse were coaxed out of children by therapists. The memories were later proved to be false, hence the name "false memory syndrome," and some individuals were found to have been wrongly accused of child sexual abuse and/or Satanic ritual abuse (SRA). The subject became very controversial with one side, backed by quite eminent psychiatrists, explaining that there is no such thing as an accurate recall tape recorder in the brain and that memories are altered and merged with others in subsequent retelling. All memories are to some extent fabricated, and counseling itself can implant ideas that later present themselves as "true memories." The other side, mostly comprised of professional therapists, asserted that memories are never false, especially those of young innocent children who do not lie and that toddlers could not invent such stories! This latter group omits the word *false* and calls the syndrome "recovered memory syndrome."

In the early days of the syndrome's becoming recognized, the idea of unearthing forgotten memories escalated very fast, and soon many people believed that anyone could have a buried history locked inside his or her brain. It was not long before a sizable number of adults (mostly women) began to go into therapy believing that they had suffered childhood sexual abuse. The social repercussions were enormous and widespread. Reputations were lost as court cases were brought against aging fathers, families were split up, and children were taken away from parents. The phenomenon soon broadened and escalated until those working in the area were at a loss to know how to evaluate some of their patients' extraordinary recollections of REINCARNATION, of abduction by aliens, or of being child sex-slaves in a Satanic cult hundreds of years ago.

The existence of repressed memories and the supposed damaging effect that they could engender, goes back to Sigmund FREUD. According to his written case histories, his patients never told him voluntarily, or even as a result of routine questioning, of their childhood victimization by sexual assault or incest. He led them toward revealing their "memories" through HYPNOSIS, often using enormous pressure to extract information. He did however later replace his original concept of the suppression of childhood memories with a new, some say even more dangerous, theory of incestuous desire—the Oedipus complex, whereby the burden of guilt was passed from the parent to the child. Nevertheless his original ideas of the supposed curative value of a sudden retrieval of lost memories became in the early 1990s current for therapists in Britain and America, with many of the professionals who were behind the early child-abuse scares actively promoting hypnosis and other hypnotic therapies for adults. Today many techniques for recovering memories—like "guided imagery," "role-playing," and "visualization"—are carried out under deep relaxation to establish a state of receptivity akin to hypnosis.

In Britain many Christian therapists, believing themselves to be under the guidance of the Holy Spirit, became convinced that Satanic ritual abuse, on a huge scale, had been widespread for decades. In the mistaken belief that they were being helpful, these mental health therapists superimposed onto their patients' already fearful memories of child abuse even more fearful memories connected to a demonically inspired occultism. But more was to come: This occult group of counselors found out that most of their "satanic survivors" were also suffering from the affliction of multiple personality disorder (MPD). This category has grown in a single decade from an exceedingly rare diagnosis into an epidemic. Critics of the group believed that the MPD category had been created by the escalation of recovered memory therapy.

With the inception of the False Memory Syndrome Foundation in Philadelphia in 1992, the recovered-memory type of therapy has increasingly come under fire. Public awareness and anger grew as scientists, physicians, and psychotherapists reported the lack of any scientific or practical basis for recovered memory treatment. No one who studied FMS was questioning, or wished to underestimate, the dreadful reality of child sexual abuse or any other kind of sexual violation. There is no doubt that child incest has been underreported, and we are now recognizing that child abuse throughout history has been more common than previously thought. Recognizing all this, studies now show how easy it is for authority figures, employing hypnotic techniques to implant false memories, even entire life histories, into the minds of suggestible patients. But even if these retrieved memories are true, are the patients benefiting from their retrieval? Perhaps those who have forgotten their disturbing experiences are the lucky ones; the problem with many trauma victims who have been through floods, fires, bloody wars, and concentration camps is that they cannot forget their horrific ordeals. But if these alleged retrieved memories are *not* true but implanted, and if repression is merely a myth, then rather than contributing to healing, therapists are doing the exact opposite, leaving their patients more damaged than they found them. Those who attend the sick must first of all do no harm. Research is showing that therapy by wielding such a suggestive baton is creating pseudomemories by tapping into people's recollections, not of personal trauma, but of stories and images learned from popular culture.

It is now suggested that therapists should study how victims of proven traumatic events have learned to get over their frightful experiences and stop indulging in Freudian psychoanalytic theories of repression, but unfortunately Freudian ideology pervades the literature of the recovered memory movement. To use a concept expressed by Thomas Kuhn in *The Structure of Scientific Revolutions* (1970), Freud created a pseudoscientific paradigm in the mental health profession. His ideas were not based on any body of generally accepted evidence and could not be tested either by predicting something new or by Popper's criterion of falsifiability. In short, his approach was not empirical. Nonetheless his theories were so widely taken up by both the scientific and the medical profession that they became effectively the dominant—at the time, the only—paradigm.

For further reading: Richard Ofshe and Ethan Watters, *Making Monsters: False Memories, Psychotherapy, and Sexual Hysteria* (Charles Scribners' Sons 1994); and Mark Pendergrast, *Victims of Memory* (HarperCollins, 1996).

FATE MAGAZINE Longest-lived, broad-based popular magazine of the paranormal. Launched in 1948 in Chicago by Ray PALMER and Curtis and Mary Fuller, *Fate* came into being during the golden era of pulp magazines. But while many of the other pulps of the time—*Amazing Stories,* for example—have ceased publication, *Fate* has managed to weather the storms of changing times. In 1955 the

Fate Magazine

Fullers bought Palmer's interest in the magazine and continued to publish it until 1989, when Llewellyn Worldwide, a NEW AGE MOVEMENT and occult book publisher in St. Paul, Minnesota, purchased the magazine.

From its inception, *Fate* has focused on "true reports of the strange and unknown." Its premiere, digest-size issue contained a first-person account by Kenneth Arnold, a pilot who saw the first highly publicized UFO of the 20th century. *Fate*'s sensational cover and Arnold's account helped launch the era's fascination with flying saucers—science fiction come to life.

First-person accounts of unexplained experiences—encounters with BIGFOOT, sightings of the LOCH NESS MONSTER, psychic dreams, visits from dead loved ones, and so on—remain a strong emphasis of the monthly magazine (now full size), although it also publishes articles about fringe science, folklore, and parapsychological research. *Fate* makes no pretense of being a scientific journal and, in fact, has traditionally had a more sensational kind of appeal.

FATIMA A small village in central Portugal where there was a supposed apparition of the Virgin Mary, the first occurrence happening in 1917—making the small village one of the most famous Marian shrines in Europe. The event was so special that it became the only one of its kind that has been authenticated by the Roman Catholic Church. The story runs like this: On May 13, 1917, three young peasant school children, Lucia dos Santos and her cousins Francisco and Jacinta Marto, were tending their parents' flock when a flash of lightning lit up the sky and was followed by a vision of "a lady brighter than the sun" sitting in the branches of a tree. According to Lucia—the only one who could hear what she was saying—the lady announced, "I am from Heaven. I have come to ask you to return here six times, at this same hour, on the thirteenth of every month. Then, in October, I will tell you who I am and what I want."

Only a few people came with the children in June but thousands attended in July. On neither occasion did anyone but the three children see anything, and consequently

Fatima

the authorities accused the Church of fabricating a miracle. The Church, fearing it was a hoax, was noncommittal. Lucia's mother admitted that her daughter had, from an early age, been prone to fantasizing and making up stories to draw attention to herself, and now feared that she was leading people astray. Nevertheless, the children, led by Lucia, refused to alter their story and, at the final appearance, on October 13, 70,000 people converged on Fatima where they witnessed the so-called "Miracle of the Sun."

The day had been dull and cloudy, and many witnesses described the sky as suddenly clearing and the Sun as leaving its place in the firmament, falling from side to side and then plunging, zigzagging on the crowd below and sending out an increasing heat. Others said the Sun broke through the clouds and, looking like a swirling ball of fire, shot beams down to earth. Nobody else outside Fatima saw or heard anything unusual on that day; only the children saw the vision of the lady, and again only Lucia could hear her speak. The lady was the Lady of the Rosary, the Virgin, and what she wanted was to give the world three messages. The first was a message of peace and, according to Lucia, a vision of hell with anguished, charred souls plunged into an ocean of fire. The second was an instruction to recite the Rosary daily and for a chapel to be built there in Her name—"If you pay heed to me, Russia will be converted to Catholicism and there will be peace. If not, Russia will spread her errors through the world, causing wars and persecutions against the Church." The third message has never been revealed; it lies in a drawer in the Vatican, too horrible for public consumption and, we are told, is read by successive Popes on their accession and then put back again.

The first national pilgrimage to Fatima took place 10 years later in 1927, and a basilica was begun in 1928 and consecrated in 1953. Since then, numerous cures have been reported, and on May 13, 1967, the 50th anniversary of the first vision, a crowd of pilgrims, estimated at one million, gathered at Fatima to hear Pope Paul VI say Mass and pray for peace. A similar vision of the Virgin was reported at Garabandal, Spain in 1961, but the Church did not accord any importance to this manifestation.

The May 1996 edition of the *FORTEAN TIMES* carries a story that officials at the shrine of Fatima are trying to stem a flood of Iranian pilgrims, who believe that the spot was named after the eldest daughter of Mohammed, Fatima Zhara, when an Islamic sect lived in the Iberian peninsula in the 9th and 10th centuries. The idea was promoted by a documentary on Iranian television, which said that the vision the children saw was of Mohammed's Fatima and not of the Virgin Mary.

Scientists say that what happened at Fatima on the day of the alleged "Miracle of the Sun" could have been mass hysteria, or retinal distortion caused by looking too directly at the sun (which, incidentally, can cause blindness). Some scientists have explained that the result of observers trying to avoid fixing their gaze on the intense rays of the sun would be to move their eyes sharply to left and right, and this would have the effect of combining image with afterimage, thus distorting vision to such an extent that they would not know what they had seen. The scientific attitude to all visions, APPARITIONS and such manifestations, especially when they are not seen by the rest of humanity, but only by one person or group, in heightened expectation, is that the thing seen must be insubstantial, nonphysical, and therefore of the imagination. People who see visions are not telling lies. They honestly believe they have seen something, but because it is subjective it cannot be subjected to any objective scientific study.

See also GARABANDAL VIRGIN; LOURDES.

For further reading: Sandra L. Zimdars-Swartz, *Encountering Mary* (Princeton University Press, 1991).

FIREWALKING The act of walking barefoot through fire, or, more commonly, over a bed of hot coals, without being burned, an act often associated with religious rites. Firewalking has been a traditional practice in many societies

Firewalking

around the world, notably in India, Polynesia, Japan, and New Zealand. Usually, it is either part of a ritual in which many people participate, not as firewalkers necessarily, but as chanters, prayers, and observers. Occasionally it is performed by a single individual as a demonstration of religious faith. The *fakirs* of India perform in public places activities such as firewalking, lying on beds of sharp nails, and provoking poisonous snakes, to demonstrate their own faith and to encourage devotion in others. In recent years firewalking has become a part of NEW AGE or Human Potential Movement practice, as participants demonstrate empowerment gained through participation in a Universal Consciousness or their self-release from fear.

In most firewalking rituals, 20–60-foot beds of extremely hot coals are prepared. Participants then chant, pray, meditate, and/or shout encouraging slogans for some time before they step onto the coals. In some cultures mental preparation for the ritual lasts for days or weeks and includes praying, fasting, and meditation; in some, preparation is a matter of a few hours under the guidance of a compelling leader. Once the event begins, some people rush across the coals, while others move more slowly; some even stop and stand, sit, or kneel on the coals for a few moments. Occasionally participants are severely burned, but an amazing number cross the coals with no apparent harm.

One explanation for this phenomenon is psychological—the power of mind over matter. This explanation suggests that participants enter something akin to an ecstatic trance which sensation does not penetrate. (Often participants who are burned say afterward that their concentration was interrupted.) Yogis have demonstrated the ability to put themselves into a trance state in which they can prevent themselves from bleeding or experiencing pain.

An alternative explanation is that participants engage in self-HYPNOTISM. Some studies suggest that self-hypnotized individuals, like yogis, can exert remarkable control over physical reactions.

There are several possible physiological explanations, as well, for the ability to firewalk. Speed is one; supposedly some participants skim across the coals so quickly that they barely touch them and thus are not burned. Another is that nervous and heat-caused perspiration sets up a physical barrier between the skin and the fire. This has been compared to what happens when a drop of water is released onto a hot iron or skillet: The drop skitters around apparently unaffected by the heat for a brief time. A few researchers have unsuccessfully attempted to test this method by purposely creating moisture barriers.

Another theory is related to chemical changes in the brain. Much as an individual sometimes gains extra strength through the release of unusual amounts of certain brain chemicals in an emergency situation, the brain might release extra chemicals when someone experiences heightened sense of excitement, fear, or will in anticipation of an act usually considered impossible.

Researcher Bernard J. Leikind asserts that successful firewalkers simply have physics on their side. All materials possess varying degrees of heat conductivity. This can be affected by adding or subtracting porousness, or air. More porous substances (feathers) conduct heat less well than denser substances (metal). Coal beds are never smooth, dense surfaces; they are uneven, filled with air pockets, and porous. Thus, unless the firewalker's foot lingers too long on a piece of coal, its chances of getting burned are small, says Leikind.

While mystics and New Agers still acclaim the mystery of firewalking, most scientists are confident that explanations can be found in the rules of physics.

For further reading: Bernard J. Leikind, "Fire Immunity," *The Encyclopedia of the Paranormal,* edited by Gordon Stein (Prometheus Books, 1996).

FISHBEIN, MORRIS (1889–1976) An Illinois physician who attacked NATUROPATHY and other forms of ALTERNATIVE MEDICINE. He authored *Fads and Quackery in Healing* (1932).

The case of American Socialist Party leader and presidential candidate Eugene V. Debs concerned Fishbein greatly. In poor health after completing a prison sentence, Debs went to the Lindlahr Naturopathic Sanatorium in Elmhurst, Illinois, for a rest. While there, he became unconscious; two days later, Fishbein was called by Debs's brother to check on his condition and found Debs to be suffering from an undiagnosed brain disturbance. The clinic's subsequent treatment proved inadequate; Debs died soon afterwards. Fishbein was strongly of the opinion that if Debs had received conventional treatment from the outset, he would probably have survived—in short, that naturopathy had done him more harm than good.

Fishbein attacked many other medical cults as unscientific and potentially harmful: HOMEOPATHY, IRIDOLOGY, zone therapy, orificial therapy, osteopathy, and CHIROPRACTIC among them.

FLAT EARTH SOCIETY An organization that promotes the belief that the Earth is flat and pancake-shaped and not an oblate sphere—the orthodox view for many hundreds of years.

The society describes itself on its Internet site as a nonpartisan, nonprofit, and nondenominational membership organization that is dedicated to improving understanding of the nature of reality through paraphysical inquiry, empirical investigation, and the exchange of ideas.

Flat Earth societies date back many years but seem to come and go with fair frequency. The present revival, founded in 1993 by Lee H. 'O'. Smith, EMF KYTP (EMF, an acronym for *Established Major Figure;* KYTP, not explained), has bases in three areas: North America, the British Isles and Australia. EMFs in each area supervise its activities, and a Chairperson named Dei Grata, who is based in California, is responsible for global administration. Dei Grata means "an acceptable God" but could be meant to read *Dei Gratis* ("by the grace of God").

The society claims that its members, who are elected, come from all walks of life and are "active in many areas of culture, science and public life."

Its tenets are that:

1. Earth is flat and has five sides.
2. All places in the Universe that are named Springfield are merely links in higher-dimensional space to one place (the atlas lists five, all in the United States).
3. All assertions are true in some sense; false in some sense; true and false in some sense; true and meaningless in some sense; false and meaningless in some sense; and true, false and meaningless in some sense.

See also VOLIVA, Wilbur Glenn.

FLETCHER, HORACE (1849–1919) Businessman, author, and traveler who advocated a dietary regimen that stressed careful and prolonged mastication and reduced food intake. Born and raised in Massachusetts, at age 15 Fletcher embarked upon a life of travel and adventure that included whaling, extensive tours of the Orient, and even some time on a pirate ship. Later he received some formal education at Dartmouth, but in most subjects he was self-taught. In his 30s he married and settled in San Francisco, where he made a fortune as an ink manufacturer and importer of Japanese goods. He would later live in New Orleans, Venice, and Belgium and continue to lecture and travel around the globe until his death in Copenhagen at the age of 70.

Fletcher's career as a popular writer began in the 1890s when, upon reaching middle age, he became concerned about his excessive weight and deteriorating health. An excellent athlete in his youth, Fletcher resolved to regain the vigor that had propelled him through his demanding early life. Drawing liberally from the New Thought tradition and his own experiences, Fletcher produced a series of books advocating what he called "menticulture," a positive-thinking variation that claimed that personal obstacles could be overcome by purging the mind of such negative emotions as anger and worry. Concurrently, Fletcher made a quick study of various works on physiology and began to formulate the theories of nutrition that would make him famous.

His basic philosophy, called FLETCHERISM, started from a belief in an inerrant "Mother Nature" whose laws were to be strictly obeyed if proper health was to be maintained. The key to health was nutrition, which in turn depended upon efficient, or in Fletcher's terms "economic," digestion. He put particular emphasis on proper chewing of foods, which he felt should be done slowly and methodically to ensure proper and thorough absorption. If done properly, the body, which acted as a "Food Filter," would need less food and retain more of its natural energy. The measure of success, which Fletcher detailed exhaustively in his various books, was to be taken from an individual's stool—ideally dry, proportionately small, and as infrequent as once a week.

Trying out his theories on himself, Fletcher was able to lose more than 40 pounds in a matter of months and claimed to have regained his lost vitality. Demonstrating remarkable physical prowess for a middle-aged man, Fletcher undertook 100-mile bicycle rides and scored high on muscle endurance tests. Detractors, however, scoffed at his dreary chewing regimen and physiological claims. In contrast to his notions of alimentary efficiency, most nutritionists later emphasized the importance of dietary roughage. Nevertheless, Fletcher became a hero for a number of progressive-era activists who wanted to promote better public nutrition and eating habits. He is also credited with convincing physiologists to reduce their estimates of daily requirements of certain foods, such as proteins, to levels that are still widely accepted.

See also COUÉ, EMILE.

For further reading: Horace Fletcher, *Fletcherism, What It Is* (F.A. Stokes, 1913); and James C. Whorton, *Crusaders for Fitness* (Princeton University Press, 1982).

FLETCHERISM See FLETCHER, HORACE.

FLEW, ANTONY G. N. A professor of philosophy at Reading University, England. Flew is a member of the SOCIETY FOR PSYCHICAL RESEARCH (SPR), the COMMITTEE FOR THE SCIENTIFIC INVESTIGATION OF THE CLAIMS OF THE PARANORMAL (CSICOP), the Committee for the Scientific Examination of Religion, and a contributing editor to the journal *Free Inquiry.* For over 40 years he has brought to the fields of the paranormal, PARAPSYCHOLOGY, EXTRASENSORY PERCEPTION, and similar inquiries the careful analytical approach of the philosopher, neither accepting nor rejecting a priori claims but subjecting them to very scrupulous examination and discussion. Flew's numerous publications are mostly philosophical, but there are several on psychical research.

Two of his works, *A New Approach to Psychical Research* (1953) and *The Logic of Immortality* (1987), particularly the final chapter of the latter on "The Significance of Parapsychology," are essential reading for any serious student of parapsychology. In *New Approach,* Flew explains what he means by psychical research and deals with spontaneous phenomena, mediumship, and survival after death. He then analyzes the more recent work of

Joseph Banks RHINE and others on experimental paranormal behavior. He finds evidence for spontaneous phenomena and mediumship both unconvincing and unlikely. The possibility of survival after death fascinates him but poses great philosophical problems. Some experimental work in the 1930s and 1940s convinced him that there were real ESP effects, but he is critical of inadequate controls in much of the work and has many suggestions for improvement. Flew is not impressed by PSYCHOKINESIS or PRECOGNITION, but approves of Soal and Goldney's work with Shackleton and of Soal's similar work with Stewart on TELEPATHY. Two case studies were appended to the book: *The Adventure* and John William DUNNE's experiments with time. The former is an account of retrocognition by the Misses Moberly and Jourdain in 1901–02 (both were principals of St. Hugh's College, Oxford); this receives fair but skeptical consideration. Dunne's time of infinite dimensions receives short shrift.

The Logic of Immortality is largely a philosophical discussion of survival and immortality from Plato's time to the present. The final chapter is of most concern. By 1987 Rhine's work was no longer respected; Hansel had shown his crucial experiments to be flawed, no longer providing incontrovertible proof of extrasensory perception. Soal had been exposed as a fraud, the Shackleton and Stewart experiments revealed as trickery. Flew looks back to the early days of SPR, remarking on the eminence and respectability of its members: Balfour, Bateson, Dodgson (Lewis Carroll), Ruskin, Stephen, and many others "seemingly of impeccable high Victorian respectability." He points out, though, that neither eminence nor respectability are guarantees of authenticity; eminent and seemingly respectable persons can be duped and some may even prove to be crooks. Flew's convictions on the later experimental work have been dented: ". . . we still appear to be as far away as ever from any repeatable demonstration of the reality of any psi-phenomenon."

FLIESS, WILHELM (1858–1928) A Berlin surgeon and an early associate of Sigmund FREUD who believed that there are two cycles that govern the behavior of living organisms: the 28-day female cycle and the 23-day male cycle. Some organisms, including humans, followed one or other of these cycles or some multiple of them: 46, 69 . . . or 56, 78 . . . days. Others, again including humans, had cycles that were combinations of both the basic 23- and 28-day periods, for example 23 plus 28, that is, 51 days.

See also BIORHYTHMS and CYCLES THEORIES.

FLOURNOY, THEODORE (1854–1920) Professor of psychology. Born in Geneva, Switzerland, Dr. Flournoy taught at the University of Geneva from 1891 until 1919. In 1901 he founded the journal *Archives de Psychologie* and wrote numerous medical and psychological works. Interested in mediumship, Flournoy became well known for his studies of Helene Smith, who while in trance narrated three distinct "dramas" in AUTOMATIC WRITING and speaking, including "glossolalia" or "speaking with tongues." Miss Smith first manifested these abilities in séances with a spiritistic group, passing along to its members supposed messages from the dead.

In his book *Des Indes à la Planète Mars* (From the Indies to the planet Mars; 1900), Flournoy recorded and analyzed the Smith "dramas," two of them telling of her supposed past lives as a Hindu princess and as Marie Antionette. The third concerned the son of one of her séance circle's members who had purportedly been abducted to Mars and who reported on people, language, and customs of the planet. Smith's dramas and messages from beyond were related through her "control," an entity named Leopold. Flournoy's analysis saw the Marie Antionette and the Mars stories, as well as the "control" entity, as what would today be called "repressed" aspects of Mlle Smith's personality. The tale of the Hindu princess, however, was much more puzzling to Flournoy. He uncovered obscure references, from books that were apparently inaccessible to Smith, confirming persons and place names described by Smith who had, additionally, spoken Hindi phrases and written in Arabic script during her trances. Flournoy concluded that Mlle Smith displayed knowledge that would be difficult if not impossible for her to acquire by normal means but rejected spiritistic explanations. While reaching no absolute conclusion, he suggested telepathy might account for the phenomenon.

Flournoy's other books on parapsychology include *Métapsychique et Psychologie* (Parapsychology and psychology; 1890); *Nouvelles observations sur un cas de somnambulisme avec glossolalie* (New observations on a case of somnambulism with glossolalia; 1902); *Esprits et Médiums* (1911), and *La Philosophie de William James* (1911).

FLYING SAUCERS, CRASHED Spacecraft that are thought to have been operated by extraterrestrials and to have crashed into Earth. There have been many reports of crashed flying saucers worldwide, particularly in the last half of the 20th century, but few of these alleged extraterrestrial craft have left any solid evidence behind them. Some have subsequently been identified as meteors, weather balloons, lightning strikes, and other manmade or natural phenomena.

Probably the most notorious and controversial crashed flying saucer incident is that said to have occurred near Roswell, New Mexico, in July 1947. Details vary, depending on the source, but most commonly it is said that from one to three spaceships crashed in the New Mexico desert. Some reports also assert that one or more living or dead extraterrestrial beings were found in the remains of the disaster. Some sources allege that the U.S. government acquired the alien(s) or their bodies and secretly kept them at an air force base in Ohio or New Mexico.

What seems fairly certain is that something crashed into Earth sometime during the first week of July 1947, only a few days after pilot Kenneth Arnold's famed UFO sighting near Mount Rainier in Washington State. Local ranch foreman William (Mac) Brazel and a teenage boy accompanying him discovered some mysterious wreckage, which they claimed was made of a metal completely unknown to them that possessed "memory"—that is, they could bend it and it would spring back to its original form without creases. They also said it was marked with some type of strange symbols, called hieroglyphs by some researchers. The Roswell *Daily Record* for July 8 reported that a flying saucer had crashed. The report at first seemed to be confirmed by a spokesperson at the nearby air base; the next day, however, the military issued a statement saying that there had been a mistake, that there had been no flying saucer crash. The military said that the debris had come from a downed weather balloon. This report ended the speculation for many people, but others were not satisfied.

In the intervening years, other witnesses have stated that they saw debris or saw the saucer crash and that government officials told them to keep quiet. The idea of the saucer crash was kept alive by some ufologists, including Kevin D. Randle, a retired U.S. Air Force captain, who concluded that the government was conducting a coverup of an incident that could have important ramifications for people throughout the world.

Until 1995, Randle and a few others had to fight to keep the Roswell story alive. Then something happened to bring the Roswell incident back into general public consciousness: British filmmaker Ray Santilli began to show and distribute a film he said had been taken in 1947 by an air force photographer of an alien body being autopsied by military doctors. The film was widely shown and debated on television, although most expert viewers, including dedicated ufologists, concluded that the film was a fraud.

Around the same time, a UFO organization succeeded in getting the General Accounting Office (GAO) of the U.S. government to investigate whether the air force or any other government agency had covered up a UFO crash at Roswell. The GAO was able to locate few documents bearing on the issue and came to no conclusions, thus allowing the controversy to continue. Skeptics say there is no solid evidence that a UFO ever crashed near Roswell, but ufologists, some former government workers, and some scientists remain convinced that the United States and other governments are covering up the evidence for this crash and other UFO-related incidents.

There have been other reports of crashed flying saucers as well (for example, Spitzbergen, Norway, 1952; Socorro, New Mexico, 1964; Varghina, Brazil, 1994), but none has been as fully explored or attained the notoriety of Roswell.

For further reading: Jerome Clark, *The UFO Encyclopedia,* 3 vols. (Apogee/Omnigraphics, 1990–96); Philip J. Klass, *UFOs: The Public Deceived* (Prometheus, 1983); *A History of UFO Crashes* (Avon, 1995); Kevin D. Randle and Donald R. Schmitt, *The Truth About the UFO Crash at Roswell* (M. Evans, 1994); Jenny Randles, *UFO Retrievals: The Recovery of Alien Spacecraft* (Blandford, 1995).

FORD, ARTHUR A. (1897–1971) One of the most influential spiritualist MEDIUMS of the 20th century. In 1917 he had entered Transylvania College, a school of the Christian Church (Disciples of Christ) to prepare for the ministry, but his education was interrupted by World War I. While in the army, he began "hear" names of people who would soon appear on the list of war casualties. After the war, as a Disciples pastor he began to explore spiritualist phenomena, then the subject of intense interest by psychical researchers. By the mid-1920s, he had developed as a trance medium and founded the first Spiritualist Church in New York City.

The most famous incident in Ford's long career occurred during his stay in New York. In 1929, Ford claimed that he was in contact with the mother of the recently deceased magician Harry HOUDINI. Ford contacted Houdini's widow, Bess, and gave her the one-word message "Forgive." She confirmed that that was the message that the couple had kept as their secret. Shortly after delivering that message, Ford was invited to hold a séance in Bess Houdini's home where he gave a cryptic message, "Rosabelle, answer, tell, pray, answer, look, tell, answer, answer, tell." This message, in the code of mentalism, which Houdini and his wife had practiced in their early life together, spelled out the world *believe,* the message they had agreed upon before his death.

Ford claimed throughout his life that his revelation of the coded message was a triumph for SPIRITUALISM and a vivid incident of spirit contact. Bess signed a statement confirming Ford's accomplishment. Others, however, were quick to charge that matters were not so simple. It appears that the confidential message "Forgive" had been published in an article in the *Brooklyn Eagle* and that the mentalist code used by the couple had been published in 1928 in a biography of Houdini by Harold Kellock. In later years, Mrs. Houdini would deny that she had received the correct message from Ford. Skeptics who have looked at the situation have branded it a hoax, though their case is not as definitive as other incidents.

In 1936 Ford founded the International General Assembly of Spiritualists, in which he could freely express his belief in REINCARNATION, an idea generally eschewed by the main body of U.S. spiritualists at the time. During his mature years, Ford emerged as a prominent spokesperson for spiritualism and contact with the dead through mediumship. His individual trance work attracted many people who were not otherwise affiliated with spiritualism. In 1956 a number of Protestant ministers and laypeople whom he had directly influenced founded Spiritual Frontiers Fellowship as a church-related organization to pursue their interest in psychical phenomena and the evidence of survival after death.

During the last years of his life, Ford conducted a televised séance with Episcopal bishop James A. Pike, whose subsequent life was greatly altered by what he felt was convincing evidence of spirit contact delivered in that encounter. Pike died a few years later in 1969 and Ford in 1971. Subsequently, in going over Ford's papers for a biography, literary executor William V. Rauscher and writer Allen Spraggett found conclusive evidence that the séance had been faked. This evidence caused many to reevaluate Ford's entire career and further substantiated questions about his relationship to the Houdini code.

For further reading: Allen Spraggett and William V. Rauscher, *Arthur Ford: The Man Who Talked with the Dead* (New American Library, 1973).

FORMATIVE CAUSATION A theory concerning the origin and growth of forms and characteristics in nature that was originally proposed by biochemist Rupert SHELDRAKE. Generally, biologists speak of what is termed "morphogenetic fields" to indicate the as-yet poorly understood factors that influence the development of growth in plants and animals. These factors are assumed to operate through the chemistry of the DNA molecules. Sheldrake proposed a more literal understanding of the morphogenetic field as structures that exist independently of space and time.

To Sheldrake, genes define the parameters within which development of an individual organism occurs, but they do not determine form. This is determined by the morphogenetic fields. Each organism is influenced by all the previous forms assumed by the previous examples of the organism; all the past fields of a given type are available to the new organism. Various morphogenetic fields influence one another by a process that Sheldrake calls "morphic resonance."

Sheldrake applied his theory to inanimate crystalline forms as well as to plants and animals. He suggested that once crystallization takes place, a new morphogenetic field is created. Subsequent crystals are influenced by the initial one. All subsequent crystallizations reinforce the field and definitively influence future crystal formations to follow the pattern of the past. In like measure, the future developments in a species are influenced by the establishment of morphogenetic fields from the past that tend to give a species some stability over time.

The theory has certain parapsychological implications that were not developed by Sheldrake. Clearly the fields could act as channels for the transmission of information in a manner similar to what is claimed for TELEPATHY and CLAIRVOYANCE.

Sheldrake's theory was met with hostility from many of his scientific colleagues. The reviewer of his book *A New Science of Life: The Hypothesis of Formative Causation,* in *Nature* (1981) denounced it as pseudoscience and nominated it as the year's best candidate for burning. Even those who decried the emotional outburst of this review offered little support for the idea, which seemed to lack the kind of falsifiability which science demands of new hypotheses.

For further reading: Rupert Sheldrake, *A New Science of Life: The Hypothesis of Formative Causation,* in *Nature* (Blond & Briggs, 1981).

FORT, CHARLES HOY (1874–1932) Writer and speculator about the paranormal. Fort was the second of three sons born to a family of Dutch ancestry in Albany, New York. His mother died when he was four and the children were raised by his father, who was a strict disciplinarian.

Fort never went to college. For a time, he worked as a reporter and wrote and published short stories and a novel, *The Outcast Manufacturers* (1909). In 1916 he came into a small inheritance, which gave him freedom to pur-

Fort

sue his own interests. For the rest of his life he conducted his own private study—acquainting himself with the current facts and theories of scientific subjects and collecting reports on mysterious occurrences and anomalies, which he referred to as the "damned" or "excluded" because they did not accord with established scientific orthodoxy. He made notes and collected press clippings from old magazines and newspapers and filed them in shoe boxes in his study. At first he worked in the New York Public Library, and then, living in London in the 1920s, continued his inquiries in the British Museum Library.

Fort saw himself neither as a realist nor as an idealist but as an "intermediatist," holding that all phenomena are approximations between "realness" and "unrealness." His attitude was to believe nothing because he said, to believe is to take a position on one side or another. Consequently his opinion toward science was highly skeptical; after ascertaining the normal scientific position on a subject, he would adopt the opposite point of view almost on principle. He wrote: "Now there are so many scientists who believe in dowsing that the suspicion comes to me that it may be only a myth after all." He often commented on how scientists argued according to their own beliefs rather than according to the rules of evidence, ignoring, discrediting, or suppressing inconvenient data. He thought that even in strictly run scientific experiments where data is being collected on inanimate objects, the experimenter can become biased—the experimenter, after all, would prefer one kind of result over another because it confirmed what the scientist wished to prove. A good example of this was the 1881 Michelson–Morley experiment (an attempt to measure the velocity of the Earth in its orbit through the hypothetical ether, by measuring the effect that such a velocity would have upon the velocity of light. The experiment gave a zero result, to account for which Einstein postulated the special theory of relativity in 1905). Fort observed that the failure of scientists to find variations in the speed of light could lead to either of two conclusions: firstly, that proposed by Einstein that the velocity of light is an absolute, regardless of the Earth's motion; and secondly, that the Earth is not moving at all. He then added: "Unfortunately for my own expression, I have to ask a third question: Who, except someone who was out to boost a theory, ever has demonstrated that light has any velocity?"

Fort was heavily criticized by the scientific establishment. His admirers, on the other hand, pointed out that some of his more colorful speculations were put forward in a spirit of outlandish fun, as well as to draw attention to the pomposity of eminent scientists who forget that no scientific theory is beyond doubt. But Fort forgot that, despite the fact that absolute certainty is unattainable, conclusions can be drawn. He could not see that it is possible to distinguish between different categories of theory, those with a high and those with a low degree of probability. Obviously there are borderline cases in which we cannot yet say whether a theory is or is not probable. But when we can, that differentiation marks the border between science and pseudoscience.

Over a decade before the flying saucer craze, Fort coined the term "TELEPORTATION" and was one of the first to speculate that mysterious lights seen in the sky might be from craft from outer space. His most bizarre explanation of these objects was that we were owned by a higher intelligence who periodically checked us out.

Fort set down his many ideas and speculations in four books. *The Book of the Damned* (1919) attempted to rehabilitate those views rejected by orthodox science. *New Lands* (1923) attacked what Fort termed the scientific "priestcraft" (and eventually led to formation of the FORTEAN SOCIETY). *Lo!* (1931) was so named because, Fort said, astronomers are always making calculations, pointing to the sky where they believe something should be, and saying "Lo!" but there was never anything there to see. *Wild Talents* (1932), published a few weeks after his death, makes the point that talents such as WITCHCRAFT and DOWSING can contribute to knowledge on an equal footing with science.

The spirit of Charles Hoy Fort lives on today in the publication *FORTEAN TIMES*.

For further reading: *The Complete Books of Charles Fort* (Dover, 1974); Martin Gardner, *Fads and Fallacies in the Name of Science* (Dover, 1957); and Damon Knight, *Charles Fort: Prophet of the Unexplained* (Doubleday, 1970).

FORTEAN SOCIETY Founded in 1931 to promote the science-debunking ideas of Charles Hoy FORT by two of Fort's longtime friends, well-known novelist Theodore Dreiser and Tiffany Thayer. Other members included renowned literary men of the time, such as Alexander Woollcott, Booth Tarkington, Ben Hecht, Burton Rascoe, and John Cowper Powys. Fort himself opposed the idea of founding a society, and it is said that he was brought to the foundation dinner on January 26, 1931, by deception. A brochure about the society describes it in these terms:

> *The Fortean Society is an international association of philosophers—that is of men and women who would live no differently if there were no laws; of men and women whose behavior is not a sequence of reflex jerks caused by conditioning, but rather the result of some cerebration, or of some mystical whimsicality of their own. . . . Eminent scientists, physicists and medical doctors are members—likewise chiropractors, spiritualists and Christians—even one Catholic priest. . . .*
>
> *The society provides a haven for lost causes, most of which—but for our sympathy—might become quite extinct. . . . A good many adherents of a flat earth are members,*

anti-vivisectionists, anti-vaccinationists, anti Wasserman-testers, and people who still believe disarmament of nations would be a good thing. . . . These members embrace the only "doctrine" Forteanism has, that of suspended judgment, temporary acceptance and eternal questioning. . . .

In 1937 Thayer, at his own expense, began to issue *The Fortean Society Magazine,* renamed *Doubt* in 1944. The publication's main purpose seemed to be to embarrass scientists by asking questions they could not answer and printing embarrassing articles about their private lives. Thayer died in 1959 and the society disbanded the following year.

The INTERNATIONAL FORTEAN ORGANIZATION (INFO), located in Arlington, Virginia carries on the tradition of the Fortean Society and publishes the *INFO* journal.

FORTEAN TIMES A modern upmarket monthly journal, founded in 1973 to continue the ideas and preconceptions of the U.S. author Charles Hoy FORT. It is published in Britain by Mike Dash for Viz and edited by Bob Rickard and Paul Sieveking, who call themselves "cosmic clerks." The magazine prints a wide selection of strange phenomena, unusual beliefs, and experiences that have been reported to it via letters, facsimiles, and E-mail by "a global network of weird-watchers" (their phrase). *Fortean Times* can now be reached on the Internet.

See also SKEPTICS' MAGAZINES.

Fortean Times

FORTUNE TELLING See DIVINATION.

FOUNDATION FOR RESEARCH ON THE NATURE OF MAN PARAPSYCHOLOGY research organization continuing the work of the former Parapsychology Laboratory founded in 1935 by Joseph Banks RHINE at Duke University. Rhine was the pioneer advocate of laboratory research on psychic phenomena. In 1927 he joined the Psychology Department faculty at Duke, then under the chairmanship of William McDougall, a former president of the SOCIETY FOR PSYCHICAL RESEARCH. Through the early 1930s, Rhine worked out the initial program of research and the new language that would be required for it. The Parapsychology Laboratory became the scene of numerous parapsychology experiments, the results of which were reported in Rhine's pioneering texts *New Frontiers of the Mind* (1937) and *Extrasensory Perception after Sixty Years* (1940). With Joseph G. Pratt, Rhine and McDougall also founded and coedited the *Journal of Parapsychology,* still the premier journal for the new discipline.

Rhine's positive findings on psychic phenomena were widely reported in the media and gave Rhine and the laboratory celebrity status. He found himself the center of two controversies, the first among psychical researchers whose methodology and research orientation he was challenging, and the second among his psychological colleagues who were hostile to the subject of his investigations. His research greatly disturbed fellow faculty members, some of whom were embarrassed by Duke's association with such a questionable endeavor. Even fellow faculty member Karl E. Zener, who had originally designed the card deck used for testing ESP, turned on Rhine and worked to have him removed from the department. Rhine's publications set off an academic controversy over ESP that has continued to the present.

The Parapsychology Laboratory became autonomous in 1950, and following Rhine's retirement in 1962, the department withdrew any formal support of it. The laboratory and the *Journal of Parapsychology* were transferred to the newly founded Foundation for Research on the Nature of Man, an independent research facility that Rhine continued to head for the rest of his life. The foundation is currently headed by Dr. K. Ramakrishna Rao.

FOURTH DIMENSION The concept that there is another dimension beyond the three that we use to describe or visualize solid objects—or describe mathematically in solid geometry. The idea entered the popular realm when explanations of Einstein's special theory of relativity appeared in the media. Einstein's theory made time the fourth dimension, inextricably connected with the three spatial dimensions, constituting the four-dimensional space-time continuum. The implications of this were negligible for everyday life but of great importance for some elements of physics and cosmology. To further complicate things, mathematicians and theoretical physicists had no difficulty in using four-dimensional mathematics and formulating five-, six-, seven- or more dimensional geometries (a good explanation of this for the layman is in Antony G. N. FLEW's *A New Approach to Psychical Research,* Appendix II). The fourth dimension is time, which in our everyday existence is independent of the three space dimensions; the interconnectedness of the three dimensions of space and the fourth dimension of time does not matter. To a particle traveling at about the speed of light (186,000 miles a second), the connection between space and time does matter. There isn't any scientifically accepted evidence for another dimension.

The multidimensional idea was seized upon by some exploring the supernatural, notably by John William DUNNE, to explain spiritualism, and pre- and retrocognition.

FOWLER, ORSON (1809–1887) Author, editor, public lecturer, and prominent American advocate of PHRENOLOGY during the 19th century. Fowler was born near Cohocton, New York, and grew up in a semifrontier setting under the watchful eye of a stern, Congregationalist father. Originally slated for a career in the ministry, Fowler became interested in phrenology while a college student at Amherst. On graduating in 1834, he and his younger brother Lorenzo began to lecture on the new "mental science" and related subjects around the country. Eventually the two settled in New York City and, with their brother-in-law, Samuel Wells, set up a combination museum, publishing house, and head-reading salon that became the center of the American phrenological movement. Besides giving individual personality readings according to the shape of the head, Fowler and Wells published the very popular *American Phrenological Journal* (in print from 1838 to 1910) and set up a lucrative mail-order business in phrenological literature and paraphernalia. Though Fowler's written work is now forgotten by all but collectors and historians, in his day his books almost always sold well, and a few went through as many as 40 editions. Most were priced low and appealed primarily to artisans and the rapidly growing middle classes. In addition to his phrenological interests, Fowler also wrote books on architecture, edited another journal (on WATER CURES), lectured on human sexuality, and set up a marriage-counseling business.

During its heyday in the 1840s and 1850s, the firm of Fowler and Wells also became a clearinghouse for a variety of reform efforts including vegetarianism, water cures, the antitobacco crusade, temperance, and the movement against tight corsets for women. Fowler supported all zealously at his frequent lectures on the Chautaqua circuit and in his voluminous writings. He was also sympathetic to the nascent women's rights movement and the antislavery cause.

Fowler received some scientific instruction while in college but never any in-depth medical training. In many subjects upon which he wrote extensively, such as anatomy and physiology, he was largely self-taught. In contrast, the founders of phrenology, Franz Joseph GALL and Johann Gaspar SPURZHEIM, had been trained physicians with considerable skill as anatomists and dissectors. They earned for phrenology most of the scant respectability it was able to garner among scientists and served as the movement's primary intellectual sources. For his part, Fowler often invoked such scientific ideals as empirical verification and the disinterested pursuit of facts, but he did little to apply these to phrenological claims. When put on the defensive, he usually resorted to polemics, dismissing antiphrenological writers as closed minded or biased. Fowler was also extremely selective in his use of evidence, and generally ignored the ever-increasing number of clinical studies that disproved key phrenological claims about the brain.

Critics, and even some other phrenologists, were quick to point out Fowler's scientific shortcomings and likewise snubbed his journalistic treatments of complex subjects such as the relationship of the brain to antisocial behavior. Judging by his considerable audience, however, his message that the sciences were both accessible to the general population and a boon to social reform fit the American temperament well during the antebellum era. It was perhaps this buoyant optimism, which Fowler brought to his phrenological activities in particular, that explains why the pseudoscience could sustain such a high level of public interest long after most medical and scientific experts had turned against it.

For further reading: Madeline Stern, *Heads and Headlines: The Phrenological Fowlers* (University of Oklahoma Press, 1971).

FREUD, SIGMUND (1856–1939) Pioneer in the development of PSYCHIATRY and psychoanalysis. Freud was born in what is now Pribor in the Czech Republic. He entered the University of Vienna in 1873 as a medical student, with growing interests in science and philosophy rather than medical practice. However, falling in love with Martha Bernays and needing an income sufficient to support a family, he changed direction and in 1882 entered the General Hospital of Vienna so that he could

Freud

qualify and start a private practice. Then followed a period of study in association with a number of eminent physicians, neurologists, and psychiatrists, including Theodor Meynert, Josef Breuer, Ernst Brücke, and Jean-Martin Charcot. In 1895 Freud published, with Breuer, *Studies in Hysteria,* the first of many groundbreaking books in psychiatry. He was early attracted to HYPNOSIS, but soon moved to analysis by free association. His theories were developed and explained in his many publications, chief among them being *The Interpretation of Dreams* (1899), *Totem and Taboo* (1913), *A General Introduction to Psychoanalysis (1915–16),* and *The Ego and the Id* (1923). In 1902 Freud invited Alfred Adler, Max Kahane, Rudolph Reitler, and Wilhelm Stekel to meet with him regularly in what became first the Psychological Wednesday Circle, then the Vienna Psycho-Analytical Society, and finally the International Psycho-Analytical Association. He formed a close friendship with and was greatly influenced by Wilhelm FLEISS.

For most of his life Freud regarded himself as a materialist and was scornful, for example, of Carl Gustav JUNG's interest in the paranormal. In his later years, however, he became a corresponding member of the SOCIETY FOR PSYCHICAL RESEARCH, with a particular interest in TELEPATHY. He wrote a paper "Psychoanalysis and Telepathy" in 1921, but was persuaded by his friend and biographer Ernest Jones not to publish (it was not published until after Freud's death). At that time he was invited to be a coeditor of a psychical journal but declined, saying, "If I had my life to live over again, I should devote myself to psychical research rather than psychoanalysis."

It has become fashionable to show that the heroes of the past have feet of clay. Freud has not been exempt, and a series of books have been published attacking both his theories and his character. That his work and his theories are not written in stone, never to be improved—nor disproved—is not surprising. What was notable is that he opened up a whole important new area of inquiry and treatment and pointed the way to continuing improved understanding—the foundations on which others could build and have built. *The Encyclopaedia Britannica* sums up very well: "These criticisms do not in any way diminish the stature of Freud, whose brilliant theories, therapeutic techniques, and profound insights into the submerged areas of the human psyche opened up a whole new field of psychological study. He was an original thinker who radically altered prevailing views of human nature."

Freud was a heavy cigar smoker and in 1923 developed cancer of the jaw. Repeated operations failed to cure the disease. Though very ill, when the Germans occupied Austria in 1938 he was persuaded to move to London, when he died in 1939.

See also PSYCHOANALYTIC THEORY.

FUTUROLOGY A term coined in 1949 by the German historian Ossip K. Flechtheim to designate the then new science of prognosis. It has often been described as an organized activity, explicitly devoted to systematic and normative interpretation of potential future histories. The new science has since been applied to a wide spread of long-range forecasting in, for example, political, sociological, economic, ecological, and other fields. Many scientists reject the notion of futurology being a science, while others quibble at the use of the unscholarly word "futurology." All are justified to some extent in their skepticism because the attractions of discussing the future are sufficiently great to attract a considerable number of cranks. Nevertheless the importance of taking cognizance of how our present actions can affect future generations is generally recognized. We *must* look to the future. The difficulties of the subject are, however, formidable enough to repel many sound scholars.

Three terms are repeatedly used in defining futurology: (1) a *conjecture,* meaning an intellectually disciplined speculation; (2) a *forecast,* which is based either on a continuing trend and its extrapolation or on some defined probabilities of occurrences, and (3) a *prediction,* which is a prognosis of a specific event, called somewhat humorously by Bertrand de Jouvenel a *proference,* the action of carrying forward a preference.

Unlike those who look into the future by pseudoscientific or paranormal methods of DIVINATION, the futurologist does not believe that the future is preordained but rather that it is being made and altered by actions that we take today. The future depends on hundreds of thousands of interactions: human, animal, climatic, which are all natural—none supernatural. There are far too many variables and contingent happenings for anyone to know exactly what is going to happen. Pseudoscience has a tendency to attribute to special individuals, for example the crystal-ball gazer or the astrologer, miraculous prophetic powers of peering into the future and seeing what is written there, or of reading the future in the stars. Most futurologists agree that there is no way we can magically foresee the future; it is not written in the stars or passed down to us from on high but is forged through a chain of causality by ourselves through the activities we engage in today. Because our present actions are always changing, projections of the future are always changing. Futurology is a very difficult subject because there is no way that one can formalize rules for forecasting the future to guarantee accurate results. All knowledge is fallible and open to modification in the light of new, more accurate data and/or theories. We can never know ultimate reality, but present observations plus past experience allow us to make causal inferences and so to develop probabilities that can guide forecasts.

The quest for certainty, said the 20th-century American philosopher John Dewey, has always proved to be elusive. He believed that the methods of scientific enquiry should be continuous with critical thinking in ordinary life and that this should be laced with generous amounts of skepticism. Scientific futurologists attempt to obtain fuller, more precise, up-to-date data and to feed this into computer simulations to obtain more and more accurate forecasting. But scientists are always aware of contingencies, the unforeseen happenings, and so usually reject projective modes of thought that extrapolate too far along a present trend; they prefer observational ones. The ability to predict well is now the test for their theories. In forecasting the future, scientists have to rely on probabilities rather than absolutes; science uses probabilities, whereas pseudoscience claims certain knowledge.

In their day, people such as Thomas MALTHUS and Karl MARX were futurologists, although they would never have so described themselves. Today it is the Ehrlichs, the Commoners, the Tofflers, and the like who fill that role.

G

GADDIS, VINCENT H. (1913–) A popular writer on paranormal and anomalous phenomena. As a teenager he had read Charles FORT's book *Lo!* (which had been serialized in a science fiction magazine) and became a devoted Fortean. He accumulated a library of material on mysterious phenomena, which he mined for articles and books over the years.

Gaddis's most famous article appeared in the popular men's magazine *Argosy* in 1964. "The Deadly Bermuda Triangle" described the area of the ocean between Florida and Bermuda in which reportedly an unusual number of disappearances of people, ships and other craft have occurred. That article, which coined the term BERMUDA TRIANGLE, launched the controversy about events in that area and led directly to the best-selling book by Charles BERLITZ a decade later. Gaddis incorporated the article in his 1965 survey of oceanic mysteries, *Invisible Horizons*. Out of his many articles, he produced several books highlighting the mysterious aspects of life, including *Mysterious Fires and Lights* (1967), and *American Indian Myths and Mysteries* (1977). While remaining a devoted Fortean all his life, Gaddis has in recent years expressed some embarrassment about his role in launching the Bermuda Triangle myth.

GAIA HYPOTHESIS Theory that Earth is not a dead mass of rock but instead a living organism. The Gaia hypothesis was first proposed by Timothy Zell in 1970 and refined by James Lovelock and Lynn Margulis in 1972. According to Lovelock, Earth escaped the fate of Venus and Mars because it was occupied about 3 billion years ago by a life form that began to transform the planet itself into its own matter. All life forms that have evolved on Earth, according to the Gaia hypothesis, are part of the Gaia organism, just as cells make up a human body. Like these cells, the different forms of life that make up Gaia interact to contribute to the health of the whole organism.

Lovelock developed the Gaia hypothesis from information gathered by the unmanned Viking probes sent by NASA to Mars in the 1970s. He theorized, based on the atmospheric composition of Mars, that life could not exist on the planet. The Viking landers confirmed that life as we know it does not exist on the Martian surface, although scientists have come to believe that at one time in the distant past simple forms of life probably did live on the planet. They reject Lovelock's thesis on this basis. Lovelock applied his ideas to the Earth, reasoning that, had a Viking lander arrived in Antarctica, it would have found few signs of life there as well. Looking at the world as an alien in search of signs of life would do, Lovelock concluded that the Earth itself was not so much a planet inhabited by living species as a holistic self-sustained and self-regulating system. In other words, the Earth was itself a living organism. Lovelock named this organism Gaia, after the earth-mother who gave birth to the Titans in Greek myth.

Scientists have difficulty accepting Lovelock's Gaia hypothesis because it is very difficult to test its basic principle. Is life on Earth sustained by the interaction of millions of different life-forms, or are these millions of different life forms only tiny parts of a much larger organism? No one has managed to devise an experiment that could answer such a question. Additionally, Lovelock believes that Gaia may have a lifespan of billions of years. Given the relatively short time that human beings have been measuring Earth scientifically, it would be very difficult to discover the changes in Earth that would prove Gaia's existence.

See also SUPERORGANISMS.

GALENISM The theory that dominated Western medical thinking during the Medieval and Renaissance periods. Galen was a second-century Roman physician who was best known for his role in establishing the science of physiology through dissection. Although Roman laws prohibited the dissection of human beings, Galen was able to learn about the internal structure of African apes and theorized that the two species were much alike. Galen established the fact that arteries carry blood (not air, as was previously believed), determined the difference in function between veins and arteries, and accurately described the valves of the heart. His reputation during the medieval period was so high that he was regarded as one of the foremost medical authorities.

Galen's theory about the causes of sickness, however, has been disproved by modern science. Galen suggested that the human body contains four major substances, or *humors,* that were excreted by various organs. He believed that, for a body to be healthy, these humors had to be in balance. Overabundance or lack of a certain humor could be diagnosed by analyzing a person's temperament. *Black bile* was supposed to be secreted by the kidneys or spleen, and too much of it in a person's system caused gloominess. An irritable person was believed to have a liver that produced too much *yellow bile.* An overabundance of *phlegm* caused a person to be cold, moist, and sluggish. *Blood* was the fourth humor and caused a person to be choleric and angry. Galen's ideas continued to be popular well into the 18th century when they were displaced by more modern theories of anatomy, physiology, infection, and diagnosis.

For further reading: Guido Majno. *The Healing Hand: Man and Wound in the Ancient World* (Harvard University Press, 1975).

GALL, FRANZ JOSEPH (1758–1828) Viennese medical doctor whose studies of the brain and cranium helped launch modern biological psychology and provided the basis for the pseudoscience of PHRENOLOGY. Born in Germany and trained in medicine in Vienna, Gall later claimed to have initiated his lifelong study of the structure and function of the brain as a young man when he noticed that persons with good memories commonly possessed bulging eyes. Building upon this insight, and knowledge gained through the study of comparative anatomy, he concluded that mental ability and behavior could be correlated with specific areas of the brain. Gall lectured on his theories in Vienna until civil and religious authorities accused him of promoting materialism and immorality and obliged him to leave the city in 1805. He subsequently embarked upon a tour of other major European cities, where his ideas were generally well received.

Gall eventually settled in Paris, where he set up a successful medical practice and indulged his tastes for gardening, mistresses, and collecting skulls. France also gave him the freedom to teach and publish his ideas. With his assistant, Johann Gaspar SPURZHEIM, he produced his major work, *Anatomie et physiologie du système nerveux en général et du cerveau en particulier,* four volumes (1810–19). In it, Gall revolutionized the study of the relationship of the mind to intelligence and behavior by departing from both humoral and mechanical philosophies in favor of a thoroughly biological approach that stressed the structure and organization of the brain and nervous system. Gall's work had many admirers, including writer Johann Wolfgang von Goethe (1749–1832), but much of the scientific establishment in France and elsewhere never fully accepted him. His attempts to lecture at the prestigious Institut de France, as well as his candidacy for the Académie de Sciences, met with failure. Undeterred, he continued to write, teach, and collect data in support of his ideas until his death in Paris in 1828.

Gall's basic theories held that the brain was actually a composite of 27 regions, or "organs," which localized all mental functions. He organized these organs into two general groups: (1) those common to animals and human beings, including the reproductive instinct, self-defense, guile, and sense perception; and (2) exclusively human faculties, including the religious sentiment, comparative sagacity, wit, and benevolence. Gall considered the organs to be innate and looked to their size and organization to explain such things as intellectual acumen and moral disposition. He would often use examples of extreme behavior and mental illness to illustrate his basic ideas and proposed reforms for the treatment of criminals and the insane.

Though Gall's principal goal was to develop a new understanding of human psychology based upon anatomy and physiology, it was his multiplex analysis of the brain that captured the public imagination. In particular, his belief that the various categories of organs could be read from the shape of the skull became the basis of the pseudoscience of phrenology. Popularized by Spurzheim and others, interest in Gall's ideas swept across Europe and the United States during the mid-19th century. Phrenologists revered Gall as their founder and great savant and often compared him to Galileo and Sir Isaac NEWTON in terms of scientific stature. This fame was something of a double-edged sword, however, as the subsequent disrepute of phrenology has often overshadowed assessments of Gall's more enduring contributions to science.

See also GALENISM.

For further reading: Erwin H. Ackerknecht and Henri V. Vallois, *Franz Joseph Gall, Inventor of Phrenology and His Collection* (University of Wisconsin Medical School, 1956); and Robert M. Young, *Mind, Brain and Adaptation in the Nineteenth Century: Cerebral Localization and Its Biological Context from Gall the Ferrier* (Clarendon Press, 1970).

GALTON, FRANCIS (1822–1911) Amateur scientist, traveler, author, and mathematician who was the intellectual spark for the modern EUGENICS movement. Galton was the seventh and youngest child of a wealthy banking family from Birmingham, England. He was related on his mother's side to the famous naturalist Charles DARWIN. A prodigy in various subjects, admirers have estimated his childhood IQ as high as 200. Galton later studied medicine at King's College and mathematics at Cambridge, but because of health problems he withdrew early and earned passing marks in only a few classes. In 1845, on the death of his father, he received a substantial inheritance that supported him comfortably for the rest of his life and effectively ended his practical need for formal or professional education. As a young man, he traveled extensively in Africa and the Middle East, gaining a reputation as an explorer that eventually earned him entry into such scientific organizations as the British Royal Society. He married Louisa Butler in 1853 and lived most of his remaining life in London. Ironically, the couple never had children.

Because of his wealth, Galton never had to work or teach and thus pursued science largely at his own leisure. In spite of this, his intellectual reputation in Victorian Britain was considerable. During his long life, he published a number of influential works on subjects as diverse as meteorology, statistics, psychology, criminology, and heredity. A firm believer in qualification, his most important work dealt with the application of mathematical analysis to the life and human sciences. He is generally credited with originating the statistical concepts of regression and correlation, giving birth to psychological testing in England, and providing the conceptual basis for the science of fingerprinting.

By far Galton's most famous work however, concerned eugenics, a term he coined in 1883 to promote allegedly scientific attempts to improve human stock by selective breeding. He had long been fascinated by the relationship of heredity to human behavior, beginning with his 1869 book *Hereditary Genius*. Using evidence taken from the histories of wealthy and accomplished English families, Galton argued that talent and intelligence are largely inherited, with education and environmental factors playing a relatively minor role. Galton eagerly applied his mathematical and statistical acumen to proving the validity of his ideas and eventually came to advocate a twofold approach to refining human nature. The first, positive eugenics, sought to increase the number of births among the more able professional classes, largely through tax incentives. The second, negative eugenics, sought to discourage high birthrates among the poor and those with below average mental ability.

Initial reaction to Galton's eugenics scheme was rather cool, if not hostile, but during the latter decades of the 19th century, interest slowly increased. This was abetted by a steady stream of writings and promotion by Galton and talented followers like Karl Pearson. In the first decade of the 20th century, the two founded the National Eugenics Laboratory, the Eugenics Education Society, and the *Eugenics Review* to popularize and add political muscle to their ideas. Similar organizations formed in Germany and the United States, and the movement peaked in the two decades after Galton's death in 1911. Organized eugenicists emphasized negative eugenics. Eugenics has been associated generally with reactionary or conservative political forces, but in the early years many socialists and other leftists (like Pearson) were attracted to eugenics as part of a larger strategy to bring scientific management to bear on social problems.

Perhaps the most basic and persistent criticism of Galton's eugenic theories was that they were founded on an unreasonable disregard for the influence of social and environmental factors and on intelligence and ability. Many felt that Galton erroneously interpreted social differences as the product of inheritable natural qualities, when in fact political factors created such distinctions. Thus, in spite of his often impressive marshaling of statistical evidence, Galton never escaped the suspicion that his work was motivated by personal or class interests rather than purely scientific ones. Some have conjectured that for Galton, whose wealth was inherited, eugenics were a way to justify his social position and overcome disappointment stemming from his own failure to produce children. In the 20th century, the Nazis' use of eugenic ideas to justify forced sterilization and racial genocide destroyed any credibility Galton's work may have had. But insofar as scientists and the public continue to debate the relationship of inherited qualities to behavior and intellectual achievement, his legacy is an enduring one.

For further reading: D. W. Forrest, *Francis Galton: The Life and Work of a Victorian Genius* (Taplinger Publishing Co., 1974); and Daniel J. Kevles, *In the Name of Eugenics: Genetics and the Uses of Human Heredity* (Alfred A. Knopf, 1985).

GALVANISM Known alternately as bioelectricity or electrophysiology, the study of the electromagnetic nature of living organisms. It stems from the work of Luigi Galvani (1737–98), 18th-century physician and professor of anatomy who indirectly discovered animal electricity. When in 1792 Italian physicist Alessandro Volta (1745–1827) proved that Galvani's famous kicking frog-legs experiment was the product of bimetallic conductivity, and not of true animal electricity, Galvani touched the skinned leg of a dissected frog to the animal's exposed spinal cord, causing the other leg to twitch. Galvani, believed that he had proved the vitalist theory of the *anima*, an electromagnetic and quasi-spiritual force that pulsed through living material. He had actually discovered the phenomenon known as the current of *injury*, which is the electric nervous signal that pulses from damaged tissues.

The electric nature of the body was largely ignored or disbelieved for long periods following Galvani's experiment, especially in the wake of further biological discoveries supporting the mechanistic standpoint of a purely chemical physiology.

Thus, even the scientists of very recent years dismissed the experiments (and successes) of maverick researchers who manipulated simple animals such as hydras, medusae, and flatworms into regeneration on reserved polarity. For example, scientists could make heads sprout from the tail ends of severed body segments.

Nevertheless, after a few strategically published papers, acceptance of electrobiology has widened. Researchers have proved that stimulation by either direct current or electromagnetic fields affects behavioral and emotional patterns, circadian rhythms, and bone growth. Bone is now considered piezoelectric; that is, if stressed or broken, it emits electricity. This has had a major influence upon how we now view skeletal growth.

Galvanism has revolutionized research on complete nerve, organ, and limb regeneration, cancer prevention and excision, anesthesiology, and bacteriology. Electric pulses from the front to the back of the head of humans or other animals can render them unconscious and throw them into deep sleep/hypnotic states, only to revitalize them in seconds with a counter signal. One can experience the refreshment of a full night's sleep in only two hours with an electrical signal administered to the brain. Although human testing has been minimal and has been used in the medical sphere only as last resort, galvanic therapy has proved its almost miraculous validity again and again.

Due to its equal potential to harm, as well as heal, electrophysiology was, and still is, a matter of national security. The invidious effects of electromagnetic fields must be studied, both for military considerations, and for the more domestic fact that our day-to-day environment waxes increasingly electric.

With its origins reaching back to the dawn of life and its implications branching innumerably into the future, galvanism will continue to be a field of scientific research.

See also ELECTROMAGNETIC FIELDS.

GANZFELD PROCEDURE German for "total field"; a test for psi (parapsychological) communication that attempts to neutralize all known possibilities of sense perception of a mentally transmitted target. Pioneered in psychology by Herman Witkin, the ganzfeld was modified for parapsychological use and introduced to that field by Charles Honorton and William Braud in their 1974 and 1975 studies, respectively.

The standard ganzfeld procedure involves one or several experimenters, who instruct and monitor the subjects, and a sender-and-receiver pair, each of whom is sequestered in a soundproof room. The sender is shown, either tachistoscopically or continually, pictures or film sequences to be communicated mentally to the receiver, who is in a nearby and separate room. The receiver is in a near-total state of sense deprivation, as white or pink noise (a loud hissing which blocks out other sounds) is broadcast to him or her through heavily padded headphones. The receiver's eyes are covered with Ping-Pong ball halves, which fill his or her field of vision with only uniform white or red light, cast from a lamp or lamps about four meters distant.

During a session, the receiver should enter an ALTERED STATE OF CONSCIOUSNESS induced by the ganzfeld and, prospectively, be able to perceive the sender's psi "transmissions." The receiver is admonished to voice his or her thoughts, in a stream-of-consciousness manner, as they arrive. After the session, the receiver and an experimenter may review the receiver's narration, and the former will choose, from three controls and the actual target, what he or she believes the subject being sent was.

A survey of most of the ganzfeld studies shows that the probability of the receivers' correctly selecting their targets at the recorded rate is millions, if not billions, to one—even when conservative figures are incorporated. The full implications of this apparent success, critically and statistically, have yet to be fully explored. Marked relationships exist between sender/receiver acquaintance, personality introversion and extroversion, gender, and belief or disbelief in extrasensory powers. The studies of Charles Honorton are perhaps the most comprehensive and potentially conclusive.

Other studies, following Honorton's and others' increasingly stringent standards, conducted at independent institutions, have yielded even more significant results, pointing to the existence of psi. Critics such as Ray HYMAN argue strongly that errors in randomization, documentation (unremarkable studies often remain unpublished), and test security have skewed results toward the positive, but most agree that the ganzfeld procedure is PARAPSYCHOLOGY's most promising prospect for a repeatable, scientific experiment.

GARABANDAL VIRGIN An appearance of the Virgin Mary reported in the small Spanish town of Garabandal. After morning Mass on Sunday, June 18, 1961, two young girls, Conchita, aged 12, and Marie Cruz Gonzales, 11, were joined by their two friends Loli and Jacinta, both 12, in the schoolmaster's garden where they were stealing apples. On the way home, probably feeling guilty, they began picking up stones and throwing them towards their left sides in the belief that this would ward

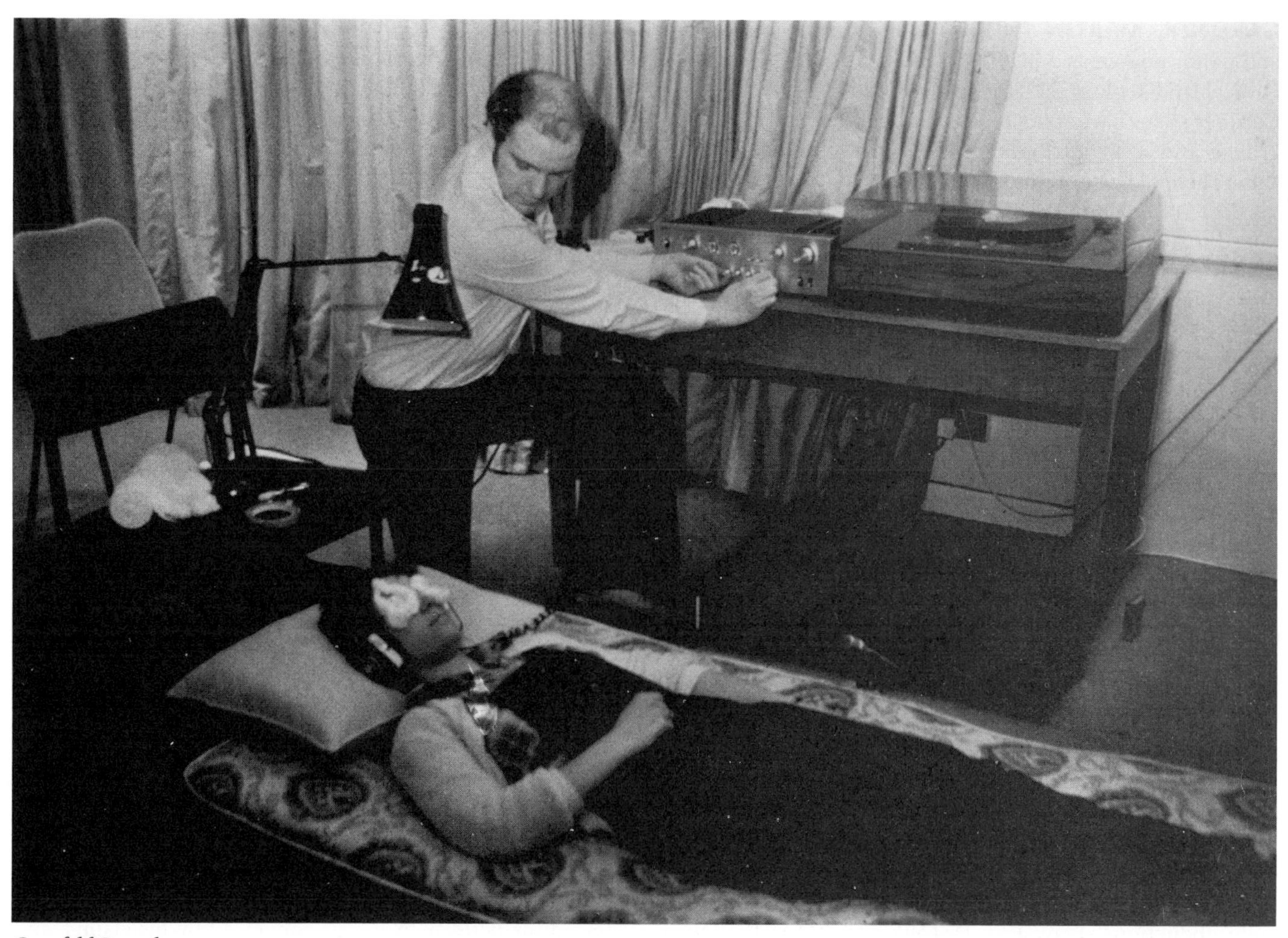

Ganzfeld Procedure

off the devil. Sitting down to rest, Conchita saw "a very beautiful figure . . . shining brightly." When she drew the other girls' attention to the figure, they too were persuaded that they saw it. Conchita said it wore a long, seamless blue robe, had big pink wings, and looked about nine years old. Returning to the same spot the following day, they all claimed to have seen the figure again, and it is reported that the children began to fall into trances when they were in church. On July 1 an angel appeared and told them that the Virgin would appear the next day; this and subsequent events were recorded in Conchita's diary. When the Virgin appeared as promised, Conchita described her this time as being very beautiful and 18 years old, and said that she had an angel on each side of her, and in addition on her right side a very large eye, which Conchita maintained was the eye of God.

When the news spread, crowds began to come to Garabandal; it is said that they all believed the girls' story because they seemed to be able to communicate with each other from afar; moreover they claimed that they could walk backwards, arm in arm, along dangerous mountain tracks where they might have been expected to fall. On July 18 Conchita fell to her knees in the street and a luminous white communion wafer appeared to place itself on her tongue. This incident was filmed. The girls are said to have had as many as 2,000 visions, and Conchita announced that the Virgin told her that many cardinals, bishops, and priests were "on the road to perdition"; she promised a miracle would follow a worldwide proclamation of the truth of the vision. The Catholic Church was unimpressed and has declined to give any credence to the manifestations. A claim of similar Marian appearances was made by young girls at FATIMA, Portugal, in 1917. This earlier claim, which included an extraordinary aerial phenomenon, was upheld by the Roman Catholic Church. The Garabandal visions were later branded hoaxes.

For further reading: Sandra L. Zimdars-Swartz, *Encountering Mary* (Princeton University Press, 1991).

GARDNER, MARTIN (1914–) American journalist and writer known as one of the premier authors of mathematical and logical puzzles and conundrums. Although Gardner spent part of his career as a reporter for the *Tulsa Tribune* and as a contributing editor to the children's magazine *Humpty Dumpty,* his greatest exposure came through his 1957–82 column in *Scientific American,* which introduced many people to the joys of mathematical and logical puzzles. Many of his puzzles and columns from the magazine were so popular that they were collected and published separately in book form.

Gardner is also a professional science writer and ranks with the late Isaac Asimov (1920–92) as one of the greatest explainers of science to nonscientists. *Relativity for the Millions* (1962), for example, looks at Einstein's concept of the relativity of time, space, and matter in a way that is easily understandable by most readers, and *The Ambidextrous Universe* (1964; revised edition, 1990) discusses symmetry and other aspects of astronomy. Other volumes explore literary history, providing sidelights on famous literary works such as *The Wizard of Oz and Who He Was* (1957), *The Annotated Alice: Alice's Adventures in Wonderland and Through the Looking Glass* (1960), and *The Annotated Night before Christmas* (1991).

In his position as an explainer of science, Gardner has launched many investigations into what he views as pseudosciences. Of these, perhaps the best known is his first book, *In the Name of Science* (1952; published as *Fads and Fallacies in the Name of Science,* 1957), which takes a debunking look at such phenomena as flying saucers and alien abductions, DOWSING, LYSENKOISM, ATLANTIS AND LEMURIA, the curious case of BRIDEY MURPHY, and many others. Gardner has continued to publish examinations of pseudoscientific cults and phenomena, including *Science, Good, Bad, and Bogus* (1981), *How Not to Test a Psychic* (1989), and *On the Wild Side* (1992). Gardner is also a founding member of the executive council of the COMMITTEE FOR THE SCIENTIFIC INVESTIGATION OF CLAIMS OF THE PARANORMAL.

GAUQUELIN, MICHEL (1928–1991) French psychologist who attempted to use statistics to prove the validity of ASTROLOGY. Gauquelin, who earned scientific degrees at the Sorbonne, became interested in astrology at an early age, and wished to discover whether it could be substantiated scientifically. While still at the Sorbonne, he began to apply statistical analysis to astrological principles and found no scientific support for astrology. Then he began to use famous physicians' birth data and discovered an astrological correlation. Follow-up studies of the birth dates of more than 30,000 people in a dozen occupations (physician, military leader, athlete, and so forth) reinforced his findings that certain planetary positions at the time of birth correlated with certain professions. Gauquelin's research and personal beliefs did not support most of the tenets of astrology—only this one aspect: that the stars under which one is born affect that portion of one's personality that leads to certain career aptitudes.

Gauquelin claimed to be able to replicate his results, one of the basic requisites for scientific proof of a theory. His findings came to be called the Gauquelin Effect, or the Mars Effect (he found a larger proportion of superachievers than was statistically likely to be, born near the rise or culmination of Mars). If Gauquelin was correct, long-held scientific skepticism about the impossibility of distant physical bodies—the stars—having a concrete effect on the lives and personalities of those living on Earth would be overturned. Although some other scientists claimed to have replicated Gauquelin's findings, these results were not found to be legitimate by the mainstream scientific community. Ultimately, unable to attain the scientific legitimacy he sought, he committed suicide and requested that all of his work be destroyed. Today the Gauquelin Effect is not accepted by mainstream science.

GELLER, URI (1946–) Israeli self-styled psychic.

Soon after serving an obligatory period in the Israeli army, Geller developed a MAGIC act. He claimed that his apparently magical powers were not produced by the usual stage illusionist's methods—trickery, misdirection, and/or deception—but were evidence of ESP—that he was psychic and had paranormal powers. Andrija Henry PUHARICH witnessed his performance, was convinced by his claims, and took Geller off to his PARAPSYCHOLOGY laboratory in California as impressive evidence of the reality of ESP.

Geller's performances in the laboratory and on television, particularly in the United States and Britain, attracted much attention, but professional and amateur magicians remained unconvinced, seeing him as not a particularly good stage illusionist. Some scientists, John G. TAYLOR and Jack Sarfatti, for example, were initially persuaded that Geller was psychic and paranormal and could be used as evidence for the existence of ESP.

Geller was the subject of two separate investigations in 1974, one by the Stanford Research Institute (SRI), the other by the *New Scientist* in collaboration with colleges of the University of London. The SRI inquiry dismissed most of Geller's claims but accepted that he displayed evidence of TELEPATHY and CLAIRVOYANCE. The *New Scientist* team concluded that SRI's controls had been inadequate and that Geller was "simply a good magician." In 1978, the *New Scientist* published an interview with Yasha Katz, at one time Geller's manager, who described some of the trickery Geller used.

See also DERMO-OPTICAL PERCEPTION; METAL BENDING; THOUGHTOGRAPHY.

Geller (seated)

GENDER THEORY The theory that entails investigation of the relationship between sex and gender. *Gender* comes from the Latin verb *generare,* "to beget," and the Latin stem *gener-,* meaning "race" or "kind." *Gender* signifies sort, kind, or class and has been used in this sense continuously since at least the 14th century. The term *gender* adheres closely to the concepts of sex, sexuality, and sexual difference. The terms *male* and *female* usually refer to biological differences, while the terms *masculine* and *feminine* usually refer to social and cultural roles.

Gender is seen as the product of social interactions rather than a set of traits derived exclusively from biological sex. Gender classification is one of our primary ways of understanding ourselves and others and of giving meaning to human behavior. Gender is not something that one has; it is something that one does—specific cultural contexts produce gender. It is not a by-product of biological sex. Thus "doing" gender means achieving one's gender through a complex set of perceptual, social, and political activities. In other words, gender is an emergent feature of social arrangements. (Some scholars also maintain that sex, too, is a socially constructed category. However, not all gender theorists hold this position.)

Historically, our views on gender have depended on our belief that women and men naturally belong to distinctly different categories within the species. We believed that women and men exhibited different emotional and behavioral patterns and that these patterns could be predicted on the basis of their respective reproductive functions. For example, we have historically viewed male bodies as more muscular and as physically stronger than female bodies. From this view came the belief that males should therefore be less emotional and more aggressive than women. Because women physically bore and had the capacity to suckle their young, we have connected women with the role of nurturer of children. Our perception of their biological capacities has also fostered the belief that women were inherently demure and compassionate.

These differences between males and females extended into divisions in work, in the family, and in expected roles, behaviors, and responses in society; further, these divisions appeared to be a natural reflection of an objective and indisputable biological reality. Gender theorists argue, however, that although size and strength are actually quite comparable between men and women, societies encourage pairings in which the male will be visibly larger, stronger, and older than the female.

It was in the mid-1930s that social scientists first began to explore the thesis that gender is socially constructed. The first expression of this thesis was the theory of gender roles. Gender roles are socially manufactured guidelines dictating socially acceptable expressions of gender identity. Often called role theory, such research explored how we learn gender behavior, and was a radical departure from the earlier notion that gender traits were straightforward extensions of biological traits. Role theory emphasized the extent to which a society encourages the construction and enactment of gender roles. It was an important development in gender theory because it examined systematically the notion that gender is not irreducibly derived from sex.

However, in the 1970s role theory came under criticism. A new generation of theorists argued that seeing gender as a *role* is inadequate to account for the influence of gender on the many other roles each person plays in society and, furthermore, that it obscures the power relationships that influence the enactment of gender.

A newer theory, generated in the 1970s and continuing today, is what is now meant by *gender theory.* Like role theorists, contemporary theorists argue that we cannot regard a world of "two sexes and two genders" as an irreducible fact, as previous *scientific* work on gender has done. To do so is to ignore both individual responsibilities for creating the world we live in, as well as the complex power structures that shape social interactions. So contemporary theorists, moving beyond role theory, maintain not only that social interactions are the basis for gender, but also that gender, in turn, is the basis for all scientific work on gender and sex; two naturally occurring sexes and two corresponding genders are neither natural nor indisputable.

The sense of objectivity and reality associated with the work of traditional science in fact arises through social interaction and not, as previously thought, because it is a mirror of reality. Traditional science has viewed gender as ultimately based on the biological imperative of reproduction. By contrast, gender theorists argue that until we realize that all aspects of gender, including the physical, are social constructions, we cannot reform these propositions. They believe that reform is essential because traditional attitudes are the basis for gender assignments and gender roles, which are unenlightened at best and oppressive at worst. These include, but are not limited to, societal attitudes toward work, family, emotional responses, and moral worth.

Traditional conceptions of gender have been androcentric, for example, centered on the ideal of maleness. This androcentrism arises not only from traditional science, but also from the predominant creation myth of the West: the Genesis story in the Hebrew Bible. We have therefore come to regard females as the second gender and consequently have not valued the qualities associated with being female as strongly as male qualities.

Herein lies the conflict: On the one hand, traditional science, seen by gender theorists as pseudoscientific, and Western tradition, mutually reinforcing, underpin a male-dominant society; on the other hand, gender theorists maintain that the recognition that gender is socially constructed will liberate men and women alike from a situation that privileges male characteristics and attributes, while ignoring others entirely. Contemporary theory reveals new possibilities for rich, full, and equal lives for men and women, informed by new conceptions of gender and engendering.

For further reading: Donna Haraway, *Primate Visions: Gender, Race, and Nature in the World of Modern Science* (Routledge, 1989); and Carol MacCormack and Marilyn Strathern, eds., *Nature, Culture, Gender* (Cambridge University Press, 1980).

GEOCENTRISM Belief that Earth is at the center of the universe. The geocentric theory of the universe dominated European astronomical thinking for more than two millennia. The original geocentric theory—that Earth is stationary and that the entire universe revolves around it—is sometimes attributed to the Pythagoreans, an ancient community established in Croton in Southern Italy by Pythagoras in the sixth century B.C.E. But the idea is probably much older, perhaps dating back to ancient Egyptian or Babylonian times. Pythagoras believed that the entire universe could be described in terms of numbers and used his famous theorem about the relationship of the sides of a right-angled triangle to combine numbers and geometry. His work formed the basis for trigonometry and allowed later philosophers to chart the positions of stars.

Although some ancient thinkers challenged the geocentric view of the universe—philosopher Aristarchus of Samos, for instance, proposed a heliocentric (Sun-centered) model in the third century B.C.E.—most accepted the view of Earth as the center of the universe. Most influential among Pythagoras's followers were the fourth-century B.C.E. philosopher Aristotle, the second-century B.C.E. astronomer Hipparchus, and the second-century C.E. astronomer Ptolemy. Aristotle was a polymath who studied and wrote about topics ranging from biology to literature. He was also the tutor of Alexander the Great of Macedon. Because his views were so widely spread through Alexander's conquests, Aristotle was accepted as the voice of authority by most of the European and Mideastern world.

The Greek astronomers Hipparchus and Ptolemy used Pythagoras's trigonometry to develop star charts that would, they hoped, help sailors navigate out of sight of land. Hipparchus developed a new method of mapmaking called stereographic projection to chart the position of the stars in the heavens. Ptolemy observed irregularities in the motion of the planets and explained them with a series of tiny circles within circles, called epicycles. He also expanded on Hipparchus's star chart in his *Almagest,* which set the standard for celestial observation for over a thousand years.

See also HELIOCENTRICISM.

GERM THEORY The theory that diseases are caused by minute living organisms—germs—that attack the body's normal functioning. In the 19th century, Louis PASTEUR showed that fermentation and putrefaction are caused by airborne microorganisms. Joseph Lister (1827–1912) in the 1860s used carbolic acid to keep airborne microorganisms from infecting wounds, causing putrefaction. In the 1880s Robert Koch (1843–1910) identified the agents that cause tuberculosis and cholera. Despite what now seems to make sense, the medical profession, always conservative, was slow to accept the germ theory, and its resistance to antiseptic conditions in operating theaters and to sterilizing instruments persisted until the turn of the century and beyond.

Counterarguments to the germ theory have surfaced from time to time throughout this century. Naturopaths have led the fight against the germ theory in the 20th century. They believe, instead, that disease is caused by a violation of the natural laws of living, and they offer different treatments from those of conventional medicine.

The naturopaths are not alone. Several religious organizations, including CHRISTIAN SCIENCE and Jewish Science, reject the germ theory; several notable persons also shared this view, including George Bernard Shaw. The general opinion of these dissenters is that germs do not *cause* disease but are a *product* of disease. Cells which have ceased to function properly—those that have become diseased—produce germs that then multiply. Some believe that, at that point, germs become infectious, can be transmitted to other people and cause them to be diseased.

Opponents of the germ theory also do not accept the use of drugs to kill bacteria. Drugs only suppress the symptoms, they hold, and do not cure the underlying ailment.

GERSON THERAPY An alternative form of treatment for cancer based upon the principles of NATUROPATHY. The treatment process was developed in the 20th century by German physician Max Gerson, who had begun his career treating tubercular patients. Working during years prior to the use of penicillin to treat the disease, he observed that a diet of fresh fruits and raw vegetables seemed to lead to a cure. He then observed the efficacious nature of the diet on other diseases. He concluded that natural foods assisted the body in mobilizing its own healing powers.

Gerson's theorizing coincided closely with that of naturopathy, an approach to medicine that largely rejects the approach of modern medicine which attacks disease-causing agents (bacteria, viruses) with specific healing agents (medicines), or attempts to remove diseased tissue through surgery. Naturopathy understands specific diseases as manifestations of the general illness of the body as a whole.

Gerson saw the replacement of potassium with sodium in the modern diet as a major factor in the development of cancer. The overabundance of sodium started a chain reaction that led a malfunction of the liver and a resulting build-up of the toxins that produce cancer. Cancer is cured by restoring the body's natural balance. Specific treatment includes the consumption of fresh fruit, vegetable juices, and herbal fortified soups, coffee enemas, and the withdrawal of meat and milk products.

Gerson published his results on treating cancer, shortly before his death in 1959, in his book *A Cancer Therapy—Results of 50 Cases* (1958). It was not well received by the medical community except by some of his naturopathic colleagues. In the 1960s, Gerson's daughter moved to San Diego, Calif., to promote his ideas through the Gerson Institute. Because the treatment is banned in the United States, the institute works closely with the Hospital de Baja, in Tijuana, Mexico, where various alternative cancer therapies are legal.

See also ALTERNATIVE MEDICINE.

GHOST LIGHTS Strange balls or patches of light whose existence and behavior are sometimes difficult to explain. They are usually white or yellow, but other colors have been seen. Sometimes they appear in the same locality, off and on, for years; and sometimes they appear only once or for a short time. In the United States there are three long-established and persistent sets of lights. The MARFA LIGHTS were first observed near Marfa, Texas, by Robert Ellison in 1883. Another set of lights near Joplin, Missouri, appear regularly. A third set, the BROWN MOUNTAIN LIGHTS, has been seen near Morganton, North Carolina, since 1913.

Rosemary Ellen Guiley in *The Guinness Encyclopedia of Ghosts and Spirits* (1994) describes the common features of ghost lights:

1. They appear in remote areas.
2. They are elusive and can only be seen from certain angles and distances.
3. They react to noise and light by disappearing.
4. They are accompanied by hummings, buzzings or outbreaks of gaseous material.
5. They are associated with local folklore surrounding a haunting due to a terrible accident or tragedy that took place at the site, involving loss of life. For example, a person loses his head in an accident, and his headless ghost returns to the site to look for it. The ghost light supposedly is the ghost's lantern.

Various attempts have been made to explain these lights, some citing normal causes, others invoking paranormal explanations. The Ghost Research Society, in Oak Lawn, Illinois, favors paranormal origins, at least for those cases that are not explained as natural phenomena. The U.S. Geological Survey decided that automobile headlights, fires, and the like were the probable causes. In 1970 the Oak Ridge Isochronous Observation Network showed that some of the Brown Mountain lights could be refracted light from sources beyond the mountain. Other ghost lights have been attributed to spontaneous ignition of ionized gas escaping from geological faults. But, inevitably, many observations remain unexplained by such natural causes. Such unexplained happenings attract paranormal explanations: ghosts, the lights of spaceships from other planets, and so on. On the other hand, because some phenomena remain unexplained, it does not necessarily follow that they must have supernatural causes, however fascinating it may be to suppose so.

See also EARTHLIGHTS.

GHOSTS Spirits or spiritual beings conceived as having survived bodily death and manifesting themselves to the living. They are also called spooks, specters, shades, haunts, phantoms, or wraiths. Ghosts may appear as living beings, nebulous—even luminous—likenesses, or strange lights; they may manifest themselves through the displacement of material objects, sounds played on musical instruments, screams, footsteps, ringing bells, or clanking chains. Sometimes they are fairly specialized: The VAMPIRE, for example, is an evil ghost that rises from its grave by night to renew its unnatural existence by sucking the blood of living persons. Oriental folklore abounds in big-headed hungry ghosts. The German POLTERGEIST is a ghost that manifests itself with noisy tantrums.

Our modern word *ghost,* like *aghast, ghastly, barghest* (a goblin), and *poltergeist,* derives from the Anglo-Saxon *gast,* which itself has been traced back to the ancient Indo-European root *gheis-,* expressive of fear or amazement. Undoubtedly, the notion of ghosts antedates history, getting its start in dreams, perhaps, or in human imagination consciously striving to make sense of the world, overly exercised by mysterious happenings and superstitious dread, or simply engaged in the fashioning of entertaining tales. Because they are based on the ancient idea that the human spirit and body are separable, ghosts figure prominently in many religions, but, as James RANDI has observed, they are also "the favorite subjects of scary stories designed to impress children and some adults."

In primitive planting societies, awareness of and dependence on natural cycles encourages the belief that ancestors are active in their descendants' lives. Thus, ghosts are usually viewed as benign and are assigned important roles in both ceremonial and daily functions. In hunting societies, which depend upon the arts of killing, ghosts are dreadful beings, to be avoided or exorcised if possible and placated if necessary. Although in both views the dead may deliberately return to communicate with the living, where ideas about death and violence are inextricably linked, ghosts are regarded, in Joseph Campbell's words, as "resenting their dispatch to the other world and now seeking revenge for their miserable state on those still alive." This durable view informs all the countless stories, from *Hamlet* to allegedly factual accounts of hauntings, in which the spirits of murdered persons return to cause trouble for their killers.

Belief in ghosts and haunted houses, widespread in Europe during the Middle Ages, began to decrease during Renaissance times but enjoyed a resurgence due to the growth of SPIRITUALISM and PSYCHICAL RESEARCH during the 19th century. The belief persists, albeit in watered-down form, even in the New Age. Here, ghosts are often identified as the astral bodies of people who died without adequate spiritual preparation. They do not realize that they have died and in their confusion they cling closely to the familiar physical world. Such ghosts are more frightened than frightening—so upset as to be pitiable objects. Accordingly, some spiritualist churches have established "rescue circles" exclusively devoted to earthbound spirits. Through a combination of logic, understanding, kindness, and firmness, the rescue circle attempts to persuade the earthbound spirit that it is, in fact, dead and ought to move on.

This approach, too, partakes of ideas that are old and widely held; for example, among the Ainu of northern Japan, the deceased is enjoined not to think about its still-living relatives and friends; moreover, it is given many presents and told to hasten with them to its ancestors, who are eagerly awaiting the gifts. Still more elaborate preventive measures may be taken to ward off the attentions of ghosts, such as removing a coffin from a house of mourning, not through a door, but through a hole made in the side of the house. While burial proceeds, the hole is repaired; thus, even should it ignore injunctions not to yearn for the company of the living, the ghost cannot find its way back into the house. The 19th-century German ethnologist and explorer Leo Frobenius described numerous ghost-proofed corpses from Africa and antiquity: To keep their ghosts from roaming, corpses had been bound in ropes, nets, or bandages, and heavy stones has been heaped on the graves; in some instances, body orifices had been stopped to trap ghosts inside corpses. Among the Aranda of Australia, a village where someone has died is razed, none may speak the name of the deceased, and surviving family members undergo ritual ordeals, all to placate the ghosts and ensure friendly relations with it—if relations there must be.

Stubbornly malign ghosts such as vampires require particularly drastic handling. According to Central European tradition, a wooden stake must be driven through a vampire's body, the head cut off, and the mouth filled with garlic; then the remains and the coffin should be buried.

See also ASTRAL/ETHERIC BODY.

GIANTS Beings of folklore with human form but of superhuman size and strength. Giants occur in many stories and creation myths of the ancient world; they are numerous in Hindu literature and figure prominently in Siamese, Hindu, Persian, and Mongol legends, among others. The ancient Chinese believed the primordial colossal divine man P'an Ku gave heaven and Earth their names, while his tears formed the rivers, his breath the wind, and his speech the thunder. Upon his death, parts of his body formed the five holy mountains of China, as well as the Sun and Moon. In Norse mythology, the giant Ymir and his progeny were overcome by Bor, Odin, Vili, and Ve, who then created Earth from Ymir's body.

Greek mythology tells of earthbound giants who were born of blood from the wounded Uranus, including Alcyoneus, Pallas, Enceladus, and Porphyrion; who renewed the battle against the Olympian gods begun by the Titans, Cyclopes, and the Hecatoncheires; or hundred-handed monsters (metaphysical giants themselves, the offspring of Uranus, or heaven, and Gaia, or Earth). Atlas, who carries Earth on his back, and Chronos, or time, an ogre often depicted devouring his own children, are portrayed as gigantic in Greek legend. The Greeks, along with the Native Americans of the Great Plains and the Scandinavians, believed that giants were the first race to inhabit Earth.

Biblical references to giants are found in the story of David and Goliath, and in Genesis (6:4) and Numbers (13:32–33). An account in the book of Enoch tells of cannibalistic giants, born from the union of heaven and Earth, and vanquished by God through his Archangels, though not before they had taught humankind the arts and sciences. Carl Gustav JUNG suggests that these giants symbolically "enlarged the significance of man to 'gigantic proportions,' which points to an inflation of the cultural consciousness at that period" (Jung, *Psychology*). In medieval European literature, giants are a favorite feature and appear in the tales of Saint Christopher, Beowulf, Sir Gawain and the Green Knight, Tristan and Isolde, Lancelot, and in other Arthurian tales.

Many accounts of "real" giants have been recorded. Magellan claimed he saw the "Giants of Patagonia" among the Tehuelche people of the New World, though no scientific evidence supports this sighting. To the small-framed European explorers, settlers, and conquistadors, tall, muscular peoples such as the Watusi of Africa or the Mississippians of the American Southeast must have seemed to be giants. Some reports of giants are no doubt linked to gigantism, a genetically transmitted disease caused by the oversecretion of growth hormone by the acidophilic cells in the anterior lobe of the pituitary, frequently accompanied by mental retardation. Persons suffering from this disease exhibit extreme height—seven feet and more—and are often short-lived. Historically they have served as soldiers or bodyguards to kings and potentates or have been displayed in freak shows and elsewhere as "monsters."

GISH, DUANE T. (1921–) A leading exponent of CREATION SCIENCE; vice-president of the Institute for Creation Research, based in California, and author of several books, including *Evolution? The Fossils Say No!* (1979), and its later version *Evolution: The Challenge of the Fossil Record* (1985).

Gish believes that the evolutionist's search for the origin of the world is hopeless because God created the world by processes which lie outside our understanding. Gish sees evolution as undermining religion and religious teaching, which are essential, in his view, to giving human beings—and in particular, children—proper moral and ethical standards. In his opinion, the story of creation in the Book of Genesis must not be challenged and must be accepted as literally true; otherwise religion will be in danger.

Gish's Institute for Creation Research (ICR) is active in providing scientific proofs of the Genesis account and of finding flaws in, or disproofs of, evolution. The Institute has a graduate school, offers Master's degrees, publishes pamphlets, books, produces and sells videos, and so on—all to further these two objectives.

The case advanced by Gish and his fellows in the ICR is an outstanding example of CONVICTION SCIENCE: They believe that the Genesis story is true, and they set out to find evidence to back up that belief, while marshaling arguments that will discredited their perceived rivals, the evolutionists. Gish's followers claim that theirs—creation science—is the true science, and demand that, at the very least, it should be taught alongside and be and of equal standing with the evolutionary account.

For further reading: Christopher P. Toumey, *God's Own Scientists* (Rutgers University Press, 1994).

GOAT GLANDS Just one of the many pseudoscientific "cures" that form part of the long history of rejuvenation medicine. Rejuvenation medicine usually deals with methods used to overcome erectile dysfunction. These methods can include certain foods, potions, and magical spells.

Like all the other supposedly sexually stimulating substances, there is no evidence whatever that goat glands or glands from other animals can be successfully transplanted into the human male to improve his youthful vigor and potency. Nevertheless credulous men came forward in the 1920s to have the operation performed by "Doc" John Romulus BRINKLEY at his clinic in Milford, Kansas. As the caprine testicle is relatively small, Brinkley transplanted it whole into his patient's scrotum and anchored it near the spermatic cord. For his rejuvenation transplant, Brinkley charged a minimum of $750; if the glands of a very young goat were used, the charge went up to $1,500, the cost in those days of a small family house.

The operation was performed in four phases. First came the grafting phase. The second phase was the Steinach operation to inject mercurochrome into the cut ends of the vas deferens. The urine would have been colored for quite a few days by this antiseptic, and the patient was told to watch out for this as a sign that things were going well. The last two phases entailed detaching a small artery and a nerve from the graft and hooking them onto the patient's own testis. Brinkley claimed that the last phase of the operation nourished and energized the implanted gland, but there is no evidence that nerves are necessary for hormone production. There is also no evidence that blood could have passed through the glandular tissues of the implant.

Brinkley became a millionaire and owned a fleet of cars, several yachts, and a private plane. Realizing the power of radio advertising, he built his own powerful, far-ranging station to send his message into every home. Along with his daily chats about the benefits of goat glands, he promoted country-and-western music and fundamentalist religion. He insisted that his patients give up alcohol and tobacco, which must have accounted for some of his successes. Nevertheless as time went on, the "Doc" came under increasing attack from the medical establishment, and in 1929 the Kansas Medical Board, pressed by Dr. Morris FISHBEIN, revoked Brinkley's medical license. Although the scam was exposed, Brinkley still had a considerable following, and in the 1930s he ran three times (all unsuccessfully) for the governorship of the State of Kansas. But his opponents were closing in on him and soon his principal means of publicity, his radio station, was taken from him.

Undismayed, Brinkley eventually moved to Arkansas. There he operated on patients with prostate trouble, using the Steinach operation in the hope of stimulating the Leydig cells and so preventing further growth of the prostate. The operation was medically useless but very profitable. Feeling that he had the situation well under control, Brinkley brought a legal suit for libel against Fishbein, his main adversary, and lost, being declared bankrupt by his own petition. The following year the U.S. Postal Service filed a complaint against him for spreading false claims about his goat-gland operation through the mail. Following his indictment for mail fraud, Brinkley moved to Mexico, where he ended up pretending to perform prostate operations and sold worthless medicine made up of a blue dye and a little hydrochloric acid.

See also APHRODISIACS.

For further reading: Roger Gosden, *Cheating Time: Science, Sex and Aging* (Macmillan, 1996).

GOBINEAU, COMTE JOSEPH DE (1816–1882) A French nobleman whose race theories influenced the Nazis. Gobineau believed that some races were superior to others—the tall, blond, blue-eyed Nordic type was the most superior; the brown Negroid type was the least.

Composer Richard Wagner (1813–83) was attracted by Gobineau's theories; they were also seized on and developed by his son-in-law Houston Stewart CHAMBERLAIN, who turned them towards German nationalism and anti-Semitism in his *Foundations of the Nineteenth Century* (1899). During the next half-century, these ideas and distortions of these ideas became the ideology of the Nazi party, culminating in World War II and the Holocaust.

See also NAZI RACISM; RACIAL THEORIES.

GOETHE'S COLOR THEORY The theory of color as proposed by Johann Wolfgang von Goethe (1749–1832). In addition to being an eminent German playwright and poet of the late 18th and early 19th centuries, Goethe also made contributions to science. One of these was an attempt to overthrow the orthodox Newtonian account of how normal white light is actually compounded out of the color rays of the spectrum. Using a prism and seeing rainbow hues only at the edges of objects, Goethe concluded that Sir Isaac NEWTON's theory, as he vaguely remembered it, was wrong. Newton had been *torturing* that holy thing "Light," he had been "putting Nature on the rack."

Newton, he went on to argue, had been misled by the excessive abstraction of mathematical forms of thinking. His own theory, following Aristotle, assumed the primacy of natural human sensation, as in the doctrine that white is a simple sensation and cannot be compounded out of colors. Goethe claimed that colors are produced by the interaction of light and dark, as dark mountains appear blue in the distance when seen through stray white light in the air and as the setting sun appears red when seen through the relatively dark sky. Goethe's powers of literary expression gained him many followers, but Newtonian scientists regarded this as a good example of the self-delusion of an individual made too self-confident by the popularity of his literary works and who would have benefited from a judicious guide.

For further reading: J. W. von Goethe, *Theory of Colours,* translated by C. L. Eastlake (MIT Press, 1987).

GOULD, STEPHEN JAY (1941–) A major contributor to the pseudoscience debate. His various books and articles, all aimed at the general reader, make clear what is acceptable as legitimate science and what is not. At the same time he shows that scientific understanding is open to amendment as our knowledge and comprehension grow. So, for example, the whole new range of understanding that the fossils of the Burgess shale opened up is very clearly explained in his book *Wonderful Life* (Penguin, 1991). Gould is an evolutionist, adamantly opposed to a creationist explanation of our world.

Gould holds a dual professorship at Harvard University in geology and zoology and is also an historian of science. In this latter capacity, he is critical of the tendency to write off past theories as pseudoscience.

In *The Mismeasure of Man* (Norton, 1981), Gould gives a detailed account of how ideas about human intelligence developed, through craniometry and phrenology, to present-day theories. He shows how the pressures and prejudices of each time influenced the current thinking—and continue to do so. Such psychological theories evolve in response to current requirements: a misogynist viewpoint, immigration policy, recruitment, selection for education, and so on. To the extent that theories are constructed to satisfy some prior social constraint, often to the neglect of the data—or by treating the data selectively, or even producing or supporting data fraudulently—Gould argues that much, if not all, of intelligence theory and intelligence testing comes under the heading of pseudoscience.

Gould is also a contributor both to the *Skeptical Inquirer* and the *Skeptic*, magazines that debate the range of pseudoscientific issues. He is a member of the editorial board of *Skeptic* and a Fellow of the COMMITTEE FOR THE SCIENTIFIC INVESTIGATION OF CLAIMS OF THE PARANORMAL (CSICOP).

See also EVOLUTION, PROGRESSIVE.

GRAHAM, SYLVESTER (1794–1851) American dietary reformer, minister, and founder of Grahamism. Sylvester Graham was born in West Suffield, Connecticut, into a family with a long history of service in the ministry and as physicians. He worked in a variety of occupations, including farmhand, clerk, and teacher, before contracting tuberculosis in the 1820s. He attended Amherst Academy in 1823 to study for the ministry and was eventually ordained a Presbyterian minister. By 1830, he was serving in the Presbytery of Newark, New Jersey. In that year he was also made general agent of the Pennsylvania Temperance Society, with the aim of promoting abstention from alcohol.

By 1830 Graham had combined his ideas about sickness, temperance, and health into a system of health maintenance and personal hygiene that came to be called Grahamism. Graham advocated a diet including rough, stone-ground whole wheat and other cereals, fresh vegetables and fruits, and pure water. He also recommended cold showers, loose clothing, sleeping on hard mattresses, fresh air, and abstention from sex and alcohol. A practiced orator, Graham drew huge crowds to his lectures partly because of his speaking ability and partly because of his frankness in speaking about bodily functions. At one point, while lecturing in Boston, he was attacked by angry bakers who believed he was ruining their business with his opposition to factory-made breads.

Graham was also ridiculed and lampooned by the popular press, although he was celebrated by Ralph Waldo Emerson as the "poet of bran bread and pumpkins." Graham's published works included *A Lecture on Epidemic Diseases General and Particularly the Spasmodic Cholera* (1833), *The Young Man's Guide to Chastity* (1834), *The Aesculapian Tablets of the Nineteenth Century* (1834), *A Treatise on Bread and Bread-Making* (1837), and the two-volume *Lectures on the Science of Human Life* (1839). Today he is best remembered as the developer of the Graham cracker. He died in 1851 after undergoing a dose of stimulants to combat his own poor health.

For further reading: Richard W. Schwarz, *John Harvey Kellogg, M.D.: Father of the Health Food Industry* (Southern Publishing Association, 1970).

GRAPE CURE The widespread belief for centuries throughout the grape-growing areas of Europe that grapes, grape juice, and grape products generally have curative properties. It has not been a dominant belief but one of those examples of folklore in the background of all cultures, a cure in the same category as goat's milk, honey, and much else. Its advocates might well point to the recent evidence that a moderate intake of wine has beneficial effects on health as validating their belief.

In the early 20th century, Bernarr Macfadden (1868–1955) claimed that a grape diet would cure cancer. The publisher of a number of health magazines and author of the massive five-volume *Encyclopedia of Physical Culture: a work of reference, providing complete instructions for the cure of all diseases through physcultopathy* (1912), he offered a prize of $10,000 to anyone who could prove that a diet of grapes did not cure the disease. Proving a negative is notoriously difficult, and the prize remains unclaimed to the present.

GRAPHOLOGY The study of handwriting, especially to assess the writer's personality, character, and/or abilities. The first attempt to analyze handwriting formally was probably that of Camillo Baldi, an Italian scholar and physician. His book *Treatise on a Method to Recognize the Nature and Quality of a Writer from His Letters* was published in 1622 while Baldi was a professor at the University of Bologna. Early scholarly and professional interest in graphology was minimal simply because very few people could read and write. As education became more widespread in the 19th century, interest in handwriting analysis rapidly increased, though at first more as an art than as a science. Johann Wolfgang von Goethe, Edgar Allen Poe, Honoré de Balzac, Charles Dickens, and other figures of this period in history dabbled in graphology.

The serious study of handwriting was undertaken in France by Abbé Jean Hippolyte Michon. He published the most scholarly work on handwriting up to that time, *The Practical System of Graphology* (1875) in which he coined the generic term for handwriting analysis. Michon's system identifying hundreds of graphic signs that were supposed to indicate specific personality traits came to be known as "the school of fixed signs." In Germany, graphologist Ludwig Klages coined the term "expressive movement" to refer to all motor activities performed habitually and automatically, without conscious thought. Klages's system of graphology, however, was esoteric and subjective, intuitive in the extreme, and therefore of dubious validity to serious scholars.

Modern graphoanalysis was founded by U.S. scholar M. N. Bunker in 1929. In the general procedure of graphoanalysis, a "perspectograph" is constructed to analyze the first hundred upstrokes that appear in a sample of writing. This sample is desirably a full page of spontaneous writing on unruled paper, performed without the subject's knowledge that it is being taken for analysis. Each upstroke is measured by a specially constructed gauge, determining degrees of slant from far forward to far backward. These measurements are recorded on a worksheet listing some 100 "primary" personality traits (those which can be determined from a single-stroke formation) and around 50 "evaluated" traits (those inferred from two or more other traits). Primary and evaluated traits are then rated as to intensity and divided into trait groups, including emotions, mental processes, and social behavior. The graphoanalyst then combines this data to yield a unified picture of the writer's personality.

Handwriting analysis has historically been plagued with the label of "pseudoscience," primarily due to the atomistic, intuitive, and loose systems of early graphology. The question of its validity continues to be raised, though graphoanalysis now has certified practitioners in all states and in most countries around the world. The discipline has a wide variety of applications, in business and industry; education; mental health clinics and hospitals; and in forensic work. Numerous validating research efforts have been conducted, findings of which are available from the International Graphoanalysis Society based in Chicago.

GREAT CHAIN OF BEING Hierarchical classification of all life on Earth. It is one of the most persistent conceptions about the nature of life and the universe in Western history. Originally presented by Roman philosopher Plotinus in the third century B.C.E., the theory states that all life—indeed, everything in the known universe—is connected in a single hierarchical structure, like a ladder or a chain. The great chain begins with the smallest single-celled amoeba and continues through increasingly complex animals, including sponges, worms, invertebrates, insects, fishes, amphibians, reptiles, birds, and mammals. Humankind represents the culmination of the great chain, the link between the material and the spiritual world.

Some historians believe that the great chain-of-being theory is a precursor to evolutionary theory. In its earliest incarnation, however, the chain of being was strictly hierarchical, static, and antievolutionary, assigning every creature a particular place in a world created and ordained by God. This chain of being represented the eternal order of things. It attracted conservative thinkers because it described a complex world that was in accord with the Bible. It also appealed to the human ego by placing humankind at the top of the hierarchy, giving humans pride of place at the top of the chain. Other thinkers expanded the chain to include biblical and spiritual creatures such as angels.

The chain-of-being theory does not fit the evidence. If life as a whole is conceived of as a chain, then all of life must be connected by individual links. In nature, however, there are large gaps between different systems—between plants and animals or between invertebrates and vertebrates, for instance. There are no known forms of life that connect some separate lines of the chain. In addition, the chain-of-being theory does not describe variations within a basic pattern, like breeds of dog or races of humans.

Some conservative thinkers went to great lengths to support the chain-of-being theory; for instance, taxonomist Charles Bonnet suggested that asbestos lay somewhere between a mineral and a plant because it has a fibrous structure that looks vaguely plantlike. Charles White, an 18th-century physician and biologist, placed the freshwater hydra as a transitional form between plants and animals. White suggested that animals and plants were related because some plants shed their leaves and some animals or birds shed their hair or their feathers. He also defined a hierarchy among the races of humankind, with his own group (Europeans) on top and black Africans at the bottom.

The great chain of being was discredited after Charles DARWIN produced his theory of evolution by means of natural selection in *On the Origin of Species* (1859). Evolutionary theory did not present the same difficulties as the chain of being. Evolution does not view species as individual links in a chain or ladder, and it does not need intermediate forms between different forms of life. Instead, it presents life as more of a bush than a ladder, with varieties of life related to and branching off from one another, but without the hierarchy implicit in the great chain of being.

See also EVOLUTION.

For further reading: Stephen Jay Gould, *The Flamingo's Smile: Reflections in Natural History* (W.W. Norton, 1985); and Arthur O. Lovejoy, *The Great Chain of Being* (Harvard University Press, 1936).

GREW/MALPIGHI Nehemiah Grew (1641–1712), an English physician, active in plant anatomy and comparative anatomy; Marcello Malpighi (1628–94), an Italian, a pioneer in embryology, plant anatomy, histology, and comparative anatomy. Both were Fellows of the Royal Society of London, the British organization of the scientific élite. In the 17th century, before specialization became common, it was possible for these men, polymaths, to play an influential role in a wide range of disciplines; science had not yet fragmented. Two things connected them: first, there was their interest in the anatomies of plants and animals and their similarities. Grew believed that animals and plants must possess similar structures, as they were "Contrivances of the same Wisdom." Malpighi's researches embraced the physiology of plants and animals. Malpighi is sometimes credited with comparing the movement of sap in plants to the circulation of the blood in animals, but this is a parallel he did not draw. Both published books on plants: Malpighi, *Anatome Plantorum* in two volumes in 1677 and 1679; Grew, *The Anatomy of Plants* in 1682. Secondly, they were connected through the Royal Society. Grew's predecessor as Secretary of the Royal Society, Henry Oldenburg, had corresponded with Malpighi and arranged for the Society to publish many of his books, a link which Grew continued after Oldenburg's death in 1677.

Both men also had ecclesiastical connections. Grew published *Cosmologia Sacra* in 1701, "to demonstrate the truth of the sacred writings"; Malpighi became personal physician to Pope Innocent XII in 1691, gave up his scientific work and moved to Rome. Jointly these men represent an important step in the development of biological science—the move from accepting the classic texts of Aristotle and Vesalius to proceeding by experiment and analysis, based on careful microscopic observation. It was they who first observed and identified the cell as the building unit of plants. Malpighi first observed the movement of blood in the capillaries of the frog's lung and so supplied the missing link in William HARVEY's proof of the circulation of the blood. In this context it is worth noting that Grew's fellow Secretary of the Royal Society was Robert Hooke, responsible above all others for the development of microscopy, and the author of *Micrographia* (Royal Society, 1665; reprinted Dover, 1961), the fundamental classical work on the subject. Moving from blind faith in others' theories to experiment and analysis was a step away from pseudoscience, and a step toward embracing the scientific method.

For further reading: H. T. Pledge, *Science Since 1500* (Peter Smith, 1990).

GUNTHER, HANS FRIEDRICH K. (1891–1968) A professor of anthropology at the University of Jena and the most distinguished academic advocate of the racial superiority of Nordics on which Hitler built his pro-Aryan policies.

Gunther accepted that not all Germans were Nordics—tall, fair-haired, of fair complexion, and blue-eyed—but maintained that there were more pure Nordics than members of other races and of mixed stock. He attributed to Nordic men and women those characteristics thought to be good, and the bad characteristics to the other races, in particular the Jews.

Gunther's apparently scientific justification made Hitler's anti-Semitic ideas more acceptable. He claimed that Nordics were cleaner in their personal habits and appearance, were more athletic, very brave, adventurous, and that they showed a preference for blues and light greens; their women were also very modest—they kept their legs together when sitting in streetcars—this as distinct from women of other races.

He condemned modern art and music as degenerate. During the Nazi era, Germany's museums were systematically purged of most of their impressionistic works, as well as works of cubism, fauvism, expressionism, dada, constructivism, and surrealism. An exhibition of purged art was held in Munich in 1937 to convince the German people of its degeneracy.

GURU Spiritual teacher or guide who claims firsthand, detailed knowledge of the invisible spirit world. *Guru* is a Sanskrit word referring specifically to a Hindu teacher, but in a general sense, it can be used to mean any teacher who teaches about matters of primary or ultimate importance—matters relating to a spiritual reality, for instance. Gurus claim to gain this knowledge through personal experience and traditionally pass their spiritual wisdom down directly from master to pupil. Many guru traditions are said to have begun with the Hindu god Siva. The techniques that the gurus teach to their students are called *yogas*.

In the United States during the 1960s and 1970s, many Indian gurus led young men and women in exploring Hindu spirituality, and different yogas won popularity and established themselves in U.S. society. Transpersonal psychology has been built on the study of gurus, and their ability to alter their states of consciousness has been extensively documented.

Some gurus also have a direct effect on their students' spiritual growth. In the discipline called Siddha YOGA, for instance, a student is taught to collect "cosmic energy," also called Shakti or KUNDALINI energy, at the base of the spinal cord. When this cosmic energy is released, the student feels a surge of physical and spiritual pleasure, or bliss. However, Siddha Yoga requires a guru in order to awaken the cosmic energy, either through a word, a touch, a look, or a thought. Because of this need for personal contact, modern gurus are sometimes accused of abusing their relationships with their students.

Although gurus who use yogic practices help many people both spiritually and physically, the scientific community has difficulty accepting their results. Scientists are unable to detect or measure the energies that gurus help students manipulate. In addition, a basic scientific principle states that results must be duplicatable—and some results obtained by gurus can only be obtained by those gurus. Science is not tied to the charisma and personality of a guru.

HAECKEL, ERNST (1834–1919) German zoologist and influential advocate of organic EVOLUTION. Haeckel grew up in Merseburg, Germany, and initially trained to be a doctor, earning a medical degree in Berlin in 1857. Along the way, however, he decided on a scientific career and studied botany, comparative anatomy, and embryology before settling upon zoology as a discipline. Appointed full professor and director of the Zoological Institute at the University of Jena in 1865, he remained at both positions until his retirement in 1909. During his long career he published numerous important scientific works, principally on the subject of marine zoology, and received memberships and honoraria from more than 90 learned societies around the world.

Haekel's academic career was just beginning when Charles DARWIN published *On the Origin of Species* in 1859. Haekel was an early convert to the principle of species transformation. His enthusiasm for evolutionism, however, was colored by philosophical and religious considerations that were not shared by Darwin and, as his critics would note, were not necessarily borne out by the content or evidentiary support for the theory. A monist, Haeckel believed that the world and all it contains, including nature, human beings, and social institutions, were part of a unified and consistent system. He also believed in panpsychism, or a world soul. These commitments in turn led him to promote a number of highly speculative and often controversial scientific concepts, which were later rejected by most mainstream biologists. These included LAMARCKIANISM, the biogenetic law, the Gastraea theory (common ancestry for all metazoans), and SPONTANEOUS GENERATION. All of these stressed notions of continuity among living organisms or between inorganic matter and organic life that complemented Haeckel's monistic philosophy. They also supported his view of nature as being intrinsically progressive and hierarchical.

Haeckel's influence in German intellectual life was considerable and was not limited to science. Intensely nationalistic, his work toward a more synthetic view of biology, religion, and society for the German Monist League extended his naturalism into social and political arenas. In this context, many of Haeckel's later works lent scientific prestige to RACIAL THEORIES that pervaded European thought in the late 19th and early 20th centuries and contributed to the racial ideology of National Socialism.

See also NAZI RACISM.

For further reading: Daniel Gasman, *The Scientific Origins of National Socialism* (Macdonald, 1971).

HARDY, SIR ALISTER (1896–1985) A biologist, he was the chief zoologist on the ship *Discovery* during the British 1925–27 expedition to the Antarctic. The experience furthered his interest in marine biology, which led Hardy in 1938 to found the Oceanographic Laboratory in Edinburgh and the journal *Bulletins of Marine Ecology* and to write many books and articles. It was an article of his, "Was Man More Aquatic in the Past?" (*New Scientist*, March 17, 1960, 642–45), that led ELAINE MORGAN to write on the aquatic ape theory of EVOLUTION.

Hardy was a practicing Christian, who believed that religion and the doctrine of natural selection could be reconciled. In 1969 he founded the Religious Experience Research Unit at Oxford, and in 1975 he published a book *The Biology of God: A Scientist's Study of Man the Religious Animal*. He claimed that a comprehensive biology of humankind should not ignore religious experience. The question he asked himself was: What is it in our makeup, in our biological system, that makes us need to seek out a religious experience continually? In trying to answer the question, he started with the emergence of early mental life in humans, of consciousness, and ultimately of esthetic, ethical, and spiritual awareness. He looked at many evolutionary theories including behavioral selection, at new discoveries in animal behavior, and also at developments in psychology and PARAPSYCHOLOGY.

Hardy saw a parallel between a dog's devotion to a human and a human's devotion to God, commenting that both we and dogs are animals who have transferred their loyalty and obedience from their former pack leader to a new master. This idea came from Hardy's belief that the very personal relationship between a dog and its human master has a biological origin; so just the devotional love relationship we have with God must also have some very fundamental biological significance.

Hardy was writing his book in the early 1970s when immanentism, the idea of a God within us (as opposed to transcendentalism, the idea of a God outside us) was very much the new theological thinking at Oxford. A convinced evolutionist, he became equally convinced of the belief that man's experience of divinity must be just as much a part of the biological system as the feeling of human affection, and indeed of love, for his own kind. These ideas remained part of Hardy's personal belief system but were never taken up seriously by the scientific community.

For further reading: Alister Hardy, *The Biology of God: A Scientist's Study of Man the Religious Animal* (Taplinger, 1975); and John A. T. Robinson, *Honest to God* (SCM Press, 1963).

HARVEY, WILLIAM (1578–1657) An English physician principally remembered for his demonstration of the circulation of the blood and the action of the heart as a pump. He was educated at the universities of Cambridge and Padua, where he was greatly influenced by Fabricius. The conjoined problem of blood circulation and heart action puzzled him, as it had many others. In the fourth century B.C.E., Empedocles had put forward the idea of the ebb and flow of the blood. Galen in the second century C.E. supported this theory, which was accepted by the medical profession for the ensuing 1400 years.

By Harvey's time Galileo's experiments and the problems of the mining industry and of land drainage had led to an interest in the way pumps functioned. There was also a growing reliance on experiment rather than on blind acceptance of hallowed texts. Anatomical dissection had established that there are one-way valves in the main veins, casting doubt on the ebb-and-flow theory. By 1616, Harvey, backed by careful dissection and experimentation both on animals and humans, showed that the heart pumped blood through the lungs, back to the other side of the heart, and then out through the arteries, returning via the veins. He published his findings in *De Moto Cordis et Sanguinis in Animalibus* (On the motion of the heart and blood in animals; 1628). As Pledge writes: "'De Moto Cordis' is a classic of science. Negatively, it was unencumbered by metaphysics, a virtue then exceedingly rare. Positively, it stands for the entry at once of the quantitative and of the comparative methods into biology."

Harvey had substituted mechanical and chemical explanations for spiritual theories of vital action. In doing so he was challenging the hitherto established authority and disturbing the beliefs of his peers. This inevitably attracted virulent attacks. Gradually, Harvey's ideas became mainstream science. Yet vitalist theories (the doctrine that life cannot be explained in purely physical terms) surfaced from time to time after his death and, occasionally, still do. George Ernst Stahl (1660–1734) put forward vitalist theories resembling those of Aristotle two thousand years earlier. And Hans Driesch (1867–1941) hypothesized a vital principle in the development of the embryo.

For further reading: H. T. Pledge, *Science Since Fifteen-Hundred* (Peter Smith, 1990).

HASTED, JOHN B. (1921–) A physicist and for many years chairman of the Physics Department of Birkbeck College, University of London, who carried out sophisticatedly instrumented experiments on METAL BENDING by children. He found small effects that he could not explain by established physical processes and published several papers describing his work.

See also PSIONICS.

For further reading: J. B. Hasted, "Paranormal metal-bending," *The Iceland Papers: Experimental and Theoretical Research on the Physics of Consciousness,* ed. by A. Puharich (Essentia Research, 1979).

HELIOCENTRISM Theory that the Sun rather than Earth is at the center of the universe. The heliocentric theory of the universe was originally proposed by the Greek philosopher Aristarchus of Samos. Around the year 280 B.C.E., Aristarchus suggested that Earth revolves around the Sun in common with the other planets—an idea that had been considered by the Pythagoreans a few centuries before as a feasible alternative to GEOCENTRISM. The competing geocentric theory, that Earth formed a stationary center and the rest of the universe revolved around it, was given precedence by such major philosophers as Aristotle and Ptolemy. When the Christian church accepted Aristotle's ideas as truth, it also accepted his geocentric model of the universe. This meant that any future thinkers who questioned the truth of the geocentric model would face possible condemnation by the church authorities.

If the geocentric model of the universe had been without practical application, the theorizing might have stopped there. However, seagoing merchants used Ptolemy's *Almagest,* a series of star tables, to calculate the positions of their ships. The church itself used the stars to regulate its calendar, which contained some important feast days, such as Easter, that had to be calculated and recalculated each year. Ancient astronomers and mathematicians had modified the original geocentric theory to account for observed variations in the motion of the planets and the stars. Ptolemy himself introduced a theory of epicycles to explain why some planets appeared to move backwards at certain times of the year.

Despite the threat of punitive action by the Catholic Church, by the 16th century some astronomers were questioning the validity of the geocentric theory. Nicolaus Copernicus (1473–1543), a clergyman from what is now Poland, found that it was much easier to calculate the church's calendar if he began with the assumption that the Sun, not Earth, was the center around which all the planets, including Earth, revolve. His book on the subject, *On the Revolutions of Heavenly Bodies* (1543), inspired other astronomers to challenge both Aristotle's geocentric theory and the church's championship of it.

Chief among the supporters of the heliocentric theory were the Italian Galileo Galilei (1564–1642) and the German Johannes Kepler. Galileo's observations of the planets supported Copernicus's theories. In 1616, however, the Catholic Church banned publication of the heliocentric theory, fearing that it might be connected with the Protestant Reformation. Galileo ignored repeated warnings from the church and in 1632 published a defense of the Copernican heliocentric system. The following year the Inquisition summoned him to stand trial for heresy. Galileo was forced to recant his views and spent the rest of his life under house arrest. Not until 1979 did Pope John Paul II, speaking for the Catholic Church, admit that the Inquisition might have been wrong in condemning Galileo.

Kepler made some important modifications in Copernicus's original theory, pointing out that planetary orbits were not true circles but ellipses. He also devised three new laws of planetary motion, which inspired Isaac NEWTON in the late 17th century to devise his theory of universal gravitation.

For further reading: John Banville, *Doctor Copernicus* (Panther Books, 1984); John Banville, *Kepler* (Panther Books, 1985); Arthur Koestler, *The Sleepwalkers* (Penguin Books, 1988).

HELLERWORK A complex form of bodywork/massage named for its founder Joseph Heller. Heller began as a student of Ida P. Rolf, who had developed a form of massage distinctive for its manipulation of the fascia, the connective tissue between bone joints. Heller became the first president of the Rolf Institute for Structural Integration in 1976. However, he was also attracted to the variation on ROLFING developed by Judith Aston that is generally termed STRUCTURAL PATTERNING. This stresses the natural asymmetry of the body and applies less force in the massage. Heller was also interested in the theories of emotional energy awareness developed by psychologist Brugh Joy. In 1978 he left the Rolf Institute to found The Body of Knowledge, an institute to train Hellerwork practitioners, located in Mount Shasta, California.

Hellerwork fits within the larger perspective of holistic healing, which attempts to treat the body as a whole and looks for continued health, in large part, in keeping the body fit and toned. In this regard, Hellerwork offers sessions featuring rolfing and body-movement exercises in an atmosphere that calls attention to the feelings experienced by the patient as the bodywork is pursued.

See also ALTERNATIVE MEDICINE.

HERBAL MEDICINE Also called botanical medicine and known in Europe as phytomedicine or phytotherapy—the prevention and treatment of disease by the use of herbs. In this context, the word *herb* means a plant or a leaf, flower, fruit, stem, bark, seed, root, or any other plant part that is used to make medicine, aromatic oils for fragrances and soaps, or food flavors. Herbal medicine plays an important part in several ancient but highly developed medical systems, such as traditional Chinese Medicine and Ayurveda from India. Eighty percent of humanity still relies on these for primary health care.

Aloe vera, cayenne, garlic, ginger, ginseng, goldenseal, and many other herbs have long histories of use in the practice of medicine. Use of at least some of these surely predates recorded history. Herbals, manuals facilitating the identification of plants for medicinal purposes, were compiled in India, China, Greece, and elsewhere before the Christian era; the ancient Chinese compilation of almost 1,900 herbal remedies is still considered to be authoritative. The manuscript herbals common in medieval Europe drew heavily on Greek and Roman sources; after the introduction of printing, herbals began to feature woodcuts as botanical illustrations (though these were generally copies of copies, tended toward stylization, and often incorporated odd mythological ideas.) Some works record departures from legitimate herbal medicine into dubious medical theory, such as the doctrine of signature, which held that human ailments could be cured by plants on the basis of supposed anatomical resemblances. Nevertheless, many excellent and accurate articles on herbals were published during the 16th century, and herbal medicine provided a firm foundation for botany, the science of the vegetable kingdom.

Herbal Medicine

Herbs, trees, and shrubs are the source of approximately one-quarter of all modern prescription drugs. The broad spectrum of plant medicines with important medical applications includes chamomile, colchicine (derived from autumn crocus), digitalis (from foxglove), morphine (from the opium poppy), peppermint, and reserpine (from Indian snakeroot). The list of ailments that are amenable to herbal remedies includes headaches, menstrual cramps, sore muscles and other minor aches and pains, symptoms of the common cold and flu, stomach upsets, constipation, diarrhea, skin rashes, sunburn, insomnia, and dandruff.

Although about three-quarters of plant-derived pharmaceutical medicines are used in modern medicine in ways that correspond with their use in traditional medical systems, pure herbs have lost ground in much modern medical practice, particularly in the United States. (In Europe, the tradition of herbal medicine is more deeply ingrained and so more widespread.)

Americans have come to depend largely upon synthetic compounds designed to mimic natural plant compounds. Because these drugs—unlike herbs—are patented, the exclusive rights to sell them are held by drug companies, and there is correspondingly less commercial incentive to invest money in collecting and preparing herbs or testing and promoting the compounds extracted from them. Thus, as little as 1 percent of all plant species on Earth today—estimates range between 250,000 to a half-million species—has been extensively studied for medicinal applications.

See also ALTERNATIVE MEDICINE; AYURVEDIC MEDICINE.

HERMETIC ORDER OF THE GOLDEN DAWN Victorian organization dedicated to the investigation and spread of occultism. Drawing on the theosophist and spiritualist movements, the Hermetic Order of the Golden Dawn (1887–1923) brought together into a single system a wide body of occult material. Under their leaders MacGregor Mathers, Dr. William Wynn Westcott, and Dr. W. R. Woodman, the Magicians of the Golden Dawn tried to organize MAGIC based on a mix of esoteric knowledge, including the legends of Christian Rosenkreuz (founder of the Rosicrucians), the rituals of Freemasonry, and the Jewish KABBALAH. The Golden Dawn and its inner circle, the Order of the Rose of Ruby and the Cross of Gold, attracted many adherents from upper-class and literary circles. Among followers were the Irish poet W. B. Yeats, his friend and supporter Annie Horniman, George Bernard Shaw's mistress Florence Farr, and the self-proclaimed "wickedest man alive," Aleister CROWLEY.

As magicians, the members of the Order of the Golden Dawn were practicing pseudoscientists. Magic is a pseudoscience because it does not rely on experimentation and the scientific method to work out its results. The order was based on one of the most famous frauds of the late 19th century. Dr. Westcott, the founder, claimed to have discovered a manuscript of unknown origin, written in a peculiar script. He deciphered it and said that it gave a brief outline of five magical rituals. Westcott asked his friend Samuel Liddell MacGregor Mathers, an experienced occultist, to complete the rituals. In the meantime, he wrote to a German, Fraulein Sprengel, whose address he found along with the manuscript, asking for her permission to start an English society to perform the rituals. In 1888, Westcott, Mathers, and Dr. Woodman formed the Isis-Urania Temple of the Hermetic Order of the Golden Dawn.

Mathers soon took over most of the management. He spent money freely, was forced to borrow often from fellow members, but demanded absolute loyalty to himself from the rest of the order. Mathers made many enemies among members who resented his high-handed ways. One of these was Crowley, who helped intensify the conflict between the rather meek but popular Westcott and Mathers. In 1900, Mathers accused Westcott of having forged letters from Fraulein Sprengel about founding the society. A year later in England, Mathers himself admitted to the order a couple of charlatans, a Mr. and Mrs. Horos, who were arrested in September 1901 for having raped one of their occult "students." Although small groups continued to follow the rituals of the Golden Dawn for years afterward, the public ridicule following the arrest of the Horos couple destroyed most of the order's credibility.

HEROIC MEDICINE Predominant medical theory and practice in the United States in the early 19th century that stressed PHLEBOTOMY, purgatives, and emetics. Based upon the belief that most diseases caused the overstimulation of various bodily systems, heroic therapy was designed to lower dangerously excited states to more natural, balanced ones. The term "heroic" referred to the aggressive intervention of doctors that was thought necessary to effect these cures. Heroic practitioners often employed harsh medicines such as opium and mercury-based calomel, and their therapeutic techniques included controlled bleeding, leeching, blistering the skin, purging, reduced diets, and induced vomiting. The various therapies were designed to cleanse the body of any offensive matter that might inhibit its normal function, lower fevers and inflammations (which were regarded as dangerous stresses upon the heart and circulatory system), and stimulate the restorative powers believed to reside in internal organs.

The theoretical perspective that informed heroic medicine was widely accepted by both the medical profession and the lay public. Patients and their families often insisted upon heroic therapy in spite of its intrinsic harshness and the many horror stories that circulated about overzealous practitioners who accidentally bled or poisoned their patients to death. Therapies usually produced an immediate and dramatic change in the patient's symptoms, for example, a lowering of body temperature or heart rate that corresponded to the expectations set forth by the theory. Deaths and relapses could be explained away by asserting the implacable strength of the affliction and were perhaps less suspicious in an era with high disease-mortality rates.

By the 1830s heroic practice began to be challenged, although it would be decades before it was displaced as the standard therapeutic philosophy. The popular success of rival healing systems, such as HOMEOPATHY and WATER CURES, spurred on the reform process by cutting into the medical market. Likewise, a new generation of doctors who stressed the self-limiting character of diseases began to speak out against heroic excesses. They advocated milder forms of therapy and championed the natural recuperative powers of the body. In response, many practitioners began to scale back their use of bleeding and severe medicines. Other physicians, however, retained the basic heroic perspective and merely switched their emphasis to a new class of alcohol-based stimulants. By midcentury a new emphasis on clinical observation and advances in microbiology further challenged traditional medical attitudes. In the ensuing decades, this new body of knowledge, termed scientific medicine, consolidated its influence over the medical profession and effectively drained away most remaining support for heroic medical practice.

For further reading: William G. Rothstein, *American Physicians in the Nineteenth Century* (Johns Hopkins University Press, 1972); and John Harley Warner, *The Therapeutic Perspective: Medical Practice, Knowledge, and Identity in America, 1820–1885* (Harvard University Press, 1986).

HEUVELMANS, BERNARD (1916–) "Father of CRYPTOZOOLOGY" and a pioneering writer on the subject of unknown animals.

Inspired by the work of Jules Verne (*Twenty Thousand Leagues Under the Sea*) and Arthur Conan DOYLE (*The Lost World*), Heuvelmans has devoted his life to the discovery of references to unrecognized, mysterious animals in the writings of travelers and scientists. His most famous work, *Sur la piste des animaux inconnus* (1955), sold more than a million copies in English translation as *On the Track of Unknown Animals* (1958). *On the Track of Unknown Animals* looks at reports of unknown animal life ranging from reported sighting of the great American ground sloth in 16th-century South America to sightings of the ABOMINABLE SNOWMAN in 20th-century India. In each case, Heuvelmans brings together the reports and suggests possible explanations for them.

Heuvelmans has also written other volumes on cryptozoological topics, including *In the Wake of the Sea-Serpents* (1968), *L'Homme de Néanderthal est toujours vivant* (Neanderthal man alive, 1974), *Les derniers dragons d'Afrique* (The last dragons of Africa, 1978), and *Les Bêtes humaines d'Afrique* (The human beasts of Africa, 1980). In 1975 he founded the Center for Cryptozoology in southern France, and in 1982 he became president of the newly established INTERNATIONAL SOCIETY OF CRYPTOZOOLOGY. He frequently contributes to the society's journal, *Cryptozoology*.

For further reading: Bernard Heuvelmans, *In the Wake of the Sea-Serpents,* (Hill & Wang, 1968); and Bernard Heuvelmans, *On the Track of Unknown Animals* (Hill & Wang, 1958).

HEYERDAHL, THOR (1914–) Anthropologist and explorer, who made his reputation in 1947 when he sailed the primitive balsa-wood raft *KON-TIKI* from Peru to an island in the Tahiti chain. The *Kon-Tiki* expedition proved that pre-Columbian South American Indians could have reached and settled islands in the South Pacific. It also won the anthropologist a worldwide reputation and brought him justification for his views. Since that time, Heyerdahl has participated in many additional expeditions, ranging from trying to cross the South Atlantic in an African reed boat to researching and excavating sites on the Maldive Islands in the Indian Ocean.

Heyerdahl, a diffusionist, believes that different cultures only rarely develop similar ideas and artifacts. In all his expeditions, he seeks to show that people usually thought to be isolated from other cultures were in fact open to contact. His 1955 expedition to EASTER ISLAND, Aku-Aku, showed that the lonely island had been inhabited from at least 380 C.E. and had experienced at least three waves of migration. In the 1960s, Heyerdahl constructed papyrus reed rafts in Africa and sailed them 3,270 nautical miles across the South Atlantic in what became known as the Ra Expeditions, showing that ancient Egyptians could have traveled to the Americas.

In the 1970s, Heyerdahl took reed boats 4,200 miles down the Tigris River in Iraq, through the Persian Gulf, and across the Indian Ocean to the African port of Djibouti. By doing this, he showed that ancient Mesopotamians could have had sea contact and trade with civilizations as far apart as Egypt and the Indus River valley. In the 1980s Heyerdahl traveled to the Maldive Islands at the request of the government. Excavations there proved that humans occupied the Indian Ocean archipelago in prehistoric times and that they had been an integral part of international trade.

For further reading: see Bernadine Bailey, *Famous Modern Explorers* (Dodd, Mead, 1963); and Arnold Jacoby, *Señor Kon-Tiki: The Biography of Thor Heyerdahl* (Rand McNally, 1967).

HIERONYMOUS MACHINE A machine invented by Thomas G. Hieronymous in 1949 that was intended to combine psychic phenomena with electronics. The first in a continuing line of such inventions, it was to analyze the eloptic radiation of minerals, a radiation hitherto unknown. The machine consisted of a box containing some tunable electronic circuitry of the thermionic valve type and a plate on which the specimen was placed. It could be used either on the mineral or on a photograph of the mineral. The original Hieronymous design was improved by John Campbell, Jr., editor of *Astounding Science Fiction,* who founded a new division of parapsychology, PSIONICS. He claimed that his electronics worked satisfactorily with or without the power supply. The user tunes the circuit while stroking a plastic plate. When the correct setting, peculiar to each user, is reached, the plate feels sticky. Something is being detected that is outside our normal sensory experiences.

Hieronymous's original machine and Campbell's improved version have disappeared, but psionics, is still around. Several electronic machines have been devised in recent years that employ electronic circuitry (not without a power supply) to explore some psychical phenomenon. None of these machines and experiments, old or new, have so far convinced the scientific community, though it would be fair to say that some of the recent results have puzzled some orthodox scientists.

HOLLOW EARTH DOCTRINE *(HOHLWELTLEHRE)* One of the many pseudoscientific ideas about Earth's form—size and shape. Some hollow Earth theories suppose Earth to be flat like a flapjack, and others suppose it to be spherical. But the most bizarre theory must be that we humans, along with the flora and fauna, are living on the inside shell of a huge sphere that looks inwards toward the whole universe, which occupies the central hollow. Most *Hohlweltlehre* are of the 19th century.

In early history before global travel, many people envisaged the world to be flat and worried about going too far afield for fear of falling off the edge. Today in the United States there are still adherents to this theory, probably followers of Wilbur Glenn VOLIVA of the Church of Zion, who propounded the idea that the Earth was shaped like a flapjack, with the North Pole in the center and the South Pole spread around the circumference. If this can be believed today—and it still is—how much more difficult must it have been in the past to abandon notions of what seemed to be quite obviously a flat earth, in order to think of it as a giant ball, with we mortals living all around its surface. It must have been like replacing an apparently plausible theory by an impossible one. In the 16th century, Martin Luther argued that the "underside" of such an Earth could not be habitable, because God would not have allowed so many people to be disadvantaged, by hiding from their view Christ's descent from heaven at his Second Coming.

Later, at the end of the 17th century, the eminent English astronomer Edmund Halley put forward the idea that Earth had an 800-kilometer–(500-mile)–thick outer shell, two inner shells the size of Mars and Venus, and finally a solid inner sphere the size of Mercury. Each sphere, he thought, was capable of supporting life, as daylight could be provided by an inner luminous atmosphere. When a display of the northern lights occurred in 1716, Halley thought it was because these interior glowing gases were escaping.

Similar ideas to those of Halley were held by a U.S. captain, John Cleves SYMMES. When he left the army in 1814, Symmes spent his remaining years publicizing his view that Earth was made up of five concentric spheres with openings many thousands of kilometers in diameter at each of the poles. He thought the seas flowed through these openings and that plants and animal life lived on and around the surfaces of the spheres. By mid-19th century, various Hollow Earth theories were alive and flourishing in Europe and America, and many novels were written on the subject. Probably the best known at the time was by French novelist Jules Verne's *A Voyage to the*

Center of the Earth (1864). Others include Edgar Allen Poe's *MS. Found in a Bottle* (1833) and *The Unparalleled Adventure of One Hans Pfaal* (1835).

Later in 1869 a pamphlet was distributed in the United States by Cyrus Reed TEED. *The Illumination of Koresh: Marvelous Experience of the Great Alchemist at Utica N.Y.*, described a vision in which he received a message from Koresh telling him that he was the new messiah whose mission was to spread the news about the true structure of the universe. Teed later published a more detailed account in *The Cellular Cosmogony* (1870), which stated that the whole cosmos is contained in the hollow of Earth. There is nothing on the outside, just a void. We are living on the inside of its 150-kilometers (100-miles)–thick shell, which is made up of 17 layers, five of which are geological strata, followed by five mineral layers and then seven metallic ones. Rotating at the center of the open interior is an invisible sun, half dark and half light; what we see is not the actual sun but a reflection of it rotating from dark to light, making it appear to rise and set to give us day and night. The moon is not material but a reflection of Earth, and the planets are also not material but reflections of "mercurial discs floating between the laminae of the metallic planes." Teed attacked orthodox scientists as quacks and humbugs, who were not proper scientists but evil personal enemies who were blocking his work from being accepted.

In 1913, American Marshall B. Gardner published *Journey to the Earth's Interior,* which advocated the Hollow Earth theory. Many of his ideas were similar to those of Symmes and Teed, except that he pictured Earth to have an outer shell that was 1,300 kilometers (800 miles) thick. Inside the great hollow was a Sun 1,000 kilometers (600 miles) in diameter giving light to the interior, and both poles had openings that were each 2,300 kilometers (1,400 miles) across. The other planets were thought to be constructed in the same way. The so-called ice caps of Mars being presumed to be openings from which a gleam of light is sometimes seen to shine, in the same way as light was said by Halley to have streamed out of our northern opening.

In the 1930s, the German aviator Peter Bender expounded the Hollow Earth Doctrine that followed closely Teed's ideas. He also claimed to have corresponded at length with the Koreshans. Bender's ideas provided the foundation for a book on the subject called *Geokosmos* by Karl E. Neupert.

For further reading: Martin Gardner, *Fads and Fallacies in the Name of Science* (Dover, 1957).

HOLMES, OLIVER WENDELL (1809–1894) Father of the famous jurist of the same name. Holmes was a physician, poet, and humorist, notable for his research into puerperal fever, which was a fever that sometimes occurred during childbirth, for his work against HOMEOPATHY, an alternative system of medical treatment, and for his lectures at Harvard Medical School, where he eventually became dean.

As author of what was known as the "Breakfast-Table" series of essays, much of his humor was directed toward medical charlatanism, then prevalent in the United States.

HOMEOPATHY System of medical treatment that relies on extremely small quantities of various mineral, plant, and animal materials that, in larger quantities, are thought to produce symptoms similar to the disease being treated. Homeopathy emerged in the 19th century as a challenge to orthodox, or "allopathic," medical practice, which relies on much more substantial doses of medicines. In spite of frequent criticism of its scientific soundness, homeopathy continues to enjoy a significant following.

Homeopathy was first developed by German physician Samuel Christian Hahnemann toward the end of the 18th century. Distressed by his inability to cure patients using standard medical techniques of the day, Hahnemann turned to experimental pharmacology. By testing the effects of various substances such as cinchona ("Jesuit's bark") on patients and on himself, he eventually arrived at the basic principles of homeopathic medicine. The first was the "law of similars," which held that substances that produce symptoms similar to particular diseases are the keys to inducing cures. Hahnemann theorized that the body's natural "vital powers" were often unable to battle disease alone. By activating an artificial form of the disease, however, the body's vital resources could be rallied against the illness. Hahnemann's second principle was the "law of infinitesimals," which held that extremely dilute doses of the similars, sometimes as small as 1 millionth of a grain, were sufficient to effect cures. He argued that careful preparation and dilution of various substances led to their "homeopathic dynamization," wherein the liquid retained a memory of the active ingredient that could excite the patient's healing powers, even though the substances in solution were chemically undetectable.

Hahnemann soon put together elaborate lists of ailments and the corresponding similars. His homeopathic practice primarily consisted of taking elaborate medical histories from his patients and then prescribing solutions based upon the similars. Eventually, he published his ideas in a series of medical manuals, the most important being *The Organon of Homeopathic Medicine* (1810). His system quickly spread throughout Europe and America.

As the popularity of homeography increased, homeopathists set up training colleges and national organizations, providing perhaps the most significant challenge to the medical orthodoxy of the era. Advocates included persons from all walks of life, but particularly women, who constituted about two-thirds of all the patrons of this alternative therapy. Also, though subject to some discrimination, women were welcome in most homeopathic training institutions, and hundreds were licensed as practitioners over the course of the century.

Homeopathists' success during the 19th century was the result of various factors. First, they provided a much milder system of therapy than the HEROIC MEDICINE most doctors practiced. Second, their theories about the law of similars and their pharmacology were plausible enough, in light of the scientific knowledge of the time, to attract many trained physicians. Third, homeopathists often demonstrated better than average cure rates for certain diseases, such as cholera. Finally, their ideas about the body's vital force and its role in the healing process could be readily linked to contemporaneous religious attitudes, particularly beliefs about the relationship of the soul to the body.

With success, however, came criticism, particularly from practitioners of standard medicine, who saw both their time-honored notions of proper therapy and their economic livelihoods threatened by homeopathy. In the 19th century, the classic antihomeopathic work was Oliver Wendell Holmes's *Homeopathy and Its Kindred Delusions* (1842), which set much of the tone for subsequent scientific and medical attitudes. In it Holmes grudgingly conceded that many patients improved under homeopathic care, but he was unwilling to grant that Hahnemann's system had much to do with the healing process. He instead attributed any improvements to the natural cycle of disease. Even less plausible was homeopathic theory, which Holmes found scientifically unsound and therefore an improper foundation for medical therapy.

Singled out for particular ridicule was Hahnemann's law of infinitesimals, which Holmes joked would require patients to consume entire oceans of solution to produce any tangible effects. On a more serious level, he (and almost all other critics of homeopathy) found no warrant for the claim that the effects, or "energies," of the various substances could be retained by the homeopathic solutions. For him, homeopaths dispensed nothing more than water, and all alleged benefits should be considered placebic. Because of the risks involved in delaying or avoiding more orthodox therapies, Holmes ultimately advocated the exclusion of homeopathic practitioners from the medical profession. He also strongly urged that homeopathic therapy be kept away from patients. In accordance, in the late 1840s, the newly formed American Medical Association (AMA) adopted strict policies hostile to homeopathy. Though much of this initial animosity would be tempered over time, strained relations between homeopaths and orthodox physicians still persist today.

In spite of AMA censure, homeopathy continued to thrive over the course of the second half of the 19th century, supported by a network of national organizations, colleges, hospitals, journals, and therapists. This occurred despite an internal schism between so-called "eclectic" practitioners, who felt that homeopathic care could be supplemented by other types of therapy (including allopathic medicine), and "traditionalists," who held steadfastly to Hahnemann's original system.

The willingness of the eclectics to use some allopathic therapies helped them establish better relations with the orthodox medical establishment in the 1880s and 1890s. Some physicians even advocated the eventual integration of the two camps insofar as the early training for both was grounded in the disciplines of anatomy and physiology. Acceptance came at a price for homeopaths, however, as these mergers led to the loss of their distinctive professional identity and alternative medical perspective. For much of the 20th century, homeopathic care seemed to be on the decline as orthodox medicine, buoyed by the development of such successful allopathic medicines as penicillin and antibiotics, consolidated and extended its public influence.

The late 1960s witnessed the revival of homeopathy. Spurred on by a general questioning of establishment institutions by the counterculture, a variety of social movements sprang up that called for a renewed emphasis on "alternative" healing methods. This renewal likewise tapped into a dissatisfaction with orthodox medical therapeutics that seemed to discount or disregard patient initiative, as well into fears that some doctors recklessly overprescribed certain medicines. Much as before, a network of national organizations, health stores, practitioners, and journals supports homeopathic practice.

With this resurgence, homeopathic beliefs about the infinitesimals and the alleged properties of their dynamized solutions have again become the focus of considerable scientific criticism and skeptical ridicule. Given the advances in biochemistry, pharmacology, and physiology over the past century, these critiques are often even more pointed than their predecessors. Homeopaths, however, seem prepared to weather the criticism and at times have countered with elaborate, ostensibly scientific, explanations of their basic beliefs. These include explanations of the medical effects of the infinitesimals based on quantum physics, or analyses of homeopathic solutions by means of nuclear resonance spectroscopy, which allegedly show that they have distinctive properties. Critics are rarely swayed, but these efforts at the very least demonstrate the ingenuity with which homeopathy has been tailored for contemporary audiences.

Other scientists and public health professionals have been willing to set explanations aside and run clinical trials on the general effectiveness of homeopathic care for a variety of ailments, including hay fever, flu, insect bites, and arthritis. The evidence from these studies has been mixed, with some indicating that homeopathy can elicit cure rates that are better than placebos, and others showing negligible or inconclusive results. Partisans on either side are predictably selective about which studies they believe are authoritative.

The legacy and prospects of homeopathy are perhaps more substantial than might be assumed if one only considers the perspective of mainstream science and medicine in the present day. In the 19th century, by virtue of the popularity of its more lenient therapeutic techniques, homeopathy contributed to the important transition away from heroic medicine. It also played a significant role in the early development of professional medical organizations. In more recent times, homeopathy has demonstrated the ability to endure and even flourish in the face of a very powerful and skeptical medical establishment. The reasons for this endurance are varied but include the apparently widespread opinion that, in spite of its implausibilities, homeopathy seems to work for many of the people who try it.

See also ALTERNATIVE MEDICINE; HOLMES OLIVER WENDELL.

For further reading: Samuel Christian Hahnemann, *The Organon of Homeopathic Medicine* (Academical Bookstore, 1843); Martin Kaufman, *Homeopathy in America* (John Hopkins Press, 1971); and Norman Gevitz, ed., *Other Healers: Unorthodox Medicine in America* (John Hopkins Press, 1988).

HOMUNCULUS Diminutive of the Latin *homo,* meaning "man"; a term adopted by embryologists in the 18th century to conjure up the idea of a very small human being. In this period, doctors were engaged in an argument that had its roots in a dichotomy that went back beyond Rome to ancient Greek science. On one side, the *preformationists* argued that embryology must represent an unfolding of preexisting structure; in other words, a tiny being complete in all its essentials, a homunculus, lay in reproductive organs concealed from view or from the magnifying power of then available microscopes. Two sites for the homunculus were disputed: The *ovists* expected to find it in the female egg, while the *spermatists* thought it more likely to be in the male sperm. On the other side of the argument was a group called the *epigeneticists* headed by Pierre Maupertuis, an 18th-century French scientist, who disliked the idea of a homunculus and thought it more likely that embryos contained coded instructions rather than preformed parts.

It is easy to see epigeneticists as the unbiased empirical scientists and the preformationists as pseudoscientists blinkered by theological prejudice, trying to uphold the centuries-old belief that St. Anne became pregnant with her daughter the Virgin Mary by a sexless conception. But these things are seldom as simple as they seem, and according to the Harvard science professor, Stephen Jay GOULD, the preformationists were the ones who were actually putting forward a view of science closer to that of our own by requiring a material cause for all phenomena. Given the science of the day, how could they believe that a complex human body could develop out of nothing? In retrospect, the epigeneticist view of causality was far too close to that of VITALISM—the idea that there was an external, nonmaterial force that could impose a design upon the unformed egg.

Maupertuis's arguments against preformation were of course the conventional rationalists' responses of the time. To the ovists, who put the homunculus in the female, he asked if the male existed only to trigger off the development of her egg, and if this was so, why we have laws of inheritance running through the male line. To the spermaticists, who placed the homunculus in the sperm cells, he raised the question of why so many little homunculi were made when so few were let out to enjoy life. To both sides he raised the issue of encapsulation, the notion of each homunculus having to contain a smaller homunculus within it and so on *ad infinitum.*

As the debate on embryology rumbled on, Maupertuis, recognizing stalemate, shifted his research elsewhere. When he was observing inheritance in polydactyly (humans having an extra finger), he found that the deformity passed on through both male and female lines. If this was so, preformationism, either in the ovist or the spermaticist version, could not be supported. The offspring must inherit its characteristics from both parents, some from the mother and some from the father.

The unfolding of the 18th-century scientific argument with all its scientific twists and turns is a good example of how science progresses. Today's scientists support a hereditary process in which coded instructions are donated by both the male and the female at conception, through DNA. The homunculus idea has been consigned to history, but in its time it was the focus of investigation and discussion, which led eventually to our present understanding. Today the term "homunculus fallacy" is often used to condemn theories of psychological processes of infinite regression that ascribe to some internal system the very psychological properties that are being investigated in the first place. For example, Sigmund FREUD is often accused of seeing the unconscious as an homunculus. A corresponding fallacy has also crept into "artificial intelligence theory," and here the term "homunculus obscurantism" can be found in occultism, where a whole other order of mystical and supernatural powers is brought into place between our seeking of knowledge about the world and the actual world outside our heads.

For further reading: Stephen Jay Gould, *The Flamingo's Smile: Reflections in Natural History* (Norton, 1985).

HÖRBIGER, HANS (1860–1931) A Viennese mining engineer and co-author in 1913 of a massive tome,

Glazial-Kosmogonie. The thesis of this magnum opus was that Earth, because it was first formed, has had several moons—six or more—in succession. Each of the moons had a catastrophic effect that accounts for Earth's geological features. During the time human beings have been on Earth, the Earth has had two Moons, the present one and the one before it. The earlier Moon's effects on Earth are described in our myths and legends. The earlier Moon had been smaller than the present one and had been much closer to Earth, which it had orbited rapidly, six times a day. The attraction of this close, rapidly orbiting Moon had distorted Earth, making it bulge at the equator, leaving huge areas away from the equator covered in ice. To get away from the ice areas, humans had concentrated in high mountainous regions of the equatorial belt. Earth's gravitational forces on the Moon had caused it to disintegrate, showering Earth with water and rocks. In consequence, Earth, no longer subject to gravitational distortion, had sprung back into near spherical shape, causing earthquakes, volcanic eruptions, and floods. These were the floods of our mythologies: the stories of Gilgamesh, Noah, and others.

Hörbiger believed that our present Moon has a surface layer of ice 200 kilometers (140 miles) thick and that other planets have a similar ice covering, that the Milky Way is a huge ring of gigantic blocks of ice, and that the moon is steadily spiraling in toward Earth, with more catastrophes yet to come. The Nazi movement seized on Hörbiger's theories as a preferred Aryan alternative to traditional astronomical theories, dismissed as Jewish science. To believe in Hörbiger's theories was essential to being a good Nazi.

After Hörbiger's death, Englishman Hans Schindler Bellamy adopted his theories and produced more mythological and folklore evidence in their support. The Hörbiger Institute in Germany continued to flourish in the post–World War II Western world.

See also DILUVIALISM.

HOUDINI, HARRY (1874–1926) The most famous escape artist in the history of stage MAGIC. He is also the key figure in the conjuring profession's long history of antagonism to SPIRITUALISM's claims to demonstrate spirit contact. Born Ehrich Weiss on March 24, 1874, he grew up in Wisconsin, joined the circus, and began adult life as a trapeze artist, but was really interested in stage magic, especially escape tricks. He took his stage name from Robert Houdin, a 19th-century French magician who had also had a history of exposing performers who worked fake miracles.

Houdini grew up in the era of what was termed physical mediumship. Many spiritualist mediums claimed that the messages they facilitated from the spirit world were accompanied by a wide variety of extraordinary phenomena, MATERIALIZATIONS, APPORTS, floating trumpets, and disembodied voices. Some MEDIUMS claimed feats equal to Houdini's stage illusions.

Houdini

Throughout his public career, Houdini took time to visit mediums and expose their tricks to the public. His enthusiasm for this nonprofessional use of his skills was a result of several motives. He genuinely desired to establish contact with his deceased mother, if that was possible, and was angry at people who would try to deceive him on such an important personal matter. He was also influenced by the ethical code of the magicians' community against claiming any "supernatural" powers. Finally, he felt competition by Houdin, the greatest magician of his day. Sir Arthur Conan DOYLE, creator of Sherlock Holmes and a dedicated Spiritualist, once claimed publicly that Houdini's feats must have been accomplished by psychic or supernatural help, a statement which cost him Houdini's friendship.

Houdini's enthusiasm for exposing mediums led to his being named to the committee established in 1924 by the *Scientific American,* which examined Mina Crandon, who under her public name "Margery" had become the most famous physical medium of the 1920s. The controversy set off by the committee would divide the U.S. psychical research community for a decade. While it is generally conceded that Crandon was a fake and accomplished her remarkable feats with the assistance of her husband and others, detection was difficult, even for Houdini. So anxious was he to prove fraud, he was himself caught tampering with Crandon's equipment.

Houdini died on October 31, 1926, following a blow to his midsection by a fan, testing Houdini's claims to take such punches without negative side effects. Distracted, Houdini was not prepared to receive the impact.

Houdini's interaction with spiritualism did not end with his death. He left a coded message with his widow and a $10,000 reward for any medium who could prove his own survival by bringing the correct message to her. That message was delivered to her in 1929 by Arthur A. FORD. The message was "Rosabelle believe," and was written in a secret code used by the Houdinis in an old mindreading act. Mrs. Houdini signed a letter confirming Ford's claim. However, two days later a New York newspaper claimed that a hoax had been perpetrated by Ford and the widow. Both she and Ford denied the claim, but the reporter could produce no confirming evidence to back up his charge.

For further reading: Harry Houdini, *A Magician Among the Spirits* (Harper, 1924); Christopher Milbourne, *Houdini: the Unknown Story* (Thomas Y. Crowell, 1970); and Arthur A. Ford and Marguerite Harmon Bro, *Nothing So Strange* (Harper, 1958).

HOXSEY TREATMENT One of many cancer cures that became available in the United States in the 20th century. It was devised and marketed by a medically unqualified licensed NATUROPATHY practitioner, Harry Mathias Hoxsey and practiced in his Dallas, Texas, Hoxsey Cancer Clinic from 1936. Hoxsey's father had devised a cancer tonic, and his son, realizing that there was room for exploitation, expanded the operation. Harry Hoxsey sold his preparations throughout the United States through a network of osteopaths. He came into frequent conflict with the regulatory authorities; the Food and Drug Administration stopped his interstate trade and warned the public that some of his claimed curing methods might actually be harmful. In 1956 one of his supporters, John J. Haluska, a Pennsylvania state senator, opened a second Hoxsey clinic in Portage, Pennsylvania, convinced that Hoxsey treatments were beneficial.

HOYLE, SIR FRED (1915–) An eminent British theoretical astrophysicist, Fellow of the Royal Society from 1957, and Cambridge University's Plumian Professor of Astronomy and Experimental Philosophy from 1958 until his retirement. He has been extraordinarily productive in several areas of science, in particular in his work on nucleosynthesis, the processes by which the heavier elements have been formed in the stars during the genesis of the universe from elementary matter. He is also a playwright and a novelist.

During World War II, Hoyle headed a small team, which included Tom Gold and Hermann Bondi, at the Royal Navy's Admiralty Signal Establishment; the team researched the propagation of electromagnetic waves, aerial design, and similar problems. Soon after the war, Hoyle was elected to a Cambridge professorship and was invited to give the 1949 Reith Lectures by the BBC. That lecture series, on radio, aroused great public interest in astronomy and considerable controversy among his professional colleagues. In 1948, together with Bondi and Gold, he had advanced an interesting explanation of the generally accepted belief that the universe is expanding: They proposed that matter was constantly being created—out of nowhere—throughout the universe and at a very slow rate—not greater than 10^{-43} kg per m^3 per second—so slow as to be undetectable, but sufficient to push stars and galaxies outward and maintain the universe so that it looked much the same throughout time. This is also referred to as the Steady State theory—an infinite universe, extending forever both in space and time. Hoyle's opponents preferred the big-bang theory: That there had been an explosion—the cause of the expansion—sometime between 10 and 20 billion years ago, the products of which form the universe today. Again the explosive material came out of nowhere. Both theories pose their difficulties, particularly finding experimental evidence for one or the other, but the Steady State theory has lost out and the Big-Bang idea is currently favored.

In 1977, Hoyle and Chandra Wickramasinghe expanded on an idea first advanced by Swedish chemist Svante Arrhenius 80 years before. They proposed that primitive forms of living matter had first been formed in the interstellar dust clouds during the first 10 billion years after the Big Bang and that when, much later, such material fell on planets where conditions were favorable, life flourished. On this theory, which its proposers advanced very forcefully, all life wherever it thrived would come from a common stock, and we might therefore expect to find life forms elsewhere in the universe similar to those on Earth. Hoyle and Wickramasinghe further claimed that those primitive interstellar life forms contained disease germs and viruses and that it was the occasional contact by Earth with disease clouds that accounted for sporadic epidemics of influenza, plague, and the like. The medical profession soon proved them wrong, but that long shot, although it failed, does not detract from the original idea, which is still stimulating research. In the last few years, astronomers have identified many organic compounds in the spectra of interstellar dust clouds and also showed that polycyclic aromatic hydrocarbons in sooty grains found in meteorites must have been formed outside the solar system. In an article in the journal *Science* (April 1996, p. 249), scientists report finding complex organic molecules, buckminster-fullerenes, in a huge crater in Ontario, Canada. The most probable explanation for their presence is that they came with the meteor that formed the crater and had been produced when the solar system was forming 4 or 5 billion years ago. In 1988 Hoyle and Wickramasinghe posited a *cosmic life force,* with "total control of the universe, with power to communicate across immense distance scales." The two researchers, in *Archaeopteryx, the Primordial Bird* (1986), also challenged the current belief that archaeopteryx was the ancestor of the world's birds.

Hoyle has never been afraid to go out on a limb. As early as 1940, he had maintained that many molecules would be common in space, contradicting the eminent astronomer Sir Arthur Eddington; the evidence supporting Hoyle's theory is now undeniable (see for example *The Chemically Controlled Cosmos* by Thomas Hartquist and David Williams, Cambridge University Press, 1995). In Hoyle's book *Frontiers of Astronomy* (Heinemann, 1955) he dismissed the idea of CONTINENTAL DRIFT, which is now well accepted. In 1977 he published *Stonehenge* (Heinemann), in which he gave a detailed account of how the stones of that prehistoric megalithic monument could have been erected and used as a sophisticated observatory to predict eclipses, especially of the Moon. He and Alexander THOM are widely credited with stimulating interest and research in this subject.

Martin Gardner sums up Hoyle very well: "Hoyle likes to beat loudest on the drums of his own inventions, but it doesn't matter because his speculations are never boring. It must have been a tragic experience for him to watch his beloved steady-state theory of the universe go down the drain as the big-bang theory became accepted, but this seems not to have diminished his mental energy, or his fondness for funky theories."

For further reading: Fred Hoyle, *Home Is Where the Wind Blows* (Oxford University Press, 1997); Fred Hoyle, *The Small World of Fred Hoyle* (Joseph, 1986); Fred Hoyle, *Ten Faces of the Universe* (W.H. Freeman, 1977); *Cosmic Life Force* (Dent, 1988); and Fred Hoyle and Chandra Wickramsinghe, *Lifecloud: The Origin of Life in the Universe* (Dent, 1978).

HUDSON, THOMAS JAY (1835–1903) U.S. journalist, lecturer, and amateur psychologist. Hudson developed a two-level theory of mind that not only opposed spiritualist claims, but later provided theoretical underpinnings of a rival religious movement, New Thought. Hudson was born on February 22, 1835, in Windham, Ohio. He studied law and was admitted to the bar in his home state, but after a brief career gave up law for journalism, becoming editor of the *Detroit Evening News.* He left journalism around 1880 and took a job with the U.S. Patent Office, eventually becoming principal examiner. During his years looking at patent applications, he had become interested in experimental psychology, a field then very much in its infancy. In 1893 he quit his job and wrote his first book, *The Law of Psychic Phenomena,* which became popular and for which he was awarded an honorary degree by St. John's College in Annapolis, Maryland.

Hudson argued that human beings have two distinct minds: the objective, the one which operates in their daily waking life, and the subjective, that most visible during the hypnotic state. The key to understanding humanity is greater knowledge of the subjective mind, which is commonly dormant but which records every experience of life. The objective mind is distinguished by its ability to reason both inductively and deductively. The subjective mind can reason only deductively with the material fed into it (thus, the off-behavior of a hypnotized person).

When published in the 1890s, Hudson's theory was received as an attack upon popular SPIRITUALISM. His theory of an infallible second mind provided a means of explaining most spiritualist phenomena. Hudson, while believing in survival after death, also believed that communication with the beyond was impossible. Attempts to communicate simply lead one into an encounter with the subjective mind which, with its vast storehouse of knowledge, leads one to believe that a spirit has been contacted. Hudson would expand upon his ideas in several later books including: *Scientific Demonstration of the Future Life* (1896), *Divine Pedigree of Man* (1900), *Law of Mental Medicine* (1903), and *Evolution of the Soul and Other Essays* (1904). Hudson died on May 26, 1903.

Hudson's idea concerning the two levels of mind was picked up by British metaphysical teacher Thomas Troward, whose popular books circulated widely in New Thought circles in both Europe and North America early in the 20th century. These books were an especial favorite of Ernest Holmes, who through them integrated Hudson's theories on mind into his teachings, known to this day as RELIGIOUS SCIENCE and science of mind.

HUMORAL THEORY The ancient doctrine that elemental body fluids or humours (blood, phlegm, yellow bile, and black bile) are the physiologic and pathologic basis of health and disease. Also called humorism, fluidism, humoralism.

Stressing the unity of the body and the strong interaction between mental and physical processes, humoral theory suggests that, provided one can discover the proper balance of humors within an individual, one can both restore the body to health and devise means of maintaining health for as long as possible. In humoralism, preventive and curative approaches are of equal relevance.

Humoral theory may have originated in Greece, around the sixth and fifth century B.C.E. By the second century C.E., it had become dominant throughout the Roman Empire. The seventh-century Arab conquests and subsequent translation of Galenic and Hippocratic texts into Arabic in the ninth century guaranteed humoral theory's primacy in the Muslim world, where it continues to play a major role in medical theory and practice. The introduction of translated Arabic medical texts into the Latin West from the 11th century onward secured the central position of humoral theory in the West. It remained the basis of Western medicine until the 19th century.

In the original Greek, the word *humor* means simply any juice or fluid. Sap in plants, blood in animals, and ichor in the gods are all humors. At the same time, other natural fluids, such as excessive mucus or bile, make their appearance only when sickness is present and may disappear again on recovery. Thus, even before the earliest medical records of Greece, it is possible to conceive of theories and therapies involving these fluids or suggesting some form of equilibrium among them. Sicilian healer-philosopher Empedocles of Acragas, who lived in the fifth century B.C.E., posited a universe created out of four basic elements: earth, air, fire, and water. Each material object was made from a proportional arrangement of these elements. For example, bone consisted of two parts earth, two parts water, and four parts fire; blood consisted of equal parts of each. The stability of objects rested on the maintenance of the proper proportional balance of these elements. It is against this wider cosmological background that the Hippocratic Corpus was formulated and written.

From the 16th century onward, many of the anatomical and physiological theories of humoralism's chief Greek exponent, the second-century physician Galen were overthrown. The apparent therapeutic advantages of humoral explanation, however, ensured its continuation with modification. When William HARVEY discovered blood circulation in the 17th century, many of the qualities that Galen had ascribed to the other humors were transferred to the blood, while the idea of health as a form of equilibrium continued to inform many 18th-century theories of health and disease. A combination of the new cellular pathology, focusing on tissues rather than on body fluids, and new discoveries in chemistry may seem to put an end to scientific acceptability of humoral theory, though the lay person still conceives of health and disease in terms that would have been familiar to Galen and Hippocrates.

HUNDREDTH MONKEY A modern myth of the NEW AGE MOVEMENT. The story of the hundredth monkey was first told by Lyall Watson in his book *Lifetide: A Biology of the Unconscious* (1979). According to Watson, four primatologists were observing monkeys living on the islands off Japan. They began to leave food for the monkeys to find and eat to keep them away from local farms. In 1953 they watched an older female monkey wash a potato to get the dirt and grit off it. She taught the process to others. Slowly the process spread, but then suddenly at one point, the practice exploded: where before only a few were washing potatoes, suddenly almost all of the monkeys were doing it. Up to a certain point—Watson supposed up to 99 monkeys washing potatoes—the behavior grew slowly, step by step. But when the next monkey joined in—the *hundredth*—almost all of the remaining monkeys immediately adopted the practice.

Watson used the story as a simple illustration of the operation of the paranormal, but one reader, Ken Keyes, saw in it a parable for the new age. If a critical mass of people attain to a new consciousness, then the idea will quickly appear throughout the human race. Keyes wrote a book applying this idea to the issue of world peace. In his book *The Hundredth Monkey,* he argued that if a critical mass of people could attain to peace consciousness, then miraculously that consciousness would be multiplied around the world. Within a few years several books appeared that further explored the implication of the hundredth-monkey idea in various realms. It was especially attractive to New Age thinkers as it supplied what appeared to be a mechanism by which the "New Age" could be brought into existence in spite of the disbelief of a great majority of the population.

As early as 1983, psychologist Maureen O'Hara attacked the story as fictitious. Then in 1985, Ron Amundsen, in an article in the *Skeptical Inquirer,* challenged the original story. He argued that nothing extraordinary occurred in the spread of potato washing among the original monkeys. Watson immediately accepted Amundsen's analysis of the story and admitted that for him the hundredth monkey was a useful metaphor based on slim evidence and much hearsay.

The hundredth-monkey concept lay behind a number of New Age events, such as the Harmonic Convergence of August 16–17, 1987, when a particularly powerful cosmic force was believed to be peaking and a goal of assembling a symbolic group of 144,000 people was projected to receive and use these energies in the cause of universal cultural transformation. Something like the hundredth-monkey concept is also at work in the TRANSCENDENTAL MEDITATION movement which believes that having a set number of meditators in a particular location has a salutary effect in preventing antisocial actions.

HUNZA PEOPLE An immortal race of people living in a peaceful paradise high in the mountains of the Himalayas, as recounted by ancient Asian legends at the end of the 19th century. Novelist James Hilton based his book *Lost Horizon* on these legends. During the period of the British Raj in India, explorers in the far northeast of the subcontinent discovered a valley colonized by people whose life expectancy exceeded that of all other known races. According to the stories, men and women living in the valley of the Hunza River in what is now northeastern Pakistan regularly lived to 100 or 120 years; some were believed to reach an age of 140 or more.

These stories slowly spread to the Western world—helped by the British takeover of the Hunza valley in the 1890s. British scientists were intrigued by the reports of centenarian Hunzakuts. In the 1930s, health specialists first launched investigations of the phenomenon. They suggested that the Hunzas' long lives were due to a combination of their diet, their spirituality, and their environment.

In 1958, the television program *People Are Funny*, hosted by Art Linkletter, sent an optometrist named Allen Banik to visit the Hunza and to determine, if possible, why the people lived such long and healthy lives. Banik reported on Linkletter's show that, in his opinion, there were two major factors that explained the Hunzakuts' longevity. One was spiritual: Banik noted the lack of violence in Hunza society and decided that a stress-free environment added years to Hunzakut lives. The other was diet: a varied intake of naturally raised, unprocessed foods including apricots, unprocessed grains, and raw vegetables. The important thing, he stressed, was that the fruits and vegetables be "organic"—that they should be raised in a natural environment and eaten with a minimum of additives.

Further research, however, showed that diet and environment were not defining factors in the search for longevity. Although the Hunzakuts led a peaceful life during the 20th century, previously they had earned reputations as fierce warriors. Their diet seemed to have little effect on others who tried it. Finally, additional enquiries showed that Hunzakut elders sometimes exaggerated their ages by as much as four or five decades in order to impress outsiders. The search to extend human life moved in other directions.

HYMAN, RAY (1928–) A psychologist member of the COMMITTEE FOR THE SCIENTIFIC INVESTIGATION OF CLAIMS OF THE PARANORMAL (CSICOP). He was very skeptical of much of PARAPSYCHOLOGY methodology. In particular, he criticized Puthoff and Targ's method of control of experiments in telepathic communication of local views of the Stanford area. By enabling "viewers" to eliminate possible "targets" one by one, they were increasing the probability of scoring a subsequent "hit." Hyman took the position that before allowing the existence of any paranormal phenomena, any possible explanations by established normal means must be totally eliminated. Acting in that role, Hyman appears in much of the skeptical literature on paranormal phenomena.

HYNEK, J. ALLEN (1910–1986) Astronomer and leading exponent of UFO research in the decades following the publication of the CONDON REPORT on UFOs in 1969. After completing his Ph.D. at the University of Chicago in astronomy, in 1935 Hynek joined the faculty at Ohio State University. He gained some fame in the mid-1950s in the large community of amateur astronomers after being assigned to train a group of them to assist the government in the tracking of the manned satellites that the government was planning to launch.

Quietly, beginning in the late 1940s, Hynek was also brought into the study of UFOs through the request of the U.S. Air Force to look at the reports and help eliminate the ones that were best explained as misidentified astronomical phenomena. He was initially doubtful that any of the reports would yield unusual data or evidence of extraterrestrial visitation, but became concerned for what he felt was the Air Force's incompetent handling of them. In 1960 Hynek became the chairman of the department of astronomy at Northwestern University and through the early 1960s stayed involved with the UFO question through his dialogs with a graduate student, Jacques Vallee, and others who continually challenged his earlier conservative statements.

During the 1960s Hynek advocated a serious inquiry into UFOs. Such a study was established in 1968–69 under the leadership of Edward U. Condon at the University of Colorado. However, the results were negative, and the final report appeared amid charges that it had been so designed from the beginning. The Condon Report killed most scientific interest in UFOs, and in the early 1970s Hynek emerged as the major spokesperson for the importance of continued investigation. His next book, *The UFO Experience* (1972) brought together his appraisal of the Air Force's incompetence with his reflections upon the Condon Report's inadequacies. He also proposed a schema for classifying the variety of UFO reports, along with some interesting examples of each. His schema was largely adopted by the UFO research community.

The following year Hynek led in the formation of the Center for UFO Studies, which initially operated out of his home in Evanston, Illinois. He gathered a small cadre of interested researchers and spent much of his time trying to raise funds to carry on a proper research program. Those funds were never found, though the center was responsible for some of the finest publications in the field. In the end Hynek, much to the consternation of many of his colleagues, began to move toward more parapsychological explanations of the otherwise interesting and unexplained UFO reports. He died in 1986 from a brain tumor. Following his death, the Center was renamed the J. ALLEN HYNEK CENTER FOR UFO STUDIES.

For further reading: J. Allen Hynek, *The UFO Experience* (Henry Regnery Company, 1972).

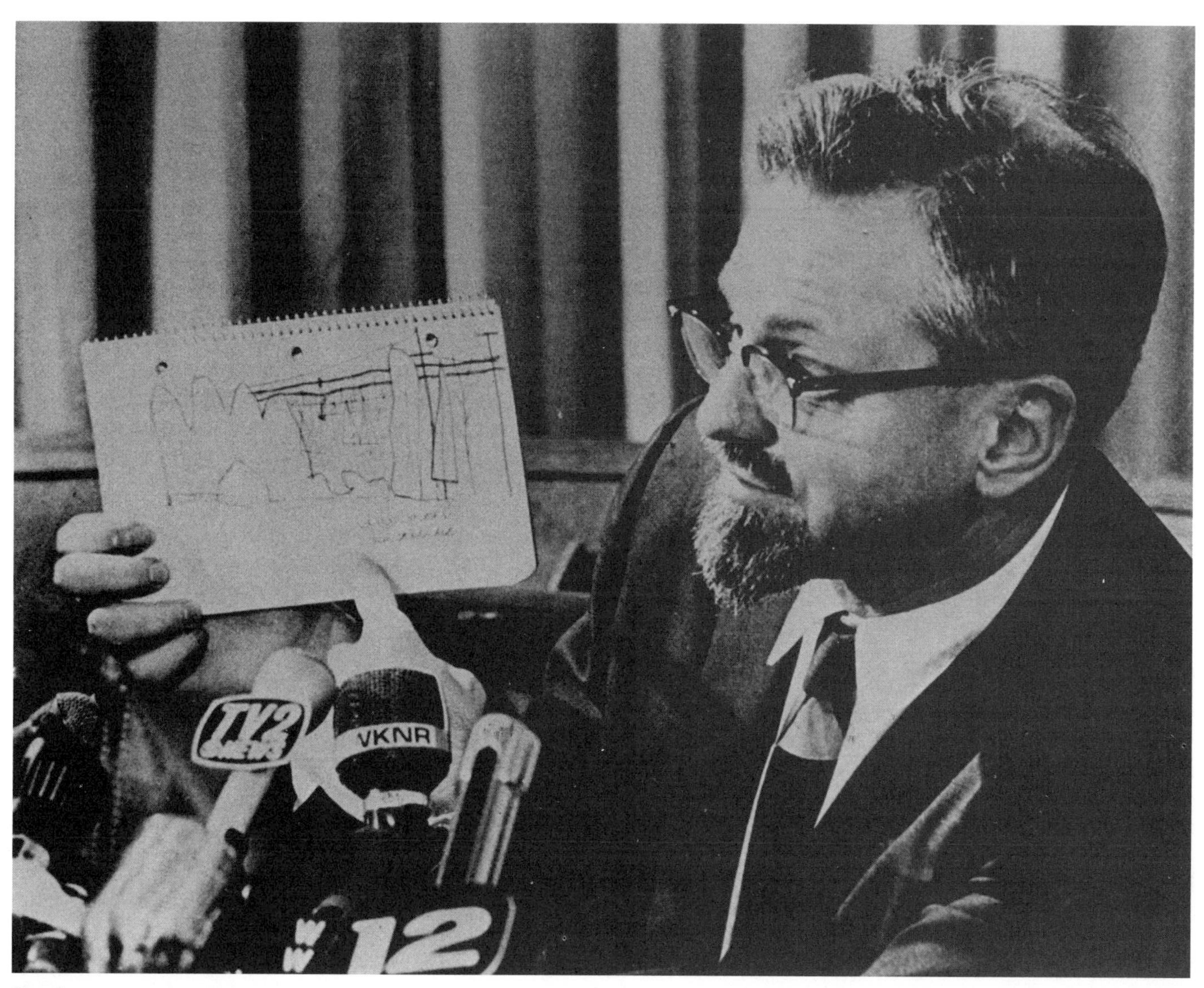

Hynek

HYPERSPACE Theoretical fourth-dimensional space that allows space travelers to circumvent the absolute speed of light. In 1887, U.S. scholars Albert A. Michaelson (1852–1931) and Edward Morley (1838–1923) showed that the speed of light in a vacuum was a constant. In 1905, Albert Einstein (1879–1955) further demonstrated that the speed of light was an absolute in terms of the three-dimensional Newtonian universe in which we live. According to Einstein's general theory of relativity, space and time are curved, and their curvature is directly related to the amount of energy and matter in the vicinity. In 1908, the mathematical physicist Hermann Minkowski (1864–1909) theorized that, under certain conditions, it should be possible to take "short-cuts" through a fourth-dimensional "wormhole" that would quickly cross what would be huge distances in four dimensional space–time.

Early science fiction writers seized on the concept of fourth-dimensional space travel as a means to move their heroes around the universe. Einstein's general theory of relativity declares that nothing can move faster than the speed of light in the universe we know; furthermore, objects traveling near the speed of light become extremely massive. Time also changes for the fast-moving objects: Hyperspace allows science fiction writers to maneuver around the problems involved in manipulating space–time. On the television series *Star Trek,* for instance, the "warp drive" used by starships gives the *Enterprise* the power to move between star systems and to return to its home base in a single lifetime without approaching infinite mass.

Scientists have confirmed Einstein's theory that matter does in fact curve space–time. The existence of hyperspace, however, is harder to prove. Some scientists believe that wormholes through hyperspace are related to black holes, but this has not been proved. Even if wormholes do exist, travel through them would be impossible because the gravitational forces involved would tear a person apart.

HYPNOSIS Derived from the Greek *hypnos* meaning "sleep"; a sleeplike state in which a range of behavioral responses may be induced by suggestion. The person being hypnotized appears to heed only the communications of the hypnotist and sees, smells, feels, and tastes as directed, apparently with no will to do otherwise. There is a school of opinion that asserts there is no such thing as hypnosis, that it is all either stage trickery or pseudoscientific rubbish. Another school just as vehemently asserts that hypnosis has paranormal roots in the occult and the magical. Still others argue for a normal psychological explanation for the phenomenon, centered round the use of suggestive imagery by powerful authority figures.

Priests in ancient Egypt, Persia, and Greece used types of hypnotic practices, and they played a spiritual role in early Christianity. But it was not until the 1840s that James Braid (1795–1860), a Scottish surgeon practicing in Manchester, England, began a scientific investigation into the phenomenon. Braid recognized that he was looking at a problem in the field of psychology and attempted to dispel the whole superstitious aura that surrounded the practice, then called MESMERISM or ANIMAL MAGNETISM; he coined a new name "neurohypnotism" which was later shortened to "hypnotism." Braid saw hypnosis as a kind of sleep induced by fatigue, resulting from the intense concentration necessary for staring fixedly at a bright object. Looking into the future, Braid hoped that the practice would be used to cure "nervous" diseases and also to alleviate the anxiety of surgery. Because of its previous taint of mysticism and charlatanism, the medical profession did not look favorably on his ideas. When Braid died, interest rapidly faded. It was not until the 1880s, when French neurologist Jean-Martin Charcot (1825–93) came upon some of Braid's papers, that interest in the subject revived. Sigmund FREUD studied with Charcot, who by then was using hypnosis in his inquiry

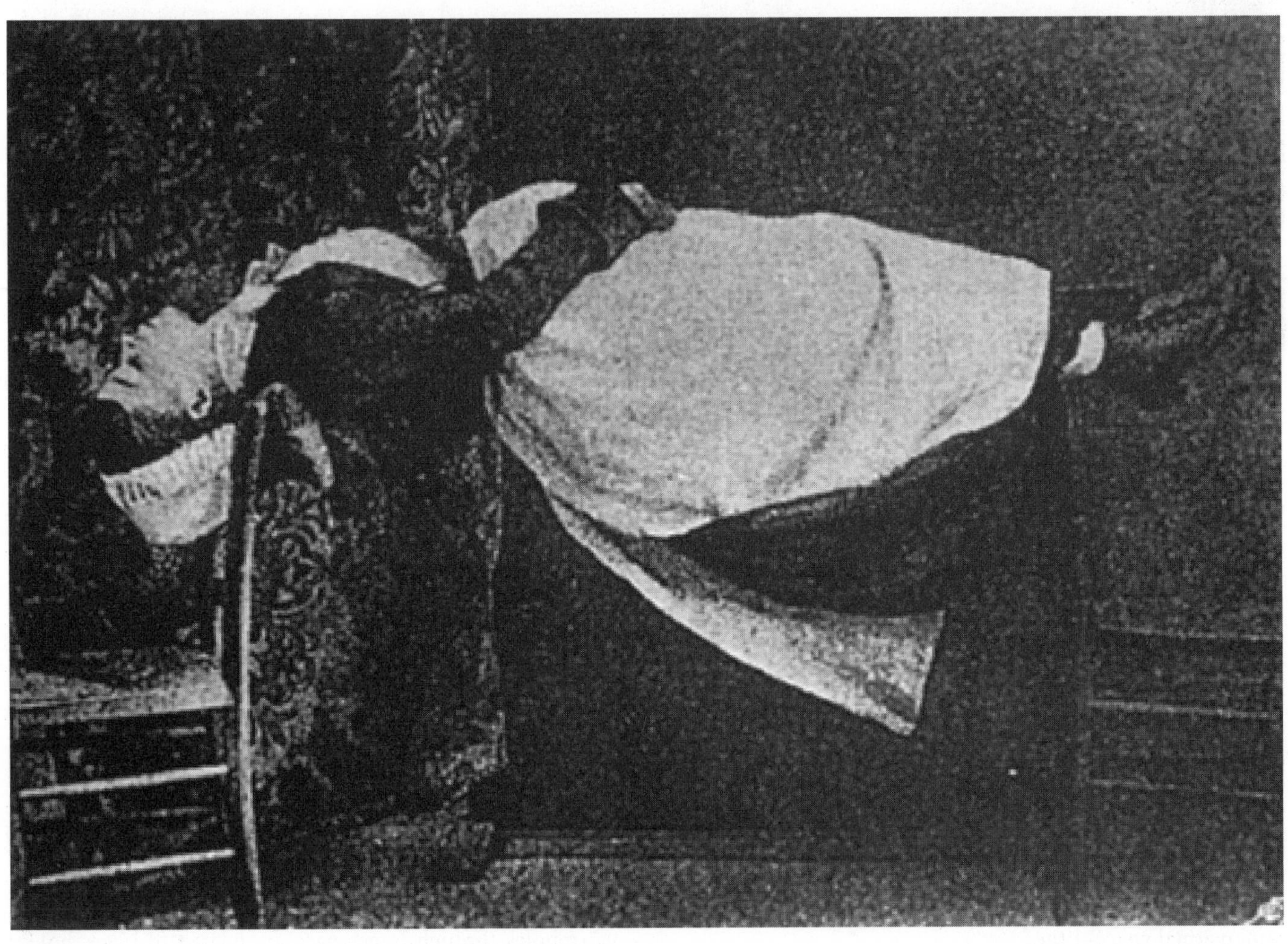

Hypnosis

into and treatment of "hysteria," a supposed women's disease then thought, quite erroneously, to be caused by a disturbance of the uterus. Sometime later Freud, along with others, developed a procedure using hypnosis that is usually regarded as the precursor of PSYCHOANALYSIS, the understanding of the self.

The most important factor in hypnosis is suggestibility, a poorly understood trait. It is thought that a subject's suitability for hypnosis is frequently determined by the degree of suggestibility that can be attained under normal conditions. There is also the strong belief that the control of the hypnotist is never absolute and that hypnotic suggestions that run counter to a person's beliefs and inhibitions will not be acted upon, but this is strongly disputed. Whatever the truth is, people do react differently to hypnosis. Some patients respond well to hypnotherapy, others do not, some do not respond at all. The use of this form of therapy for such addictive habits as alcoholism and smoking involves strengthening the willpower, thereby enhancing the patient's own determination to stop the habit. Phobias, such as fear of spiders and fear of flying, are also treated by hypnosis, but these conditions also respond well to aversion therapy. It has been observed that hypnotic suggestions to bolster willpower fade over time and that some patients revert to previous bad habits, but again this opinion is strongly disputed.

Hypnosis was once thought to help people recall forgotten memories, but courts are now banning its use as evidence, or even as a tool of criminal investigation, because they have found that memories that surface under hypnosis are less reliable than those recalled without it. There is also a likelihood that the beliefs of the hypnotist, however carefully he or she proceeds, will be communicated to the subject. False memories can be implanted even in minds that do not believe themselves vulnerable or uncritical. Because of the sometimes extraordinary power to control another person, hypnotism can have the most sinister implications.

In surgery, hypnosis has been used in place of anesthetics, and for some patients and some operations, it has proved very successful. But it has fallen out of favor as a therapeutic tool in psychoanalysis. Many psychiatrists distrust its use in the treatment of neurosis because they believe that it only tackles the symptoms of the trouble, leaving the underlying causes untouched. Nevertheless, within alternative medicine, hypnotherapy is frequently used to treat many kinds of emotional and physical problems.

Neither a professional board nor the government regulates the practice of hypnotherapy, and there is some concern about how it is taught and practiced. Practitioners do not have to know anything about orthodox medicine; consequently, they may attempt to treat ailments that are physically based rather than stress-related. By alleviating the pain, hypnotherapists can remove the body's warning of a possible serious dysfunction.

For further reading: Robert A. Baker, *They Call It Hypnosis* (Prometheus Books, 1990).

I

I CHING An ancient Chinese text, one of the five classics of Confucianism. Devotees believe it to be a book of life, containing within it a system of cosmology that gives an explanation of the entire laws of the universe by which everything is governed. It also contains explicit instructions on how humans should conduct themselves to remain continually in harmony with those laws.

For many years the *I Ching* has been used for DIVINATION (fortune-telling), the cryptic text allowing the user leeway in interpreting its significance. Today it has enjoyed a popular revival as a book of guidance with members of the NEW AGE MOVEMENT. Carl Gustav JUNG, the analytic psychologist and cult hero of many New Agers, was very influential in promoting the modern usage of the book. The procedure of consultation is very complex and like a ritual, taking many hours or even days to answer a single question. Questions must be put one at a time or ambiguity will arise.

Simply described, the system comprises 64 symbolic hexagrams, each of which is built up line by line from the bottom by casting lots using dice, coins, or sticks. A hexagram may emerge with entirely whole lines, entirely broken lines, or some combination of the two. Whole lines represent yang, the male principle, while broken lines represent yin, the female principle. These two principles form a different pattern within each hexagram, and as the total number of hexagrams is built up and put together, they are said, if properly understood and interpreted, to contain profound meanings applicable to daily life and, some say, capable of controlling future events.

Some scholars are troubled that the *I Ching* is included among the Confucian classics because Confucius deliberately avoided creating esoteric doctrines. The inclusion happened in the second century B.C.E., during the Han dynasty, when followers of Confucius, influenced by the Taoist quest for immortality, justified their use of the *I Ching* text by attributing certain of the commentaries to Confucius. Whether true or not, the original *I Ching* is said to have predated Confucius in China. It was only three centuries ago that Jesuit missionaries brought it to Europe, where it has been used ever since, along with Virgil's *Aeneid* and the Holy Bible, as an oracle book. In the modern era, scientific orthodoxy and many mainstream churches have marginalized this form of occultism as speculative and perverse. The usage and practices of the *I Ching,* whether resulting in benign or malign consequences, are classed as pseudoscience.

For further reading: Max Maven, *Max Maven's Book of Fortunetelling: The Complete Guide to Augury, Soothsaying, and Divination* (Prentice Hall, 1992).

INCORRUPTIBILITY Freedom from decay of a corpse that has been reported to extend for weeks, months, and even years in spite of the absence of any artificial preservative process such as embalming, freezing, or placement in a sealed metal coffin. In the Western Christian tradition, incorruptibility has been reported in the case of several saints, most prominently St. Francis de Sales, who died in 1622 and whose body remained undecayed for a decade; St. Francis of Paula, whose body was preserved for 55 years; and St. Francis Antonius of Florence, whose body was reported uncorrupted 130 years after death.

Accounts of such events raise several questions. Science has yet to investigate and explain such incidents, the very difficulty of obtaining such a corpse (which takes on the characteristics of a sacred object) for investigation being a significant problem. Theologically, the argument that such incidents are evidence of miracles has its own problems. While there are numerous incidents of incorruptibility, some of the most notable saints—St. Francis of

I Ching

Assisi or St. Catherine of Siena, for example—have not been granted such a grace. On the other hand, while not as well known, incorruptibility has also been observed among leaders of other religious traditions, as has the related experience of an unnatural perfume issuing from such bodies instead of the smell of decay.

In Eastern Christianity, incorruptibility carries a very different evil connotation. It is a sign that the Earth will not receive the body of an especially evil person and that such a being is likely to become a vampire. Bodies which do not decay as expected are often mutilated (beheaded, staked, etc.) and abused.

INCUBUS AND SUCCUBUS A male spirit or DEMON that visits a sleeping woman for purposes of having sexual intercourse with her; a female demon who likewise visits a sleeping man. This is a particularly vivid manifestation of two very basic and common ideas in human history: The first is that sexual intercourse can take place between mortals and gods or supernatural beings; the second is that NIGHTMARES are the result of an external will, such as that of a lewd demon. The word INCUBUS comes from the late Latin *incubus* and *incubo,* meaning "nightmare." These derive from the classical Latin *incubare,* meaning "to lie upon" (with *succubare* meaning "to lie under"). The second entry under "incubus" in the *Oxford English Dictionary* defines it as "a feeling of oppression during sleep, as of some heavy weight on the chest and stomach; the nightmare."

The incubus may remain invisible to the woman he is visiting, but if he appears to her, he normally does so in the form of her lover. Despite the familiar form, the woman finds the sex unpleasant, although the erotic element makes it an ambiguous experience. Often the incubus has been thought to subject the woman to sexual depravity, lust, and terrifying nightmares. Incubi were generally deemed sterile, but they were able to impregnate women with semen collected from nocturnal emissions of men. Reports of incubi visitations were always more common among women than men and among virgins and widows than among married women; they were most prevalent among nuns.

It is useful to distinguish between interpretations of the incubus and the succubus that believe that an external will is involved and those that assume that there is nothing more than a DREAM to be explained in scientific terms. For the mythic incubus, the medieval Christian version of the phenomenon offered the most thoughtful interpretation, emphasizing the element of sin. In St. Augustine's *City of God* (XV, Ch. 23) and St. Thomas Aquinas's *Summa Theologiea* (I, Q. 51, A. 3–6), incubi were interpreted as demons sent to tempt humanity. Incubi and succubi were thus understood as a test of one's strength or virtue in the face of temptation; therefore women who were susceptible to them were associated with depravity or WITCHCRAFT.

For science, the incubus is something that appears in a dream and dreams are explained entirely with reference to psychology, broadly defined. The association of demons with dreaming has almost entirely lost its hold on the modern mind since the 19th century and the emergence of psychology. People are now perfectly willing to believe that what they experience while asleep has to do with their physical minds—influenced by external factors to be sure, but not by external entities with conscious intent.

The Freudian interpretation focuses on the erotic element in the encounter with the incubus or succubus. In the dream, there is a tension between pleasure and pain, desire and fear. This is interpreted within the doctrine of intrapsychical repression, where the negative elements of the experience result from the attempt to repress the erotic desires. This repression takes place through the mechanism of projecting the erotic desires onto an external agent or demon. It is argued by Freudians that the tendency to make such projections was reinforced by Christian theology in the Middle Ages.

Because some scientists also consider Freudian psychology mythic, recent psychology proposes to understand dreams and nightmares in more biological and neuropsychological terms. The experience of what may still be called an incubus consists of nothing more than these three factors: dreaming at the onset of sleep, a sense of impending doom, and the perception of paralysis. The traditional content—the vivid imagery of demonic intercourse—is no longer of pressing concern because this content has faded from the modern imagination. At work in psychological terms is an hallucination, something that by definition occurs just as one is falling asleep and therefore still partially conscious. The hallucinations that occur are the result of imagery from the rapid eye movement (REM) or dreaming state of sleep intruding upon the conscious mind during the transition from wakefulness to sleep. This may also occur during the transition from sleep to wakefulness.

See also FREUD, SIGMUND; and POSSESSION.

INSTINCT A biological term meaning the inborn tendency to behave in a way characteristic of a species. Over the years, scientific concepts of what constitutes instinct and whether it can apply equally to humans as well as to animals has undergone many changes.

Some early philosophers held that human behavior is controlled by rational thinking, that actions can be accounted for by knowledge of and desire for certain consequences. In short, humans have free will. The corollary of this is that animals are driven by blind instinct, and therefore their actions are determined by their nature. However, modern science, beginning with Charles DARWIN, views humans as just another species of animal. Darwin treated instincts as complex reflexes that were inherited and therefore subject to natural selection. In his view such instincts evolve alongside other aspects of the animal such as their physical form and their behavior.

In the first half of the 20th century, two diverging views of instincts emerged. Konrad Lorenz, an eminent researcher of animal behavior, epitomized one position: His basic argument was that humans, like the rest of nature, are "red in tooth and claw," but because of their superior inventive powers, human beings are now escaping from the Darwinian laws of selection and adaptation and will shortly be left at the mercy of their own aggressive instincts. At the other extreme, the behaviorists, notably B. F. Skinner, postulated that humans were mere stimulus-responsive machines molded mainly by rewards and punishments, the only exception being a few basic learned rules. In an attempt to reconcile the two positions, psychologists sought to identify an instinctive "force" or "drive" for every aspect of behavior—a hunger drive, a thirst drive, a sex drive, and so on. This proved to be an impossible premise when it resulted in deriving a drive for thumb-sucking and another for nail-biting, for example.

More modern views retreat from the extremes of the NATURE-NURTURE DEBATE and abandon the notion of instinctive "forces" by recognizing that the extent to which an animal's nature or nurture determines its behavior varies from species to species and activity to activity. Even what was once thought to be inbred activity has been shown to be influenced to some extent by experience within a certain environment. Currently, scientists are studying how genes affect behavior. It appears logical to abandon this strict nature/nurture dichotomy and acknowledge all behavior to be influenced both by the animal's genetic make up and by the environmental conditions that existed during its development.

For further reading: Richard L. Gregory, *The Oxford Companion to the Mind* (Oxford University Press, 1987).

INSTITUTE FOR CREATION RESEARCH The primary organization presenting scientific creationism. In 1970, Dr. Henry Morris left his career in electrical engineering to help found and become academic dean of Christian Heritage College; the college was to be built on an understanding of creationism, biblical authority, and a conservative Protestant, but transdenomenational, Christianity.

At the same time, Morris founded the CHRISTIAN SCIENCE Research Center at the college and became its director. However, two years later, Morris reorganized it at the Institute for Creation Research. It remained a division of the college until 1981 when it became an independent research facility.

The institute, in espousing scientific creationism, one of a variety of creationist perspectives, has argued that the creation model is at least as scientific as the evolutionary model and can stand on the same nonreligious scientific basis, quite apart from biblical revelation. Thus, the institute holds, it is a viable alternative to be presented in the public schools, and local school boards should be free to adopt creationist literature and teach it to the children and youth in attendance.

In 1981, at the time when formal ties were broken with Christian Heritage College, the institute founded a graduate school to offer master's degrees in astro/geophysics, geology, biology, and science education.

The institute's staff presents its positions in public lectures; a weekly radio show, *Science, Scripture, and Salvation;* and two periodicals, *Acts & Facts,* a newsletter, and *Days of Praise,* a daily devotional quarterly. The institute, which has also published several textbooks, is the most active and visible of several creationist organizations.

See also CREATION SCIENCE; GISH, DUANE T.

For further reading: Henry M. Morris, *What Is Creation Science?* (Creation Life Publishers, 1982); Henry M. Morris, *A History of Modern Creationism* (Master Book Publishers, 1984); and Ronald Numbers, *The Creationists* (University of California Press, 1993).

INSTITUTE FÜR GRENZGEBIETE DER PSYCHOLOGIE UND PSYCHOLHYGIENE (INSTITUTE FOR BORDER AREAS OF PSYCHOLOGY AND MENTAL HYGIENE) Leading German parapsychological organization founded in 1950 by Hans Bender. The institute has been a symbol of the rebirth of German parapsychological research.

Bender had become interested in psychic research and did research on AUTOMATIC WRITING at the College of Paris. He finished his doctorate at Bonn just before the Nazis suppressed PARAPSYCHOLOGY. After World War II, Bender taught at Albert-Ludwig University at Freiburg, where he began to introduce parapsychological material into his lectures. In 1954 he was named to the newly established chair in parapsychology at Freiburg. Bender took the lead in developing instrumented research in parapsychology. He had a particular interest in poltergeist cases and developed the use of video cameras to record phenomena that had previously only been described in the literature.

Bender chose the name of the institute carefully. He viewed the institute as having two roles—scientific research and education. The education program deals on the one hand with those holding irrational and superstitious beliefs, and on the other hand with those whose understanding of the paranormal is purely negative and/or skeptical. The institute's journal, *Zeitschrift für Parapsychologie und Grenzgebiete der Psychologie,* has been a major instrument in its educational endeavor.

INTELLIGENCE TESTS Attempts to measure the quantity and quality of human intelligence. One of the principal themes in the study of human behavior has been "biological determinism": that people's behavior, as well as their social and economic circumstances, come from inherent qualities that are passed from parents to their children. The idea that the phrase expresses is as old as the Greeks: Plato credits Socrates with developing it. Some 19th- and 20th-century scientists concluded that almost all human qualities could be traced to inherited characteristics. One of their primary aims was to devise a way to measure people's worth and group them on the basis of their intelligence. To do this, however, they made some erroneous assumptions. They concluded that intelligence was a single quality that could be determined through a simple test and, sometimes, that intelligence was directly related to a person's race.

Most of the biological determinists working in the 19th century tried to group intelligence testing into two general categories. One group believed that intelligence was directly related to brain size and could be predicted by measuring the skull of an individual, a study known as *craniometry.* In his *Crania Americana* (1839), Samuel George Morton gathered craniometrical data that, he believed, proved the racial inferiority of Native Americans to white Europeans. Craniometry lost its status as a respectable science when a later examination of the skulls Morton used showed that he had falsified his data. Later examiners, including Frenchman Paul Broca (1824–80), perhaps the best-known and most respected adherent of craniometry, also tried to find a link between brain size and intelligence in humans, but their results were inconclusive.

The other group believed that intelligence could be measured through certain types of testing. Chief among these were the IQ TESTS developed by Frenchman Alfred BINET and adopted by such Americans as H. H. Goddard, Lewis M. Terman, and R. M. Yerkes. Binet originally intended to use the tests to identify students who might require special attention because they were not performing well in regular classrooms. However, Goddard, Terman, and Yerkes, who brought the Binet test to the United States, corrupted Binet's original intentions to support their own prejudices about society.

Although the tests and measurements were by themselves fairly harmless, they were taken over by politicians and used to support social programs intended to "improve" society—attitudes that were linked to racist and anti-immigrant feelings prevalent in the country during the same period. Racists in the antebellum South cited cranial measurements to prove that blacks were mentally inferior to whites. Terman classified poor people and what he called "inferior races" as low performers not recognizing that his tests were biased in favor of wealthy white children. Intelligence testing developed by Yerkes—again with a bias in favor of middle-class and wealthy white Americans—spread widely during World War I as testers tried to group soldiers and identify potential officers by measuring their intelligence.

In the United States of the 1920s, researchers such as Goddard and Terman cited the results of IQ tests to argue that potential immigrants with a low score should be excluded. Congress passed the Immigrant Restriction Act of 1924 based on information derived from these intelligence tests. The government reduced the number of immigrants from southern and eastern Europe because people from these areas scored lower on the tests than did people from northern and western Europe.

Local and state governments used the data gathered from the intelligence tests in their own social programs; for example, between 1924 and 1972 the State of Virginia forcibly sterilized more than 7,500 men and women who tested mentally deficient on standardized intelligence tests. The state believed that any children born to these people would inherit their parents' feeblemindedness and become a burden to society—dependent on state support.

For further reading: Paul Ehrlich and S. Shirley Feldman, *The Race Bomb: Skin Color, Prejudice, and Intelligence* (New York Times Book Co., 1977); Stephen Jay Gould, *The Mismeasure of Man* (W.W. Norton, 1981); and Ashley Montagu, ed., *Race and IQ*, expanded ed. (Oxford University Press, 1996).

INTERNATIONAL FORTEAN ORGANIZATION (INFO) An organization, founded in 1965 by Ronald J. Willis, devoted to research and study of unexplained natural phenomena. INFO filled the vacuum left by the demise of the FORTEAN SOCIETY five year previously. The books of Charles Hoy FORT and the efforts of the society had created a community interested in exploring strange, anomalous phenomena. A number of Forteans believed that in its later years the society had lost its original focus, and they observed that many of those attracted to it had been people also drawn to various alternative social movements and beliefs. The society's energy was drained in support of their extraneous causes.

INFO emerged as an organization more closely focused on unexplained phenomena. INFO's *Journal* publishes articles on strange and unexplained phenomena and updates information on items discussed in Fort's books (many of which have, in the intervening decades, been more closely investigated and thus the subject of more conventional explanations). INFO, currently headquartered in Arlington, Virginia, sponsors an annual convention that supplies a platform for the many researchers and writers in the field.

INTERNATIONAL SOCIETY OF CRYPTOZOOLOGY An organization founded in 1982 (one of its founders was Roy MACKAL) to promote the study of CRYPTOZOOLOGY. It publishes *Cryptozoology* and *ISC Newsletter* and acts as the discipline's professional body. Its president is Dr. Bernard HEUVELMANS, recognized as the founder of the discipline, a professor of zoology and the author of *On the Track of Unknown Animals* (1995).

The society now initiates studies and expeditions, researching as many as 150 possible animals. To dismiss their findings as mythological, pseudoscientific, or merely folkloric legends would be mistaken. The work of cryptozoologists in the last 50 years has resulted in acceptance of the existence of several animals hitherto thought to be either extinct or mythical. It is reasonable to expect that more will be discovered. There is a certain urgency to this work because so much of the relatively inaccessible areas of the world, where such animals might be found, are being destroyed to meet the demands of the expanding population and its expectations.

For further reading: Bernard Heuvelmans, *On the Track of Unknown Animals*, 3d ed. (Kegan Paul, 1995); and Karl Shuker, *In Search of Prehistoric Survivors: Do Giant "Extinct" Creatures Still Exist?* (Blandford, 1995).

INVISIBILITY The process through which a person or a thing can be rendered unseen. Science fiction has long offered the idea that invisibility is viable. One of the most famous works of British science fiction writer H. G. Wells is *The Invisible Man* (1897). The book tells of the adventures of a misanthropic scientist who discovers the secret of making himself invisible. Since that time, the idea of invisibility has been a staple theme of science fiction.

A more recent use of invisibility occurs on the television show *Star Trek*. Some starships on the program use a "cloaking device" that renders them invisible to sensors—including light-sensitive devices. Lawrence Krauss, in his book *The Physics of Star Trek*, suggests that the starships may use a device that is powered by antimatter and is related to the "warp drive" to bend light around themselves. Another example of reputed "invisibility" is the PHILADELPHIA EXPERIMENT, in which the U.S. government reputedly rendered the battleship USS *Eldridge* invisible.

Physicists see two problems with the science-fiction idea of invisibility. The first is that an invisible person, by definition, would not be able to see anything. If light is being bent around a person, then no light can penetrate to that being's eyes. Anyone trying to operate a "cloaking device" on a spaceship would be entirely in the dark, unable to perceive anything about his or her own position or surroundings.

The second problem has to do with the type of device needed to create the cloaking effect. It may be possible to bend light around an object, but the process requires a very powerful gravitational field; in fact, the presence of such a field would be just as revealing as the light itself. BLACK HOLES are by nature invisible—they pull light rays into their own gravitational field—but they are not undetectable. A personal gravitational field such as that generated by a black hole would render a person technically invisible, but the apparatus needed to generate such a field would be impossible for anyone to carry. In addition, the gravitational field would tear the user apart.

IVY LEAGUE POSTURE PHOTOGRAPHS Nude or seminude photographs, in front, side, and rear positions, required of freshman students at most Ivy League and Seven Sisters schools from the 1940s through the 1960s. At Harvard the practice started as early as 1880. While students were told that the photos were taken to assess posture, some participating schools and an archive in Washington, D.C., made the photos available to university and corporate researchers for a variety of purposes without subjects' permission. Although Yale reports destroying its collection of posture photographs, hundreds of photos from Yale and other institutions are thought still to exist, and rumors of lost or stolen posture photos have circulated widely.

News of the posture photographs received national attention in 1992, when Naomi Wolf, author of *The Beauty Myth,* wrote about them in an op-ed article in *The New York Times.* Shortly afterward, Yale art history professor George Hersey wrote a letter to the *Times,* in which he suggested that the photos were used primarily for anthropological research similar to that conducted by the Nazis. According to Hersey, participating institutions were taking these photos in order to draw conclusions about the subjects' intelligence, temperament, moral worth, and probable future achievement for eugenic purposes. Hundreds of nude posture photos from Harvard illustrate the pages of W. H. Sheldon's book on body types, *Atlas of Man.*

IQ TESTS Ratio of a person's "mental age" to their physical age. In 1904 Alfred BINET, director of the psychology laboratory at the Université de Sorbonne, devised a series of tests for the French government. The tests were intended to identify students who might require special attention because they were not performing well in regular classrooms. Originally Binet's tests were arranged in order of increasing difficulty, but in 1908 he sorted them according to age level, dividing them according to the youngest age at which a child could complete the test successfully. He suggested that children who needed help could be identified by subtracting their chronological age from their tested "mental age." In 1912, a German psychologist suggested that the identifier should be computed by dividing the mental age by the chronological age, and not by subtracting. Dividing allowed measurement of the relative difference between the two qualities, whereas subtracting did not. From this comes the term *intelligence quotient,* or *IQ.*

Binet developed his test solely as a means of identifying children who needed help in schooling. However, three scientists brought the Binet test to the United States and corrupted Binet's original intentions to support their prejudices about society. H. H. Goddard, who believed that intelligence was passed from parent to child, interpreted the tests as a method of measuring this inborn intelligence. L. M. Terman did not recognize the effects that poverty or culture might have on test results and believed in the inferiority of non-Western races. He refined the original Binet scale and suggested that IQ tests could be used to assign people to the jobs they were best suited for. R. M. Yerkes gave the IQ tests that Terman developed to U.S. soldiers during World War I—again in a form that was biased in favor of white Americans of western European origin. The U.S. Congress used the data that Yerkes compiled to pass the Immigration Restriction Act of 1924, which discriminated against poorly educated people who did not perform well on IQ tests.

See also INTELLIGENCE TESTS.

IRIDOLOGY A method of diagnosing illnesses from the appearance of, and changes in, the iris of the eye. It is a diagnostic technique and not a treatment. A sectional map is made of the iris with precise positions located that refer to parts or functions of the body or head. If a spot or color change is found in a certain position on the iris, the practitioner will be able to locate the exact seat of the illness. The system was invented by Ignatz Peczely, a Budapest doctor, who published a book in 1880 that at the time inspired homeopaths practicing in Germany and Sweden. The practice of iridology was introduced into the United States by Henry E. Lahn, who in 1904 wrote the first book published in English on the subject. Later, in 1917, a pupil of Lahn's, Henry Lindlahr, produced a definitive study of naturopathic diagnosis called *Iridiagnosis and other Diagnostic Methods.*

Lindlahr wrote that Peczely made this great scientific discovery when, at the age of 10, he caught an owl and accidentally broke its foot. At the precise moment that the bone snapped, he was looking into the bird's eyes and noticed the sudden appearance of a black spot in the lower central region of its iris. Taking the bird home for a pet, the boy found that as the leg healed the black spot developed a white border, which was, according to Ignatz, related to the formation of the scar tissue around the bone.

According to today's iridiagnosticians, the iris is divided into sections like the portions of a cake and also into six rings or zones. The innermost zone relates to the stomach, the second to the intestines, the third to the blood and lymph system, the fourth to glands and organs, the fifth to muscles and the skeleton, and the sixth and outermost zone to the skin and the elimination of waste products. The left iris corresponds to the left side of the body, and the right to the right side, but many functions are reflected in both irises. Generally the top part of the iris will relate to organs in the upper body, and the bottom part to lower body organs. Dark marks indicate the part of the body that is not working properly, and white marks indicate stress. A dark rim round the edge of the iris shows poisons are accumulating in the skin because of inefficient elimination of waste products.

Although medical practitioners have always looked into their patients' eyes for clues to gauge their well-being, such as yellowing, dullness, or unnatural brilliance, few would accept the detailed linking of areas of the iris to parts of the body. The iris is not a map of the body. Most traditional doctors consider iridology to be a pseudoscience that is dangerous because it is more than likely to fail to diagnose a condition that may need urgent surgery or drug control. Many iridologists give only diagnosis, but some also devise treatments suitable to their diagnoses, and in cases like this, a second opinion from an orthodox doctor should always be sought.

See also HOMEOPATHY; NATUROPATHY.

J

J. ALLEN HYNEK CENTER FOR UFO STUDIES A scientific research center, the J. Allen Hynek Center for UFO Studies grew out of the reaction of J. Allen HYNEK to the devastating effects of the CONDON REPORT (1968–69) on the study of unidentified flying objects (flying saucers). As an astronomer at the Ohio State University, Hynek had been drawn into the UFO controversy as early as 1948 when the U.S. Air Force asked him to examine and help them understand the strange reports they were receiving. However, in the 1960s he became drawn more intensely into the controversy and in 1966 appeared at a hearing before the U.S. House of Representatives Armed Services Committee arguing for a civilian examination of the Air Force files. That study was held at the University of Colorado under the leadership of Edward U. Condon. The Condon Report argued that the study of UFOs was of no scientific value. In the years after the report, Hynek emerged as the leading scientist who would speak positively on the UFO issue. In 1972 he wrote his most important book, *The UFO Experience,* in which he argued that the Air Force had been incompetent in handling UFO reports and had treated some scientifically interesting material with indifference. He also began to create a network of scientists and scholars to consider the UFO problem.

Then, in November 1973, during a conversation with his friend Sherman Larsen, Hynek expressed his frustration at being unable to stimulate further interest in the UFO question. Larsen suggested that an organization might help. He actually still had control of a small organization that had grown out of the demise of one of the early UFO study groups, the National Investigations Committee on Aerial Phenomena (NICE). That corporation became the seed from which the Center for UFO Studies (CUFOS) was generated. The first center was opened in Evanston, Illinois, near Hynek's office at Northwestern University in 1973. It quickly emerged as the major scientific organization looking at UFO phenomena.

During the late 1970s the program developed to its fullest. The more substantial *International UFO Reporter* replaced the earlier newsletter. A series of conferences was organized, a journal was issued sporadically, and a variety of monographs on investigations of important sightings were published. In 1985, the *Reporter* was turned over to Jerome Clark, under whose editorship it focused on the discussion of critical issues and cases in UFOLOGY.

In 1985, Hynek moved to Arizona. He died there a year later. He was succeeded as head of CUFOS by sociologist Mark Rodeghier. The center assumed its present name a short time later. Under Rodeghier's leadership, amid the waxing and waning public interest in UFOs, the center has been able to expand and has become the major intellectual focus of study of the UFO phenomenon, though it has never been as large at the MUTUAL UFO NETWORK.

JAMES, WILLIAM (1842–1910) Outstanding philosopher and psychologist, who was also a pioneer in PSYCHICAL RESEARCH in the United States. James began his long and distinguished career following his graduation from Harvard Medical School in 1869. Two years later, he became an instructor in anatomy and physiology at Harvard College, where he was deeply involved in the emergence of psychology as a science distinct from both philosophy and physiology. In 1890 he published his classic text, *The Principles of Psychology,* which established psychology as a lab science subject to experimentation. He then turned to philosophical concerns and wrote his important essays on pragmatism,his career culminating in his influential work, *The Varieties of Religious Experience* (1902).

James became interested in psychical research out of both psychological and philosophical concerns. He participated in the formation of the SOCIETY FOR PSYCHICAL RESEARCH (1882) and served as its president for the 1884–85 term and vice-president from 1896 to his death. He was also active in the formation of the AMERICAN SOCIETY FOR PSYCHICAL RESEARCH (1885). In 1885 he became aware of Leonora Piper (1859–1950), a MEDIUM in Boston. Greatly impressed by his initial visit, he conducted extensive observations of her for a period of 18 months and felt that he had found in her significant evidence suggestive of human survival of bodily death. Piper, who lived into her 90s, would later be studied by a number of prominent researchers, and while many of the mediums would be caught in fraudulent activity, Piper's reputation has survived as one of the more extraordinary subjects of psychical research. She was the occasion of one of James's most famous quotes: "If you wish to upset the conclusion that all crows are black, there is no need to seek demonstration that no crow is black; it is sufficient to produce one white crow; a single one is sufficient . . . My white crow is Mrs. Piper."

James authored a number of important essays on psychical research including "What Psychical Research Has Accomplished" (1897). Toward the end of his quarter-century of work, he noted his belief that most of the phenomena of psychical research were rooted in reality. He also concluded that the spiritistic hypothesis (that certain phenomena are due to the activity of spirits) was unproved, and opted instead to explain them as due to various faculties in the human mind and consciousness.

Less known than his work in psychical research were his explorations in various forms of alternative healing, though some are discussed in *The Varieties of Religious Experience*. James was himself a victim of insomnia for a period of his life and in seeking a cure visited Lydia E. PINKHAM, the developer of a famous NOSTRUM for women.

JANUARIUS, ST., MIRACLE OF Miracle attributed to the blood of the bishop of Benevento and patron saint of Naples, believed to have been martyred in 305 C.E. Januarius's exact identity is unknown; in fact, there is no evidence that he was an historical person. His fame rests solely on a relic: a vial of what is alleged to be his dried blood that is kept in Naples Cathedral and frequently shown to the public when, after varying intervals of time, it is observed to liquefy. There is no historical record of the "blood" prior to 1389, over one thousand years after the presumed date of Januarius's death. There are some twenty other saints' bloods, with similar properties, to be found in and around the Naples area, which probably indicates the existence of a local artificial blood manufacturing technique in the 14th century. This was a period when the mass manufacture of relics was prevalent throughout Italy. It was not until the Council of Trent (1545–63) that reforms were brought about and rules laid down to clean up the trade and ensure the authenticity of relics.

No detailed scientific examination of the "blood" has ever been allowed, and descriptions of what happens vary. Some say beeswax has been mixed with the dried blood; others believe that the mixture is of olive oil and beeswax, with some red pigment such as dragon's blood added. Whichever is the case, it is thought that it is the beeswax that melts when the vial is brought out into the heat. But at the same time as the contents of the vial turn to liquid, an increase in weight and volume has been noticed; this would not happen to a blend of beeswax and blood or olive oil. Taking this increase in weight and volume into account, it is more likely to be caused by deliquescence. There are many substances which deliquesce—that is, have the property of picking up moisture from the air to such an extent as to dissolve in it; in other words, they become liquid when exposed to humid air (air high in water vapor). Scientists do not know what substance is in the vial, but whatever it is it cannot be blood, as blood is not deliquescent. When blood is released from the body it oxidizes, turns brown, and is attacked by fungi, going moldy.

Recently members of the Italian COMMITTEE FOR THE SCIENTIFIC INVESTIGATION OF CLAIMS OF THE PARANORMAL (CSICOP) took a new look at the phenomenon. This team, as with other scientists, has not been allowed to inspect the relic (nor other relics in Italy for which similar claims have been made). They experimented with materials that could replicate the behavior of the contents of the vial. Their efforts produced vials of a dark brown solid, resembling dried blood which, when tapped or shaken, turned to liquid. The property of such solids is called "thixotropy" and it has a number of applications, the most familiar of which is in non-drip thixotropic paint. Whatever the "blood" comprises, it is now thought not to be a genuine relic of the saint—if in fact he did exist.

For further reading: Joe Nickell and John F. Fischer, *Mysterious Realms: Probing Paranormal, Historical, and Forensic Enigmas* (Prometheus Books, 1992).

JENNY HANIVER A term for an artificial monster made out of the remains of certain fishes, especially skates or rays. The nostrils, mouths, and gills of these fishes in their natural state look rather like an evil, grimacing face. With judicious cutting, partially separating the pectoral fins from the body, they can be made to resemble monsters: winged snakes, dragons, or whatever other creature the manufacturer thinks credulous visitors will buy. In medieval times, Jenny Hanivers were often passed off as basilisks—mythological creatures who were so ugly and poisonous that their very breath could kill. Buyers believed that the basilisk's hide, when made into a medicine, would prove a potent remedy for poisoning.

Sixteenth-century Swiss naturalist Conrad Gesner, who was the first writer since classical times to compile an encyclopedia of animal life, described a Jenny Haniver in his *Historia Animalium* (1558). Gesner published a drawing of the Jenny Haniver, along with a description of one that he saw in a shop in Zürich. Gesner stated that the shopkeeper—who also sold quack medicines—believed that the monster was a small basilisk. Gesner himself, however, explained that the Jenny Haniver had in fact been made from a ray, and he warned his readers not to mistake such forgeries for winged serpents or small dragons.

Gesner was only the first of many writers about natural science to tell the truth about Jenny Hanivers. Henri-Louis Duhamel du Monceau's *Traité générale des pesches* (1777) also exposed Jenny Hanivers as frauds. Yet, despite this well-publicized debunking, Jenny Hanivers continued to be made and passed off as genuine monsters until modern times. Jenny Hanivers are preserved in collections throughout Europe. In the early 20th century, a couple of them were manufactured in the United States, but they were regarded as curiosities rather than monsters.

JENSEN, ARTHUR (1923–) A prominent U.S. educational psychologist who has made important contributions to the study of intelligence from his graduate student days to the present. He maintained in an article published in the *Harvard Educational Review* in 1969 that there is an hereditary difference in the average Intelligence Quotient (IQ) of U.S. whites and blacks—that blacks are of inferior intelligence on average. He has researched this subject ever since, and though his thesis is hotly contested, has continued to insist on the validity and applicability of his findings. In 1979 Jensen reinforced his earlier article with the book *Bias in Mental Testing* (Free Press). Jensen extends his intelligence scale to include every form of living being from amoeba to man and even posits that higher extraterrestrial intelligence is in his hierarchy.

Jensen joins a long line of intelligence measurers—Francis GALTON, Alfred BINET, and Sir Cyril BURT—and picks up where they left off. Their shared and evolving belief is that intelligence can be measured on a simple unidimensional scale and that it is both a measure and a predictor of performance. In the development of intelligence testing, it became apparent that people perform differently in different areas—artistic, mathematical, literary—and so specific tests were devised. Jensen believes that from these different measures, a general measure, for which the symbol *g* is adopted, can be derived. This is overriding and universal: *general intelligence,* which can be measured and used as a scale on which to rank humans and groups of humans. Jensen extends his theory to include life forms that are supposedly less or greater than human. His faith in g is clear in the following quote from his book *Straight Talk about Mental Tests* (1981):

> *Although psychologists can devise tests that measure only one group factor, they cannot devise a test that excludes g. . . . The ubiquitous common factor to all tests is g, which has been aptly referred to as the primary mental ability . . . and the same g permeates scholastic achievement and many types of job performance, especially so-called higher level jobs. Therefore, g is most worthy of our scientific curiosity.*

There are both scientific and social problems with Jensen's theories. 1) They presuppose an anthropocentric outlook, that is, one in which humans are taken to be the ultimate purpose and pinnacle of evolution, conceived as a progressive process. Actually we are relative newcomers on one branch of the evolutionary tree, a branch that may quite possibly be short-lived. Other life forms at the ends of other evolutionary branches are as well or better adapted to their particular niches and are neither higher nor lower than human beings but are equally the products of their species' evolutionary history. Many of those life forms have survived successfully under dramatically varied conditions for much longer than have humans, some as many as hundreds of millions of years. 2) Scientists have demonstrated that if you perform the mathematical process to abstract the measure of general intelligence from the several separate measures of abilities, what you are left with is a number that has no real significance. 3) The results of Jensen's measurements and calculations are used to support culturally and socially generated racist and sexist prejudices. Blacks in the United States, for example, suffer severe handicaps in social and economic life. To argue on the basis of these highly suspect measurements that their have-not position in U.S. society is an inevitable consequence of their lower general intelligence is to justify social and economic inequity.

See also INTELLIGENCE TESTS; IQ TESTS; SOCIAL DARWINISM.

For further reading: Arthur Jensen, *Straight Talk about Mental Tests* (Free Press, 1981); Stephen Jay Gould, *The Mismeasure of Man* (Norton/Penguin, 1981); Raymond E. Fancher, *The Intelligence Men* (Norton, 1987); R. J. Hernstein & C. Murray, *The Bell Curve* (Free Press, 1994).

JIN SHIN DO A form of healing bodywork developed as a variation on JIN SHIN JYUTSU by Ron and Iona Teegarten in California in the 1970s. Jin Shin Jyutsu is a form of ACUPRESSURE in which various points on the body are massaged in a set sequence. Following their introduction to Jin Shin Jyutsu in the 1960s, the Teegartens began to explore a means of addressing what they felt were the emotional states underlying pains and body tensions. They taught those who came to them for bodywork the techniques of YOGA breathing, exercise, and MEDITATION.

The practice of Jin Shin Do continues to draw on the Chinese understanding of the self, which pictures the body as permeated with an invisible network of pathways (called meridians) that allow for the flow of universal energy (*chi*) through the body. Health is coincidental with the free and balanced flow of *chi* to all parts of the body. There is little evidence that the *chi* is a healing force that is capable of affecting specific diseases. However, that one may attain an overall positive effect from practicing Jin Shin Do with its exercise and mediation components is readily understandable.

JIN SHIN JYUTSU A form of healing practice developed in Japan in the 1940s by Jiro Mirai. Literally the "Art of Universal Energy in Compassionate Man," Jin Shin Jyutsu is based on ACUPUNCTURE and is a direct refinement of ACUPRESSURE. It draws on the perspective of the Chinese idea that the human body is enlivened by *chi* (universal energy), which flows through the body along a set of invisible paths called meridians. Various points on the meridians are directly connected to the different organs of the body. Illness is caused by the unbalanced or thwarted movement of *chi*. That movement can be restored by treatment that is directed at the points along the meridians. In acupuncture, needles are inserted into the skin at these points. In acupressure, massage with the fingertips replaces the stimulation provided by the needles. Jin Shin Jyutsu differs from acupressure in that the massage pressure along the meridians is applied in a set sequence of pairs of acupoints. A complete treatment may involve as many as 40 to 50 pairs of points.

In the 1950s, Mary Bermeister brought Jin Shin Jyutsu to the United States, where it joined other forms of acupressure such as SHIATSU. One U.S. variation on the practice is called Jin Shin Ryoho. While there is little evidence that Jin Shin Jyutsu and related acupressure procedures have any immediate and direct effect on specific diseases, Jin Shin Jyutsu has not been shown to do any harm to those who utilize it.

JOSEPH OF COPERTINO (1603–1663) Ecstatic Franciscan priest called the Flying Friar. He was born near Brindisi of very poor parents and in his youth was sickly and slow-witted, earning the nickname "the gaper" for his habit of wandering around with his mouth open. He joined the Franciscan order, working as a servant, and was eventually admitted as a novice, although he remained very backward in his studies. His life in the church was marked by a long succession of reported mystic happenings: He was most famous for his ability to levitate, of which 70 instances were recorded in 17 years. One of the most spectacular was a flight high above the altar to make secure images that had been placed there. On another occasion he helped workmen to erect a cross,

Joseph of Copertino

36 feet high, by flying in midair and lifting it into place "as if it were a straw." In 1767 he was canonized, not for his levitations, which were said to have disturbed his superiors, but for his extreme patience and humility. These feats were recorded long after Joseph of Copertino's death and are based on legend.

Scientists believe that persons or inanimate objects cannot levitate because this would entail the suspension of the natural law of gravity which anchors everything to the Earth. Without gravity we would all float around like the objects in a space ship.

See also LEVITATION.

JUNG, CARL GUSTAV (1875–1961) Swiss psychologist born in Kesswil, the son of a pastor of the Swiss Reform Church. Jung studied at both the universities of Basel and Zurich, receiving his degree in 1902, and had a background in biology, zoology, paleontology, and archaeology. He eventually formulated a psychological system based on broad realms of study and partly on the occult beliefs of his day. Two of his basic doctrines—that of the archetypes and that of SYNCHRONICITY—have been shown to be without the clinical basis that Jung claimed for them.

In 1907 Jung met Sigmund FREUD and there began a close association, with at first Jung defending Freudian theories. But in 1912, with the publication of Jung's *Psychology of the Unconscious,* Jung split with Freud over his very different concept of libido; Jung views this as total psychic energy, disagreeing with Freud that libido is restricted to sexuality. Biographers of both Jung and Freud say that the relationship was doomed from the start because both men were prone to dogmatic assertions. Neither were strictly scientifically orientated; each started from different preconceptions and their similarly stubborn characters drove them further and further apart. To explain neuroses, Freud tried to establish causal links going back to childhood that implied a somewhat mechanistic view of human nature. Jung's notion of analytical psychology instead placed humans into an historical context, which he said gave their lives meaning and dignity. This view, diametrically opposed to Freud's, implied a place in a purposeful universe, and so was metaphysical. While Freudian analysis dominated the academic world, Jung was adopted as a GURU by the NEW AGE MOVEMENT.

By the time of the split with Freud, Jung had spent two years on independent investigation into the symbolic significance of the content of Western and Eastern religions, myths, legends, literature and folktales. His best-known contribution to psychology was his notion of the collective unconscious (a reservoir of universal thoughts, feelings, and memories shared by all humanity) and the archetypes that he believed to be its manifestation. Six examples of archetypes we are thought to live by are: the Hero, the Orphan, the Wanderer, the Warrior, the Martyr and the Magician. Jung wrote of quite unsophisticated patients whose dreams exhibited parallels with archetypes from myths found in various traditions. In addition, patients would describe many pictorial elements they encountered in dreams, including: the snake, the Sun, the star, and particularly the mandala—a diagrammatic representation of the cosmos or some aspect of it that is used in Eastern religions as a focus for contemplation. Because Jungian therapy involved a thorough understanding of the human cultural preoccupation with such archetypes, he came to see the mandala as a symbol of individuation, or the sought-for unity of self, in which the parts are held together within a unity of the whole. Jung's overriding objective was neither metaphysics nor the philosophy of human nature so much as to be one who cures souls.

A central theme in Jungian theory is his distinction between introversion and extroversion, which he saw as two generalized orientations, one with the libido focused inwards on private experiences and feelings (introversion), the other with the libido focused outwards on external events (extroversion). Although he believed one of the two types dominates in any individual and thought this constitutionally determined, Jung also believed per-

Jung

sonalities can alternate between types, since each orientation, while it exists on the conscious level also subtends its opposite at the unconscious level. Therefore Jung believed that it would be unwise to label a person as wholly one type or the other, and saw the key to mental health as giving each orientation its proper role.

From his doctrine on archetypes, Jung extrapolated that each of us carries in our psyche an unconscious image of ourselves as the opposite gender. The female element in a man he called the *anima* (Latin, "spirit"), and the male element in a woman, the *animus* (Latin, "mind"). Myths in which the archetypical hero rescues the female in distress were, he thought, illustrations of the male need to rescue the female in himself. The animus was personified by Psyche, who in the Greek myth was loved by Eros but forbidden to look directly at him; she disobeyed and was abandoned, only to regain his love through hardship and much suffering. Jung thought this story shows the negative animus rebuilding itself, but said this does not always happen because the male element in women masks a very destructive mode, that, if left to develop unchecked, can completely take over the female psyche. Jung thought that the only cure powerful enough to correct this violent urge, once identified, might be seduction, a beating, or rape. Unfortunately, Jung sometimes tried out some of his most bizarre ideas, without foundation in science, on his patients, thus betraying his ambition to heal the soul.

Jung's ideas helped to form much of the early thinking of the New Age Movement. Although he equivocated with his usual wary erudition on some pseudoscientific ideas, Jung was a committed believer in ASTROLOGY, TELEPATHY, LEVITATION, and MATERIALIZATION. In fact, he coined the word SYNCHRONICITY to describe a paranormal force that creates coincidences. Furthermore he claimed to have himself seen PSYCHOKINESIS in action and to have actually dreamt future events including both World Wars.

In a recent book, *Against Therapy: Emotional Tyranny and the Myth of Psychological Healing* (1988), historian and critic Jeffrey Moussaieff Masson argues that any psychotherapeutic relationship, because of its unavoidable imbalance in power, is little more than a modern version of the insane asylum. In Masson's view, Jung was one of the more reprehensible practitioners of this psychological tyranny. Besides being an anti-Semite and an admirer of Hitler, Jung, under the guise of the "wise old man archetype," offered wild interpretations of his patients' experiences and dreams that had more to do with mythological voodoo than with science or healing.

For further reading: Richard L. Gregory, ed., *The Oxford Companion to the Mind* (Oxford University Press, 1987); Frank McLynn *Carl Gustav Jung* (Bantam, 1996); Jeffrey Moussaieff Masson, *Against Therapy: Emotional Tyranny and the Myth of Psychological Healing* (Atheneum, 1988); and Richard Noll, *The Jung Cult* (HarperCollins, 1996).

KABBALAH Reputed source of secret Jewish wisdom. Although the Hebrew word *kabbalah* actually means "tradition" and referred originally to the legal practices of Judaism, it has become associated with mystical Judaism; in particular, it refers to a system of obscure and mysterious practices that were formulated in the medieval period. The most famous of the kabbalistic texts, known as the *Zohar,* was composed by the scholar Moses de Leon around the year 1290. Jewish students of the kabbalah in southern Europe stressed speculation about the nature of the universe, while those in northern Europe applied the practices to social, ethical, and even magical themes. The study of the kabbalah spread during later medieval and Renaissance times, especially among the Jewish communities that were expelled from Spain in 1492 and Portugal in 1495. Jewish mystics such as Isaac ben Solomon Luria in the 16th century and Sabbatai Zevi in the 17th century used the kabbalah as the foundation for new schools of thought and the basis for personal messianic status. Luria's work with the kabbalah formed the core of Hasidic Judaism.

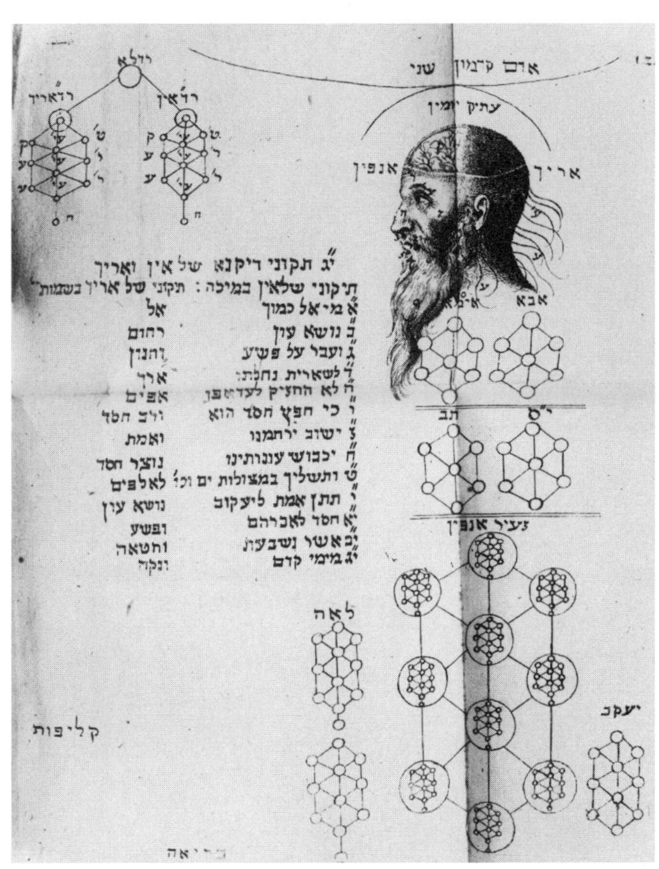

Kabbalah

Medieval Jewish thinkers derived the kabbalah from the Old Testament, using symbolic techniques based on combinations of letters and numbers. From the material revealed by these techniques, the kabbalists worked out some singular beliefs about the creation of the world and the redemption of humankind. They theorized that the world radiated from an unknowable God in a series of creations. They believed that there had been four of these creations, each one interlocking with the next. When Adam sinned, the immanent spirit or divine presence of God was exiled from the current creation. This divine presence, which the kabbalists regarded as female, would bring about harmony and redemption to the universe when reunited with the transcendent aspect of the divinity.

KACHINAS The spirits who guided the Hopi Indians' ancestors from the previous world to this one and who continue to guide these peoples' lives. The Hopi believe that we today live in the Fourth World; the previous world was underground, and the kachinas—spirit people who are intermediaries between humans and the Great Spirit—guided the people through a hollow reed up onto the surface of Earth. The kachinas then taught the people ceremonies, gave them guidance in living in the new world with dignity and moderation, and showed them the interconnections between nature, the land, and the people. The kachinas also bring rain, a vital element in the Hopi's stone mesa and desert environment.

The kachinas said they would live part of each year with the Hopi, but, being spirit people, they could not stay with them all the time. Their home is thought to be in the nearby San Francisco Mountains. In midwinter, the kachinas return to the people, and Hopi ceremonies show them interacting with the people, as Hopi in kachina costume demonstrate through dance the guidelines and methods of the spirit people who taught them how to grow crops, capture game, and live together peacefully. The Hopi also carve kachina dolls, wearing costumes and masks similar to those the ceremonial dancers wear. These are not idols nor are they meant to be worshiped; they are used to teach children the ways of the people.

The Hopi still practice their ceremonies, and tourists and other collectors purchase finely detailed kachina dolls for their collections.

KAMMERER, PAUL (1880–1926) An Austrian biologist who claimed early in this century to have produced experimental evidence of the validity of the Lamarckian genetic mechanism for the inheritance of acquired characteristics. He worked on sea squirts, salamanders, and midwife toads. For example, the midwife toad is land dwelling, and the male lacks the dark and rough nuptial pads on the knees of similar species that breed in water. Kammerer claimed to have bred nuptial pads into his (male) midwife toads by forcing them to live and mate in water. Other biologists could not check his findings, for they were not as skilled as he in keeping such animals in abnormal conditions.

Kammerer's work was disrupted by the First World War and by subsequent Austrian inflation. His claims were vigorously attacked by William Bateson in 1913. Debate with English geneticists continued after the war. In 1926, it was found that the knee pads on a preserved specimen of a midwife toad had been darkened with ink. Kammerer denied all knowledge of the apparent fraud but soon after committed suicide. People drew their own conclusions. The case was the subject of a book by Arthur Koestler which claimed Kammerer was a misjudged martyr for LAMARCKIANISM.

For further reading: Arthur Koestler, *The Case of the Midwife Toad,* (Hutchinson, 1971); and R. G. A. Dolby, *Uncertain Knowledge* (Cambridge University Press, 1996).

KEEL, JOHN A. (1930–) Popular writer on UFOs and anomalies who believes that such phenomena are demonic entities he calls "ultraterrestrials." Keel has asserted that such things as UFOs, BIGFOOT, and LAKE MONSTERS are inhabitants of another, paranormal, order of existence. Because they easily slip out of normal "consensus" reality, it is not surprising that scientists have difficulty verifying their existence and collecting specimens for laboratory investigation.

Keel was born March 25, 1930, in upstate New York and grew up in the small town of Perry. He began a career in writing as a teenager, having developed an early interest in the writings of Charles Hoy FORT. He produced five books on UFOs and a series of articles in various UFO and men's adventure magazines. Keel castigated the more conservative ufologists who were probing for evidence of extraterrestrial life and argued instead for the existence of more mysterious masses of paranormal energy from an unimaginable *other* reality that have been appearing in

Keel

our world to assault and manipulate humans since time began. The UFOs and other well-known anomalous phenomena are thus viewed as negative and hostile toward humanity. Keel thought that UFO reports especially mirror older supernatural stories of angels and DEMONS.

Keel's ideas gained some support in the 1970s from ufologist Jacques Vallee, who moved toward an occult interpretation of the paranormal. Nevertheless, they have generally been more at home among those interested in psychic and occult phenomena. For the last decade, Keel has written a regular column for *FATE MAGAZINE*.

See also UFOLOGY.

KELLOGG, JOHN HARVEY (1852–1943) U.S. medical doctor, inventor, author, and spa administrator who promoted dietary and other health reforms. Kellogg was born into a large family of Seventh Day Adventists in Battle Creek, Michigan, where he grew up listening to the health-reform teachings of prophetess Ellen White. Adventist leaders pushed Kellogg toward a medical career, and he began to study the works of health and hygiene reformers Sylvester GRAHAM and Russell Trall. Later he earned a degree from the prestigious Bellevue Medical School in New York City, where he studied surgery and recent European advances in microbiology and physiology.

In 1876 Kellogg returned to Michigan to take over the moribund Western Health Reform Institute, a health spa run by the Adventists, which he revived to spectacular success. Renaming it the Battle Creek Sanitarium and revamping its health and hygiene programs along the lines of his theories of "biologic" living, Kellogg made the "San," as it came to be known, one of the most popular and prestigious institutions of its kind in the United States. Thousands, including presidents, business magnates, and celebrities, flocked to its halls to dine upon nut-based meat substitutes, to exercise vigorously, and to submit to Kellogg's harangues against meat diets, sexual excess, alcohol, and tobacco. In his spare time, Kellogg practiced surgery, invented new foodstuffs and therapeutic devices, traveled and lectured around the world, published the journal *Good Health* and *Modern Medicine and Bacteriological World*, and wrote more than 50 books on health and nutrition.

Kellogg's basic philosophy grew out of the vegetarian, temperance, proexercise, and whole-grain orientation of Grahamism, but it often appropriated more recent scientific advances to bolster its claims. Although many of Kellogg's basic ideas were sound, he was also prone to exaggeration, if not outright quackery; for example, he argued against eating meat both for moral reasons and because it allegedly caused dangerous levels of toxic bacteria to build up in the colon—a condition known as autointoxication. Critics pointed out that the human digestive system could effectively filter out or neutralize such bacteria and was well equipped to handle a partial meat diet. Kellogg's withering array of exercise and therapeutic devices, which variously employed muscle-toning paddles, hot and cold water, electricity, and radioactive gases, were likewise criticized for their faddishness and dubious health benefits.

Yet, some of Kellogg's ideas have proved beneficial: He developed breakfast cereals such as cornflakes to make basic grains more digestible, and his nut butters, developed as a protein substitute for meats, have become a standard part of the American diet.

For further reading: Gerald Carson, *Cornflake Crusade* (Rinehart & Company, 1957); and John Harvey Kellogg, *The Natural Diet of Man* (Modern Medicine Publishing Co., 1923).

KENSINGTON STONE Runic inscription discovered in Minnesota and supposed to represent the extent of Norse exploration in North America. The Kensington Stone is a slab of rock that was found embedded in the roots of a large tree by a Swedish farmer named Olaf Ohman on his Minnesota farm in November 1898. It measures 91 centimeters long by 38 centimeters wide by 14 centimeters thick (about 36 inches long by 15 inches wide by $5\frac{1}{2}$ inches thick) and weighs 86 kilograms (230 pounds). The stone is covered with runic writing that tells the story of about 30 men on a exploratory journey in 1362 westward from Vinland (northeastern North America). The stone records that a third of the company was killed while the rest of the men were out fishing, and prays to the Virgin Mary to be spared from a like fate. The stone is generally regarded as a forgery, but this has not been proven to the satisfaction of all scholars.

The historicity of the Kensington Stone was originally challenged because scholars felt that the language it used was not contemporary with the 1362 date and because it mixed characters from the Latin alphabet with the runic signs. However, the earliest scholars to investigate the stone (from the University of Oslo) drew no conclusions about who might have perpetrated the forgery. The first to accuse Ohman, the stone's finder, of forging the inscription was former university professor and newspaper editor Rasmus B. Anderson. In 1910 Anderson claimed in his newspaper that three men—Ohman, defrocked Lutheran minister Sven Fogelblad, and Swedish worker Andrew Anderson—had colluded to produce the stone. Andrew Anderson and Olaf Ohman both denied Rasmus Anderson's charges and defended the authenticity of the stone.

Writer Hjalmar R. Holand challenged the interpretation of the stone as a forgery. Holand pointed out that the original interpretation of the stone was based on a faulty idea—that the stone was contemporary with Leif Ericsson's original voyage of discovery to Vinland around the year 1000. Instead, Holand suggested that the carving on the stone was done by members of an expedition sent out by King Magnus of Norway to the Norse colony in Greenland, which he had heard had fallen away from Christianity. According to Holand's interpretation, the members of Magnus's expedition sailed west of Norway into Hudson's Bay and then down a chain of lakes and rivers into south-central Minnesota. Holand suggested that it was members of this expedition who were attacked by natives at their camp, lost a third of their number, and recorded the disaster on stone.

In 1959, excavations at L'Anse aux Meadows in Newfoundland proved that Norse voyagers had visited North America, removing one of the major arguments against the authenticity of the Kensington Runestone. Although many scholars still do not accept the authenticity of the stone, its genuineness has been defended in several recent monographs, including Robert Anderson Hall's *The Kensington Rune-Stone, Authentic and Important: A Critical Edition* and Rolf M. Nilsestuen's *The Kensington Runestone Vindicated.* The stone is currently held in a local museum in Minnesota.

KICKAPOO OIL A NOSTRUM that was first made famous by its sale in one of the largest medicine shows in the 19th-century United States. Kickapoo Oil was later immortalized in the 20th century as Kickapoo Joy Juice in the *Little Abner* comic strip drawn by cartoonist Al Capp. The Kickapoo show was founded in the early 1880s by John E. Healy (better known as Colonel Healy) and Charles H. Bigelow (better known as Texas Charlie). During the show, a "tribal medicine man" would prepare the nostrum, which was originally called Sagwa. Healy and Bigelow called their headquarters in New Haven, Connecticut, the Principal Wigwam, and advertised that from there, Native Americans would scout the local woods for the ingredients with which to make Sagwa and other remedies. For more than 30 years, the company successfully sent out shows to tour the land.

The success of Healy and Bigelow has been ascribed to the quality of the entertainment they offered, their honesty and reliability in running their shows, and the quality of their remedies. The remedies, of course, had none of the ingredients advertised (roots, berries and other woodland products) but were carefully made preparations using a variety of ingredients, such as molasses and Jamaican rum. Products for external use had camphor, myrrh, and other odoriferous ingredients. Each show had a professionally trained orator to promote the medicines, which were sold in the audience by Native Americans during intermissions. The showmen also left quantities with druggists, who distributed them throughout the year.

Early in the 20th century, the Food and Drug Administration (FDA) charged that Kickapoo Cough Syrup did not possess properties necessary to treat any lung ailments. The company was fined $25 and court costs. In 1912 Healy and Bigelow decided to retire and sold the business for $250,000.00. The Kickapoo remedies, needless to say, soon disappeared from drugstore shelves as the FDA moved against "medicines" without medicinal properties.

KILNER, WALTER JOHN (1847–1920) British physician and author of a famous book documenting his research on human AURAS. He was born on May 23, 1847, at Bury St. Edmunds, Suffolk, England, attended St. John's College at Cambridge University, and pursued his medical studies at St. Thomas's Hospital in London. In 1879 he was appointed head of electrotherapy at St. Thomas's. He joined the Royal College of Physicians in 1883 and spent most of the rest of his life in private practice.

Kilner had become aware of Karl VON REICHENBACH's reports of having seen a luminous radiation around both the poles of magnets and human hands. Various psychics had reported radiations coming from human bodies, but there was little hope at the time that scientists could verify these reports or reproduce the phenomena. Kilner reasoned, however, that if the aura existed, it might merely need the proper filter to make it visible.

Around 1908 Kilner began his experiments. He eventually developed a filter composed of a layer of dicyanin, a coal-tar derivative, pressed between two pieces of glass. Kilner first looked at bright daylight through the filter. Then, putting the filter aside, he turned to look at a human body against a dark background in a dimly lit room. He claimed that the technique revealed a radiation with three layers, all lying in the ultraviolet end of the spectrum. The first layer followed the contour of the body and was about 0.6 centimeters ($\frac{1}{4}$ inch) thick. He related this layer to theosophical concepts and called it the etheric double. The second layer also followed the body's contour and ended approximately 8 centimeters (3 inches) from the body. The third layer was approximately 30 centimeters (1 foot) from the body.

According to Kilner, the aura was affected by magnets and/or electric currents near the body, changed colors as the body was afflicted with different diseases, and dissolved away at death. It was Kilner's hope that his discoveries could lead to the use of aura analysis in medicine. He published his findings in his book, *The Human Atmosphere* (1911).

Kilner died in 1920 in London. Although his research generally met with indifference from his colleagues, several pursued it further and used his techniques for diagnostic purposes. In the 1930s British biologist Oscar Bagnell experimented with various dyes that, like dicyanin, could be used to sensitize the eyes to the auric radiations. In 1937 Bagnell reported on his own findings in *The Origin and Properties of the Human Aura,* a sequel to Kilner's early work.

Kilner and his followers failed to convince many of their colleagues of the significance of their discoveries. Some who tried the colored screens obtained quite different results in their attempts to see the auric radiations. However, they provided the psychic community with what many believed were scientific underpinnings for belief in the aura and related occult realities such as the etheric human double. Periodically, individuals have created goggles with dicyanin-colored lenses and have marketed then through the psychic community's publications. Occasionally, both Kilner's and Bagnall's books have been reprinted.

During the 1970s, a photographic process utilizing an electrical discharge onto a photographic plate generated images of an auralike corona around the human body. However, ultimately KIRLIAN PHOTOGRAPHY also proved to be a dead end.

See also ASTRAL/ETHERIC BODY.

For further reading: Oscar Bagnall, *The Origin and Properties of the Human Aura* (London, 1937/rev. ed., University Books, 1970); and Walter J. Kilner, *The Human Atmosphere* (London, 1911, reprinted as *The Human Aura* [University Books, 1965]).

KINDERHOOK PLATES An archaeological hoax purporting to show Egyptian influence on Native Americans. The story of the Kinderhook Plates began in 1843, when a set of six brass plates were unearthed by amateur archaeologist Robert Wiley, a merchant in Kinderhook, Illinois. Each plate had a hole at the top through which an iron ring had been passed, binding the plates together. When the plates were cleaned, they revealed writing that resembled Egyptian hieroglyphics. The discovery became the subject of much discussion throughout the state, in large part because the Mormon movement was growing in Illinois. Smith claimed that the *Book of Mormon,* the group's new scripture, had been translated by him from a "reformed" Egyptian text. The plates were shown to Joseph Smith in 1844, but he was assassinated before he could study them and offer an opinion.

The plates disappeared shortly afterward, and it was not until 1920 that one of them resurfaced. It was given to the Chicago Historical Society. In the meantime, however, the hoax had been exposed. In 1855, a letter revealed how Wiley had manufactured the plates with the help of two coconspirators, Wilbur Fugate and Bridge Whitton. Whitton, a blacksmith, cut out the original plates and Fulgate created the hieroglyphics in wax. The plates were etched with acid, artificially aged, and buried to be discovered later.

In the 1980s, tests were run at Northwestern University on the plate owned by the Chicago Historical Society. Scientists discovered that the brass in the plates was of a purity unlike that of ancient times and that the markings had in fact been produced with nitric acid.

KING SOLOMON'S MINES Legendary site of the wealth of biblical King Solomon. When the 19th century German explorer Karl Mauch discovered the ruins of ZIMBABWE in southern Africa in 1871, he believed he had uncovered the remains of an ancient biblical settlement—perhaps, he believed, the remains of the Queen of Sheba's home or the site of Solomon's fabulous mineral wealth. Mauch reported that local natives, when questioned, declared that the great walls had been built in the distant past by a foreign race of white people. In his report on the site, Mauch dated the site to before the birth of Christ. He also believed he had found evidence that the ancient inhabitants of the lost city practiced Jewish rites of sacrifice.

The identification of Great Zimbabwe with Solomon had more to do with Mauch's 19th-century racism than with biblical figures. Great Zimbabwe, later archaeologists showed, was constructed by native Africans using local materials, but the idea of an ancient settlement lying somewhere in Africa sparked European imaginations. In September 1885, British novelist and civil servant H. Rider Haggard (1856–1925) published his romantic adventure story *King Solomon's Mines,* which told the story of a small band of English adventurers fighting their way through the wilds of Africa and finally securing a fantastic cache of diamonds. *King Solomon's Mines* was a best-seller and defined Africa for millions.

Haggard's novel had little respect for the facts of African history. Unlike King Solomon's mines, Great Zimbabwe had exported gold, not diamonds. Later archaeological research placed the site of Solomon's wealth in the Arabian peninsula—probably in modern Saudi Arabia. The exact site is unknown.

For further reading: William Minter, *King Solomon's Mines Revisited: Western Interests and the Burdened History of Southern Africa* (Basic Books, 1986).

KINSEY, ALFRED C. (1894–1956) Founder of the Institute for Sex Research, Indiana University, in 1942, where he and his colleagues published two important reports: *Sexual Behavior in the Human Male* (1948) and *Sexual Behavior in the Human Female* (1953). They revealed wide variations in sexual behavior and created something of a furor at the time. They were criticized for scientific inadequacies: sampling shortcomings and the problems of using material gathered by personal interviews as objective data. Some of the criticism was probably oblique opposition to the outspoken nature of the contents, since the subject was surrounded by taboos.

The two reports are statistical analyses of 5,300 white U.S. males and 6,000 U.S. females. Neither was a statistically representative sample of the U.S. population, but both drew on broad groups and produced information that has yet to be bettered. Nine types of sexual behavior and eight characteristics of the subjects were identified. Many of Kinsey's findings surprised and shocked the U.S. public, but the reports were undoubtedly milestones on the road of sexual understanding. Perhaps the most shocking statistics were the proportions of those admitting to homosexual acts (one-third of men and one-fifth of women) and as to being exclusively homosexual (4 percent of men, 2 percent of women).

The Kinsey reports attempted to produce a map of male and female sexual behavior, in that time and place, by examining a large group of male and female subjects, something Henry Havelock ELLIS had attempted to do in Britain 40 years earlier, on a much smaller scale, based only on his observations of his patients. However imperfect the sampling and allegedly unscientific the method, the reports provided a basis for further work in the field. They brought the subject out of its Victorian closet; there was no going back.

For further reading: Gardner Lindzey, Calvin S. Hall, and Richard F. Thompson, *Psychology*, 3rd ed. (Worth, 1988).

KIRLIAN PHOTOGRAPHY A technique of photographing the AURAS of living things. Developed by Russian scientists Semyon Davidovich Kirlian and Valentina Khrisanova Kirlian, the method is based on earlier efforts to render the supposed psychic energy fields that surround living beings visible. The Kirlians were inspired by the 19th-century work of such scientists as Karl VON REICHENBACH and Walter John KILNER, a British physician who sought ways to prove the existence of the auras. Kilner discovered that if he sealed two pieces of glass together, with a small amount

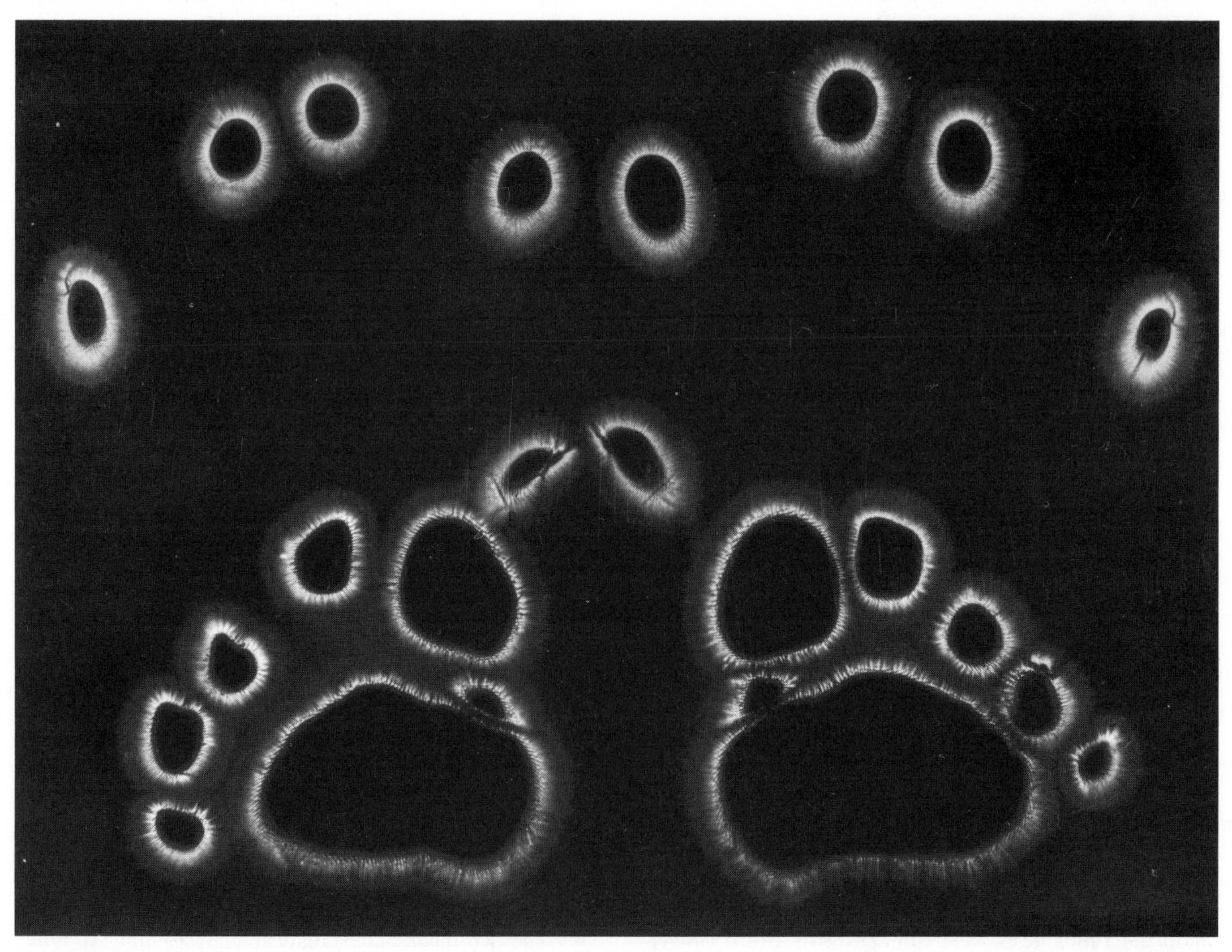

Kirlian Photo

of a bluish dye called dicyanin in between, he had a lens that allowed him to view a kind of vaporous energy that extends from living beings. Kilner's methods were improved on in the 20th century by Harry Boddington, among others, who found that he could accomplish the same end by using glass dyed to the same spectroscopic color as dicyanin.

In the 1940s, the Kirlians discovered that they could photograph auras (without physically seeing them) by transmitting a brief shock of high-voltage electricity through the photographic subject. The subject had to be in direct contact with photographic plates while the shock was being administered, so until recently, it was only practical to photograph the supposed auras of small things—most often a person's fingertip or a small living, or recently living, aspect, such as a leaf. Early Kirlian photos were black and white, but film and equipment advances have made it possible to film the aura in color and to photograph much larger objects, without direct contact with photographic plates.

Kirlian photographers assert that the photos can yield important information about a person's physical and mental health. The size, brightness, patterns, and colors of various parts of the aura provide the clues.

Although some scientific testing was done of earlier Kirlian photography, no strong, supportable conclusions were reached as to its efficacy. In the 1970s, Thelma Moss and Kendall Johnson at the University of California's Center for Health Sciences, Los Angeles, performed several interesting experiments. One used leaves: Aura photographs taken of a leaf growing on a plant were remarkably different from photos of the leaf being approached by a human hand and being picked. These experiments gave credence, some researchers thought, to the idea that plants, like animals, have emotions. Nevertheless, the results weren't strong enough to convince most scientists of the importance of aura photography or, indeed, of the existence of auras. Today's aura cameras are most often seen at psychic fairs and are used by some alternative medical practitioners.

The largest question surrounding Kirlian photography is whether psychic or bioenergetic fields are actually being photographed or whether the images are merely of electrical discharges caused by the voltage sent through the subject.

See also SPIRIT PHOTOGRAPHY.

KIYOTA, MISUAKI (1962–) Japanese psychic who produced paranormal photography (nengraphy) and claimed extraordinary psychokinetic powers at METAL BENDING and nengraphy. Kiyota emerged as a teenage star in 1977, in the wake of the extensive international coverage of psychic Uri GELLER. He was tested and filmed by Japanese scientists and his phenomena appeared on both Japanese and American television. Japanese parapsychologists put him through further tests at both metal bending and producing images on film. Two U.S. researchers, Walter and Mary Jo Uphoff, went to Japan to study him and wrote a book about their observations and findings in 1980. In 1982 several U.S. parapsychological laboratories tested him. In response to the paranormal photographs Kiyota produced, some Japanese scientists formed the Japan Nengraphy Association.

That same year, the parapsychological community was shocked by Kiyota's confessions that at least some of the effects he had produced were produced by fraud and that the ability of parapsychologists to detect trickery by a skillful conjurer was inadequate.

In the wake of Kiyota's confessions, some parapsychologists have tried to argue that at times, including during the 1982 experiments, Kiyota did produce genuine paranormal effects under tightly controlled and photographed conditions. They have suggested that Kiyota resorted to fraud only when placed in the position of having to produce phenomena on demand in repeated experiments. However, Kiyota has not been involved in any experiments designed to replicate any of his earlier experiments.

For further reading: Arthur S. Berger and Joyce Berger, *The Encyclopedia of Parapsychology and Psychical Research* (Paragon House, 1991).

KLASS, PHILIP J. (1921–) One of the foremost and most forceful critics of UFOLOGY. He graduated from Iowa State University in electrical engineering, worked during World War II as an aircraft electronic engineer for General Electric, and then for many years was editor of *Aviation Week* (subsequently *Aviation Week and Space Technology*).

In the mid-1960s, he began to investigate the reports of flying saucers and UFOs. Klass was soon convinced that they could be explained as hoaxes, publicity stunts, optical illusions, or possibly as electric discharges akin to BALL LIGHTNING in places where the electric field was exceptionally high (a theory later shown to be scientifically unsound). In succeeding years, Klass became the scourge of the ufologists, writing several books—*UFOs—Identified* (1968), *UFOs Explained,* and *UFOs—The Public Deceived* (1983)—and publishing the *Skeptics UFO Newsletter.* He appeared on talk shows attacking UFO claims. He became associated with Paul Kurtz, editor of the *Humanist,* and several other skeptics in the COMMITTEE FOR THE SCIENTIFIC INVESTIGATION OF CLAIMS OF THE PARANORMAL (CSICOP). In the 1980s, he investigated claims of alien abductions and published *UFO Abductions: A Dangerous Game* (1989). Klass attributed these highly dramatic accounts of abduction, rape, dissection, and so on to hoaxes, fantasies, and in particular to FALSE MEMORY SYNDROME, since he thought memories elicited

Klass

under HYPNOSIS were a mixture of fantasy, recalled DREAMS, dim memories of science fiction, and suggestions by the hypnotist.

Klass came under attack in 1983 when it was claimed that he had persuaded the University of Nebraska to refuse permission for a UFO meeting on campus, a change he denied. There was also some confusion when it was discovered that a Philip Klass, under the nom de plume of William Tenn, had written many science-fiction stories in the 1940s, 1950s, and 1960s—the themes of which were strangely at odds with Klass's professed and very public skepticism. But it turned out to be another Philip Klass. For more than 30 years, the real Philip Klass (who thereafter styled himself Philip J. Klass) has been, and continues to be, highly skeptical of all accounts of flying saucers, unexplained celestial sightings, UFOs, alien abductions and similar stories.

See also ABDUCTIONS, ALIEN.

For further reading: Thomas E. Bullard, "Klass Takes on Abductions; Abductions Win," *International UFO Reporter* (November/December 1987); Jerome Clark, "Phil Klass vs. the UFO Promoters," *Fate Magazine* (February 1981); and Jim Schnabel, *Aliens, Abductions, and the UFO Obsession* (Hamish Hamilton, 1994).

KNOSSOS (CRETE) Largest archaeological site of a famous pre-Classical Mediterranean civilization. In 1900, Sir Arthur Evans (1851–1941) began excavations at a site in the north of the island of Crete. Evans was following up on several examples of hieroglyphic writing that he had discovered in an Athens flea market. Learning that the samples had come from Crete, Evans resolved to look there for the beginnings of ancient Greek civilization uncovered in the previous two decades by such archaeologists as Heinrich Schliemann (1822–90). Evans visited a native Cretan named Minos Kalokairinos, who had excavated at Knossos 20 years before; his work had been stopped by the Cretan parliament, who feared that their Turkish overlords would confiscate any treasures he uncovered. Kalokairinos had explored the site sufficiently to know that extensive ruins lay under it. Based on this evidence, Evans came to the conclusion that the undiscovered Cretan world reflected the Homeric world that was recently uncovered by Schliemann.

Like Schliemann, Evans had little background in archaeology, but he did have a lifelong interest in antiquities. Beginning in 1883, he served as the keeper of the Ashmolean Museum at Oxford. However, he was unprepared for the riches he discovered within weeks of beginning his excavations at Knossos. The fire that had destroyed the palace more than three millennia before had left frescoes and other artifacts—including a throne, bathtubs, and extensive plumbing—almost untouched. Some of these artifacts reminded him of the few ancient Greek myths that had been passed down about Crete: many-roomed Knossos resembled the labyrinth of King Minos, while the frescoes of young men and women leaping over bulls brought to mind the legendary Minotaur.

Evans had in fact discovered an entirely new civilization, seemingly unrelated to the contemporary cultures that had been uncovered on mainland Greece, in Asia Minor, and on the islands of the Cyclades. Like Schliemann, Evans chose to interpret Knossos according to his own vision of how the palace was built and how it ran. Evans invested his own money not only in excavating but also in *restoring* the site. He rebuilt the palace as he continued to excavate, using stone, steel, and concrete where the original wood had rotted away. He also restored damaged frescoes and shored up architectural features such as the grand staircase that dominates the eastern side of the palace's great central court.

Evans became so convinced of the superiority of his Minoan culture that between 1921 and 1936, in his four-volume work *The Palace of Minos,* he declared that the Minoans had dominated the Aegean Sea and mainland Greece between Neolithic times and the late Bronze Age. The artifacts discovered by Schliemann at Troy, Mycenae, and Tiryns, including bodies masked with gold lying in elaborate graves, were only offshoots of Minoan culture, produced by Minoan artists employed by barbarian kings. The Minoan world was non-Greek-speaking, and the Trojan War was only a revamped Cretan myth. Evans's theories were proved wrong—an English architect named Michael Ventris proved conclusively in 1954 that the Linear B tablets from the Cretan palaces were written in a primitive Greek—but his contribution to Minoan scholarship and the history of the early Aegean remains unchallenged.

For further reading: Maitland A. Edey, *Lost World of the Aegean* (Time-Life Books, 1975); and Michael Wood, *In Search of the Trojan War* (Facts On File, 1985).

KOCH, WILLIAM F. (1885–c. 1950) A medical doctor who claimed in 1919 to have found a cure for a great variety of illnesses including cancer, tuberculosis, and leprosy. He was awarded a Ph.D. in chemistry in 1917 from the University of Michigan; in 1918 he received a degree in medicine from Wayne University. Koch taught histology and embryology at Michigan from 1910 to 1913 and physiology in the Medical College at Wayne from 1914 to 1919.

In 1919 Dr. Koch claimed to have synthesized a substance that he called glyoxylide which cured four out of five patients, no matter what their illness, by stimulating the patient's own ability to manufacture remedies. He published two books explaining the claimed scientific basis for his cure: *Cancer and its Allied Diseases* (1929, revised 1933) and *The Chemistry of Natural Immunity* (1938). In 1943 and 1948, the U.S. government twice took Koch to court to stop him from performing what it saw as a potentially dangerous procedure. Government chemists testified that glyoxylide ampules were indistinguishable from distilled water. The trial was unsuccessful, and Koch continued to sell thousands of glyoxylide ampules. The Christian Medical Research League, Detroit, still gives the injections.

Although in 1942 the Federal Trade Commission prevented Koch from advertising his product in the United States as a cure, word-of-mouth kept him in business; in addition, he then started an aggressive and lucrative business in South America.

KOESTLER, ARTHUR (1905–1983) Journalist and author one of whose claims to fame occurred in 1985 when the Koestler Chair of PARAPSYCHOLOGY was established in the Department of Psychology at the University of Edinburgh, Scotland. He and his wife Cynthia left a bequest in their wills to make this possible. On Koestler's death from leukemia and Parkinson's disease, his wife, several years younger than he and apparently in good health, took her own life. A note explained that they had made an agreement to die together. They were a devoted couple and presumably wished to be together after death. This raises a question: Was the Koestler chair intended to initiate research that might enable communication with them on the other side?

Koestler was born in Budapest and after studying in Vienna became a journalist and then an author who dealt with some of the central issues of his time. He became a communist, fought against Franco in the Spanish Civil War, was imprisoned under sentence of death, and wrote a book drawing on his own experience—*Spanish Testament* (published as *Dialogue with Death*, Penguin, 1938). Koestler was British by naturalization and wrote in English. One of his books is *The Sleepwalkers: A History of Man's Changing Vision of the Universe* (1959), in which Koestler claimed that scientific discovery occurred more by accident than design and likened scientists to sleepwalkers stumbling upon their knowledge while not being fully awake. In a more recent book, *Insight and Outlook*, Koestler speaks of Joseph Bank RHINE's work as having ushered in a new Copernican revolution.

In *The Roots of Coincidence*, published in 1972, he ascribed some sort of psychic significance to examples of unusually rare chance happenings. Koestler seemed unable to believe that if these were mathematically possible, as predicted by probability theory, then they were also inevitable. He began to believe in ESP because he thought he saw in some ideas of modern science, particularly quantum physics, absurdities that made the absurdities of parapsychology a little less preposterous.

The first occupant of the Koestler Chair is Robert Morris who, in his own words, "has attempted actively to blend research on what's not psychic but looks like it, with research aimed at examining some of the procedures that appear to provide strong evidence for psi, to see if valid effects can be obtained with sufficient strength and consistency that they can lead to systematic research to uncover their nature."

In addition to *The Sleepwalkers*, Koestler wrote several important books, among them *The Case of the Midwife Toad* and *Creativity*. Although he had no special claim to authority in these subjects, and although the professionals have often been dismissive, he brought to them a fresh, critical eye that encouraged debate.

For further reading: Iain Hamilton, *Koestler: A Biography* (Secker and Warburg, 1982); Paul Kurtz, *A Skeptic's Handbook of Parapsychology* (Prometheus, 1985).

KON-TIKI Craft used in a voyage across the Pacific Ocean led by Norwegian anthropologist Thor HEYERDAHL (1914–) to prove that Polynesian inhabitants of the South Pacific could have been influenced by South Americans. The craft, named after the legendary Incan sungod *Kon-Tiki*, was a primitive balsa-wood raft designed along the lines of ancient rafts used by South American Indians. Heyerdahl, a diffusionist, believed that Polynesian culture showed influences that came directly from South America. Most anthropologists believed that all Polynesians and their culture had come to the islands eastward across the sea from Asia. In a trip lasting 101 days, Heyerdahl's expedition traveled 7,000 kilometers (4,300 miles) across the South Pacific. On August 7, 1947, the *Kon-Tiki* ran onto a reef in the Tuamoto Archipelago, proving that it was possible to reach Polynesia from South America using only Stone Age technology. The craft was later rescued and preserved in an Oslo museum.

The *Kon-Tiki* voyage from Peru to Tahiti showed that contact between the American mainland and the South Pacific was possible, but it did not prove that it had taken place. It also did not prove that South Pacific islanders were the descendants of South American Indians. However, Heyerdahl's record of the trip, published as *Kon-Tiki* (1948), brought him worldwide fame and won justification for his views. In 1961, the Tenth Pacific Science Congress in Honolulu, Hawaii, declared that South America had fathered some of the inhabitants and the cultures of the Pacific Islands.

For further reading: Thor Heyerdahl, *Kon-Tiki: Across the Pacific by Raft* (Rand McNally, 1950); and Thor Heyerdahl, *American Indians in the Pacific: The Theory behind the Kon-Tiki Expeditions* (Rand McNally, 1953).

KRAKEN A giant squid—one of the few unseen animals whose existence is not in doubt. It was originally described by Erik Pontoppidan, a Norwegian bishop, in his *Natural History of Norway* (1752–53) as immense; he declared it was so large that it could drag ships under water. Pontoppidan's opinions about the kraken were ignored by biologists until 1847, when a Danish naturalist named Johan Japetus Steenstrup lectured on the subject to the Society of Scandinavian Naturalists. Steenstrup published a scientific description of the kraken in 1857 and gave it its scientific name of *Architeuthis*. A series of squid carcasses washed onto Canadian beaches during the 1870s helped popularize Steenstrup's work.

The modern controversy over the giant squid rests not on whether it exists, but on how large it grows. Because most krakens live in deep water, they never reach the surface. One way to measure the size of these creatures is to look at the scars they leave on their major predators—sperm whales. The whales have shown sucker scars that measure as much as 45 centimeters (18 inches) in diameter. If the scars are proportional to the size of the squids, the kraken may measure up to 27 or 30 meters (90 or 100 feet) in length.

Some scientists dispute this conclusion, suggesting that the scars may have been made when the whales were smaller. The scars would then have increased in size as the whales grew. Many modern cryptozoologists, however, reject this claim. They state that baby whales would be unlikely to come in contact with giant squids, and if they did, they would be unlikely to survive the encounter.

KREBIOZEN A substance used for the treatment of cancer. It was originally prepared as a serum extract from horses that had been injected with a deadly fungus. Krebiozen was first promoted by Yugoslavian physician Dr. Steven Durovic and by Dr. Andrew Ivy of the University of Illinois. Krebiozen therapy grew in popularity during the 1950s and early 1960s. Proponents claimed that Krebiozen produced a high level of activity against a variety of cancers that were considered incurable, with few, if any, side effects. Its contents were at first kept secret from the scientific community, for fear of sabotage by the medical establishment. Thus Krebiozen was shielded for a time from the conventional research standards of thorough and timely disclosure and peer evaluation and critique. In 1963, the federal government completed a comprehensive examination of Krebiozen. Tests of Krebiozen showed that it contained mineral oil and creatinine, a substance normally excreted by the body; neither agent has any proven anticancer activity. The federal investigation included a review of the records of more than 500 patients who had reportedly benefited from Krebiozen therapy. After evaluating these cases submitted by the Krebiozen Research Foundation, an expert committee concluded that there was no evidence to suggest Krebiozen had any therapeutic effects whatsoever on cancer patients.

KUHN, THOMAS S. (1922–1996) U.S. historian of science. He graduated in physics from Harvard in 1943. After working on radar countermeasures in Europe during World War II, Kuhn returned to Harvard, where he obtained a Ph.D. in solid-state physics. While a postgraduate student, he was invited by James B. Conant to teach case studies in the history of science. That early experience convinced him that the prevalent view that the science of the past was bad—wrong and misguided—was mistaken. Scientists tended to think of the history of science as the history of past mistakes; Kuhn realized that the science of the past was not pseudoscience but the knowledge of its time—the best that scientists could then do with the information available to them. That realization was a turning point both in his professional life and in the history of science.

Kuhn left Harvard in 1957 to start a history of science program at the University of California at Berkeley. In that same year he published *The Copernican Revolution: Planetary Astronomy in the Development of Western Thought* based on his case studies course. This, his first book, marked an important step in thinking about science. Conant, in his foreword, wrote, "I wish to register my conviction that the approach to science presented in this book is the approach needed to enable the scientific tradition to take its place alongside the literary tradition in the culture of the United States. . . . Professor Kuhn's handling of the subject merits attention, for, unless I am much mistaken, he points the way to the road which must be followed if science is to be assimilated into the culture of our times."

Kuhn's next book, *The Structure of Scientific Revolutions* (1962), proved a revolution in thinking about science. In it he distinguished between normal science, the activity that engages the scientific community for most of the time, and revolutionary science, which occurs in a science when there is a major change of outlook, what Kuhn termed "a paradigm shift." Most scientists, most of the time, work on problems within normal science—that is, refining and enlarging the prevailing overview or paradigm. As inconsistencies and unsolved problems arise within normal science, some—very few—scientists start to explore alternative frameworks, looking for one that, although satisfactorily accounting for all that the prevailing view explains, also resolves the inconsistencies and problems, or at least promises to do so—a paradigm shift. This is what happened when HELIOCENTRISM supplanted GEOCENTRISM, or when relativistic physics subsumed Newtonian physics. Such upheavals inevitably meet with opposition and take time to be accepted, but nevertheless occur in a relatively short time and are followed by long periods of stability—until another paradigm shift takes place. These paradigm shifts are the scientific revolutions referred to in the title of Kuhn's book.

Kuhn was not attempting in any way to undermine the authority, the value, or the validity of science, but to show the manner in which scientific understanding develops. At the same time, he worked to show how scientists, science, and scientific theories are influenced by the social and cultural milieu, the belief systems, in which they function. Science does not stand apart from society but is an integral part of the culture of its time. Kuhn's strategy might be summarized thus: To understand the philosophy of science, the way in which it functions, and how it affects us, it is necessary to examine its history, which, when understood and set in context, points us toward the answers.

For further reading: Barry Barnes, *T. S. Kuhn and Social Science* (Macmillan, 1982).

KUNDALINI Religious experience espoused by Hindu groups in the *shakti* tradition. In Hatha YOGA, *kundalini* energy, also called *shakti*, is a form of "cosmic energy" that collects at the base of the spine. This *kundalini* energy is generally visualized as coiled, a potential energy that can be released and used. When it is released, it creates pleasure or bliss throughout the body. Some yogic practitioners and GURUS use the release of *kundalini* as a way to relieve stress. Others use it during meditation as a way to further their spiritual growth. Still others recommend using it during sex to heighten the physical and spiritual intimacy between partners.

Western yogic practitioners say that *kundalini* energy is a real product of the body, not a visualization of the mind. Alexander Lowen, founder of bioenergetics, declared that a human body is in a constant state of vibration, linked to the rhythmic contractions of all the cells within the body. When such vibrations are in harmony or are very intense, they can spark powerful emotions. Wilhelm REICH, a researcher into theories of bodily energies, stated that muscle tension, when relaxed, can result in an increase in pleasurable sensations that he called "streaming." Margo Anand, in her popular book *The Art of Sexual Ecstasy* declares that the release of *kundalini* energy stimulates an "Ecstatic Response. It is an intense sensation of joy, a feeling of being cleansed and relaxed."

Some yogic practitioners, such as Gopi Krishna in his book *The Biological Basis of Religion and Genius* (1972), claim that *kundalini* energy marks an evolutionary distinction between humans and animals. Gopi Krishna helped establish the Kundalini Research Foundation in Connecticut to continue research into the *kundalini* phenomenon. Scientists are frustrated by claims like these because they have found no way to measure the *kundalini* energy itself. They can measure muscle tension that, according to the yogis, blocks *kundalini* energy, but they can only detect normal muscle activity—nothing that resembles the description of *kundalini* energy, and nothing at the base of the spine. With no way to detect and measure it, the idea of *kundalini* energy is considered pseudoscientific.

L

LAETRILE An extract from apricot stones or almonds, sometimes also known as bitter almonds, vitamin B17, or amygdalin. In the mid-1970s, Ernst Krebs, Jr., promoted it as an alternative treatment for cancer. He maintained that it killed or inhibited cancer cells while having no adverse effects on healthy cells. In ALTERNATIVE MEDICINE, it was the primary anticancer agent. Trials on its efficacy have proven inconclusive and inconsistent, and it has had unwelcome side effects for some patients. There is controversy in medical circles about its use. It appears to be most effective if used in conjunction with a modified diet, controlled lifestyle, and so on; then it becomes difficult to know how much of any improvement is due to laetrile. Laetrile contains, or releases in reaction with the body's enzymes, small amounts of cyanide, which is a powerful poison and can be dangerous if the levels build up from overuse. Its use is illegal in some U.S. states. Laetrile's popularity has diminished since the 1970s.

LAKE MONSTERS Anomalous creatures of various types and nicknames said to inhabit lakes around the world. Among the best known lake monsters are Nessie, the LOCH NESS MONSTER of Scotland; the OGOPOGO, of Lake Okanagan in Canada; and CHAMP, of Lake Champlain. Other, less well-known monsters haunt smaller bodies of water. Tessie inhabits Lake Tahoe, while Sandy lives in Big Sandy Lake in Minnesota. An unnamed water monster lives in Lake Vorota in eastern Siberia, and a bunyip inhabits Lake Modewarre in Australia. Among the hundreds of reports from around the world are those from people who claim to have seen lake monsters even in artificial lakes.

Scientists suggest that many reports are based on misperceptions and preconceptions. An odd-shaped log, a patch of seaweed, shadows of waves and sunlight, or a glimpse of an everyday item may all easily be seen as something else. This may happen accidentally because of poor visibility and bad conditions. Additionally, people often see what they want to see or expect to see instead of what is actually there. Countless tests of human observation skills show that misperception and preconception are very common.

Some lake monsters have turned out to be frauds. In 1993, the last survivor of the Loch Ness expedition that produced the famous 1934 photograph confessed that the monster was a fabrication. Among the other lake monsters clearly acknowledged as frauds is the Silver Lake (New York) monster, first reported in 1855 near the shore village of Perry. Several sightings that summer led to monster hysteria and brought many tourists to Perry and the surrounding area. In 1857 firefighters found a strange contraption in the attic of a burned hotel. The hotel's proprietor, A. G. Walker, fled to Canada, but later investigation strongly suggested that the bizarre canvas-and-wood object was the "lake monster" and that Walker had conspired with an itinerant photographer and three or four others to create the economy-boosting monster of Silver Lake.

To date no lake monster has been fully verified through scientific inquiry. Most cases rely on often unverifiable eyewitness accounts, photographs of such poor quality that no conclusions can be drawn, and physical evidence that later proves to have another explanation. The occasional mystery corpse more often than not proves to be the partially decomposed body of a known creature. Cryptozoologists are still awaiting the day that a live or dead lake monster is captured for study.

LAMARCK, JEAN BAPTISTE (1744–1829) French naturalist and formulator of a distinctive theory of organic EVOLUTION. Lamarck was born in the Picardy region of France to a noble family of somewhat declining fortunes.

Originally slated for a religious career, Lamarck's education began in a Jesuit school where he studied ancient languages, logic, mathematics, and scholastic thought. After his father died, however, Lamarck turned away from the religious life and joined the army. Eventually, he distinguished himself for bravery during the Seven Years' War and was promoted to the officer's rank, which afforded him enough leisure time to study music and botany. After an injury mustered him out of the military, Lamarck went to Paris, where he embarked on the scientific career that would make him famous.

In Paris Lamarck began his career as an associate of the famous naturalists Bernard de Jussieu and Georges-Louis Buffon. Initially interested in plants, Lamarck worked as a collector and classifier at the royal botanical garden. In 1789, he produced his first major scientific work, *Flore française,* which was well received and prompted his election to the prestigious Academy of Sciences. In the midst of the French Revolution, he helped organize the Natural History Museum (Musée d'histoire naturelle). Lamarck focused his studies on invertebrate zoology and cultivated side interests in meteorology, hydrogeology, fossils, and conchology. His most important works were produced between 1800 and 1822, including *Zoological Philosophy* (1809) and *The Natural History of Invertebrates* (1815–22). In them Lamarck outlined a systematic theory of nature that included his evolutionary theories. These stressed the inheritance of acquired characteristics as a mechanism of organic change. The merits of the latter, dubbed LAMARCKIANISM, would be debated by biologists and philosophers well into the 20th century.

Though admired by many Republicans, with whom he openly sympathized, Lamarck's reputation in French scientific circles suffered under the Restoration. He was disliked by Napoleon, and his theories often clashed with those of the comparative anatomist Baron Georges CUVIER, one of the most famous and influential figures in French science. At the time of his death in 1829 Lamarck had only a small number of followers. After the publication of *Origin of the Species* (1859), however, interest in Lamarck's work revived in the context of the wide-ranging social, scientific, and philosophical debates stimulated by Darwin's evolutionary theory.

For further reading: Madeleine Barthelemy-Madaule, *Lamarck the Mythical Precursor,* translated by M. H. Shank (MIT Press, 1982).

LAMARCKIANISM Theory of EVOLUTION that stresses the inheritance of acquired characteristics as the mechanism of organic change. Over time somewhat different species of Lamarckianism have emerged, but at a basic level all share the belief that the physical changes, structural enhancements, and sometimes even mental qualities achieved by an organism in its lifetime can directly effect the genetic makeup of subsequent generations. The concept was first put forth in a rigorous scientific fashion by French naturalist Jean Baptiste LAMARCK in the early 19th century.

Lamarck's belief in the inheritance of acquired characteristics was actually part of a more comprehensive theory of natural history and organic development. Although he never used the term *evolution,* his ideas were one of the earliest systematic attempts to explain the development of complex organisms from simple ones. In his theory, Lamarck accounted for the origin of life via SPONTANEOUS GENERATION. Once the simplest organisms (both plants and animals) had been produced, environmental influences and time affected them in such a way as to produce the variety of complex organisms now found in the world.

The heart of the theory was a four-part process that governed organic development and change. First, driven by a natural tendency, the general size and complexity of living organisms tends to increase up to certain naturally imposed limits. Second, new organs or specialized limbs develop from new wants or needs that arise and are continuously felt in the life of the organism. Here Lamarck invoked what was to become one of his most famous and widely misunderstood concepts, the *sentiment intérieur,* to explain these changes. In the course of later popularization of his work, this idea was often reduced to such familiar examples as the giraffe who stretches after food on high branches and thereby develops a longer neck. These kinds of illustrations can be somewhat misleading, however, as the actual theory was much more complex. Lamarck observed that, in addition to reflex action, higher animals also engaged in a variety of voluntary motions in response to physical needs (such as hunger or reproduction) that in time became habitual. He then conjectured that, in response to such needs, the brain of an animal directed a "nervous fluid," with properties somewhat akin to electricity or magnetism, that could satisfy those needs, to the appropriate muscles or tissues. If the stimulus was repeated over and over again, in time a new or modified organ would be produced by the action of the fluid on the original anatomical structures.

Lamarck's third major principle held that the development of organs and their potency in the life of a particular organism was proportional to their use. Often called the use–disuse principle, its corollary held that unused organs and limbs gradually atrophied and disappeared. Finally, Lamarck's fourth postulate maintained that all the changes that occur in the life of an organism are preserved in its offspring—the familiar inheritance of acquired characteristics. In historical context many of these concepts had been discussed and accepted by many naturalists, particularly laws three and four, although few had been inclined to interpret them in such an explicitly evolutionary context.

During his own life, however, Lamarck's theories were controversial and little accepted, even in his native France. Intellectually, his arch-nemesis was fellow naturalist Baron Georges CUVIER, (1769–1832) whose resistance to the idea of species transformism and whose powerful position within the French scientific community proved to be formidable obstacles. After Lamarck's death in 1829, however, an increasing interest in evolutionary ideas during the course of the 19th century prompted a series of reassessments of his ideas. Particularly after the publication of Charles DARWIN's *Origin of Species* in 1859, many scientists and intellectuals adopted a version of Lamarckianism that championed the concept of the inheritance of acquired characteristics. Often called neo-Lamarckianism, its advocates included British philosopher Herbert Spencer, author Samuel Butler, playwright George Bernard Shaw, and a host of European and American scientists representing a number of different disciplines.

In scientific terms, neo-Lamarckianism challenged weaknesses in Darwin's somewhat murky theories of inheritance (as yet not combined with Mendelian genetics) and reflected a widespread dissatisfaction with Darwin's emphasis on chance variations and natural selection as the mechanisms of evolutionary change. Little supported by either experimental evidence or practical experience (critics constantly pointed to offspring that had obviously not inherited their parent's acquired strength, skill, or learning), neo-Lamarckianism can be viewed as an intrusion of wish, ideology, and religious belief into scientific discourse. Often, neo-Lamarckianism had close ties with persons and political movements that stressed historical progress and saw in the theory a potent metaphor for their larger social vision (as was the case with LYSENKOISM in the Soviet Union). It also attracted persons who, although compelled to accept the broader concept of organic evolution, were resistant to Darwinism because it seemed to deny purpose and direction to natural history. Religious modernists, for example, were often attracted to Lamarckianism because it seemed to retain a strong teleological orientation, in ironic contrast to critics earlier in the century who had accused Lamarck of promoting materialism.

Within a few decades, however, most scientists abandoned neo-Lamarckianism. Particularly damaging were a series of experiments in the 1880s and 1890s by German biologist August Weismann (1834–1914) that showed how mice that bred after their tails were cut off continued to give birth to offspring with tails. Since the reassertion of Darwinian evolutionism in the early 20th century (now combined with genetics and population biology), variation and natural selection have eclipsed the inheritance of acquired characteristics as the focus of evolutionary theory for most mainstream biologists. Arguably, though, Lamarckianism retains a certain level of popular support.

For further reading: Peter J. Bowler, *Evolution: The History of an Idea*, rev. ed. (University of California Press, 1989); and Jean Baptiste Lamarck, *Zoological Philosophy* (1809), trans. by Hugh Elliot (reprinted by Hafner, 1963).

LAPLACIAN PHYSICS The hypothesis that all nature is in principle explicable by the laws of theoretical physics—celestial and terrestrial mechanics—and subject to experimental tests.

Pierre Simon Laplace (1749–1827) made major contributions to theoretical physics and celestial mechanics. In their time he and Adrien-Marie Legendre (1752–1833) were the outstanding figures in these fields, often developing each other's ideas. Laplace lived through tumultuous times in France, held public office under several regimes, was appointed a Count of the Empire, later a Marquis, and in 1916 elected one of the Forty Immortals of the Académie française. He is said to have discussed astronomy with Napoleon Bonaparte on the field of battle. He published both mathematical and explanatory books on celestial mechanics and on the theory of probability. His work on probability laid the foundation for all subsequent work in the subject.

Laplace researched the Moon's motion, tides, the variation in the orbital velocities of Jupiter and Saturn, the stability of the solar system, the eccentricity and planes of planetary orbits, and the shape and rotation of Saturn's rings. His investigation of a rotating fluid in equilibrium led him to suggest the NEBULAR HYPOTHESIS of the formation of the solar system. Although his output of original and synoptical works in a wide range of subjects was enormous, he said on his death bed, "What we know is minute; what we are ignorant of is vast."

Laplace's more famous aphorism was his reply to Napoleon when he asked him whether God had created the planetary order described in his celestial mechanics: "Sire, I have no need of that hypothesis." It is that attitude, the rejection of any approach to science not rooted in the testability/falsifiability thesis—long before Sir Karl POPPER (1902–94)—that exemplifies the antithesis to pseudoscience that is the essence of Laplacian physics.

For further reading: Stephen G. Brush, *History of Modern Planetary Physics* (Cambridge University Press, 1996).

LASER BEAMS Narrow beams of intense radiation, produced at a fixed frequency, that depend on the molecules in the cavity of the laser. The word "laser" is an acronym for Light Amplification by Stimulated Emission of Radiation. Lasers may function continuously or in a pulsed fashion. The pulsed laser transmits small bursts at a very high energy level; indeed, one of the first uses of the beam was for cutting and welding in heavy industry. It was first introduced sometime between 1957 and 1960 in California.

In medical practice, lasers are now used for cutting tissue; the localized heat that they generate can coagulate blood, arresting bleeding. By passing the laser through fiber optics, the energy can be transmitted to vessels of the heart from some distance away—for example, from an incision in the thigh—without affecting other surrounding tissues. In ophthalmic surgery, the small diameter of the laser beam allows for the delicate welding of a detached retina back onto its base, thus restoring sight. Birthmarks like the disfiguring port-wine mark can be taken away without the overlying skin been broken or burnt.

It was once thought possible to introduce lasers into nuclear fusion by bombarding a volume of hydrogen with an intense pulse of laser light, but the development of a laser sufficiently powerful for fusion to take place and to maintain the right amount of hydrogen has proved too difficult to achieve.

President Ronald Reagan tried to introduce a pseudo-scientific idea using lasers as a wonder weapon, the Strategic Defense Initiative (or Star Wars project, as it came to be called). The theory was that a high-energy laser beam could burn holes into each incoming ballistic missile before it crossed the U.S. coastline, thus rendering each in turn useless. It is doubtful whether this could have been done, and the project is now on hold—not least because there is no super-power left in the world to challenge the United States with ballistic missiles from afar. Knowing what we do now about Soviet technology, there probably never was such a power.

A United Nations conference in Vienna in 1995 called for an international ban on laser weapons that were designed primarily to blind enemy troops. The main opponent to the regulation banning such weapons was the United States, but negotiations are continuing. Nevertheless, advances in laser-weapon technology continue. Weapons experts predict that laser weapons will soon be held in the armories of most nations and will become commonplace on battlefields by the year 2025. As Jones points out: "The speed of light opens a whole new dimension." For example, a conventional anti-aircraft gunner must predict where his target will be when his shell arrives on target—a notoriously difficult operation. Jones predicts that because a laser beam traveling at the speed of light arrives at its target virtually instantaneously, "Neither side will be able to operate aircraft, missiles, or satellites above the horizon of the other's speed-of-light weapons." In other words: If you can see it, you can shoot it down; all visible targets will be vulnerable.

LAWSON, ALFRED A. (1869–1954) Developer and promoter of the science of LAWSONOMY. Born in London, he grew up in Detroit, and worked at various menial tasks, until he ran away from home in his teens, riding freight cars for several years. Between 1888 and 1907, Lawson became first a professional baseball player with a number of teams, later a manager. During this period he wrote his first book, a work of science fiction, *Born Again* (1904). Then followed a period in aeronautics, first publishing and editing popular magazines, *Fly* and *Aircraft*, and subsequently designing and building airplanes. The Lawson Aircraft Corporation, Green Bay, Wisconsin, built the first passenger planes in 1919 and 1920 and was for a while a successful business.

In his early fifties, Lawson began to formulate and promote theories in both the social and physical sciences, which were the basis of his Lawsonomy. He promulgated these theories in *Manlife* (1923), which sets out Lawson's Laws of Penetrability and Zigzag-and-Swirl motion. In the 1930s, Lawson became the leader of a popular economic reform movement, the Direct Credits Society (DCS), which attacked financiers' stranglehold on the economy, supported the abolition of interest, and advocated loosening credits to end the Depression.

LAWSONOMY A collection of ideas promoted by Alfred A. LAWSON, which he believed provided an overarching theory for virtually everything. Lawson dismissed current orthodox theories as the work of incompetents and replaced them with an entirely new set of principles and concepts. He thought that mechanics were governed by penetrability, suction and pressure, and "zigzag-and-swirl"; for Lawson gravity was nonexistent and its effects were attributed to Earth's suction. The functions of the human body were attributed to the actions of thousands of little suction and pressure pumps. The universe is occupied by regions of different density, and movement takes place by pressure and suction, governed by Lawson's Law of Penetrability. Bodies do not move in straight lines or simple curves but along complicated paths according to the principle of zigzag–and–swirl. The activities of the human brain are the work of billions of mental organizing particles, "menorgs" in conflict with disorganizing particles, "disorgs." When thinking, a human being marshals the vast army of menorgs to battle against the disorgs. Lawson expounded his ideas in many books, including the three-volume *Lawsonomy* (1935–39), and *Penetrability* (1939).

In 1942 Lawson opened the Des Moines University of Lawsonomy. As its name implies, it taught Lawsonomy and only Lawsonomy, to the exclusion of anything not sanctioned by Lawson and recorded in his writings and speeches. The teachers were Knowlegians, top teachers were Generals, and Lawson himself was Supreme Head and First Knowlegian. Students paid no fees but were expected to work on campus.

Until his death Lawson remained convinced of the ultimate triumph of Lawsonomy, predicting that it would replace currently accepted wisdom by the year 2000.

LEARY, TIMOTHY (1920–1996) U.S. psychologist and proponent of sensory expansion. Leary's self-described "illumination" occurred in Mexico in August 1960 as a consequence of eating "sacred mushrooms." In 1962, Leary, then a professor at Harvard, was introduced to the hallucinogen LSD by a biologist who had been studying the drug's effect on spiders. Leary subsequently began to promote LSD as a social cure-all. He set up the Castalia Foundation, coined one of the defining buzz phrases of the 1960s, "Turn Off, Tune In, Drop Out," and cowrote *The Psychedelic Experience* (1964) with Richard Alpert, a like-minded Harvard colleague. Leary and Alpert believed LSD to be a sacramental chemical that could induce spiritual revelations, and they accordingly modeled *The Psychedelic Experience* after the Tibetan *Book of the Dead.* They intended their book as a manual for mind expansion that would impart a mystical, even religious, frame of reference to the LSD experience.

Leary's activities attracted the disapproving attention of the FBI, leading to a conviction on charges of possession of marijuana, a sentence of imprisonment for ten years, and a jailbreak engineered by a radical group called the Weathermen. Leary escaped to Algeria, was recaptured in Afghanistan, was returned to the United States, and remained incarcerated until 1976. He later toured on the lecture circuit with Watergate conspirator G. Gordon Liddy. Leary also wrote about himself (*Flash Backs,* 1983) and extensively about such subjects as exopsychology, neurologic, neuropolitics, neurogeography, and rejuvenation.

LEE, SAMUEL, JR. and **LEE SAMUEL H. P.** See BILIOUS PILLS.

LEMMINGS Rodents closely related to mice and voles. Lemmings are small (about 8 to 12 centimeters [3 to 5 inches] long) and brown or red in color. They have short tails, live underground, and eat grass, roots, and plant products. They breed often and quickly; females are capable of producing a litter of 9 young only 20 days after mating. Species of "true lemmings" (genus *Lemmus*) live in open, grassy areas of Norway, Canada, and the United States; other lemming genera include the Arctic (or collared) lemmings, the red-backed (wood) lemmings, and bog lemmings.

Many people believe that lemmings commit mass suicide by trekking to the sea and throwing themselves over cliffs to drown in the water; this is not true. True lemmings (genus *Lemmus*) have a population cycle that, every few years, explodes. When food supplies begin to run short, lemmings will migrate in large numbers in search of new feeding grounds. The Norwegian species of lemming will cross rivers and lakes in search of food, even if many of them drown in the process. They do not necessarily stop when they reach the sea, which may have given rise to the suicide legend.

Many more lemmings are killed by predators than die in these migrations. Owls and foxes, their most common predators, have a reproductive cycle that follows that of the lemming population. Because of their "boom-and-bust" population cycle, lemmings form an important part of the food chain in the Northern Hemisphere.

LEVITATION Raising the human body or any object into the air without mechanical aids, apparently defying gravity. History is full of levitation reports. Mystics of many religions have been said to have the power to levitate themselves. Saint Teresa of Avila (1518–82) described raptures in which her body was lifted from the ground, and Saint JOSEPH OF COPERTINO (1603–63), called the Flying Friar, was believed to have flown in the air 70 times in 17 years, the most spectacular occasion being when he helped workmen erect a Calvary cross 11 meters (36 feet) high by flying upward and lifting it into place "as if it were a straw."

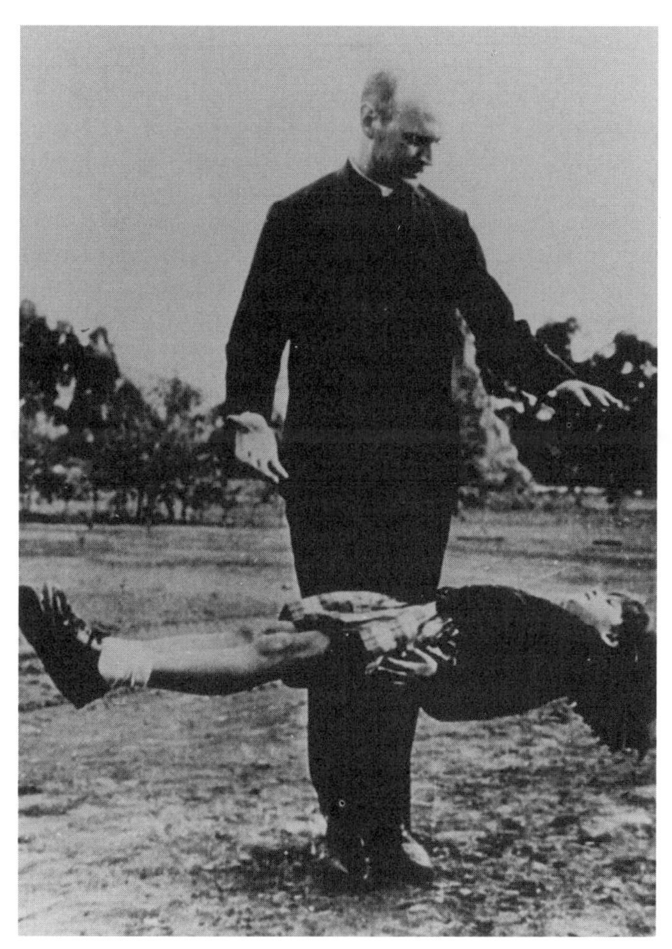

Levitation

Throughout the ages, witches, who were believed to be possessed by the devil, were thought to have the ability to fly on broomsticks by the application of flying ointment to their bodies, but nobody now takes this claim seriously. In the early 20th century, it was occasionally reported that Indian fakirs had performed an amazing feat called the Indian rope trick. A fakir's young assistant was said to have climbed up a rope that remained in the air without any visible means of support, and then on a sign from the fakir, the young lad had disappeared from view. As these reports were always second- or thirdhand, the British Magic Circle in 1919 offered a large sum of money for anyone who would perform the trick under controlled conditions. There were no takers, and it is now thought that reports of such happenings were hearsay and that the rope trick may never have existed.

Many spiritualist MEDIUMS have claimed the power to levitate, and some have been seen during a séance to take the table with them, so that all those holding hands around it had to stand up. The celebrated and very clever 19th-century medium Daniel Dunglas Home (1833–86) was reported in 1868 by a group of English lords to have achieved a spectacular levitation. Home apparently floated out of a third-story window 25 meters (80 feet) above the ground and back in at another window, in Ashley House, London. One of those present said that he knew what was going to happen through telepathic communication. All present described the windows as having wrought-iron balconies outside each, with a 2-meter (7-foot) gap, and only a 10-centimeter (4-inch) ledge between the two. The building is no longer standing, but before it was taken down some of these measurements were disputed, especially the gap between the balconies which was said to be only 1.2 meters (4 feet). There is no way that the matter can be resolved now, except to say that Home was caught in fraud many times, but he was always clever enough never to be exposed publicly.

Another famous levitator was Carlos Mirabelli, whose 1934 signed and validated photograph of himself in his white lab coat standing in midair close to the ceiling, is now in the annals of photography. The negative has recently been scrutinized and found to have been doctored. In fact, Mirabelli was not levitating but standing on top of a stepladder that did not show in the photograph because it had been erased on the negative. At the time, when people thought the camera could not lie, no one thought it necessary to check the negative. This finding questioned other photographs of Mirabelli, levitating in a chair and, on another occasion, in an automobile together with several other passengers.

Today courses in TRANSCENDENTAL MEDITATION claim to teach the discipline of levitating in the lotus position, about a meter (3 feet) off the ground, and eventually of flying through the air horizontally. People who have taken the course say the floor is very well padded, and that, after a few months, they have developed the concentration and muscle power to jump 2 or 3 centimeters (an inch or so) into the air, for a few seconds. They feel that with practice this can be improved upon, and although no one has yet achieved true levitation some are convinced that it can be done.

Numerous very clever stage magicians' acts are built around levitation. A typical act starts with the curtains being drawn back to reveal a scantily dressed young woman appearing to lie rigidly balanced on the points of swords. The swords are removed and she is left suspended in midair as if in a cataleptic trance, while hoops are passed over and under her to show that there is no trickery. She then rises up very slowly to the applause of the audience. A parlor game demonstrating levitation, which used to be very popular at parties, involves four people at each corner of a chair lifting it with only their index fingers, while a person is sitting on it. This cannot be done without the four lifters first pressing down for a few minutes on the top of the head of the person seated. Then the levitation can proceed, but for only a few inches off the ground. An explanation that has been put forward is that the initial pressing down causes a resetting of the natural level of muscle tension in the lifters' arms and blocks the pain so that a greater force than usual can be applied to the lift.

Magicians and parlor games aside, what would levitation as a genuine phenomenon entail? It would mean that the gravitational field is somehow inactivated either universally, thus ceasing to act on all persons and things everywhere, or just locally, that is, ceasing to act on just one person or a single object. The universal option would have widespread and disastrous consequences, with objects usually kept in place by the downward pull of gravity suddenly starting to take off like the unanchored contents of a spaceship. How local would the local option be? Nobody has ever suggested that a levitating person could suspend gravity to the extent that he or she flies off into space. So how could the lift be limited to a decorous raising of a few feet off the ground? And why does it not always include the bystanders, as was the case in the medium's table? If this is a real physical possibility, it should be applied to more serious and important ends. When we begin to consider the phenomenon rationally, and examine all its implications, it is clearly absurd. If, instead, we suspend thinking and consider it irrationally then anything is possible. This is the difference between science and pseudoscience.

The word "levitation" is also used to describe a technological process which reduces friction in a great variety of transport systems, in particular in rail transport. The technique uses repulsion between like magnetic poles to raise the vehicle clear of its supports, thus effectively eliminating mechanical friction and the resultant energy loss. The first magnetically levitated trains (MAGLEV) use superconducting electromagnets, thereby reducing electrical energy losses, to lift, guide, and propel the train. The first public MAGLEV line went into operation in Berlin in 1988. Similar systems are planned in the United States, Britain, and Japan. This form of levitation is not, as defined above, "without mechanical aids, apparently defying gravity," but is a technological development, well explained by the laws of physics, and therefore in no sense pseudoscience.

See also MAGIC and WITCHCRAFT.

For further reading: see Richard P. Brennan, *Levitating Trains and Kamikaze Genes—Technological Literacy for the 1990s* (Wiley, 1990).

LEVY, WALTER J. (1947–) Famous perpetrator of fraudulent experiments in PARAPSYCHOLOGY. Following graduation from medical school, Levy joined the staff of the Institute of Parapsychology headed by Joseph Banks RHINE in Durham, North Carolina. Levy ran what appeared to be a successful series of experiments that seemed to indicate ESP in animals, especially mice and chicken embryos, from which he published a number of papers. In 1973 Rhine appointed him director of the institute.

Levy's very success suggested to some that he was falsifying data. Suspicions arose in 1974, when some of his colleagues at the institute noticed him hanging around the computer during an experiment. There was no need for him to be there, as all of the equipment was running automatically without the need of any human intervention. Suspicions seemed confirmed by the special significance of this particular experiment. The workers in the laboratory rigged the equipment to bypass the computer at the point where they suspected Levy was manipulating the machine. If two different sets of data were produced, then he had manipulated one set. When two divergent sets of data appeared, it was hard evidence of his faking positive results. The results were presented to Rhine, who in turn presented them to Levy. He admitted what he had done and was shortly thereafter fired from his position. He dropped out of the parapsychological community.

The incident called all of Levy's work into question. Most affected was James Terry, who had worked with Levy on some of the earlier experiments. He began to repeat the affected experiments but was unable to produce positive results. As a result, Rhine withdrew all of Levy's work, and Terry reported on his unsuccessful replications at the Parapsychological Association.

The Levy deceptions have been widely discussed as one of the two major incidents of prominent fraud in parapsychological research, the other being the case of Samuel G. SOAL. It is to the credit of the community that it uncovered the fraud, confronted Levy with the findings, and announced what had occurred to the world, embarrassing as this had been.

LEY LINES An invisible network of alignments connecting sites of sacred or ritualistic significance. Leys (pronounced *lays*) were first named and defined by Alfred Watkins, a Briton, in the 1920s when he discerned notable patterns linking ancient burial sites, beacon hills, churches built on early pagan sites, stone circles, holy wells, and other landmarks. Watkins's books, *British Trackways* (1922) and *The Old Straight Track* (1925), and his public lectures did much to establish his theories in the popular imagination. Watkins asserted that ancient sites were not situated by chance; early people carefully planned their trades routes, city sites, and sacred places according to very particular straight-line alignments with astronomical connections. He found that England was covered with a vast network of straight-line connections—sacred and other important sites lined up with important sun alignments (such as midsummer sunrise). Watkins's supporting evidence included the observation that the names of many sites, both ancient and more modern, that were aligned on particular paths contained variations of the word *ley: lee,*

Ley lines

leigh, ley, lea, and so on. (The Old English word from which *ley* is ultimately derived means "grassland.")

Others began to discover similar significant alignments in other countries; in fact, some ley hunters, as they call themselves, believe that the entire Earth is covered with a vast, significant gridwork.

During the rise of the NEW AGE MOVEMENT in the 1960s and 1970s, the concept of leys evolved into the idea that leys are paths of a supernatural energy grid that covers Earth. The intersections of leys are thought by some occultists to be strong energy vortices that provide power to those who know how to access and use it. These vortices are also said to be frequent sites of such anomalies as UFOs and EARTHLIGHTS.

Skeptics maintain that one can take a map and connect any number of sites with straight lines without these having any significance at all. They also point out that sites allegedly located on leys often come from extremely different time periods, shedding doubt on the idea that ancient people planned the leys. (Believers suggest that the locations from different time periods do not disprove the ley theory; quite the contrary, they show that modern people, too, are drawn inexorably to follow these ancient and perhaps mystical earth patterns.) In general, archaeologists and other scientists do not accept the validity of leys, finding no evidence in their favor.

For further reading: Paul Devereux, *Shamanism and the Mystery Lines: Ley Lines, Spirit Paths, Shapeshifting, and Out-of-Body Travel* (Quantum, 1994); and John Michell, *The New View Over Atlantis* (Thames & Hudson, 1983).

LINDORMS Giant serpents reported to exist in Scandinavia. In 1885, Swedish scientist/folklorist Gunnar Olof Hylten-Cavallius published some 50 accounts of sightings of the lindorm, which he described as being 3 to 6 meters (10 to 20 feet) in length. These reports gave verification to the earlier discussion of lindorms in *The Natural History* of Norway by Erik Pontoppidan (1698–1764), the Bishop of Bergen, in the middle of the 18th century. Pontoppidan had related the belief that the lindorms were youthful SEA SERPENTS that leave the land for life in the sea when they grow to the point that they can no longer move easily around the rocks.

According to Hylten-Cavallius's summary, the lindorm is a black snake with a yellow underside, whose body is as thick as a man's thigh. Its neck shows a growth of hair similar to a horse's mane, and it possesses a forked tongue and a mouth full of teeth. In general, it was reported to have a foul temper and could raise itself up and attack any prey. It was generally reported in unpopulated areas. Encounters tended to have a traumatic effect on the people involved.

Following publication of his book, Hylten-Cavallius offered a reward for a specimen. None was forthcoming, and none was produced in the following century. Throughout the 20th century, the number and frequency of reports have steadily declined. There is no physical evidence that lindorms exist or ever have existed, and explanations of the many reports have to be sought in psychology and folklore.

LOCH NESS MONSTER The most widely publicized and most widely believed survivor of the dinosaurs. The first recorded sighting of this animal, possibly a living plesiosaur, was by Saint Columba in 565 C.E. There were a very few further sightings reported until 1933. "Nessie" has since become an international celebrity, being photographed and reported in newspapers and on radio and television across the world, though remaining very elusive. Eminent scientists have lent their support to Nessie's "incontrovertible" existence; well financed and equipped expeditions have pursued it, hoaxers have admitted to their pranks, but the search continues.

Over the years, many photographs, allegedly of the monster, have been produced, none very clear and unambiguous. An expedition in 1972, using an underwater camera, produced photographs that were grainy and indistinct but which, when enhanced by computer, showed a 1.22-meter (3-feet)-long fin. In 1975 two more pictures were published, one showing the monster's head, neck, and body, and the other showing a close-up of its face. In 1984 *Discover* magazine accused the 1972 team of having retouched its picture. Adrian Shine, of the Loch Morar Project (which later became the Loch Ness Project), charged that the 1975 head photograph actually showed a tree stump, and Tim Dinsdale, another Nessie researcher, suggested that the photograph was actually of an engine block discarded by a local boatman. (Loch Morar, near Mallaig on the west coast of Scotland, is another landlocked lake. For many years, there were claims of sightings of sea creatures, "Nessie's cousins," in Loch Morar; explorations of the two lochs—although some 60 kilometers (40 miles) apart—were linked. That possible connection seems to have been dropped.) The 1972 expedition also obtained sonar traces of two large objects chasing a school of salmon.

Another expedition in 1987, Operation Deepscan, obtained no photographs but did report 10 sonar echoes from large mobile objects. The most recent photograph was printed in the *Nottingham Evening Post* in 1995. In that year, the search had been intensified and three sightings/soundings were reported in the December issue of the *FORTEAN TIMES*. In April 1995, sonar on the boat the *Royal Scot* indicated a strange object in Loch Dochfour, a shallow lake at the north end of Loch Ness—the first time that Nessie or some relation had been reported in Loch Dochfour. In July 1995 a submarine exploring Loch Ness recorded strange "grunts" 140 meters (450 feet) down, close to the west bank, and a member of the scientific

Loch Ness Monster

team hazarded the opinion that "perhaps the sounds are the monster's mating call."

Although the existence of Nessie remains in doubt, about one thing there is no doubt: The Loch Ness Monster has made a massive contribution to the Scottish tourist industry.

See also DINOSAURS, CONTEMPORARY; LAKE MONSTERS; MORAG.

For further reading: Henry H. Bauer, *The Enigma of Loch Ness: Making Sense of a Mystery* (Johnston, 1991); Stuart Campbell, *The Loch Ness Monster: The Evidence* (Birlinn, 1996); Tim Dinsdale, *Loch Ness Monster*, 4th ed. (Routledge, 1982); Nicholas Witchell, *The Loch Ness Story* (Corgi, 1989).

LOCH NESS INVESTIGATION BUREAU Originally called The Bureau for Investigating the Loch Ness Phenomena Ltd., a bureau that was established to serve a dual function: to receive reports of observations and investigations and to promote actively research concerning the loch and its famous inhabitant (or inhabitants)—the LOCH NESS MONSTER. The bureau was established in 1962 at Achnahannet on the shore of the loch. It recorded random sightings, initiated projects investigating the loch—principally concerned with locating and identifying the monster(s) beyond reasonable doubt—and helped other organizations in their explorations of the loch. It fulfilled that role until it closed in 1972.

In 1967 Professor Roy MACKAL became a director of the bureau and garnered substantial support from the United States for its activities. The bureau's headquarters at Achnahannet also served as an exhibition for the eccentric and unusual connected with the loch. In 1971 alone, 50,000 visitors viewed its small display of photographs, maps, and charts. The bureau recorded 200 eyewitness sightings of the monster between 1963 and 1972, after scrupulously rejecting any that it felt were doubtful. It assisted investigating teams from as far afield as Europe (including Britain), the Americas, Australia, and Japan. When it closed in 1972, lacking both sufficient funds and planning permission to continue on its present site, it could rightly claim to have done a thorough and conscientious job. Nevertheless, it had failed to locate and identify the monster "beyond reasonable doubt."

For further reading: Nicholas Witchell, *The Loch Ness Story* (Corgi, 1989).

LOMBROSO, CESARE (1835–1909) A 19th–century Italian physician, who believed that there is a gene that predisposes its possessors to criminal or aberrant behavior. He claimed to have identified features that characterize criminals: "the enormous jaws, high cheek bones, prominent superciliary arches, solitary lines in the palms, extreme size of the orbits, handle-shaped ears." He maintained that these characteristics were shared by criminals, savages and apes, which led them to exhibit "insensibility to pain, extremely acute sight, tattooing, excessive idleness, love of orgies, and the irresponsible craving of evil for its own sake, the desire not only to extinguish life in the victim, but to mutilate the corpse, tear its flesh and drink its blood."

Lombroso thus distinguished between criminal behavior by normal people and the behavior of born criminals, people who had evolved with characteristics that predisposed them to criminality. It is a variation of the familiar NATURE-NURTURE DEBATE: All humans have the capacity, presumably small, to behave criminally (nature), and whether or not they do so depends on their circumstances (nurture); but there is a distinct group who have evolved differently with a much larger criminal nature, and this latter group can be easily recognized by their ape-like features. Its members are a throwback to our common ancestry with the apes.

Lombroso

Lombroso had to show the association between animals and criminal behavior to support his argument. If our criminal classes resembled animals, then one could only predict criminality if animals were by their nature disposed to evil. So he proceeded to establish that link in his chain of argument. He described cases in a great variety of animals and insects that proved his point: ants, birds, beavers. It's not hard to add similar cases to his catalog from the newspaper accounts of our day: domestic cats killing birds, foxes destroying all the occupants of a henhouse, dogs attacking children.

Lombroso then added another step in the argument: He believed that the more primitive groups of humans lagged behind the civilized groups in the evolutionary process. They were nearer to the apes and therefore more criminal; they behaved more savagely and had physical features resembling those of born criminals. Lombroso wrote a treatise on the Dinka tribe of the Upper Nile to justify his theory.

Lombroso also used the moral shortcomings of children, particularly the children of the lower classes, to support his ideas.

All the features that Lombroso draws upon to make his case are distributed continuously among the population. Nevertheless Lombroso's theories, while alienating many, were attractive, and were often adopted in a modified form by others. Italian educator Maria Montessori and Henry Havelock ELLIS explicitly supported Lombroso. Bram Stoker in *Dracula* bases the title character on Lombroso's theories. In the 1930s Earnest Hooton, professor of anthropology at Harvard, made an exhaustive study correlating body types with different forms of criminality and was convinced of a significant relationship. The idea of inborn criminality still shows up from time to time today.

Lombroso appears in another role in the history of pseudoscience. He, a man of standing in the world of science, was approached in 1888 by the physical medium Eusapia Palladino's sponsor, who requested that he should test her scientifically. In 1990 Lombroso, known to be skeptical about such matters, did examine the medium's claims when in Naples and, much to other scientists' surprise, he pronounced her genuine. That endorsement was important in setting Palladino on the road to fame and fortune.

See also CRIMINAL GENES.

For further reading: Stephen Jay Gould, *The Mismeasure of Man* (Norton/Penguin, 1981).

LOST TRIBES OF ISRAEL Ten tribes of the northern Hebrew nation of Israel taken captive by the Assyrians in the eighth century B.C.E. They never returned to Israel.

The fate of the Lost Tribes became a matter of much theological debate, especially after the discovery of the Americas in the 15th century. Some clerics advanced the opinion that Native Americans were the descendants of the Israelites, settled by divine Providence in the New World. This idea was later adopted by Joseph Smith, founder of the Mormon Church.

Some theorists suggest that one of the lost tribes removed to Ireland, where they became the ancestors of the Irish. Most historians now believe that the Lost Tribes of Israel became part of the Diaspora, the great worldwide Jewish community that began to emigrate after the destruction of Jerusalem by the Romans in 70 C.E. Many of the Israelites probably settled in Mesopotamia (modern Iraq), although ancient Jewish communities also existed in Egypt, India, and throughout the Roman Empire.

LOURDES A place of pilgrimage for millions of devout Christians each year, where visions of the Virgin Mary were seen by a local girl in the mid-19th century. Lourdes, of Roman origin, is situated in southern France, at the foot of the Pyrenees. On the right bank of the River Gave, which flows through Lourdes, stands a medieval castle, a state prison from Louis XIV's reign until the beginning of the 19th century. Lourdes has two main areas: above lies the old fortified city with its castle, residential, and commercial area; below lies Massabielle with its grotto, sanctuary, chaplains' residences, hotels, and shops.

In 1858, Bernadette Soubirous—a 14-year-old peasant girl—claimed to have received visions of the Virgin Mary in the grotto of Massabielle. Since her father's milling business had failed when the millstream ran dry, the family of six lived in impoverished conditions in a one-room hovel. Bernadette suffered from asthma, was ill-nourished, illiterate, and spoke only the local dialect, akin to Basque. Though not unintelligent, she was constantly chided by her teachers for ignorance of her religion.

On February 11, 1858, while gathering firewood with her sisters on the banks of the Gave, Bernadette claimed that she saw a beautiful young woman, standing in a niche in the grotto of Massabielle. Bernadette instinctively knelt down and began to pray. When her sisters returned, they found her in a trancelike state and feared that she had died. She later described "Aquero" (that one) or "the Lady" in detail to her sister Marie and was reproved by her family for her foolish fancies, and forbidden to go again; but neither threats nor persuasions could keep her away from the grotto. The Lady appeared 18 times between February and July, and on her third appearance, she invited Bernadette to visit the grotto every day for a fortnight, adding, "I cannot promise you happiness in this world but in the next."

No one else saw the vision or heard the words spoken to Bernadette, but news spread, and the young girl was ridiculed by many who came to scoff. At first, the attitude of the parish priest was hostile and he forbade his fellow priests to visit the grotto. Bernadette was harshly interrogated by the Commissioner of Police and Imperial Procurator, in an effort to stop the place attracting thousands and disturbing the peace. Believing the girl was deranged, they sought the opinion of the local Dr. Dozous—a non-believer—hoping to disprove her claims in the name of science and medicine. He visited the grotto and examined the girl for signs of catalepsy or hysteria. By chance, this was the occasion of the sixth apparition and the doctor made careful notes and observations of Bernadette in a state of ecstasy. He concluded that her condition was due less to loss of normal consciousness than to extreme concentration. Moreover, he witnessed one of the first sudden and unexplained cures, that of a dying, two-year-old child—his own patient, Louis-Justin Bouhohorts—who was suffering from cachexia (paralysis and wasting). The day after being plunged into the icy waters of the grotto's spring, the child could walk.

Even the skeptical were moved by the transfigured face of Bernadette, as she gazed on "the Lady." At the ninth appearance, Bernadette was instructed to "Go to the Spring yonder and drink and wash yourself." Seeing no spring, Bernadette moved towards the Gave, but "the Lady" said, "Not to the Gave, please!" Bernadette scratched the earth under the rock where she was directed, coming at last to muddy water with which she smeared her face, to the shame of her family and the laughter of the crowd, who thought her demented. Before long, the underground spring of Massabielle began to flow and has flowed ever since, at the rate of 20–40 liters per minute. At the 13th manifestation on March 2, Bernadette reported that "the Lady" said, "Go and tell the priests to come in procession and build a chapel." At the 16th apparition, "the Lady" revealed her name, speaking in a local dialect. "Qué soy era l'immaculada councepciou" (I am the Immaculate Conception). Bernadette did not understand these words, which she repeated to the priest, Father Peyramale. He had previously dismissed her account of the apparitions, but subsequently wrote to the Bishop of Tarbes, who created a Commission to investigate Bernadette's claims.

The authorities of both Church and State were reluctant to accept the manifestations at the grotto, although local people had no doubts. They had witnessed cures caused, they believed, by the healing qualities of the spring; for example, a blind man who washed his eyes in the spring's water regained his sight. They set up a little shrine with candles and flowers and proposed the construction of bathing establishments. An eminent chemist, Professor Filhol of the Faculty of Science at Toulouse, analyzed the water and reported that it had no special medicinally curative constituents. Therefore, the extraordinary effects claimed could not, "in the present state of science, be explained by the nature of the salts (chemicals) of which the analysis shows the existence."

Only in 1862 after four years, first of opposition, then of stringent examination—during which Bernadette tranquilly but resolutely held to her claims—were the visions declared authentic by the Catholic Church. By order of the bishops, the spring water was channeled to fill a reservoir with three taps and to fill *piscines* to receive the sick for bathing. Since then Lourdes has become a major international pilgrimage center. About 5 million pilgrims, some 50 thousand of them sick or disabled, go there annually.

The busy Rue de la Grotte is now noisy, jostling, and full of shops filled with cheap Lourdes souvenirs. By contrast the Domaine, which early on was bought by the Church to keep it unsullied for the worship of God and intercession of Our Lady of Lourdes, is open to pilgrims day and night as a place of prayer and meditation. There are now four basilicas in Lourdes: the Upper Basilica, completed in 1871, was built over the Crypt—the first chapel open for worship—begun in 1863. The Rosary Basilica, opened in 1889, became inadequate to hold the increasing number of pilgrims, so an immense Underground Basilica of reinforced concrete, holding 30,000, was consecrated in 1958. Six entrances without steps allow easy access for disabled pilgrims in wheelchairs. St. Bernadette's Church, inaugurated in 1988, seats 5,000 and has a sanctuary, meeting rooms, and a chapel.

Sick pilgrims are cared for by an army of helpers under the guidance of the Association of the Hospitallers of Our Lady of Lourdes, established in 1884. In early days most carers were from noble families, but today's volunteers, from all walks of life, include members of the forces and religious orders, professionals, manual workers, housewives, and students. They work without payment as nurses, helpers at the baths, and *brancardiers* (stretcher bearers) paying their own expenses, and adhering to strict rules of conduct. Four medical centers, accommodating up to 2,400, receive sick pilgrims who visit Lourdes to partake in the Masses, the baths, the afternoon procession of the Blessed Sacrament, and the evening torchlight procession.

Alleged cures are rigorously investigated, first by doctors and then by the Canonical Commission, and must satisfy seven criteria before being declared miraculous: the disease must be organic, grave, incurable, without effective treatment; the cure must be instantaneous, complete, and permanent. The Medical Bureau is responsible for examination of patients when an alleged cure takes place. Scrutiny is so severe that only a fraction of cases are referred. Patients are examined and cases discussed exhaustively. If a record seems significant, the patient is kept under observation by a physician in his/her own region and a second examination at the end of the first year is required before judgment by the International Medical Commission of Lourdes. Lastly, the Canonical Commission in the patient's own diocese, consisting of five people, two of whom must be specialists in the disease, considers the case. Cures, which can only be declared miraculous by the bishop, are never announced in under two years' time—usually much longer.

An international hydrologist, Abbée Richard, investigated the source of the spring, and reported on its geography and geology in 1859. In the 20th century Norden, Moog, and Lepage—eminent scientists—searched for the possible existence of radioactive elements, but the content was insignificant, "only 0.7 millicuries of radon per liter at the point of emergence." Therefore, "the water from the Spring at Massabielle is not magic water, miraculous water or holy water, but natural water of good quality. It is devoid of therapeutic properties, without anything which could be bactericidal or bacteristatic, or any other curative agent." Nevertheless, a bacterial analysis of a sample of the water, taken *after bathing* by sick pilgrims, was made at the Pasteur Institute in Paris in 1948, and revealed the presence of many microbes. However, guinea pigs injected with these microbes survived, whereas those inoculated with the same microbes in water from the Gave died. Lourdes water, it is claimed, even when polluted with billions of bacilli, has at no time caused harm or infection to any one.

Full details of the miraculous cures accepted by the Catholic Church—only 65 in 126 years—are recorded in the Archives. The nature of pilgrims' illnesses has changed with advances in medicine. In the 19th century, sick and dying patients with tuberculosis, poliomyelitis, and rheumatic heart disease were brought to Lourdes in hope of a cure. These diseases are rarely seen today. Pilgrims with neuromuscular diseases, congenital disorders, cancer, kidney disease, and severe injuries from traffic accidents are more common now. The number of authenticated cures has decreased in recent decades, since many referred do not meet Lourdes's exacting criteria, having already received all the advanced treatment of modern medicine.

Four people were declared miraculously cured between 1965 and 1978, two French and two Italians. A soldier, Vitorio Micheli, aged 22, was admitted first to Verona and then to Trente Military Hospitals, in April 1962, suffering from severe sciatic pain, a swelling of the left buttock, and limited hip movement. After X rays and biopsy, an advanced pelvic sarcoma, a cancerous tumor of embryonic connective tissue was revealed. The only treatment available was to encase his leg and pelvis in plaster. Though unable to walk, he made a pilgrimage to Lourdes in May 1963, where he took several baths in the pools. He was convinced that he was cured, but the Lourdes and Italian doctors did not believe it possible. One month later he was able to walk. When the plaster cast was removed in February 1964, his bone structure and hip configuration were normal.

The last Lourdes cure to be classed as exceptional was that of Delizia Cirolli, an 11-year-old Sicilian girl diagnosed, by means of X rays and bone biopsy in May 1976, as having a malignant tumor of the right knee. Her parents refused permission for amputation and her condition worsened. By December she was close to death, but regained full health and strength after her mother, who had deep faith in Lourdes, prayed and gave her Lourdes water every day. Her case was examined by the Medical Bureau in 1977, and accepted as an extraordinary cure of Ewing's Sarcoma of the knee in 1980. In 1982, the International Medical Committee declared the cure "a completely exceptional event in the strictest sense of the term, contrary to all known information and expectation in medical experience, and hence inexplicable."

Bernadette herself retreated to a life of prayer and penance in the Convent of Nevers in July 1866. She died of tuberculosis on April 16, 1879, aged 35. She was canonized on 8 December 1933, by Pope Pius XI, when 77-year-old Louis-Justin Bouhohorts was present. Her family resolutely refused to benefit at all from the remarkable experience of their daughter.

What are we to make of the extraordinary apparitions and cures associated with Lourdes? There has been a steady growth of interest by doctors since 1882, when five doctors visited, to 1953 when there were 1,556, and the number grows. Some are interested, curious, and unbiased; others are perplexed, incredulous, or hostile; but all respect the scientific procedures and strict practices of the Medical Bureau. Though the church has only authenticated a small number of miraculous cures, and does not demand belief by its members, many thousands of sick people have attributed their recovery to miraculous healing powers in Lourdes. Many hopes for a cure are unfulfilled, yet most leave Lourdes with an inner serenity and sense of well-being that is hard to explain. Neither Bernadette nor "the Lady" ever mentioned cures; their requests were always for prayer and penance. The water of Lourdes, therefore, is seen by believers—like so many examples in scripture—as a sign, symbolizing a gift from God to pilgrims, for their spiritual and physical good. Perhaps the Protestant Free churchman John Oxenham was correct, when he wrote in 1923, after visiting Lourdes: "For myself, I believe Lourdes to be a genuine revelation of the goodness of God to a world which, every day, stands more and more in need of it."

See also FAITH HEALING.

For further reading: Ruth Cranston, *The Miracle of Lourdes* (Doubleday, 1988).

LOWELL, PERCIVAL (1855–1916) American astronomer best remembered for his popularization of the existence of Martian "canals," first reported by 19th-century Italian astronomer Giovanni Virginio SCHIAPARELLI. Through his early career, Lowell was successful in business, finding opportunities to develop cotton mills and electric companies. Between 1883 and 1893 he traveled extensively in the Far East, serving for an extended period as counselor and foreign secretary to the first U.S.-Korea mission. Lowell became an accomplished astronomer later in life and correctly predicted the existence of a planet that caused changes in Neptune's orbit. That planet was discovered by Clyde Tomabaugh in the 1930s and named Pluto.

Lowell's fascination with the so-called Martian canals began with a misconception: In 1877, Schiaparelli reported the existence of *canali,* an Italian word meaning "channels," not "canals." What Schiaparelli had actually seen were thin, dark lines, supposedly connecting broad dark areas. Lowell decided that the markings were canals that were meant to carry water to and from the thin polar ice caps and between the dark areas (thought to be seas) on the planet's surface. He concluded that such canals were evidence of intelligent life on Mars. He founded the Lowell Observatory in Flagstaff, Arizona, in 1894 to observe the Martian surface, and for more than 10 years he charted the dark markings.

Although Lowell's claims about the Martian canal system were challenged by other scientists, his ideas about extraterrestrial life remain in the popular imagination. Only after the Mariner space probes of the 1960s, and the Voyager landings on the planet in the 1970s was Lowell's canal concept finally disproved. Although few credit the possibility of intelligent life on Mars, some scientists believe that there may at one time have been life on the planet, perhaps in the distant past when climatic conditions may have been more favorable. Others think that life may exist in environments deep underground.

See also MARS CANALS.

LUCKY/UNLUCKY NUMBERS One of the most abiding superstitions throughout history, that some numbers are lucky and others are unlucky. In our society, 3 and 7 are commonly held to be especially lucky; why this should be so is uncertain. Perhaps 3 is from the Holy Trinity, or it may have an earlier origin, such as the three-legged stool of the Greek prophetess Pythian. Seven may come from the seven days of the biblical creation, the seven ages of man, the seven deadly sins or the seven wonders of the world.

Thirteen as an unlucky number is even more puzzling, the more so because some communities regard it as lucky. Thirteen's unlucky character may originate from the 13 seated at the table at the Last Supper: Christ and his 12 apostles, one of whom betrayed him.

The lucky and unlucky numbers are most usually those in single digits, with a few exceptions.

1. is lucky; in France 1 magpie is unlucky. Does this derive from the Sun or from our monotheism?
2. is also lucky, and in France 2 magpies are lucky.
3. especially lucky, as above—but very occasionally is seen as unlucky, there being a possible association with Peter denying Christ three times.
4. is unlucky in Far Eastern cultures, representing death.
5. is neither strongly lucky or unlucky.
6. also is mixed, except that a child born on the sixth day of the month will have prophetic gifts.
7. as already noted, is especially lucky; the seventh son of the seventh son will have the gift of second sight. We speak of being in seventh heaven but, to the contrary, in the United States, France, and the United Kingdom, breaking a mirror brings seven years bad luck. The French ask pardon 77 times 7 times.
8. has no special connotations.
9. is lucky, associated with the nine months of pregnancy and, most important, being 3 squared, 3 X 3.

Once out of the single digits, the associations are few. **10** is generally good, perhaps because there are ten commandments. **11** is mildly unlucky. **12** the number of Christ's disciples, is therefore lucky. **36** has several superstitious associations for the French. **40** is a dangerous age. Last, but not least, **666** is the number of the Beast, the Devil's number—definitely a bad omen.

The practices that believers in numerical superstitions engage in are even more mind boggling. Oral Roberts needed a total of around $100 million to finance the building of a medical complex (The City of Faith—14 letters) near the Oral Roberts University. The hospital was to have 777 rooms, donations were solicited in sums of sevens: $7.77, $77.77, and so on. Building work was started in 1977 (the year 7777 being too long to wait), and 77 doves were released on that occasion. Is this obsession with seven because 777 has some Christian significance? Or because Roberts has seven letters? In 1980 a gambler at a craps table in Las Vegas placed a bet of 777 thousand dollars and won. MADAME BLAVATSKY, the Theosophist, published *The Divine Secret,* in which she rewrote the history of the universe in terms of the ROOT RACES: 7 root races, each of which had 7 sub-races, each of which had 7 branches.

The unlucky 13 is taken seriously by sufficient people for there to be no 13th floor in many tall buildings in the United States, the elevators going straight from the 12th to the 14th floor. In many streets in France there is no house numbered 13; sometimes straight from 12 to 14, or 11 to 15, sometimes 12A between 12 and 14. Airlines are even more of a puzzle, there being no obvious rationality to their behavior. Some American airlines have a row 13 in their planes, some don't, the same being true of British, Canadian, Australian and Japanese airlines. Almost all other European airlines, and all Asian airlines, do not have a row 13.

If there was anything in inherently lucky or unlucky numbers, it would soon show up in the returns from the casinos.

See also NUMEROLOGY.

For further reading: *The Cassel Dictionary of Superstitions* (Cassel, 1995).

LYCANTHROPY The act of transforming from a human to an animal. *Lycanthropy* comes from the Greek words for wolf (*lúkos*) and human (*anthropos*). Many cultures believe this concept. Even today some Native American tribes believe in shapeshifters. Most commonly, when Western people think of lycanthropy, they think of werewolves, the human–wolf beings who slash their way through horror movies and who used to be common in European folk beliefs and anecdotal reports. In Africa, India, and other parts of Asia, the lycanthropic image is more likely to be a tiger.

Lycanthropy is described in many historical writings, including the Greek classic *Satyricon,* Saint Augustine's fourth-century work *The City of God,* and many medieval

Lycanthropy

accounts. Detailed accounts of incidents have been reported even in the 20th century.

Folk belief has it that lycanthrops can either willfully transform themselves magically into animals or that people can be turned involuntarily into animals through black magic. Many witches during the notorious witch hunts of the Middle Ages confessed to lycanthropy or were accused of it. Often, animals that were thought to be witches in animal guise were executed. Some historians think that belief in lycanthropy developed out of unexplained shocking crimes—violent murders in which people were savagely slashed, bitten, and even cannibalized. The only explanation appeared to be that such a crime was the work of an animal; yet, other evidence suggested human involvement.

Today, psychologists and medical experts offer several explanations for lycanthropy. These include drug-induced hallucination, in which a person believes he or she has become an animal during the course of drug-taking. This explanation also fits many WITCHCRAFT accounts in which the alleged witch claims to have applied a special ointment to himself or herself to become an animal; records of recipes for witch ointments often include hallucinogenic substances such as deadly nightshade or henbane. Mental illness also incorporates delusions of animal personae, as do the diseases porphyria (a rare genetic disease whose symptoms include skin lesions and discoloration and excessive sensitivity to light), and congenital generalized hypertrichosis (a rare genetic disease that causes extraordinary hairiness on the face and upper body)—all these may help explain the phenomenon.

For further reading: Richard Noll, ed., *Vampires, Werewolves, and Demons: Twentieth-Century Reports in the Psychiatric Literature* (Charles C. Thomas, 1983); Charlotte F. Otten, ed., *A Lycanthropy Reader: Werewolves in Western Culture* (Dorset Press, 1986); and Ian Woodward, *The Werewolf Delusion* (Paddington Press, 1979).

LYELL, SIR CHARLES (1797–1875) Geologist who vigorously attacked the theories of CATASTROPHISM. Lyell believed instead in uniformitarianism—the uniformity of geological forces that requires an Earth old enough for these forces to account for its present state. Lyell was a powerful influence on Charles DARWIN, providing for him the stage on which he could set his theory of EVOLUTION and natural selection. In 1830–33, Lyell published *Principles of Geology, Being an Attempt to Explain the Former Changes of the Earth's Surface by Reference to Causes Now in Operation* (three volumes).

Lyell was so passionate about the uniformity of the past that he refused to accept the validity of any catastrophe, abrupt change, or progression in the natural history of the Earth. The debate went on for many years between Lyell and the great champion of catastrophe, Baron Georges CUVIER. Scientists accepted a more balanced form of uniformitarianism following the publication of Darwin's theories.

For further reading: Stephen Jay Gould, *Time's Arrow, Time's Cycle* (Harvard University Press, 1987/Penguin, 1990).

LYSENKOISM Marxist approach to agricultural science. Trofim D. Lysenko (1898–1976) attempted to reform Soviet agricultural practices during the 1930s and 1940s by utilizing a version of Jean-Baptiste LAMARCK's theories. Lamarck's version of EVOLUTION rested on the idea that life starts out simple and then becomes more complex as environmental forces act on it. In the late 19th century, evolutionists looking for an alternative to Charles DARWIN's theory of evolution through natural selection adopted some of Lamarck's ideas and formed a new theory, generally called neo-Lamarckianism. Neo-Lamarckianists suggested that an environment imposes shape and function on an otherwise passive organism; in other words, the creative element in evolution came from the environment and not from an individual animal or plant.

In the 19th century, Austrian monk Gregor Mendel (1822–84) demonstrated through his research with plants that organisms actually pass on their characteristics through genetic mutation. Lysenko rejected Mendel's theory in favor of a neo-Lamarckian approach. He proposed to force desired changes that could be passed from one generation to the next by altering the environments of his subject plants. When his experiments failed to produce superior crop plants, he falsified data and solicited political support from Communist Party officials, including Josef Stalin. In 1940, Stalin named Lysenko to the directorship of the Institute of Genetics of the Soviet Academy of Sciences.

Lysenko used his director's position to consolidate his political and scientific power. Soviet scientists who opposed his views were imprisoned or banished to Siberian prison camps. In addition Lysenko refused to allow any alternate theory of evolution to be taught in Soviet schools. In 1948, in an address to the Lenin Academy of Agricultural Sciences, Lysenko claimed victory for his ideas by declaring that they had been approved by the Central Committee of the Communist Party. After Stalin's death in 1953, however, Lysenko's career fell into decline. He lost his position in 1965, and his career and methods were censured by the scientific community.

MACKAL, ROY (1925–) One of the founding members of the INTERNATIONAL SOCIETY OF CRYPTOZOOLOGY, an explorer searching for the fabled long-necked monster MOKELE-MBEMBE of Lake Tele in the Congo and one of the many investigators into the legend of the LOCH NESS MONSTER in Scotland.

In 1980 Dr. Roy Mackal accompanied by James H. Powel, a crocodile expert, ventured deep into the wildest part of Likouala country around Lake Tele to reconnoiter, gather, and collate reports on sightings of the Mokele–Mbembe. The local people generally described it as being red-brown in color with a long neck, topped by a small reptilian head and having a thick body and a long, whiplike tail. They believed that it lived in deep pools close to the bank and usually emerged early in the morning or late in the afternoon, making little noise and leaving behind three–clawed footprints.

In his book *A Living Dinosaur?* (1987), Mackal said that the descriptions that he had collected did not fit any known living animal. He suggested that with its long neck and tail, four legs, vegetarian diet, and length of 4.5–9 meters (15–30 feet), the Mokele–Mbembe sounded very similar to a small sauropod dinosaur, perhaps a surviving relative of the giant diplodocus or the brontosaurus. Fearing that the Mokele–Mbembe was perhaps withdrawing to an even more inaccessible part of the swamp or heading for extinction, Mackal mounted a second expedition the following year. Since then, expeditions from all over the world have looked for the monster. Some even brought back film footage. One sequence purported to be a close-up of Mokele-Mbembe and was screened in the 1980s by the U.S. TV show *That's Incredible*. Roy Mackal later learned that the monster was a balsa-wood model and that the film had been taken in close-up with a camera strapped onto the shoulders of a swimmer.

Mackal also investigated the persistent claims of sightings of the Loch Ness Monster with equal thoroughness and persistence. Again unconvinced, he is nevertheless credited with providing the definitive account of sightings and research on Nessie, as the legendary animal is affectionately named, in his book *The Monsters of Loch Ness*.

See also CRYPTOZOOLOGY and DINOSAURS, CONTEMPORARY.

For further reading: Roy Mackal, *A Living Dinosaur?* (Brill, 1987); Roy Mackal, *The Monsters of Loch Ness* (Swallow, 1976); and Roy Mackal, *Searching for Hidden Animals* (Doubleday, 1980).

MACROBIOTICS A dietary system derived from the methods of a late 19th-century Japanese doctor, Sagen Ishizuka. The term *macrobiotics* comes from two Greek words meaning "large" and "life" and indicates that this way of eating will make adherents healthy enough to enjoy life to its fullest. The diet is semivegetarian, using whole-grain cereals, and vegetables with no white rice or refined sugar.

Macrobiotic diets are not based on the nutrient content of the food but on the amount of *yin* and *yang* (alleged energy modes) they contain. A diet is designed to meet the circumstances in which individuals find themselves. Everyone must balance the amount of *yin* and *yang* in the body, as an excess of either will result in illness. *Yin* is the more inactive, cooler, and darker side of nature, and *yang* is the dynamic stronger side, characterized by heat, movement, light, and energy. Someone who wishes to have a relaxing day should eat *yin* foods, and a person who lacks energy and has a busy day's work ahead should eat *yang* foods.

Moderate *yin* and moderate *yang* foods are always recommended over extreme *yin* or *yang* foods, such as sugar, spices, and alcohol (very *yin*) or meat, eggs, and cheese (very *yang*). Generally, raw foods are considered more *yin*, and the longer the cooking time the more *yang* food becomes. The upper parts of plants are thought to be more *yin*, so vegetables are cut from top to bottom or diagonally to get some *yin* and some *yang* in each piece. Using the correct cooking utensils is also very important. Cast-iron pots and pans are preferred to copper and aluminum because they, especially the latter, leave behind traces of metal that may contaminate the food.

The macrobiotic way of life takes into account other factors, including climate. Both temperature and humidity are said to unbalance the *yin* and *yang* in the body. *Yin* foods are seen as cooling in hot weather, and *yang* foods are recommended as warming in cold weather. The amount and type of exercise taken can alter *yin* and *yang* content of the body. For instance YOGA is considered to have a *yin* effect, and aerobic exercises to have a *yang* effect.

Some macrobiotic practitioners claim that this diet is effective in treating, and even preventing, cancer, AIDS, and other serious diseases. Diagnosis can come from pulse reading, and recommendations for treatment are based on one or more of the following: ancestral diagnosis; astrological diagnosis; aura and vibrational diagnosis; environmental diagnosis, which includes influences from tides or celestial bodies; and spiritual diagnosis, including past memories and future visions.

Although there is little scientific evidence to support any of these medical or diagnostic claims, the whole-food diet has proved to be popular and beneficial to many in the West. But dietitians warn that even though some macrobiotic diets contain sufficient amounts of nutrients, others are totally inadequate and could cause serious malnutrition. Mainstream doctors warn that using the diet as an alternative to modern medicine to try to cure a serious disease is putting the patient at risk.

For further reading: Patrick C. Pietroni, ed., "Eating Your Way to Inner Harmony and Health," *Reader's Digest Family Guide to Alterative Medicine* (Reader's Digest, 1991); and J. Raso, *Mystical Diets* (Prometheus, 1993).

MAGIC From the Greek *magika,* meaning "what wizards do"; any ritualistic practice intended to produce results without using the causal processes of the physical world. In its less elevated form, it could be any sleight-of-hand trick used by conjurers or any demonstration of mind reading or TABLE RAPPING by stage entertainers. The concept behind the serious belief in magic is that physical actions performed in one context can have an effect in a totally different context, perhaps miles away or years apart; for example, nature is thought to be controllable by rain dances, or evil spirits are warded off by carrying a talisman, or good luck is brought about by the carrying of amulets or charms. The two contexts—the action and the result—are either seen to be directly connected, or somehow linked together, by a third external, mystical force.

Because magic depends on a worldview that believes that there are supernatural forces at work outside the ordinary laws of nature, it is, in its broadest sense, at the core of most religious systems. But many Western religious thinkers make a distinction between, on the one hand, the external manipulation of magic symbols with the emphasis on technique, such as water when blessed acquiring curative qualities, and, on the other hand, the effect exerted by personal inner grace on an outside force for good. But even in the latter case, religious ritual is not always seen to be very distinct from magic as, for example, when the contemplation of icons is thought either to heal directly or to help evoke a state of grace, which then promotes a process of healing.

Social anthropologist James Frazer (1854–1941) in his 12-volume book *The Golden Bough* (published in installments between 1890 and 1915) described how, in certain climates of opinion where magic is the only available concept, magic has been used to influence behavior and therefore future events. Even in our scientific age, the "self-fulfilling prophecy" has been proved to be very effective.

In the mid-20th century Claude Lévi-Strauss (b. 1908) has examined magic in structuralist terms and come to the conclusion that, because all human beings have a tendency to order and classify phenomena, belief in magic, belief in religion, and belief in science are all ways of understanding the world, each having its own total structure. Furthermore he thinks that each belief system has its own internal logic, within which we have to take on faith certain basic principles that cannot be verified.

Very few scientists would accept Levi-Strauss's view of magic being another way of making sense of the world that is as valid as science. Scientists argue that there are significant differences between themselves and magicians, shamans, and prophets: The latter group straddles the supernatural and the natural world making statements that do not have to be verified, while scientists work solely in the natural world, always questioning the work that has gone before. Scientists observe, categorize, and then formulate a theory to explain those categories and observations. Then—and this is the crucial difference—they must test this theory by further experiments. If the tests fail, the theory must either be abandoned or modified. Magic undergoes no comparable processes or tests. Furthermore, no scientist could conduct a cause-and-effect experiment if work were being continually interrupted by a supernatural happenings or statistical data upset by miracles.

For further reading: James Frazer, *The Golden Bough: A Study in Magic and Religion* (Penguin Books, 1996); A. Lehmann and J. Myers, *Magic, Witchcraft and Religion,* 2d ed. (Mayfield, 1989); J. Skorupski, *Symbol and Theory* (Cambridge University Press, 1976); and Claude Lévi-Strauss, *The Savage Mind,* 2d ed. (Weidenfeld and Nicholson, 1972).

MALICIOUS ANIMAL MAGNETISM The notion of an evil force permeating nature that was put forward by Mary Baker Eddy founder of Christian Science. She had borrowed the idea from 18th-century Austrian physician Franz Anton MESMER, who believed that he had identified a sort of magnetism associated with animals and plants, an idea he soon realized to be without foundation. In Eddy's revived form, it was supposed to be a scientific version of casting evil SPELLS: One person could do harm to another by transmitting something damaging to his or her well-being through ANIMAL MAGNETISM, thus "malicious" animal magnetism. Although there is no evidence for this, it remains a part of the doctrine of Christian Science.

See also HYPNOSIS.

MALTHUS, THOMAS ROBERT (1766–1834) British economist, statistician (now he would be called a demographer), and clergyman in the Church of England, who is best known for his *Essay on the Principle of Population as It Affects the Future Improvement of Society,* first published in 1798. A second edition in 1803 included information collected on trips to the other main European countries. His central idea, although not original, was that population growth would always exceed the growth in food supply.

Malthus pointed out that populations tend to grow at a geometric rate (1, 2, 4, 8, 16, 32), while the food supply tends to grow at an arithmetic rate (1, 2, 3, 4, 5, 6). So he foresaw populations growing by unrestricted reproduction and food supply restricted by the law of diminishing returns. He was firmly convinced that the difference between the two growth curves, with the population growth inevitably outstripping the food supply, would result in poverty and hunger for the masses. A solution to the problem, according to Malthus, would be celibacy, late marriages, and moral restraint—yet he was not optimistic about anyone taking his advice. It was because of this pessimism of Malthus, and others who agreed with him, that economics came to be termed "the dismal science." But a clergyman of the English church at the time could suggest none other; he was writing before the development of BIRTH CONTROL devices, although induced abortion and infanticide were practiced. He was also writing before industrial dominance of the land had developed or was thoroughly understood, and his education and his thinking were probably influenced by a group of economic philosophers called the "Physiocrats," who believed in the primary importance of the land and agriculture. He therefore underestimated the scope of technological progress on agricultural productivity.

Since the days of Malthus, at least in the West, output of food has increased enormously, and birth control is now available, although we have not reduced population increase to near-replacement level. Nevertheless the consequence has been that we have a temporary respite from the two main Malthusian concerns: unrestricted population growth and restricted food supplies. But by the middle of the 20th century it became obvious that all was not well. Today the United Nations envisage the world population soon surpassing 7.5 billion, and perhaps reaching 14 billion by 2025, with most of the growth in less developed countries where, for various reasons (including crops being grown to sell to the West for cash), good arable land on which to grow food for local consumption is scarce. Already disease and war are breaking out in parts of Africa and emigration has begun, with developed countries tightening their border controls in anticipation.

Some politicians in the third world are arguing that their population growth has no real importance when set against the consumption patterns of highly developed countries which are, for example, creating the greenhouse effect that will cause massive problems for the environment, with consequent difficulties for humanity and all wildlife. The argument for a per capita count of carbon emission gives only a snapshot of an effect which is hard to assess. Who is going to have the greater impact—a 70-year-old in Britain with 2 children, 3 or 4 grandchildren and 6 or 7 great-grandchildren, or a 70-year-old in Kenya who might have eight children, 50 grandchildren and 180 great-grandchildren? The greenhouse effect, although enormously important, is not the only detrimental environmental change. There are also soil degradation, overuse of water supply, and the destruction of forests, rare species, and wild plants. In addition, there is the conversion of valuable agricultural land into towns and roads to house and serve the growing human population.

Malthus, using the best information he had at the time, produced his famous predictive model. As his dire predictions failed to materialize, his theories fell into disfavor and were seen as pseudoscientific. The industrial revolution, with its mechanized farming, and its improved fertilizers and insecticides, increased the amount of food produced. In addition, improved transport opened new granaries, and refrigeration and canning techniques increased the availability of perishable foodstuffs. Anti-Malthusian models began to appear, using all the new information on agriculture and food supplies, with the result that the expectation grew that humans would not have to control their own fecundity. Today, in light of the detrimental impact technology has had on the environment, and sophisticated projections of population increases, neo-Malthusians are producing another, even more worrying, model than Malthus's original. Now it is the optimistic model that refuted Malthus's pessimistic predictions that occupies the position of a pseudoscience. This time technology will not solve the problem; in this latest Malthusian model, technology is seen as part of the problem.

See also MALTHUSIANISM.

For further reading: Paul R. Ehrlich, *The Population Bomb* (Pan/Vallantine, 1972).

MALTHUSIANISM The theory of population—having both scientific and political dimensions—based on the writings of Thomas Robert MALTHUS that views a check on the rate of population growth as both desirable and essential. These checks are either positive—including famine, war, and disease—or preventive—involving moral restraint: late marriage, premarital abstinence, and celibacy. Malthusianism is generally associated with the operation of positive checks on population growth and, if as is likely, these fail to achieve a balance, the inevitability of disastrous shortages.

MARFA LIGHTS A phenomenon of lights reported near Marfa, Texas, a small town in the southwest part of the state. Accounts of the phenomenon go back to when the town was first settled and continue to the present. The lights may be seen at several sites between Marfa and Alpine, Texas, when looking in a southwest direction toward the Chinati Mountains. The lights resemble car headlights and are seen to move mysteriously across the desert floor. They have been photographed on numerous occasions.

The Marfa lights fall into a larger group of phenomena collectively termed GHOST LIGHTS. At present, the nature of the Marfa lights is unknown.

MARGERY CONTROVERSY A major controversy in the mid-1920s that split the psychical research community for a generation. The Margery controversy centered on MEDIUM Mina Crandon. Crandon had emerged as a medium in 1923. Within a short time, she began to go into trances in which a spirit control named "Walter" spoke through her. A short time later, she began to manifest direct voice communication and AUTOMATIC WRITING.

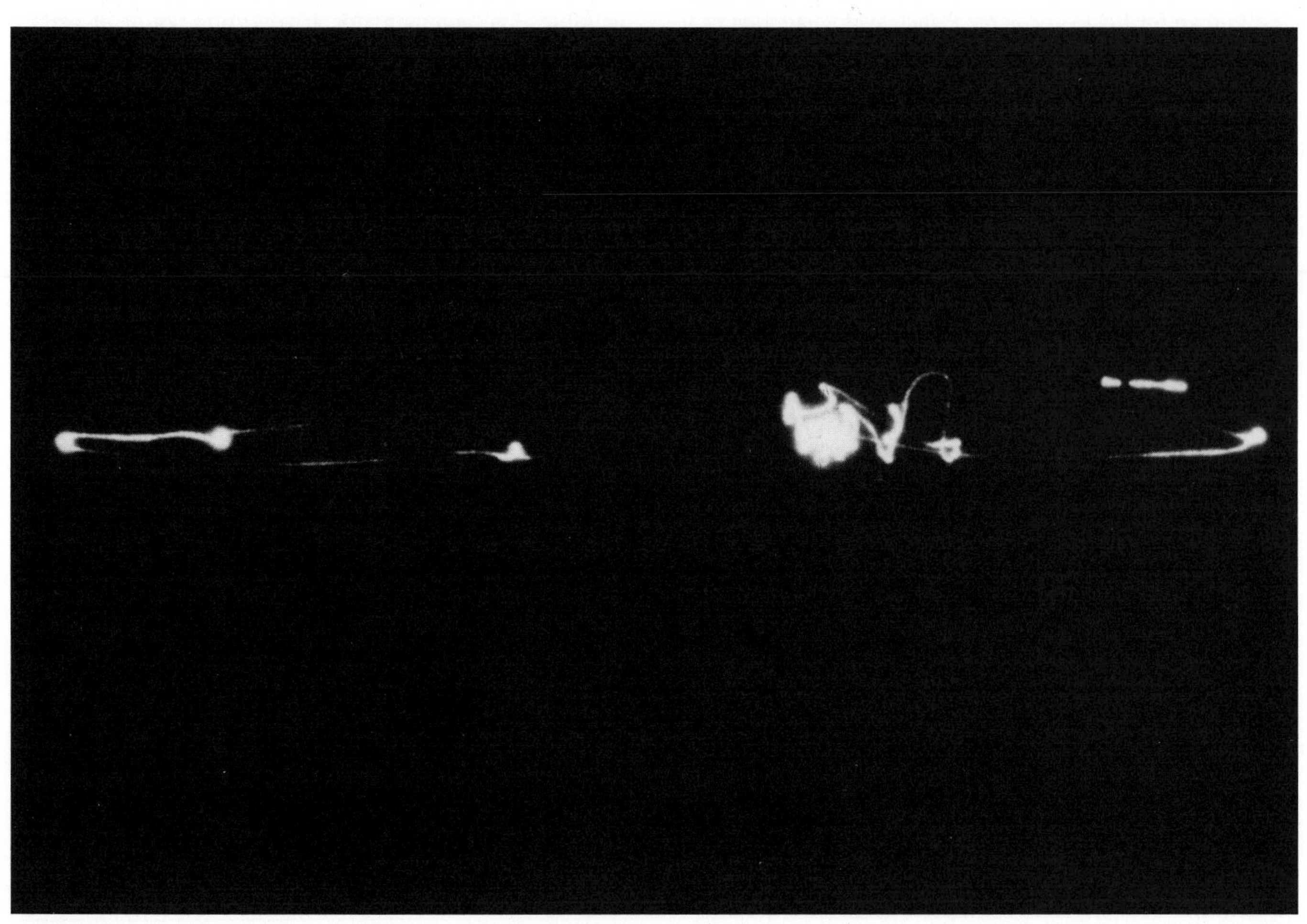

Marfa Lights

A group from Harvard conducted an initial investigation of Crandon with inconclusive results. She then allowed a number of psychical researchers on both sides of the Atlantic to study her. To allow her some personal privacy, she adopted the name "Margery," which was used in all published reports. Researchers assessed her talents favorably.

In the spring of 1924, the controversy over Margery's mediumship grew when a committee organized by the magazine *Scientific American* could not reach a conclusion about her reported powers. Psychical researcher Hereward Carrington pronounced Margery a genuine medium, but fellow researchers Walter Franklin Prince and William McDougall remained noncommittal. Only stage magician Harry HOUDINI openly denounced her as a fraud, but his observations were countered by charges that he had himself tampered with the committee's experiments in order to predetermine their outcome. A second investigation by a group from Harvard and a report by the British SOCIETY FOR PSYCHICAL RESEARCH also reached no conclusions. The *Scientific American* committee continued to hold sittings. By early 1925, the committee secretary Malcolm Bird had privately concluded that the phenomena were genuine, while Prince had become equally convinced that they were fraudulent.

Prince joined with Joseph Banks RHINE to denounce Margery's mediumship. Prince won over McDougall and Carrington, but not Bird. The board of the AMERICAN SOCIETY FOR PSYCHICAL RESEARCH (ASPR), for which Prince worked, supported Margery. They signaled their disapproval of Prince's skepticism in 1925 by appointing Bird as the society's second research officer with Prince. This appointment split the society as Prince, McDougall, and others left the ASPR and formed the BOSTON SOCIETY FOR PSYCHICAL RESEARCH.

The controversy remained alive for another decade, while evidence of Margery's fraud accumulated.

In 1930, Malcolm Bird suddenly resigned from the ARPR and disappeared from the psychical research community. His resignation came soon after his submission of a report in which he admitted that he had become convinced that Margery was a fraud early in the investigation. At one point, he revealed, he had been approached by her to join in a conspiracy to produce fraudulent results for Houdini. The ASPR suppressed the report for a number of years, and only after the Margery controversy and those involved in it passed into history were the facts made public.

For further reading: Thomas R. Tietze, *Margery* (Harper & Row, 1973).

MARS CANALS Thin black lines stretching across the surface of Mars, resulting in speculation by some that intelligent life existed on the plant. In 1877, Giovanni Virginio SCHIAPARELLI observed a fine network of lines on the planet's surface. In spite of being a committed believer in life on other worlds, he presented his description with scientific caution. However, the neutral Italian term *canali* (which just means "channels," either natural or human-made) was translated into English as "canals," implying that the lines were artificially constructed waterways, requiring either past or present intelligent life on Mars. Committed believers in the plurality of inhabited worlds, foremost among them Percival LOWELL, a leading U.S. astronomer, soon built up an elaborate picture of Martian life in the irrigated strips on each side of the canals, which carried water from the Martian polar regions to the drier equatorial areas. Lowell's popular books on the subject were picked up by science-fiction writers of the early 20th century and brought to many thousands of readers through their fantastic stories.

In 1892, when observers again looked at the planet, they did not see canals in the same places as Schiaparelli's, encouraging speculation about changing patterns of seasonal crops growing along water channels. Schiaparelli later commented that the Mars discussions had attracted many charlatans.

Photography was still not sensitive enough for astronomy, so there was no way to eliminate the problems of human observation.

By the end of the 19th century, the idea gained currency that the lines seen on Mars were optical illusions. Yet orthodox astronomy moved toward describing Mars as a dry and dusty planet with many transient marks that were changed by widespread storms, until the 1970s popular culture retained the image of a dying planet kept alive by global irrigation schemes. In 1969 the *Mariner 6* and 7 spacecraft sent back detailed pictures of the planet's surface that showed no canals, finally putting to rest that chimera.

We now have detailed maps and pictures of Mars that show that there are channels—not canals—on its surface. Valles Marineris, a highland region, covering about ten percent of the Martian surface, is cut by many deep channels stretching several thousand kilometers. They may have been generated by rifting of the planet's crust and/or by eruptions of huge amounts of subterranean water. If the latter happened, the resulting flooding would have swept rocks and other material through the Valles Marineris, forming lakes which subsequently drained, or more probably evaporated, leaving the swept material behind. Present theories about the Martian geology must necessarily be speculative.

For further reading: M. J. Crowe, *The Extraterrestrial Life Debate 1750–1900* (Cambridge University Press, 1986).

MARX, KARL (1818–1883) Known for his profound influence on social and economic thought in the late 19th and the 20th centuries. He saw capitalism as a stage in history, which he understood, in characteristic 19th-century terms, as progressive. Capitalism followed feudalism, which followed ancient slave-based societies. In its turn capitalism would be followed by communism. The driving force of these historical changes lay in the economic base of society, that is, the means and relations of production, which determined the allocation of political power.

In the case of capitalism, the exploitation of wage-earning workers by the owners of the means of production would inevitably increase as competitive pressures squeezed the profit margins. With the rise of working-class consciousness of exploitation, revolution was increasingly likely.

Subsequent events did not turn out as Marx had anticipated, partly because of the influence of his ideas, especially in the political movements he supported. Capitalism grew from the wealth gained from geographical expansion and the greater efficiency of machinery in which much capital had been invested. Wages rose (union bargaining helped), and workers also became consumers who helped sustain demand. Although the system continues to exploit workers in the underdeveloped world, a majority of workers in the West had too much to lose in revolution. In the years since Marx's death, there have been many socialist or communist states professing some debt to Marx and marxist ideas, mainly in the less-developed countries.

The 19th century was an age of scientific enthusiasm, and to lay claim to be scientific carried great weight. Marx claimed to have founded the science of economics, by which he meant that he had exposed the underlying structure of economics, particularly the economics of capitalism, in a way which gave people the understanding with which to gain control. His use of the word "science" did not imply a set of laws beyond our control but an understanding that empowered humans to intervene. Nor did Marx see his theories as immutable but rather as general principles that were open to constant reassessment and revision. Friedrich Engels (1820–95), Marx's friend, collaborator, and popularizer, pushed the claims of marxism much further toward positivism and scientism. At Marx's funeral he stated "Just as Darwin discovered the law of the development of organic nature, so Marx discovered the law of development of human history." Later, in *Anti-Dühring* (1894), Engels claimed marxism to be "the science of the general laws of motion and development of nature, human society, and thought." To this day, the many versions of marxism are claimed by their proponents to provide insights into almost every realm of thought—and to be essentially scientific.

As a science (and also as a political program), marxism never presented a single unified image. Thinking marxists have internalized their own version of how Marx would see the problems they wish to deal with. Marxism has a tendency to fall into factional divisions because modern problems now take a quite different form from those of the mid-19th century, because marxism is often learned in a derivative form from a diversity of present-day movements, and because it invokes combinations of values on which people do not readily agree.

Marxism is probably best thought of not as a science but as a set of values developed into a plethora of rival frameworks, some of which can act as frameworks for social science. Marxism does not provide authority for answers to our social, political, and economic problems; rather it is a resource for critical thinking about these matters, an irritant to the complacent assumptions of orthodoxy. Some marxist-inspired challenges to intellectual orthodoxy have taken hold in the wider culture.

For further reading: Maurice Comforth, *The Open Philosophy and the Open Society: A Reply to Dr. Popper's Refutations of Marxism* (Lawrence & Wishart, 1968); David McLellan, *Marx* (Fontana/Collins, 1975); and Karl R. Popper, *The Open Society and Its Enemies* (Routledge & Kegan Paul, 1966).

MARY CELESTE U.S. cargo ship that was found sailing erratically between the Azores Islands and the coast of Portugal, with cargo intact but with no signs of life aboard. The ship was found in December 1872, one month after it had started its voyage from New York City to Genoa, Italy. All aboard, including Capt. Benjamin Spooner Briggs, his wife and young daughter, and the crew of eight, had disappeared without a trace. Although evidence on the ship suggested that some kind of violent or hasty event may have occurred, no human traces were found either at the time the ship was recovered or later. Discovered by the *Dei Gratia,* another cargo ship that had left New York about the same time as the *Mary Celeste,* the ship was boarded by members of the *Dei Gratia*'s crew, who found a stained sword beneath the captain's bed, a smashed glass in the box containing the ship's compass, two broken barrels among the ship's crude alcohol cargo, two hatch covers awkwardly thrust aside from their normal position, and no lifeboat. The ship's log remained aboard (the last entry was 11 days before the ship was found), but the navigation instruments and ship's papers were missing. The food and water stores remained on board, as did the captain's and the crew's sea chests and other personal belongings.

Dei Gratia crew members sailed the abandoned ship to Gibraltar with the *Dei Gratia* accompanying them. An official investigation ensued, with many possible explanations offered for the puzzling disappearance of captain and crew. These included suggestions of mutiny by the crew, an insane massacre by the captain who then jumped into the sea himself, conspiracy on the part of the two captains—Briggs and the *Dei Gratia*'s Capt. Morehouse—to commit insurance fraud, and an unwarranted hasty abandonment of the ship when water was found in the holds. The British Admiralty court in Gibraltar, for the first time in its history, was unable to reach a conclusion about the fate of the *Mary Celeste*'s crew, but it did award a salvage fee to the *Dei Gratia* crew, and it returned the ship to its surviving owner (Briggs was part-owner) in New York. He quickly sold the ship, which resumed its unlucky history. (It had been involved in several accidents before the 1872 voyage, including a collision and a fire, and its history continued in this vein. Ultimately, the ship's career ended in 1885 when its owner scuttled it on a reef as part of an insurance scam.)

The *Mary Celeste* became a part of mystery folklore when a new young writer Arthur Conan DOYLE wrote a fictionalized account of the ship (misidentified as *Marie Célèste*) that was published in 1882 in the British *Cornhill Magazine*. Doyle's popular story inspired other imaginative accounts of the *Mary Celeste*'s fate. Maritime experts today generally concur that the *Mary Celeste* was probably abandoned in panic and that the people aboard, ill prepared to face the ocean in a small lifeboat with no supplies, died at sea.

For further reading: Charles Eden Fay, *Marie Celeste: The Odyssey of an Abandoned Ship* (Peabody Museum, 1942); and Colin Wilson with Damon Wilson, *The Encyclopedia of Unsolved Mysteries* (Contemporary Books, 1988).

MASSAGE Body therapy technique used by ALTERNATIVE MEDICINE practitioners. Massage has been used since ancient times. As long ago as the fifth century B.C.E., the Greeks used massage for religious as well as medical purposes. Greek philosophers, including Plato and Socrates, promoted the blessings of massage to relieve pain and prolong human life. Alexander the Great found Indian kings being massaged in the fourth century B.C.E. In ancient China masseurs developed a theory of relations between the skin and internal organs that led to the development of ACUPUNCTURE. Famous Romans, including Julius Caesar, the great orator and statesman Cicero, and the famous lawyer Pliny, claimed that massage had cured them of ills such as neuralgia and speech defects. The second-century physician Galen prescribed massage in combination with regular exercise to help patients keep fit.

Massage fell out of favor in the West during the medieval period, perhaps because of the influence of the Catholic Church, which frowned on manipulation of the body. It was rediscovered and applied by Renaissance physicians. When Mary Queen of Scots fell ill with typhus, fainted, and was given up for dead, her doctor, a man named News, massaged her until she recovered consciousness.

Since the Renaissance, massage has continued to be used as a way to improve personal health in modern times. Victorians developed machines to manipulate the muscles. After the Second World War, massage and other physical medicines had something of a revival because of serious war injuries and the poliomyelitis epidemic. Doctors found that massage could relieve pain, assist in the circulation of blood, and help keep muscles supple. Today massage therapy is used in treating diverse problems such as diabetes, effects of stroke, nerve and spinal cord injuries, birth defects, and arthritis.

MATERIALIZATIONS The solid-looking spirit faces, body parts (such as ghostly hands), or complete spirit figures that are supposedly brought into being by a MEDIUM during a séance. The materialized parts are alleged to be made of a milky white substance called ectoplasm, the consistency of which varies from medium to medium. Sometimes it is described as cold and leathery, and sometimes fluid and slimy. Mediums claim that ectoplasm is made inside the body and during a séance is extruded out of one of the orifices, often from the mouth, nose, or ears but sometimes from the vagina. Materializations are mostly associated with ectoplasm and, we are told, must not be confused with the related phenomena known as pseudopods, which are purported to be temporary growths from the medium's body.

Manifestations by mediums were popular entertainment during the second half of the 19th century, and they usually followed a simple routine: The medium would enter a "cabinet," a confined wooden structure used as a working space where she (mediums at this time were usually women) kept her props and disguises, which included wigs, luminous gloves, life-size photographs of recently deceased, well-known people, silk and chiffon scarves, and various musical instruments. From a curtained door, apparitions would appear, trumpets without visible means of support would sound off, ghostly hands would clash cymbals, and skulls would let out piercing screams.

Fraud, especially in the case of materializations, permeated mediumship from the very beginning, and after cabinets had been investigated and secret recipes for the production of ectoplasm had been circulated in the newspapers, this form of entertainment became discredited; cabinets are now rarely used in modern physical mediumship.

Today materializations take the form of small objects being conjured up out of thin air. Strictly speaking, this kind of phenomenon where an object appears in the presence of a medium is called APPORT from the French *apporter*, meaning "to bring." An Indian religious leader Sai Baba, who claimed to be the reincarnation of many religious leaders including Jesus Christ, was so convincing with his demonstrations of materializing expensive gold trinkets, which he then gave to wealthy followers at public ceremonies, that his actions caused a public controversy. Rich Indians were accused of using the ceremony as a scam to obtain gold that otherwise would be prohibited under India's Gold Control Act. The Indian authorities were pressed to prosecute Sai Baba for the manufacture of

Materializations

gold objects without license. But the Indian government quite wisely took the position that the materialization method of bringing an object into existence did not constitute manufacture. Sai Baba narrowly avoided another conflict with the authorities when he materialized a prototype of a Seiko watch not yet on sale but complete with its serial number. There are numerous videotapes available showing Sai Baba producing his materializations and, although shot from strange angles and very skillfully edited, viewers familiar with sleight-of-hand conjuring movements will be able to recognize how he creates the illusions and so misleads his audience.

For further reading: James Randi, *Flim-Flam!* (Lippincott and Crowell, 1980).

MATHER, COTTON (1662–1727) Puritan clergyman, investigator of WITCHCRAFT, and early proponent of vaccination. Cotton Mather is probably best remembered in history for his role in the Salem witch trials of 1692–93. Although he publicly supported some of the executions, Mather opposed admitting the spectral evidence brought against many of the accused. Instead, he recommended prosecution only for cases based on solid evidence, and execution only in extreme cases. Mather investigated some witchcraft cases personally; in 1688 he took a witch-child into his own house, recording his observations of her behavior in *Memorable Providences, Relating to Witchcrafts and Possessions* (1689). Despite his relatively liberal views on the subject, Mather supported the Salem verdicts in books such as *Wonders of the Invisible World* (1693). He also supported the Salem judges against one of their harshest critics, Robert Calaf, who expressed skepticism about the existence of witchcraft with his book, *More Wonders of the Invisible World* (1700).

Mather was also a corresponding member of the Royal Society of London and an early proponent of vaccination. In 1721, the city of Boston was ravaged by a great smallpox epidemic. Mather, exposed to the idea of vaccination through his Royal Society associates, interested Dr. Zabdiel Boylston in the practice. However, most of the community of Boston, the majority of the town's physicians, and some of the town's clergy opposed the process. Their fears were in part grounded in fact: At that time, a person was inoculated by introducing material directly from an infected wound into an open wound on another person. This process, however, made the inoculated person potentially as contagious as anyone who was not infected with the disease on purpose. Despite ferocious opposition, Mather and Boylston convinced the selectmen of Boston to support the practice, in part through Mather's publications *Sentiments on the Small Pox Innoculated* (1721) and *An Account . . . of Inoculating the Small-Pox* (1722). Of nearly 5,000 people who were inoculated, only 46 died; of 700 who contracted the disease naturally, 124 died.

MEDITATION A way to bring about total relaxation, to slow down all mental processes until cares and strains are shed and inner peace is attained. Ideally, the mind becomes released from all disturbing memories of the past and from all worrying decisions to be made in the future, and is liberated into the present by staring into space and drifting. No drugs, alcohol, tranquilizers, antidepressants, or sleeping pills are used. Advocates claim that stress-related conditions such as migraine and insomnia can be alleviated through meditation.

Meditation has been practiced for thousands of years in India and most of Asia, but was virtually unknown in the West until the 1960s, when it attracted popular attention with the growth in interest of Indian customs and music. The culture centered around the Beatles (an English pop group) and their guru, the Maharishi Mahesh Yogi, who taught just one type of meditation, TRANSCENDENTAL MEDITATION (TM). The technique involves sitting twice a day for 20 minutes each, clearing the mind of all distractions through the repetition of a secret word, or *mantra.* Some 500 studies on meditation have been published since the mid-1970s; most of these relate to TM.

In the late 1960s, a scientific research project headed by the eminent Harvard cardiologist, Herbert Benson, confirmed that meditation was beneficial, but he concluded that TM was not the only or necessarily the best system. Benson was surprised to find that simply sitting quietly and giving the mind a focus produced similar physiological changes to that of the more complicated TM. Using his simple procedure, heart and respiratory rates slowed down and brain waves assumed a distinctly quieter pattern. His conclusion was that benefits were not only associated with TM but with all kinds of relaxation techniques, such as deep breathing, YOGA, repetitive chanting, and muscle relaxation.

For further reading: Herbert Benson and William Proctor, *Beyond the Relaxation Response* (Berkeley Publishing Co., 1985); Herbert Benson and Miriam Z. Klipper, *The Relaxation Response* (Avon, 1976); and John Kabat Zinn, *Full Catastrophe Living* (Piatkus, 1996).

MEDIUM In the context of pseudoscience, an intermediate agency in the world of SPIRITUALISM—a person who mediates between the spirit world and the material world. Mediums perform at a séance or sitting, where they enter into a trance—a halfway state between sleeping and waking—from where they claim they can connect with the dead (or in their words "those who have passed over")

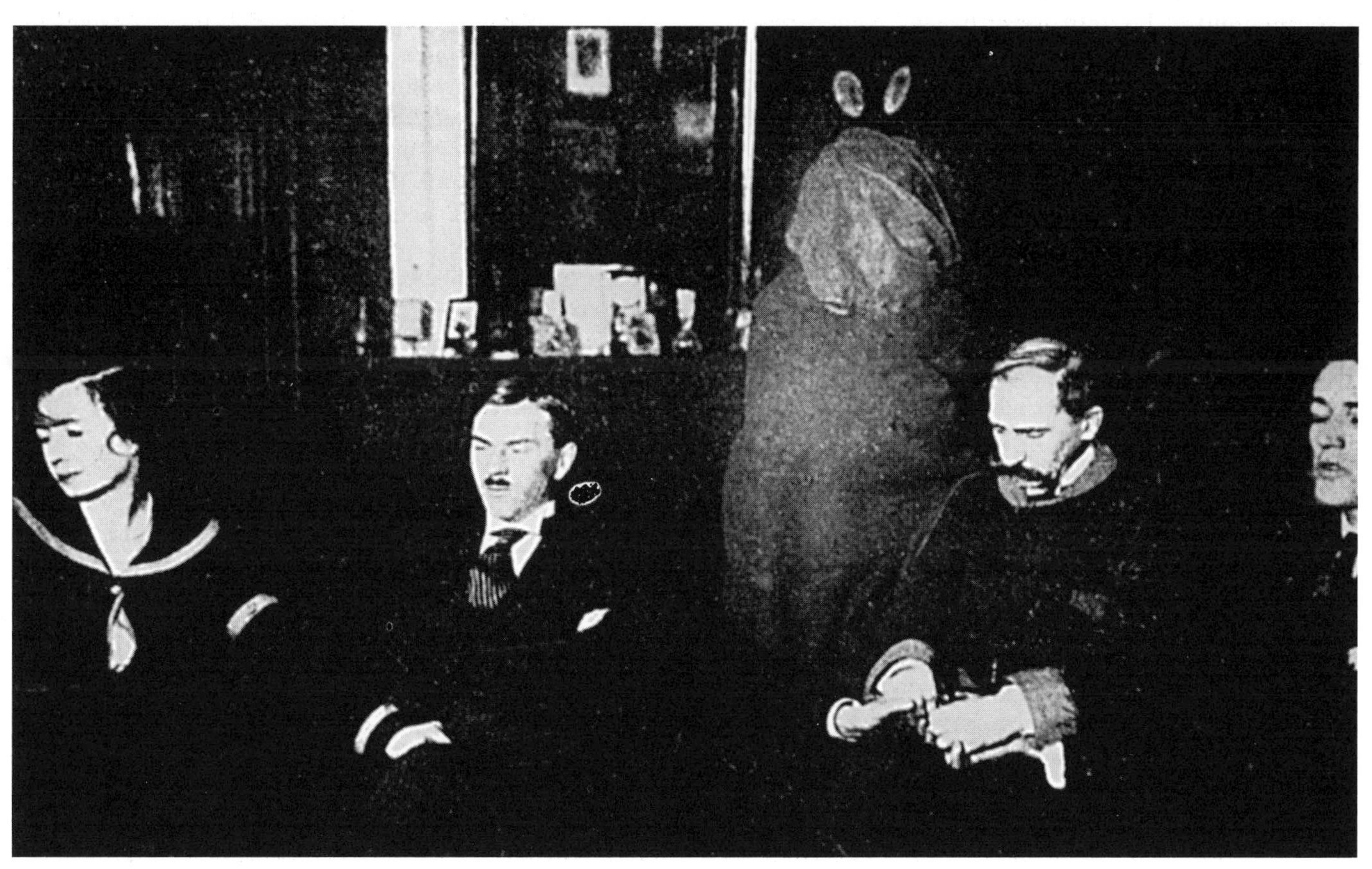

Medium

and transmit their messages to living relatives or loved ones here on Earth.

Mediums have always been keen to test their skills against the scrutiny of science, and in the 19th and early 20th centuries, scientists have not been backward in coming forward to do the testing. British scientist and early worker in wireless telegraphy, Sir Oliver Lodge, was an agnostic in most subjects and became interested in PSYCHICAL RESEARCH after his youngest son Raymond was killed in World War I. In 1916 Lodge published a book called *Raymond,* which was an immediate success. The very common loss of a loved one in that war no doubt contributed to it selling so well.

Sometime later, Lodge thought that his scientific approach and his knowledge in signaling through space without wires could be put to good use to find out the truth about mediumship. To this end he devised a set of cards with a circle, a triangle, and a square, which were "sent," one at a time in random order, by his fellow researcher calling himself the transmitter. The receiver (note the scientific terms Lodge used), a psychic medium in another room, had to guess the card that was being sent by thought transference and at the end of each run had to have an overall better-than-chance score. The idea was later adopted by Joseph Banks RHINE. These were early days in this type of experiment, and nothing conclusive came out of the tests. Nevertheless Lodge persisted and he became much in demand as a tester of the claims of many mediums. He remained convinced that what he was doing was important even when he had caught a well-known medium, Eusapia Palladino, cheating.

But something very definite came out of Lodge's final experiment, which was set up in his will. He promised to be available after his death to any medium who tried to contact him, and then he would divulge to them a series of messages that he had previously placed with his solicitor in a package containing seven sealed envelopes, one inside the other, with a message in each one and a special message in the last. The details of the experiment were approved and judged fair by a panel of scientists and mediums. After his death in 1940, more than 100 séances were held between 1947 and 1954, and many outside mediums contacted the executors, all claiming contact with the dead man. None, however, could describe the contents of the packages or get anywhere near revealing the final message—a child's short piano exercise. The case was not proved, and so, according to scientists, mediumship must remain a pseudoscience.

There is no doubt that Sir Oliver Lodge's reputation in the scientific world was damaged by his association with spiritualists, but Lodge always remained fairly skeptical; he just thought that there was enough merit in the idea for him to give it his attention and possibly find a scientific explanation. He never believed that there was anything supernatural involved, but that if investigated and confirmed, it would be found to fall within the natural order.

For further reading: W. P. Jolly, *Sir Oliver Lodge* (Constable London, 1974); and Oliver J. Lodge, *Raymond, or Life and Death, with Examples of the Evidence for the Survival of Memory and Affection after Death* (Methuen London, 1916).

MEGALITHIC YARD Standard unit of measurement used by Neolithic British monument builders and discovered by Alexander THOM. Thom, a former professor of engineering and science at Oxford University, derived the megalithic yard during the 1960s from his studies of megalithic monuments. After surveying more than 600 different MEGALITHS, Thom came to two conclusions: that the megalithic sites were all used as astronomic observatories, and that they were all based on a standard unit, which he labeled a megalithic yard. The megalithic yard measures about 2.72 feet, almost exactly as long as the traditional Spanish *vara,* or rod. Thom suggested that the rod was passed down from the original builders of the megaliths and has survived to the present day.

Thom's theory of the unit of the megalithic yard also helps explain the shapes of some of the stone rings of the British Islands. Thom showed that the builders actually erected the stones in noncircular shapes—ellipses or ovoids—to avoid complex fractions in the diameters and circumferences of their rings. Some of Thom's theories are not universally accepted, but the megalithic yard is so pervasive that most scholars accept it as evidence that the monument builders of Stone Age Britain shared a common culture. Thom's ideas have also sparked a reevaluation of megalithic sites in Europe and Britain. If Thom's conclusions are correct, they imply that ancient Europeans had a sophisticated geometry long before the cultures traditionally credited with developing mathematics. Some theorists even suggest that classical mathematics and the theory of number derive directly from the work of the ancient monument builders.

MEGALITHS Giant boulders used by Neolithic cultures (4500–1500 B.C.E.) for building monuments, including tombs. Megaliths are found on Malta island in the Mediterranean Sea, in Germany, in Spain, and in Greece, but their greatest concentration is in northwestern France and in the British Isles. Because they were built so long ago, all memory of their original functions has vanished.

All megaliths can be divided into three general types: (1) *menhirs,* or single stones that stand upright and alone; (2) groups of menhirs, sometimes arranged in circles, semicircles, or parallel lines; and (3) *dolmens,* three or more menhirs that support a capstone like a roof. If the groups of menhirs are arranged in circles and surrounded by banks of earth, they are called henge monuments, after the most famous megalithic site, STONEHENGE. Menhir groupings also occur in parallel lines, as at AVEBURY; in multiple parallel rows, as at Carnac in southern Brittany, France; and in fan shapes, as in northern Scotland. Dolmens were occasionally used as tombs, and they also fall into three basic patterns: the single chamber tomb; the passage grave, which has a long corridor that ends in a chamber; and the gallery grave, in which the chamber is enlarged and then subdivided into a number of smaller areas. If the tombs are covered by earthen mounds, they are known as barrows. Some scholars believe that most dolmens were covered at the time of their construction but blame ages of weathering—and digging by local residents—for wearing away their outer layers. Others believe that many dolmens were never meant to be covered at all.

Historians and antiquarians have long debated why and how megaliths were built. During the Middle Ages writers speculated that the great standing stones were the work of the Romans. The medieval authors drew connections between the imposing Roman ruins scattered across Europe and the other great stonework in the same area. Other, more romantically minded writers suggested that the standing stones were the remains of one of Caesar's legions, petrified by a great Celtic magician, or that they were giants who had been turned to stone magically. During the Renaissance, scholars investigating the megaliths suggested that they had been raised by Egyptian refugees from the Persian invasions of 525 B.C.E., by Celtic Druids, or by barbarian invaders.

Most modern researchers believe that the megaliths of Western Europe were erected by a neolithic folk called the Beaker People, who moved into the area during the third millennium B.C.E. Archaeologists do not know where the Beaker People originated, but they do know that they brought with them the ability to smelt copper.

The function of the megaliths is perhaps even more mysterious. Some, such as the grouping at Stonehenge, were apparently used as astronomical observatories. Others, including most of the dolmens, were used as burial places, although this may not have been their original purpose. The most popular theory has been that the megaliths were used to mark special holy places in the religion of the persons who erected them. Some menhirs and dolmens, especially in Brittany and Ireland, are decorated with symbols that may be religious in origin: zigzag lines representing sacred serpents, mystical spirals, and pictures that may represent the sun or the great mother goddess. During the Renaissance, one popular theory held that the megaliths were sites where the Druids practiced human sacrifice. Archaeologists have found human remains at some sites, including Woodhenge and Avebury, but most megaliths show no evidence that such sacrifices were ever offered there.

Modern thinking suggests that the megaliths were erected with simple workmanship, using simple tools: ropes, ramps, and rollers. Scientists believe that to erect some of the larger menhirs or the capstones that cover dolmens, may have required workforces of hundreds of people at a time. However, Norwegian anthropologist Thor HEYERDAHL, while working with descendants of the people who raised the great statues of Easter Island, found that a group of 30 islanders could raise a 27 metric-ton (thirty ton) statue four meters (12 feet) up onto its base, using only levers and stones for propping. The carvers who originally quarried or shaped the stones probably used equally simple tools: stone mallets or mauls with wooden handles, or by alternately heating and cooling the rock.

The megaliths also demonstrate the mathematical powers of their creators. Alexander Thom, a trained engineer, showed in his books *Megalithic Lunar Observations* (1971) and *Megalithic Sites in Britain* (1967) that many megalithic circles are not true circles but are in fact complex geometrical shapes. They could only have been created, Thom suggests, by people with a sound grasp of basic geometry. Thom also believes that the megalithic builders had evolved a theory of triangular proportions that is usually credited to the Greek philosopher Pythagoras—who lived about 2,000 years later.

Because of their imposing appearance, the megaliths have attracted much speculation about their origins. Numerologists have found complex relationships in the measurements of the giant stone monuments, hiding special esoteric knowledge. This knowledge, according to John Michell, author of *The View Over Atlantis,* includes answers to such pressing questions as the time of the Second Coming of Christ. Some people have suggested that the megaliths were in fact erected by members of the lost civilization of Atlantis, or by extraterrestrial visitors who had the ability to erect structures to predict such events. Others believe in an "Old Straight Track"—special places on Earth's surface where magnetic lines of force run just below the surface. These theorists believe that the megaliths serve as collection stations for these magnetic forces and that the ancient megalith builders knew how to tap the energy, using it to send messages over great distances and to levitate heavy objects.

See also ATLANTIS AND LEMURIA.

For further reading: Paul Caponigro, *Megaliths* (Little, Brown, & Co., 1986); Kevin Crossley-Holland, *The Stones Remain: Megalithic Sites of Britain* (Rider, 1989); and Glyn Daniel, *Megaliths in History* (Thames & Hudson, 1972).

MEGAVITAMIN THERAPY Practice of taking large concentrated doses of vitamins to prevent disease. Megavitamin therapy was a side development of the rash of newly discovered vitamins that took place during the

1930s. Scientists had recognized the importance of diet in a healthy life for centuries. Scottish physician James Lind showed during the 1750s that sailors could reduce their risk of contracting scurvy (caused by a lack of vitamin C in their diet) by drinking lime juice. The term "vitamin" itself, however, was not developed until 1912, when Polish chemist Casimir Funk isolated vitamin B_2, or thiamine. In 1937, scientists in the southern United States were investigating outbreaks of pellagra, a disease caused by a deficiency of a vitamin called niacin. Pellagra victims suffered from some physical symptoms—reddened skins, sore mouths—but they also showed signs of mental disorders, including confusion and depression. These scientists discovered that when their patients were treated with niacin, their mental problems disappeared along with their physical symptoms.

In the 1950s, scientists began to treat mental problems specifically with niacin. Two Canadian doctors, Abram Hoffer and Humphrey Osmond, tried a niacin-based therapy on schizophrenic patients. Their results, published in 1962, showed that administering large, controlled doses of niacin had relieved some of the symptoms of the patients' schizophrenia. Popular medicine picked up on this "megavitamin" approach and introduced it to the public. People began to believe that many physical diseases, as well as mental disorders, could be prevented through large doses of vitamins. Large amounts of vitamin B_1, it was thought, could prevent mosquito bites. Some scientists, including Dr. Linus Pauling, advanced the theory that the common cold could be prevented through megadoses of vitamin C. As a result of this kind of thinking, nearly half the population of the United States takes some form of vitamin supplement each day to ward off disease.

No solid evidence exists to prove that huge doses of vitamins beyond the U.S. government's minimum Recommended Daily Allowance (RDA) contribute to general health. In special cases, doctors may prescribe megadoses of vitamins in order to treat certain diseases. Some doctors, for instance, prescribe niacin to reduce cholesterol. A number of scientists suggest that these large doses of some vitamins may have side effects, but these are usually not harmful. Vitamin B_6, for instance, if taken in large quantities, can cause numbness in fingers and toes. Other vitamins, such as Vitamin A, can be very dangerous if taken in large quantities.

MENTALISM A form of stage MAGIC; a pseudopsychic magic that imitates demonstrations of psychic powers used by spiritualists in séances. Combined with other forms of stage magic, mentalist practices are used to produce psi effects in SPIRITUALISM that are then attributed to the spirits. In the case of mediumship, and fortune-telling, mentalism includes the psychological skills of cold reading clients to assess what they want to hear. This is done in many ways, but a very frequent ploy is to ask questions that sound like general statements and then to watch for subtle clues revealed by the client's body language.

In the past stage, mentalists had to invent ingenious codes, such as the famous time-delay code arranged between sender and receiver where an accomplice can transmit bleeps by touching a switch in his cuff, or even in his shoe. Today, miniaturization technology has produced much easier forms of communication.

Scientists are often called in to adjudicate on alleged cases of mentalism. But science writer Martin GARDNER believes they are the last people who should be invited to investigate such cases, because they are used to working in a universe that always plays fair, so they never expect trickery. The best psi investigators are such magicians as James Randi, who are themselves skilled in mentalist techniques.

For further reading: Uriah Fuller, *Confessions of a Psychic* (Karl Fulves, 1975); Uriah Fuller, *Further Confessions of a Psychic* (Karl Fulves, 1980); and M. Lamar Keene, *The Psychic Mafia* (St. Martin's Press, 1976).

MENZEL, DONALD H. A Harvard astronomer and astrophysicist who summarized the scientific argument against the ideas of Immanuel VELIKOVSKY.

Menzel also attacked the idea of flying saucers. He argued that these discs or saucers were a sort of mirage caused by unusual weather circumstances, which in different times would not attract much, if any, interest. But, he suggested, the public's imagination had been stimulated by the growing belief that there was a real possibility of some sort of invasion, benign or malign to taste, from outer space: from Mars or from a planet from another star system. With their minds receptive to these ideas, many people were more than ready to invest the objects or lights that they observed with an exotic significance. He published his interpretation of these flights of imagination—as he saw them—in *The Truth About Flying Saucers,* 1956.

MERLIN Legend Celtic magician and prophet. The sorcerer Merlin, also known as Myrdin and Myrddin, is an important figure in Arthurian legend. He is best remembered as the enchanter who makes possible King Arthur's accession. According to the 12th-century chronicle *Historia Regum Britanniae* (completed c. 1135–39) by Oxford clergyman Geoffrey of Monmouth (c. 1109–54) Merlin was the child of a British princess and a demon. The British king Vortigern, fleeing from the invading Saxons, learned of the fatherless child while building a stronghold in the Welsh mountains. He had Merlin summoned by the advice of his wise men, who claimed that Vortigern's tower would stand if he killed a fatherless child and

Menzel

sprinkled its blood as a sacrifice over the stones of his fortress. Merlin shamed Vortigern's wise men by proving that he himself had the gift of prophecy. He later became an important part of the story of King Arthur. He engineered Arthur's conception and supported the young man in his position as war leader of the Celts. Existing sources depict Merlin as a mysterious character who differs from all other mortals in his powers and abilities.

One of the major questions asked about Merlin is whether he really existed. The best evidence for Merlin's existence lay in the book known as the *Prophecies of Merlin,* which Geoffrey of Monmouth claimed to have translated from an ancient British language into Latin. During the Middle Ages, the story of Merlin and the book he left behind was accepted without question. Medieval mystics regarded Merlin's *Prophecies* in the same way that Renaissance mystics regarded the prophecies of Nostradamus. Later critics denied the authenticity of the *Prophecies,* believing that they were invented by Geoffrey of Monmouth.

Recently, however, another critic has rejected this interpretation. According to Norma Lorre Goodrich's *Merlin,* a study based on examinations of French, Scottish, Welsh, and English sources, the original Merlin was probably born in Scotland around the year 450 C.E. and died in the year 536. Goodrich suggests that the *Prophecy* is in fact a history of Merlin's own time rather than a series of predictions of the future. She shows that the historical Merlin was a member of the Romano-Celtic British aristocracy. The *Prophecy* uses traditional Celtic totem imagery which, Goodrich believes, demonstrates that the writings date from sixth-century Britain.

MERMAIDS Legendary humanlike sea-dwelling mammals. Most reports of mermaids can be traced to one of two sources: legends represented by the stories of the ancient Roman writer PLINY THE ELDER and the tales of the *selkies,* or seal-people of northern Europe; or mistaken identity, in which sailors confused dugongs, manatees, or other sea mammals for mermaids. In addition some mermaids are frauds. Fishermen, especially in Japan during the 19th century, learned to create artificial mermaids by sewing fish tails to the upper torsos of monkeys. Yet, despite these frauds, people continue to report seeing mermaids and mermen. As recently as August 1949, fishermen off the coast of Scotland claimed to have seen merfolk at a distance.

In some cases, the being in question is unmistakably a manatee or dugong. Christopher Columbus reported seeing mermaids during his voyages of discovery in the late 15th century, but he concluded that "they are not so fair as they are painted." In 1927 the British naturalist W. P. Pycraft identified a mermaid sighting in the Red Sea as a dugong. In the mid-20th century reports of a *ri,* a mermaidlike animal, off the coast of Papua New Guinea brought about an expedition by U.S. cryptozoologists; through underwater photography, they discovered that the animal was in fact a dugong.

Many of the sightings, however, refer to very humanlike qualities: large breasts, long arms, and human eyes and faces. In 1830, fishermen in the Hebrides Islands off the Scottish coast accidentally killed a mermaid, which was examined by the local sheriff. He described it as being the size of a four-year-old child, but with fully developed breasts. Its tail was like a fish's, but without scales. In 1978, Filipino fisherman Jacinto Fetalvero reported that a mermaid with blue eyes and red cheeks helped him in his nightly catches.

The question of the mermaid remains a matter of individual belief. Most sightings can be explained as unclear views of dugongs or manatees. However, some people suggest that the mermaids may in fact be an unknown species of sea mammal or even an aquatic primate. Perhaps more acceptable, however, is the explanation that mermaid sightings are optical illusions or hallucinations. Two Norwegian scientists have shown that atmospheric inversions that precede storms can explain some eyewitness testimony.

For further reading: Richard Carrington, *Mermaids and Mastodons: A Book of Natural and Unnatural History* (Rinehart & Co., 1957); Gwen Benwell and Arthur Waugh, *Sea Enchantress: The Tale of the Mermaid and Her Kin* (Citadel Press, 1965).

MESMER, FRANZ ANTON (1734–1815) Austrian physician and inventor of a technique of HYPNOSIS known as mesmerism. Mesmer has often been dismissed as a fraud who manipulated credulous people for his own personal gain. He was intrigued by the physical phenomenon of magnetism, and he formulated a theory that animals and plants as well as inanimate objects give off magnetic fields. He believed that diseases, both physical and mental, could be treated by manipulating these fields. Although Mesmer used natural magnets and large quantities of iron in his experiments, most of his results came from hypnosis and massage techniques that have since been used successfully by physicians and psychotherapists. The theory of ANIMAL MAGNETISM, on the other hand, has been discarded by modern medical workers.

Mesmer popularized his ideas in the salons of prerevolutionary Paris, France. He paid great attention to details of atmosphere. His rooms in which the magnetism sessions were held were exotically equipped and decorated with stained glass, mirrors, incense, and Aeolian harps. Patients sat holding hands in a circle around a tub filled with iron shavings and rods, intended to pass the magnetism from person to person. At a certain point in the proceedings, Mesmer himself appeared, dressed in exotic robes. He "magnetized" the patients by massaging them along the head, neck, and back and through hypnosis, staring directly into their eyes. Patients later testified that they could feel the magnetic force moving through their bodies, energizing and healing them.

Mesmer's techniques caused a great furor among the upper classes and intelligentsia of 18th-century France. Detractors fought a bitter battle against the doctor's practices, accusing him of being a fraud, of defrauding his patients, and of being in league with the devil. Admirers—including Queen Marie Antoinette—praised Mesmer's results, often in extravagant language. The debate raged in pamphlets for several years until a group of doctors exposed Mesmer's "animal magnetism." Mesmer left the country, taking with him a fortune of 340,000 francs.

MESMERISM A medical treatment—the precursor of hypnotism—exploiting a universal magnetic fluid that 18th-century Austrian physician Franz Anton MESMER claimed to have discovered. Attacked by physicians for practicing MAGIC, Mesmer left Austria and set up in Paris where the therapy became a craze for a few years in the early 1780s. Mesmer claimed to be able to produce convulsions in the afflicted parts of his patients by drawing magnetic fluid through them, either with magnets or with several passes of his hands. The technique quickly developed further, including in 1784 the production by the Marquis de Puységur of a state of artificial sleep—somnambulism.

However, Mesmer again attracted hostility from the medical profession, and in 1784 King Louis XVI appointed a commission of inquiry, among its members Ben Franklin and Antoine Lavoisier, to examine and report on mesmerism. The commission could find no trace of the magnetic fluid and concluded that the effects were produced by what the patient *imagined* had happened.

Interest in the controversial phenomenon continued to spread. In the mid-19th century, it was proposed by James Braid (1795–1860) that what is now called HYPNOSIS is a subjective phenomenon involving a special form of suggestibility that is readily produced by fixation of the attention on a single object.

Although his methods were attacked both in Vienna and in Paris, denounced as using magic and being based on false theories, and therefore being pseudoscientific, Mesmer was neither a fraud nor a charlatan. He had developed a method, derived from ancient practices, of treating a number of symptoms with some success. As so often with treatments discovered empirically, he did not understand how it operated. But others continued to explore the method, notably Braid and later Jean Charcot and Sigmund FREUD. Its derivative, hypnotism, for which there is still no satisfactory explanation, is regularly used today.

For further reading: H. G. Ellenberger, *The Discovery of the Unconscious* (Fontana, 1994); and Stuart Gordon, *The Book of Hoaxes* (Headline, 1995).

METAL BENDING The ability to bend metal through psychic powers. In the mid-1970s, a young Israeli, Uri GELLER became an international media celebrity apparently by displaying a psychic ability to bend metal objects such as forks and spoons in front of witnesses and even while being filmed.

Geller provoked controversy, in part because he persuaded many people, including a few scientists, that his powers were genuinely psychic, while others, such as the professional magician and debunker of psychic fraud James Randi, insisted that conjurors could easily duplicate Geller's feats. Randi pointed out Geller's use of standard techniques of misdirection and his insistence on the presence of a close associate, Shipi Strang. In 1974, *New Scientist* discredited his ability.

For further reading: John Hasted, *The Metal-Benders* (Routledge & Kegan Paul, 1981); J. Randi, *The Magic of Uri Geller* (Ballantyne, 1978); and J. Taylor, *Superminds* (Picador, 1975).

METEORITE Small particles of stone or iron from space that survive passage through the atmosphere and then fall to Earth.

Meteorites have been observed since antiquity and have been known to come from the sky. Their celestial origin and fiery entry to Earth gave rise to the belief that they came from the gods and various myths, miraculous and religious, were built around them. In the 18th century, scientists, particularly those of the French Academy, began to question meteorites' celestial origin, thinking it impossible that stones could fall from the sky. However, in 1794 Ernst F. F. Chladni, a German physicist, confirmed their extraterrestrial origin. A substantial fall of meteorites at L'Aigle in France in 1803 clinched the matter.

There are several types of meteorite, and one, carbonaceous chondrite, contains organic material. This is difficult to explain and has added fuel to the belief by a small number of 20th-century scientists, including Sir Fred HOYLE, that life entered Earth from outer space, perhaps carried by meteorites. Until recently, careful studies had concluded that there was no sound experimental evidence to support this supposition. In 1996 a group of scientists reported what they believed to be convincing evidence that small meteorites from Mars contained the microscopic remains of organic material. If confirmed, this would imply that there had been life on Mars. Their work has since been challenged, and resolution must await further research.

See also ORGEUIL METEORITE.

MICROBE KILLER A NOSTRUM developed in the 1880s to exploit the newly posited germ theory. The Microbe Killer was concocted by William Radam, a gardener from Texas. He argued that although some doctors believed that some diseases were caused by germs, he had discovered that *all* diseases were so caused, and he offered his compound (later shown to contain 90 percent water and minuscule amounts of red wine, hydrochloric acid, and sulfuric acid) as a cure-all. He suggested that killing germs was similar to killing garden bugs. The product reportedly had a variety of uses: It supposedly could be poured into drinking water as a general preventative, and, taken internally, it was said to release vapors that attacked worms and cured a variety of diseases from tuberculosis to leprosy. The product gained popularity and was successfully marketed long after Radam's death.

A significant attack on the Microbe Killer began in 1889 when pharmacist and physician R. G. Eccles published his analysis of the contents of the product. Eccles also noted that the manufacturer made a 6,000 percent profit for a product that was largely water. Writing in a professional journal, the *Druggists' Circular,* he argued that any universal microbe killer would necessarily kill almost all living things. His critique largely fell on deaf ears because his fellow druggists were making money from the popular medicine, which had no documented harmful effects.

It was not until 1913 that an effective effort to remove the product from drugstore shelves was launched. In 1912, the Sherley Amendment to the Pure Food and Drug Act defined a product as fraudulent if the label claimed false curative effects. Dr. Carl L. Alsberg, then head of the U.S. Government's Bureau of Chemistry, charged with enforcement of the Pure Food and Drug Act, arranged for a shipment of the Microbe Killer to be sent from its manufacturer in New York to Minneapolis. At the trial that followed, Alsberg argued that the Microbe Killer could have a bad effect on public health. People might be misled into taking the medicine when ill instead of something that might actually have an effect on their disease. The jury ruled that the manufacturer marketed the product with clear knowledge that it had no effect on the various diseases it claimed to cure. The product was destroyed, and through the case the bureau demonstrated the value of the Sherley Amendment as an enforcement tool.

MIDWIFE TOAD See KAMMERER, PAUL.

MILITARY USE OF PSYCHICS The interest in using psychics for intelligence and planning operations by the three U.S. armed services during a period of 20 years was closely involved with work at the Stanford Research Institute (SRI) and the Science Applications International Corporation (SAIC). They were used to try to hunt down Muammar Gadhafy prior to the 1986 U.S. bombing of Libya, to look for plutonium in North Korea, and to locate the kidnaped Brigadier General James L. Dozier in Italy. In none of these ventures were the psychics successful and, during 20 years and for $20 million, the return on the investment of time and money was virtually, if not actually, zero.

The military used six allegedly proven remote viewers, notably Ingo Swann and Keith Harary at SRI and Major Ed Dames and Sergeant Mel Riley. According to a BBC television program, Major Dames's psychic powers were used in the attempt to locate Gadhafy, prior to the bombing raid on Libya. Testing of the remote viewers' reliability ran into the problems that beset so much of PARAPSYCHOLOGY. In particular, because the program was primarily concerned with a virtually infinite number of possible visual targets and not with a limited number of cards, the scope for interpretation was great. For example, a viewer sees a skyscraper at the designated location and draws a rough picture; the analyst, whose mind is not focused on skyscrapers, interprets the picture as a rocket in a silo. As always with such programs, the perception of success depends strongly on expectations. Professional psychologists are well aware of this and take steps to protect their work from such undue influence (see Alfred BINET for example). But the amateur enthusiast may be misled.

MILLENARIANISM From the Latin word *millenniu,* meaning literally "a period of a thousand years." Millenarianism, and another more modern word, "Millennialism," each stem from a statement given in Chapter 22 of the Bible Book of Revelation that claimed to know the actual date of the Day of Judgment. Similar claims are said to be found in a miscellaneous collection of Jewish apocalyptic literature called the KABBALAH.

At the end of the first millennium, certain Christians thought that something momentous was going to happen at that time. Chief among these were two sects which in the second century had been declared by the early Christian church to be heretics, namely the Gnostics, who believed that only by concentrating on the spiritual element in mankind would there be redemption, and to this end they declared the body corrupt; the other was an extreme ascetic sect called the Montanists. But the mass of the peasantry and the country priesthood would have been ignorant of such predictions. The chronicler Rudulfus Glaber (985–1047), a well read monk and considerable traveler, was well aware of the millennium and observed that there seemed to be little interest in the event. Today, as we approach the year 2000, many more people are aware of the ancient sources and fears are rising that something might happen at this millennium.

Ancient millenarian doctrine promised that a thousand years' rule of righteousness on Earth will begin with the "Second Coming" of Christ, who will reappear in the guise of a warrior, defeat Satan, and throw him bound into a bottomless pit, where he will be held prisoner. This period is the millennium, the thousand-year rule. During it, Christian martyrs of the past, those who died for their faith, will be resurrected and will rule with Christ. At the end of this period, Satan will be released for a time, during which he will collect his followers in preparation for the great final battle between good and evil at Armageddon. There will then be a decisive victory over evil, followed by a general resurrection of all the dead. The last judgment of all souls will take place, and those judged to be without sin will receive redemption; the rest will be damned.

There are many variations on the above story, but all support a sudden dramatic or cataclysmic intervention of a God into history and describe eschatological or end-time events that are apocalyptic. Clearly, this kind of thinking cannot be contained within atheistic thought, and history shows that certain types of religion are more conducive to millenarianism than others. It is mainly worldviews, whose adherents believe that their God is working through history toward a preordained end, that provide a framework for millenarian beliefs. Religions in which history has no meaning and those that have a cyclic conception of time will be less open to millennial ideas. Nevertheless, whatever the religion, all millenarian literature deals with the divine judgment of all people, salvation of the faithful or elect, and the eventual unity of the saved with God in heaven. The early Judeo-Christian genre flourished mostly between 200 B.C.E. and 200 C.E. Zoroastrianism also falls into this category. Historians believe that outbursts of millenarianism have usually taken place in periods of social transition and against a background of disaster:

plagues, pestilence, droughts, floods, fires, and calamitous wars. The literature is thought to have been written primarily to give hope, within a religious framework, to groups in stress because of natural or cultural upheavals or social persecution.

Millenarian movements date not only from our ancient past or the middle ages; modern movements such as the Shakers, the Seventh-Day Adventists, the Jehovah's Witnesses, and the Mormons hold similar views, along with many individuals within older mainstream religions. The term "millenarian" is not used now in its strictly specific historical sense when describing groups such as the Branch Dravidians in Waco, Texas, but topologically to designate those that expect, via any path, to attain collective salvation. In the Western world, new kinds of inflated expectations, rather than grinding poverty as in the past, have been important conducive factors. People who lack status and power, who are on the bottom of the social pile, are particularly attracted to the myth of an elect, and the notion of reversal of roles, such as believing that the meek will inherit the earth, are important aspects of all millenarian ideology. Anthropologists have seen similar belief systems, which they have called "cargo cults," arising in deprived parts of the world; for example in Papua, New Guinea, where the indigenous population had been exposed to missionary teaching about the coming of the Messiah, alongside the introduction of Western material culture. Cult leaders promised that dead tribal ancestors would arrive on their shores with a cargo of European goods that would make them immediately more prosperous than the white population.

As we approached the year 2000, many irrational doomsday theories took hold in the West, with people believing that the end of the world was nigh. The tragic death in San Diego, California, of 39 "Heaven's Gate" disciples in March 1997 was connected to millenarian themes that had become mixed up with Star Trek ideas about the comet Hale-Bopp. It is interesting to see how Marshall Herff Applewhite, the leader of the group, had taken his theology from the ancient Gnostics, one of the second-century sects mentioned earlier, who believed that the human body was a corrupt and disposable vessel. Applewhite had had himself castrated, and his disciples had described their bodies as "vehicles" or "containers." In 1976 Applewhite preached the desirability of a "release" from the body, a typical neo-Gnostic solution. As always, modern millennialism shows the predisposing element of frustration and powerlessness arising from the feeling that those in control of our affairs are corrupt and evil, and this is associated with an upsurge of Satanism. The material causes are many and are often connected with fears, and sometimes misunderstandings, of science and technology: the fear of high technology warfare, especially nuclear war; fear of pestilence—for example, the current fear of the ebola virus or AIDS; fear of pollution, global warming, widespread unemployment, and economic breakdown. For many, all these fears add up to a conviction that rational thinking and science is not going to save them, and so they turn to irrational thinking and pseudoscience for answers. Some believe that another Great Master will arrive in a spaceship from a faraway planet and that all the unbelievers will be transported to a certain planet X. Others believe that the Great One, arriving in a UFO, will be the anti-Christ and he will mark all Christians, using a laser beam, with the special mark of the beast; those who are marked will then be led away for extermination. Another belief holds that kidnaping has already started, and that UFO watchers have observed aliens spiriting people off into their spaceships.

Modern scholarship, seeking a more accurate historical dating of Christ's birth, calculates that, because Herod died in 4 B.C.E., his census, which required the Holy Family to travel to Bethlehem, must have taken place in 4 B.C.E. or before. So even if we still wished to become excited about the Second Coming, a more-accurate bimillennial date would have been 1996 or before.

For further reading: Norman Cohn, *The Pursuit of the Millennium* (Secker and Warburg, 1957); Krishan Kumar and Stephen Bann, *Utopias and the Millennium* (Reaktion Books, 1993); and Peter Worsley, *The Trumpet Shall Sound: A Study of Cargo Cults in Melanesia,* 2d ed. (Shocken, 1968).

MIND CONTROL Also called Coercive Persuasion or Brain Washing; a technique that aims at the systematic erosion and reversal of a person's habits or convictions. Such persuasion has many purposes: political indoctrination or interrogation, religious proselytizing, educational or reeducational programs, or just the personal domination of one individual over another as in the master-slave relationship.

The technique varies with the circumstances and the perpetrator, but in most cases the basic classic scenario progresses as follows:

1. to control the physical and social environment of the subject to destroy old loyalties, either by forced incarceration, especially solitary confinement, or by social ostracism that causes loneliness and deprivation;
2. to show by various means, either by the stick or the carrot, that the old attitudes, beliefs, or behavior are not acceptable, are therefore incorrect, and must be changed; and
3. to substitute new loyalties, beliefs, and so on to take the place of the old.

Many different agents are used to facilitate the control of another's mind. The most obvious is the use of drugs. Certain substances can bring about a physicochemical action within the brain that can affect the ability to think, remember, and plan so that the mind does not have the ability to look after its own body, let alone make complicated ideological decisions. In addition, drug dependency brings the fear of substance withdrawal, this can change a person's motivation, principles, and priorities. An alternative, very drastic, but crude form of mind control is brain surgery. Another technique using direct interference is electrical-current injection. But any dictatorial government or religious sect that seeks to control masses of people would find the old fashioned methods of controlling the means to information more effective than brain implantation or mass drug administration.

Brain manipulation has often been bracketed with education. Control what people read and leave the rest to humankind's wish to conform, and all but a very few will fall into line. The best of education exists in parallel with the wide outside world and respects the subject's conscious mental process. The very worst takes the subject out of external influences, disrupting the balance between self and outside world, and allows only certain doctrinal material to be taught.

Complete milieu control not only seeks to influence all that the subject sees, hears, reads and experiences, but finally seeks to penetrate the subject's inner life, what we might call inner conscience, so that the individual can then become self-controlling. At this stage, the new ideology is totally mandatory and the person believes herself or himself to be the "chosen one" in the vanguard of a mystical higher purpose, or law of social development. This great mission divides the world into the "absolutely good" and the "absolutely evil" and so demands that ends should always justify means. Thus messiahs such as Jim Jones could lead 900 of his followers into mass suicide at his commune in Guyana in 1978.

MIND MAPPING The ability to locate areas in the brain that control specific functions. Stephen J. GOULD writes (in *Eight Little Piggies*):

> *Modularity pervades all neurological organization, right up to what Darwin called "the citadel itself"—human cognition. This principle of breaking complexity into dissociable units does not disappear at the apex of known organization. . . . The brain does a great deal of work by complex coordination among its parts, but we have also known for a long time that highly particular aptitudes and behaviors map to specific portions of the cerebral cortex.*

At its very simplest, mind mapping identifies the control of the actions of the right side of the body by the left hemisphere of the brain and vice versa. There is a school of thought, mainly in popular psychology, that asserts that the left hemisphere dominates quantification and analysis and the right is responsible for holistic and integrative thinking: in general, mathematicians, engineers, and scientists are left-brain people; artists and writers are right-brain people.

Since the early observation from brain-damaged patients that there was at least some localization of function within the brain, a vast amount of information has been accumulated by neurologists, both by experiment and by observational correlation. Experiments have been carried out on animals, principally on rats but also on primates, by deliberately disconnecting a part of the brain and then testing the animal's functions, both bodily and mental. In this way clues have been obtained, but the human brain is distinctly different from any other animal brain, and mapping the areas that control specific functions has been slower and more difficult, but still continues. At first, it was thought that each control area would be independent of the others, but it was soon realized that this is not the case; some specific functions are usually located in certain areas of the cerebrum but, except for a very few areas, if any moderate-sized portion of the cortex is destroyed, other segments gradually take over the lost functions. For example, stroke victims suffer function loss, but some function, sometimes all, returns, providing there are no further strokes. It is possible to make a map of the brain showing the following: (a) the sensory areas—visual, auditory, olfactory, and so on; (b) the motor areas—both to stimulate and inhibit the many muscles; and (c) the association areas—the mental processes: memory, intelligence, learning, and so on. Recent developments in noninvasive techniques such as Positron Emission Tomography (PET) have added to our ability to identify and locate brain function.

It is, however, important to emphasize the interdependence of these locations. There are voluntary, involuntary, and learned motor reactions. Examples of each are the heartbeat, withdrawal from a painful touch, and kicking a football—and the latter two involve communication between motor, sensory, and association areas.

The danger in mind mapping is to oversimplify, and it is in doing so that it becomes pseudoscientific; the human brain is a huge and complex organ with many active components and pathways, both in use and dormant. John von Neumann, the pioneer of computing science, expressed the caveat well, responding to the numerous facile parallels made between computers and biological organisms, particularly the brain, when he said: "It is dangerous to identify the real physical (or biological) world with the models which are constructed to explain it. The problem of understanding the animal nervous action is far deeper than the problem of understanding the mechanism of a computing machine. Even plausible explanations of nervous reaction should be taken with a very large grain of salt."

See also BICAMERAL MIND.

For further reading: Tony and Barry Buzan, *The Mind Map Book* (BBC Books, 1993); B. Edwards, *Drawing on the Right Side of the Brain* (J. P. Tarcher, 1979); and R. M. Reystak, *The Mind* (Bantam, 1988).

MINNESOTA ICEMAN Hairy hominid completely enclosed in ice exhibited at midwestern country fairs in the 1960s. Two prominent cryptozoologists, Ivan T. SANDERSON and Bernard HEUVELMANS, traveled to Minnesota to see the creature. After careful examination—or as careful as could be done considering that it remained encased in ice in a cramped sideshow trailer—they concluded it to be real and a previously unknown human ancestor. Sanderson proposed that the Smithsonian Institution in Washington, D.C., undertake further examination of the creature, but it disappeared before the institution had an opportunity. The exhibitor claimed that the owner had taken it away to an undisclosed location. He claimed to have replaced the original with a model, which he continued to exhibit for several more years. The creature that Sanderson and Heuvelmans believed they examined was never seen again.

The creature resembled a very hairy human, about 1.8 meters (6 feet) tall and sturdy, with large feet and hands and a broken arm and a damaged eye socket. Others who saw the creature or who examined Sanderson's drawings and notes concluded that it was more likely a well-done latex model created specifically for fair sideshows.

See also CRYPTOZOOLOGY.

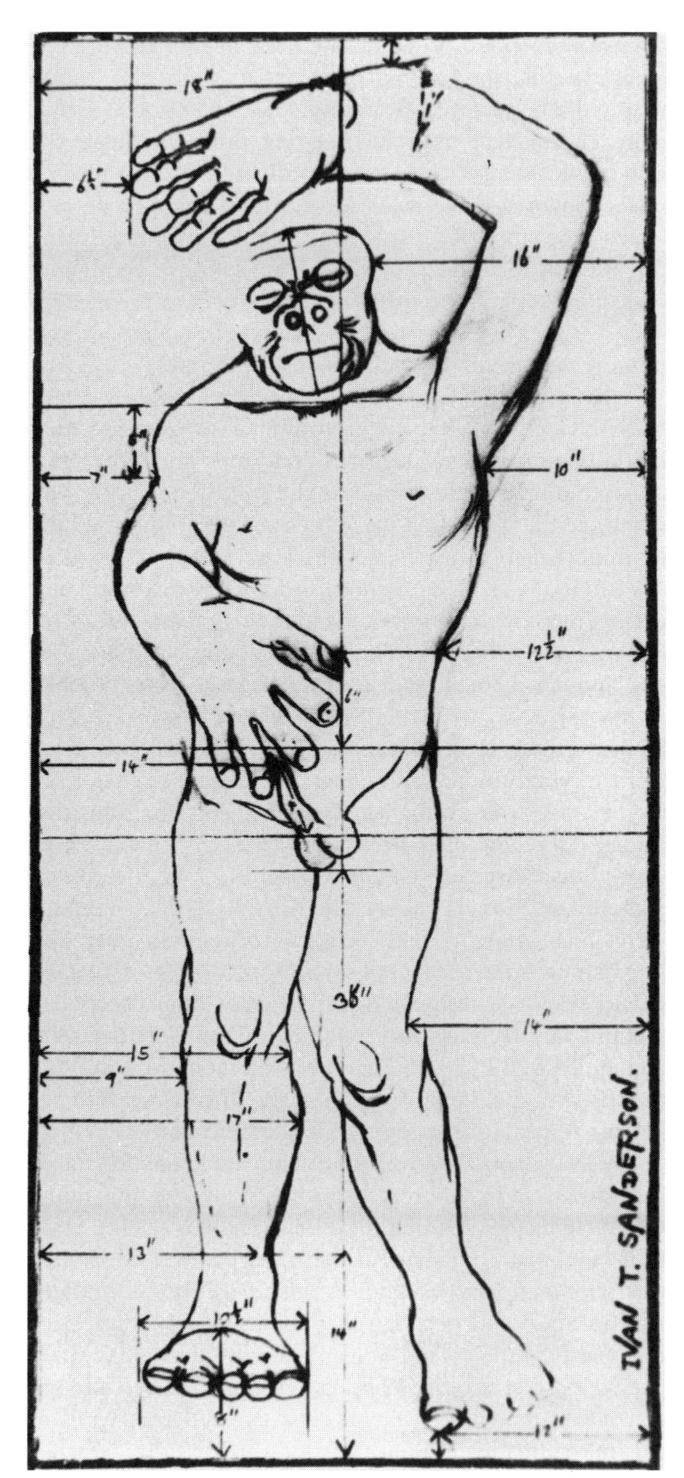

Minnesota Iceman

MISSING DAY IN TIME The assertion that National Aeronautics and Space Administration computers had discovered a missing day (actually 23 hours, 20 minutes) at a point in time that corresponds to an incident recorded in the biblical book of Joshua 10:12. The Hebrews had just defeated the Amorites, and Joshua commanded the sun to stand still so that the victors could revenge themselves on the Amorites. Analogous to the Joshua story is the account of Isaiah making the sun move backward for King Hezekiah in II Kings 20:8–11. These stories have continually been produced to question the accuracy of the Bible on matters of scientific fact.

From the 1970s to the present, however, a story circulated among Evangelical Christian groups that at one point in the 1960s, the staff at NASA were checking the positions of the planets by computer, and they discovered two missing periods of time. These periods corresponded in time with the two incidents involving Joshua and Hezekiah.

Harold Hill, former president of the Curtis Engine Company, a Baltimore firm that did some work for NASA, apparently heard the story and repeated it to the Evangelical Christian community. In 1980 writer Tom McIver traced the story to Hill, who explained that he used it as a means of convincing them that science really supports the truth of the miraculous occurrences in the Bible. Hill justified the story in that it produced converts to his particular version of Christianity. Hill's story was repeated within the community and was used as evidence of the scientific accuracy of the Bible.

Further investigation revealed that the story was based on changes in the way NASA measured time. In the early days of the Apollo program, the mathematics required to compute the motion of Earth relative to the Moon required a more sophisticated timekeeping system. Originally, the system used for keeping time had been the Epheneris sidereal time, but Coordinated Universal Time (also known as Greenwich Mean Time) came to be preferred. The story of the missing day was generated during the attempt to resolve the timing in the move from sidereal to universal time. Although there was some discrepancy between the two timing systems, at no time was a missing day or missing periods of time found.

MISSING LINK Extinct protohuman, or hominid, that marks a transitional stage between apes and other primates and humankind. Finding the so-called missing link was once the goal of all paleontologists. The hunt for the hominid began in the mid-19th century with the discovery in Germany of the first Neanderthal skeletons. Later discoveries of more-modern human remains at Cro-Magnon in France underlined the conclusion that modern human beings had ancestors with some apelike characteristics. In 1891 and 1892, Dutch surgeon Eugène Dubois (1858–1940) excavated a partial skull and some teeth mixing humanlike and apelike characteristics from the bed of a river on the island of Java. He named his discovery *Pithecanthropus erectus,* the "erect ape-man."

Dubois's discovery sparked a series of discoveries of protohuman fossils across the world. Similar remains were uncovered in Heidelberg, Germany, and near Beijing, China. Another Dutch excavator working in Java, G. H. R. van Koeningswald, uncovered more complete remains like Dubois's *Pithecanthropus.* Van Koeningswald concluded that *Pithecanthropus* was more modern-looking than it was apelike and renamed the form *Homo erectus,* placing it in the same genus as modern humans.

In 1924, South African university Professor Raymond Dart discovered an extremely apelike skull from the Taung valley. He recognized that the animal had too small a brain to be a human, but that it stood upright. Dart christened the specimen *Australopithicus africanus,* the "southern African ape," and declared that it was a human ancestor that was nonetheless not human—a link with humankind's apelike past. A Scottish colleague of Dart's, Robert Bloom, discovered that there were at least two different types of Australopithicines in South Africa: a lightweight "gracile" form, and a large "robust" form with huge teeth and jaws.

Australopithicus was not universally accepted as a human ancestor. The "discovery" of PILTDOWN MAN in southern England suggested that humankind had developed in a different direction: that humans developed big brains before they began to walk upright. Piltdown Man was later proved to be a forgery. It was not until 1959 that Louis and Mary Leakey, a husband-and-wife team working in Olduvai Gorge in Tanzania, discovered Australopithicine fossils in conjunction with primitive stone tools. The fossils, which they placed in a new species, *Australopithicus boisei,* were the oldest protohuman remains then known, dating back about 1.75 million years. A year later, the Leakeys discovered another hominid skull dating to the same time as the *boisei* specimens, but with much more human characteristics. Leakey named the new find *Homo habilis,* identifying it with the stone tools scattered around the gorge and making it a direct ancestor of modern humans. *Homo habilis* pushed back the date for a missing-link ancestral to both apes and humans and threw the australopithicines out of the direct line of human ancestry.

Since the Leakeys' finds, paleontologists have uncovered still more hominids. Richard Leakey, son of Louis and Mary, found evidence of *erectus* and australopithicines near Lake Turkana in Kenya. In 1974, U.S. anthropologist Donald Johanson discovered an almost-complete australopithicine skeleton at Hadar, Ethiopia. He placed the remains, which he called Lucy, in a separate species, *Australopithicus afarensis,* dated them to about 3.5 million years ago, and suggested that they represented a link, a form ancestral to both australopithicines and humans. Some paleontologists do not accept Johanson's reasoning, but they all recognize that the lines of apes and humans split long before the date he gives for Lucy. Scientists now believe that humans and apes separated between 12 and 8 million years ago.

Paleontologists no longer consider the discovery of the missing link the primary goal of their profession. The idea of a single missing link began with the discarded chain of being theory, which stated that all life on Earth is connected in a hierarchy stretching from the simplest single-celled amoeba to humans. However, there were several transitional stages missing, in particular that between humans and apes. Under Charles DARWIN's theory of EVOLUTION by means of natural selection, apes and humans are not linked in a direct chain. Instead, they represent two different branches of primate development. Journalists still occasionally use the term *missing link* loosely to describe newly discovered protohuman species.

For further reading: Michael H. Day, *Guide to Fossil Man,* 4th ed. (University of Chicago Press, 1986); Eric Delson, ed., *Ancestors, the Hard Evidence: Proceedings of the Symposium Held at the American Museum of Natural History April 6–10, 1984 to Mark the Opening of the Exhibition "Ancestors, Four Million Years of Humanity"* (A. R. Liss, 1985); Maitland A. Edey, *The Missing Link* (Time-Life Books, 1972); Maitland A. Edey and Donald Johanson, *Lucy: The Beginnings of Humankind* (Simon & Schuster, 1981); Brian M. Fagan, *Snapshots of the Past* (AltaMira Press, 1995); and Paul Mellars, ed., *The Emergence of Modern Humans: An Archaeological Perspective* (Cornell University Press, 1990).

MOKELE-MBEMBE A dinosaurlike animal whose existence has long been reported by native and colonial sources in the swamplands of west-central Africa. *Mokele-mbembe* means "one who stops the flow of rivers." Stories about the animal showed up in reports filed by colonial officials dating back to the 18th century. Englishman Alfred Aloysius Smith recorded encountering the mokele-mbembe's tracks in Gabon in his 1927 book *Trader Horn*. Popular science writers Willy Ley and Bernard HEUVELMANS both recounted legends of the animals in their respective books *The Lungfish, the Dodo, and the Unicorn* (1948) and *On the Track of Unknown Animals* (1958). Natives from the Cameroon, Equitorial Guinea, Gabon, Congo, and the Central African Republic all agreed that a large crocodilelike beast lived in the swamps. Although the natives had different names for the monster (in Gabon, for instance, it was called *n'yamala*), they agreed that it was water loving and dangerous to approach. When the natives were shown pictures of various animals, they chose the diplodocus, a sauropod dinosaur, as being the animal most like the mokele-mbembe.

During the 1970s and 1980s several expeditions traveled to the Congo to investigate reports of the mokele-mbembe. Herpetologist James H. Powell, Jr, and biologist Roy MACKAL both made several trips to the area and collected descriptions of the animal from eyewitnesses. In 1981, Mackal saw signs of a large animal with habits like the mokele-mbembe's along the Likouala River. Around the same time U.S. engineer Hermann Regusters saw a large animal matching the description of the mokele-mbembe swimming in Lake Tele in northern Congo. He was unable to take pictures of it because the high humidity had damaged his cameras. In 1983, an expedition led by Congolese official biologist Marcellin Agnagna sighted a mokele-mbembe in Lake Tele. Agnagna at first claimed he was unable to film the animal but later produced ambiguous pictures. In 1986, his reputation was attacked when

Mokele-mbembe

four Britons accused him of stealing their film and supplies on still another mokele-mbembe sighting mission.

The identity of the mokele-mbembe is a mystery. Although many descriptions and eyewitness accounts have been collected, no one has produced any physical evidence of the animal's existence. Most cryptozoologists agree that it is not impossible that a species of sauropod dinosaur survives in modern times in equatorial Africa. However, until someone produces flesh and bones, the animal remains a question mark.

See also CRYPTOZOOLOGY; DINOSAURS, CONTEMPORARY.

For further reading: Willy Ley, *The Lungfish, the Dodo, and the Unicorn* (Viking, 1948); Bernard Heuvelmans, *On the Track of Unknown Animals* (Hill & Wang, 1958); and Roy P. Mackal, *A Living Dinosaur? In Search of Mokele-Mbembe* (Brill, 1987).

MONBODDO, LORD JAMES BURNETT (1714–1799) Scottish lawyer, judge, and pioneer anthropologist who explored the origins of language and society. His book, entitled *Of the Origin and Progress of Language* (6 vol. 1773–92) is typical of Enlightenment thinking in 18th-century Scotland, containing a large body of curious ideas, as well as sober anthropological learning, on the manners and customs of primitive peoples. He believed, contrary to the received opinion of his time, that human history is not a decline from primeval perfection, as portrayed in Genesis, but a slow and painful ascent from imperfection.

Following the 17th-century pioneer of comparative anatomy, Edward Tyson, Monboddo was one of several thinkers who anticipated the principles of Darwinian EVOLUTION—notably he related the human to the orang-utan, and then traced human development from these origins towards a social state. As part of his evolutionary theory, he believed that children were born with tails, an idea that he never thought necessary to investigate. The notion just grew out of his general idea that orangutan were from the same genus as a human, but a type of human who had failed to develop.

The notion that higher apes are not animals at all, but either primitive or retrograde humans, had been held by a number of post-Renaissance naturalists and was common currency among the literati of the French as well as Scottish Enlightenments. At this time apes were rarely seen by Europeans, so naturalists looking for information were obliged to rely on travelers' tales, which were a mixture of fable and fact, from which they speculated wildly.

Monboddo is seen by some as anticipating DARWIN, but Darwin, writing a century later, was, in contrast to Monboddo, the consummate scientist, meticulously observing everything, making drawings, documenting his evidence, and coming to conclusions from years of accumulated observation. Monboddo saw no need to validate his ideas against observation and experiment. However, Monboddo must be assessed in the context of his times. Living on the threshold of the scientific revolution, he could not have understood the nature of scientific evidence, but he did have the courage and initiative to question ancient shibboleths and to push knowledge forward the only way he knew how—by imagining the outrageous.

For further reading: E. L. Cloyd, *James Burnett, Lord Monboddo* (Clarendon Press, 1972).

MONTGOMERY, RUTH (n.d.) Leader of the New Age Movement.

Montgomery met MEDIUM Arthur A. FORD at a 1958 conference on SPIRITUALISM and the two became friends. Encouraged by Ford, Montgomery began to practice MEDITATION and soon found that she had the ability to do AUTOMATIC WRITING, a form of CHANNELING in which one enters a trancelike or deeply meditative state, allowing spirits to guide the hand and write messages. Using a control named Lily, Montgomery would communicate with an anonymous group of entities known as Guides.

Lily and the Guides dictated their philosophy to Montgomery. According to them, all souls are "sparks of the Creator" that exist naturally on higher vibrational planes of existence that our own. Souls either advance spiritually to ever higher and purer planes, or choose to work out past karmic debt in flesh incarnate.

Montgomery depicted her new spiritual doctrine in *A Search for The Truth.* She elaborated on it in *Here and Hereafter* and *A World Beyond,* for which she came close to attributing authorship to Arthur Ford, by this time dead, and the Guides.

Ruth Montgomery is now known primarily for her leadership and innovation in the New Age movement. She speaks on tours regularly and is a perenially esteemed spokesperson for New Age spiritualism.

MONUMENTS OF MARS Objects on the surface of Mars that resemble pyramids, a human face, and a city. To some, they suggest that Mars is or has been at some time inhabited. During the 1960s and 1970s, a series of U.S. space probes took thousands of pictures of Mars, all indicating that there was no life on the planet. However, the two photo analysts—Vincent DiPietro and Gregory Molenaar of Goddard Space Flight Center in Maryland—came across what appeared to be a cluster of pyramids and several other structures. Although NASA attributed the images to optical illusion, DiPietro, Molenaar, and others began to search the *Viking* photos for more evidence of artificially created structures. Science writer Richard Hoagland promoted the idea that civilized beings had either lived on or visited Mars long enough to build these fantastic monuments. In *Monuments of Mars,* he included the results of experiments by an electrical engineer and image-processing expert, Mark M. Carlotto, who enhanced the images. Carlotto believes his enhancements, which gave hints of roadways and entrances to the structures, prove that the structures are artificial. Not everyone is convinced, however. NASA continues to deny that anything except natural structures were found on Mars.

More recently, Hoagland claims to have found evidence of artificial structures on the moon as well. He and his colleagues continue to urge NASA to explore Mars more fully, preferably with a manned landing party, to discover once and for all whether the monuments are artificial or natural.

See also MARS CANALS.

For further reading: Richard C. Hoagland, *Monuments of Mars: A City on the Edge of Forever,* rev. and enl. (North Atlantic Books, 1992).

MOON, LIFE ON THE Historical reports of extraterrestrial life on the Moon. The concept of life on the Moon has a long history. The first person to suggest that the moon was, like Earth, inhabited was Greek philosopher Anaxagoras (c. 500–c. 428 B.C.E.). Second-century writers Plutarch and Lucian of Samosata both wrote about the Moon as an inhabited world. Lucian's story, *A True History,* tells about a war between the king of the Moon and the king of the Sun. Sixteenth-century Italian poet Ludovico Ariosto told of a trip to an inhabited moon in his epic poem *Orlando Furioso.* Similarly, Cyrano de Bergerac (1619–55) in *Voyages to the Moon and the Sun* predicted travel to the moon by rocket. Scientists such as Johannes KEPLER (1571–1630) and Galileo Galilei (1564–1642) depicted the moon as a world with intelligent life.

By the early 19th century, scientists were convinced that life—and certainly not intelligent life—could not exist on the Moon. In 1835, however, the *New York Sun* sponsored a "Moon Hoax," claiming that the British astronomer John Herschel (1792–1871)—then on an expedition to Cape Town, South Africa—had discovered life on the moon using an extremely powerful telescope. The perpetrator of the hoax was Richard Adams Locke (1800–71), an immigrant writer. On August 25, 1835, Locke published a fantastic story claiming that Herschel had seen flowers, lakes, and trees on the surface of the moon. Locke elaborated on the story in future installments; he later stated that the scientist had also seen large animals resembling bison and unicorns and finally concluded that Herschel had seen intelligent winged humanlike beings.

Locke's hoax was soon revealed for what it was—an irresponsible piece of science fiction masquerading as journalism. Scientists knew that no telescope available to Herschel at the time was capable of the resolution that Locke had claimed, making objects of 45 centimeters (18 inches) in diameter visible. Nonetheless, the story did achieve its objective of boosting the circulation of the newspaper for several days. Thousands of people were deceived by Locke's work; they and their heirs continued to believe in a Moon populated by living beings. The case was not definitively resolved until the Apollo missions of the 1960s and 1970s brought back moon soil that showed no signs of life.

For further reading: Isaac Asimov, *Extraterrestrial Civilizations* (Crown, 1979).

MOORE, ANN (n.d.) A woman who claimed the ability to live without eating. According to Moore's own account, in 1807 she decided that she could not stand to eat any more.

Moore became the subject of some curiosity. Pamphlets written about her speculated that she had learned to live off air. A few suggested that she was eating secretly when no one was looking. In September 1809, a group of neighbors decided to test her. They watched Moore for 16 days, changing shifts hourly. At the end, they all testified that she had received no food. Moore gained a set of loyal supporters and her fame spread. In 1813 further tests revealed Moore as a fraud. She finally confessed the hoax she had been living, describing all the means she and her family had devised in order to pass food to her. The observations of Moore called claims of extended fasting into question and also help account for the lack of such claims in the last century.

MORAG One of several reported LAKE MONSTERS similar to the LOCH NESS MONSTER that have been reported in nine of the larger Scottish lakes. Morag is the name given the creature that has been sighted in Loch Morar, which lies approximately 110 kilometers (70 miles) southwest of Loch Ness. It was named for Mhorag, the traditional spirit of the loch, who was, according to local folklore, a mermaid. There have been sightings of a beast in the lake since at least the late 19th century, though little notice was taken of it until 1969. That year, two men reported a beast having hit their boat. Wire services picked up the newspaper accounts of the incident and alerted the investigators at Loch Ness.

The following year, several people from the LOCH NESS INVESTIGATION BUREAU formed the Loch Morar Survey. The group had an immediate success when a biologist, Neil Bass, spotted what he described as a "hump-shaped" black object near the northern end of the loch. He called the others to see it, but it quickly disappeared. They did see the disturbance in the water where the object had submerged. In the early 1970s, in the wake of the interest in the Loch Ness monster, Elizabeth Montgomery Campbell and folklorist E. Macdonald Robertson collected stories of Morag.

Among cryptozoologists, Loch Morar is seen as one of the more likely homes of an as yet undiscovered large aquatic species (in the absence of any specimens and assuming that lake monsters are all of the same species). Some suggest that the Morag is a relative of the plesiosaur, though the coldness of the loch's water argues against it. Biologist Roy MACKAL suggests a modern relative of the zeuglodon, a primitive whale that dropped out of the fossil record some 20 million years ago. More skeptical voices have suggested a range of mundane explanations, including misidentification of everyday objects and hoaxes.

See also CRYPTOZOOLOGY; DINOSAURS, CONTEMPORARY.

MORGAN, ELAINE (1920–) British author and lecturer on evolutionary theory who holds scientifically unorthodox opinions, and who started to write at the end of the 1960s at the height of the women's movement. She saw the story of creation and similar male myths of the past, as well as evolutionary explanations of the development of humankind to have all, in their various ways, marginalized the contribution of females.

Morgan asserts that a high proportion of scientific thinking about our past has remained androcentric. The impression given by the common usage of the male pronoun for both sexes has confused "man as species" with "man as male"; the mental image portrayed is still very much a male one. The prehominid male has been made the great hero of the evolutionary story. "He," the Tarzanlike figure of the Mighty Hunter, was preeminent, while "She" was portrayed as the Hunter's mate. Everything worthwhile about our species was held to have developed out of the male lifestyle: We walked erect because the Mighty Hunter had to stand tall to scan the horizon for his prey; bipedalism enabled him to race after game while carrying a weapon; we learned to use our hands because the making and use of weapons was necessary for hunting; we lived in caves because the Mighty Hunter needed a base to come home to; we learned to speak because the Mighty Hunter needed to communicate the happenings of his day and to plan strategy for the next safari with others.

Morgan, who was not a scientist, started her researches believing that evolutionary patterns might just as likely have followed women's requirements and lifestyle as men's, for it was in *her* that the future of the species lay. It was no use inventing a superior male figure if *she* and the babies inside her did not thrive—a scenario that could easily have come about when the forests of Africa dried up during the heat and drought of the Pliocene—a 10-million-year gap in fossil and skeletal evidence for which scientists have been singularly unsuccessful in unearthing the vital missing links in our ancient past. In Morgan's readings, she came upon an idea that was much more sympathetic to woman's survival and that postulated that, during the many changes in the Pliocene, a branch of the primitive ape-stock might have moved to coastal regions in search of food on the beaches and in the shallow waters.

The idea was first put forward by Professor Sir Alister HARDY, and published in 1960 in the *New Scientist* under the title "Was Man More Aquatic in the Past?" Hardy had been toying with this concept for many years. Although he stressed that it was only a hypothesis lacking any hard evidence, he nevertheless believed it was scientifically possible. He reminded us that the ancestors of the mammals we see in the seas today must have initially come out of the water and adapted to the new terrestrial life, learning how to breathe and reproduce on land. Then later, for reasons of overpopulation, climate change, or shortage of food, they reversed the process and returned to the water. Hardy believed this might have happened to mankind's ancestors for a period, pointing to possible evidence of our maritime heritage in many of our bodily features—for example, the loss of our body hair, as had happened to other mammals that returned to the sea. The hair on our heads was not lost but remained as a protection from the direct glare of the sun's rays, an adaptation process very necessary for a wading and swimming animal, as opposed to one that lived entirely in the water, only coming up to the surface occasionally to breathe. Examining the tiny remnant hairs on the surface of the trunk of the human fetus, he found them to be different from those on apes, particularly noting the hair tracts on the back, which he found to all point downwards and inwards towards the spine. This might be an indication that, before disappearing completely, our body hair first modified to follow the flow of water passing over a swimming body. Hardy also noted we are unique among the primates in possessing a thick layer of subcutaneous fat—equivalent to the blubber on whales or seals—an important feature for insulating a warm-blooded creature from the water's cold.

After consulting the eminent scientist, Morgan immediately saw that an aquatic past for our species could answer other questions. For example: How did humans progress to the erect posture and evolve deft hands? The chimpanzee, our nearest evolutionary relative, has knuckle-walking hands. Our sensitive hands and fingers could not have developed at the same time as they were being used to sustain body weight while mastering the upright position on land, whereas sea wading with water supporting the body weight would have allowed early humans an erect stance, freeing the hands to be used as delicate instruments. Both sexes, even pregnant females, would have been able to join in the search for food, groping around rock pools and the seabed, feeling for shellfish, and then learning how to pry them open.

The hypothesis has still to be confirmed; the evidence is still indirect and speculative. Morgan needs hard scientific proof—for example, hominid fossils or some kind of remains found in the seabed. Meanwhile, in further published work, she poses many new questions: Why do we cry salt tears? Why are our nostrils shaped the way they are? Why are the human female's breast and buttocks so different from those of other primates? Why do humans copulate face to face? Although answers to these questions cannot be established scientifically, Morgan points out, for example, that since practically all land mammals use the rear approach to sex, and practically all aquatic mammals use the frontal approach, the connection must be more than fortuitous. The manatee and the dugong, for example, mate face to face, and the female's genital organs are modified accordingly. The same is true of the whale, dolphin, porpoise, and also the sea otter—and, although it swims in fresh water rather than salt—the same is true of the beaver.

Morgan's ideas have still to find some basis for acceptance; meantime, they must remain pseudoscientific.

For further reading: Elaine Morgan, *The Descent of Woman* (Souvenir Press, 1972); Elaine Morgan, *The Aquatic Ape* (Madison Books, 1982); Elaine Morgan, *The Scars of Evolution: What Our Bodies Tell Us about Human Origins* (Penguin Books, 1990); Elaine Morgan, *The Descent of the Child: Human Evolution from a New Perspective* (Penguin Books, 1994); and Machteld Roede, Jan Wind, John Patrick, and Vernon Reynolds, eds., *The Aquatic Ape: Fact or Fiction?* (Souvenir Press, 1991).

MU Legendary lost land whose existence was proposed by 19th-century French physician and explorer Augustus Le Plongeon. Le Plongeon translated a rare surviving Mayan text, the *Troana Codex*, in which, he claimed, lay evidence of a lost land whose inhabitants were ancestors of both the Mayans and the Egyptians. He suggested that the continent of Mu had been destroyed by an earthquake and had sunk into the Pacific Ocean. Author and adventurer James Churchward wrote several books in the 1920s and 1930s, adding to the Mu legend. Much of his evidence, he claimed, came from ancient stone and clay tablets he found in India, although he never showed the tablets to anyone else. He also asserted that legends and artifacts of many cultures around the world supported his claims of Mu's existence.

Churchward claimed that Mu sank 13,000 years ago, destroying an advanced civilization of some 60,000 people, and leaving behind the Polynesian islands. Le Plongeon and Churchward also suggested that the inhabitants of Mu had been responsible for the ancient, mysterious statues on Easter Island. Churchward's four Mu books detailed the lost civilization.

If the continent of Mu existed, no convincing physical evidence of it remains today. Most authorities consider it mythical.

See also ATLANTIS AND LEMURIA.

For further reading: James Churchward, *Children of Mu, Cosmic Forces of Mu, The Lost Continent of Mu,* and *The Sacred Symbols of Mu* (reprinted by The Paperback Library, 1968).

MULTIPLE PERSONALITIES When two or more distinct personalities exist within the same person. Sometimes one or more of these personalities are aware of each other; sometimes not. One personality tends to be dominant, the primary personality, except during occasions brought on by specific conditions or stress. Commonly but inaccurately called schizophrenics, people with multiple personalities are considered to have a mental disorder (multiple personality disorder, or MPD).

Multiple personalities have been known throughout history, although until the rise of psychiatry in the late 19th century, their symptoms were usually attributed to demonic POSSESSION. MPD cases can have any number of personalities; some have been documented in the dozens.

One of the most famous MPD cases of modern times was that of "Eve," about whom both a book and a movie were made (*Three Faces of Eve,* 1957), bringing the problem and its causes to widespread attention. Since that time MPD has been widely documented and has also been used as dramatic device—usually specious—in many fictional works.

MPD usually occurs as a result of severe trauma, most often aggravated childhood abuse. The child who is badly battered, whether emotionally, physically, or sexually, often recedes into an involuntarily self-made shell during which one or more new personalities develop, some to protect the child from knowing of the abuse and thereby being able to cope with day-to-day events, some to act as companions to the child, some for unknown reasons. Even after the individual escapes the abusive environment, some or all of the personalities tend to remain, sometimes causing notable problems for the individual: People with MPD may appear erratic or even insane to those around them and even to themselves; they may not be able to form meaningful attachments with others.

Although MPD has been looked on with suspicion, it is viewed as a genuine mental illness by the American Psychiatric Association. There have even been court cases in which secondary personalities have been allowed to testify. Problems in authentication often arise because of the question of whether the therapist has led the patient to the diagnosis. Also, the secondary personalities often remain hidden until provoked by a specific condition.

Once diagnosis is made, the aim of treatment is also controversial. Traditionally, the aim has been to eradicate all the secondary personalities, but some patients and therapists believe the correct route is to allow the personalities to remain and to help them integrate and work with each other in a way that is practical and positive for the individual in whom they reside.

MUTUAL UFO NETWORK Largest of the several UFO organizations to survive into the 1990s. The Mutual UFO Network (MUFON) was founded in 1969 as the Midwest UFO Network. The name was later changed to reflect its growth into an international organization. MUFON holds an annual symposium that provides the largest regular gathering of ufologists and their supporters, and its program has become a good indication of the shifting emphases in research and debate over UFOs. Though MUFON has endorsed the extraterrestrial hypothesis of UFO origins, a range of opinions, including skeptical ones, is expressed in its program. The lectures are published as the annual *Mutual UFO Symposium Proceedings.* MUFON also publishes a *Manual* for its field investigators that attempts to bring some order to the chaos of UFO reports received on a regular basis. Its *MUFON UFO Journal* (formerly *Skyhook*) is the largest circulating journal in the field.

See also UFOLOGY.

MYERS, FREDERIC WILLIAM HENRY (1843–1901) A member of the Committee on Thought Reading (CTR) of the British SOCIETY FOR PSYCHICAL RESEARCH. Myers and his colleagues made a determined effort to assess many claims of paranormal phenomena, especially thought reading. Some of his books are still referred to today. In his most notable work, *Human Personality and Its Survival after Bodily Death, Vols. 1 & 2* (1903), he presents what he believes are authenticated accounts of hundreds of cases of various paranormal events.

The first report of the CTR in 1882 was warmly received, having clearly established the reality of thought reading by six young girls. In 1888, the girls admitted that they had tricked their observers. Yet, Myers continued to be convinced of the reality of thought transference.

Myers played a part in paranormal phenomena even after his death. In the early part of the 20th century, communicating with the dead by AUTOMATIC WRITING was in vogue. The spirit of a dead person would take control of the medium's hand and write messages to the living. One of the spirits frequently invoked in this process announced itself as "Myers."

For further reading: Antony Flew, *A New Approach to Psychical Research* (Watts, 1953); and C. E. M. Hansel, *ESP: A Scientific Evaluation* (Scribner's, 1966).

MYSTERY HILL American structure of mysterious origins possibly predating the Columbian contact. Outside Newport, Rhode Island stands a round tower built of field stones. It is about 25 feet in diameter and stands about 25 feet high. Referred to as the Old Stone Mill, it is supported by eight round columns, which in turn support Romanesque arches. It has two stories, with a very few small windows and three small slits that may have been used by defenders to fire through. The roof and the floors of both stories are missing, but otherwise the building is fairly complete. Although the structure has been called a windmill, a fort, a church, a watchtower, and a temple, its origins remain a mystery.

In 1677, the tower stood on land owned by the colonial governor Benedict Arnold, who called it "my stone-built windmill" in his will, dated December 24th of that year. However, it may not have been built by Arnold; it is not mentioned elsewhere until February 1677, during the middle of King Philip's War, and it is overbuilt for the purpose of milling. Some historians suggest that Arnold adapted an existing structure for his own purposes. The building may also have been constructed around 1640 as a watchtower, but some reject this, claiming that the English would not have used precious resources building a stone watchtower with so few and such tiny windows.

A more romantic view is that the tower was the rotunda of a Norse church, built by an expedition from Norway to the colony in Vinland in 1355 to 1364. The scholar Hjalmar Holand argues that the building was constructed by a team dispatched by King Magnus Eriksson to locate emigrants from the Icelandic/Norwegian colony in Vinland. He believes that the tower was intended both as a church and as a headquarters for the expedition.

This Viking theory was first propounded by a Danish historian, Charles C. Rafn, who sought evidence of early Scandinavian settlements in North America. An 1839 supplement to his 1837 tome, *Antiquatatis Americanae,* asserted that the tower had been built by a Greenland bishop as a church. He pointed out the style similarities to several European 12th-century churches. This theory took hold with many people who believe the Vikings were the first Europeans to visit the New World.

Several 20th-century scholars found evidence to support Rafn's contentions. The tower is built (according to Arlington Mallery, an archaeologist who under the sponsorship of Harvard University's Peabody Museum did a dig around the tower in 1948) in a Celtic manner (the mortar in particular) but in the style of a Norse church. Supporting this dual Irish-Norse style was the fact that by the late tenth century, the Vikings had Irish slaves who often traveled with them and who might have been responsible for the tower's construction. Mallery also found round nails like those used by the Vikings during his dig around the tower. The building has intriguing niches that could have been used as shrines, and its structure is similar in many respects to several Norse round churches, including the 13th-century St. Olaf's Church in Tonsberg, Norway.

For further reading: Hjalmar R. Holand, *Explorations in America before Columbus* (Twayne Publishers, 1956); and Barry Fell, *Saga America* (Times Books, 1980).

N

NATURAL LAW The concept that there is a system of justice inherent in the world that is more certain, more just, and superior to the written laws of any society, which are always specific to time and place and made by imperfect human beings, and therefore fallible. The idea that there is an objectively discoverable set of principles of right moral conduct was originally postulated by the early philosophers in the classical period and has run through Western thought ever since.

In ancient Greece, Plato (c. 429–347 B.C.E.) conceived of law as a disposition or arrangement of reason, which was one of his "Universals" or "Forms." Later, Aristotle (384–322 B.C.E.) thought that the permanent and changeless universal law was more important than laws written by states, which were the product of individuals like himself and therefore liable to errors of judgment. The Roman Stoic conception of universal reason led Cicero (106–43 B.C.E.) to believe that there are natural laws that have been built into the universe by a rational deity. The medieval Christian philosopher/theologian St. Thomas Aquinas (c. 1225–1274 C.E.) synthesized Catholic theology and Aristotelian metaphysics. He defined natural law in relation to God's eternal law and held that the eternal law is God's reason, which governs the relationship of all things in the universe to each other. Thomists (Aquinas's followers) said that this eternal law is conveyed to human beings in part through revelation and in part by their own reason, introspection and dialogue, and it is the latter, the part that can discerned by reason and which relates to their own behavior, that is natural law. According to this thinking, the principles of natural law must not be understood to be immutable but could be allowed to be variously developed at different times and in different places. Protestant theologians like Karl Barth (1886–1968) and many others criticized this notion, holding that sinful and sinful and fallen men and women cannot have any direct knowledge of God's intent without the special aid of revelation.

Today the notion that there is a set of ideas and principles adding up to a system of natural law that can be tapped into persists in various NEW AGE MOVEMENT religions, whose adherents feel the need for the reassurance of a permanent set of values and meaning in a world of conflicting global ideologies and fast-changing standards of behavior. Skeptics would say that the whole notion of natural law is a projection of human morality on an essentially meaningless universe.

There are now natural law parties in British and American politics. In the U.S., John Hagelin, once a particle physicist but now a follower of Maharishi Yogi (see TRANSCENDENTAL MEDITATION), has twice run for president of the United States. He alleges, without going into too many details, that through transcendental meditation he, in unison with others, can create a "coherent field" through which he can tap into natural universal morality and cause crime to vanish, end drug abuse, cleanse the inner cities, and bring about world peace. More amazingly, he claims he can make the national debt disappear. He claims that the whole concept can be proved by particle physics, and if the Superconducting Supercollider had been built, he could have worked out the details and been able to give proof of his "coherence creation" theory. By using the ideas and terminology of particle physics, Hagelin is trying to legitimize his pseudoscientific theories on how to activate natural law and in a flash give the answer to all questions and so put the world to rights.

Most skeptics agree that there is no independent universal law of nature relating to a superior conception of human justice. Consequently the extraordinary claim of activating it for our benefit not only violates the laws of physics but flies in the face of logic and common sense. In actuality it is wishful thinking.

For further reading: J. M. Finnis, *Natural Law and Natural Rights* (Oxford University Press, 1980).

NATURAL THEOLOGY The belief that religious knowledge can be obtained by human reason by working upon information gained by observing the world and all its wonders, as opposed to the more usual revelation by faith or reading the Scriptures. It was Thomas Aquinas (c. 1225–74) who formulated the distinction between natural and revealed theology, as opposed the older Augustinian view that there is no knowledge of God except through revelation.

The subject of theology is not the usual area in which issues of science and pseudoscience can be contrasted. But when a theology claims that fundamental knowledge, even the knowledge of God, can be sought by rational reflection without the use of revelation, then science is immediately involved. The world "science" comes from the Latin "scientia," meaning knowledge, and one of its characteristics is its universal quality: Its truth does not depend on adhering to a specific form of revelation, so, at least in theory, Natural Theology and Science should be able to engage in a dialogue. In fact, a dialogue was the result.

In the 17th century, following an interest in astronomical and physical systems, there arose a religious position known as Deism, which inferred God's existence from looking at the natural systems of the universe. One of the most compelling theories it advanced was "the Argument from Design." The argument runs thus: The order, the pattern, and the beautiful structures of the universe form a grand design, and such a design must have a designer, just as houses and watches have a human designer. Therefore God, the great designer, must exist. The basic principle employed here is one of cause and effect, with the Deity as the great watchmaker—the First Cause who created the universe and put it into motion. Deists were quick to point out that although we can infer God from this argument, we can infer nothing else, neither his moral nor his aesthetic nature. Critics of Deism were outraged by the idea of a God who did not take part in the affairs of the world. God is more than just a powerful first cause, they argued; in addition to being omnipotent he is an all-loving father. In England under autocracy of the Anglican Church, many Deists suffered persecution and imprisonment for their beliefs. Nevertheless, the Church of England could not stop the philosophers and religious thinkers being attracted to Deism, and by the 18th century it had spread to the rest of Europe and America.

The arguments around the "watchmaker God" and the divine purpose of the universe, especially the ethical problem of how an all-good and all all-powerful God could allow evil to exist in the world, remained hotly contested issues throughout the 18th century. The Scottish Enlightenment philosopher David Hume (1711–76), in his *Dialogues Concerning Natural Religion*, showed how both the churchgoer of faith and the natural theologian, in their arguments, gradually moved towards skepticism. In their desire to defeat each other, each side eventually destroyed the other's argument.

The next important step in the Natural Theology argument came when Charles Robert DARWIN (1809–82) published his *Origin of the Species by Natural Selection* in 1859. After the initial shock engendered by the book had died down, theologians on both sides began to move away from the notion of species being individually created by God toward accepting the evolutionary process of natural selection. The actuality was that the post-Darwinian world view, where nature is a place of competition and struggle rather than a place of God's love and harmony, had shaken those who believed in the loving God of revealed theology. On the other hand, Darwinism had destroyed completely the Deist's belief in the world as a mechanism that needed a first cause. With the death of Deism, Natural Theology had to change its form, and a new idea called Process Theology, which saw God as an agent within evolution rather than producing each species ready made, was advanced by A. N. Whitehead (1861–1947). This position has been attacked by some as encouraging pantheism (the belief that God is present in everything); others, weary of all the twists and turns, drifted away from theology and towards science and its postulation that the evidence of EVOLUTION reveals a universe developing without a predetermined design.

The above was a very English development, but there was another cultural stream that emerged in the United States in the 19th century that can be associated more easily with American individualism and American reaction against European institutions. This was *transcendentalism*, and it is the history of this movement that formed the basis of the natural theology of the NEW AGE MOVEMENT. Important to the acceptance of transcendentalism were the writings of Ralph Waldo Emerson (1803–82) and Henry David Thoreau (1817–62), who exalted individual spiritual awareness over conventional mainstream religions and subscribed to pantheistic Hindu ideas of the divine being present in all nature. Although Emerson and Thoreau renounced the Protestant ethic of the 19th century, they nevertheless still worked within the original Protestant spirit of Luther and Calvin, who claimed direct access to God against the mediation of the Catholic clergy. Both Emerson and Thoreau saw all conventional churches as standing unnecessarily between the believer and the divine. Of course, those who follow a personal New Age religion without intercessors are very much by themselves and have to accept the anxiety of the increased responsibility of their own spiritual health. Some adherents of New Age thought use psychedelic drugs to find ways of attaining a higher divine realm or reaching a higher state of consciousness. Although the New Agers cannot be defined by a strict canon of principles, they nevertheless tend to hold a combination of religious beliefs, such as REINCARNATION, karma, Taoism, Zen, spiritism, animism, and shamanism, yet they generally reject Christian fundamentalism, Catholicism, and Judaism

For further reading: Frank J. Tipler, *The Physics of Immortality: Modern Cosmology, God, and the Resurrection of the Dead* (Doubleday, 1994); Paul Davies, *God and the New Physics* (Touchstone Books, 1984); and Richard Dawkins, *The Blind Watchmaker* (Norton, 1986).

NATURE-NURTURE DEBATE A term first coined by Sir Francis GALTON (1822–1911) for a controversial question: whether humans' nature—their genetic make-up—*or* nurture—the influence of their environment, both physical and social—affects them most. Galton vigorously opposed the belief, then widely held at the end of the 19th century, that all babies were born equal and subsequently diverged in response to their upbringing and surroundings. He was convinced that humans had different inherent potential and that these differences could be affected but not eradicated. Nineteenth-century philosopher John Stuart Mill, on the other hand, while not denying that some part of the variation between humans may be innate, believed strongly that education played the much bigger part.

There are human characteristics that are clearly and unarguably genetically transmitted, the physical differences between Asians and Caucasians, for example. But other characteristics are not so obviously uncontroversial. For example, a major matter of concern in the nature–nurture debate has been about intelligence: Are we born with a fixed value for our intelligence quotient, or can IQ be raised or lowered by experience, education, and training? It is an important question because its answer can influence the educational policy of a nation or authority, and often has done. The effects may spread even wider. Galton, acting on his belief, originated the EUGENICS movement to improve the quality of the nation's men and women by selective breeding and sought to find a way of measuring intelligence to be able to select the breeding stock. Today, a century later, argument continues, with the two sides as polarized as ever. But there is now a third position in the debate: Some participants now believe, to quote Raymond E. Fancher, "that interactions are so variable and so ubiquitous as to make futile *any* discussion of heredity and environment as separate factors." He stated, "Thus most investigators now recognize that there is not, and never can be, any universal or final answer to the nature-nurture question with respect to intelligence."

See also INTELLIGENCE TESTS.

For further reading: Raymond E. Fancher, *The Intelligence Men: Makers of the IQ Controversy* (Norton, 1987).

NATUROPATHY The idea that, given the right set of circumstances, the body has the power to heal itself through its own "inner vitality." Like many other systems of ALTERNATIVE MEDICINE, naturopathy claims that disease is brought about by an imbalance in the working of the whole body. Naturopaths concentrate on the person rather than the disease, believing that the success of the system depends on their patients' ability to heal themselves.

The origins of naturopathy go back more than 2,000 years probably to the Greek physician Hippocrates (460–370 B.C.E.), known as the "Father of Medicine," who it is believed was the first to recognize nature's own healing power. He laid down as a first principle of healing—"first do no harm"—which seems to fit well with the noninvasive ideas of the naturopaths. Modern naturopathy was pioneered in Germany in the early 19th century when various WATER CURES, basically hydrotherapy, were developed. Patients came to outdoor baths that were built in the woods; they were encouraged to eat a simple diet and drink pure water to cleanse their kidneys by flushing out toxins. The treatment also included enemas and plunging into first hot and then cold baths to stimulate circulation. Since then, naturopathy has developed slightly differently in different parts of the world, with European and United States practitioners emphasizing slightly different ways to detoxify the body to bring it back to the point where it can start to heal itself. Europeans use more hydrotherapy and HERBAL MEDICINE, while in the United States more homoeopathic treatments are used. But on both sides of the Atlantic, naturopaths recommend exercise, fasting, and special "natural food diets" with vitamins, and in addition some form of osteopathic and such CHIROPRACTIC techniques as manipulation and massage.

Medical science believes that illness is caused by germs and viruses that must be attacked and eliminated. Naturopathy teaches that symptoms are the body's efforts to overcome these external agents and that they should not be suppressed unless they pose a threat to life. Letting the body itself tackle the toxins helps it to build up a future resistance to the disease. Many of the treatments put forward by naturopaths form the basis of good modern health recommendations.

Most naturopaths believe that virtually all diseases are within the scope of their practice, and it is this belief that causes the most concern to orthodox physicians. Although naturopathy forms the basis of modern healthy eating, it is not safe for everyone, and some of its ideas, such as fasting, are dangerous for all.

For further reading: K. A. Butler, *Consumer's Guide to Alternative Medicine* (Prometheus, 1993).

NAZCA LINES Amazing ancient tableau of gigantic figures and lines drawn on the desert floor about 402 kilo-

Nazca Lines

meters (250 miles) southwest of Lima, Peru; brought to prominence with the 1970 publication of Erich VON DÄNIKEN's best-selling book *Chariots of the Gods?* The several hundred drawings of birds, animals, humans, geometric shapes, and straight lines cover about 500 square kilometers (200 square miles) and are believed to have been inscribed by the Nazca people between 500 B.C.E. and 500 C.E. They were created by people who systematically removed surface rocks and scraped off the surface of the desert in the forms of their designs. Stone cairns, fire pits, and evidence of animal sacrifice remain at some of the sites. How and why they were created has been the subject of controversy.

One of the most amazing aspects of the Nazca drawings is that they can only be seen fully when viewed from the air. This fact has led to fantastic speculation, most notably by von Däniken, who theorized that the drawings are a large airfield or, he later amended, at the very least navigation aids for visiting extraterrestrials. Von Däniken asserted that the drawings could not have been made without overhead supervision, and who better than the extraterrestrials themselves?

Von Däniken was not the first to voice such speculation, but his book achieved such widespread notoriety that his theories fully captured the world's imagination. Scientists immediately attacked von Däniken.

In the 1970s, members of the International Explorers Society (IES) popularized another theory suggested by artifacts found in some Nazca graves. They claimed these showed that the Nazcans had the technology to build a balloon and thus could have supervised the desert etching—or could have taken ritual skyrides to view the finished works.

Maria Reiche, a German mathematician who came to Peru in 1946, studied the drawings for decades and spent most of the last years of her life trying to save them from the ravages of tourism and climate. She dismissed the IES theory. She believed that they could have been—and most likely were—created entirely on land, using a very precise proportional scaling method: The designers would lay out the patterns on a small scale and then, using sticks and ropes of designated unit lengths, the pattern would be enlarged. Most archaeologists believe that Reiche's theory is sound, although there is no agreement yet on the purpose of the drawings. They could have been created for religious or astronomical purposes, or for some purpose unimagined by us today. Since few Nazcan artifacts survive, only speculation remains.

For further reading: Evan Hadingham, *Lines to the Mountain Gods: Nazca and the Mysteries of Peru* (Random House, 1987); and Jim Woodman, *Nazca: Journey to the Sun* (Pocket Books, 1977).

NAZI RACISM An increasingly extreme racial policy that was carried out by the Nazi regime (1933–45), culminating in genocide. Nazi ideologists produced a coherent and systematic body of thought that drew on a number of older racist themes. Right-wing political theory had earlier emphasized the dangers of mixing the blood of different races. The Aryan race, which predominated in northern Europe, was to be the focus of German national pride. Other racial groups, in particular the Jews, were singled out as sources of Germany's recent problems. Popular feelings of anti-Semitism, widespread in central and eastern Europe, were turned into state policy.

The Nazi state was established after a century of German pride in the quality of its science. The Nazis accepted this, though they made a distinction between the scorned Jewish science, which tended to be abstract and theoretical, and the more respected Aryan science, which was more experimental and practical.

Were their ideas of race, which appeared to their foes to be an especially dangerous example of pseudoscience, good science by the standards of Nazi culture? The theoretical underpinning of their main racial distinctions were not in accord with the physical anthropology or the biology of other European societies of the time and have no basis in more recent genetics. Yet their views were the expression of a coherent cultural viewpoint. On the practical side, their racist "science" systematically developed the potential of eugenical ideas, based on Charles DARWIN's theory of EVOLUTION and the ideas of German zoologist Ernst HAECKEL. They established a program of selective breeding of captive populations (especially from Poland) with valued racial characteristics (positive EUGENICS) and the systematic killing of those they regarded as racially inferior (negative eugenics). Their actions put eugenics in lasting disrepute, for it illustrated dramatically how difficult it is to establish objective values to which we may appeal in judging biological superiority and inferiority in human populations.

In spite of its systematic cultural basis, Nazi racism cannot stand as science, even as "Nazi science." Nazi racism operated in terms of unquestioning acceptance of the kind of simplistic stereotypes that are so often produced by the pressures of political polarization of thought into black-and-white extremes. People in Germany just before 1945 did not dare publicly question the assumptions underlying official Nazi racist views, fearing that they would then be subject to the even stronger pressures of thought control in a totalitarian society. Within the Nazi party, only Hitler was in a position to rethink and modify the racist ideas, and the final trend of his thought was ever more extreme.

See also GOBINEAU, COMTE JOSEPH D.; RACIAL THEORIES.

For further reading: D. Gasman, *The Scientific Origins of National Socialism* (Macdonald, 1971); and R. G. A. Dolby, *Uncertain Knowledge* (Cambridge University Press, 1996).

NEAR-DEATH EXPERIENCE (NDE) Mystical experience that occurs to some people who have been near death.

Most often people who have an NDE say they felt themselves leave their body, and many report floating above it, observing what's happening to themselves and observing other events in the vicinity. They also report perceiving themselves traveling through a dark, narrow tunnel toward a lighted opening. Once they near or arrive at the light, they may meet "beings of light," in many cases their own deceased relatives and friends; in others, mystical beings. These beings are usually encouraging and supportive. Often the experiencers are told by the beings to return to their bodies; in some cases, they believe they have a choice of returning to their bodies or remaining with the beings of light; those who return to life believe they have chosen to do so.

Most people who experience an NDE think of themselves as having visited an afterworld paradise, a sort of heaven, or perhaps a waystation between Earth and heaven, and some resent being brought back to life. A small percentage feel that they have visited a hellish place where terrifying things happen to them. A person's character—virtuous or sinful—seems unrelated to whether he or she has a positive or negative experience. At this time, no factors are known to specifically influence the nature of the experience specifically.

Not much was known about NDE until Dr. Raymond C. Moody's book *Life After Life* was published in 1975. Moody's research revealed that NDE is not uncommon; in fact, a Gallup poll conducted in 1982 showed that more than 8 million American adults had undergone an NDE. (Later studies focused on children, who were found to have similar experiences when close to dying.)

To many people, Moody's findings and those of researchers who came after him proved that there is life after death. But not everyone agrees that NDEs prove life after death. Even Raymond Moody says his studies only relate to *near* death. Some say the NDE is not a mystical but a physiological experience. Some researchers have found similar experiences in users of certain types of drugs, and some theorize that the brain itself produces a chemical that under the stress of a deathlike experience causes an NDE hallucination.

Still others say the NDE is a psychological phenomenon caused by fear of death: Those who think they're dying may induce paradisiacal hallucination to reduce their fear. They recall the things they've been taught about the afterlife, and they create their own comforting scenario of what will happen to them when they die.

There is no empirical evidence for NDEs; they cannot be safely induced, after all. It is clear that many people experience *something* when they nearly die, but the exact nature of that experience is open to interpretation.

See also ASTRAL/ETHERIC BODY; OUT-OF-BODY EXPERIENCE.

NEBULAR HYPOTHESIS A theory of the formation of the Sun's planetary system advanced by Pierre Simon Laplace (1749–1827) in his *Exposition du Système du Monde* (1796). He theorized that gravity would condense a rotating disc of matter with most of its mass near the center into a massive central body and a number of much smaller orbiting planets. The theory gave a plausible explanation of why all the planets orbit in the same direction and why their orbits are in approximately the same plane.

In 1873, James Clerk Maxwell (1831–79) showed theoretically that radiation would exert a pressure on any surface on which it fell. This was confirmed experimentally in 1900. Sunlight, for example, is calculated to exert a pressure of about 2 pounds per square mile, so small as to be difficult to measure. But in the interior of stars and galaxies where the radiation intensity is much higher, radiation pressure is a significant factor. In 1900 two scientists in the University of Chicago showed that the Maxwell radiation pressure in a disc of the size required in the nebular hypothesis would counterbalance the gravitational forces and prevent condensation; they advanced an alternative theory (see CHAMBERLIN-MOULTON HYPOTHESIS). Sir James Jeans later showed that the radiation-pressure difficulty only held for astronomically small aggregations such as the solar system and that the nebular hypothesis could explain the condensation of stars from a galactic disc.

However, in 1949 the nebular hypothesis was given a new lease of life when first Carl von Weizsäcker in 1944 and Gerard Peter Kuiper in 1949 put forward theories of the formation of the solar system that incorporated a form of the nebular hypothesis. Weizsäcker postulated a flattened ring of gas, either the matter remaining after the Sun formed or produced by a near collision, that first formed cellular vortices and then condensed into the planets. His theory gave predictions that fitted the known planets' constituents, their total mass, and the radii of their orbits. Kuiper proposed a cloud with about 10 percent of the Sun's mass that first broke up into what he called protoplanets, which subsequently formed the planets and their satellites. His theory accounts well for the planets' distances from the Sun, the direction of rotation of both the planets and their satellites, and the closeness of the orbital planes; Neptune is still a problem, its orbit's angle to the ecliptic being too high.

More recently Orson FOWLER and Sir Fred HOYLE produced a nebular hypothesis requiring 1 percent of the Sun's mass to separate from the Sun in its early stages when it was spinning rapidly and then to form very small protoplanets—microprotoplanets—that finally condensed into the planets. In the process about one-tenth of the material, most of the lighter atoms, escaped into space. This theory explains well the differing chemical element composition of the planets and the meteorites.

All explanations of the formation of the solar system make highly improbable assumptions, and none are seen as satisfactory. The nebular hypothesis is still a candidate but has many problems as yet unsolved.

What does the nebular hypothesis have to do with pseudoscience? Only that it adopts the methods of orthodox science to find an explanation of the formation of the solar system by natural processes to set against the mythological accounts adopted by all societies from that of the ancient Egyptians to the present. It offers a scientific alternative to the pseudoscientific.

See also LAPLACIAN PHYSICS.

For further reading: John Charles Duncan, *Astronomy* (Harper, 1955); and Jay M. Pasachoff, *Astronomy,* 4th ed. (Saunders, 1995).

NEDERLANDSE VERENIGING VOOR PARAPSYCHOLOGIE Leading parapsychological organization in the Netherlands. The Nederlandse Vereniging Voor Parapsychologie (Dutch Society for Parapyschology—NVVP) was formed in 1960 as a result of an internal controversy in the STUDIEVERENIGING VOOR PSYCHICAL RESEARCH (SVPR).

In the early decades of the 20th century, the Netherlands had been the single most-receptive country in continental Europe to the investigation of paranormal phenomena. Parapsychological work made a relatively rapid recovery in the late 1940s, following its suppression by the Nazis. The revived parapsychological community was led by Wilhelm Heinrich Carl TENHAELF, who headed the SVPR. Tenhaelf had founded the *Tijdschrift voor Parapsychologie,* the primary Dutch parapsychological journal, and was gaining additional prestige with his successful research using psychic Gerard Croiset. However, Tenhaelf had a dictatorial leadership style that was not appreciated by many of his colleagues. In 1960, led by George Zorah, who was second in stature only to Tenhaelf himself, they withdrew to form the NVVP. Shortly thereafter, Tenhaelf's work was discredited.

NEOASTROLOGY An updated version of ASTROLOGY, claiming to put the subject on a scientific basis. Traditional astrology has been under attack for a long time. In Shakespeare's *Julius Caesar,* Cassius says:

> *Men at some time are masters of their fates:*
> *The fault, dear Brutus, is not in our stars,*
> *But in ourselves, that we are underlings.*

In recent years the main attacks have come from science.

In the face of these attacks, those who firmly believed in astrology turned to science and used the critics' own weapons to counterattack. François and Michel GAUQUELIN were the leading exponents of the scientific approach. They carried out a massive analysis, running into many tens of thousands of samples, of the recorded characteristics of individuals and their birth times and dates. They then compared these to the positions of the planets at these times and carried out a statistical analysis of the correlation. Conventional astrological readings of the subjects were also made. The statistical analyses showed a significant correlation between the position of Mars, for example, and the conventional "martian" characteristics, but no correlation between the planet's position and traditional astrological characteristics; in some cases, there was a significant negative correlation. On this basis, Gauquelin and the neoastrologists claimed there is a scientific case for a new style of astrology but no scientific validation of traditional astrology. Astrological symbolism appears to be statistically demonstrated, at least for planets previously observed as having some influence on personality—such as the Moon, Mars, Venus, Jupiter, and Saturn. But this claim needs careful qualification. Birth has to be natural for the association of planet position and personal characteristics to hold. The fetus, so it is said, knows when it is the right time to emerge into the world, and if it has its way, the neoastrological association holds good to a statistically significant extent. If there is any interference with the natural process—induced birth, artificial rupture of the membranes (amniotomy), forceps delivery, Caesarian section, or anything else that alters the time of birth from the natural—then the astrological bond is broken or substantially diminished.

Artificial satellites exploring the solar system in the 1960s showed that the planets trail disturbed magnetic fields behind them, called magnetic tails. In his book *Cosmic Influences on Human Behavior* (1974), Gauquelin suggests that "magnetic tails may perhaps provide a 'psychological' explanation for a large number of recent puzzling observations." He was proposing a possible scientific mechanism for how the planets could affect human beings. A very similar idea is advanced by Percy Seymour, astronomer and director of the William Day Planetarium at the University of Plymouth in England, in his book *Astrology: The Evidence of Science* (1988). John Anthony West, in *The Case for Astrology* (Viking Arcana, 1991), maintains that the impasse between the old and new astrologists and the skeptics is probably insoluble, for "it does not rest upon proof or disproof, the desire for objectivity, or even upon deliberate bad faith and stupidity on the part of astrology's opponents, though there is no shortage of either. It rests ultimately upon psychological factors, upon levels of understanding, and upon different—and diametrically opposed—ways of viewing the world."

Astrology has persisted for over four thousand years and is turned to today by millions in the less developed and industrialized worlds. There are more astrologers than priests in France, more astrologers than astronomers in the United States. When people feel out of control of their lives, they turn to anything that may help restore their confidence. Neoastrology is an attempt to buttress astrology against rising criticism, by making more modest claims than those made by conventional astrology and by demonstrating some scientific basis. In the mid-1970s, however, a statement entitled "Objections to Astrology" was published by 186 leading scientists in *The Humanist*. It stated: "Those who wish to believe in astrology should realize that there is no scientific foundation for its tenets," and added: "We must all face the world, and we must realize that our futures lie in ourselves, and not in our stars," thus echoing Shakespeare three hundred years earlier.

For further reading: R. B. Culver and P. A. Ianna, *Astrology: True or False? A Scientific Evaluation* (Prometheus, 1988); Michel Gauquelin, *The Truth about Astrology* (Blackwell, 1983); Percy Seymour, *Astrology: The Evidence of Science* (Lennard Publishing, 1988); and John Anthony West, *The Case for Astrology* (Viking Arcana, 1991).

NEOPAGANISM A modern form of Paganism, adapted to the beliefs and mores of the 20th century. The term "pagan" comes from an Ancient Roman word for a civilian or villager, "paganus" as opposed to "miles," a soldier. After the Roman emperor Constantine (274–338 C.E.) converted to Christianity, his soldiers came to the faith with him, leaving the civilians of the Empire to follow their diverse polytheistic religions. Thus, those of the Christian faith, from the third century onwards, labeled every practice and belief system that was not based on the worship of a single God, as was their own, "pagan." Those of the Jewish faith, and later, in the early seventh century, those of the Islamic faith, were spared the label that by then had come to have a pejorative connotation.

As in paganism of the past, the term "Neopaganism" covers a whole variety of belief systems. But today, probably because those in mainstream religions are more tolerant, those outside have managed to take what was once a derogative name, with all kinds of supposed negative implications like "unenlightened" or "followers of false gods," and turn their beliefs into a positive philosophy encompassing many NEW AGE MOVEMENT social, political, and economic concerns. Neopaganists have gone back nearly 2,000 years and adopted the original meaning of the word "paganus"—one of the people—and not a soldier who was then a representative of the Roman ruling hierarchy. Some Neopagan groups are against hierarchies in any form, but most are especially against hierarchies as used in the established churches, with their ecclesiastical orders of priests, bishops, cardinals, etcetera. The whole system, they feel, is elite and patriarchal, reeking of male power, that has done its best over the centuries to marginalize all ancient pagan beliefs. In neopaganism "he" and "she" are equal, not identical, and have equal standing in its ceremonies.

Most varieties of Neopaganism do, however, recognize the teachings of Jesus, who they claim was not divine but an ordinary person. They say His divinity was claimed later by the Church in order to give credence to its teachings, which at that time, were being distorted to create a scriptural orthodoxy in order to elevate early Christianity into a monarchistic religion with views of God as a king or overlord to whom homage has to be paid, and who has to be praised, worshiped and venerated as though he was a feared potentate to be appeased. In Neopaganism there is no belief in, or worship of, such a potentate. Neither is there any worship of Satan; that is a case of mistaken identity, the neopagan Horned God is a God of the woodlands who is actually the protector of life and the living. They see the Devil as a creation of Christian dualism that needed an evil as well as a good god. Neopagans in the United States feel that their government's policy of freedom of worship with no state religion gives small group worshipers like themselves a better chance of survival than similar groups in many countries in the world.

Neopagan groups have various codes of conduct that, they believe, are rooted in ancient tradition. Most revere the Earth and see Her as a Mother who gives life, and cares and feeds them. Therefore they feel the need to care for Her in return. Many rituals and festivals are attuned to seasonal cycles, venerating the wealth of the Earth's harvest. These types of issues have become attractive to many young people today because they reflect our current concerns for the ecology of the Earth. Furthermore, the equality of men and women in neopaganism has attracted many feminists who feel the need to believe in something more sympathetic to their sex than many of the assumptions of the male-centered monotheistic religions. Neopagan groups have a variety of pseudoscientific beliefs and practices including much magical lore, the raising of energy, holistic and psychic healing and WITCHCRAFT.

See also ASTROLOGY, DIVINATION, CRYSTAL GAZING, I CHING, MAGIC, NUMEROLOGY, REINCARNATION, and TAROT CARDS.

For further reading: Janet Farrar, Stewart Farrar, and Gavin Bone, *The Pagan Path* (Phoenix Publishing Inc., 1995).

NEPTUNISM The name given to a late 18th-, early 19th-century theory of geology that proclaimed the aqueous creation of all rocks; this theory was opposed to Plutonism or Vulcanism, which argued that granite and many other rocks were of igneous origin. The disagreement between the two schools of thought was one of the great geological arguments of the day. The name "Neptunism" was taken from the ancient Roman god of water and of the sea, Neptune; Pluto was the Greek god of Hades, the underworld, and Vulcan was an early Roman deity—a fire god associated with volcanoes.

The geologist who founded the Neptunist school was Abraham Gottlob Werner (1750–1817), a lecturer at Freiberg School of Mining. He taught that Earth was originally an irregular solid body covered completely with a heavy, saturated water solution. Over time, solids precipitated out of the water in regular succession. First came the primitive rocks such as granite and gneiss, then the transition rocks such as slates and some limestones, and finally the recent rocks such as sandstone and other types of limestone. As the water lost its chemical content, it gradually became the salty oceans that we know today. In this scenario volcanoes were local phenomena caused by burning coal seams, and there was no theory concerning Earth's inner heat.

The rock basalt became a specific issue between Werner's followers and other geologists, particularly the Vulcanists who had identified it as cooled-down volcanic lava. Werner explained the basalt structure of the Giant's Causeway on the northern coast of Ireland as being formed by a local volcano that had melted pre-existing basalt; then it had cooled down into the formation we see today. In his theory, there were other flaws of a very basic kind: Namely, he never explained where the surplus fluid that had covered Earth had gone, and although he allowed for erosion, he never allowed for the formation of new rocks.

In 1785, James Hutton, a Scottish geologist, published *Theory of the Earth* in which he proposed an alternative hypothesis to Werner's, giving great prominence to the interior heat of Earth—hence the name Plutonism, which was given to the theory. Hutton thought, quite wrongly, that Earth's heat would cause sediments in the ocean to rise up and become new continents, and in 1787, he thought that he had identified some sedimentary material that had been changed this way. He had actually found metamorphic rocks.

Although both Werner and Hutton did get things wrong, they were both practical men and early adherents to systematic observation in the new field of geology. During the industrial revolution, with the increase in prospecting for coal and ores, geology changed from an amateur study to become a professional occupation. But it was the differences in their overall theories of the developmental history of Earth that became important for the future of geological study, and it was Hutton's explanation that led him to be called the father of modern geology. Whereas Werner's ideas added up to an Earth that had one creation point and then followed a preordained path toward a universal desert, Hutton thought geological EVOLUTION had been a continuous process. This view he called uniformitarianism, and it stated that the natural processes now at work on and within Earth have been operating in the same general manner throughout the ages. It was from uniformitarianism that Sir Charles LYELL later developed the general principle of small changes taking place over long periods of time, the basic geological view that still holds good today.

For further reading: Stephen Jay Gould, *Time's Arrow, Time's Cycle: Myth and Metaphor in the Discovery of Geological Time* (Penguin, 1987).

NEURAL ORGANIZATION TECHNIQUE A treatment that claimed to correct various learning and neurological disorders. Neural Organization Technique was developed by Carl A. Ferreri, a New York practitioner of CHIROPRACTIC medicine and of APPLIED KINESIOLOGY. Ferreri suggested that such conditions as dyslexia, learning disability, minimal brain dysfunction, and related conditions that become noticeable during school years were caused by the misalignment of certain bones of the skull. He believed that the misalignment could be corrected by physically adjusting the bones.

Ferreri made his proposal for correcting these disorders in two books: *Dyslexia and Learning Disabilities*

Cured (1984) and *Breakthrough for Dyslexia and Learning Disabilities* (with R. B. Wainwright, 1985). Ferreri and Wainwright argued that as we breathe the bones of the skull move because of a relationship of connective tissue bands. He suggested, for example, that dyslexia is caused by a faulty motion of the sphenoid bone (the center skull bone) and one of the temporal bones. Other conditions are caused by the faulty movement of other specific bones. The technique advocated manual manipulation of the cranial bones, as well as the more familiar manipulation of the spine.

Ferreri's ideas entered a new phase in 1987 when the Del Norte Unified School District of northern California agreed to participate in a study of Neural Organization Technique. A psychologist associated with the school district had attended a seminar given by Ferreri in San Francisco in November 1985 and brought Ferreri's work to the attention of the district. In the seminar, Ferreri noted that his technique had not yet been subjected to a formal clinical study (though there had been several studies of applied kinesiology that had proven negative). Following a presentation by the psychiatrist, the school board accepted the idea of a study using children from their special-education program. Meanwhile, Ferreri's technique was encountering some legal problems. In Utah a court restrained a chiropractor from promoting the technique for use with learning disorders, and in Texas a chiropractor had had his license suspended for practicing the technique.

In the fall of 1987, 17 of the 48 parents whose children participated in the study filed formal complaint with the California Board of Chiropractic Examiners, claiming that their children had suffered both emotional and physical harm. A lawsuit was filed against Ferreri, the school district psychologist, the chiropractors who participated, the research consultant, and the school district. A jury awarded a half-million dollars in damages in 1992. The chiropractors settled for lesser amounts out of court.

For further reading: C. A. Ferreri and R. B. Wainwright, *Breakthrough for Dyslexia and Learning Disabilities* (Exposition Press, 1985).

NEUROLINGUISTIC PROGRAMMING (NLP) A theory and therapy for achieving greater health, emphasizing the mind–body connection. "Neuro" refers to the workings of the brain and to consistent, detectable patterns of thinking. "Linguistic" refers to verbal and nonverbal expressions of the brain's thinking patterns. "Programming" implies that these patterns are recognized and understood by the mind and that they can be altered. The NLP practitioner's goal is to uplift a person's state of health and well-being by helping to "reprogram" that person's beliefs about healing and about themselves. By "reading" automatic body changes such as skin color changes, muscle tension, eye movements, moisture on the lips and eyes, as well as other physiological responses, the NLP practitioner discerns how a client perceives and relates to issues of identity, personal beliefs, and life goals. NLP practitioners attempt to help clients replace false or negative perceptions with positive, life-affirming beliefs. As an adjunctive therapy and as a separate methodology in its own right, NLP has been used in cases of AIDS, cancer, allergies, arthritis, Parkinson's disease, and migraine headaches, as well as to enhance general well-being, self-esteem, and interpersonal skills.

NLP was developed in the early 1970s by John Grinder, professor of linguistics at the University of California at Santa Cruz, and Richard Bandler, a student of psychology and mathematics at Santa Cruz, in an effort to define the qualities of excellence in several accomplished individuals. Grinder and Bandler studied the thinking processes, language patterns, and behavioral patterns of Fritz Perls, father of Gestalt therapy; Virginia Satir, a renowned family therapist; Milton Erikson, M.D., notable hypnotherapist; and Gregory Bateson, a leading anthropologist and communication theorist. They concluded that many of the behavioral and psychological elements that enabled these individuals to excel were unconscious and intuitive and that the participants could not describe their own exceptional qualities.

Grinder and Bandler analyzed the speaking patterns, voice tones, word selection, gesticulations, postures, and eye movements of these individuals, finding numerous correlations between body language and speaking patterns. They then related this information to the internal thinking process of each participant. According to their findings, eye movements, posture, voice tone, word choice, and breathing changes reveal unconscious patterns affecting a person's emotional state. For people experiencing emotional difficulties or physical illness, Grinder and Bandler suggest that once these unconscious patterns are discovered, the client can be assisted in adopting new, healthy patterns of thinking that trigger positive immunological responses and guide the mind and body toward greater health and well-being.

NEW AGE MOVEMENT An umbrella term covering a variety of beliefs and practices, but all having the common denominator of stemming from a mystical paranormal mental set. The thinking that distinguishes this mind-set is the belief that there is no difference between material, physical reality and imaginary reality. Some go so far as to deny the existence of material reality completely, believing that psychic experience is all there is, so they claim to be able to invent reality with their own minds. These ideas are pseudoscientific.

The first to use the name "New Age" was Alice Bailey who wrote in the 1920s about her telepathic communications with a Tibetan spirit guide. But "New Age" is really a misnomer; most of Bailey's ideas were not new but derived from 19th-century ideas from two continents, all of which were in turn borrowed from many ancient worldwide traditions. From Europe she used the THEOSOPHY of Madame BLAVATSKY and the anthroposophy of Rudolph STEINER; from the United States, she drew on the transcendentalism of Ralph Waldo Emerson (1803–82) and Henry David Thoreau (1817–62). These 19th-century writers had one thing in common: They all renounced orthodox Protestantism while still holding on to the Protestant claim of a direct access to God, with only the self determining that relationship. Before the 16th century, Catholicism was the main Christian belief in the West, and the church in the form of popes and councils stood between humankind and God, interpreting the scriptures for the people rather than letting them read the Bible and decide for themselves.

New Age thinking today has been summed up by writer and psychologist Maureen O'Hara in her article "Science, Pseudoscience, and Mythmongering" (1989) as a system of beliefs focusing on the intuitive spiritual consciousness within human beings. This consciousness is thought by its adherents to be in the process of transcending and expanding, until the power of the mind will at some time in the future, perhaps for a few at the millennium, arrive at a point where it undergoes a paradigm shift away from all the old ways of knowing. These old ways are identified as being left-brained and rational, scientific, materialistic, competitive, and so on. New ways of thinking are said to be all notions of psychic phenomena, ESP, mediumship, REINCARNATION, karma, Taoism, Zen, animism, shamanism, SPIRITUALISM, and its updated version, CHANNELING. Rejected religious beliefs are Catholicism, Judaism, and some forms of fundamental Protestantism. Lifestyles are based on the counterculture of 1960s, with the use of LSD and other psychedelic drugs to attain the higher consciousness. Connected with this movement were Timothy LEARY (1920–96), Baba Ram Dass, various rock bands such as the Grateful Dead, and authors such as Carlos Castaneda and Allen Ginsberg.

Many of those who are writing today about New Age thinking employ the language of science and technology to explain their psychic ideas, employing, for example, the paradoxes of quantum mechanics and the Heisenberg "uncertainty principle" or the language of electronics as used by José Arguelles in his book *The Mayan Factor* (1987). But once more, there is nothing new in this ploy; one of the revered predecessors of the New Age movement, Emanuel Swedenborg (1677–1772), appropriated the scientific language of his day to explain his mystical theology, but he had a scientific training, and when he claimed to converse with angels on his astral journeys, he at least knew the geography of the solar system and what lay beyond. Many New Age writers today do not understand the scientific words and phrases that they employ.

Most New Age theories of health and disease start from the assumption that illness comes from a separation from the divine. The ill person has drifted away from the correct spiritual path, so sickness and sin are consequence and cause. We are told that we can eliminate illness from our lives by how we correctly apply our minds to the divine, and a daily diary of the state of one's spiritual health is strongly recommended. Those who practice spiritual healing use prayer, MEDITATION, and visualization to link themselves to divine or mystical healing forces that are then channeled to patients, activating their power to heal themselves. Some New Agers believe that all, even children, who are ill or had accidents have themselves brought on their pain and suffering by inheriting a bad karmatic debt. Believing in the Hindu and Buddhist doctrines of karma and reincarnation, they explain that a stricken child must have been evil in one of his or her previous lifetimes. Its karma must be balanced; treatment by modern medicine and surgery prevent the child earning spiritual points toward wiping out the debt. Is there anything new in this?

Skeptics in the scientific community, trying to explain the attraction of the New Age, point to the believers' upside-down thinking where, even in the title New Age, there is a slip back into an Old Age when people lived in a nonliterate culture. In the New Age paradigm, individuals can invent their own science, their own religion, and their own medicine. There is no need to study and no need to find proofs for statements. Solutions are anything you imagine them to be, dreams come true, delusion is reality, anything is possible, and the foundations of Newtonian physics are up for grabs.

For further reading: Robert Basil, *Not Necessarily the New Age: Critical Essays* (Prometheus, 1989).

NEWMAN ENERGY MACHINE The invention of Joe W. Newman who designed what he claimed was a free-energy machine—a massive direct-current motor driven by a large set of batteries. The idea of a "free" source of energy is a chimera that has attracted many inventors for obvious reasons. Newman said that if his idea were implemented, "there would be no more pollution, no more Ethiopias (for example, mass starvations). Deserts will become oases. People will work only one hour a week and have all the material goods they need," and so on. There are echoes here of many similar claims. It was initially claimed that nuclear power would be so cheap that it would not be worth metering; COLD FUSION had similar attractions. However, neither Newman's nor any other invention has yet given us free energy—nor are they likely to.

See also PERPETUAL MOTION MACHINE; ORFFYREUS'S WHEEL.

NEWPORT TOWER See MYSTERY HILL.

NEWTON, SIR ISAAC (1642–1727) Mathematician and physicist whose ground-breaking work revolutionized the study of the physical world. Born at Woolsthorpe, England, Newton graduated from Trinity College, Cambridge, in 1665. That autumn, Newton retired to Woolsthorpe for some 18 months to avoid the plague, during which time he formulated the basic features of his greatest works in mathematics, mechanics, astronomy, and optics, captured in his theories of "fluxions" (calculus), motion and gravitation, and the composition of light.

In 1667 Newton returned to Cambridge and was appointed Lucasian professor of mathematics. It was not until 1669 that he communicated any of his theories born of the Woolsthorpe period. Throughout his career, Newton was notoriously possessive of his ideas, reluctant to publish, and inordinately sensitive to criticism. In response to criticism, he often became resentful and vindictive, threatening to withdraw from science entirely or using the power of his reputation or office to denigrate others. Be that as it may, when some of his works in mathematics began to circulate in 1669, his genius was quickly recognized.

In 1687, Newton's *Philosophiae Naturalis Principia Mathematica* was published, meeting instant and universal acclaim. The *Principia,* presented in three books, constitutes the founding of "mechanical explanation," though numerous successors would correct and refine Newton's original theories. In 1689, he was elected to Parliament as a representative of the university, serving for about one year. He was appointed warden of the mint in 1696, became its master in 1699, and was elected president of the Royal Society in 1703, holding both positions until his death. In addition to his monumental works in science and mathematics, Newton spent considerable time and effort throughout his life exploring history, alchemy, and theology. Quarrels and controversy marred his later years, most notably in the dispute as to whether Newton or Gottfried Leibnitz had been first in formulating calculus. Though Leibnitz was accused of stealing Newton's ideas, it is now generally accepted that the two arrived at calculus independently.

By the second quarter of the 18th century, "Newtonian mechanics" had triumphed over René Descarte's physical theories and had begun to influence areas outside the field of science. Mechanics took a central place in human knowledge, regarded as the ultimate explanation for phenomena of any kind and extended to all manner of human endeavor. The modern pre-eminence of scientific thought reflects the Newtonian image of the world as a perfect machine governed by simple mathematical laws, and of Newton himself as the perfect scientist.

NIGHTMARES Terrifying DREAMS. Throughout history, many superstitions have attached themselves to nightmares, which have been regarded as punishments, previews of frightful future events, or the effects of visitations by DEMONS or other unwholesome characters. During the Middle Ages, many Europeans believed that bad dreams were caused by a demon squatting on a sleeper's chest, creating a sense of suffocation—thus "nightmares," from the German *mara,* a word meaning "crusher." The feeling of having a crushing weight upon the chest is common among victims of nightmares, as is the paralysis that persists briefly following awakening.

The precise nature of the beings who menace people in their sleep has varied over time, but their purposes are nearly always malign. Sometimes the nocturnal attackers have been supposed to be VAMPIRES, subsisting on the victim's blood or life force. The Roman INCUBUS (from the Latin term for nightmare, *incubo*) was a demon in male form that lay with women and occasionally fathered children by them; its ostensibly female counterpart, the succubus, lay with men.

Scientists now know that nightmares may have either psychological causes or physical causes, such as indigestion or poor circulation or breathing problems.

NOETICS A contemporary term, derived from the Greek *nous* ("mind"), for the science of consciousness and its alterations. The term was first used in this manner by Madame BLAVATSKY, co-founder of the Theosophical Society, in an article in a theosophical magazine in 1890. She compared "noetic" to "manasic" (from the Sanskrit *manas* or "mind") in her attempt to contrast the materialistic psychology of her day with both ancient Hindu and occult concepts of human beings. She suggested that there was a higher character, a divine consciousness, of the Mind than the individual human ego.

Unused for a century, the term "noetics" was picked up in the 1960s by Charles Musés. Musés was looking for a word to describe the new focus on the nature of consciousness by contemporary humanistic and transpersonal psychologists. As defined by Musés, "noetics" referred to the study of the nature, alteration, and potentials of consciousness. Musés' definition has parapsychological implications, especially for results obtained with instruments such as the electroencephalograph. It also has value because of public interest in states of consciousness identified with the production of psychic phenomena.

In 1973, former astronaut Edgar Mitchell founded the Institute of Noetic Sciences to give focus to his own interest in parapsychological research. Mitchell had become interested in psychic phenomena in 1967 as part of a general crisis of faith that led him to search for meaning in mysticism and the psychic. In 1971 he conducted an ESP test while on a mission to the moon—the trip during which he became the sixth person to walk on the moon. Mitchell later expressed the goal of transforming humans into "psychonauts," or travelers in inner space. In his opinion, explorations of inner life bring an awareness of spirituality and a sense of unity with one another and the universe. According to him, PARAPSYCHOLOGY is a tool to assist in the transformation of world consciousness.

Although it was originally established to study the relationship of mind and body and the nature of human consciousness, the Institute of Noetic Sciences has ranged over the traditional fields of parapsychology. In part because of this, noetics is often used simply as another name for parapsychology.

For further reading: Charles Arthur Musés and Arthur M. Young, eds., *Consciousness and Reality: The Human Pivot Point* (Dutton, 1972); and Edgar D. Mitchell, ed., *Psychic Explorations: A Challenge for Science* (G. P. Putnam's Sons, 1974).

NOSTRADAMUS (1503–1566) Sixteenth-century French physician and mystic known today for his cryptic poems that purported to foretell many future events. Born in to a well-educated Jewish family who later converted to Roman Catholicism, Nostradamus had a wide and deep education in languages, science, religion, and ASTROLOGY. A gifted healer, he saved many patients from the ravages of the great plague, and he also developed his own remedies, which were said to be more benign than the bleeding and blistering techniques that were popular during his day.

After he lost his own wife and two children to the plague, he drifted around Europe for a time and became more involved in the occult. He married a wealthy widow and began auguring, summoning mystic visions in a bowl of water. Astrologer to French kings Henry II and Charles IX, he became known for the accuracy of his prophecies. Around 1550, he began to record his prophetic visions in a great series of poems he called "centuries"—each century contained 100 four-lined verses or quatrains. Because of the era's ambivalent attitude toward the occult and its virulence toward anything that could be considered heretical, he couched his prophecies in mystery, mingling a variety of languages and ambiguous images. Most of these poems were published during his lifetime in two editions of *Les Prophéties de M. Michel Nostradamus*. He composed 10 centuries (the seventh had only 42 quatrains, for some unknown reason) and was planning two more at the time of his death. He told one of his sons he had made forecasts up to the year 3797.

Since Nostradamus's death, scholars and other interested people have spent very many hours trying to decipher the poems and to determine whether he did indeed predict future events. Many proponents believe that he was the greatest prophet of the modern era, and they point to numerous successful predictions. They say he accurately predicted the deaths of kings, troubles of the Catholic Church, the rise and fall of Hitler, the troubles in the Middle East, and many other events that have occurred during the four centuries since his death. Some also point out that this successful prophet appears to have predicted the end of the world in the year 3797. (Others say he did not predict the end of the world, but a transition to another kind of world or civilization.) More skeptical analysts say that, like many prophecies, Nostradamus's are so ambiguous that they can be interpreted to mean anything the interpreter wants them to mean—particularly in light of the purposeful obfuscation that Nostradamus used to protect himself from persecution—or from challenge.

See also DIVINATION.

NOSTRUMS Quack medicines, also referred to as patent or proprietary medicines. Nostrums emerged in the modern world of commerce from the folk remedies of the past. As early as 1692, a Boston newspaper advertised a product called "Aqua anti torminales" which was supposed to cure the "Griping of the Guts and the wind Cholick" and prevent the "Dry Belly Ache." Early in the next century, ads appeared for a product called "Tuscarora Rice" that was supposed to cure tuberculosis (then called consumption). In 1733 the *New York Weekly Journal* published an advertisement for "Dr. Bateman's Pectoral Drops," a medicine which had a patent from England's King George.

It was from the issuance of such royal patents for medical preparations that the name "patent medicines" began to be applied to substances. After the opening of the U.S. Patent Office, a few medicinal products were patented, but it was quickly discovered that patents did not protect the most valuable part of the product. Patents expired after 17 years, and when the patent expired, anyone could make a competing product. The issue was clearly demonstrated in the case of "Pitcher's Castoria." Immediately after the Pitcher's patent expired, a number of competing castoria clones appeared, and "castoria" became a generic name in medicine. Most manufacturers moved to gain trademarks over the names and slogans associated with their product and copyrights over the advertising text. Their products are more properly termed "proprietary medicines," though the name "patent medicines" has remained much more popular.

The rise of patent medicines coincided with the spread of literacy because they were dependent on newspaper advertising. The patent medicines in turn had a noticeable effect on the emergence of newspapers as a popular medium because they were the first major nonlocal advertisers. They also bought large blocks of newsprint to reprint testimonials from people praising the effectiveness of their product.

The first generation of nostrum peddlers can be divided roughly into three types. First, natural healers offered herbal medicines, reportedly drawn entirely from natural extracts. Their sale depended on public faith in folk wisdom. In popular imagination, Native Americans were believed to possess a storehouse of knowledge about various healing plants and herbs. A second group drew on whatever healing substances they observed, including herbs and chemicals. A third group based their remedies on mechanical devices. Some of the more popular of these used static electricity in different ways. Body trusses were also widely marketed.

Among the earliest and most successful patent-medicine manufacturer and salesman was Dr. Elisha PERKINS, a Connecticut physician who in 1796 received a patent to manufacture "metalic tractors." His colleagues had little faith in his work. In the following year, they kicked him out of the Connecticut Medical Association, labeling him a quack in spite of his successful practice as a surgeon. Perkins's tractors used a mild electrical charge and became a fad over the next generation. His son opened an office in London and spread the practice through England. Perkins later died of yellow fever—the same disease he had earlier supposedly cured.

Among the nostrums available in the early 19th century were a set collectively termed "opodeldocs." The term originated with a Dr. Steers, who advertised an all-purpose medicine called "Dr. Steers Chemical Opodeldoc." The term was satirized in a poem by Robert Browning and the "old Opodeldoc" became a standard character in dramas, providing the comic relief. By the 1820s there was a wide range of all-purpose nostrums, as well as numerous products for specific ailments.

Among the most successful of the medicine manufacturers was British-born Thomas W. Dyott, who settled in Philadelphia in the early 19th century. He had been an apothecary's apprentice in London, and as soon as he had saved enough money, he opened a drugstore in his new home. In the first decade of the 19th century, he developed a line of "family remedies," which he began to market throughout the Northeast. The remedies covered most of the major problems afflicting the population, ranging from pains accompanying the menstrual cycle to the then-dreaded and incurable tuberculosis. Dyott's success led to his buying his own glass factory for the production of medicine bottles and his building a model village for his employees, complete with a library, singing school, rules against liquor, and an ordered lifestyle. The successful business only came to an end when the industrious Dyott went into the banking business as a sideline. His bank failed in 1836 and he was tried for fraud and sentenced to prison. Although he was later pardoned, his release did not occur in time to save his medicine-manufacturing operation.

The decades following the Civil War introduced a new era in the history of patent medicines. The growing population, the opening of the West, and the rise of newspapers provided ever-expanding opportunities for drug merchants. The leaders in the country's unregulated marketplace were the SARSAPARILLAS, herbals based upon the extract of the vine *Smilax* and a few related aromatic roots. Of the sarsaparillas, Ayer's Extract of Sarsaparilla led the market. James Cook Ayer was the first to understand the potential of the rural Western market, where a scattered population had to rely on family members for medical assistance. Ayer's success was soon followed by the discovery of the healing properties of celery. In 1872 Wells & Richardson, a company founded by a small group of Civil War veterans, began the manufacture of a nerve tonic called Paine's Celery Compound. The success of Paine's led to numerous similar medicines, all exploiting the reputed, if vaguely described, virtues of CELERY. Joining the racks of successful medicines was the famous elixir for female disorders developed by Lydia E. PINKHAM.

The widespread and unregulated use of nostrums began to be attacked effectively late in the 19th century. The American Medical Association, founded in 1847, made public health a concern and set itself against harmful and ineffective cures. Their effort was aided by the work of Harvey W. Wiley (1844–1930), an official for the Department of Agriculture, who in 1883 began a campaign to force manufacturers of products to label their ingredients properly. The watershed in the battle with patent medicine did not come until the beginning of the 20th century. In 1905, *Collier's Magazine* became the instrument of the first systematic crusade against patent medicines. Reporter Samuel Hopkins ADAMS attacked what he called a "Great American Fraud" perpetrated by skillful advertising. He called on newspapers and magazines to stop accepting advertising from manufacturers of patent medicines.

Adams was contemptuous of alcohol content, the major active ingredient in the most successful nostrums. He blamed the nostrums for causing many people to sink into alcoholism and drug addiction. He also attacked the practice of selling medicines for five or more times the cost of production. He challenged people, including clergymen and temperance leaders, for endorsing such products and called on the government to stop the sale of liquor disguised as medicine. Adams was particularly harsh on those who offered a cure of tuberculosis. Some of the medicines, which included opium, chloroform, and prussic acid, not only did no good but actually contributed to the ill health of the user.

Collier's 10-part series took everyone involved in the conspiracy to sell patent medicines to task. The Pure Food and Drug Act had largely been tabled due to the efforts of the manufacturers of patent medicines and food processors. Although the AMA had earlier lobbied to push the act through Congress, many historians attribute the bill's passage to the effectiveness of Adams's writing.

In the early 20th century, Dr. Arthur J. Cramp built a coalition between the government and the Better Business Bureau, which effectively destroyed patent medicines as a significant part of pharmaceutical stock. Cramp is best remembered as the author of three very detailed volumes entitled *Nostrums and Quackery*, which discussed in detail the many products examined and branded as worthless by the AMA. Cramp's efforts were bolstered by the activities of Morris FISHBEIN, editor of the *Journal* of the AMA who authored a number of articles and one important book, *Fads and Quackery in Healing* (1932). Cramp's efforts were somewhat blunted by Prohibition (1919–33), when those nostrums with a high alcohol content became very popular while alcoholic beverages were illegal.

As medicine improved through the 20th century, so did standards for medical products. Fishbein lamented the problem of faddish legislation passing state legislatures, while at the same time he opposed too much federal intervention in dictating medical practice. However, a major step in dealing with nostrums came during the late 1930s. The AMA merged its interests with those of the newly formed Consumers Union (founded in 1936) and fought for the passage of the Food, Drug, and Cosmetic Act of 1938. Then in 1940 the Food and Drug Administration was established as an enforcement arm to implement the provisions of the act. The role of the FDA was strengthened in the late 1960s as consumerism became a popular social movement. The FDA has continually had to fight powerful manufacturing interests opposed to the continued expansion of regulations. A few nostrums, having passed some basic tests determining the limits of their usefulness, remain on the market even at the end of the 20th century.

For further reading: Stewart H. Holbrook, *The Golden Age of Quackery* (Macmillan Company, 1959).

NOTED WITNESSES FOR PSYCHIC OCCURRENCES Important compilation of anecdotal accounts of paranormal events collected in the 1920s. The volume was compiled by Walter Franklin Prince during his years as research officer for the BOSTON SOCIETY FOR PSYCHICAL RESEARCH, a rival organization to the AMERICAN SOCIETY FOR PSYCHICAL RESEARCH (ASPR). This data collection was designed to counter the idea that psychic occurrences were confined to particular groups, especially the uneducated, the religious, or members of certain ethnic groups.

Beginning with the CENSUS ON HALLUCINATIONS in the 1880s, various surveys showed that a significant percentage of the most educated populations of modern urban society reported paranormal experiences. In *Noted Witnesses for Psychic Phenomena,* Prince countered the idea that reports of paranormal occurrences were usually offered by people whose credibility and trustworthiness could not be verified.

Prince brought together 170 scientists, lawyers, doctors, military officers, politicians, diplomats, religious figures, and others from different walks of life. Among the people whose stories were told in the volume are Elizabeth Barrett Browning, Luther Burbank, William Gladstone, Oliver Wendell Holmes, Rudyard Kipling, and Mark Twain. While not proof of psychic phenomena, *Noted Witnesses* was effective in driving home the point that having psychic experiences did not make one crazy or dysfunctional.

For further reading: Walter Franklin Prince, *Noted Witnesses for Psychical Occurrences* (University Books, 1963).

N-RAYS A new type of ray (following soon after the discovery of X rays, alpha, beta, and gamma rays), which René Blondlot, an eminent French physicist at the University of Nancy, believed he had discovered in 1903. This was the year in which Henri Becquerel and Pierre and Marie Curie were jointly awarded a Nobel Prize for the discovery and investigation of radioactivity. Wilhelm Röntgen had recently discovered X rays. Physicists across the world were alert to the possibility of undiscovered rays. At the same time the instrumentation for discovery and analysis of such phenomena was very crude and this made such work difficult and liable to error. Blondlot had been investigating the nature of X rays—particles or electromagnetic waves—when he found another type of ray and began to sort out its properties. N-rays, as he called them—*N* for *Nancy*—were emitted by X ray generators, some materials such as tempered steel, and by Nernst glowers. They passed through wood, paper, mica, quartz, and thin sheets of some metals and were absorbed by water and rock salt. They could be bent by aluminum and quartz prisms and dispersed into a wide spectrum. Another Nancy scientist, medical physicist Augustin Charpentier, showed that contracting muscles, active nerves, and the central nervous system all emitted N-rays. Other physicists, excited by the appearance of a new field of research, joined in the research. Some reported success, and soon a series of papers confirming and extending Blondlot's discovery appeared. Others failed to find any evidence of N-rays, some attributing their lack of success to inadequate technique, some questioning whether the rays existed. Most of the successful experimenters were in and around Nancy and had the benefit of Blondlot's advice and guidance. Another group in Paris, led by Jean Becquerel and A. Broca, also turned to Blondlot for help. All those based in countries other than France were unsuccessful; some were skeptical and others felt that they lacked the necessary expertise. In 1904 the French Academy of Sciences honored Blondlot for the whole of his works, with a special mention of his new ray at the end of the citation.

When R. W. Wood, a physics professor at Johns Hopkins and an authority on physical optics, failed to reproduce Blondlot's observations, he visited Nancy to check out Blondlot's work. Blondlot's detectors were either spark gaps or phosphorescent strips. An increase in brightness detected either by eye or on a photographic plate indicated the presence of N-rays. Wood could not detect the claimed effects. He then removed the aluminum prism in one experiment and replaced the N-ray emitter by a nonemitter in another, in both cases undetected by Blondlot and his coworkers. In neither case were the Nancy scientists aware of any change. Wood believed that Blondlot and other N-ray enthusiasts had deluded themselves and published his conclusions in *Nature,* September 1904. Despite this, N-rays still had a short vogue among some, but by no means all, French physicists.

Eventually the French journal *Revue Scientifique* proposed a test to Blondlot that would settle the matter: Let him perform his experiments using two superficially identical boxes as N-ray sources, one of which contained an alleged N-ray emitter and one that did not. After some delay Blondlot in 1906 declined to participate. Blondlot died in 1930 and with him the last believer in the reality of N-rays.

For further reading: R. G. A. Dolby, *Uncertain Knowledge* (Cambridge University Press, 1996).

NUMEROLOGY The belief in the magical power of numbers. This belief is probably universal in all sophisticated cultures. In the West it is based on the Pythagorean idea that all things can be ultimately reduced to a relationship with numbers. Pythagoras (sixth century B.C.E.), a mystic, regarded mathematics as a spiritual discipline leading to the discovery of abstract general truths. Numerology was used in China in the first century, and Hindu and Buddhist teachings also embody number patterning and number symbolism. Jewish students of the KABBALAH have their own numerological system *gematria.* In the Middle Ages music, geometry, and astronomy were connected by numbers. People's lives were thought to be the microcosm of the greater macrocosm of the universe.

Today numerology is prominent in many forms of fortune-telling and character reading (see DIVINATION, OCCULTISM, and MAGIC). A frequently used system allocates numbers to letters of the English alphabet. But this leads to difficulties because in ancient numerology there was no English language, so practitioners have to correlate the English letters to a Hebrew alphabet, with some help from the Greek, thus:

1	2	3	4	5	6	7	8
A	B	C	D	E	U	O	F
I	K	G	M	H	V	Z	P
Q	R	L	T	N	W		
J		S			X		

The numerologist then inserts the letters of a person's name and the corresponding numbers are added together. If the total comes to double digits, they are then added together to yield a single-digit result. So, for example, if the first total comes to 13, the 1 and 3 are added together to yield a final total of 4. This first simple total gives what is known as the Key Number. Next the consonants' numbers only are added together to give the Heart Number, and finally the vowels' numbers are added to give the Personality Number. These three numbers reveal different aspects of the named person: The Key Number describes the whole person, the Heart Number the inner personality, and the Personality Number the outward impression. In modern psychological terms, the last two might be the introvert and extrovert aspects of the individual. Furthermore, each of the numbers is held to have a range of associated characteristics. For example, the Key Number 2 denotes what were thought to be the essentially feminine qualities: sensitivity, tact, kindness, calmness, etc., indicating a follower rather than a leader. But a person with a Key Number 2 can turn up a Heart or Personality Number indicating other qualities thought not to be feminine, like that of leadership.

There is a further complication. People are often known by more than one name: the one recorded on their birth certificate; a family name; a name used by friends; or a professional name. These different names usually yield a quite different set of Key, Heart, and Personality Numbers, with different resulting analyses. So the British biologist Richard Alan Baker does not equate with Sandy Baker, and Norma Jean Baker (no relation) is not the same numerological character as Marilyn Monroe. In short, the numerologist is left with a wide range of discretion as to which interpretation to use for a given client and, as so often with divination, much depends on the numerologist's ability to make maximum use of clues he or she picks up about the client—from dress, posture, behavior, accent, preliminary conversation, and so on.

For further reading: Francis X. King, *The Encyclopedia of Fortune-Telling* (Hamlyn, 1988).

O

OCCAM'S RAZOR Philosophical rule first put forth by the medieval English scholar William of Occam or Ockham (c. 1285–1349). "Occam's Razor," sometimes called the "law of parsimony," states that explanations for a phenomenon should be simple and should not be multiplied without a very good reason: "Entities must not needlessly be multiplied." In other words, the simplest explanation is to be preferred over more complex explanations. When faced with two hypotheses that explain the observations equally well, choose the simpler. Observers should try to explain unusual happenings as the result of known phenomena, rather than assuming that they occur from new or exotic causes.

Occam's Razor is a useful tool for philosophical thinking because it places logical limits on the possible number of explanations for any given happening. One example might be the space theories of Immanuel VELIKOVSKY: they offer complex ideas that might be true, but only if a very complex chain of events had taken place. Occam's Razor suggests that a simpler explanation would be more likely.

Occam himself was not a scientist. He was a monk of the Franciscan Order and a lecturer at Oxford University from 1309 to 1319. Occam believed that the theories of the Scholastic thinkers, who believed that the Christian religion could be understood through human reason, were wrong. He maintained that religion was mostly a matter of faith, not reason. In part because of these views Occam was tried for heresy in 1324 by Pope John XXII and fled to Germany to escape persecution. His "razor" was a weapon he used against the scholastic religious scholars he debated: it pared away excess or confusing complications, leaving only the simplest form of an argument. It was only after the beginning of the Scientific Revolution in the 17th century that Occam's Razor began to be applied to scientific theories.

However, Occam's Razor is only a tool. It cannot provide answers itself; it can only help a thinker choose between possibilities. It can lead to false conclusions, especially in cases where some phenomena were not observed or were observed incorrectly—as, for instance, among the 19th-century scientists trying to link brain size with human intelligence. False conclusions can also come from the expectations of the observer: the prehistoric remains of PILTDOWN MAN were accepted as genuine for decades because they matched the conception of human evolution at the time they were discovered.

See also INTELLIGENCE TESTS.

OCCULT CHEMISTRY A revised version of chemistry written in the early 20th century by Annie BESANT and Charles Webster Leadbeater (1854–1934) under the guidance of the theosophists' spirit masters. The chemistry revealed in this manner ranged from the structure of atoms and molecules to the chemical properties of materials. It is said that some of the writing was done while Leadbeater was sitting on a bench in Finchley Road, London, England. On other occasions, when the material under spiritual research was not on hand, Leadbeater had to make astral visits to glass cases in museums where they could be found. It is not clear why the authors felt it necessary to rewrite the subject except insofar as it formed part of a larger exercise. Besant was, at this stage in her life, convinced of the reality and authority of the theosophists' spirit guides and strongly influenced by Leadbeater. It is doubtful whether Leadbeater shared Besant's convictions: He was probably making the most of the status that his role as the conduit to the spirit masters gave him. The masters dictated to Leadbeater, he told Besant, and between them they recorded the truths so revealed. In this way, they rewrote texts on geology, the history of the world, the continents of ATLANTIS AND LEMURIA, the history of Christianity (in which Christ was born in 105 C.E.), and chemistry. Seen as part of this larger exercise, it is not surprising that this version of chemistry—never accepted by professional chemists or outside the small circle of the theosophists and their adherents—has not withstood the test of time.

OCCULTISM Any doctrine, principle, or practice associated with unknown forces or spirits, whose existence is cut off from most of humanity and available only to an initiated few. The word occultism comes from the Latin *occultus,* meaning hidden, concealed, or cut off from view by interposing some other body between the eye and the object. In this sense astronomers talk about the sun, the moon, or a planet "occulting" each other. As well as being used in the sense of objects not being readily observed by the eye, the term has also been used subjectively to mean not being readily understood, therefore mysterious.

Devotees study very early esoteric writings like the Jewish KABBALAH and the Chinese *I CHING* (see also VIRGIL) in the belief that ancient people had a facility to access mystical secrets, a facility that is believed to have been lost to modern people in the process of acquiring technical skills. Occult practices can be connected to any section of the hidden supernatural world and include for example: MAGIC, ALCHEMY, ASTROLOGY, DIVINATION, THEOSOPHY, Satanism, and WITCHCRAFT. None of these practices or rituals fall readily within the province of the major religions in the West. Christianity in particular speaks out against the dangers of meddling in the occult.

There are, of course, many things still unknown and in this sense hidden from orthodox science, but none of these are in principle unknowable. Occultism assumes the existence of things beyond the reach of natural science, known only by those who have the secrets of absolute knowledge. The intellectual history of the Western world, since the seventeenth century, has rejected such assumptions along with notions of any other way of knowing about the world than by empiricism, that is experience through the senses. Scientists believe such other forms of knowledge are supernatural, and occultism is therefore not seen as science but as pseudoscience.

For further reading, Patrick Grimm, ed., *Philosophy of Science and the Occult,* 2d ed. (New York Press, 1990).

OD A subtle radiation that, according to 19th-century scientist Baron Karl von Reichenbach (1788–1869), flows from every substance in the universe, especially stars, planets, crystals, magnets, and the human body. Reichenbach derived the term from Odin, the Norse deity. He postulated the existence of the odic force based on the writings of the mesmerists and the reports of sensitives who, Reichenbach believed, could feel and, on occasion, see the force.

Reichenbach began his study of the od in the 1830s. By that time, he had already developed a reputation as a brilliant scientist. He had spent two decades in industrial chemistry; had discovered paraffin and a variety of coal-tar products, including creosote, eupion, and pitch; and had also published a series of papers on meteorites, which were little understood by scientists of the day.

To pursue his study, Reichenbach gathered a set of people who he believed were sensitive to the od. He distinguished these people by the reactions they had around other people and by their sense of polarity. They also were sensitive to certain colors, foods, metals, and mirrors.

Reichenbach ran his special group through a series of tests. The group demonstrated to his satisfaction the ability to see auras around human beings and emanations from crystals and metals in total darkness. Later Reichenbach applied the concept of od to psychokinetic feats such as table-tipping. Lacking the variety of equipment used by scientists today, he attempted to measure the force by the only means available to him—generally, accepting the reports of sensitives, who were the only recorders of direct contact with the force.

Reichenbach's papers on his studies were widely circulated in the 1840s—English translations of his collected papers appeared in 1850 and 1851; they were not well received. They challenged the existing mechanized model of contemporary physics and chemistry. In addition, Reichenbach's colleagues tended to dismiss his work, claiming that it was spoiled by bad methodology and that his results were due to the power of suggestion. They equated his work with MESMERISM, even though Reichenbach made a large attempt to separate his ideas from those of the animal magnetists.

Throughout the 20th century, various people have tried to discover, define, and measure a subtle force in living organisms, claiming Reichenbach as their inspiration. The SOCIETY FOR PSYCHICAL RESEARCH, for instance, briefly attempted to revive Reichenbach's theories, but the organization had little success in reproducing his reported results. To date, the existence of such a force remains a matter of controversy, though some of the experimental data on psychokinesis is certainly suggestive of its existence.

For further reading: Karl von Reichenbach, *The Odic Force: Letters on Od and Magnetism* (University Books, 1968); and Karl von Reichenbach, *Researches on Magnetism, Electricity, Heat, Light, Crystallization and Chemical Attraction in Relation to the Vital Force* (University Books, 1974).

OGOPOGO LAKE MONSTER allegedly found in Lake Okanagan in British Columbia and related by name to several other Canadian lake monsters, including Manipogo of Lake Winnipeg and Igopogo of Lake Simco. Early Native legends told of a demon that possessed a human who then murdered an old man named O-Kan-He-Kan. In honor of the murdered man, a beautiful lake was named for him, later giving its name to the entire valley. The gods changed the murderer into a lake serpent, which came to be called Naitaka, or "lake demon." This monster, condemned to suffer for his crime, also made others suffer. He devoured fishing people and others who came too near the lake. For many generations, Indian people avoided the lake or brought gifts for Naitaka when travel on the lake was unavoidable. Indian tales also inspired caution in European settlers of the 19th century. Settlers' stories filtered back to England, where Ogopogo became a music-hall mainstay and acquired its present name.

Ogopogo has been described as dark-colored, long-necked, narrow, and humpbacked—like many other lake monsters. It is said to be somewhere between 12 and 21 meters (40–70 feet) long. In various accounts it has also been described as having horns, a mane, a beard, and a forked tail.

In 1914 a blubbery, partially decomposed animal corpse washed up on the shores of Lake Okanagan. Many people thought it was Ogopogo even though it didn't fit the most common descriptions of the legendary beast. An amateur naturalist identified the body as that of a manatee, but that ocean animal would be no more likely to be found in an inland lake than would a sea serpent.

In the past century, many sightings of Ogopogo have been reported, many of them by people considered reliable witnesses. One 1974 report claimed the monster looked something like a whale, only long and snaky. Other witnesses have also provided whalelike details. Some cryptozoologists suggest that the animal could be a Basilosaurus or zeuglodon, a possible leftover from the age of the dinosaurs. However, there is no evidence beyond the anecdotal to support this idea.

In 1968 and again in 1980, vacationers took movie footage of what they believed to be the lake monster. Unfortunately, neither film is sharp enough to positively identify its subject. Both show dark shapes moving on the water. They could be interpreted as sea serpents, but alternative explanations include bobbing logs and strange wave and shadow formations. Another film, taken in 1989, shows a shape of a distinctly smaller size than the first two films. All of these films provide fodder for those who believe in lake monsters, but are not convincing to skeptics.

See also CHAMP; LOCH NESS MONSTER; MORAG.

For further reading: Mary Moon, *Ogopogo, The Okanagan Mystery* (J. J. Douglas, 1977).

OMEGA POINT Supposedly the point in time when the evolutionary development of human culture theoretically reaches such an advanced stage of humanization that personal consciousness will merge with God. The idea was put forward by 20th-century French Jesuit philosopher and paleoanthropologist Pierre TEILHARD DE CHARDIN in *The Phenomenon of Man,* a book which, because of the disapproval of the Roman Catholic Church, had to be published posthumously. His theory blended science and Christianity so closely that he saw human development resembling "nothing so much as a way of the Cross." Teilhard coined words in order to express his ideas: "cosmogenesis" was the development of the world in which humankind was central and especially favored; "noogenesis" was the growth of the human mind, our emergence from the apes; "hominisation" and "ultra-hominisation" were the stages of our humanization toward a hyperpersonal consciousness and from there to a Godlike state.

Although in recent years Teilhard's ideas remained purely of academic interest and had largely dropped from public consideration, there has been of late a resurgence of interest in connection with reports of NEAR-DEATH EXPERIENCES and OUT-OF-BODY EXPERIENCES, and also UFO abductions. University of Connecticut psychology professor Kenneth Ring has been researching the psychology of people who report these experiences, has compared them with those who are merely interested, has found them to be of a distinct type, and suggests that they may be people pushing toward the omega point. Furthermore, he speculates that they may be giving a message to the world that we must change our lifestyle if we are to move toward the Omega Nirvana. From a Christian point of view, the main criticism of Teilhard's thesis, before and after his book had been published, was that of blasphemy, with many objecting to the fusion of God and humankind at a real point in time on Earth. But today the objections tend to come from the scientific community, especially from those who study Darwinian theory. They say that the idea of an omega point gives an intention and a meaning to EVOLUTION, which is misleading, and furthermore that the preselected development and eventual apotheosis of the human makes Darwinism essentially homocentric, which it is not. To give evolution a purpose is a wrong reading of Darwinian theory, which is essentially a contingent one in which chance favors the development of each and every species including our own. There is no preordained design to evolution, it happened—and still happens—in response to the many changes in the environment and not to things that were predestined to happen, such as the progressive climb of humankind up the evolutionary ladder to reach the summit, the Omega Point.

For further reading: Stephen Jay Gould, *Hen's Teeth and Horse's Toes* (Penguin, 1984); and Kenneth Ring, *The Omega Project: Near-Death Experiences, UFO Encounters, and Mind at Large* (William Morrow, 1992).

OMEN Any phenomenon or circumstance purporting to portend good or evil. In order to believe in omens a specific prior belief is absolutely necessary—that the future *is* knowable, in other words, that everything that is to be has been foreordained. Thus an omen is an event that presupposes destiny. The chief feature of an omen is its fortuitous and unsought nature: the black cat runs across your path, the owl hoots as you pass by. From far back in recorded history many changing aspects of nature have been noted as harbingers of good or ill; it is interesting that most forebode ill. The probable origin of superstitions of this kind was mankind's attempt to know the future and perhaps, through fore-knowledge, to avoid disaster. Omens have always formed a basis for action rather than an indication of inexorable fate.

In the ancient world an omen did not have to be an extraordinary happening, like a comet foretelling misfortune. Quite ordinary events observed, or heard, to happen in the sky were noted: thunder and lightning, or the flight or song of birds, could be deemed to be auguries. Very important was the direction from which a sign first came; it was this that indicated whether it predicted good or evil. In Ancient Greece an owl hooting on the left was an unlucky omen while one hooting on the right was lucky; in Roman society the values were reversed. The anomaly could have arisen from Rome's early national struggles for power in the Mediterranean, leading to the belief that an omen that was bad for Greece was necessarily good for Rome.

In the Western world belief in omens has survived down the centuries and is still with us today, linking us to our distant magical past. Many find it very hard to shrug off belief in prognostic signs that for thousands of years were part of the survival strategy of human beings and have now become ingrained—so much so that in today's scientific world omens are frequently presented in a pseudoscientific guise. Red skies are said to be omens of bad weather according to an old verse learned as children:

Red sky at night, shepherd's delight.
Red sky in the morning, shepherd's warning.

These omens are likely to be more correct than not when we realize that "sky at night" means sunset in the west, and "sky in the morning" means sunrise in the east, and that, in Britain and the northwestern United States, the prevailing winds come from the west. So, at sunset if the sky is sunny and bright in the west, the clear, moisture-free air between the observer and the horizon is scattering predominantly red light towards the shepherd. This means the weather will remain dry for many hours whilst the wind moves towards the east. At sunrise, however, when the red sky is in the east, there is no guarantee that these conditions will remain for long, because the prevailing winds will soon blow the clear air away and the next weather front carrying rain clouds will likely blow in from the west. But rain clouds might not come in from the west, so red sky in the morning is not as reliable a shepherd's warning as red sky at night.

ONTOGENY AND PHYLOGENY Theory that relates the development of an individual to its biological past. The phrase "ontogeny recapitulates phylogeny" was, during the 19th century, one of the main arguments of scientific racism in the Western world. It was first stated by German zoologist Ernst HAECKEL in *Über Arbeitstheilung in Natur und Menschenleben* (1869) as a way of understanding EVOLUTION. Because the fossil record was very incomplete, Haeckel proposed that it might be possible to read the evolutionary history of species—its "phylogeny"—through the development of an individual, or "ontology." He believed that every individual passes through stages that represent the adult development of its ancestors—or, as Stephen Jay GOULD puts it in *The Mismeasure of Man,* "an individual, in short, climbs its own family tree."

Individuals of a species do seem to develop characteristics associated with other animals. Human embryos, for instance, develop (and lose) gill slits on their necks, a three-chambered heart (which expands to four chambers), and a tail. These characteristics seemed to prove the 19th-century idea that evolution was a ladder of progress with humans at the top of the ladder. Recapitulation theory even placed human races in a biologically determined hierarchy. Some races were ranked superior to others; white Western Europeans and Americans were placed on top. Scientists theorized that other human races could be understood by studying the behavior of white children. When 20th-century biologists replaced the model of evolution as a ladder with one of a many-branched bush, they no longer had a place for recapitulation theory, and it was discredited.

ONZA A large feline, unrecognized by zoologists, whose main habitat is reportedly the Sierra Madre Occidental range of northwest Mexico. Among the animals of folklore being considered by CRYPTOZOOLOGY but yet to be officially recognized by biologists is the onza. Accounts of the onza go back to the Aztecs, who called it *cuitlamiztl.* They clearly distinguished it from both the jaguar and the puma, the better-known and recognized large cats from the same area. Europeans first saw the animal in the Aztec emperor Montezuma's zoo, where it was distinguished from the puma by its wolfish appearance. In the mid-18th century, Jesuit missionary Ignaz Pfefferkorn gave a more complete description of the onza: He likened it to a cheetah with a long narrow body and long thin legs. The onza, he added, was also notable for its ferociousness.

Onza

Modern consideration of the onza began in 1930s when two hunters, Dale and Cecil Lee, and a client, Joseph H. Smirk, killed a strange cat on a jaguar-hunting expedition in Mexico. When they described the animal to U.S. zoologists, their story was ridiculed. They received little attention until the story was included in the 1961 book *The Onza* by Robert Marshall. The story of the onza was picked up again in 1982 by the INTERNATIONAL SOCIETY OF CRYPTOZOOLOGY (ISC). Within a very few years, several skulls had been located in various collections, and in 1986 a complete specimen was obtained for dissection by ISC secretary J. Richard Greenwall. The specimen fit the traditional description of the onza, with distinguishing long ears, body, and legs. What remains to be established is whether the onza is a totally new species or merely a local, if extreme, variation of the puma.

The proof of the existence of the onza has been one of the major accomplishments of cryptozoology. It clearly demonstrated that the descriptions of indigenous people can be quite literal and accurate. The story of the onza has given cryptozoologists hope that specimens of some of the other mysterious animals reported to exist in remote corners of the world also will eventually be located for study.

ORACLE A person, object, or shrine through which humans can communicate with the gods to learn the future; also, the divinely inspired forecast itself. Many ancient cultures (and a few remaining ones today) revered male or female holy people who were considered oracles and were consulted before many important events. Usually the person consulting the oracle brought some kind of sacrifice, often an animal. The human oracle then communicated with the divine, sometimes by examining the sacrifice and "reading" it, sometimes by entering a trance state through MEDITATION, self-hypnotism, drugs, or frenzied dancing. Some oracles also read ritualistic items rather than the sacrifice.

Among the best-known ancient oracles was that of Delphi, in Greece. According to legend, Jupiter sent two birds in opposite directions—where they met would be the center, or navel, of the Earth; the place they met was Delphi. Here people erected a temple, which included a white marble "navel." Female priests, or sibyls, dedicated themselves to the temple and to Apollo, and became the MEDIUMS between gods and humans. At first, the oracle was chosen from among virgins, but at some point this became impractical, and older, virtuous women, past their childbearing years, became the oracles. A petitioner would bring a question and a sacrifice to the oracle; she would burn the question, enter a trance, and speak the god's answer. Many political and military leaders consulted the oracle. Although Delphi has remained the most famous, Greece had other well-known and oft-consulted oracles as well.

Egyptians in the Old and Middle Kingdoms (2680–1786 B.C.E.) put great faith in oracular dreams, especially those of designated priestesses. In the New Kingdom (1570–1342 B.C.E.) the people consulted oracular cult statues, usually dedicated to Amun, god of fertility, agriculture, and the breath of life. With the help of priests, the statues spoke or nonverbally indicated answers to petitioners' questions.

Andean Indians in Peru up to recent times consulted oracular elements of nature (certain rivers, trees, and so on) imbued with divine spirits. They offered sacrifice to and consulted these *huillcas* on both major and everyday matters. Even today in Tibet oracles (generally not priests, but psychic laypeople), after careful preparation, enter trance states and respond to petitioners' questions.

Many other people and cultures have consulted oracles over the centuries. In fact, many people consider most DIVINATION methods forms of oracles. When a medium or prophet engages in CRYSTAL GAZING, TAROT

CARDS, I CHING, or PALMISTRY, he or she is, in essence, consulting an oracle.

Is there any scientific evidence for oracular powers? The ancient oracles cannot, of course, be tested. Their persistent existence points to popular faith in them, but there is little but anecdotal evidence in their favor. People and devices that allegedly foretell the truth today can be tested, but rarely is this done. Believers resist having matters of faith subjected to scientific analysis. Organizations such as the COMMITTEE FOR THE SCIENTIFIC INVESTIGATION OF CLAIMS OF THE PARANORMAL (CSICOP) annually report on the success rates—or, more accurately, the lack-of-success rates—of popular psychics who forecast upcoming events, but even this is done in an informal manner. There is little hard or systematically gathered evidence one way or the other about the efficacy of oracles.

ORANG-PENDEK A mysterious humanlike creature reported to inhabit the Indonesian island of Sumatra. The orang-pendek is usually contrasted to other Sumatran animals such as the gibbon, the orangutan, and the sun bear. It is described as being between 76 and 152 centimeters ($2\frac{1}{2}$–5 feet) tall, covered with short hair, and possessed of a bushy mane on its back. Its arms are shorter than those of most anthropoid apes. It is most often reported on the ground rather than the trees. Its footprints are very humanlike but slightly broader. Although the orang-pendek has been reported by Dutch settlers throughout the 20th century, the lack of a viable specimen has led many primatologists to discount the species' existence. They have attributed the stories to misidentifications or hoaxes.

The issue of the existence of the orang-pendek was raised anew in 1989 by British travel writer Dorothy Martyr. While visiting southwest Sumatra, she was told about the orang-pendek and began to collect stories of sightings. She was shown footprints that resembled those of a child but were much too broad. She took a cast and gave it to the local authorities. It was sent to the Indonesian National Parks Department but later disappeared. Indonesian zoologists reject the existence of the orang-pendek, but cryptozoologists are still trying to establish its existence.

ORCHIS EXTRACT Early 20th-century product claiming the ability to revive male sexual prowess. At the turn of the century, a number of remedies tried to help men who could not hold a penile erection. During the 1920s, perhaps the most successful of these remedies, which were usually sold through the mail, was manufactured by the Packers Product Company of Chicago. Packers, whose remedy was called Orchis Extract, was owned by Fred A. Leach, who had previously owned a business that sold a vacuum apparatus which, he claimed, men could use to lengthen their sex organ. Orchis Extract was named for, but had no material relationship to, orchids—flowers with a prominent stamen resembling an erect penis. The company's floral logo quickly communicated the claims for the product.

Packers also used a picture of the stockyards of the nearby Armour & Company in its advertising. The advertising implied, but did not clearly state, that the company was taking the sexual organs of the recently slaughtered animals to make a serum, capitalizing upon the latest scientific discoveries. The literature mentioned in passing that the extract was made from a substance derived from the testicles of rams.

Eventually, the picture on the Packers' literature was brought to the attention of the owners of Armour & Company. They filed suit with the United States Post Office and the post office in turn prosecuted Leach as a perpetrator of mail fraud. He was put out of business, but he reappeared a few months later as head of the Organo Product Company. Organo simply transferred the text of the Packers' literature to a new letterhead without the picture of the stockyards that had caused the lawsuit. In August 1919, the post office again shut down his operation.

See also APHRODISIACS.

ORFFYREUS'S WHEEL A PERPETUAL MOTION MACHINE (also known as Orffyreus) invented by Jean Ernest Elie Bessler (1680–1742). The idea of a large overbalanced wheel, delivering more energy from a series of falling weights than is required to return the wheel to its original state and therefore gaining energy as it rotates, was first suggested by Villard de Honnecourt (c. 1225–c. 1250) in the 13th century. A large, heavy wheel, once turning, has great momentum and will continue to turn for some time. This attractive idea was explored by many in the ensuing centuries, Bessler (Orffyreus), following in their footsteps with several attempts, finally believed himself to have succeeded in achieving perpetual motion. A famous Dutch mathematician, philosopher, and physicist, Wilhelm Gravesande (1688–1724) remembered today for his method of measuring Young's Modulus—the extent to which a substance stretches under tension—of a material in the form of a wire, was impressed with this last version and described its appearance in a letter to Sir Isaac NEWTON. Bessler was unwilling to reveal to Gravesande the secret of its internal construction, afraid that others might steal the idea. The secret died with him.

The French Academy of Sciences ruled against any further consideration of claims to have devised perpetual motion machines in 1775. Both the British and the U.S. Patent Offices have refused to examine patent applications for such machines, in the latter case unless accompanied by a working model or other convincing evidence—there have been no takers so far.

ORGONE ENERGY A discovery, more a revelation, by 20th-century Austrian physician Wilhelm REICH of the vital life force that explains everything. It is blue in color and is responsible for the color of the sky, sea, lakes, and much besides. It is the mechanism by which gravity acts and is responsible for the formation of galaxies at the big end of our universe. On a smaller scale, it explains illness and health and is involved in nuclear energy.

ORGUEIL METEORITE A shower of stones, of meteoric origin, that fell near the village of Orgueil, near Toulouse in the south of France, on the night of May 14, 1864. The stones were of the carbonaceous chondrite type, and about 20 were collected for examination. They were found to contain hydrogen, carbon, and oxygen in a material that was like peat and lignite—organic substances that occur naturally on Earth. How such substances came to be in this meteorite (and are also found in other carbonaceous chondrite meteorites) is still a puzzle.

So far there is nothing in the above to single out the Orgueil meteorite from the thousands of others that have fallen on Earth. Its importance is twofold. First, at that time Louis PASTEUR had started a furious debate in France by disproving the current belief that life can arise from inanimate matter: SPONTANEOUS GENERATION. One version of this was that microbes entered Earth's atmosphere in meteorites and when given the right conditions developed into different organisms. Careful examination of several of the fragments failed to find any microbes. The idea that microbes enter our atmosphere from space has persisted nevertheless and has recurred several times in this century. Lord Kelvin supported the idea as did two Swedish scientists—Svante Arrhenius and Louis Bachman. Another Swedish professor, Knut Emil Landmark, favored the meteorite theory, and University of California professor Charles B. Lipman claimed to have cultured microbes from pulverized meteoritic material. That experiment has eluded repetition, and his cultures are thought to have been the result of inadequate sterilization. More recently still, Sir Fred HOYLE and Chandra Wickramasinghe, eminent astrophysicists, promoted the theory of invasion of Earth by bacteria from outer space, which again failed to stand up to scrutiny.

The second cause of the Orgueil meteorite's notoriety is that a hoaxer attempted to use some of its material to mislead the proponents of spontaneous generation. He mixed some seeds and coal into a sample of the meteoritic material and replaced it. However that particular sample was not examined, and the hoaxer lost the opportunity to discredit Pasteur's opponents. Many years later, in 1964, the sample was examined and the attempted hoax was disclosed.

ORTHOGENESIS Suggested mechanism to explain how the process of EVOLUTION works. Literally, "orthogenesis" means "straight line generation," from two Greek words: *orthos,* meaning "a straight line," and *genesis,* meaning "origination." Orthogenetic organisms have their entire evolutionary possibilities preprogrammed into their genes. Proponents of orthogenesis believed that some uncontrollable evolutionary tendencies can lead to the extinction of species; for instance, 19th-century scientists believed that the Irish elk became extinct because its huge antlers (a secondary sexual characteristic) grew so large that the animals were no longer able to lift their heads.

Another example was given by British paleontologist A. E. Trueman, who published a paper in the 1930s stating that the coiled shell evolved by oysters of the genus *Gryphaea* led to their extinction. Trueman cited fossil evidence to support his theory that the oysters were unable to stop their shells from coiling. Eventually, the scientist concluded, the *Gryphaea* oysters had coiled so much that they were unable to open their shells, and so they became extinct.

For a time in the 19th and early 20th centuries, there were four major suggested evolutionary mechanisms: Charles DARWIN's *natural selection;* the inheritance of learned characteristics, or *Lamarckism;* sudden genetic change, or *mutation;* and *orthogenesis.* During the 1930s, natural selection emerged as the preferred mechanism to explain how evolution works. Few scientists now believe that organisms are preprogrammed to follow an evolutionary destiny.

OUDEMANS, ANTOON CORNELIS (1858–1943) Prominent Dutch zoologist and pioneer cryptozoologist best remembered for his study of SEA SERPENTS. Oudemans was born into a family known for its scientific and intellectual accomplishments. He developed an early interest in the study of animals. He completed his doctorate in zoology in 1885 and shortly thereafter became the director of the Royal Zoological and Botanical Gardens at The Hague. A decade later, he moved to the University of Sneek before beginning a 27-year tenure at the University of Arnheim.

As a youth, Oudemans became interested in the question of sea serpents and began to assemble a collection of reports. He reached an early tentative conclusion that the reports referred to a variety of primitive whale only known from fossil records. His ever growing interest in the sea serpent question led to a book, *The Great Sea Serpent,* published in 1892. By that time, he had concluded that the sea serpents were probably a form of long-necked seals. The book met mixed reviews, but although it was denounced by some, it succeeded in renewing consideration of the reports of a giant ocean animal, which had persisted through the century—in spite of the general consensus among his colleagues that no such beast existed. In 1933, a decade after his retirement, Oudemans suggested that the LOCH NESS MONSTER, which had at that time first become well known internationally, was also a variety of long-necked seal. He expressed hope that its identification would solve the sea-serpent question once and for all.

Although honored for his contribution to modern CRYPTOZOOLOGY, few scientists agree with Oudemans's conclusion about the identity of the sea serpent. One exception is writer Peter Costello, who suggests that such seals might be behind the sightings of the Scottish LAKE MONSTERS.

OUIJA BOARD A modern version of an ancient instrument for DIVINATION and communication with supposed spirit entities. The modern Ouija Board was invented by Elijah Bond in 1892. Its name was created from the French and German words for yes, *oui* and *ja*. It represented an improvement over the PLANCHETTE, a similar spirit communication instrument that had been in use since the 1850s. The planchette consisted of a heart-shaped piece of wood supported by two wheel-castors and a pencil. The pencil point rested on a piece of paper, and the hand of the operator(s) rested lightly on the planchette. Many people found that the planchette would write out messages either spontaneously or in response to questions.

The Ouija Board replaced the sheet of paper with a board upon which were written the letters of the alphabet, the numerals 0 through 9, and the words "yes" and "no". The wheel-castors and pencil of the planchette were replaced with three felt-tipped legs. One end of the instrument became a pointer. Usually, two people placed their fingertips on the instrument which would, like the planchette, move at its own direction or in response to questions. By pointing to the letters, it spelled out messages.

Bond sold the idea of the Ouija Board to William Fuld of the Southern Novelty Company (later the Baltimore Talking Board Company), who marketed it as a mystifying toy. It attained a level of popularity during and immediately following World War I, amid general grief and anxiety over loved ones who had become war casualties. It has since enjoyed waves of popularity as interest in the occult has waxed and waned, and it has remained a steady seller. In 1966 the rights to the board were sold to Parker Brothers. Though marketed as a entertaining toy, a number of modern MEDIUMS, most notably Betty White, the medium of the *Betty Book* literature, and Jane Roberts, who channeled *The Seth Material,* reported that their career began with the use of the Ouija Board. In each case, however, they soon moved beyond it to trance speaking. British medium Hester Dowden was one of the few mediums who made extensive use of the Ouija Board.

The Ouija Board became quite controversial both within the psychic community and among parapsychologists. In spite of the many attestations of its usefulness, psychic critics claimed that it led to an unmonitored access to the unconscious resulting in the release of negative material and the development of obsessional behavior (although very few accounts of such occurrences have been reported in the literature). Parapsychologists have dismissed the Ouija Board, noting that it did not address any of the critical objections directed earlier against the planchette—that is, that it produced information only from the unconscious and operated through subtle muscle movements. They claimed that it was of no value in exploring paranormal phenomena. More hostile critics have seen the Ouija Board as a harmful instrument spreading superstitious belief under the guise of harmless entertainment.

The Ouija Board had a number of historical precursors, including a moving table that was reportedly employed by the philosopher Pythagoras in his mystical school in the sixth century B.C.E. Analogous instruments have been reported in use among the ancient Chinese, the Mongols, and Native Americans. Roman historian Marcellinus Ammianus described an instrument consisting of a metal disc around the edge of which the Greek alphabet had been engraved. It stood on a tripod and was operated by a person standing above it with a pendulum consisting of a string, at the end of which was a ring.

For further reading: Gina Covina, *The Ouija Board* (Simon and Schuster, 1979).

OUT-OF-BODY EXPERIENCE (OBE) According to author H. J. Irwin, "an experience in which the center of consciousness appears to the experient to occupy temporarily a position which is spatially remote from his/her body." No scientific theories (for example, those based on psychology and physiology) have been able to explain these mysterious experiences; in fact, the above definition is necessarily based on reports by people claiming to have had OBEs. These reports have not as yet been scientifically verified.

OBE reports have several common features. In a great number of cases, the person having the experience observes his or her body lying motionless or corpselike as the "mind" or consciousness floats above it. Observations and sensations of the OBE are usually described as more vivid than those of a DREAM. Often a subtle, ghostlike ("astral") body is perceived to accompany this straying consciousness. The astral body is frequently attached by a silvery ethereal cord—an astral umbilical cord—to the physical body.

It is not uncommon for the person having the OBE to report traveling through walls and other solid objects. A bright, white light is often associated with the experience, and often the experient is in a long dark tunnel with a white light at the end of it. The OBE is more often than

Out-of-body experience

not reported to be a pleasant and even liberating experience. At the same time, the unexpected observation of his or her detached body occasionally shocks and frightens the person experiencing out-of-body travel. This shock usually brings the experience to a sudden end. There is frequently a sensation, sometimes painful, during the process of reentering the body.

Out-of-body experiences occur more often during consciousness than during sleep. They seem especially prone to occur during periods of high stress, serious illness, or acute trauma. Exactly what these conditions have to do with the phenomenon remains a matter of dispute, although significant progress has been made by psychologists in recent years. Research has found the experience to be fairly widespread: Somewhere between one-quarter and one-third of the adult Western population admits to having had at least one OBE. Gender and age do not seem to be factors in the likelihood of occurrence.

OBEs have been recorded throughout human history. Passages in the Bible as well as from classical writings make reference to what may be classified as OBEs; they are particularly prevalent in Western literature since the early 1800s. In fact, any belief in or concept of consciousness as separable from the body is the essence of OBE.

It is possible to distinguish two types of out-of-body experiences. In the first, the exteriorized consciousness observes a scene of normal reality, either the body and its immediate surroundings or something further afield. In the second type, the person travels to what he or she regards as another plane of existence, normally a spirit or "astral" realm of some kind.

There is little reason to doubt that people do have out-of-body experiences. At issue, therefore, is not whether people genuinely experience leaving their bodies, but whether, in the form of some nonphysical mind or soul, they really *do* leave their bodies. In scientific terms, it is a matter of whether the OBE is purely a function of psychology and physiology—something that ultimately takes place in the imagination (and hence within the physical brain)—or whether it is instead an actual separation between mind and body. Either way, it is necessary to explain what happens during the OBE and what function it serves.

Perhaps the simplest, and to many the most appealing, explanation of what happens during an OBE is that a nonphysical "mind" or soul in fact separates from the body. These theories are generally called separationist. They rely on the existence of a nonphysical level of reality, something to which the mind, consciousness, spirit, or soul belongs and in which it is able to travel. This, however, is extremely difficult if not impossible to demonstrate scientifically, as are REINCARNATION, apparitions of the dead, POSSESSION, and other theories of a spirit realm.

This claim necessarily raises questions. How does this nonphysical mind retain the powers of observation and reflection that belong to the body? Furthermore, what is the relationship between this nonphysical consciousness and the normal consciousness to which we are accustomed? Such questions present the separationist's position with a challenge. This results in an effort by separationists to prove in rational terms what is apparently nonrational and so easily leads to false science.

Separationist theories do not satisfy the scientific mind partly because they are unable to present any testable predictions about out-of-body phenomena. When are OBEs most likely to happen? Which individuals are most likely to experience them? What function do they serve? Aside from the questionable character of the answers they do provide, separationists have special difficulties with these larger questions.

From a scientific perspective, out-of-body experiences take place in the imagination for either primarily physiological or psychological reasons. Thus, these are called imaginal theories based on imaginal experiences. One prominent imaginal explanation interprets the OBE as a mechanism to protect the threatened self. In physiological terms, the body experiences a sudden and threatening shift of state due to drugs, trauma of either emotional or physical type, or some kind of sensory deprivation. The body then reestablishes its equilibrium through an OBE.

Psychologically, the astral body can be understood as fulfilling the need to base the self on a bodily form. Similarly, the birthlike imagery of the dark tunnel with bright light at the end and the astral umbilical cord can be interpreted as providing a necessary feeling of security. In fact, the general approach of explaining OBEs in terms of the self can become entirely psychological; such a line of explanation usually focuses on the concept of body image. Generally, strict physiological explanations are not entirely satisfying; physiology needs rather to play a supporting role in any broader theory.

Many psychological explanations tend to focus on such predictable factors as the fear of death. The OBE in this way can be interpreted as a mechanism to assure consciousness that it will survive death. Another explanation that emphasizes fantasy points to the birth imagery common in OBEs. In addition to the dark tunnel and astral cord, floating sensations and roaring noises are frequently reported.

In more concrete terms, scientists have noted certain fundamental elements of the OBE as a psychological phenomenon falling under the category of ALTERED STATES OF CONSCIOUSNESS. As a unique variety of an altered state of consciousness, OBEs involve vivid and detailed imagery combined with minimal sensory input from the body and a minimal attention to reality (or what is called reality testing). Added to these is the continued capacity to perceive and think clearly. This approach suggests a largely physiological precipitation of the OBE, beginning with either reduced or ignored sensory input that allows consciousness to temporarily ignore or drift away from attending to the body. This is a framework in which much of the data relating to the OBEs can be explained. It is only complete, however, if one is willing to ignore the claims of paranormally acquired information (unavailable normally) during an OBE.

While some psychologists prefer to supplement their imaginal theories with reference to CLAIRVOYANCE, PSYCHOKINESIS, and other paranormal phenomena to explain the occasional mystery, and thus step beyond the boundaries of standard scientific research, others are content to dismiss the validity of the alleged knowledge and to pursue the scientific explanation as comprehensive. The trend in psychological research, at any rate, is toward the investigation of altered states of consciousness and mental imagery. One line of inquiry, for instance, concerns whether or not people who experience out-of-body travel exhibit fantasy-prone personalities. Such approaches need to be filled out with empirical research, but again this is difficult due to the primarily spontaneous nature of the OBE.

It is common for the out-of-body experience to be confused with other experiences of seeing oneself. Some of these experiences are set within the widespread belief that individuals have doubles, called by the German word *Doppelgänger.* Closely related is autoscopy, or the experience of seeing oneself as an apparition. The distinction between seeing such a double or apparition of oneself and an OBE is that in the OBE one leaves the real body and experience takes place in the double or astral body.

Another related phenomenon is depersonalization, the feeling of being estranged from oneself and of not being a real person. The connection between this and OBE is the concept of body image, which would appear to be disturbed in both cases, though they should be distinguished. The categories of individuals who experience depersonalization are not correspondingly susceptible or unsusceptible to OBEs. Also, the OBE does not involve the lack of identification or alienation from the body—

even though it is temporarily separated—that occurs in depersonalization.

There are also important distinctions to be made between OBEs and dreams. OBEs most often occur when awake. They tend to be more vivid and lucid, and they happen in what is experienced as a state of full consciousness. Furthermore, preliminary physiological testing suggests both that there is no discrete physiological state in which OBEs occur and that they usually do not coincide, physiologically, with the *REM* or dreaming state. The chief weakness with these observations is that they are from laboratory experiments while most OBEs happen spontaneously.

See also ESP; NEAR-DEATH EXPERIENCE.

For further reading: Susan J. Blackmore, *Beyond the Body: An Investigation of Out-of-the-Body Experiences* (Granada, 1982); Glen O. Gabbard and Stuart W. Twemlow, *With the Eyes of the Mind: An Empirical Analysis of Out-of-Body States* (Praeger, 1984); and H. J. Irwin, *Flight of Mind: A Psychological Study of the Out-of-Body Experience* (Scarecrow Press, 1985).

P

PALEY, WILLIAM (1743–1805) British theologian and moral philosopher. Paley studied mathematics at Christ's College, Cambridge, in 1759, was elected a fellow of his college in 1766, and ordained an Anglican priest the following year. Paley taught at Cambridge for nine years, until his marriage, and rose in the church to be the Archbishop of Carlisle. Paley authored three books, all of which were widely read and accepted as textbooks. His *Principles of Moral and Political Philosophy* (1785) is more a description of duties and obligations than it is a philosophical treatise. To Paley, the will of God is for mankind to find everlasting happiness; thus one should carry out actions that promote general happiness and avoid those that diminish it. God's will can be found either in the scriptures or in "the light of nature," both of which lead to the same conclusion. Paley's *Principles* contains his famous satire on private property, in which he describes the effects of private property on the social order of a flock of pigeons—for which he earned the nickname "Pigeon Paley."

Paley's other two books defend Christian belief. In *A View of the Evidences of Christianity* (two volumes, 1794), he argues that the miracles recorded in the scriptures are genuine, proven by the steadfastness of the early Christians who bore witness to miraculous events in spite of risks to comfort, happiness, and life. The book is essentially an apologetic essay, its thesis resting on the reliability of historical witnesses who reported on events that necessarily only they could have experienced. Nevertheless, it was a huge success, and following its publication Paley was made a prebendary of St. Pancras in the Cathedral of St. Paul's and the subdean of Lincoln.

Paley's third book, *Natural Theology, or, Evidences of the Existence and Attributes of the Deity, Collected from the Appearances of Nature* (1802) is, as the title implies, an essay on natural theology. In it Paley argues for the existence of God, for where there is mechanism, instrumentality, or contrivance, there must be an intelligent designer. Mechanized nature is, to Paley, the work of a divine mechanic. That God is, moreover, the sole creator and a beneficent one is proven by the global uniformity of natural law, by the usefulness of natural objects, and by the experience of pleasure.

PALINGENESY The resurrection of plants. Palingenesy was a concept held by some scientist-philosophers of the 18th century that offered a chemical understanding of ghosts and apparitions, an alternative to either supernatural or hallucinatory explanations. The idea of palingenesy derived from Greek philosopher Lucretius (c. 98–55 B.C.E.), who thought that ghosts were in fact thin filmlike products analogous to the dead skin of a snake.

The idea was revived in the 17th century among alchemists, who at a meeting of Royal Society of England burnt a plant and extracted a salt from its ashes. They then mixed the salt with an unnamed substance and subjected the mixture to heat. The particles began to move and in the midst of the powder the image (an APPARITION) of the plant emerged. When the heat was taken away the form disappeared, and when it was reapplied, it reappeared.

The hypothesis the alchemists derived from this incident was that the substantial form of a living object resided in its salt. They concluded that the form of human beings resided in the human body's salts and that the heat of fermentation caused the salts to form in the shape of the body. Several additional reports of experiments carried out in Paris to test the palingenesis hypothesis have been reported in the literature, including a rather unbelievable story involving alchemist Robert Fludd (1574–1637) described by John Webster in his volume *The Displaying of Supposed Witchcraft* (1677).

PALMER, RAY (1910–1977) Internationally known writer and publisher of science fiction, sensationalized science fact, and serious science, especially during the heyday of the pulp magazines. Palmer edited such popular magazines as *Amazing Stories, Fantastic Adventures,* and *Flying Saucers.* He also founded several popular publications, including FATE MAGAZINE. A Wisconsin native, Palmer began his professional publishing career with Ziff-Davis, a large popular magazine publisher, in the 1930s and ended it with *Space World,* a serious publication about space exploration. A prolific writer as well as editor, he published many stories, both fiction and nonfiction, under an array of pseudonyms.

Palmer is particularly remembered for the Shaver Mysteries, a series he published in AMAZING STORIES in the 1940s, purportedly the true accounts of Richard Shaver, a Pennsylvania working man, who told of his experiences with a malevolent race of beings who lived inside Earth. He called these beings *deros* and said they were responsible for many of Earth's catastrophes. What started out as a tantalizing, potentially true tale eventually became so far-fetched that *Amazing Stories*'s subsequent editor, Howard Browne, killed the series off in 1949.

Palmer was the proponent of many unusual and not widely accepted ideas. These included the idea of an inner-Earth–dwelling race, as described in the Shaver Mysteries, and the idea that the North and South Poles were the sites of hidden UFO bases. Palmer led a controversial life and career, but his impact on science fiction and the myths of modern-day fringe science cannot be denied.

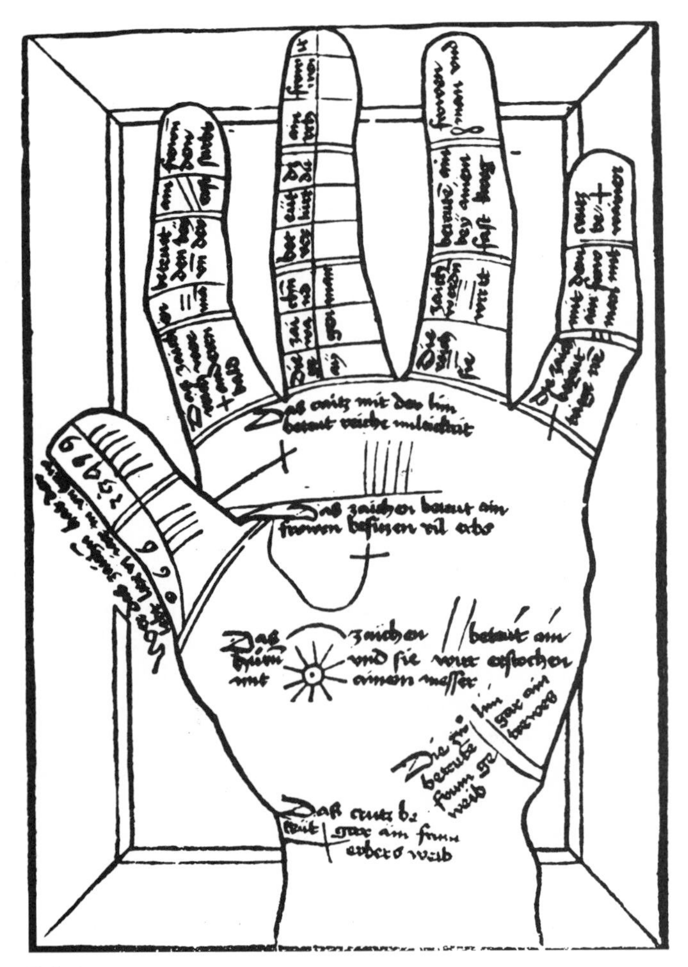

Palmistry

PALMISTRY The practice of telling a person's character and fortune by interpreting the lines and undulations on the palm of the hand, also known as chiromancy or chirosophy. A version of this art is called podoscopy, where fortunes are read from the soles of the feet.

The origins of palmistry go back to antiquity, there are allusions to it in ancient Egyptian hieroglyphics and in the Bible. It had an unsure history prior to the early 13th century when the first known manuscript on the subject appeared. The first book on palmistry was published in Augsberg about 1475. In general the system has not changed much since then, although there have been some additions and variations; for example, modern palmists do their readings from palm prints. They find the print, done with water-soluble ink, shows up the fine lines in greater detail than does direct observation. Although palmists look at both hands they usually concentrate on the client's dominant hand.

Notwithstanding the specifics of reading the lines of the hand, the practitioner usually wishes to see the client in person to talk to and to make observations about their age, wearing apparel, and so on, at the same time looking at the backs of their hands and noticing how he or she holds them at rest—relaxed or otherwise. The color and the texture of the skin and the condition of the nails provide clues to the client's vocation, work patterns, and temperament. This overall assessment of a client's lifestyle is called cold reading and is used in all methods of DIVINATION. With assessments gleaned from cold reading, palmists can then formulate generalized positive statements about their client, which are usually accepted as a reasonably accurate evaluation of their character. Experiments have been done on students (Forer, *Journal of Abnormal and Social Psychology* 44, 1949) giving every person in a class an identical personality assessment. They were all told the profile was based on a personal evaluation and that they were not to confer with each other. Every profile was rated to be highly accurate.

For further reading: F. Gettings, *The Book of the Hand: An Illustrated History of Palmistry* (Hamlyn 1965).

PALUXY TRACKS Some curious fossilized footprints that were discovered during the 1930s in the bed of the Paluxy River near Glen Rose, Texas, by Roland Bird, a field researcher for the American Museum of Natural History. Some of the prints were dinosaurian and dated to the Cretaceous period about 100 million years ago. Others looked almost human. They showed signs of heels and arches in the appropriate places. However, the tracks ranged between 40 and 50 centimeters (15–20 inches) in length—roughly twice the length of the average human foot—and measured as much as 20 centimeters (8 inches) across.

Some creationists seized on the Paluxy tracks as proof that dinosaurs and human beings existed at the same time. Creationists, who reject the theory of EVOLUTION and insist on a literal interpretation of the Christian Bible, stated that the tracks proved that giant humans existed in ancient times, just as the Book of Genesis said. They believed that these humans and the dinosaurs died together in a Great Flood around 4000 B.C.E. Writers such as Brad Steiger, who do not hold creationist views, declared that the Paluxy tracks were evidence of an advanced ancient civilization of giant humans. The giants were wiped out through a combination of natural disasters and atomic war.

In the 1980s, with the resurgence of interest in creationism, paleontologists began to look closer at the Paluxy tracks and related sites in Texas. In some cases, they were able to show to the satisfaction of the creationists that the tracks were forgeries, carved in recent times by human beings; in others, the scientists theorized that the tracks were made by dinosaurs, but for some reason the toe marks had been lost. It was dinosaur enthusiast Glen Kuban, however, who finally solved the Paluxy tracks mystery. His close observations of the tracks revealed that the prints *were* those of a dinosaur. The dinosaur had left toe prints, but they had been filled in at a later time by a differently colored sediment. After similar tracks were discovered in New Mexico, Kuban was able to convince both paleontologists and creationists of the truth of his theory.

See also DINOSAURS, CONTEMPORARY.

For further reading: Philip Kitcher, *Abusing Science: The Case Against Creationism* (MIT Press, 1982).

PARACELSUS (1493–1541) German chemist who worked to reform medieval medicine. Philipus Theophrastus Aureolus Bombastus von Hohenheim, better known as Paracelsus, had trained as a physician and surgeon and held degrees from several universities in Italy. He theorized that illness was not caused by an imbalance in an internal system of humors, as medieval practitioners believed (see GALENISM) but was due to outside forces. He believed that the way to attack these invading forces within the body was through the use of chemicals rather than herbs or techniques such as bleeding. Paracelsus planned to create these chemical cures using the techniques of ALCHEMY. Previously, the major goal of alchemists had been to find a way to produce gold from base metals; Paracelsus changed that to the production of medicines, setting the basis for modern chemistry.

Paracelsus

Paracelsus was, by all accounts, a difficult man to associate with, having little tolerance for human frailties or for disagreement with his own ideas, roundly condemning practitioners of traditional medicine for their failure to adopt his ideas. Perhaps because of this, he failed throughout his life to find a permanent job or a university position. He died without ever seeing his ideas win favor among the medical and academic establishments. Although his work was instrumental in establishing the tenets of modern medicine, Paracelsus was not himself a scientist. He attacked the theory of humors on mystical, religious grounds rather than on scientific grounds.

PARACRYPTOZOOLOGY Name given to the consideration of the most paranormal CRYPTOZOOLOGY reports and to the paranormal world posited to contain them. Modern cryptozoology is largely confined to accounts of sightings of unknown species that exist in exotic environments, especially remote jungles and mountains or the middle of oceans and deep lakes. The major subjects of cryptozoological research are animals that modern zoology suggests are unlikely to exist but which are by no means impossible. The production of an actual specimen of a LOCH NESS MONSTER or an "abominable snowman" would cause some revision but no major revolutions in the field.

However, along with improbable sightings of BIGFOOT in the rural northwestern United States and sauropods in remote areas of Africa, at the margin of cryptozoology are many extraordinary reports of hairy hominids roaming around in more populated areas, not to mention even more exotic creatures such as flying hominids or even classical mythological creatures—dragons, satyrs, and centaurs. The study of these reports is known as paracryptozoology.

The reports of such exotic creatures occasionally include paranormal elements. These creatures suddenly appear or disappear. People threatened by such strange apparitions have shot at them, but bullets seem to have no effect. Some have reported being in telepathic contact. In some cases, their appearance has been tied to reports of UFOs. Although most cryptozoologists dismiss these anomalous reports and take pains to separate their work from them, some writers have championed the cause of these more exotic creatures and have attempted to explain them by reference to the existence of an alternative universe or reality.

Paracryptozoologists range from the more conservative such as Loren Coleman and Janet and Colin Bord to the more extreme such as Jon Erik Beckjord and F. W. Holiday; for example, in his book, *The Dragon and the Disc* (1973), Holiday suggested that the creatures who inhabit Loch Ness are in fact true dragons, supernatural demonic creatures and the UFOs are counterbalancing forces of good.

See also INTERNATIONAL SOCIETY FOR CRYPTOZOOLOGY.

PARAPHYSICS The subsection of physics (not recognized by the professional institutions) to which paraphysicists subscribe. Paraphysicists are professionally qualified physicists who investigate claims of the paranormal in the belief that their expertise as physicists gives them a special ability to examine such claims. For approximately 100 years, there have been a number of physicists who could be so described: Sir William Fletcher BARRETT, Sir William CROOKES, John B. Hasted, and Sir Fred HOYLE among them.

It is difficult to generalize about such a group when physics itself has been subject to such great change. Today, paraphysicists fall into two groups: those who believe that paranormal phenomena, such as ESP, are real and have an explanation within the known scientific range—for example, John TAYLOR, who concluded by a process of elimination that paranormal phenomena must be electromagnetic; and second, those who believe that paranormal phenomena are real but that the explanation for them must lie in some hitherto unknown aspect of quantum mechanics—for example, that there is an infinite series of quantum levels of which we as yet only know the uppermost and that paranormal events take place in lower levels.

These and other speculations, of course, are just that—pure speculation. These paraphysicists may reply that any advance must require speculation as a first step. But the next step to validate such speculations is to seek the evidence and to find it. That is, so far, missing.

See also TAO OF PHYSICS.

PARAPSYCHOLOGY The study of anomalous experiences and the beliefs arising from them, using the tools of psychology. The term implies an aspiration that the field will eventually be accepted and incorporated into mainstream psychology. "Parapsychology" literally means beside or alongside psychology. Regrettably, there is no universally agreed definition of parapsychology, but as many definitions as there are definers. To believe in the validity of parapsychology is to subscribe to a dualist model of the nature of humans, that there are two separate components: a physical or mechanistic one, and another nonmaterial one—the mind, the spirit, or the soul. If this were proved beyond doubt it would corroborate the Platonic-Cartesian view of humankind.

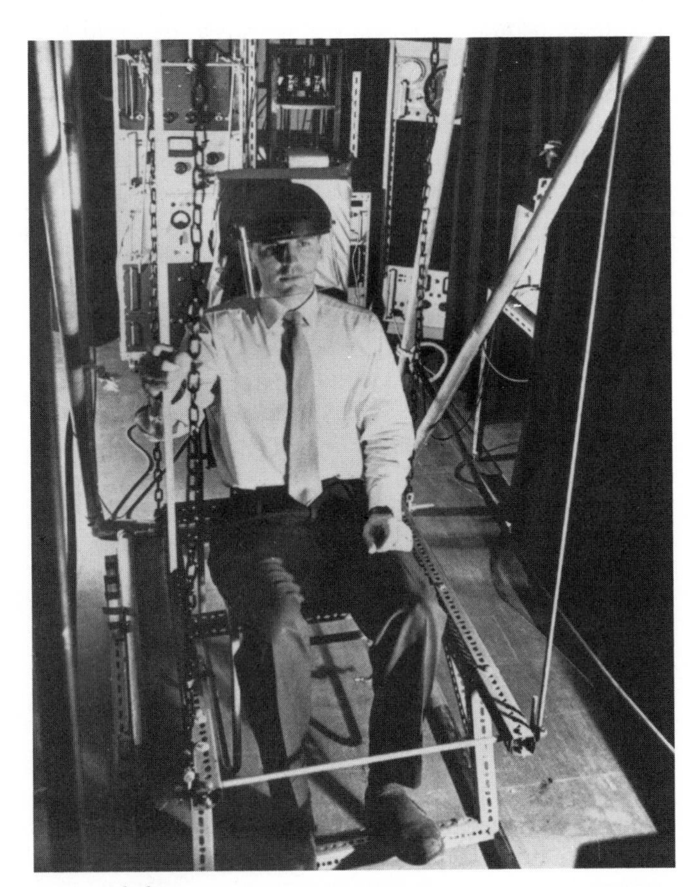

Parapsychology

Parapsychology has four main, but not entirely independent strands: TELEPATHY, CLAIRVOYANCE, PRECOGNITION, and PSYCHOKINESIS (PK). The first three together constitute ESP and all four form psi. The parapsychologist's job is to investigate one or more of these under properly controlled conditions to find out whether they happen and, if so, how, without making any a priori assumptions about whether the explanation is a new discovery in physics and/or biology or some new property of mind, outside conventional physics and/or biology. Parapsychology became established in the late 1920s and 1930s. Prior to that, PSYCHICAL RESEARCH, a term that some still prefer, had described and evaluated largely uncontrolled events but had also carried out some laboratory work. Parapsychology turned more toward controlled laboratory experiments (again not exclusively) but set its face against ASTROLOGY, UFOLOGY, CRYPTOZOOLOGY, and the like. While initially concentrating on card and dice experiments, which are open to statistical evaluation and therefore expected to be respectable scientifically, its range of experimental methods has been extended in recent years. Controls have been steadily tightened, and both skeptics and proponents have been involved in their monitoring. This increasingly scientific approach has been reflected in some relaxation of the traditional hostility of the orthodox science community. In 1969, the Parapsychological Association was admitted as an affiliated organization of the American Association for the Advancement of Science, and in 1996 Professor Robert Morris, professor of parapsychology of the University of Edinburgh, was elected president of the Psychology Section of the British Association for the Advancement of Science. Neither development would have been possible only a few years earlier; even today very few respected refereed scientific journals will accept papers on the subject of parapsychology.

ESP comprises three different types of investigation, with increasing degrees of control: spontaneous cases, field investigations, and laboratory experiments, attention turning most toward the last. ESP is treated essentially as a communication problem, an approach first adopted by Sir Oliver Lodge a century ago. The source/transmitter sends information or a message by an unknown channel (all known channels having been blocked as far as possible by the experimenter) to a receiver/participant. The results are then analyzed to determine to what extent the transmission and reception have been successful. The information/message can be in the form of ZENER CARDS, pictures, or even emotional states, and the abilities of both the sender and participant are tested.

Likewise, PK can be explored in three different ways, but now the sender is the only participant, whose transmission is to affect the receiving system: experiments with (1) dice or random event generators, (2) shifting otherwise stable systems, and (3) moving static objects (where the experimenter has to beware of professional magicians). All have been investigated.

The field is still a minority occupation within psychology, staffed by a small band of committed scientists who are convinced that there is something real to be discovered, described, investigated, and brought within the fold of orthodox psychology, and thus within reach of science.

Although, according to British philosopher Antony G. N. FLEW, there is as yet no acceptable evidence of any parapsychological phenomena, that view is not universal. The belief that they exist persists; their existence needs only to be established to the satisfaction of the scientific community. Despite his caution, Robert Morris—and many others—continues to work to shift parapsychology from the status of pseudoscience to what would render it acceptable as legitimate science.

See also PSYCHICAL RESEARCH FOUNDATION.

For further reading: Gordon Stein, ed., *The Encyclopedia of the Paranormal* (Prometheus, 1996).

PARAPSYCHOLOGY FOUNDATION Foundation cofounded in 1951 by the well-known trance MEDIUM Eileen Jane Garrett (1892–1970) and Congresswoman Frances Payne Bolton (1885–1977), who was its financial backer. It was to be a "nonprofit educational organization to support impartial scientific inquiry into the total nature and working of the human mind and to make available the results of such inquiry."

The Parapsychology Foundation gives financial support through grants for both qualitative and quantitative scientific research in both field and laboratory settings. A parapsychology research scholarship is annually awarded to a college-level student for studies in parapsychology. In 1953, the foundation began to sponsor an annual international and interdisciplinary conference during which invited participants would present formal papers and discuss some parapsychological theme. The foundation publishes the *Proceedings* of these conferences as well as a series of *Parapsychological Monographs*. The foundation published the bimonthly *Parapsychology Review* (which replaced its *Newsletter of the Parapsychology Foundation* and *The International Journal of Parapsychology* [1959–1968]) between 1970 and 1990 and the magazine *Tomorrow* from 1952 to 1962.

The foundation's concern is not only with parapsychology itself but also with its relationship to other disciplines that study the nature of the mind. These include psychology, PSYCHIATRY, philosophy, physics, medicine, religion, and psychopharmacology. The foundation has no membership and does not now conduct research of its own. It maintains an extensive reference library (9,500 volumes) on parapsychology that is made available to serious students and researchers.

PASTEUR, LOUIS (1822–1895) Famous French chemist who made many notable discoveries in chemistry, biology, and medicine.

In 1848, Pasteur separated tartaric acid into two forms, both having the same chemical formula; however, one form rotated a polarized light beam to the right (dextro-rotatory) and the other to the left (levo-rotatory). Pasteur's discovery of molecular dyssymmetry opened up the field of stereo-chemistry. In 1856, Pasteur showed that fermentation is caused by yeast, not by chemical agents; some years later, he showed that similar micro-organisms were responsible for causing wine to go sour and beer to deteriorate. In 1862, Pasteur stirred up opposition by ridiculing the theory of SPONTANEOUS GENERATION—that living organisms can arise from inanimate matter. In 1877, Pasteur noted that some bacteria produce substances that kill other bacteria, the basis of antibiotics—that made possible the use of antibodies in the 20th century. In 1885, he uncovered the mechanism for immunization through VACCINATION. Pasteur put forward the germ theory of disease in 1880.

Pasteur's theories were attacked as unscientific, and their acceptance by the medical profession was a slow and painful process. He was denounced by the medical community as a mere chemist who was not qualified in their discipline. Yet by the 20th century, scientists had accepted Pasteur's germ theory and relegated the idea of spontaneous generation to pseudoscience.

Pasteur

For further reading: J. B. Conant, *Harvard Case Histories in Experimental Science,* cases 6 & 7 (Harvard University Press, 1957); René Dubos, *Louis Pasteur: Freelance of Science* (Da Capo, 1986); Émile Duclaux, *Pasteur: The History of a Mind* (Scarecrow Press, 1973); and R. B. Pearson, *The Dreams and Lies of Louis Pasteur* (Sumeria, 1994).

PAST-LIFE REGRESSION A technique, most often using HYPNOSIS, to help people recall a previous life. For many years, hypnotism has been used as a tool to help people remember things that happened in their childhood. A few high-profile cases suggested that it could also be used to regress people back beyond their present life into a life lived sometime in the past. In the latter half of the 20th century, it was probably the case of BRIDEY MURPHY that brought hypnotism as a regression tool into prominence.

An amateur hypnotist named Maurey Bernstein in Pueblo, Colorado, published *The Search for Bridey Murphy* in 1956. In the book, Bernstein details his hypnosis sessions with a young woman named Virginia Tighe, who began slowly not only to recall the events of a life she lived in 19th-century Ireland but to speak in the voice and accent of Bridey Murphy. Tighe (called Ruth Simmons in the book to protect Tighe's privacy) recalled many obscure details of life in Country Cork during the time she claimed to have lived (1798–1864). Bernstein was able to verify some of these details through careful research, and he believed Tighe had had no access to any information about Irish life of the earlier time.

Bernstein's book caused a sensation. Independent researchers were also able to verify some of Tighe's facts, although they were not able to find any birth or death records for a woman named Bridey Murphy in County Cork during the appropriate time period. Some evidence also suggested that Tighe may have known an Irish woman (a relative or neighbor) who may have told her tales of the homeland during Tighe's childhood; Tighe may not have consciously remembered these tales, some skeptics said, but she might have recalled them under hypnosis.

The truth of Virginia Tighe's account of a past life was never completely determined. Many people thought her case offered proof of REINCARNATION, while others felt the case was a fraud, whether purposeful or not. In any event, it did lead to more experimentation with hypnosis and past-life recall.

Additionally, as time went by, hypnosis, once looked upon as primarily a tool for entertainers or charlatans, became more accepted as a way to get at individuals' repressed memories. It became a relatively accepted tool for therapists, some of whom claimed to uncover past lives in their patients; in fact, some of them found the causes of their patients' present-day problems in their past lives.

In the 1960s and 1970s, psychologist Helen Wambach did a detailed 10-year study of past-life regressions. She claimed to have regressed more than 1,000 subjects to live lives during the past 4,000 years, and she felt that her study showed that their regressions were genuine.

Today, hypnotism is considered a difficult tool at best. Many cases relating to crime, sexual abuse, abductions by extraterrestrials, and other matters have shown that it is relatively easy for the hypnotist, consciously or completely unconsciously, to lead the subject. Extreme care must be taken by the hypnotist not to suggest a direction for the subject to follow. For that reason, many skeptics do not consider past-life regression through hypnotism to be legitimate proof of reincarnation.

PAULING, LINUS C. (1901–1994) One of the 20th-century's leading chemists, born in Portland, Oregon, and educated first at the Oregon State Agricultural College (now Oregon State University) and then at the California Institute of Technology (CIT), from which he received his Ph.D. in 1925.

His main work in chemistry was in determining the nature of chemical bonds and molecular structure. In 1931, he published the milestone work on the subject, *The Nature of the Chemical Bond, and Structure of Molecules and Crystals.*

Pauling suggested that the benzene ring owed its stability to resonance; in 1946, he suggested that enzymes worked by lowering the reaction barrier by binding to a transitional state as atoms in a compound moved about the central core; in the early 1950s, he was in the race with Maurice Wilkins, Rosalind Franklin, James Watson, and Francis Crick—and many others—to find the structure of DNA and reported his work and theories freely. All these and other ideas were explored by other researchers and many proved to be fruitful. Some of the successes were his. In 1951, with Robert B. Corey, he showed that proteins had an alpha-helix structure, an advance in the understanding of complex organic molecules, which laid the foundation for later discoveries, including the structure of DNA. In 1954, he was awarded the Nobel Prize for Chemistry for his work on chemical bonds.

In the 1950s, Pauling became a very active campaigner against nuclear weapons and nuclear weapon testing, publishing *No More War* in 1958. In that same year, he presented a petition to the United Nations pressing for an end to nuclear weapon testing, signed by 11,000 scientists from many countries. In 1963, he was awarded the 1962 Nobel Peace Prize for his opposition to atmospheric nuclear-weapon testing and his other activities in the cause of peace. He was only the second person to be awarded two Nobel Prizes (Marie Curie being the first). In 1963 he left CIT and spent the next six years at the Center for the Study of Democratic Institutions at the University of California at Santa Barbara. He left that institution in 1969 in protest against the educational policies of Governor Ronald Reagan of California, moving to Stanford University. In his later years, still very full of ideas, he became convinced that massive doses of Vitamin C were effective in staving off colds.

See also VITAMIN C.

For further reading: Linus Pauling, *Vitamin C: the Common Cold and the Flu* (Freeman, 1976).

PEARY, ROBERT EDWIN (1856–1920) Arctic explorer usually credited with leading the first expedition to reach the North Pole in 1909.

Peary began his career in the U.S. Navy in 1881 and early on showed an interest in Arctic exploration. During his first expedition in 1886, Peary and Matthew Henson, his former black servant, traveled over the Greenland ice sheet for 161 kilometers (100 miles). Five years later in 1891–92 he returned to Greenland with seven others, including his wife and Frederick Albert COOK, a surgeon, this time sledging 2,092 kilometers (1,300 miles), finding evidence that Greenland is an island. The region of northern Greenland that he was the first to explore, Peary Land, is named after him. He also made a study of an isolated Eskimo tribe, the Arctic Highlanders, who later helped him in his unsuccessful first attempt to reach the North Pole in 1893–94. Peary's second attempt, in 1905, was similarly unsuccessful.

Nevertheless, accompanied by Henson and four Eskimos with sledges and dogs, Peary claimed success on his third attempt in 1908–09. On returning to the Greenland settlement of Etah, Peary learned that Frederick Cook, the surgeon on his 1891–92 expedition, had already claimed the Pole five days earlier.

Some time later Cook's assertion was found to be false, but during the disputation, Peary's records were scanned by a National Geographic Society subcommittee. His diary entries were minimal and only attested to by Peary himself, as those with him did not have the education to check his calculations. To have gained the Pole on Peary's calculations, his party must have traveled 690 kilometers (429 miles) in polar conditions between April 2 and April 9, an average of 86 kilometers (53 miles) a day—a feat questioned by many accustomed to trekking with dogs in this kind of habitat.

Despite opposition, in 1911 Congress passed a bill crediting Peary with being the first to reach the North Pole and retiring him from the navy with the rank of rear-admiral. His published works included *Northward over the "Great Ice"* (1898), *The North Pole* (1910), and *Secrets of Polar Travel* (1917).

For further reading: Dennis Rawlins, *Peary at the North Pole: Fact or Fiction* (Luce, 1973).

PERKINS, DR. ELISHA (1740–1799) Medical doctor who developed the Perkin's Patented Metallic Tractor, based on his belief that metals had the property of drawing out pain and diseases from the body. He patented the device in 1796. The tractor consisted of two 7.6-centimeter (3-inch)-long rods—one an alloy of copper, zinc, and gold and the other of iron, silver, and platinum. When the tractor was drawn over the affected part of the patient, it drew out the disease. Perkins sold his tractors at a fairly high price and had many impressed clients, including George Washington. His son Benjamin published a book of testimonials in 1796 and pushed the sale of his father's tractors both in the United States and abroad, particularly in England. Though some were impressed, orthodox doctors ignored the device.

PERPETUAL MOTION MACHINE (PERPETUUM MOBILE) A hypothetical device that remains in continuous motion without any sustaining force or added energy.

Long a goal of both tinkerers and inventors, the perpetual motion machine has been contemplated and attempted since classical antiquity. Designs varied widely, and employed various strategies including falling weights, unbalanced wheels, gears, pulleys, belts, sponges, and running water to achieve their goal.

From the 13th through the 19th century, perpetual motion machines exercised considerable influence on the imaginations of scientists and engineers who seemed little deterred by repeated failures. A few persons became famous for their devices, including Edward Somerset, the Marquis of Worcester (1601–67) and Jean Ernst Elie-Bessler, known popularly as "Orffyreus" (1680–1745). Working in the 17th and 18th centuries, both devised large "overbalanced" wheels with shifting weight near the rim that remained in motion for impressive intervals but always eventually slowed to a stop. During the 17th century, Robert Fludd (1574–1637), and Georg Andreas Böckler drew up plans for "closed mills," which employed standard milling apparatuses whose supply of falling water was constantly replenished by the machine itself from its own runoff. These too could be made to work for a time but eventually would grind to a halt.

Some of the more famous names in the history of science engaged in the pursuit of perpetual motion. Leonardo da Vinci (1452–1519), for example, sketched a number of designs for overbalanced wheels in his notebooks, although these were probably based upon earlier designs of other inventors. More original were the musings of 18th-century Swiss mathematician Johann Bernoulli (1667–1748) and 17th-century Irish chemist Robert Boyle (1627–91), who both had expertise in hydrodynamics. Bernoulli believed that solutions of dissimilar densities could be brought into dynamic interaction, resulting in constant flow of liquid. His rather simple device never succeeded because of the difficulties involved in producing a necessary filter that would separate the lighter from the denser liquid.

Boyle thought that perpetual motion might be possible using capillary action. Nevertheless, he never developed a workable device, believing that human artifice could not match the immense scale involved in the natural elevation and seepage of water. Later seekers did develop machines based in part upon capillary action. Many of these employed a circle of linked sponges that would absorb water during one part of a cycle, only to have it squeezed out during the next. The constant imbalance between the wet and dry sponges, in theory, would maintain the motion of the chain. In practice, however, these designs could not stir from static equilibrium without external help and could not remain in motion for very long.

The search for a perpetual motion machine ended in the mid-19th century with the articulation of the first and second laws of thermodynamics. These laws express the principle of the conservation of energy and the ability to translate heat energy into work. They make it clear that mechanical systems cannot operate continuously without external help. Some energy is always lost in the process of doing work, and therefore no closed system can overcome the forces that tend to return it to a static state. Interest in perpetual motion machines, however, has not entirely subsided. In the late 19th and the 20th centuries, claimants have continued to press their case. In this period, as in earlier ones, elements of hucksterism and flimflam can be found alongside genuine attempts at innovative engineering. Many devices represented in public demonstrations as genuine perpetual motion machines, in fact, had hidden motors or compressed air systems that maintained motion.

In spite of the failure to produce bona-fide perpetual motion machines, the search itself no doubt contributed to the craft of engineering by refining ideas about mechanical efficiency and the reduction of friction. Similarly, scientific theorizing about the nature of heat, energy, and work benefited from some of the practical knowledge gained by the inventors who pursued what is now thought to be an impossible dream.

See also OFFYREUS'S WHEEL.

For further reading: Arthur W. J. G. Ord Hume, *Perpetual Motion: The History of An Obsession* (Allen & Unwin Ltd., 1977).

PERUNA The single best-selling NOSTRUM at the beginning of the 20th century. Peruna appeared on the market at some unknown date, probably in the late 1880s. The formula was discovered (or concocted) by S. B. Hartman of Columbus, Ohio, who believed that the bane of the human race was catarrh, an obsolete term referring to an inflammation of any mucous membrane accompanied by an increase of mucus. The most well-known example of catarrh is, of course, a common head cold with a runny nose. Hartman claimed that catarrh of all parts of the body afflicted humanity and that his product, Peruna, cured catarrh.

At some point in the early 1890s, as Hartman struggled to have Peruna recognized among the hundreds of competing panaceas, he received a surprise order from Waco, Texas, for a full rail car of his concoction. While his staff worked overtime to fill the order, Hartman went to Texas to meet Frederick W. Schumacher, the man who had placed the order. Hartman later hired Schumacher to develop his advertising program. Under Schumacher's guidance, sales of Peruna rapidly increased, seemingly based upon the growing amount of newspaper advertising. It claimed among its supporters a variety of the rich, famous, and influential—including a number of governors, congressional representatives, and senators.

While Peruna had its supporters, it gained its enemies as well. *The Ladies Home Journal* was editorially opposed to nostrums in general and Peruna in particular. In the fall of 1905 it ran a story of its research on Peruna that included the results of a test showing the alcohol content relative to beer. The *Journal* labeled Peruna no better than cheap whiskey. Hartmann did not escape from the attention of the famous series of articles also written in 1905 by Samuel Hopkins ADAMS (1871–1958) for *Collier's*. Adams discovered that the government had forbidden the distribution of Peruna on Native American reservations. He informed his readers that it consisted of nothing but alcohol, water, some flavoring, and burnt sugar (for color) and challenged Hartman either to put some real medicine in his product or simply to open a bar.

In 1907 Hartmann responded to his critics and the imminent passage of the Pure Food and Drug Act by dropping the alcohol content of Peruna from 27 percent to 20 percent. He also added a quantity of two herbs, senna and buckthorn and limited the claims for his medicine's healing properties. Sales plunged and never recovered. In an attempt to recover his profits, Hartman began to manufacture Peruna under a new name, Ka-Tar-No, and sold it as an alcoholic drink. Ka-Tar-No never did well, but Peruna continued at a reduced popularity for several decades; it gained a new audience when radio arrived in the 1920s.

Both Hartman ad Schumacher died wealthy men. The latter left his fortune to the Columbus Gallery of Fine Arts.

PHILADELPHIA EXPERIMENT A mythological experiment in which the U.S. Navy beamed a ship and its crew between two naval dockyards. The 20th-century myth was put about in the United States after World War II. The story goes that during the war the U.S. Navy carried out top secret experiments using very advanced technology. There are various accounts but the general gist is that they involved one or more highly improbable processes: making ships and their crews invisible, TIME TRAVEL, EXTRATERRESTRIALS, and/or beaming up a ship and its crew from one naval dockyard in Philadelphia and materializing them in another—Norfolk, Virginia. There is no hard or reliable evidence that this, or anything like it, ever occurred but, nevertheless, the myth has lived on with extraordinary persistence.

In 1956, Carl Allen approached Morris K. Jessup, author of *The Case for the UFO* (1955), claiming that the navy had conducted an experiment in 1943 rendering a ship—which much later he identified as navy destroyer DE173—invisible, and teleporting it from Philadelphia to Norfolk and back minutes later. The crew was very badly affected, so much so that the navy abandoned further experiments along these lines. Despite its implausibility, the story continues to have some currency to this day, sometimes embroidered in its later versions. Allen himself has admitted more than once that it was a hoax and subsequently retracted his admissions.

In 1984 the story was the basis for a movie, *The Philadelphia Experiment,* in which time travel was added to the earlier version. Later still, in 1990, Al Bielek claimed to have been an officer, Edward Cameron, on board DE173 in 1943, transported forward in time to 1984, and then back to 1943. His memory of the events was erased and was only recently recovered, hence his very late support for Allen's story.

There seems little doubt that the whole idea of the Philadelphia Experiment is a product of Allen's fertile imagination, accepted in various forms by those open to such stories.

For further reading: Ted Roach, *The Physics of a Flying Saucer* (Roach Industries, 1997); and Douglas Rushkoff, *The Ecstasy Club* (Hodder and Stoughton, 1997).

"PHILIP" An artificial GHOST produced in one of the most important experiments in survival. In September 1972, eight members of the Toronto Society for Psychical Research formed a séance circle. Instead of waiting for whatever spontaneous manifestations might occur, they created a fictional character, "Philip," for whom they supplied a history and description. A. R. G. Owen and Joel Whitton sat in on the group as observers. After initially negative results, the group adopted the suggestion of British psychologist Kenneth Batcheldor—to use the various methods of traditional spiritualism to create an atmosphere in which manifestations seemed natural—sitting around a table, singing, etc. Within a few weeks "Philip" began to respond with raps and table movements and emerged as an independent personality ready to respond to the group's inquiries. "Philip" filled out the story of his life and acted like any other purported spirit entity. The story, described in full in a book by the two major experimenters and videotaped for a television show, called into question the conclusions of paranormal investigators about survival after death. It cast into doubt their hypothesis of the existence of a spirit entity as the most likely explanation of séance room phenomena.

For further reading: Iris M. Owen and Margaret Sparrow, *Conjuring Up Philip* (New York: Harper & Row, 1976).

PHILOSOPHY OF SCIENCE ON THE DEMARCATION OF SCIENCE FROM PSEUDOSCIENCE The matter of how pseudoscience is differentiated from science. This is difficult question to address because neither term has a precise and universally accepted definition. A trawl through some dozen books on the philosophy of science shows that only one had considered this question, which is a clear indication that this is not a matter with which philosophers concern themselves. The outstanding exception is Sir Karl POPPER (1902–1994). By his own account, he enunciated the principle of falsifiability as a young man in response to his disillusion with the spurious claims of personality psychology and Marxist political economy. This principle says that a claim—proposition or theory—can only be considered scientific if a test can be devised that could, if failed, show it to be wrong. As Popper explained in *The Logic of Scientific Discovery* (1934) and in *Conjectures and Refutations* (1963): It is falsifiability, not verifiability "that is to be taken as a criterion of demarcation" between scientific and nonscientific (or pseudoscientific) theories. "A theory which is not refutable by any conceivable event is non-scientific; . . . every genuine *test* of a theory is an attempt to falsify it." In Popper's view, there were many areas being given attention and claiming to be scientific that defied contradiction; their supporters were convinced of their truth, and no test could be devised which would shake that belief. If opponents produced what they thought was evidence that discredited the theory, its supporters shifted ground, shouted foul, explained that the conditions were unfavorable, or in some way sidestepped the objection. They *knew* they were right. There was no test that they would accept which would shake their belief. To Popper, such theories were pseudoscience and their advocates pseudoscientific. Yet, such people claimed that they *were* scientists because they were always verifying their theories in the way approved by the "positivists" of the Vienna circle. Also, Popper's principle had another important significance: It was essential to the progress of science—science develops through scientists' conjectures and nature's refutations. For the young Popper, *real* science was shown by Einstein's invitation to the astronomers to refute his general theory of relativity with the results of the forthcoming eclipse of 1919.

This was an interesting and persuasive argument, and other philosophers gave it some consideration—a whole new area to be explored. They soon showed it to have problems, principally that the history of science showed that in practice scientists didn't work like that. Maybe the science community *ought* to reject immediately any theory that failed the agreed test, but the evidence showed that the community *doesn't* (or, at any rate, *hadn't*); many respectable and respected scientists had doggedly clung to their theories long after they had failed the early test or tests and had behaved in the way Popper had described as pseudoscientific. Only after some time and repeated and refined testing had their theories triumphed. For example, both classical physics (Copernicus and Newton) and relativity physics (Einstein) would have been summarily rejected if Popper's falsifiability criterion had been ruthlessly applied. Furthermore, it is not always possible to get agreement on an acceptable falsification test. And adherents and skeptics disagree about whether a particular result is a falsification, and theories that the science community is adamant are pseudoscientific are sometimes open to falsification, for example Rupert SHELDRAKE's hypothesis of FORMATIVE CAUSATION.

Chief among Popper's critics were Thomas S. KUHN (1922–96) and Imre Lakatos, each offering different criteria for defining what in practice was scientific behavior. But both were primarily concerned with reconciling the observed behavior of those who are or were unquestionably scientists with some revised version of Popper: rational/irrational, sustained defense, and so on. Lakatos proposed shifting the issue of demarcation away from Popper's concentration on hypotheses onto research programs: If a program is progressive, getting steadily stronger, and obtaining more and more testable results, then it is on the science side of the line; if it is degenerating, running into more and more problems, having to invent new hypotheses, and reducing its claimed achievements, then it is pseudoscientific. But that runs into trouble too. All the work in the 19th century that tried to detect motion through the ether—surely by reputable scientists and undoubtedly science, though doomed to failure—would be classed on this criterion as pseudoscientific. Philosophers of science have continued to gnaw away at this problem, seeking to find a generally agreed solution, without much success.

A focus of attention has been with what happened at the times that one scientific paradigm gave way to another and how scientists in both the old and the new camps behaved—with less attention turned to pseudoscience. Part of the difficulty has been that very often different sorts of nonscientific matters have been lumped together. Philosopher Mario Bunge distinguished thus:

Pseudoscience is any cognitive field that is nonscientific and yet is advertised as scientific; it is this *claim* to be scientific that separates the pseudoscientific from other nonscientific matters.

Despite any shortcomings, Popper's falsifiability principle remains the philosophy of science's best rule-of-thumb means of distinguishing science from pseudoscience. Science progresses by a process of trial and error; pseudoscience does not. In broad terms, that seems clear enough. It is in the fine detail and in the heat of controversy that it becomes difficult, but it is a job that needs to be done. As Marcello Truzzi writes in *The Encyclopedia of the Paranormal:* "There is surely bunk in the world in need of debunking, but there is always the potential for premature dismissal of a protoscience in the name of fighting pseudoscience." However, it should be accepted that not all the onus falls on either the philosophers or the scientists. The main responsibility for establishing a claim to be scientific is with the claimant—the need to be open to new ideas must be balanced against the need to produce extraordinary proof for extraordinary claims.

Another approach to defining pseudoscience has been suggested by philosopher Jerry Ravetz in connection with quantitative methods. He has observed that in a variety of fields, which are accepted as scientific because they use quantitative data and mathematical methods, there is no control of the uncertainties at any stage of the process. He takes over the American acronym GIGO (Garbage In, Garbage Out) from computer practice and defines "GIGO-Science." This is a field claiming to be scientific where the uncertainties in inputs must be suppressed, lest the outputs become indeterminate. A symptom of GIGO is that precision of numerical outputs goes up as accuracy of quantitative inputs goes down. Although the criterion has not yet been applied systematically, candidates for such status are fields where computer models are used for the analysis of social and natural phenomena.

PHLEBOTOMY The practice of opening a vein as a therapeutic measure, also called venesection, bloodletting, or "bleeding." A historically standard treatment for a vast array of ailments and considered the chief remedy for some, phlebotomy did not fall entirely out of fashion until the late 19th century. A symposium conducted by the Philadelphia County Medical Society in 1860, discussed reasons for the rapid decline of bloodletting. Physicians noted the change in types of disease and in the constitution of patients, the proliferation of propaganda by homeopaths and other vitalistic thinkers, and empirical substitution of other remedies for bloodletting as among the causes of its decline. The abandonment of bloodletting as a remedial agent entailed a shift in therapeutical thinking, from the idea that disease could be forced out of the body to the realization that gradually "building up" and conserving strength were essential for recovery from illness.

PHLOGISTON The invisible material believed for many years to be the source of heat. To heat a body was to add phlogiston—phlogistication; to cool it was to take phlogiston away—dephlogistication. Today that seems pseudoscientific, but at the time, it was an explanation of what was observed in heating and cooling. So, for example, Lancelot Hogben in *Science for the Citizen* (George Allen and Unwin, 1938) says:

> *The doctrine of phlogiston, which was the last attempt to sustain the elemental nature of fire, [the four elements of which everything was supposed to consist of were earth, air, fire, and water] was concocted towards the end of the seventeenth century. It provides an instructive example of the way in which facts may be used to* illustrate *instead of to* test *the truth of a theory. The argument runs as follows. It is self-evident that if things burn, they must contain the fire principle. A combustible substance is, therefore, a combination of a calx or non-combustible material with the fire principle phlogiston. The escape of phlogiston when a combustible substance burns is accompanied by production of incombustible material which actually weighs more than its predecessor. It is therefore self-evident that phlogiston must be endowed with the opposite of weight,* levity, *or the power to make a body weigh less. Much valuable time was wasted in disproving a theory with nothing to commend it but the elegance of flawless reasoning from premises which have no foundation in fact.*

Isn't hindsight wonderful!

By contrast J. D. Bernal in *Science in History* (1983) writes:

> *We are apt, looking at it from the point of view of its immediate successor—the theory of combustion as oxidation—to treat the phlogiston theory as absurd; in fact it was an extremely valuable theory and it co-ordinated a large number of different phenomena in chemistry. It proved a good working basis for the best chemists of the mid-eighteenth century, and was firmly adhered to till the end by many of them, including the man whose experiments were to destroy it, Joseph Priestley.*

The theory had its problems—this became more evident as chemists developed quantitative methods. Some substances, wood for example, lost weight on burning, leaving a dephlogisticated residue—ash; others such as metals gained weight. Priestley's discovery of oxygen provided the key. Lavoisier showed between 1770 and 1790 that it was invariably involved in any combustion, that burning was the addition of oxygen. So if the wood ash and the combustion gases were both weighed, there was an increase in weight, that of the added oxygen. When a metal was burnt, oxygen was added—oxidation; when iron ore was "burnt" in a furnace to give shining metallic iron, oxygen had been removed from the ore—reduction. Careful quantitative experiments showed that the sums added up, and by 1800 Lavoisier's oxygen theory was generally accepted and the phlogiston theory abandoned.

The foregoing is an abbreviated and simplified account of the problem that dominated much of chemistry during the 17th and 18th centuries and rumbled on into the early 19th century. With hindsight, it all seems so readily understandable and the confusions and complications so unnecessary. But without a clear knowledge of what constituted a chemical element and what those elements were, with only a dawning realization of the nature of chemical reactions and a limited appreciation of quantitative measurements, disentangling what was happening must have been very difficult. The phlogiston theory was an important step in that process.

For further reading: R. C. Olby, ed., *Late Eighteenth Century European Scientists* (Pergamon Press, 1966); and David Steele, ed., *The History of Scientific Ideas* (Hutchinson Educational, 1970).

PHOTOGRAPHIC EVIDENCE Film developed by Semyon Kirlian in the 1940s that had been exposed while in a powerful electric field, showing auras around human skin and other living matter. It was reported in the 1970s in the West in *Psychic Discoveries behind the Iron Curtain.*

Photographs have also been produced to support other claims of strange and supernatural happenings, the most well-known being the many photos purported to be of UFOs and the photos of the COTTINGLEY FAIRIES. One example of the former were photos that Ed Walters of Gulf Breeze, Florida, claimed to be of a UFO that he had seen. They were pronounced absolutely genuine by Dr. Bruce Maccabee of the Fund for UFO Research. Later a neighbor, Tommy Smith, disclosed that he had helped Mr. Walters to fake UFO pictures and a model flying saucer was found in Walters' attic. The Cottingley Fairies were for many years forcefully asserted by many, including Sherlock Holmes author Sir Arthur Conan DOYLE (1859–1930), to be genuine photos of fairies, without doubt. These photos were taken in 1917 by two young girls, Frances Griffiths, aged 10, and her cousin Elsie Wright, aged 16, who maintained that they had seen the fairies at the bottom of the Wrights' garden in Cottingley, Yorkshire, England. Elsie had photographed Frances with four of the fairies dancing before her. Later the two girls produced other photos showing fairies in their company. Skeptics, Frances's father among them, were unable to shake the girls' story, and spiritualists and occultists seized on it as incontrovertible proof of the existence of the little people. Attempts by the two girls four years later, in 1921, accompanied by the clairvoyant Geoffrey Hodson, to obtain similar photos were unsuccessful. Doyle founded the Society for the Study of Supernormal Photographs. In 1926 he published *History of Spiritualism,* in which he maintained that the fairy pictures were genuine. Many years later, in the 1970s, the women were interviewed on two occasions on television programs. When asked about the photographs Elsie said, "I've told you they're photographs of figments of our imagination and that's what I'm sticking to," and Frances supported her.

See also KIRLIAN PHOTOGRAPHY.

For further reading: Martin Gardner, *Science, Good, Bad and Bogus* (Prometheus Books, 1989); Stuart Gordon, *The Book of Hoaxes* (Headline, 1995); and Edward L. Gardner, *Fairies: The Cottingley Photographs and Their Sequel* (Theosophical Society Publishing House, 1966).

PHRENOLOGY The belief that the brain is actually a composite of various "organs" that localize various social, moral, and intellectual qualities; also known as craniology, organology, and bumpology. By measuring the surface of the skull, the relative size and strength of the organs can be estimated. This information can then be used to assess an individual's personality, moral capacity, and intellectual aptitude.

As a historical phenomenon, phrenology can be related on one hand to various DIVINATION techniques that purport to read the inner character and fate of a person from various external markings on the body—for

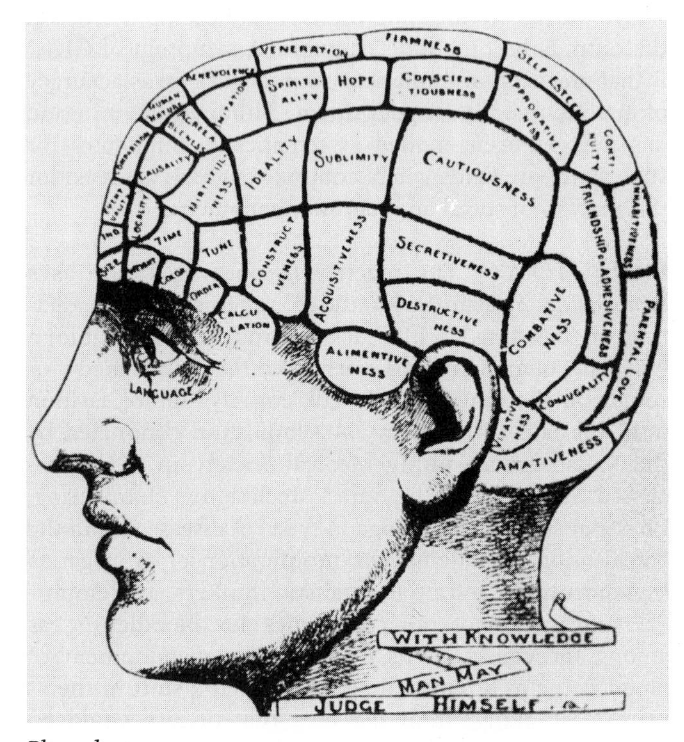

Phrenology

example PALMISTRY, PHYSIOGNOMY, and the Renaissance practice of metoposcopy (reading foreheads). On the other hand, it can be related to long-standing speculation about the precise relationship of the brain to human behavior. Often, as in the medical teachings of GALENISM, this resulted in theories of brain localization, which related particular areas of the brain to particular mental attributes like imagination and memory.

Phrenology as it has come to be known in the modern sense stems from the work of two 19th-century Viennese doctors, Franz Joseph GALL (1758–1828) and Johann Christophe Gaspar SPURZHEIM (1776–1832). Building on Gall's anatomical studies and measurements of various heads, the two began to lecture on the relationship of the brain to human behavior to appreciative audiences in Vienna in the 1790s. Quickly denounced by authorities for being subversive to religion and morals (a charge they vehemently denied), they were obliged to take their lectures to other cities in Europe, where they were generally well received. Eventually they settled in Paris, where they produced the key foundational work of phrenology, *Anatomie et Physiologie du système nerveux en général et du cerveau en particular,* four volumes (1810–19). Revolutionary insofar as it advocated a thoroughly biological (as opposed to philosophical, HUMORAL, or mechanical) approach to the study of the mind, the work stressed the importance of the structure and organization of the brain and nervous system to any understanding of human behavior.

The scientists' most central claim was that the human brain was actually a composite of 27 distinct measurable organs that accounted for all mental activity and behavior. Later phrenologists would add to this total, but most followed them in organizing the mental organs into two general classes: (1) animalistic traits such as the sexual instinct and self-preservation, and (2) qualities exclusive to human beings such as the religious sentiment and comparative wisdom. General considerations looked to the size of the organs in this second "moral and intellectual" group, particularly in proportion to the more "animal" organs of the first group, as an index of intelligence, moral vigor, and religious devotion.

Gall and Spurzheim's scientific credentials were more than adequate for their era. Likewise, the techniques they used to develop their allegedly scientific claims, observation, and correlation were widely accepted. Both men had medical degrees and earned widespread respect and admiration for their innovative dissecting skills. In developing their original scheme of phrenological organs (which was the basis of all later claims) they combined their anatomical expertise with head measurements and personality information from a cross-section of different people. They then correlated the data to develop a portrait of the various regions and attributes of the human brain. Later critics would fault Gall and Spurzheim for basing their mental categories on less than scientific criteria (such as anecdote, philosophical taste, and personal whim), but for subsequent phrenologists their map of the cranium was an indisputable empirical fact.

After breaking with Gall in 1813 because of a perceived lack of recognition and other differences of opinion, Spurzheim began to popularize phrenology as a social reform philosophy. Phrenological ideas were subsequently disseminated throughout Europe, North America, and Australia. Such key advocates as Scotsman George COMBE (1788–1858), American Orson FOWLER (1809–75), and Spaniard Mariano Cubi y Soler (1801–75) saw phrenology as a positive new "science of the mind" and promoted it as the basis for a variety of reforms including temperance, public education, the treatment of the mentally ill, and penology. At the popular level, phrenology was best known for its "head readings," which purported to reveal personality and character through the careful examination of the shape of the cranium. Although nominal interest in phrenology developed in virtually all social strata, its core constituents were the rising middle classes and astisans who were keenly interested in both social reform and finding new ways of fashioning personal identity in the face of rapid industrialization. Despite ongoing criticism from both scientific and religious sources, enthusiasm for phrenology grew robustly throughout the first half of the 19th century; thereafter, support fell off sharply, although pockets of adherents survived into the twentieth century.

Opponents attacked phrenology on various grounds and became more vocal and alarmist as the movement grew in popularity. Antiphrenological literature often highlighted theological and philosophical issues. Religious intellectuals and clergy, for example, often accused phrenologists of promoting materialism and fatalism insofar as all human activity and abilities, even religiosity, were understood as manifestations of particular areas in the brain. Many believed, in spite of ardent protestations from phrenologists, that the "science of the mind" had dispensed with the need for a soul, self, or free will. Other detractors felt that the phrenologists' belief in innate mental categories led to deterministic attitudes about antisocial behavior and low achievement.

Scientific critics tended to disparage the specific claims made by phrenologists about their various cranial "organs." Some pointed out that external cranial measurements were not reliable estimates of the size of the brain as the thickness of skin, bone, and other tissues differed from person to person. Others doubted that the mind could be divided into the discrete categories at all, much less measured from external appearances. Perhaps the most damning evidence against phrenology came from a group of scientists who actually shared a number of similar assumptions. By the middle of the 19th century, many anatomists and physiologists who studied the brain advocated, like the phrenologists, the biologically based study of psychology and behavior as well as the localization of certain mental functions. As they began to amass evidence from pathological cases and animal experimentation, however, it became evident that the phrenologists had wildly oversold their claims. By the 1850s, phrenology had lost most of its important supporters from established scientific and medical communities and was well on its way to being classified as a pseudoscience by all but its advocates. Significant scientific work continued in related areas, however, particularly on the localization of sensory and motor functions. In this sense, phrenology contributed to the rise of the modern study of neuroanatomy and neurophysiology.

For further reading: Roger Cooter, *The Cultural Meaning of Popular Science: Phrenology and the Organization of Consent in Nineteenth-Century Britain* (Oxford University Press, 1984); John D. Davies, *Phrenology, Fad and Science: A 19th-Century America Crusade* (Archon Books, 1971); and Robert M. Young, *Mind, Brain and Adaption in the Nineteenth Century: Cerebral localization and its biological context from Gall to Ferrier* (Clarendon Press, 1970).

PHYSIOGNOMY Judging character from features of face or form of body. The idea goes back to Aristotle who argued by analogy: Those with features resembling an animal might be expected to show similar characteristics. So, for example, someone with a face like a fox would be cunning. Despite its long history and its eminent and respected originator, the idea has generally fallen into disfavor and today is regarded as not supported by the evidence and therefore pseudoscientific. Nevertheless, it has been revived many times and has had other eminent advocates. Nineteenth-century physician Cesare LOMBROSO (1835–1909) claimed that a large proportion of criminals were biologically determined and that their predisposition to criminal behavior could be seen in their facial features. In the 18th century, Swiss J. K. Lavater was an enthusiastic advocate.

To discredit the idea totally is difficult. That there are cases where facial features do have implications cannot be denied. So, for example, doctors can diagnose, almost from the moment of birth, whether a newborn child has trisomy-21 (Down's syndrome or mongolism)—the eyelids, the tongue, and the general facial conformation are the indicators. Other genetic predispositions, such as cretinism or hydrocephalus, have their facial or bodily markers. But this is very different from the greater claim of physiognomy to be able to assess a wide range of personal characteristics from physical features. One claim, often made, is that people with long fingers are likely to be good pianists; a quick trawl through great pianists shows this to be mistaken. There have been several thorough surveys during a 200-year period that show convincingly that the correlation between physical and psychological characteristics is so low as to be negligible. In today's environment, the simple physiognomic relationship has been supplanted by genetic and biochemical influences: If a person has certain genetic or biochemical abnormalities, these will produce certain behavioral effects. Biological determinism is a very seductive idea, and it will probably never go away.

PILTDOWN MAN Paleontological fraud. Perhaps the most famous fraud in the history of paleontology, the faked remains of Piltdown Man (*Eoanthropus dawsoni*) influenced theories about human origins for more than 35 years. The hoax began in 1912, when a lawyer and collector named Charles Dawson visited Arthur Smith Woodward, the keeper of the British Museum geological collection. Dawson claimed that he had found some ancient skull fragments in a gravel quarry in Sussex, England. Together Smith Woodward and Dawson, accompanied by the Jesuit priest Pierre TEILHARD DE CHARDIN searched the quarry for more remains. On one of these expeditions, Dawson uncovered a lower jaw that looked as apelike as the skull fragments looked modern. However, two teeth still embedded in the jaw showed signs of wear like human teeth. Based on this evidence, Smith Woodward described *Eoanthropus dawsoni* in December 1912 as the "missing link" in the evolutionary chain between apes and humans.

From the beginning, some scientists expressed doubts that the Piltdown jaw and the skull belonged together. Two significant parts of the lower jaw were missing—the chin and the joint where the lower jaw connects to the upper jaw. Other finds quieted the questioners. In 1913, Teilhard de Chardin discovered a canine tooth that again combined characteristics of apes and humans. In 1915, Dawson found two more skull fragments and a tooth at another site some miles away. Piltdown Man went unquestioned by most scientists until 1949 when Kenneth Oakley proved that the remains had not been in the quarry very long. Finally, in 1953, Oakley, working with colleagues, discovered that the bone fragments had been artificially stained and that the teeth had been filed to resemble human teeth. The skull fragments were in fact from a modern human skull; the jaw was an orangutan's.

One of the most intriguing questions about the Piltdown forgery is: Who was responsible for the fraud? Most investigators blamed Dawson, who made most of the discoveries and had the best opportunity. Others implicated rivals of Dawson, amateur collectors jealous of his success. Some leading paleontologists, including Stephen Jay GOULD, have suggested that Teilhard de Chardin was a coconspirator with Dawson.

Equally as interesting as the "whodunit" question is the question of why the forgery went undetected for so long. Louis Leakey recalled that the remains were kept under lock and key and that scientists were never allowed to examine them closely. Gould suggests that the remains supported what people wanted to believe: Most scientists believed that humans evolved a large brain first, while keeping other apelike characteristics. For English nationalists, Piltdown Man gave England an important place in the history of humankind, rivaling the French Neanderthals and Cro-Magnons. For Eurocentrists, it gave scientific evidence for claims of the antiquity and superiority of white European culture.

For further reading: Stephen Jay Gould, "Piltdown Revisited," *The Panda's Thumb: More Reflections on Natural History* (W.W. Norton & Co., 1980); and Joseph S. Weiner, *The Piltdown Forgery* (Oxford University Press, 1955).

PINKHAM, LYDIA E. (1819–1883) The most successful manufacturer of proprietary medicine in 19th century America. Lydia Pinkham was born and raised in Lynn, Massachusetts. Under her maiden name, Lydia Estes, she became a schoolteacher, and an activist in the causes of slavery abolition, temperance, and women's rights, and a devoted disciple of the natural diet program of Grahamism. She was also deeply impressed by the theories of Jacob Bigelow of the Harvard Medical School, who suggested that many diseases were self-limiting and would disappear more rapidly if left to the natural recuperative powers of the patient. In 1843 she married Isaac Pinkham.

During the 1870s, Lydia Pinkham began to manufacture and market an herbal remedy for "female complaints." The venture was undergirded by one of the most impressive advertising initiatives in U.S. history. At first, the bulwark of the advertising program was a small pamphlet written by Pinkham describing in plain terms the nature and remedy of female disorders. These were distributed initially by the thousands, and later in the millions, throughout the cities of New England. In 1901 the pamphlet was expanded to 64 pages and translated into five languages. Second, Lydia's grandmotherly, comforting face appeared on the cover of the pamphlet and, beginning with an early ad on the front page of the *Boston Herald,* it later appeared in newspapers across the United States. The initial front-page ad in the *Herald* produced such spectacular results that similar ads spearheaded the company's growth.

Following Lydia Pinkham's death in 1883, her son Charles, her daughter Aroline, and Aroline's husband William H. Gove (the lawyer who had trademarked the Pinkham name) formed a corporation. Under their guidance, Lydia Pinkham's vegetable compound went on to new successes and new markets in Canada, Mexico, Spain, and even China.

When broad attacks on patent and proprietary medicines began in the early 20th century, the compound was only lightly touched. Even Food and Drug Administration investigator Morris FISHBEIN had very little but praise for it. Except for the revision of its original broad medical claims and the reduction of the percentage of alcohol content (from 18 to 13 1/2 percent), the compound passed several chemical analyses that revealed it actually had healthful effects on the conditions for which it was advertised. Among its ingredients was estrogen, a female hormone.

By the beginning of the 20th century, Lydia Pinkham's face had become the most recognizable in North America. She had entered the folklore of the country and had become the subject of a several tongue-in-cheek songs that were heard in bars, clubs, and medical conventions. Only in the last generation have new products been developed that have pushed Lydia Pinkham's Vegetable Compound out of the market.

PIRI REIS MAP A map discovered in 1929 by historians in the Palace of Topkapi, Istanbul. It shows, in remarkably accurate detail, the coastlines of North and South America and the geography of Greenland and Antarctica *below their ice sheets.* The mapmaker also had accurate knowledge of relative longitudes.

The legend of this map's production is as follows: Piri Reis, a 16th-century Turkish admiral, was an avid collector of old maps at a time when all maps were somewhat inadequate and charted only coastal waters, which was not surprising. The more maps an admiral had, the more he was able to cross-check them against each other and against his experience, thus improving his navigation. Among Reis's captives taken in an early 16th century sea battle was a man claiming to have been one of Columbus's pilots on his epic voyages to the West Indies. He further claimed that Columbus had not sailed west acting only on a hunch but that he had had maps, which were still in his, the captive's, possession. The admiral appropriated them and used them, together with his other old charts, to construct the Piri Reis map in 1513—the one discovered by 20th-century historians.

Since the discovery of the map, cartographers, assisted by the U.S. Navy Hydrographic Bureau, have examined it carefully. It contains much accurate information on geographic details that are not believed to have been known prior to 1513: relative longitudes, the fact that Greenland is two islands below the covering ice sheet, the structure of Antarctica below its ice cover, and so on. We have extensive written records from ancient Egypt, Greece, and Rome, the Moors, Spain, and Portugal; none contain information about, for example, Antarctica (which was not discovered until 1818). The accurate determination of longitude came from the work of the astronomers and clockmakers of the mid-18th century.

No one knows if the Piri Reis map is an elaborate hoax perpetrated in the 1920s, the inspired work of Admiral Piri Reis putting together ancient charts that had been made by navigators in preceding centuries who possessed knowledge, lost for hundreds of years—of continents, coastlines, mountain ranges, and sub-ice detail—of which there is no other record. But the ice cover of Antarctica is millions of years old; how did these ancients know what lay beneath and what equipment had they to probe beneath the ice? Could this map have been produced in some paranormal way? Arlington T. Mallerey, an authority on ancient maps, after careful scrutiny, stated that "it was evident that there was very little ice then, at either pole. But, secondly, they had a record, for example, of every mountain range in Northern Canada and Alaska, including some ranges that the Army Map Service did not have. The U.S. Army has since found them! Just how they were able to do it we do not know."

PLACEBO EFFECT Phenomenon of curing people without administering medication. Statistical tests have shown that sick people who have taken "sugar pills"—tablets with no drugs in them—but who are told that they have taken drugs often seem to recover, even though their diseases have not really been treated. In "blind trial" experiments, some patients are given healing medication, and others are given tablets with no medical value. The patients do not know that some of the tablets have no drugs. A significant number of those receiving the placebo recover anyway, apparently just by taking what they believe is medicine. This psychological factor in the treatment of disease is known as the placebo effect. Although doctors sometimes use the placebo effect to treat a patient, the effect is difficult to reproduce and many physicians regard it as a statistical anomaly.

Doctors have long noted that a patient's attitude plays a significant role in his or her recovery. People who trust doctors and their medication stand a better chance of recovering faster and more completely than people who do not. This attitude plays an important role in the treatment of long-lasting diseases with no apparent cause, such as chronic pain. In some cases, chronic pain can be treated through the use of placebos. Pain is reduced because patients believe they will improve. As a result they become less anxious and more confident and their pain diminishes, even though they haven't taken any therapeutic drugs. Some forms of medical treatment, such as HOMEOPATHY, rely almost entirely on this result for their effectiveness.

The problem with the placebo effect is that, while it is a statistical phenomenon that can be documented throughout a group of people, its effects cannot be accurately predicted on a given individual. Its effectiveness on a person can also vary through time, depending on that person's state of mind. As a result, it is very difficult for doctors to know when the placebo effect will be an effective treatment.

PLANCHETTE An instrument created in 1853 for communication with supposed spirit entities. It was named for its inventor M. Planchette, a French spiritualist, and was used by his fellow believers for the next 15 years. The planchette consisted of a simple heart-shaped piece of wood mounted on two wheel-castors, with a sharpened pencil serving as a third leg. The planchette was placed on a piece of paper with the pencil point down. The operators placed their hands on top of the device. Spiritualists found that the instrument would often begin to write spontaneously in answer to questions. It had an obvious advantage over raps and table-tipping in that it could respond to question that were not phrased in a yes-or-no format and could even write messages that were not related to the sitter's questions.

In 1868 a U.S. toymaker produced a toy planchette, which led to the device becoming a popular party item for people wishing to experiment with SPIRITUALISM. The device was widely used among spiritualists into the early 20th century. It competed with the OUIJA BOARD, by which it was eventually replaced. During its heyday, however, it was used to write complete books of spirit messages.

While spiritualists claimed that the planchette served as a facilitator of spirit communications, critics both within and outside of the PSYCHICAL RESEARCH community quickly found alternative explanation for the speedy movement of the device. The critics suggested that the planchette moved through subtle muscle movements created by the operator's unconscious mind. The messages received through the planchette bear a remarkable resemblance to those obtained with the Ouija Board and the pendulum.

PLEIADES An open cluster of stars in our galaxy, the Milky Way, that is supposed to be the source of spirit messages that are being channeled through MEDIUMS. It is a group of about 500 stars on the border of the constellations Taurus and Perseus, 410 light years away; only about 10 of them are visible with the naked eye.

The messages that are claimed to be transmitted from Pleiades are instructions to the medium's client to meditate, to pray, and to purify him- or herself for the New Age when Earth will take its rightful place in a new galactic order. The last is sometimes coupled with the information that Earth is now in the grip of a sinister extraterrestrial conspiracy that is putting implants into our skulls. Some added credence was given to these claims when it was recently discovered that, under certain conditions, one of the main oblique passages of the Great Pyramid of Giza (the purpose of which is unknown) is aligned with the Pleiades.

See also NEW AGE MOVEMENT; PYRAMIDOLOGY.

PLINY THE ELDER (23/24–79 C.E.) Roman encyclopedic writer whose extant writing lies on the boundary between the science and the pseudoscience of his times. He was born into a wealthy equestrian family and died during the eruption of Vesuvius by breathing in sulfurous fumes while making scientific observations at too-close quarters. Pliny was a man of great industry and thirst for knowledge; he always found time to read, write, and collect information during his crowded career (first, he practiced at the bar, then he was on military service, and finally he was an administrator to the state). Most of what he observed, heard, or read was commented upon in his wide-ranging writing: on military science, oratory, grammar, biography, the German Wars, and contemporary Roman history—unfortunately now all lost. But fortunately his major work *Historia naturalis,* the greater part of which was published posthumously, has survived.

The term *historia naturalis* must be explained: *Historia* can mean either "history" or "enquiry" as in research; *naturalis* can mean "belonging to nature" or "to the nature of things." So although we may translate the title as *Natural History,* a better title might be *Researches into the World.* Thus, questions about Pliny's taxonomy, the strange way in which he seems to categorize his material, should take into account the meaning we today read into his title. Furthermore, the whole text must be seen in the context of his times, when the divisions between the arts, science, and technology were not hard drawn. For example, in the part on minerals and metallurgy, Pliny included information on their use in medicine and in the arts, with a digression at one point on the history of painting. Likewise in the section on geology, he discusses the making of marble statues and the beauty of the works of Phidias.

The unifying thread throughout Pliny's work was that of anthropocentrism—he regarded man as the central factor of the universe. As Greek philosopher Protagoras said, "Man is the measure of all things." It was part of Greek and Roman thinking that animals and plants, in addition to being useful as beasts of burden or for food or medicine, could in their life styles and habits teach us moral lessons. Accordingly, Pliny included what we now regard as superficial chitchat next to a serious discourse on the customs and practices of his times; for instance he described the variety of mattresses on beds; woolen clothes in demand; ostrich feathers worn on military helmets; the removal of wrinkles with asses' milk; and a decree issued by the senate to prevent hedgehogs' quills, used for carding (cleaning and combing wool), from being monopolized. Although not strictly scientific, such information has been enormously useful to historians in understanding the beliefs and habits of his times, information that, but for Pliny, would have been lost.

The problem that Pliny was continually grappling with was that of differentiating between science and superstition. He tried hard to rid himself of the myths and fallacies of his time, often without success, and to provide factual accounts of each subject under discussion. He was not afraid to criticize the religion of his day, abhorring the popular clamor over the Roman pantheon of gods, and being skeptical over the much used phrase "the vicissitudes of Fortune." Instead he put forward the belief that humans could help themselves more if they joined in the constant struggle with external natural forces. He often upset people by writing against the widespread Roman belief of life after death. More respectable to Roman popular thinking were his Stoic beliefs of duty, virtue, and the acceptance of the status of life in which you found yourself. In politics he was a true Roman, accepting the Roman imperial system as indispensable and praising the security offered by the *Pax Romana.* Despite his trenchant criticism of Greek doctors and Greek medicine, Pliny always retained his philhellenic outlook and in particular his liking for Herodotus—called "the father of history"—because he was the first to introduce realistic, as opposed to mythological (sometimes called poetical), accounts of events and because of his painstaking checking of sources.

Throughout his career, Pliny was intensely interested in the strange and unusual phenomena of nature, and here again he had great difficulty in sorting out fact from fiction. Nevertheless, he did try to remain skeptical. One of the many bizarre specimens that he reported was the skeleton of a SEA MONSTER brought from Joppa to be exhibited in Rome. It was purported to be the very one to which Andromeda was sacrificed before she was rescued by Persius. Included for "authenticity" were the chains by which she was bound to the rocks.

In the early Middle Ages, Pliny's material from the *Historia* proved a useful source for authors of popular books of fables and marvels, describing as it did legendary animals, magic stones, and dogs with two heads. But unfortunately none of Pliny's skepticism was included. Later on, his work began to be used more responsibly, and his conviction, derived from Plato and Pythagoras, that the world was a sphere and could be divided into zones was accepted. The Venerable Bede, an eighth-century Anglo Saxon scholar, relied heavily on Pliny's astronomical and meteorological material, adapted and corrected his work on tides, and used his compilation on gems. From this time on, Pliny became firmly established in the Christian tradition of learning and was read in all the best monasteries. At Oxford in England in the 12th century, a Christian interpretation of Pliny's Stoic beliefs enhanced his reputation even more. The *Historia naturalis* was one of the earliest Latin texts to be printed, appearing in Venice in 1469. Today one way of looking at the work is to see it as the "Encyclopedia of Science and Pseudoscience" of its period.

For further reading: Pliny the Elder, *Natural History: A Selection* (Penguin, 1991).

POLAR SHIFT The motion of Earth's magnetic poles, through geological time, over the surface of Earth. It has been standard fare in physics textbooks for scores of years to describe a movement of Earth's magnetic poles around Earth's axis of rotation. For example, S. G. Starling in *Electricity and Magnetism* (1912) explains that the magnetic North Pole describes a circle of 17° radius around the axis in about 960 years.

That, however, turns out to be a relatively minor polar shift. In the last 50 years the science of paleomagnetism—the study of magnetized rocks—has advanced dramatically. At the time rocks are laid down, they are very weakly magnetized by Earth's magnetic field. The direction and strength of their magnetization provides a record of Earth's field at the time and place that they were laid down. If the poles had remained about where they are now, as described above, we should expect the record to show a magnetic field roughly pointing north—with the north, north, and the south, south. That turns out not to be so. Rocks can be dated accurately over the last 5 million years, and they show complete polar reversals about 25 times in that period. The North and South Poles change places, abruptly in geological terms, they flip over in about 1,000 to 10,000 years. The changeovers are very erratic: steady as we now are for the last 800,000 years and then six flips in the preceding million years—and that irregular pattern is typical.

Rock dating before 5 million years ago is not reliable, but we do have an accurate chronology on the ocean floor. Along the Mid-Atlantic ridge, molten rock wells up from the ocean floor and solidifies; as more rock wells up, the last lot is pushed to either side, and so on. The solidifying rock records the magnetic field at the time and show how reversals have occurred during the last 150 million years. Beyond that, we are dependent on chance finds of well-dated rocks. The irregular pattern of reversals is now well documented with a long quiet period—no reversals—between 84 and 114 million years ago.

There is nothing in the above that today could be classed as pseudoscientific: Both polar rotation and polar reversal are established as legitimate science beyond question; only when it comes to explaining the how and why, particularly of reversals, is this subject emergent science. Current theories assert that Earth's magnetic field is generated by dynamo action in the liquid iron core that lies between the solid mantle and the solid inner core. Intensive work in the last decade has narrowed down the possible explanations of the dynamo's behavior, which could account for the observed reversal record.

There is a good account of this whole subject in an article by David Gubbins in *Physics World* (1996). He summarizes the present position thus: "Although theoretical ideas are still speculative, they are consistent with a dynamo theory that is becoming increasingly sophisticated. The main obstacle is that the parameters we can readily observe are difficult to predict from theory, and vice versa. However, theory and measurement are beginning to come together."

For further reading: J. Jacobs, *Reversals of the Earth's Magnetic Field* (Cambridge University Press, 1994).

POLTERGEIST Noisy or mischievous spirit. The term *poltergeist* comes from the German words *poltern* ("to knock") and *geist* ("spirit") and refers to unexplained incidences of noises and knocks and moved, thrown, spilled, and broken objects. Unlike GHOSTS and hauntings, poltergeist incidents usually do not involve sites of tragic, violent, or emotionally charged events or the spirits of people who died violently. Poltergeists generally are affiliated with a particular person, often an adolescent, and their manifestations typically are harmless but may be frightening and a nuisance. Their manifestations may last a few hours or more than a year, but typically they start and end abruptly. Poltergeist incidents have been

Poltergeist

recorded for many centuries. They include rocks raining from the skies; houses with mysterious rappings and other unexplained noises; vases and other fragile objects thrown and smashed; small fires with no known origin; and occasional physical attacks on persons; usually involving scratches, pinchings, bruises, and—rarely—sexual assaults.

One well-known 20th-century incident revolved around a young German secretary named Anne-Marie Schaberl. In the summer of 1967, while Anne-Marie was working, light bulbs began to burst in the offices and hallways, documents moved mysteriously from one room to another, heavy file cabinets moved out from the walls against which they had stood, and telephone records showed that an impossible number of calls was being made to a time-information number. Investigators were called in, including Dr. Hans Bender, psychologist and parapsychologist from Freiburg University, and two physicists from the prestigious Max Planck Institute for Plasmaphysics. They discovered that the incidents only occurred during office hours and only when Anne-Marie was present. The investigators absolutely ruled out the possibility that Anne-Marie was causing these events purposely. When Anne-Marie left for vacation and ultimately another job, the incidents abruptly stopped.

This case contains several classic traits: Anne-Marie was an adolescent; she was working at a boring and frustrating job toward which she harbored significant repressed hostility; the incidents started and ended abruptly and had no apparent physical cause; and there were no indications of fraud, although there were strong indications that the events were connected to Anne-Marie in some way.

Although many supposed poltergeist events have been proven to be frauds or to have natural explanations (environmental conditions, house settlings, underground streams that cause strange noises, and so forth), many have not. The most popular parapsychological explanation is that poltergeists are caused psychokinetically; that is, the focus person somehow unconsciously causes the events with his or her mental energy. Psychical researcher Nandor Fodor suggested in the 1930s that intense feelings of repressed anger, hostility, and sexual tension (all common in adolescents) can lead to the development of concentrated psychic energy, which manifests as poltergeist activity.

In the past people often believed the unexplained noisy and destructive poltergeist activities were caused by DEMONS. Today skeptics are more likely to explain poltergeist incidents as fraud, naturally caused events, or hallucination.

POLYWATER A dense, semiplastic form of water found on surfaces on which water has condensed; also known as anomalous water or orthowater. As its name implies, it is water in an extraordinary form. Claims to have discovered water with unusual properties were first made by scientists in the Soviet Union in 1968. Water condensed from vapor into very small glass or fused quartz vessels showed atypical vapor pressure, freezing point, and so on. This led to a series of investigations of this strange substance in laboratories in several countries, over several years, all apparently bearing out and extending the original claims. Polywater showed lowered vapor pressure and melting point; raised viscosity, density, and thermal stability; and Raman and infrared spectra that were abnormal.

Water—normal water—has, of course, many very unusual properties; if it were not for these properties, the present terrestrial life forms would not have been possible. This could have been yet another addition to water's existing catalog—polymerized water, two or more water molecules bonded together, perhaps? But eventually scientists became suspicious and closer examination showed that the anomalies were all accounted for by the presence of impurities.

It is an interesting example of a particular form of pseudoscience known as a bandwagon effect—scientists jumping aboard what looks like being a new a fruitful line of research. The original claims are accepted uncritically and even embroidered by the recruits, until eventually the scientific community steps back and takes a good skeptical look at the subject, challenges the evidence and shows the whole affair to have been mistaken.

See also COLD FUSION; MARS CANALS; N-RAYS.

For further reading: F. Percival and Alexander H. Johnstone, *Polywater: A Library Exercise for Chemistry Degree Students* (The Chemical Society London, 1978).

POPPER, SIR KARL (1902–1994) Philosopher of science who argued for the demarcation between true science and pseudoscience by means of a criterion he called falsifiability.

Born and educated in Vienna, Popper was trained in mathematics, physics, and philosophy during the same era that produced LOGICAL POSITIVISM. He was not, however, a member of the famed Vienna circle and in fact disagreed with it on many key issues. Anti-Semitism and the rise of Nazism restricted his career in Europe, but after the publication of his acclaimed first book, *The Logic of Scientific Discovery* (1935), he was able to get an academic position at Canterbury University in New Zealand. Later, he moved to England, and from 1949 until his retirement, he was a professor of logic and scientific method at the London School of Economics. His published work includes some of the most celebrated works of 20th-century philosophy, among them *The Open Society and Its Enemies* (1945) and *The Poverty of Historicism* (1957), as well as numerous essays.

Popper formulated his basic philosophy of science early in his career and maintained its core principles throughout his life. Since their introduction, his ideas have enjoyed widespread support among scholars of many disciplines, though not without extensive debate and criticism. Perhaps the most fundamental problem he addressed concerned the demarcation of science proper from other sorts of assertions, including pseudoscience, metaphysics, and myth. In opposition to earlier philosophies that had championed the notion that science was that which could be conclusively verified, Popper asserted that the criterion for the scientific status of a theory should in fact be its "falsifiability." At a basic level, this means that for a proposition to be truly scientific, it must be capable of undergoing testing that could potentially demonstrate its falsity. Science is that which perseveres in the face of challenges, and the scientific quest becomes, at least in part, a search for falsifying instances of promising theories.

Pseudosciences (and other nonsciences) fall short of this criterion by being unable to weather challenges or in many important instances, by being unfalsifiable, that is, untestable by scientific methods. Popper often used the examples of marxism and Freudian psychology to illustrate his argument. Though he felt both of these contained some valuable insights, they ultimately were not sciences because they were based upon flexible arguments that resisted refutation. One could hardly imagine a historical event or personality trait that adept marxists or psychoanalysts could not explain via their respective theories. Popper concludes that this comprehensive explanatory ability, far from being the mark of scientific authenticity, actually set such analysis apart from empirical science. A similar line of reasoning distinguishes science from metaphysics.

In spite of his insistence on demarcation by falsification, Popper was not as dismissive of nonscience as some of his philosophical and scientific peers. In keeping with the general tenor of his thought, which viewed certainty as elusive, he realized that boundaries were often blurred. Pseudoscientific ideas could, with the proper tempering, later emerge as fully fledged scientific theories (though they must eventually submit themselves to the falsifiability criterion). Likewise, many seemingly indispensable theories, such as the 19th-century concept of ether, sometimes have to be abandoned.

The falsification criterion was meant as a corrective to an earlier view of science that had perhaps overemphasized induction (the gradual emergence of general truths through the accumulation of data). Popper realized that science often proceeds by a process of bold theorizing, the supporting evidence for which was only collected *after* the basic concepts were conceived and related. In the brave new world of 20th-century science that was awash in new ideas about relativity, quantum mechanics, and subatomic physics, Popper hoped that his falsification criterion could be used to discriminate between various theoretical assertions.

Critics of Popper have noted that the falsification criterion by itself is perhaps too weak to differentiate all nonscience from science. Many endeavors that are widely considered to be pseudoscience can easily satisfy the falsification burden; for example, astrological assertions could be tested by correlating zodiac signs with marriage and divorce rates. Whether or not this proves that certain pairings are more compatible than others, Popper's falsification criterion is technically satisfied, thus earning astrology scientific status. In light of this and other examples, some philosophers have called for the modification of Popper's arguments with criteria that also consider the standards of established scientific communities. For these critics, falsification should be replaced or at least supplemented by criteria such as the capacity of a theory to unify and explain a broad range of phenomena, stimulate new research, and provide problem-solving strategies that are useful to ongoing studies.

See also PHILOSOPHY OF SCIENCE ON THE DEMARCATION OF SCIENCE FROM PSEUDOSCIENCE.

For further reading: Karl Popper, *Conjectures and Refutations* (Harper & Row, 1963).

POSSESSION A feature of many societies in which a person is taken over by spirits and loses control of his or her behavior. It is an important part of most African tribal religious systems. It is the MEDIUM through which their deity or ancestral spirit speaks to the tribe. In this context, possession is usually formalized into a one-to-

Possession

one relationship between the tribal god and the intermediary, usually a witch doctor. This way credence can be given to his divination and his spiritual leadership over all others so that he can exert a powerful controlling influence over alternative forms of possession. In effect, the tribe would interpret the situation as the good tribal god dealing with minor evil spirits who are possessing a member of the tribe. In some forms of "evil" possession, the person is regarded by the community as having committed a spiritual transgression, and recovery might require a sacrifice.

In Western religions such as Christianity, spirit possession is also believed to occur. But in Christianity, possession by the Holy Spirit has seldom been seen as the prerogative of any class or group. Nevertheless, spirit possession by members of the congregation mostly occurs outside the institutional churches as, for example, in the Pentecostal movement. Manifestations of possession include violent unusual movement, shrieking, groaning, and uttering disconnected or strange speech (called "speaking in tongues"). In some cases, the behavior is stimulated by drugs, but mostly it is either self-induced or brought about by chanting, drumming, or collective hysteria. Sometimes, a normally pious member of a congregation utters blasphemies or exhibits terror or hatred of sacred objects. In cases like this, the religious leader might consider the possibility of possession by evil spirits or some other malevolent transcendental cause, and exorcism is considered. Films, books, and television and radio programs have capitalized on this in their plots.

Today some space-age followers believe in alien possession. Thousands of Americans are so worried about been abducted and having their minds possessed by little green creatures that they are taking out insurance policies against it, although it is not clear how they would go about claiming. This form of possession moves away from GHOSTS, spirits, and witches and goes high-tech.

A typical scenario has a spaceship landing in the dead of night, disgorging aliens who sneak into bedrooms, abduct sleepers into their spaceship, and subject them to some form of possession before returning them to their beds. The ostensible reason for the aliens' behavior is their wish to establish a core of automatons who will obey their commands when they finally decide to invade and take over planet Earth.

Most scientific studies treat possession as a psychophysical manifestation to be dealt with medically or within the sphere of social psychology. Conditions that previously have been termed *demonic possession* have also come to be treated as a form of mental illness such as hysteria (male or female), somnambulism, or schizophrenia.

See also CHANNELING; DEMONS; FLYING SAUCERS, CRASHED.

PRECOGNITION The belief that there is a form of TELEPATHY in which the person receiving can somehow apprehend the message that the transmitting person has not yet sent and does not even know but will shortly. The theory arises from statistical analysis of the results of telepathy experiments in which there was found to be a significant better-than-chance correlation between the receiver's guesses and the subject of the transmitter's next message—before the transmitter knew what it was to be.

See also RHINE, JOSEPH BANKS.

PRECOLUMBIAN DISCOVERIES OF AMERICA Possible expeditions from Eurasia and Africa to the Americas before Christopher Columbus. Although Columbus has long been celebrated in popular imagination as the discoverer of the New World, historians now recognize that he was only one, and not necessarily the first, of many early explorers of the Western Hemisphere. Columbus—who insisted throughout his life that he had discovered a sea route to the East Indies—was preceded by Norse explorers nearly 500 years earlier. In the early 1960s, excavations at L'Anse aux Meadows at the very north of Newfoundland revealed a Viking settlement dating from about 1000 C.E., concrete evidence of a European presence in America before Columbus.

The first Precolumbian discovery of America took place in the very distant past, between about 30,000 and 10,000 B.C.E., during the ice ages, when ethnic Asians crossed a land bridge over the Bering Strait between Siberia and Alaska. They were probably neolithic hunters who followed migrating herds of animals through an ice-free corridor into what is now the United States. These neolithic hunters became the ancestors of the Native Americans and during the millennia spread from the far northwestern corner of North America to the extreme southern tip of Tierra del Fuego in modern Chile.

Some scholars trace a European presence in the New World back much further than the Viking settlement. Dr. Barry Fell, in his books *America BC* (1976) and *Saga America* (1980), puts forth a theory of contact between Europe and America dating from as long ago as around 3000 B.C.E. Fell claims to have found rock inscriptions in characters and language used by ancient North Africans and residents of the Iberian peninsula (modern Spain) in what is now New England, Iowa, and as far afield as present-day Arizona and New Mexico. Many of these inscriptions are in an ancient Celtic script called Ogam.

Fell also finds similarities between the Celtic languages of Europe and Native American tongues, including Algonquian. He draws comparisons between Egyptian hieroglyphics and characters used by the Micmac nation in Maine and Maritime Canada. European contact with America, the author declares, was maintained until the collapse of the Roman Empire. "The main outlines . . .

seem fairly clear," Fell concludes in *America BC*. "Various peoples from Europe and from northwest Africa sailed to America three thousand years ago and established colonies here. The primary evidence rests in the structures they built and in the inscriptions they wrote in letters that we now can identify as spelling phrases and sentences whose meaning we can grasp."

Not all scholars accept Fell's conclusions, but many more embrace a Norse presence in North America before Columbus's first voyage. The first written evidence in Europe of the existence of America comes from several long Norse poems, or sagas. In 982 C.E., Norwegian immigrant Erik Thorwaldson (or Erik the Red) discovered the island of Greenland and established a colony there. About four years later, according to the sagas, Bjarni Herjulfsson, a relative of one of Thorwaldson's companions, was blown off course on his way to Greenland and found himself in waters near a strange coast. Herjulfsson did not land on the coast but merely sailed north and east until he found the coast of Greenland.

However, Herjulfsson's report of the new lands to the south and west influenced Thorwaldson's son Leif to make a trip of his own. Leif Eriksson left Greenland in the year 1000. He and his crew made landfall at three points on the North American coast, places he named (from north to south) Helluland, Markland, and Vinland. Eriksson set up a winter camp in Vinland (named for the grapevines it had in abundance), spent the remainder of the year there, and returned to Greenland the following spring. In about 1010, an Icelandic explorer named Thorfinn Karlsefni followed Eriksson's trail to Vinland. Like Eriksson, he spent the winter on the continent. However, Karlsefni's group first traded and then fought with the natives, whom the Norse called *skraelings*. After only three years Karlsefni's group abandoned the settlement and returned to Iceland.

Most historians identify either Eriksson's or Karlsefni's settlements in Vinland with the archaeological site at L'Anse aux Meadows. The site includes the remains of eight huts, including a great hall where the leader of the expedition would have held meetings, and a smithy where iron was smelted and shaped. Most scholars identify Baffin Island as Helluland, Labrador as Markland, and Newfoundland as Vinland. However, historians differ over the question of whether L'Anse was in fact Vinland. Eriksson's description of Vinland as a place where grapevines grew wild seems to describe a place further south on the North American coast. Hjalmar R. Holund, in his *Explorations in America Before Columbus* (1956), suggests that Helluland is now known as Flat Rock Point in southeastern Newfoundland, that Markland is now part of Nova Scotia, and that Eriksson founded his settlement, Vinland, on the southern coast of Massachusetts. Thorfinn Karlsefni, who followed Eriksson to Vinland, may have made his own camp as far south as Virginia or, as Holund suggests, Manhattan Island.

Although Karlsefni abandoned Vinland after an attack by the local natives, Vinland may have been settled by Norse from Greenland throughout the high Middle Ages. Holund cites some sources that suggest that a refugee from the Norwegian court fled to Greenland and then to Vinland in or around the year 1047. In 1112, it is reported that an Icelander, Erik Gnupson, was named bishop of Greenland. It is further reported that in 1121 he went to Vinland and never returned—implying that there was still a Norse settlement there in need of Christian guidance. In 1342, when the Greenland settlements were threatened with starvation, settlers may have fled to the shores of North America. The Norwegians believed this to be the case and, in 1355, launched an expedition in search of the lost colonists. This expedition may have traveled as far west as Minnesota and may have left documentation of its passing in the KENSINGTON STONE. The Norse explorers apparently never found the lost Greenlanders, but 17th-century explorers of northern Canada reported the existence of two types of natives in what is now Baffin Island: the aboriginal American type and a taller sort, tending toward fair skin and blond hair. These, according to Holund, may have been the descendants of the Greenlanders.

For further reading: Hjalmar R. Holund, *Explorations in America Before Columbus* (Twayne, 1956); Barry Fell, *America BC: Ancient Settlers in the New World* (Times Books, 1976); and Barry Fell, *Saga America: A Startling New Theory on the Old World Settlement of America Before Columbus* (Times Books, 1980).

PREMONITION A warning given in advance of an occurrence; an uneasy knowing about an event before it takes place. Premonition does not involve a pictorial vision but rather an overwhelming feeling that something is going to happen—a foreboding—that comes from deep inside. The feeling is seldom a happy one; it mostly portends an unpleasant or unwelcome happening.

Premonition involves a particular kind of knowing that does not come empirically through the senses—that is, through sight, hearing, smell, taste, or touch—but according to ancient lore is believed to be connected to some form of nonphysical, spiritual, or psychic entity that is extrasensory. The information received comes at any time of the day or night and is not available, at that time, to anyone else. PSYCHICAL RESEARCH has yet to prove the existence of premonitions. There is much anecdotal evidence for its existence, but information is usually reported retrospectively, after the event has happened. To aid their study, researchers now have set up premonition registries in the United States where people can report their forebodings as soon as they occur.

Scientific tests to prove or disprove the phenomena are very difficult to devise because the premonition hypothesis seems so implausible. Scientists believe there is always a contingency element in life where things are liable, but not certain, to occur because they are dependent on hundreds of thousands of diverse happenings that are in turn dependent on hundreds of thousands of other happenings. For premonitions to be a real phenomena, the future would have to be predetermined and then laid out like a long scroll for us to read; and furthermore, acting alone, we would have to be able to avoid something happening without interfering with other contingent events.

PRICE, GEORGE MCCREADY (1870–1963) A Seventh Day Adventist and a vigorous opponent of theories of evolution. He received his college education at Adventist schools and subsequently taught at several as a geology professor, finally retiring in 1938. He published a number of books, among them *The New Geology* (1923), which describes his theory of Earth's history in great detail. He was a devout believer in the literal truth of the Genesis account of Earth's origin—it did take place in six days, ordinary 24-hour days, not days of great length as some suggest. The fossil record is evidence of the great flood, the fossils being the remains of species on Earth before the flood. The age of Earth is only a few thousand years.

Price distanced himself from those who had earlier insisted on the literal truth of the biblical account, in particular from 17th-century Anglican clergyman Thomas Burnet, chaplain to King William III. Price argued that Burnet was mistaken in using the biblical account as the *source* of his theory, thereby bringing the biblical account into disrepute. Price was insistent that the source had to be Earth's geology, that the record had to be determined from the rocks, and that the evidence thus painstakingly accumulated would confirm—and did confirm—the biblical account. In *The New Geology,* he attacks Burnet and his ilk: "Their wild fancies deserve to be called travesties alike on the Bible and on true science; and the word "diluvial" has been a term to mock ever since. Happy would it have been for the subsequent history of the sciences, if the students of the rocks had all been willing patiently to investigate the records, and had their fancies sternly in leash until they had gathered sufficient facts upon which to found a true induction or generalization."

Price dismisses the theories and arguments of the evolutionists as circular and self-justifying: Fossils are used to date levels of stratification; fossils are dated by the level in which they are found. When strata are found in the evolutionists' wrong or reversed order, they explain this as caused by folds or upheavals produced by buckling of plates or earthquakes. All this required millions of years to happen. To Price, all these fancy theories are unnecessary. Creation by God in six days a few thousand years ago, two periods in Earth's history: the ante- and postdiluvial, species killed off and buried in the flood, those remaining being God's creatures fixed for ever. No evolutionary explanation is necessary—nor has evolution happened.

Those geologists—almost all—who support the orthodox evolution theory will, of course, have nothing to do with this. They claim that there is a great deal of collateral evidence that confirms the chronology of strata and fossils, that the theory of evolution does not rest solely on a circular argument.

The geological side is, of course, only a part of the evolutionists' case. Charles DARWIN's *Origin of Species* (1859) is primarily concerned with the relation between species and the manner in which they have evolved. Price would have nothing to do with humans being descended from an apelike ancestor. Modern apes, in his view, did not exist before the flood but "are degenerate or hybridized men." He accepted that the present huge number of species must have developed from the relatively small number that survived the flood. Mongoloid and Negro types of humans, for example, must have come about by amalgamation—presumably interbreeding—of the original pure races. The apes he saw as the results of an even greater degeneration.

Price's views, backed up as they were by weighty geological evidence, were music to the ears of other fundamentalists. Price himself thought that it was only a matter of time, a few years only, before his theories became generally accepted by the scientific establishment. Three-quarters of a century later, that has not yet happened; on the contrary, they are farther from acceptance than they were in 1923.

For further reading: Martin Gardner, *Fads and Fallacies in the Name of Science* (Dover, 1957).

PRINCETON ENGINEERING ANOMALIES RESEARCH LABORATORY (PEAR) A research center exploring the interaction of consciousness with the physical world. PEAR was founded in 1979 by Robert Jahn, dean emeritus of Princeton's School of Engineering and Applied Science and a NASA and Department of Defense researcher. In the past 17 years, experiments involving millions of samples have been conducted at PEAR in which ordinary people (that is, people who do not consider themselves to be psychics) have been able to influence the performance of machines by mind alone. In some experiments, subjects have been able to change the output of equipment thousands of miles away.

Measurable effects of these mental manipulations recorded at PEAR have been minuscule; however, where sophisticated instrumentation is concerned, minuscule changes can be critical. The example of a fighter pilot in the cockpit of a modern aircraft illustrates this point. In an emergency situation, a tiny anomaly could be disastrous. If the studies at PEAR prove conclusively that human beings can mentally influence machinery even to the smallest degree, the implications are far reaching.

According to PEAR, consistent differences are found in the results of men and women. Men tend to produce a stronger correlation with their intentions than do women, but women tend to produce more dramatic effects than men. PEAR's findings further suggest that the power of what we call love is at play in these experiments of mind over matter. When couples have been tested together, results have tended to reflect the combination of the male correlation with intention and the stronger effects of the female.

Two decades ago, a Princeton senior studying under Jahn developed an original machine called a Random Event Generator (REG) for a thesis on psychokinesis. The machine is like a computer with a horizontal flat line running across the screen. Subjects were able to persuade mentally a second line, moving from left to right, to dip either above or below the horizontal line. The implications of these experiments led Jahn to found PEAR. That human beings can manipulate subjective information—create music, for example—is fundamental to our understanding of ourselves. PEAR's findings are now showing that we may also manipulate objective information.

PROJECT ALPHA See RANDI, JAMES.

PROPHECY A religious phenomenon, generally associated with Judaism and Christianity, but found throughout the various faiths of the world since the earliest times. Prophecy is derived from an inspired person believing he or she is the deliverer of a message from the deity. The message is usually ascertained through visions, DREAMS, or the casting of lots. The purport of prophecy now, as in the past, is twofold: either to declare the divine will in a set of dicta telling us what to believe and how to live our lives and behave towards each other, or to forewarn us in a set of predictions about future events.

Frequently prophetic mystics maintain that religious truths are revealed only to them alone, and their truths are the only authentic certain set of beliefs. This type of prophecy is usually instigated by a call from God. The Old Testament prophet Jeremiah's call was in the form of a vision in which Yahweh told him that he had been chosen before he was born (Jer. 1:5). The significance of such a call is that it legitimizes whatever follows, which could be a denial of a traditional belief system or the proposing of a completely new ethical social approach. In effect, given the right circumstances, prophecy could lay the foundations for the founding of new religions, as, for example, it did in the cases of Zoroaster, Jesus, or Muhammad. If a high ecstatic state is reached, prophecy may start to be proclaimed in the first person singular, making the audience believe that God is speaking through the mouth of the prophet. Other prophets, like the Old Testament Ezekiel, and later Islam's Muhammad, frequently went into ecstatic trances, throwing themselves around and completely losing control of themselves, which led some people to regard them as madmen.

In the Christian religion the predictive form of prophecy is illustrated by MILLENARIANISM, which is based on the New Testament belief in Christ's return. The basic doctrine is "chiliasm" (from the Greek *chilioi*, "thousand"). It prophesies that Christ will come to earth in a visible form and set up a theocratic kingdom over all the world, and thus usher in the millennium—the 1,000-year reign of Christ and his elect. The early and medieval Christian Church mostly opposed chiliasm because it tended towards division. Protestant Reformation churches likewise found it too radical. Nevertheless, the prophecy persisted in Anabaptist circles and later in reform theologies, such as Pietism (the elevation of individual piety into the universal priesthood of all the faithful), where it is found in some Lutheran churches, and in various revivalistic movements, including several religions of the NEW AGE MOVEMENT.

Prophecy, from whatever religion it derives, relies on paranormal forces to explain its powers, and so it is pseudoscientific. The professional iconoclast of all things paranormal, James RANDI formulated a set of rules for those who wish to be successful prophets:

1. Make lots of predictions and hope that some come true. If any do, point to them with pride. Ignore the others.
2. Be very vague and ambiguous. Definite statements can be wrong, but "possible" items can always be reinterpreted. Use modifiers whenever possible.
3. Use a lot of symbolism. Be metaphorical, using images of animals, names, initials. They can be fitted to many situations by the believers.
4. Cover the situation both ways and select the winner as the "real" intent of your statement.
5. Credit God with your success, and blame yourself for any incorrect interpretations of His divine messages. This way, detractors have to fight God.
6. No matter how often you're wrong, plow ahead. The believers won't notice your mistakes, and will continue to follow your every word.
7. Predict catastrophes; they are more easily remembered and more popular by far.
8. When predicting after the fact, but representing the prophecy as *preceding* the event, be just wrong enough to appear uncertain about the exact details; too good a prophecy is suspect.

In certain scientific subjects, as for example astronomy, predictions can be very accurate. Prophecy has never achieved similar accuracy.

For further reading: Lloyd M. Graham, *The Prophets, Deceptions and Myths of the Bible* (Bell Publishing, 1979); and James Randi, *The Mask of Nostradamus: Prophecies of the World's Most Famous Seer* (Prometheus, 1993).

PSAMANAZAR, GEORGE (1679–1766) Initially a very successful literary fraud. Almost certainly this was not his real name, and his place of birth is not known (possibly Avignon, France), but he came to London in 1701 claiming to be a native of Formosa (Taiwan) who had been recently converted to Christianity. He was taken under the wing of the bishop of London who provided him a living in Oxford where he trained missionaries to continue the good work in Formosa. He translated church texts into a language that no one else understood, purportedly Formosan, and convinced people that Formosans looked more European (like him) than Oriental. In 1704 he published an entirely fictitious *A Historical and Geographical Description of Formosa.* In it he described a weird Formosan lifestyle: centenarians, strange diet, ritual slaughter, polygamy, cruel punishments, and so on. This appealed to the public's imagination, and his book sold well, going into a second edition in 1705 despite exposures by Formosan missionaries. In 1728, fearing that he was mortally ill, he admitted that the whole thing was a fraud (which had rewarded him handsomely). He spent the rest of his life, a repentant born-again Christian of his day, using his writing and linguistic talents to produce articles for reference works, some about China and Japan, and working on his *Memoirs,* which were published after his death.

For further reading: Nick Yapp, *Hoaxers and their Victims* (Robson Books, 1992).

PSIONICS A term derived from combining *psi,* the shorthand for PARAPSYCHOLOGY, and *electronics,* to describe the application of electronics to PSYCHICAL RESEARCH. The early such instruments were the original HIERONYMOUS MACHINE and the improved version of it developed by John Campbell, Jr., in the late 1940s and early 1950s. Since then, more-sophisticated devices have been devised. Professor John B. HASTED, head of the physics department at Birkbeck College, London, experimented in the late 1970s on psychic METAL BENDING by school children, using strain gauges to detect any effect: "The signals were of a character such that they could not have been produced by any known physical force under the given experimental conditions." Hasted was convinced that the integrity of his instrumentation ruled out any possibility of error or trickery. His results could only be explained by some effect on the metal caused by the child subjects willing it.

Attention recently has focused on the work of Professor Robert Jahn (1930–). Until 1986, he was Dean of Engineering at Princeton. When he began to publish his parapsychology research, he lost his deanship and was reduced to an associate professorship. He has continued his research in a small laboratory, the PRINCETON ENGINEERING ANOMALIES RESEARCH LABORATORY (PEAR), financed by the Fetzer Institute and the McDonnell Foundation. Expecting to find very small effects—of the order of 0.1 percent, one part in a thousand—he built a very sensitive and stable apparatus, one that is protected against interference and fraud, to detect at that level. The equipment is a random noise generator, the amplitude of the signal from which is sampled a thousand times a second. If enough samples are taken, half should give a positive reading and half negative. Left alone the generator does just that. A person is then asked to think the generator into altering the balance, in some trials to push the average to the positive, some to push to the negative. Jahn claims to obtain very small but statistically significant effects, around his expected one in a thousand. The number of samples taken is huge, many millions. The trials have been monitored by representatives of the COMMITTEE FOR THE SCIENTIFIC INVESTIGATION OF CLAIMS OF THE PARANORMAL (CSICOP), an organization essentially skeptical. The equipment has been checked and cross-checked and tightened up wherever there seemed possible room for error. The results held, even when the person doing the influencing was as far away as New Zealand or Russia. Jahn believes that it is not the physical equipment, the random generator, that is being affected but the laws of statistics and advances a possible quantum explanation. Skeptics raise a whole raft of questions about the experiment and remain unconvinced. Jahn is not discouraged but says that the test will be for others to carry out similar experiments and check his data. Toward that end, he has designed, produced, and distributed a cheap version of his random generator and awaits results.

Stan Jeffers a physicist at York University, Ontario, built an optical equivalent of Jahn's experiment, in which people were required to bend a beam of light in what is essentially a simple interferometer, a very sensitive instrument in which the beam contains countless numbers of photons, the packets of radiation that make up a light beam. If the thinker can affect the beam and bend it, then that should show immediately and unmistakably. No effect was observed from trials with 80 people.

A major difficulty with testing the validity of claims in this field is that a null result is never final. The possibility always remains that the next idea, equipment, experiment, or whatever will come up trumps. Also, there is always a plausible explanation for the failure: inadequate sensitivity, a bad day, the subject lacks psychic powers, and so on. The skeptics likewise can always find room for error in those experiments claiming positive results: faulty statistics, ignoring unfavorable results, unconscious interference, and so on.

John McCrone, writing about this in the *New Scientist* (November 1994), concludes, "Recent experience suggests there may never be a simple, conclusive test of the existence of psychic powers. However, Jahn's work does seem to narrow the boundaries somewhat, for if such abilities exist, then their effects appear microscopically small. They also seem quite bizarrely resistant to the constraints of time, place and logic. Knowing what science is not looking for, at least is knowing something."

See also EXTRASENSORY PERCEPTION.

For further reading: Robert Jahn and Brenda Dunne, *Margins of Reality* (Harcourt Brace Jovanovich, 1987).

PSYCHIATRY The part of medicine that is concerned with the study and treatment of any disorder—whether it is behavioral, physical, or mental—where psychological factors are important either as causes or as clinical features. So, for example, psychiatrists deal with disorders as diverse as schizophrenia, depression, and addictions to narcotics, alcohol, and tobacco. Generally accepted as an important professional practice that is traditionally concerned with the institutional care and treatment of the insane, psychiatry has always had critics. Its claims to be science-based come from its links with medicine, especially to the claim that many forms of madness have a demonstrable brain pathology and that even more psychiatric disorders benefit from drugs that act on brain function. The arguments against are the great variation in the theories and practices of psychiatrists and the difficulty of either proving or falsifying those theories.

Criticism was especially intense in the 1970s and was focused on arguments that many forms of psychiatric care and treatment were judged successful largely because it was in the interest of those in control of them to say that they worked. Anthony Clare stated the situation thus:

> *"It is one of life's paradoxes that at the precise time when psychiatry is under intense attack for the vagueness of its subject matter and the arbitrariness of its judgments, the demand for psychiatric intervention within an ever-widening sphere of human activity continues to grow. . . . Simultaneously, fringe groups have appeared, dedicated to the provision of alternative and less orthodox interventions and united in a common antagonism towards what Michael Barnett called 'the expertise, formalized power, the cult of knowledge and the collective atrocities and idiocies of modern, mechanistic psychiatry.' The indiscriminate use of psychotropic drugs is widely deplored as a short-sighted panacea for superficial symptoms which leaves the deeper wounds to bleed unseen; yet the everyday clinical response is a pill and a platitude and, given the immense demand and the paucity of resources, it is difficult to see how it can ever be otherwise."*

Thomas Szasz argued that we should not incarcerate the insane involuntarily in institutions, saying that it was for the patient's own good—the courts would be preferable to psychiatrists in deciding on conflicts of interest between an individual and a society that wished to be rid of him. R. D. Laing argued that schizophrenia could be seen as an inward escape, a comprehensible response made by some people to impossible social and personal demands. SCIENTOLOGY is also strongly opposed to psychiatry, preferring its own approach to restoring the integrity of the individual. However, the practical need we continue to have for some way of coping with the insane means that no matter how much it is criticized, we will continue to need psychiatry.

Psychiatrists now apply their skills to a great range of psychological, medical and social problems—schizophrenia, manic-depression, psychosomatic disorders, addictions, sexual deviations, child psychiatry, forensic psychiatry, and the exploration of the contribution of social and economic factors to mental disorders. This much extended range has led to the emergence of psychiatric social workers—not medically qualified, as are psychiatrists, but trained, as the name implies—to deal with social problems causing mental disorders.

For further reading: Anthony Clare, *Psychiatry in Dissent* (Tavistock, 1980); and Thomas Szasz, *The Myth of Mental Illness* (Hoeber-Harper, 1961).

PSYCHIC ARCHAEOLOGY An approach to archaeology that uses pseudoscientific and often paranormal methodology, as opposed to modern scientific site-finding and site-surveying procedures that apply geographical, statistical, and technological skills.

Archaeology (from the Greek meaning "study of ancient things") is a comparatively young discipline dating from the rediscovery and first excavations of Pompeii in the 18th century. Initially it was little more than an organized treasure hunt, but from the early 19th century onward, the subject became less entrepreneurial and more academic. Along with its developing academic status, a much more systemized scientific approach developed; nevertheless the psychics stayed working alongside the subject, using the same age-old processes they had used from the start.

Psychics claim to find sites intuitively through their sixth sense. Some say that they contact the spirits of the dead people they are studying and gain direct information from them. Some use DOWSING techniques, and they allege to be able to locate objects buried under the earth with their dowsing rods and their pendulums. They do not even have to be on site, they can be miles away using their dowsing instrument over a map.

Traditional archaeologists are very skeptical about these paranormal claims and point out that prehistoric settlements are not so rare; in fact they are quite common in certain areas. For the reasonably intelligent nonpsychic mind, site location within these areas is not too difficult to predict as it is really just common sense. A few simple facts must be kept in mind: for example, early people always needed to live near a source of fresh running water; they needed access to fuel and building materials, preferably stone; they also needed rich soil and gentle slopes to grow their crops.

Some psychics claim to be able to reconstruct the exact, aboveground, appearance of the site before the excavation starts. They say this is possible by examining the electromagnetic photo fields (EMPFs) by walking over the site holding wires and noting how they react with local magnetic fields. They believe the magnetic fields were produced in early prehistoric times when atomic or subatomic particles from outer space bombarded the area from above and ground structures and features such as fences became magnetized. Physicists say this is rubbish, and archaeologists point out that once the site is located and a rough estimate of its period is assessed, the in-site reconstruction is reasonably predictable because it follows a similar pattern to other sites of the same period. Academic studies have led archaeologists to believe that culture is always patterned behavior.

Yet another psychic trick is the reading of an artifact to give very specific and particularized details of it; the term they use for this technique is "psychometrize." Apart from some rather general information about the culture from which the object came that could have been found in any book on the period, the other very personal details are often so exclusive and private, such as the individual history of the person who made it or the love life of the person who used it, that they are outside the realm of scientific testing. These details do however often contain common modern misconceptions of ancient ideas and beliefs of the period that are found perpetuated in historical movies and novels; for example, they talk of freedom and justice that would have been alien concepts in early times. Archaeologist Marshal McKudick in his detailed examination of psychics in archaeology calls this placing of concepts out of their period the "captive-of-his-own-time syndrome." An interesting story is told by McKusick in the *Journal of Field Archaeology* (1982) about Edgar CAYCE, the so called "Sleeping Prophet of Virginia Beach," who proclaimed that the PILTDOWN MAN discovered near Lewes, England, in 1912 was one of a group of immigrants to England from the lost continent of Atlantis. At that time the Piltdown fossil was thought to be genuine, but later, after Cayce's death, it was proved to be a fake. Despite his alleged psychic abilities, Cayce was obviously just another ordinary captive of his own time, believing the phony skull to be authentic.

The myth of Atlantis is being kept alive today by members of the NEW AGE MOVEMENT whose perspective on archaeology is decidedly psychic. In their view, ancient people were in many ways more advanced than those of the modern technological world. They had large naturally intuitive minds and were enlightened from within, spiritually manifesting a greater connectedness between themselves and the natural forces of the universe. A New Age theme, current at the time of writing, is that the guiding minds behind the building of the pyramids were not the ancient Egyptians but the people of Atlantis. In actuality, Atlantis never existed. It was an imaginary island beyond the Pillars of Hercules, postulated by Plato, in two of his Socratic dialogues *Timaeus* and *Critias*, as an argumentative device to set against his just but equally imaginary Republic.

There are many New Age themes about extraterrestrial aliens visiting the earth in ancient times and encoding their wisdom into ancient structures like STONEHENGE; the claim is that if we had the ability to understand what they were saying, we too would become equally enlightened. Yet another theme that has emerged recently is that the Maya people were intergalactic beings who traveled the universe as DNA codes.

See also ATLANTIS AND LEMURIA.

For further reading: Sprague de Camp, *Lost Continents: The Atlantis Theme in History, Science, and Literature* (Dover, 1970).

PSYCHIC DETECTIVES Stories of psychically gifted individuals using their abilities to aid people and law enforcement, find missing items and persons, solve crimes, and bring criminals to justice can be found in many societies and in most historical periods. These range from the biblical tale (Samuel 9) of Saul finding lost livestock after consulting a seer in the land of Zuph, through the many incidents recorded during the 15th and 16th centuries where victims of theft recovered their goods through the help of cunning men and wise women, to contemporary cases of psychics being consulted by police and even the Central Intelligence Agency.

Psychics who specialized in working on police cases in Europe during the 1920s and 1930s gained much notoriety, and included Germany's August Drost, W. de Kerler and George Mittelman; Austria's Dr. Leopold Thomas (who worked with a hypnotized helper known only as Megalis), Maximilian Langsner, and Raphael Scherman (a psychographologist); Hungary's Alfred Pathes and Michael F. Fischel and, especially Janos Kele; France's Madame Luce Vidi; Czechoslovakia's Frerick Marion; and the United States's Eugenie (Gene) Dennis and Florence

Sternfels. Kele and Marion are particularly noteworthy for volunteering for laboratory studies of their alleged abilities. Among these sleuths, the records left by Dennis and Sternfels are perhaps the most impressive. Eugenie Dennis, who left her native Kansas to give her "readings" on the vaudeville stage when she was 16, had many successes reported throughout the United States and obtained the endorsement of the city's police department, including testimony on her behalf from the captain of that department's Missing Persons Bureau. New Jersey's Florence Sternfels reportedly "read more people and solved more crimes than any medium who ever lived" (according to the American Psychical Institute), received many public police endorsements, and was said to correspond with England's Scotland Yard and France's Sureté.

The postwar period was dominated by two Dutch psychic detectives: Gerard Croiset, perhaps the most internationally celebrated, and Peter Hurkos, best known in the United States. Other publicized psychic sleuths include Netherlands "paragnosts" Marinus Dykshoorn and Jan Steer, Sweden's Olof Jonosson, and Britain's Doris Stokes, Nella Jones, and Robert Cracknell. Since 1980, the most prominent U.S. psychic detectives include the highly publicized Dorothy Allison, Greta Alexander, John Catchings, Nancy Czetli, Beverly Jaegers, Noreen Renier, Dathlyn Rhea, and Dixie Yeterian.

Psychic detectives have been enlisted for a wide variety of tasks that include finding or at least describing killers and other criminals; finding lost property and missing or kidnapped persons; locating pets and downed aircraft; ascertaining crime scenarios; unearthing clues and weapons, locating hostages, sunken treasure, and buried archaeological items; and helping with the selection of jurors. On some occasions, they have even been employed by the accused or his family to find evidence that might clear him and help indict the real perpetrator. Some psychics specialize or have had particularly good success in certain areas: Greta Alexander seems to have been particularly good at finding missing bodies, and Noreen Renier largely restricts her efforts to murder investigations.

Most psychic detectives are amateurs who only become involved with one or a few local cases, but others are full-time professional sleuths and investigate cases for clients all over the world. A few are licensed professional private investigators. At least one psychic sleuth, Riley G. is a retired New York City police officer. Some are flamboyant, but most are quite conservative in appearance and manner. Some are involved in other psychic areas such as healing (e.g., Groiset and Czetli) or even public performance/entertainment (e.g., Marion, Dennis, Uri GELLER, Joseph Dunninger, and Dreskin). The field is not even limited to human beings, for the exhibited "mind-reading horse" Lady Wonder (whose alleged abilities were tested 25 years earlier, inconclusively, by Drs. Joseph Banks RHINE and Louisa E. Rhine) was widely credited in both *Life* and *Newsweek* magazine stories in 1952 with correct prediction of the location and deaths of two missing children. Though the best-known psychic sleuths have been men, today the field seems dominated by women. Some modern psychics have networked and formed groups to pool their intuitions. These have included the U.S. Psi Squad in St. Louis, Missouri; John Catchings's North Texas PARAPSYCHOLOGY Association; E. S. Peters's Professional Psychics United in Berwyn, Illinois; and Stephan Schwartz's Mobius group in Los Angeles.

While some psychics claim that a supernatural element is involved with their "powers" (e.g., Alexander and Allison have spoken of their "angels") and some use occult divinatory practices in their investigation (e.g., Bill Ward uses ASTROLOGY), most contemporary psychics merely speak of their "intuition" or "extrasensory" perceptions, preferring the language of modern parapsychology. The vast majority claim help from their use of psychometry, the purported ability to obtain information by physically handling items connected with the victim or the crime. Some claim their "gift" from birth; others claim that some traumatic event produced their powers (e.g., Greta Alexander says she was struck by lightning, and Peter Hurkos said he became psychic after near-fatally falling off a ladder onto his head). Still others (e.g., Noreen Renier and Beverly Jaegers) assert that they trained themselves to develop psychic abilities; several of these "trained" psychics offer courses for others, including seminars for police officers.

Frequently, a well-publicized criminal investigation produces many communications from amateur psychics offering their services to the police. Most such contacts are dealt with politely but are otherwise ignored. If police involve a psychic, it is usually at their own initiation (not the psychic's) and is almost always done on an informal rather than official basis. Many police, like other members of the general public, believe in psychic abilities and, when finding themselves at a dead end with routine methods, may privately contact a psychic whom they know or have had recommended to them by trusted others. When psychics are used, there is generally great police concern about public or media ridicule, so typically they are employed without publicity and are based on internal police referrals and recommendations from other departments that have had some success with the particular psychic. Far more frequently, psychics are brought into cases by the families or friends of the victims, often when the police have been perceived as unsuccessful. Often such local groups obtain the cooperation of the police who, if only for good public relations, may bring the psychic at least into the periphery of their investigation. According to a 1993 skeptics' survey of urban police departments, 17 out of 48 responding (35 percent) indicated that they either had used or were currently using a psychic. In a follow-up survey of small and medium-size cities' police departments, these researchers found that only 32 out of 113 respondents (28 percent acknowledged using a psychic. Thus, urban departments indicated a greater probability that they had used a psychic. Though this study was intended to challenge the claim that psychics were more likely to be used by rural than urban departments, it demonstrated the reverse. Because rural departments greatly outnumber urban ones, these percentages actually support the challenged view: More psychics are used by rural than urban departments.

Reportedly, government intelligence agencies also have occasionally used psychics. In addition to such well-publicized instances as the search for kidnapped U.S. general James Dozier in 1982, psychics were employed to locate U.S. hostages during the Iranian crisis of 1980 and to locate Panama's then-president Manuel Noriega in the early 1980s. Following their retirement from the military, several of these army-trained psychics, led by Edward Dames, set up a company, Psy-Tech, which offers similar services to business and industry and claims to have worked with the United Nations on locating chemical plants in Iraq following the Gulf War.

In examining the claims of psychics, it is important to distinguish between *authenticity* and *validity.* The former refers to the sincerity or honesty of the people making the claims; the latter refers to whether the claims are objectively true. There certainly have been inauthentic psychic sleuths (the record indicated that Peter Hurkos and Gerard Croiset misrepresented a great deal in their claims), and some have been outright crooks, in a few recorded cases even the perpetrators of the crimes they claimed to investigate. Most psychics who work with the police, however, appear to be authentic; that is, they believe in the reality of their own abilities. Like all human beings, they occasionally distort their memories, but there is little reason to believe that very many are the con men or women that some critics have suggested they all are. Their most common error is a typical distortion of the frequency of their successes. Many claim to be helpful in 80–90 percent of their cases, but this would require a very broad definition of what constitutes being helpful. This distortion may partly result from wishful thinking, but it also may come about because psychics used by police (who may only thank them for their assistance) often get no feedback as to how useful they were.

More significant than authenticity is the issue of validity. The media frequently distorts psychic success, reporting it more frequently than failure. Reports of success are often unreliable and are frequently written for maximum sensationalism, especially in the tabloids. At best these reports are anecdotal and therefore weak evidence in terms of the strict canons of science. Despite the sometimes remarkable qualitative evidence these stories present (sometimes even acknowledged by the police and officials involved), and despite strong evidence that these tales are accurately told, there ultimately is little rational basis for judging whether such successes are the result of chance because we have no box score to tell us the ratio of psychic successes to failures, even for individual psychics. In the case of some well-known psychic detectives, literally tens of thousands of inquiries have come to them; would we not expect a few dozen correct guesses that might appear startling in isolation? At the present time, there simply is no adequate database that can be used to reach definite conclusions on the matter of validity. There is also the problem that critics and proponents do not agree on criteria for validity. If we use the strongest criterion—that a psychic must produce new evidence that leads to a conviction in a court of law—then there are probably no such cases. If we insist that psychic assistance be proved beyond a reasonable doubt (eliminating all other possible alternative explanations), as some critics insist, that too is impossible. But if we use softer criteria—whether or not the police confirm the psychic provided help or if we rest their case on a mere preponderance of evidence—the case for psychic sleuths is quite strong.

Most practicing psychic sleuths have never been properly tested (though many may give the impression they have), and in some instances where they have been tested, they have badly failed. A notable example was the negative results psychologist Charles Tart found when examining the abilities of the late Peter Hurkos. Nonetheless, a few psychic sleuths succeeded in obtaining impressive results, most notably: Gerard Croiset, Pat Price, Olof Jonsson, Dr. Keith Harary, and Dr. Alex Tanous.

The main experiments with psychic detection have been group studies. The first of these was conducted in the Netherlands in 1958 under the direction of police officer Dr. Filippus Brink. Testing four clairvoyants (including Croiset) who had previously worked with police. Brink showed them a wide variety of photos and objects for their "readings." Brink and his colleagues concluded that "the results proved nil" and did not evince "anything that might be regarded as being of actual use to police investigations." Although this report seems quite damning, Brink gives too few details about his method and analysis to properly assess the strength of his conclusions. Another experiment was conducted in 1984 by television reporter Ward Lucas and his Denver, Colorado, investigative news team. Here, a group of psychics' statements were compared to those of a control group of college students. Lucas concluded that both groups scored at a chance level. Like Brink's study, Lucas's provides us inadequate information on how these determinations were made. This is especially true in regard to how hits and misses were scored for those unsolved cases. The most impressive experimental tests to date have been those conducted by the Los Angeles Police Department's Dr. Martin Reiser and his associates. In their first experiment, Reiser tested 12 psychics by having them give information about four crimes (two solved and two not). He concluded that "the usefulness of psychics as an aid in criminal investigation has not been validated." Unfortunately, serious questions have been raised about the methodology used in this study. There is some question as to how representative these particular psychics were of those who have worked with police. Also only 50 percent of the information provided by the psychics was deemed verifiable, even though much more of it seems to have been capable of verification; most questionable, silence (no information provided in some category) was statistically scored as a failure along with actual mistaken information. In Reiser's second study, he compared the statements of two groups of psychics against two control groups, one consisting of college students, the other of homicide detectives. Unfortunately, the same criticism raised about his first study can be applied to the second. In general, the few group experiments on psychic detection have produced negative but not definitive results. Further and better-refined studies need to be conducted.

Those interested in the value of psychic detection broadly fall into three camps. Most of the writing about these sleuths has come from their supporters who are typically uncritical if not credulous. They include many biographies and autobiographies as well as such overviews as those by Fred Archer (1969), Paul Tabori (1974), W. S. Hibbard and R. W. Worring (1981), and Colin Wilson (1984). A second category of writers includes the debunkers, usually better researchers, but too often more interested in discrediting than disproving and typically uncritical of one another's analyses while engaging in hypercritical examinations of the sleuths and their records. From this standpoint, all reported psychic successes are actually the result of either misreporting, misinterpretation, coincidence, or fraud. The main value of this approach is in its elaboration of alternative explanations and detailed analysis of pseudopsychic methods such as cold-reading techniques used for simulation of the psychic, much of which casts light on what the psychic's clients may misinterpret as genuine new information that is really ambiguous or devious. The best example of this sometimes uneven approach appears in the recent collection for investigations edited by Joe Nickell (1994).

A third approach is that of anomalists who are skeptical but are concerned not to overstate their analysis in either direction. Examples of this approach would include the several investigatory articles of the late Piet Hein Hoebens and the overview by Arthur Lyons and Marcello Truzzi (1991). The latter demonstrates its caution by concentrating not on psychic powers but on the broader notion of a "blue sense" (from the police vernacular that refers to the intuition police officers develop from their work experience). The blue sense *may* include psychic abilities, but Lyons and Truzzi prefer to leave that matter open and seek to examine *whatever* it is that may be going on in these often mysterious episodes, whether it be intuition, unconscious mental processing and pattern recognition, or sensitivity to small cues.

The anomalist's perspective centers less on the issue of validity and concentrates instead on that of *utility*. This is deemed appropriate because criminal investigation remains largely art rather than science. A police officer's goal is to solve or resolve a case; the evidence he or she seeks is intended for a court of law, not a jury of scientists. Most police are quite willing to use unorthodox or controversial techniques if they feel that such methods might prove useful. Thus, police regularly use polygraph testing for lying, hypnosis for memory recall, psychiatric profiling, and numerous other highly controversial (and often legally inadmissible) tools. The anomalist also recognizes the social science issues involved which include questions of the functional value of psychics in these investigations whether or not psychic powers really exist. Psychics play many roles in their interactions with victims' families, police, lawyers, and so on. Much of their work is quasi-therapeutic and involves assuagement of anxieties. There are also instances where police have promoted the "myth of the psychic," pretending to use a psychic for proactive purposes. Culprits who believe in psychics have been known to confess once a psychic was brought in (much as is the case with polygraphs, which also depend on subjects' beliefs to extract confessions). In one successful ruse, police planted a false story in the press about a psychic being called in to the investigation because they believed this would get the culprit to return to the scene of the crime, which he did. Psychics have sometimes been credited as the source of information actually obtained from other covert sources, such as police informants or illegal wiretapping, thus using the psychic as a cover story to protect the real source. The point that many critics overlook is that even a psychic who obtains outside information (from an informant such as an elevator operator or a cab driver) whom the police otherwise might not have had, may help the police. If the psychic is merely a clever and keenly observant person, whether authentic or not, he or she may still be helpful to the police. In the final analysis, criminal investigation must depend upon a costs versus benefits analysis. If a psychic is useful, whether or not he or she possesses extrasensory abilities, that in itself may prove critical. Such hard realism and pragmatism is now frequently found in the growing professional police science and criminal investigation literature that discusses the use of psychics. Though caution is emphasized, psychics who are willing to keep a low media profile, who have performed successfully in the past, and are endorsed by other departments are now recommended for otherwise dead-end investigations. It seems that as a last resource, police authorities recognize that it may be quite rational to try even long shots like psychics who might lead to a solution.

See also MILITARY USE OF PSYCHICS.

For further reading: Fred Archer, *Crime and the Psychic World* (William Morrow, 1969); Whitney S. Hibbard and Raymond S. Worring, *Psychic Criminology: An Operations Manual for Using Psychics in Criminal Investigations,* (Charles C. Thomas, 1981); Arthur Lyons and Marcello Truzzi, *The Blue Sense: Psychic Detectives and Crimes* (Mysterious Press/Warner Books, 1991); Joe Nickell, Ed., *Psychic Sleuths: ESP and Sensational Cases* (Prometheus Books, 1994); Paul Tabori, *Crime and the Occult* (Taplinger, 1974); and Colin Wilson, *The Psychic Detectives* (Pan Books, 1981).

PSYCHIC METAL BENDING See METAL BENDING.

PSYCHIC PHOTOGRAPHY Either (1) photography of psychic or paranormal phenomena such as GHOSTS, spirits, and fairies; or of cryptozoological or mythic animal species; or of questionable technological wonders like flying saucers; or (2) photography claiming to have been produced by psychic or paranormal means, as for example in THOUGHTOGRAPHY, KIRLIAN PHOTOGRAPHY, or aura photography.

When photography was first presented to the public in the middle of the 19th century, experts were quick to exploit the processes to produce paranormal phenomena. As early as 1856, prints of ghostly looking ethereal figures sitting next to the person being photographed were being sold as joke novelties. Nineteenth-century Scottish physicist Sir David Brewster, who was famous for his law on the polarization of light, wrote explaining how this was done by removing the would-be ghost part way through the long exposure time needed in those days. Nevertheless, for those who did not care to accept this explanation and for others who found it a profitable field to exploit, there grew a belief that "the camera cannot lie," and so photographs of many alleged paranormal entities or events were presented as firm proof of their existence.

The first ghost image to be produced in the United States is usually credited to W. W. Campbell of Jersey City, New Jersey, who in 1860 produced a picture of a boy sitting in a chair. But he did not pursue this type of photography. The first to make a solid career of spirit photography was William Mumler of Boston, Massachusetts, who from 1861 onward, using a different method, made many photographs of living persons accompanied by rather flat-looking ghostly spirits that came to be called extras. A portrait of Mary Todd Lincoln accompanied by the spirit of the assassinated Abraham Lincoln became quite famous. Accusations of fraud dogged Mumler, with experts trying to prove he was using cutouts of recently dead people inserted by double exposure, but somehow none of the charges stuck. There followed an explosion of spirit photography accompanied by an explosion of fraud cases; one Frenchman E. Buguet claimed that he only resorted to fraud when the genuine spirits were too shy or forgetful to appear. When asked to explain why the extras were so flat and tended to be in the very same poses as photographs in magazines, spirit photographers put forward a special theory of "reproduction" that explained that spirits tended to forget what they had looked like on Earth and had to manufacture their images by jogging their memories from magazines and newspapers.

Similar cutout methods were used in 1917 in the making of the photographs of the COTTINGLEY FAIRIES that showed two young cousins from Bradford, Yorkshire, England, cavorting with fairies. The younger girl Francis was 10; the elder girl, Elsie, was 16 and was employed as a photo retoucher. In 1978 the writer on ghosts in photographs, Fred Gettings, found the source of the cutouts: *Princess Mary's Gift Book,* a volume for children published in 1915. In 1981, Francis at the age of 74 finally admitted the fraud. Nevertheless, there are still some who cannot accept the confession and still believe the fairy story to be true.

After the 1930s interest in spirit photography waned along with interest in SPIRITUALISM, and the genre turned toward other areas of parapsychology. Photographs of a cryptozoological nature became very popular, and the search for a LAKE MONSTER in Loch Ness continues to this day, spurred on by the art of photography showing floating driftwood purporting to be the monster "Nessie." In the 1980s, natives living near Lake Tele in the Congo region showed photographs of a supposed living dinosaur they called MOKELE-MBEMBE that looked surprisingly like a living sauropod. Many expeditions were mounted, including a high powered one from Japan, and, despite all the photographs and aerial videos taken, they were—and we are—still no closer to a positive identification.

Although we may look on monsters and their like as strictly the concern of folklore and zoology, the unidentified flying object (UFO) is no longer only a technical concern as there are many claims that these objects are paranormal and have a psychic nature. A book by John Keel, *Our Haunted Planet* (1971) covers this aspect of UFOs. The photographs of unidentified flying hardware like the photographs of ghosts, spirits, monsters, and so on always seem to land in the nonproven out-tray. As these phenomena cannot be pinned down, there always remains a space for subjectivity in the interpretation of the evidence. It is at this point, when stalemate is reached, that psi photography moves on to yet another supposed psi phenomenon to generate fresh public interest.

Another possible method of psychic photography is thoughtography, often called skotography. It was made famous in the 1960s by Ted Serios, who was said to have had the ability to project his thoughts onto the film inside a Polaroid camera.

KIRLIAN PHOTOGRAPHY and aura photography are two other kinds of photographic imaging, both claiming to be produced by other than normal means. In 1939 two Russian engineers, Semyon Davidovitch Kirlian and Valentina Kirlian, demonstrated how they made photographic images without the use of a camera, lens, or lights. The technique was not new (it had been done previously and called corona discharge photography), but the Kirlians claimed to be psychic and added a touch of supernatural artistry to their demonstration. They switched off the lights and sat their subject on photographic paper or film placed on an earthed (grounded) metal plate. An electrode placed above the subject was connected to an apparatus that generated a high voltage (15,000–60,000 volts), low-amperage, high-frequency electric current. When the generator was switched on, electrical discharges passed between the high-voltage electrode and the earthed metal plate, through the subject and the photographic material. The glow created by these discharges did the exposing. After normal photographic development, the subject showed as a silhouette with its edges glowing with beautiful fields of color and spiked with lightninglike jabs of bright light. The basic physics of the process is well understood, but the discharges are very variable.

Those who practice Kirlian photography today claim that they can reproduce the complete "aura" or "bioplasm" of any living thing, not only in its wholeness as it lies in the apparatus but even after part of its body has been lost, for example a salamander with a lost tail (a neat trick that triggered off many speculations on how it was done). Skeptics have shown that inanimate objects can produce auras just as well as plants, animals or humans and have suggested that degrees of moisture could produce different auras from the same object. James RANDI made completely different auras of his fingers by altering the pressure on the film. Two physicists, Watkins and Bickel, looked at the phenomenon and identified 22 variables that could influence the aura.

In 1992, Guy Coggins developed a similar method of producing aura photography. This time, the results were much more unfocused, looking more like professional modern abstract photography. Coggins's subject sat on the plate of a high-frequency electrical generator in front of a screen that, we are told, was made up of 10,000 pixels. Such a small number of pixels would produce a very coarse image, not the subtle gradations of color said to be necessary to produce a close diagnosis of the subject's psychic health or state of mind. Experts looking at the results say that the promoter's claim is not born out by the photographs, which show much finer detail than 10,000 pixels would allow, and look more like either simple double-exposure shooting into colored lights or specially prepared transparencies showing defocused fields of color.

Experiments with this kind of photography suggest that psychics' ideas, that only human beings have an aura and that this occurs mainly round the head are incorrect. Gifted psychics used to boast that they could identify a person by the colors they could see in their aura, but examination of changes in the photographic color field has been shown to indicate variances in pressure, humidity, and conductivity, all of which are explained by elementary physics rather than by the paranormal ability of psychics.

For further reading: Fred Gettings, *Ghosts in Photographs: The Extraordinary Story of Spirit Photography* (Harmony, 1978); and James Randi, *Flim-Flam!, Psychics, ESP, Unicorns, and Other Delusions* (Prometheus, 1982).

PSYCHIC SURGERY The surgical side of psychic healing, performed through the mind and spirit of the healer during an alleged visionary experience. The gift of healing must be given to the would-be healer by the spirits; then the skills and practice must be acquired during an apprenticeship and initiation period taking up to five years to complete. Techniques differ, and there are two main centers practicing and teaching psychic surgery: one is close to the city of Manila in the Philippines, and the other is in the village of Congonhas de Campso, north of Rio de Janeiro in Brazil. Other centers are now being opened all over the world.

The Brazilian organization was founded by José Pedro de Freitas who practiced there between 1950 and 1971, the year he died in an automobile accident. He came to be known as Ze Arigo, the peasant Brazilian surgeon-healer. Arigo used a spirit guide or control in his work. His control was Adolphe Fritz, a German doctor who had died in 1918. When Arigo was working on patients he purported to be in a trance, possessed by Fritz. In this condition Arigo always spoke in a guttural German accent and cupped his hand over his left ear, the ear through which Arigo claimed Fritz always controlled him, fortunately speaking to him in Portuguese as Arigo could not speak German.

Arigo worked quickly, seeing and treating more than 300 patients a day, diagnosing in seconds and performing operations in a minute without any anesthetics, nor did he use any antiseptics, although he used unsterilized penknives, kitchen knives, and scissors. There was no pain, no need for stitches, and no tying off of blood vessels as there was very little bleeding. It is reported that he could stop the flow of blood by a sharp verbal command. Another report was put about that on one occasion, when operating on a woman called Sonja who had cancer of the uterus, Arigo opened her up and dropped his scissors into the incision, then stood back, and watched the malignant tumor being cut out by the scissors that were, we are told, moving under a mysterious force unknown to science.

During the late 1960s and 1970s, hundreds of Americans and Canadians traveled to the Philippines for treatment by Dr. Tony Agpaoa. At first, the operations were carried out in the hotels where these patients were staying, but later he established his own healing center in Luzon. Agpaoa's work was examined by James RANDI, magician-turned-investigator-of-paranormal-claims; Randi had no difficulty in demonstrating how any stage magician could fake such operations, even with witnesses at close quarters. He also wondered why Agpaoa, when he needed his appendix removed, did not trust his colleagues but flew to a hospital in San Francisco for an orthodox operation. In 1977, British TV documentary *World in Action* visited Luzon, and the narrator Mike Scott and his team managed to acquire bits of tumors that they saw taken from patients. When they returned home. Scott had them analyzed at Guy's Hospital London, where they were found to be from cows and pigs.

In 1982, Agpaoa died of cerebral hemorrhage and another healer, Jun Labo, took over. Agpaoa rejected the notion of a personal spirit guide, especially one from Europe. Labo preferred to continue the third-world tradition of ancient spirits who concentrated power into the hands of the healer, who is in a semitrance before he touches the body of the patient. Advertising material tells us that pus, worms, stones, and diseased tissue materialize on the surface of the body and are wiped away with a cloth soaked in coconut oil. Sometimes, we are told, the spirits cause the healer's hands to enter the body and be attracted to offending tissues, which are pulled away without leaving an internal wound; then, we are led to believe, the opening closes as the hands leave the body.

Another "operation" performed by these psychics does involve actual invasion of the body. It is a medieval procedure known as cupping and is described by James Randi in his *Encyclopedia of Claims, Frauds, and Hoaxes of the Occult and Supernatural* (1995). First, a small incision is made in the body; then a piece of cotton wool soaked in alcohol is placed on a coin on top of the cut. The cotton is ignited and a small inverted glass is placed over the area; then the whole site covered with a cloth. A partial vacuum is created, causing the flesh and blood to be sucked up into the glass, which is revealed when the cloth is whisked away.

Minnesota physician William A. Nolen researched faith healing in general and wrote *Healing: A Doctor in Search of a Miracle* (1974), a book that uncovers many cases of fraud, including the first exposure in the United States of psychic surgery. In this category, Nolen found no cures of any serious organic (nonpsychogenic) diseases. If a simple orthodox surgical operation is performed when a child's spleen is ruptured, the child will recover completely. But take the child to a faith (or psychic healer) and the child is dead in a day.

For further reading: John G. Fuller, *Arigo: Surgeon With the Rusty Knife* (Crowell, 1974); Martin Gardner, "Uri" and "Arigo," *Science Good, Bad, and Bogus* (Prometheus, 1981); William A. Nolen, M.D., *Healing: A Doctor in Search of a Miracle* (Random House, 1974); James Randi, *The Faith Healers* (Prometheus, 1989); and James Randi, *Flim-Flam!* (Prometheus, 1982).

PSYCHICAL RESEARCH The investigation of the range of phenomena held to be paranormal, established as a field of enquiry in the late 19th century. In the mid-19th century, an epidemic of paranormal phenomena—that is, happenings that defy explanation by any established and understood mechanism, lying outside what is accepted as normal science—was generated by a growing middle class with time on its hands to whom this was a fascinating occupation. In the 19th century, mortality was still fairly high, and there were very few families who had not lost one or more loved ones. There was a strong belief in life after death and therefore the possibility, however remote, of communication between the living and the dead. Séances became the rage, in which the spirits contacted by physical MEDIUMS were producing spirit writing and materializing. Those contacted by mental mediums gave messages to members of the séance, often with information that no living person but the spirit's loved one could have known.

In that age of scientific enthusiasm, it seemed that it might be possible to establish an empirical scientific basis for phenomena that appeared to be closely linked to religion. In Britain the SOCIETY FOR PSYCHICAL RESEARCH was founded; in the United States it was the PSYCHICAL RESEARCH FOUNDATION. Cases were sought for which only a paranormal explanation seemed possible.

The focus of interest slowly moved from the most dramatic cases to those for which the evidence seemed strongest. In the 1920s, Joseph Banks RHINE's series of tediously repetitious studies mainly using ZENER CARDS (25 cards, 5 each of: star, cross, square, circle and wavy lines), claimed to have built up a great weight of statistical evidence for TELEPATHY, CLAIRVOYANCE, PSYCHOKINESIS, and PRECOGNITION. However, all research concerning the existence of psychic powers continued to be controversial with abounding accusations of fraud.

For further reading: R. G. A. Dolby, *Uncertain Knowledge* (Cambridge University Press, 1996); C. E. M. Hansel, *The Search for Psychic Power: ESP and Parapsychology Revisited* (Prometheus, 1989); M. Lamar Keene, *The Psychic Mafia* (Dell, 1976); and A. Owen, *The Darkened Room: Women, Power, and Spiritualism in Late Nineteenth-Century England* (Virago, 1989).

PSYCHOKINESIS (PK) Literally meaning "mind movement," sometimes known as telekinesis, the direct action of mind on matter. As a result of mental concentration directed on an inanimate object, the object is supposedly caused to move or change its shape. PK is the physical side of psychic phenomena in contrast to the cognitive side, which is extrasensory perception. Some writers on the subject include within psychokinesis: POLTERGEISTS, TABLE RAPPING and LEVITATION of one's own body. Although related activities, none of these can be strictly included because spirits, rather than a human person's mind, are involved in poltergeists and table rapping. In levitation, the influence is not on an external object but on the practitioner's own body. However, table turning or lifting, the bending of metal spoons, and the causing of dice to land with a certain number upward all fall into this category because the MEDIUM's concentration alone is used to move a real object.

The first attempt to test psychokinesis critically involved the 19th-century British scientist Michael Faraday, who was skeptical when a craze of table turning first hit his country in 1812 and thought it likely that physical force by the medium's hands, rather than any psychic influence, was the cause. With this in mind, he devised a complicated system of experiments with cards glued to each other (one over the other), with soft cement. When the medium sat at the table with her hands lightly placed on the upper cards, the table moved as he had expected. On close inspection Faraday found that the displacement of the cards showed that the medium's hands always moved in advance of the table; in fact, the table had been dragged along by her hands.

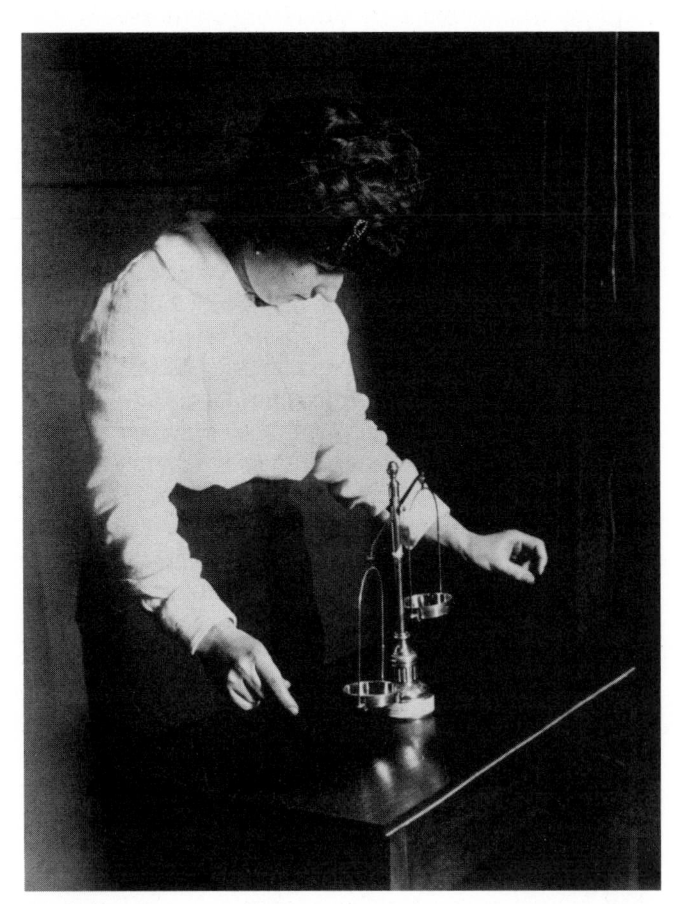

Psychokinesis

During the late 19th and early 20th centuries, committees of eminent scientists looked into the claims of a woman named Eusapia Palladino, who appeared to be able to move a balance by calling on the aid of her spirit guide "John King." Sir Oliver Lodge (see SPIRITUALISM) attended one of her performances and was impressed by her supernatural powers. When Palladino arrived in Paris, French scientists Pierre and Marie Curie were members of the investigating committee. They uncovered fraud: She was holding a fine cord between her hands, and when a screen was put between her and the balance, the balance could not be moved. Nevertheless, the committee was puzzled by some of the happenings.

In 1910, Palladino came to the United States, where she was tested by a group of scientists and magicians. In a series of controlled experiments, she was unable to produce any psychic phenomena. Later, Palladino was offered $1,000 if she could perform a trick that magicians who did not have her powers of PK could not reproduce. Palladino refused the offer and returned to Italy.

After this exposé, the scientific community lost interest in investigating the supernatural until 1934, when Joseph Banks RHINE began a series of experiments at Duke University, North Carolina, to find out whether the fall of dice could be influenced by PK. The results of these and subsequent tests have always been inconclusive.

One of the most prominent entertainers claiming to use PK in recent years was Uri GELLER who first appeared in 1973, bending forks and starting broken watches; these were good conjuring tricks. There are books that describe many methods of doing them; the basic principles can be found in Bob Couttie's book, *Forbidden Knowledge: The Paranormal Paradox* (1988).

For further reading: Bob Couttie, *Forbidden Knowledge: The Paranormal Paradox* (Lutterworth Press, 1988); C. E. M. Hansel, *E.S.P. A Scientific Evaluation* (Scribners' 1966); C. E. M. Hansel, *The Search for Psychic Power: E.S.P. and Parapsychology Revisited* (Prometheus, 1989); and James Randi, *The Magic of Uri Geller* (Ballantine, 1975).

PSYCHOANALYTIC THEORY The theory and practice of psychoanalysis as developed by Sigmund FREUD in the 1890s and the decades following. At the heart of his theory was the idea that the instinctive sexual drive goes through important stages of development in early childhood and that many circumstances can disturb its natural development to produce neuroses. Psychoanalysis as a therapy seeks to bring disturbing past experiences to consciousness to reintegrate the fragmenting personality of the patient.

Freud asserted that we are not fully conscious of the instinctive drives of the id, the mental system with which an infant enters the world, the drive for immediate gratification without regard to any constraints: concern for others, dangers, morals, or what is possible. The id, he said, is guided by the pleasure principle. Freud believed that both our DREAMS and our everyday mistakes reveal these instinctive drives (most famously the Oedipus complex, the sexual attraction of a boy toward his mother). The ego is the conscious self, operating independently of others; it develops and acts as a control, modifying the demands of the id as the growing infant encounters disapproval, punishment, and difficulty. The ego is governed by the reality principle. The superego, the third subsystem of a person's psychology, is the manner in which the ego internalizes socially required self-control, especially as learned from parents in early infancy. The ego is an individual's effective controller, finding a path between the demands of the id on the one hand and the prohibitions of the superego on the other.

Freud's theory was, and still is, heavily criticized because (a) it was not exposed to the kinds of empirical tests that are normally required of science, (b) psychoanalysis was never shown to work better than nonpsychoanalytic therapies or than natural recovery, and (c) it was not adequately integrated into academic psychology but flourished in the less-demanding milieu of popular culture at the fringes of medicine.

Freud's theories have been subjected to much criticism, especially by scientists who see them as pseudoscientific for the reasons stated. But he was a pioneer in the field of PSYCHIATRY, radically changing the way we think about ourselves.

For further reading: R. G. A. Dolby, *Uncertain Knowledge* (Cambridge University Press, 1996); H. Eysenck, *The Decline and Fall of the Freudian Empire* (Penguin, 1985); and Sigmund Freud, *Psychopathology of Everyday Life* (Penguin, 1938).

PSYCHOSYNTHESIS A branch of developmental psychology concerned with aiding people—usually business managers—to make the most of their abilities during the pursuit of their professional careers. Psychosynthesis is an approach to human development that was developed by Roberto Assagioli in 1910. It is both a theory and a practice where the focus is to achieve a synthesis of the various parts of an individual's personality into a more cohesive self. That person can then function in a way that is more life affirming and authentic. Another major aspect of psychosynthesis is its affirmation of the spiritual dimension of the person, for example, the "higher" or "transpersonal" self. The higher self is seen as a source of wisdom, inspiration, unconditional love, and the will to meaning in our lives. Psychosynthesis is founded on the basic premise that human life has purpose and meaning and that we participate in an orderly universe structured to facilitate the evolution of consciousness. A corollary is that each *person's* life has purpose and meaning within this broader context and that it is possible for the individual to discover this.

Assagioli was a pioneer of the psychoanalytic movement in Italy and a contemporary of both Sigmund FREUD and Carl Gustav JUNG. Early in his work, he observed that repression of higher, superconscious impulses (later known as repression of the sublime) could be just as damaging to the psyche as repression of material from the lower unconscious. Traditional psychoanalysis recognizes a primitive, or "lower" unconscious—the source of our atavistic and biological drives. But Assagioli supposed that there is also a higher unconscious, a superconscious—an autonomous realm from which originate our more highly evolved impulses: altruistic love and will, humanitarian action, artistic and scientific inspiration, philosophic and spiritual insight, and the drive toward purpose and meaning in life. Psychosynthesis is concerned both with integrating material from the lower unconscious and with realizing and actualizing the content of the superconscious. To this end, it uses a wide range of techniques for contacting the superconscious and establishing a bridge to that part of our being where, so Assagioli claimed, true wisdom is to be found. The superconscious is thereby accessible, in varying degrees, to each one of us and can provide a great source of energy, inspiration, and direction. Psychosynthesis helps us in our attempt to manifest this part of ourselves as fully as possible in everyday living.

Advocates and practitioners of psychosynthesis believe that each human being has a vast potential that generally goes largely unrecognized and unused. They also believe that we each have within ourselves the power to access that potential. Psychosynthesis is often seen as an unfolding process where the person actually possesses an inner wisdom or knowledge of what is needed for that process at any given time.

Although the psychosynthesis movement appears to be most active in the United States, there are centers in several other countries. It is a minor branch of psychoanalysis, and, as with the major parts of this field, it is a set of practices based on a set of beliefs that are not susceptible to the testing process that distinguishes science.

PUHARICH, ANDRIJA HENRY (1918–1995) A qualified medical doctor who began to explore PARAPSYCHOL-

OGY in the early 1950s, investigating a range of phenomena and publishing *Beyond Telepathy* (1962) and *The Sacred Mushroom* (n.d.). In 1971 Puharich met Uri GELLER, an Israeli stage magician, and was convinced of his extrasensory powers. In 1974 Puharich published *Uri: A Journal of the Mystery of Uri Geller,* a chronicle of the many and varied things that happened around Geller in Puharich's presence and of the messages from higher beings that Geller conveyed while under HYPNOSIS.

In his book, Puharich reports that while Geller was under hypnosis and regressing (i.e., tracing his steps back through his life), an unearthly and metallic voice from above said, "It is we who found Uri in the garden when he was three. We programmed him in the garden for many years to come, but he was also programmed not to remember. On this day his work begins. Andrija, you are to take care of him." Geller and Puharich remained associated for several years, Puharich acting in effect as Geller's promoter or impresario.

Prior to the Geller association, for which he is best known, Puharich had enthusiastically endorsed the activities of a Brazilian unorthodox healer, José Pedro de Freitas, known as Arigo. Arigo, who claimed that his powers came from spiritual guidance, treated a huge range of disorders by both physical and mystical techniques.

Puharich became very interested in the possibility of a link between some of the stranger aspects of theoretical physics and the various extrasensory phenomena, the reality of which he was persuaded. He was an influential actor in the symposium. "Towards a Physics of Consciousness" held at the Harvard Science Center in May 1977. He there reported on a "Geller Child" (that is, a child who exhibited powers similar to Geller's) and on materializing a tree, and he also explained that Geller's powers could be due to tachyons—particles that could in theory travel faster than light but that are as yet undetected. He subsequently edited *The Iceland Papers: Experimental and Theoretical Research on the Physics of Consciousness* (1979).

PYRAMID INCH Supposed standard unit of measurement used in the construction of the Great Pyramid. In 1864 Charles Piazzi Smyth, Astronomer Royal of Scotland and a professor at the University of Edinburgh, published a book entitled *Our Inheritance in the Great Pyramid.* In it Smyth proved to his own satisfaction that the Great Pyramid of Egypt was based on a divinely inspired unit of measurement he called a pyramid inch. Smyth's work has become a staple of PYRAMIDOLOGY, and the pyramid inch has been used to measure the distance from Earth to the sun, Earth's density, and its mean temperature.

Smyth discovered that when he divided the length of the base of the Great Pyramid by the width of a casing stone—the stones that originally lined the exterior of the pyramid—he got 365, the number of days in the year. The Edinburgh professor defined the length of the casing stone (about 63 1/2 centimeters [25 inches]) as the biblical cubit. He found that by dividing the width of a casing stone by 25, he got a unit that was only a little larger than the standard inch but was also 1/10,000,000 of the distance from the middle of Earth to the North Pole. In 1865, Smyth traveled to Egypt to confirm his conclusions by making his own measurements of the Great Pyramid. He published his results in *Life and Work at the Great Pyramid* (3 volumes, 1867) and *On the Antiquity of Intellectual Man* (1868).

Smyth's pyramid inch was disproved years later when other casing stones were uncovered. They proved to have different widths from the casing stone he had used—and therefore his pyramid inch was his own creation.

PYRAMIDOLOGY The reading of messages that are implicit in the structure of the great pyramids of Egypt and, latterly, attributing magical influences to pyramid-shaped objects.

The idea that the Great Pyramid of Giza was constructed according to a very precise quantitative plan and that its measurements encode messages about future events for posterity has been a popular point of discussion among pyramidologists for more than a century. The Great Pyramid is featured in several medieval and Renaissance religious cults and in such occult sects as the Rosicrucians. Modern interest in pyramidology originated with John G. TAYLOR in *The Great Pyramid: Why Was It Built? And Who Built It?* (1959). He believed that the pyramid had been built by the Israelites, possibly by Noah, and he discovered significance in many of its dimensions: the mathematical constant *pi,* the biblical cubit, a relationship to Earth's radius, and so on. He also read hidden references to the pyramid in many passages in the Bible. The idea had been seized on and developed earlier by others, notably Charles Piazzi Smyth, the Astronomer Royal of Scotland. Toward the end of the 19th century, a group based in Ohio urged the adoption of a system of pyramid-based measures as a counter to the evil French metric system, insisting that this would be the Lord's wish. It became a cornerstone of the creed of Jehovah's Witnesses from the end of the 19th century until 1928 when, many of its prophecies having failed to materialize, it was disowned. But many groups and individuals persisted and still persist with some version of pyramidology. An important contribution was made by an engineer from Leeds, England, David Davidson, who published a massive work—*The Great Pyramid: Its Divine Message* (1924)—on which several later works are based. Madame BLAVATSKY and the theosophists subscribed to their own version of pyramidology.

In the *FORTEAN TIMES* (May 1996) a mathematics class at University College School, London, reported its experiments with accurately scaled models of the Great Pyramid. The class reported a significantly greater growth from seed planted in the pyramid models than the growth of the control group. The report concluded "some of the findings are sufficiently strange to warrant further and more rigorous investigation, and this we invite." In this and similar magazines there are occasional advertisements for pyramid products and publications. The mysteries of the Great Pyramid seem set to fascinate for many years yet.

Skeptics say that the freedom to reinterpret the precise value of the ancient Egyptian units of measurements, and the complete freedom in choosing which measurements should be seen as significant, make it possible to read any message at all into the pyramid's structure.

For further reading: P. Lemesurier, *The Great Pyramid Decoded* (Compton Press, 1977); and M. Toth and G. Nielson, *Pyramid Power: The Secret Energy of the Ancients Revealed* (Destiny Books, 1976).

R

RABBITS Any of several soft-furred, large-eared, rodent-like burrowing mammals of the family Leporidae. The ancient Britons may have made use of rabbits for purposes of DIVINATION; it was unlawful to eat them at the time of the Gallic Wars (c. 50 B.C.E.), and in all the British Isles it was a common complaint that witches and hags changed themselves into rabbits in order to steal milk from pastured animals. Sailors avoid saying the word "rabbit" before going out to sea; a dead rabbit on board is said to bring bad weather and shipwreck. In Great Britain, the superstitious repeat "rabbit" or "white rabbit" three times on the first day of the month to bring good fortune for the duration of the month. Seeing a white rabbit near one's house is, on the contrary, an omen of death. Numerous curative powers have been attributed to the severed foot of a rabbit. It is said to lessen the pain of gout, cramps, and rheumatism, and brushing a rabbit's foot across a baby's cheek or putting it under a child's pillow guards against accidents. Carrying a rabbit's foot is said to prevent accidents for the bearer and to guard against WITCHCRAFT, and also brings good luck in acting, gambling, and thievery. Bad luck, however, comes to all those who lose their rabbit's foot.

RACIAL THEORIES To classify people on the basis of their skin, their intelligence, or some other quality that is perceived as inborn. This urge is quite ancient. According to Greek philosopher Plato, his teacher Socrates proposed that, in the ideal state, people would be raised from birth to believe that their social system reflected an inborn biological order.

Racial theories emerged during the 19th century in the United States as a justification for the enslavement of African Americans. Southern theorists sought a NATURAL LAW justification that would defend the institution of black slavery as an outgrowth of the supposed "natural inferiority" of Africans. They drew on the concept of "biological determinism"—the idea that social and economic distinctions in society are inherited and inborn rather than learned. Biological determinists sought scientific justification of the differences between human races. People at the bottom of the social level, biological determinists argued, were at the bottom because they were innately inferior. Many famous 19th-century United States statesmen, including Thomas Jefferson and Abraham Lincoln, adhered to this view.

One way to explain both the differences between races and to justify mistreating persons of a different-color skin was to deny them a common ancestry with white Europeans. This view—regarded as scientific by some because it rejected the biblical story of a human being (Adam) fathering humanity—is called "*polygenism*". If other races were not related to white Europeans, then they were not entitled to be treated like human beings. They were, the polygenists argued, separate species. British philosopher David Hume defended polygenism in the late 18th century, basing his conclusions on data gathered during service in the British Colonial Office. Charles White, an English physician, wrote an impassioned defense of polygenism entitled *Account of the Regular Gradation in Man* (1799). White explicitly rejected using racial criteria as an excuse for enslaving non-Europeans and instead claimed that humankind could be classified on aesthetic grounds—with white Europeans receiving the highest marks for beauty.

The most famous of the U.S. polygenists was 19th-century Swiss-born naturalist Louis Rudolphe AGASSIZ. Agassiz converted to polygenism after immigrating to the United States in the 1840s. He published his defense of the doctrine in the *Christian Examiner* magazine in 1850. During the Civil War, he wrote letters to Abraham Lincoln arguing against allowing African Americans social equality with white Americans. Agassiz, who supported both creationism and CATASTROPHISM, died unhappy, isolated, and frustrated, having lost most of his students and colleagues to the evolutionary theory of natural selection as proposed by Charles DARWIN.

Another famous 19th-century U.S. polygenist was the Philadelphian Samuel George Morrison. A doctor and a scientist, Morrison believed that white Europeans (and Americans) were innately superior to other races. He sought to prove this superiority through craniometry—measuring the size of the brain case of the skull—believing that brain size was directly related to intelligence. Other scientists, including Paul Broca and Maria Montessori, seized on this idea and used it as justification for their own theories. Craniometry was later rejected as both a measure of innate intelligence and a means of classifying races when the raw data was reexamined and showed no significant differences between races. Other methods of measuring intelligence, such as the famous IQ TESTS, have also failed to show any significant differences between races.

In the 20th century, the most infamous of racial theories were those proposed by the German National Socialist (Nazi) party from the 1920s through the 1940s. The Nazis believed in the innate superiority of Germans, and they tried to rationalize their racism through misapplication of scientific theories. As early as the publication of *Mein Kampf* (1925), Adolf Hitler proposed the elimination of Jews from the state as a means of solving Germany's financial and social problems. Hitler and various top Nazi officials also believed that Jews threatened the purity of the so-called Aryan race—meaning the Teutonic Germans. The Nazis passed regulations known as the Nuremberg Laws to prevent the intermarriage of Jews and Germans and to legalize the sterilization of existing "half-breeds" (children with one Jewish and one German parent) to prevent further "contamination."

The Nazis tried to bolster their racism by drawing on misinterpreted ideas of EVOLUTION and Darwinian natural selection. At the Wannsee Conference on January 20, 1942, Nazi party officials hatched what they euphemistically called the Final Solution—the extermination of Jews in Europe. According to the Wannsee Protocol, the Nazis regarded their concentration and death camps not as houses for political prisoners, but as a means of natural selection—a natural "weeding out" of weaker members of a race or species.

See also NAZI RACISM.

For further reading: Stephen Jay Gould, *The Mismeasure of Man* (W.W. Norton, 1981); and Stephen Jay Gould, *Dinosaur in a Haystack* (W.W. Norton, 1995).

RADIESTHESIA A phenomenon similar to WATER DIVINING and DOWSING in which a small pendulum is used instead of a forked twig. The pendulum is made by suspending a weight of any material, for example a finger ring, on a chain or thread. Without any conscious effort on the part of the operator, the weight when suspended is said to start swinging in one of many ways: in straight lines back and forth or sideways, or in clockwise or counterclockwise circles or ovals. Depending on the context and the questions to be answered, these motions are interpreted differently.

The term "radiesthesia" was coined by a Swiss priest, the Abbé Mermet, in the 1920s, but the pendulum is thought to have been used in Europe as early as the 18th century. The system was used by the abbé initially as a method of medical diagnosis; the term actually means "sensitive to radiation" and denotes a concern with disharmonies in energy patterns in the body. The motion of the pendulum, when hung over a patient, was believed to detect, diagnose, and even cure an ailment. The pendulum was also put to use to find the growing places of wild plants for herbal medicine. Many even wilder claims on the power of the pendulum have been made; for example, the device was used to find lost objects and the locale of buried treasure. As a follow-up to this, it was put to use in archaeology, where it was alleged not only to find sites but also to locate, describe, and date artifacts before they were even unearthed. Later, the pendulum came to be used on all occasions where previously the dowsing rod was utilized, as for example in finding underground water and mineral deposits. There were also forensic claims as in detecting criminals and in authenticating paintings.

In his book *Psychical Physics* (1949), Dr. Solcon W. Tromp maintained that the phenomena of radiesthesia, like that resulting from the use of divining rods, was due to electromagnetism (there is a connection here to LEY LINES). He thought that every substance in the world had an aura that could be detected through ELECTROMAGNETIC FIELDS and that a person sensitive to these fields could unwittingly redirect this knowledge into the movement of his pendulum. Tromp became interested in experiments to recognize the sex of unborn children; he believed the pendulum would rotate clockwise above females and counterclockwise above males, but many operators swear that the reverse is true. Tromp puts this confusion down to the occurrence of a "shadow phenomenon," where the pendulum is picking up something else, perhaps the shadow of a past event like that of an impression of where a person had previously laid or sat or even a spot on the ground where an animal had urinated. U.S. novelist, Kenneth Roberts, writing on the same theme in *Henry Gross and His Dowsing Rod* (1951), says that the pendulum should swing in circles over males and back and forth over females and that this should accurately indicate the sex of people, babies in the womb, animals, eggs, and also photographic images. He thought that there would be one exception to its accuracy: If a woman's blood contained the Rh-negative factor, the pendulum should have a male reaction. Roberts's protégé Henry Gross was tested on his ability to tell the sex of the babies of 16 pregnant women at the obstetrics department of the Maine General Hospital. Gross was right in seven cases.

Many operators of the pendulum claim that they do not have to be "in the field" to pick up information, that it works just as well over a photograph or over a map of the area. The German army under Rommel in World War II used radiesthesia over maps at their base camp during the North African campaign. Their navy did the same and swung pendulums over maps of the North Atlantic to locate enemy (British and U.S.) convoys. U.S. and U.K. naval scientists considered it nonsense and preferred to experiment on perfecting their radar.

For further reading: K.L. Feder, *Frauds, Myths, and Mysteries: Science and Pseudoscience in Archaeology* (Mayfield, 1990); Martin Gardner, *Fads and Fallacies in the Name of Science* (Dover, 1957); and David Tansley, *Radionics—Interface with the Ether Field* (Health Science Press, 1975).

RADIONICS The instrumental detection of vital energy patterns and associated diagnosis and therapy. Radionic theory suggests all living things radiate an electromagnetic field that will vary according to the health or disease of the subject. Energy patterns are given a numerical value or "rate," usually calibrated by use of a diagnostic apparatus called a black box. The first black box, sometimes called an oscilloclast, was invented by physician Albert ABRAMS of San Francisco. It consisted of several variable rheostats and a thin sheet of rubber mounted over a metal plate. A blood sample from a patient was put into the machine, which was attached to a metal plate placed on the forehead of a healthy person. By tapping the abdomen of the healthy subject, the doctor diagnosed the patient according to "areas of dullness" in relation to dial readings on the apparatus.

After the death of Abrams in 1924, his procedures were further developed by Dr. Ruth B. DROWN in the United States and by George De la Warr in Britain. De la Warr constructed black boxes and an apparatus that produced photographs relating to the condition of the patient whose blood sample was placed in the machine. De la Warr claimed the photographs captured a radiation pattern showing the shape and chemical structure of the radiating body and that, given a suitable blood sample, the camera plate could also record its pathology. The Delawarr Laboratories in Oxford manufactures radionics instruments and diagnoses and treats patients as well. A Radionics Association in the United States and also in Britain (Surrey) trains and represents radionics practitioners. In the United States, Thomas G. Hieronymous invented a machine to analyze a new type of radiation in 1949, leading to U.S. interest in radionics under the name PSIONICS. Instructions for building a HIERONYMOUS MACHINE were published in June 1956 in the journal *Astounding Science Fiction.*

RAINMAKING One of the earliest efforts at weather modification by means of sympathetic MAGIC. Early farmers, in an effort to encourage rain for their crops, used a variety of means to encourage rainfall, ranging from sex magic in Babylonian times to human sacrifice among the ancient Maya Indians of Central America. Magical means to induce rain fell out of favor in Christian times—although medieval Christians actively sought divine intervention in the weather through prayer—and the process of rainmaking declined until the 20th century.

With the coming of new scientific knowledge about the atmosphere, however, rainmaking—or, at least, control of rainfall—became more feasible. Research airplanes reported that clouds were made up largely of ice particles. During the 1930s and 1940s, scientists discovered that they could encourage the production of ice particles in clouds by seeding them with frozen carbon dioxide (CO_2—"dry ice") or silver iodide (AgI) crystals. The ice particles that formed around the crystals encourage the growth of snowflakes, which then melt and fall as rain.

There are problems with this method. The "seeding" only works with certain types of clouds—ones that have enough water in them at the right temperature to form ice crystals. The method does not work in areas that are naturally dry, such as deserts, where no clouds have formed. Also, no one has devised a technique to control where and how the rain falls. Some areas may still get too much water, while others do without. Because of these limitations, the technique of cloud seeding is only rarely used to encourage rainfall. The dream of the early farmers—rain that comes and goes on command—is still out of reach.

RAMPA, TUESDAY LOBSANG (c. 1911–1981) Supposedly a Tibetan lama (priest), Rampa claimed to have been born in 1911 in Lhasa and to have spent many years, from the age of seven, in esoteric training for the priesthood, specializing in Buddhist medicine, and then to have written an autobiography, *The Third Eye,* which was first published in 1956.

The book covers events at the lamasery under his guru, Lama Mingyar Dondup, and tells of the operation to open Rampa's "third eye" by drilling a hole into his head just above the bridge of his nose. This was purportedly to get into the center of his psychic system, which the monks thought to be associated with the pineal gland found in the limbic system of the brain. A sliver of wood treated with herbs and fire was inserted into the hole, and he immediately experienced a blinding flash and spirals of color. After this initiation there followed experiences of CLAIRVOYANCE, TELEPATHY, LEVITATION, astral travel, and a friendship with the Dalai Lama.

In 1958, the story about Lobsang Rampa and the opening up of his third eye was exposed as a scam by detectives in England, who were hired by suspicious Tibetan scholars, including Hienrich Harrer, the author of *Seven Years in Tibet.* The book pretending to be Rampa's autobiography was proven to have been written by Cyril Henry Hoskins, the son of a plumber from Plympton, England.

Undaunted by the charge of fraud and the fact that he spoke no Tibetan and had never owned a passport, Hoskins claimed that he had been bodily possessed by Lobsang Rampa. This had happened, he said, when he had concussed himself after falling from a tree when photographing an owl. At the very moment that he returned to consciousness, he remembered seeing a lama in his saffron robes drifting toward him and cutting his astral cord. Then the lama cut his own astral cord, which immediately floated back to Tibet. At the same time, he connected himself to the cord end emerging from Hoskins, taking possession of the body of the Englishman. This is all related in a book called *The Rampa Story.*

By the end of 1957 *The Third Eye* had become a bestseller, and the status of the book was affected little, or perhaps even enhanced, by the fraud accusation in 1958. By 1990, many more sequels had appeared, of which 12, including the two already mentioned, are still in print. Most people now agree that the stories, although entertaining, are just fiction. There are better sources, such as the Tibetan *Bardo Thodol* (Book of the Dead) and Hoff's *The Tao of Pooh* (1982) for the realities of Tibetan Buddhism.

For further reading: F. Freemantle and C. Trungpa, trans., *Tibetan Book of the Dead: The Great Liberation through Hearing in the Bardo* (Shambhala, 1992); T. Lobsang Rampa, *The Rampa Story* (Ballantine, n.d.); and T. Lobsang Rampa, *The Third Eye* (Ballantine Globe Comm-Inner Light, 1986).

RANDI, JAMES (1928–) Known as James "The Amazing" Randi, a professional magician and escapologist. Born and educated in Toronto, Canada, Randi moved to the United States and was naturalized a citizen in 1987. Subsequently, he became an amateur archaeologist and astronomer, author, lecturer, and founding member of the COMMITTEE FOR THE SCIENTIFIC INVESTIGATION OF CLAIMS OF THE PARANORMAL (CSICOP), where for 40 years he has investigated many unusual and pseudoscientific claims. He has appeared in a great many TV specials and documentaries and lectured and performed his sleight-of-hand tricks all over the world, always making the distinction between his art of conjuring as a form of entertainment as opposed to its frequent use as a deception for the purpose of fraud. He has taught courses in many universities and has given seminars to surgeons on pseudoscience and medicine and to police officers on charity scams and on detecting fraud and swindle perpetrated by fortune-tellers and psychic operators. He received a special grant from the MacArthur Foundation to assist him in this work, especially with reference to the claims for supernatural, the occult, and the paranormal powers of the TV evangelists and psychics. He belongs to numerous humanist and scientific organizations, is a regular contributor to *Skeptic* magazine, and in 1995 was given a doctorate, *honoris causa,* from the University of Indianapolis. Randi is still single and now resides in Florida.

Randi has ruthlessly debunked what he considers to be pseudoscientific rubbish, and, like Harry HOUDINI before him, has used any method available in his efforts to unmask psychic or mystical claims, including offering a cash prize of $1.1 million to anyone who can prove paranormal powers. The prize remains unclaimed.

Randi is best known both for his attack in the 1970s on Uri GELLER, the Israeli alleged psychic spoon bender, whose paranormal feats he demonstrated to be conjuring tricks, and for his criticism of the methodology used by parapsychologists Harold Puthoff and Russell Targ of the Stanford Research Institute (SRI), who validated Geller's powers. Randi in his turn was accused by Puthoff and Targ of distortion and factual errors in his account of the tests in *The Magic of Uri Geller* and of hypothesizing the existence of a loophole condition that they maintain did not exist. Nevertheless Randi's conclusions were publicly supported by prominent scientists Christopher Evans, Carl SAGAN, and others.

Randi has exposed CHANNELING, FAITH HEALING and PSYCHIC SURGERY. His usual method is to train someone to perform the deception and then to introduce that person into research studies to see whether researchers will uncover the fraud. For example, in his Project Alpha Experiment in the summer of 1983, he planted two young magician friends—Steve Shaw and Michael Edwards—into the McDonnell Laboratory for PSYCHICAL RESEARCH at Washington University, St. Louis, Missouri, to see whether pro-psychic researchers had enough controls to detect trickery, and, if not, to see if they would accept expert conjuring advice to put things right. The laboratory failed on both counts, lost their funding, and had to shut down, causing an enormous uproar with many accusing Randi of unethical methods.

In 1993, Randi decided to separate from CSICOP because of its acceptance of "The New Skepticism," voiced by its chairman Paul Kurtz, professor of philosophy at the State University of New York at Buffalo, and endorsed by Carl Sagan. The philosophy exhorts skeptics to be positive and constructive and tells investigators to be open-minded always about new possibilities and to be willing to question and overturn the most well-established principles in the light of further inquiry. This principle, said Kurtz, is part of the scientific method: The method never reaches any absolute truth, only approaches to the truth. Any pretense that science has the monopoly of the right approach is not constructive and creates an "us and them" situation. Scientists must recognize the human roots of superstition and pseudoscience by allowing skepticism itself to be questioned.

For further reading: The following books by James Randi: *Conjuring* (St. Martin's Press, 1992); *An Encyclopedia of Claims, Frauds, and Hoaxes of the Occult and Supernatural* (St. Martin's Press, 1995); *The Faith Healers* (Prometheus, 1989); *Flim-Flam!* (Prometheus, 1982); *The Mask of Nostradamus* (Prometheus, 1993); and *The Truth About Uri Geller* (Prometheus, 1982).

RAUDIVE TAPES Communications of the spirits of deceased people with the living via electronic tape recordings. Although he was not the first to discover these phenomena, also known as ELECTRONIC VOICE PHENOMENA (EVP), Latvian psychologist Konstantin Raudive popularized and devoted many years of research to them in the 1960s and 1970s.

A professor at the University of Riga, Latvia, and a newspaper editor as well as a psychologist, Raudive met Friedrich Jurgenson, a Russian-born Swedish painter and film producer, when Raudive moved to Sweden following the Soviet army's invasion of Latvia. In 1959, Jurgenson accidentally came upon electronic voice phenomena when he recorded a Swedish finch and found on playback a message from his deceased mother. Intrigued, Raudive took up the study of this phenomenon and during the next several years made more than 100,000 recordings of spirit voices, many of the messages fragmentary or extremely difficult to understand as human speech. Some, however, seem to be clear examples of human voices speaking words, phrases, and sentences, sometimes meaningful, often cryptic.

Raudive found that he could discern messages made on very ordinary tape recorders in recordings of music, of the white noise found on empty radio bands, or of silence. The voices are never discernible during the recording process itself; in other words, the person doing the recording does not physically hear anything unusual while the taping is going on. It is only during playback that the voices can be heard.

Several explanations have been offered for these phenomena, including both paranormal and natural ones. Paranormal ones suggest that these are indeed the voices of spirits attempting to communicate with living people or, conceivably, the voices belong to beings from other worlds or other planets. Natural explanations include the accidental recording of fragments from radio stations, the imprints of voices from not fully erased previous tapings on the same piece of magnetic tape, natural sounds (such as someone touching the tape recorder or brushing hands across clothing while recording), and psychological explanations: The person listening to the tape wants to believe and so interprets random or natural sounds as spirit voices.

While there are organizations that carry on Raudive's work today (The American Association of Electronic Voice Phenomena, for example), scientists find no evidence to support this phenomenon.

RECOVERED MEMORY SYNDROME See FALSE MEMORY SYNDROME.

REFLECTOGRAPH An instrument developed by the ASHKIR-JOBSON TRIANION GUILD to help record supposed messages from the dead. The guild was founded by two British researchers, A. J. Ashdown and B. K. Kirby after the death of their colleague, George Jobson. Prior to Jobson's death, the three had made a compact that the first to die would attempt to communicate with the other two. Ashdown and Kirby believed that such a contract had been established through a MEDIUM previously unknown to them, Mrs. L. E. Singleton. She not only gave them the agreed-on message, but through her, Jobson offered instructions for several communication devices, including the reflectograph.

The reflectograph consisted of a large typewriter that was modified in such a way that its keys were very sensitive to the slightest pressure. Even a slight depression would close a circuit and light up a corresponding letter on a letter board. It was situated just outside of the medium's cabinet. Once she went into trance, one of her hands would reach outside the cabinet and type out any messages.

Unlike the other devices developed by the guild from Jobson's messages, the reflectograph was never developed to a point that it could be operated without the mechanical pressure of the medium's hand or her presence as the operator. It was, however, the earliest of a series of devices that have appeared through the 20th century that claim to make regular spirit communication easier. The reflectograph supposedly made the process more scientific by taking the medium's conscious presence, always seen as a distorting factor, out of the way.

REFLEXOLOGY Also called zone therapy; an approach to natural healing based on the belief that each body part is represented by a zone on the hands and feet and that pressing on this specific reflex area can have therapeutic effects on that part of the body. Reflexologists claim that the body is divided into 10 zones and that each organ or body part is represented on one of these zones. The therapy also alleges a preventative/diagnostic power. By feeling in and around these zones, reflexologists claim that

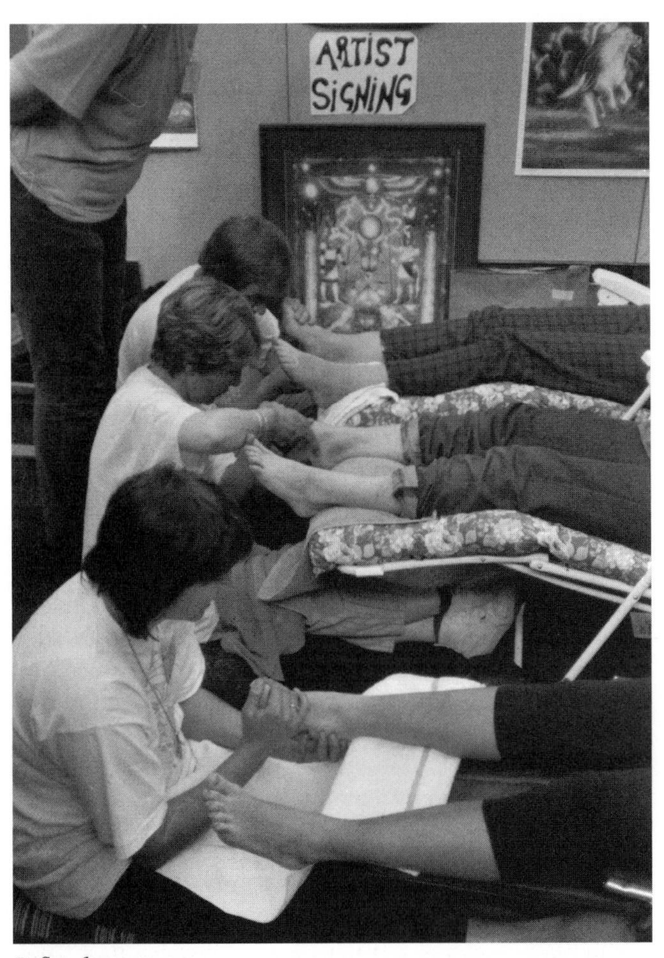

Reflexology

abnormalities in the body can be diagnosed before they actually manifest themselves in the affected area.

Reflexology is an ancient art. The Chinese in 3000 B.C.E. practiced a form of pressure therapy that is thought to have developed alongside ACUPUNCTURE. It was also used in Ancient Egypt in 2330 B.C.E., as illustrated by paintings on the physician's Tomb in Saqqara. There is then a gap in our knowledge about its usage until publications on the subject appeared in the 16th century in Europe, when there were reports of its use by a Dr. Bell of Leipzig in 1582. But it was not until the early part of this century that it was taken up seriously in the West, and then it became known as "Zone Therapy."

Modern reflexology divides the body into 10 longitudinal zones extending through the body from front to back—5 on one side of the body and 5 on the other—with a medial line separating the two sides. Three transverse zones have also been located forming a sort of grid system. This allows the accurate location of the patient's problem to be transferred to equivalent areas in the feet and/or hands for treatment. Because trouble can be located very accurately, therapy can commence immediately without recourse to damaging X rays or lengthy internal examination. Then without delay, the vital energy can be directed to flow through the right pathways back to the affected parts. Color therapy is often used in combination with reflexology, the two reinforcing each other.

Although reflexology is seen by critics to be at the respectable end of ALTERNATIVE MEDICINE, it shares with the rest the same basic assumptions that often slip the bounds of rationality. Reflexology is based on the belief that there are energy flows through the body, and treatment is aimed at unlocking the supposed channels that have become blocked, so allowing the energies to flow freely. But the pathways connecting the body parts and the zones have not been scientifically shown to exist, and no clinical trials have been conducted to prove any of their claims. Furthermore, there seems to be no scientific explanation of how it works, except that it helps the body to relax, and perhaps the first part of its name, "reflex," might indicate an involuntary or unconscious response that follows a stimulus. As with other approaches to natural healing, orthodox doctors have little objection, providing that the therapy remains *complementary* and does not become *alternative*.

For further reading: J. Rasso, *"Alternative" Healthcare: A Comprehensive Guide* (Prometheus, 1994); and Chris Stormer, *Reflexology—The Definitive Guide* (Hodder and Stoughton, 1995).

REICH, WILHELM (1897–1957) An Austrian physician and psychiatrist (an early associate of Sigmund FREUD), who fled the Nazis, settling first in Norway, then from 1939 teaching at the New School in New York, and subsequently retiring to Maine. While in Norway, he claimed to have discovered ORGONE ENERGY and later, while at the New School, he set up the Orgone Institute in Long Island and a press in Greenwich Village.

Reich made substantial contributions to PSYCHIATRY, convinced, like Freud, that many or most problems were basically sexual, but from early in his time in the United States he shifted his attention to exploring the extent and effect of orgone energy. He claimed that it was not electromagnetic but that it was the vital life force of all nature and that it was blue in color, which was the reason for the blueness of the sky, oceans, lakes, and much else besides.

Orgone energy manifested itself in human sexual intercourse, flowing out during orgasm and back in after. Reich harnessed it in a device that resembled a modern multiple rocket firer that was comprised only of empty tubes; he pointed it at clouds to bring on rain by drawing orgone energy from the clouds, causing them to break up. He invented an orgone accumulator, a metal lined wooden cubicle that concentrated orgone energy. The patient sat inside it to build up his or her energy charge, with therapeutic effects. The treatment was to be used for fatigue, anemia, early cancer (except in the brain or liver), colds, hay fever, arthritis, chronic ulcers, some migraines, sinusitis, burns, and wounds.

Reich saw orgone energy as the fundamental something that explained everything. Streams of orgone energy meet, interact, and form galaxies. The converging motion of orgone-energy streams explains gravitation, the apparent attraction all bodies have for each other: particles, Sun and planets, stars, galaxies. He saw atomic/nuclear energy and orgone energy as complementary: the one destructive and malign, the other constructive and benign. In his later year, he researched the possibility of neutralizing nuclear energy by some process using orgone energy.

Reich has never received the plaudits from the scientific establishment that he considered he deserved, becoming somewhat resentful and embittered as a consequence. He has railed against his unfair treatment in print but comforted himself with the belief that he is an unappreciated genius in good company—Copernicus, Bruno, Galileo, and others—and, like them, he will be celebrated in due course—or he may go down in history as Carl SAGAN dismisses him: "Here was Wilhelm Reich uncovering the key to the structure of galaxies in the energy of the human orgasm." Since Reich's death, his daughter Eva, a Maine pediatrician, has continued to promote his theories.

For further reading: Martin Gardner, *Fads and Fallacies* (Dover, 1957); W. Edward Mann, *Orgone, Reich and Eros* (Simon & Schuster, 1973); Wilhelm Reich, *The Mass Psychology of Fascism,* trans. by Vincent R. Carfagno (Souvenir Press, 1997); Wilhelm Reich, *Selected Writings* (Farrar, Straus and Giroux, 1979); Myron Sharaf, *Fury on Earth: A Biography of Wilhelm Reich* (Da Capo, 1994); Colin Wilson, *The Quest for Wilhelm Reich* (Doubleday, 1981).

REICHENBACH, KARL VON (1788–1869) A German 19th-century physicist and chemist who in 1833 discovered creosote—carbolic acid—by distillation from wood-tar. Baron Karl von Reichenbach, like his contemporaries Joseph Priestley in England and Alessandro Volta in Italy, was interested in the process known as destructive distillation—the reducing of an element to its basic components, usually by means of heat. Reichenbach applied destructive distillation to wood and wood products. In 1830 he discovered that some woods, when distilled, produced a flammable waxy substance that he called paraffin (now known as paraffin wax). In 1832, he found that the wood of the beech tree, when distilled, produced a slightly yellowish liquid that burned easily and gave off a strong smell. Further experimentation led Reichenbach to discover that the liquid, which he named *Kreosot* (creosote), could function as a preservative for other wooden products, keeping them safe from attack by wood-eating insects.

Creosote and paraffin quickly became important during the Industrial Revolution then taking place in Europe. Paraffin wax was cheaper than beeswax and could easily be made into candles. Like wax, too, it could be used to treat cloth in order to make it waterproof. In addition to its preservative qualities, creosote proved to be a fine antiseptic, and, as medical knowledge improved throughout the 19th century, creosote came into wide use as a means of preventing infections in hospitals, particularly in surgical procedures.

In part because of his work with distillants, Reichenbach became interested in the theories of 18th-century Austrian physician Franz Anton MESMER. Mesmer believed that all life, including both animals and plants, gave off magnetic fields. He theorized that diseases, both physical and mental, could be treated by manipulating these fields. Reichenbach accepted most of Mesmer's reasoning, but he redefined Mesmer's simple magnetic fields as being a force of a different kind, an odic force which he called OD. This odic force was shared by animals, plants, and some minerals. In certain human subjects, it manifested itself as a stream of energy coming from their fingertips. It could also be detected through the use of magnets, certain crystals, light, heat, and chemical action—including the means that Reichenbach had used to discover paraffin and creosote.

Reichenbach published several works on his experiments with the odic force, including *The Odic Force: Letters on Od and Magnetism* (1968) and *Researches on Magnetism, Electricity, Heat, Light, Crystallization and Chemical Attraction in Their Relations to the Vital Force* (1974). His theories later became an important part of the system of beliefs known as THEOSOPHY.

See also GERM THEORY.

REINCARNATION The idea that we die only to be born again as some other person or being—human, animal, or, in some versions, vegetable—and that evidence of such a process surfaces from time to time in memories of one or more of our past lives. These memories of our past(s) are believed to emerge under HYPNOSIS. A particular case that is often quoted is that of Virginia Tighe who, under hypnosis, remembered that in a former life she had been an Irish girl, BRIDEY MURPHY. For a while, this case was seen to provide convincing evidence for reincarnation, but closer inquiry revealed that Virginia Tighe was uncovering memories of an Irish girl who had been one of her neighbors in Chicago when they were children.

Another notable case is that of Kardecism, an occult sect attracting many in Brazil, combining SPIRITUALISM with reincarnation. The MEDIUM calls up a spirit mediator, a reincarnation of some appropriate person from a past life, who can then be asked to advise, provide information, or act as the intermediary between the living and the dead.

Reincarnation or the transmigration of souls is, of course, a tenet of some religions, in particular of Buddhism, which holds that behavior in this life mediates the form of the next, and that, if the good life is followed invariably, the individual moves up the ladder toward a state of nirvana in which the soul is free from all desires. A form of the belief is present in some Near Eastern religions (for example, Orphism) and in Manichaeism and Gnosticism. The Cathars, a heretical Christian sect that flourished in southern France and Italy in the 12th and 13th centuries, also subscribed to reincarnation. Reincarnation is common today in Asian religions and philosophies: Hinduism, Jainism, Sikhism—all of which, together with Buddhism, have their origins in India.

The idea has appeared in the West from time to time during the last two centuries in different forms, in particular in THEOSOPHY, and it is still current today. According to the polls, one-quarter of the people in the United States believe in reincarnation. The doctrine also has a substantial body of adherents in Britain, encouraged by the BBC series of the 1970s which was recorded by producer Jeffrey Iverson in the book *More Lives Than One?* (1977). The TV series and the book were based on a collection of tapes, assembled by Cardiff hypnotherapist Arnall Bloxham, of regressions to past lives by patients under hypnosis. The BBC described a selection of the most convincing as "the most staggering evidence for reincarnation ever recorded . . . accounts so authentic that they can only be explained by the certainty of reincarnation." The acid test of such claims is twofold: They must be accurate descriptions of the circumstances of a past life, and there must be no other possible, more mundane explanation of how the regressor came by the information. Skeptical analysts subsequently looked carefully at these "authentic accounts" and found them lacking in both respects. There were gross inaccuracies, and the subjects had had access to the information—both historically correct and incorrect!—from contemporary sources.

As with all such beliefs/doctrines, reincarnation must either be supported as a belief, regardless of the evidence or of rational argument—in which case there is nothing more to be said—or it must be open to challenge by scientists, philosophers, psychologists, and any other interested parties, some of whom will inevitably be skeptical. They ask such questions as:

- For how long has reincarnation been going on? Do past lives stretch back infinitely? Will reincarnations continue for infinite time in the future?
- Do our past virtues or misdeeds follow us; that is, do we start each life with a clean slate or not?
- If not, are some reincarnated people given a head start (or a head handicap) in their current life and, if so, does this explain infant prodigies or whatever the opposite may be? If so, should we praise or punish them for something beyond their control in their present life?
- Is not this subject open season for hoaxers? Should we not therefore be doubly aware?

Many believers in reincarnation today are convinced by the work of Ian STEVENSON, founder of the Division of PARAPSYCHOLOGY at the University of Virginia and a firm supporter of reincarnation. He has examined some 2,000 cases in all, including many studies of young children's claims of reincarnation, dismissing many as lacking corroboration, but maintained that there was a residue of cases with satisfactory confirmation. He found that children's memories of their previous lives improved until they were about five years old and had virtually disappeared by the age of eight. Stevenson's researches on the corroborated cases confirmed that the remembered people did in fact exist and many of the details fitted. Those skeptical of his evidence claim that there are more mundane explanations than reincarnation—of the Bridey Murphy type, for example—for his cases. The single case for which Stevenson is most renowned was that of Englishman Edward Ryall, who claimed in 1970 that he had an earlier life as a 17th-century West Country farmer, John Fletcher. Stevenson found Ryall's recollections so impressive that he collaborated with him in writing a book about them, *Second Time Around* (1974). The British Broadcasting Corporation put on a program in 1976 with Ryall, Stevenson, and two skeptical interrogators—London University physicist John G. TAYLOR, and Manchester University psychology professor John Cohen. Ryall coped with the questioning very convincingly, reinforcing Stevenson's support. However, there then ensued a more exhaustive examination of the detail of Ryall's book by several people, including Michael Green, an Inspector of Ancient Monuments and Historic Buildings; Renée Haynes, an editor for the British SOCIETY FOR PSYCHICAL RESEARCH; and author Ian Wilson. Ryall's remembered details did not stand up to scrutiny; for example, he located John Fletcher's house in what had been marshland at the time, but there was no record of Fletcher or his wife in the parish records. Ryall died in 1978. Stevenson continued to defend his claims, albeit with diminishing conviction that never quite evaporated.

What conclusions can be drawn about reincarnation? For the believers, those who subscribe to reincarnation without question, including those whose religious beliefs tie them into belief in reincarnation, there is nothing more to be said. For each such person, there is a particular form of doctrine that is accepted as true and not open to critical examination or doubt, but for the rest, reincarnation or *any* doctrine of reincarnation poses serious problems. First, there are no recorded cases of anyone recalling convincing detail of a previous existence—human, animal, or vegetable—that could not have an alternative explanation. All those cases that have, over the years, been heralded as "so authentic that they can only be explained by the certainty of reincarnation" have invariably been discredited. There is no accepted *evidence* for reincarnation. Second, will the arguments for belief in reincarnation stand up to skeptical examination? The answer must surely be "No"; reincarnation raises more problems than it purports to solve and gives rise to so many contradictions that there is no adequate ground for such a belief.

For further reading: Jeffrey Iverson, *More Lives Than One?* (Pan Books, 1977); Ian Stevenson, *Twenty Cases Suggestive of Reincarnation* (American Society for Psychical Research, 1966); and Ian Wilson, *Mind Out of Time: Reincarnation Claims Investigated* (Gollancz, 1981).

RELIGIOUS SCIENCE One of several New Thought religious communities developed upon an understanding of popular psychology in the early 20th century. The Religious Science Movement was founded by New Englander Ernest Holmes who was influenced by the philosophy of Ralph Waldo Emerson and the teachings of Mary Baker Eddy. He moved to California in 1912 and discovered the writings of the British metaphysical teacher Thomas Troward. In Troward's works, Holmes believed he discovered a true understanding of the human mind.

Troward in turn based his ideas on the writings of popular lecturer and writer Thomas Jay HUDSON who proposed an alternative to SPIRITUALISM. Hudson's theory was that humans have two minds: an objective mind, evident in daily waking life, and a subjective mind, most clearly seen when a person was in a hypnotic trance. The objective mind is a thinking, reasoning entity that can engage in both deductive and inductive reasoning. The subjective mind is more limited: It can only reason deductively. However, the subjective mind has some interesting powers: According to Hudson, it stores a complete record of all impressions formed throughout a person's life.

From Hudson, Troward developed a method of feeding data and suggestions into the subjective mind. Troward believed that manipulating the subjective mind would put it to work manifesting the desires of the waking conscious being. Troward's concept provided the rationale for what Holmes later called the science of mind, a term earlier used by Eddy. Holmes believed that the concept lifted religious teaching to the level of psychological science and provided an understanding of spiritual healing.

While Hudson's work has been abandoned even by popular psychologists, his ideas still inform the religious teachings of Religious Science. The Religious Science movement is now embodied in two churches: the United Church of Religious Science and Religious Science International.

For further reading: Ernest Holmes, *The Science of Mind* (Robert M. McBride, 1926); Thomas Jay Hudson, *The Law of Psychic Phenomena,* (1893 & reprint Hudson-Cohan, 1997); and Thomas Troward, *The Edinburgh Lectures* (Mead, 1901).

REMOTE VIEWING Seeing a distant place by psychic means. Just as the U.S. Air Force was attracted to the possibility of using TELEPATHY to transmit messages, the idea of being able to survey locations, installations, possible bombing targets, and much else by some form of ESP was very attractive to the military and to intelligence agencies. To be able to garner the information about some target site thousands of miles away without the expense, the labor, and the risks of satellite surveillance, spy planes, spies—the whole gamut of the intelligence-gathering world—is indeed very attractive, so the U.S. intelligence agencies and the military had to examine the possibility.

In the early 1970s, the CIA funded a program of experiments into remote viewing at the Stanford Research Institute. The two principals were Harold Puthoff and Russell Targ, both physicists and both convinced of the existence of this form of ESP. They tested many possible subjects, some drawn from military personnel, and selected a handful who showed promise. Although the CIA withdrew support at the end of the 1970s, apparently then convinced that the program could not yield useful intelligence information, the Defense Intelligence Agency (DIA) took the program under its wing and financed it until finally the agency abandoned it early in 1995. The program, in this second phase, was called Stargate and monitored foreign governments' psychic intelligence activities, supplied psychic viewers to other U.S. government agencies from its small gifted group, and continued the laboratory experiments that were started at the Stanford Research Institute (SRI) and then transferred to Science Applications International Corporation (SAIC). All this was done in secret as befits intelligence activities.

When the program ended, the Senate declassified it and put it back under the CIA with instructions to evaluate its success or lack of success. The CIA commissioned the American Institutes for Research (AIR) to carry out an investigation. The AIR employed Ray Hyman (1928–), a known skeptic psychologist who is well informed on many aspects of ESP; and Jessica Utts, a statistician of repute who was not so skeptical about the evidence for psychic phenomena. They in effect produced two reports. Hyman assessed the program as having produced nothing of value; 20 years of work and millions of dollars had yielded no useful intelligence, and the laboratory experiments showed no satisfactory evidence of remote viewing. Utts was more favorable in her assessment: Some of the laboratory experiments had statistically significant scores, better than chance, which persuaded her that remote viewing had occurred. The two assessors had limited their evaluation to the 10 best experiments carried out by SAIC near the end of the program. They agreed that these had given scores significantly above chance. But the experiments were methodologically flawed, and they further agreed that the results were still doubtful until the experiments could be repeated under satisfactory conditions.

On the basis of Hyman's and Utts's evaluation, the AIR submitted a report to the CIA, which finished: "First, . . . evidence for the operational value of remote viewing is not available, even after a decade of attempts. Second, it is unlikely that remote viewing—as currently understood—even if its existence can be unequivocally demonstrated, will be of any use in intelligence gathering due to the conditions and constraints applying in intelligence operations and the suspected characteristics of the phenomenon. We conclude that: Continued support for the operational component of the current program is not justified."

See also MILITARY USE OF PSYCHICS.

For further reading: David Morehouse, *Psychic Warrior* (Penguin, 1997); and Michael D. Mumford, Andrew M. Rose, and David A. Goslin, *An Evaluation of Remote Viewing: Research and Applications* (American Institutes for Research, 1995).

RHINE, JOSEPH BANKS (1895–1980) A botanist from Chicago who first became interested in PSYCHICAL RESEARCH after hearing a lecture on SPIRITUALISM by Sir Arthur Conan DOYLE. In 1926 he became a research assistant to William McDougall in the psychology department at Harvard, following him to Duke University in 1927. His first published paper, in 1929, was on a mind-reading horse, Lady Wonder; he was soon shown to have been tricked by the horse's owner but continued to believe in mindreading and prescient animals. In 1934 he published *Extra-sensory Perception,* by now convinced of its reality, and in 1940 he founded the Duke Parapsychology Laboratory and was named as it first Director.

In the 1930s he and colleague K. E. Zener had followed up an idea of Sir Oliver Lodge's 50 years earlier. Lodge, a member of the SOCIETY FOR PSYCHICAL RESEARCH, was interested particularly in thought transference but became impatient with the unsatisfactory, somewhat uncertain nature of evidence to that time. In the 1880s he introduced the idea of using cards carrying simple pictures—a square, a circle, and a triangle—so that the results of a series of thought transfers could be analyzed statistically. Rhine was attracted to the idea of quantifying psychical research and tried out several methods including the cardreading experiment, now with five symbols and called ZENER CARDS. Rhine's fame rests mainly on his laborious series of tests carried out on cardguessing, especially on what are known as the Pearce-Pratt and the Pratt-Woodruff series. A detailed account of this work is given in *ESP: A Scientific Evaluation* (1966) by C. E. M. Hansel, and its revised version *ESP and Parapsychology: A Critical Reevaluation* (1980). Rhine also became convinced that PRECOGNITION and PSYCHOKINESIS could be demonstrated experimentally. He reanalyzed his card-guessing data to show that the receiver had better than chance success in predicting what the next turned-over card would be; that is, the subject could tell what was coming up without anybody else knowing, without TELEPATHY being involved—knowing in advance: *precognition.* Psychokinesis is the effect upon some physical system obtained by just thinking or wishing. Rhine explored this effect on the roll of a dice and claimed better than chance results with some subjects.

Hansel in the earlier of his two books was not impressed by Rhine's experimental methods: Rhine's results could have been obtained by means other than psychic—hoaxes by some of Rhine's assistants, unconscious cuing, selective statistics. In his later book, he is less dismissive. Rhine's integrity is not at issue. Rhine was convinced from the outset that there were parapsychological effects to be found and took great pains to remove possible error and to improve his methods when criticized. The question about his work is not whether he was dishonest but whether he was mistaken. On the statistical selection, for example, Martin Gardner in FADS AND FALLACIES (1957) writes:

> *This alleged "selection" is not a deliberate process, but something which operates subtly and unconsciously. . . . let us suppose an experimenter tests 100 students . . . to determine who should be given additional testing. . . . about fifty of these students will score above average and fifty below. (Group tests, according to Rhine, nearly always show such over-all chance results.) The experimenter decides that high scorers are most likely to be psychic, so they are called in for further testing. The low scorers in this further test are again dropped, and work continued with the high ones. Eventually, one individual will remain who has scored above average on six or seven successive tests. As an isolated case, the odds against such a run are high, but in view of the selective process just described, such a run would be expected.*
>
> *A competent experimenter would not, of course, be guilty of anything as crude as this, but the illustration suggests how tricky the matter of selection is. To give a better illustration, let us imagine that one hundred professors of psychology throughout the country read of Rhine's work and decide to test a subject. The fifty who fail to find ESP in their first preliminary test are likely to be discouraged and quit, but the other fifty will be encouraged to continue. Of this fifty, more will stop work after the second test, while some will continue because they obtained good results. Eventually, one experi-*

menter remains whose subject has made high scores for six or seven successive sessions. Neither experimenter nor subject is aware of the other ninety-nine projects, and so both have a strong delusion that ESP is operating. *The odds are, in fact, much against the run. But in the total (and unknown) context, the run is quite probable. (The odds against winning the Irish Sweeptstakes are even higher. But someone does win it.) So the experimenter writes an enthusiastic paper, sends it to Rhine who publishes it in his magazine, and the readers are greatly impressed.*

Gardner goes on to develop this argument maintaining that, in the absence of any trickery, fraud, hoax, or unconscious cuing, if only chance is operating and there is no ESP, continued experiment with a successful subject will result in his scores reverting to chance. Rhine does record such dropping off in high scorers but attributes it, not to the nonexistence of ESP, but to tiredness, loss of interest, and the like.

Rhine's work, and the controversy surrounding it, continued until his death and still continues. Hansel softened his criticisms—maybe there was something in it. But Antony G. N. FLEW who had earlier been persuaded that the reality of telepathy was established beyond question (while ruling out psychokinesis and precognition), states quite categorically in his review of 1987 that there is no believable experimental evidence to support ESP and regards all Rhine's work as discredited. It is Flew's opinion that, in the absence of such evidence, there is no reason for believing that ESP exists.

Nevertheless, Rhine's work, backed by his transparent sincerity, helped to give the whole field of psychical research, including parapsychology, a certain limited respectability. Many doors are still closed to parapsychology: The National Science Foundation in the United States will not give it funds, most scientific journals will not publish its papers, and most established scientists in other fields either reject its claims or are highly skeptical. But in 1962 the Parapsychology Association was accepted into the American Association for the Advancement of Science—but not into the British Association; now one British and a few American universities have parapsychology programs, laboratories or institutions, usually within psychology departments, and/or offer courses in parapsychology.

For further reading: John Beloff, *Parapsychology: A Concise History* (Athlone, 1993), Antony Flew, *The Logic of Immortality* (Blackwell, 1987); Martin Gardner, *Fads and Fallacies in the Name of Science* (Dover, 1957); Martin Gardner, *Science: Good, Bad and Bogus* (Prometheus, 1989); C. E. M. Hansel, *ESP and Parapsychology: A Critical Reevaluation* (Prometheus, 1980); C. E. M. Hansel, *The Search for Psychic Power: ESP and Parapsychology Revisited* (Prometheus, 1989); and James McClenon, *Deviant Science: The Case of Parapsychology* (University of Pennsylvania Press, 1984).

RIGHT BRAIN/LEFT BRAIN See BICAMERAL MIND.

ROLE THEORY See GENDER THEORY.

ROLFING One of the manipulative and structural therapies, sometimes called structural integration; a system of physical manipulation aimed at loosening up the body to allow the posture to realign itself with gravity. The term "rolfing" comes from its inventor Dr. Ida Rolf, a U.S. biochemist who started to train therapists in her techniques in California in the mid-1960s.

Practitioners of rolfing claim that our bodies are constantly battling against the pull of gravity, causing poor posture and using up more of our energy than is necessary just to remain upright. The result is that we soon become tired and our emotional and physical well-being is effected. The theory is that chronic long term tension causes the *fascia,* a network of fibrous tissues covering the muscles, to shorten, and that rolfing releases the fascia so that they become longer and are no longer in strain. The assertion is that the body then becomes realigned with gravity and the individual functions better.

Rolfing does not claim to cure disease but rather to alleviate musculoskeletal problems. The angle of the pelvis is the keystone to integrating the body's weight, and the "standing pelvic tilt" can be measured by looking at the distances between the tip of the tailbone (coccyx), the front of the pubic bone, and the upper part of the pelvis. By massaging and pummeling the connective tissues around the shortened muscles of the body, these muscles become elongated and are made more flexible, blood flow is improved, and energy loss is eliminated. Those who have been through sessions of rolfing say that it is often painful. Practitioners agree that there can be pain as the fascia are released, but they point out that it should not be any more painful than any deep massage given in CHIROPRACTIC, osteopathic, or the ALEXANDER TECHNIQUE.

Orthodox medical opinion warns those who seek this treatment to choose a rolfer who works within the client's limits. Those who have experienced severe emotional trauma and also those who bruise easily or are obese should be wary of this treatment.

For further reading: K. A. Butler, *Consumer's Guide to "Alternative" Medicine* (Prometheus, 1992).

ROOT RACES The idea appears in Madame BLAVATSKY's book *The Secret Doctrine* (1888), claimed to be extracted from *THE BOOK OF DYZAN* written in Senzar (a hitherto and still unknown language) by her astral guides, the "Masters," with Blavatsky acting as an intermediary or MEDIUM and translator. The second volume of this work is in effect the spirits' version of a theory of EVOLUTION, a sort of reply to Charles DARWIN. Humankind is descended from spiritual beings on another planet, the Moon, gradually taking physical form through a succession of root races. The present phase in this development, human history, is just one step in the spirits' evolution through a series of rebirths moving from planet to planet. We, the human race at the end of the twentieth century, are now in the dark age (Kala Yuga) of the Fifth Root Race, which started when Krishna was killed by an arrow on February 16, 3102 B.C.E.

Each root race—there are seven of which only five have so far appeared—has seven subraces, each of which has seven branches. The first, located somewhere near the North Pole, was of fire-mist beings—ethereal and invisible. The second, located in northern Asia, was of beings with astral bodies that approached visibility, propagating by some sort of fission process that evolved to sexual reproduction after going through a hermaphrodite stage. The third was in Lemuria, a continent or large island in the Indian Ocean, so-called because it provided an explanation of the geographic distribution of lemurs, forming a sort of land bridge between the African and Indian continents at that time. The Lemurians were apelike giants with corporeal bodies that were slowly developing into an approximately modern human form. Lemuria disappeared into the ocean in some cataclysm after the fourth subrace had migrated to Atlantis to begin the fourth root race. The fifth, the Aryans, the present civilization, came from the fifth subrace of the Atlanteans. The sixth root race is slowly emerging from the sixth subrace of the Aryans, in southern California, where the climate most suited the theosophical ideal of paradise, to take up its home on Lemuria, risen once again, this time from the Pacific! While this happens, the American continent will sink. The seventh root race is predicted to develop from the seventh subrace of the sixth root race. When it in turn falls, the Earth cycle will have ended and the next cycle will start on Mercury.

Others took up the story where Madame Blavatsky left off (she died in 1891). In 1914, W. Scott-Elliott published *The Story of Atlantis,* expanding on the history of the subraces of that island, the home of the fourth root race. The first subrace arriving from Lemuria was the Rmoahal, beings 3–4 meters (10–12 feet) tall with mahogany black skin. The second, the Tlavatli, were copper-colored and ancestor worshipers. Then came the Toltecs whom Annie BESANT described as 8 meters (27 feet) tall with flesh so hard that it could not be cut with a knife. They survived for 10,000 years, were even more copper-colored and taller than the Tlavatli, were Grecian in appearance, had a very advanced science (traveling in airships driven by a cosmic force that is unknown to us today), and were sometimes given to drinking the hot blood of animals. Next came the Turanians, an irresponsible race but great colonizers. Then it was the turn of the Semites, with high reasoning abilities and an inner conscience, a turbulent group who often warred with neighbors. Next were the sixth subrace, the Akkadians, who were the first legislators, and finally, there were the Mongolians, who migrated to Asia, being the first subrace to develop an Atlantean culture off the island.

Rudolph STEINER, who broke away from the theosophists to found the ANTHROPOSOPHICAL SOCIETY, added more information to this saga, obtained from a secret source, in *Atlantis and Lemuria* (1913). Lemurians could not reason or calculate; they lived by instinct, had no speech, communicated by TELEPATHY, lived in caves, had high will power, and so could lift very heavy weights. The atmosphere in Lemuria was denser than now, the water more fluid, and the earth still plastic and unconsolidated.

James Churchward (1850–1936) claimed to have been shown tablets by priests in a temple-school monastery in India written in ancient Lemurian (MU) script. Between 1926 and 1934, he wrote a series of books about Mu. According to his account, this original Eden was specially created 200 million years ago; it did not evolve. It was a very advanced civilization, much better than ours, incorporating a cosmic force that could nullify gravity, the same forces as those used to walk on water by Jesus, whose teachings are all Mu-derived. An explosion about 12 thousand years ago sank Mu, killing 64 million inhabitants, leaving only Pacific islands as remaining traces.

Lemuria and Atlantis myths, the surviving remnants of the root races story as it were, still persist, encouraged by Hans S. Bellamy and later Immanuel VELIKOVSKY, each with their own versions. Magazines, books, and societies promoting these ideas flourish. A Danish group has issued stamps and currency and devised a flag, and the myths have been incorporated into science fiction.

See also ATLANTIS AND LEMURIA.

For further reading: Elsie Benjamin, *Man at Home in the Universe* (Point Loma, 1981); Martin Gardner, *Fads and Fallacies in the Name of Science* (Dover, 1957); and Peter Washington, *Madame Blavatsky's Baboon* (Secker and Warburg, 1993).

ROSENBERG, ALFRED (1893–1946) An Estonian who joined Adolf Hitler in Munich in 1919 and became editor of the party newspaper, *Völkischer Beobachter,* and a leading ideologist of the Nazi movement. In 1934, he published *The Myth of the Twentieth Century* in support of NAZI RACISM. The use of "myth" in his title is not intended to disparage; on the contrary the myth is of the purity of the German's Nordic qualities and their superiority over most, if not all, other supposed types—Jews, Catholics, and so on. His theories claimed to provide the intellectual underpinnings for the Nazi racist policies and the brutal acts of genocide and repression that the Nazis carried out. Rosenberg was continuing and extending a German tradition of Nordic supremacy and anti-Semitism that was largely developed in the 19th century from the works of Frenchman Comte Joseph D. GOB-

INEAU and of composer Richard Wagner's English son-in-law Houston Stewart CHAMBERLAIN. His pseudoscientific theories gained support from respectable contemporary scientific academics such as Professor Hans Friedrich K. GÜNTHER of the University of Jena. Anti-Semitism ran deep in the German psyche and was fertile ground for exploitation.

In 1923, Rosenberg republished *The Protocols of Zion,* a work that was presented as the record of meetings where Jewish leaders planned to control the world. The work was already known to be a fraud, a revision of an 1865 book by Maurice Joly that attacked Napoleon III and had nothing to do with Jews, but it made useful ammunition for Hitler's campaign.

During World War II, Rosenberg was minister for the Eastern European territories occupied by the Germans. He was tried at Nuremberg as a war criminal, found guilty, and hanged.

For further reading: D. Gasman, *The Scientific Origins of National Socialism* (Macdonald, 1971).

S

SAGAN, CARL (1934–1996) One of the most famous U.S. scientists of this century and an outspoken critic of claims of the paranormal. He earned a degree in physics from the University of Chicago in 1954, received his Ph.D. in astronomy and astrophysics in 1960, and taught at Harvard before moving to Cornell where he became Director of the Laboratory for Planetary Studies.

Sagan's influence extended outside the science community. He wrote, co-authored, or edited 29 books, some with his wife Ann Druyan, many written for the general reader. Among the best known of these is *Cosmos* (1981), based on his 1980 television series. He received the Pulitzer Prize for Literature in 1978 for *The Dragons of Eden: Speculations on the Evolution of Human Intelligence* and the Joseph Priestley Award "for distinguished contributions to the welfare of mankind." The National Academy of Sciences gave him its highest award in 1994.

A large part of Sagan's work was designed to expose the fallacies of various pseudosciences and mysticisms that seemed to fascinate so many in our late 20th century Western society. He maintained that scientific thinking and a healthy skepticism are necessary for all citizens for the defense of both our technology-based society and our democracies. He was concerned that the approaching millennium was fostering pseudoscience and superstition, inimical he thought to our culture and society: "Sooner or later this combustible mixture of ignorance and power is going to blow up in our faces." He was an ardent believer in the need to balance wonder at nature, skepticism and openness—not, on the one hand, to dismiss any new idea as improbable or impossible or, on the other hand, to accept too readily, to be too gullible.

This openness, the ability to consider seriously the possibility of even the apparently farfetched, extended to his professional research and sometimes brought him into conflict with some of his peers. So, for example, he collaborated with a Soviet astronomer, L. S. Shklovskii, in carefully exploring the likelihood of life outside our planet (an unfashionable idea at that time), and they co-authored *Intelligent Life in the Universe* (1966). He was a leading member of the CETI (Communication with Extra-Terrestrial Intelligence) project. He searched for coherent radio signals from outer space that would confirm the existence of intelligent beings but without success. Again his attitude was sensibly cautious: "In assessing evidence and in evaluating statistical estimates of the likelihood of extra-terrestrial intelligence, we may be at the mercy of our prejudices. At the present time there is no unambiguous evidence for even the most simple form of extra-terrestrial life, but the situation may change in the coming years. Whether we have been too optimistic or not optimistic enough, only the future will tell." That was written before the tentative discovery of possible primitive life forms in meteorite fragments from Mars, still far short of intelligent life, which he found very exciting.

Sagan helped to organize a debate between scientists on the proposition that some UFOs were spaceships at an annual meeting of the American Association for the Advancement of Science and, with Thornton Page, edited *UFOs: A Scientific Debate* (1972). He designed the engraved plate that is carried by all U.S. satellites in the very slight possibility that one may reach a remote populated planet.

In his last book, published shortly before his death, *The Demon-Haunted World: Science as a Candle in the Dark* (1996) Sagan wrote about many subjects germane to pseudoscience: Science and Hope, Aliens, The Fine Art of Baloney Detection, and When Scientists Know Sin are some of the chapter headings. He wrote:

Science is different from many other human enterprises—not, of course, in its practitioners' being influenced by the culture they grew up in, nor in sometimes being right and sometimes wrong ((both of) which are common to every human activity), but in its passion for framing testable hypotheses, in its search for definitive experiments that confirm or deny ideas, in the vigor of its substantive debate, and in its willingness to abandon ideas that have been found wanting.

Both skepticism and wonder are skills that need honing and practice. Their harmonious marriage within the mind of every schoolchild ought to be the principal goal of public education. I'd love to see such domestic felicity portrayed in the media, television especially: a community of people really working the mix—full of wonder, generously open to every notion, dismissing nothing except for good reason, but at the same time, and as second nature, demanding stringent standards of evidence; and these standards applied with at least as much rigor to what they hold dear as to what they are tempted to reject with impunity.

In the 1970s, Sagan was asked to sign a manifesto "Objections to ASTROLOGY." He declined, not because he subscribed to astrology but because he felt that the widespread addiction to it, extending from ordinary citizens to presidents, meant that it had to be treated and contested seriously and sympathetically, not dismissed by a statement from however distinguished a group.

Sagan, from his earliest work that included showing experimentally how the chemicals essential to living matter might have been produced, through more than 400 published papers, showed an extraordinary commitment to science. He also participated in an open-minded way in the examination of claims of the paranormal, as a member of the COMMITTEE FOR THE SCIENTIFIC INVESTIGATION OF CLAIMS OF THE PARANORMAL (CSICOP) and of the U.S. Air Force Scientific Advisory Board committee on UFOs. Throughout his many projects, his guiding principle was the insistence on testing the credibility of any idea, however improbable, wherever possible by experiment, and rejecting those ideas which failed the test, whether in the realm of legitimate science or of pseudoscience.

See also EXTRATERRESTRIAL INTELLIGENCE, COMMUNICATION WITH (CETI).

For further reading: Carl Sagan, *The Demon-Haunted World: Science as a Candle in the Dark* (Headline, 1996); Carl Sagan, *The Dragons of Eden: Speculations on the Evolution of Human Intelligence* (Random House, 1977); Carl Sagan and Ann Druyan, *Shadows of Forgotten Ancestors: A Search for Who We Are* (Random House, 1992); Carl Sagan and Thornton Page, eds., *UFOs: A Scientific Debate* (Cornell University Press, 1972); Carl Sagan and I. S. Shklovskii, *Intelligent Life in the Universe* (Dell, 1967); and Carl Sagan and Richard Turco, *A Path Where No Man Thought: Nuclear Winter and the End of the Arms Race* (Random House, 1990).

Sanderson

SANDERSON, IVAN T. (1911–1973) Naturalist, cryptozoologist, world traveler, and Fortean. Scottish born and well educated (with degrees in zoology, geology, and botany), Ivan Terrence Sanderson began his life work by collecting animals for the British Museum. Perhaps it was his interest in exotic animals that led him to an interest in other exotic phenomena and, eventually, to found the Society for the Investigation of the Unexplained, in 1965. Sanderson was also a prolific writer, contributing to popular magazines such as *Saturday Evening Post* and *Argosy*, as well as to scientific journals. He also wrote several books, many of them popular works on unusual animals and unexplained phenomena, and he appeared frequently on radio and television. Among Sanderson's interests were the ABOMINABLE SNOWMAN and the MINNESOTA ICEMAN, the latter an instance in which this man of science and curiosity was most likely hoodwinked.

SARCOGNOMY A new science of the relationship of the brain and the rest of the body, which was proposed by

Joseph Rhodes Buchanan (1814–99). Working from phrenological theory, Buchanan proposed the idea that the body is basically expressive of character. He concluded that each part of the body's surface had interesting psychological powers, as well as certain physiological characteristics. Thus, each part of the skin exercises a direct action on a particular part of the brain through the nervous system. Buchanan explored these relationships in an effort to discover their value for the diagnosis and treatment of disease. In the end, he was unable to influence even the majority of his colleagues who taught PHRENOLOGY, and sarcognomy died for lack of supporting evidence.

SARSAPARILLAS A tropical vine, the extract of which gave its name to a variety of NOSTRUMS that became popular in the mid-19th century. The *Smilax* vine is found throughout the American tropics and subtropics. The extract of *Smilax* known as sarsaparilla was originally introduced into Europe in the 16th century as a cure for syphilis, though it proved to be an ineffective treatment. When sarsaparilla reappeared in the 18th century, it was offered as a health tonic and purifying agent for the blood. Manufacturers imported *Smilax* plants from the tropics and used a domesticated form called *Aralia nudicaulis*.

Among the first to market sarsaparilla in the United States was C. C. Bristol. In the 1840s, Bristol's sarsaparilla led the market over a number of similar products, each of which advertised itself as a more potent form of sarsaparilla. However, Bristol's and its competitors were overwhelmed by James Cook Ayer's product, Ayer's Extract of Sarsaparilla. Ayer, a physician graduate of the University of Pennsylvania, was one of the first patent-medicine manufacturers who understood the importance of the U.S. West as a new market. After the Civil War he marketed a set of products including Ayer's Extract of Sarsaparilla. As Ayer became more successful, he expanded his operation, even buying an interest in a railroad in order to move his product from Lowell, Massachusetts, his manufacturing center, as cheaply as possible. He died in 1878 a wealthy man and honored citizen of Lowell.

Sarsaparillas were a notable segment of the nostrum market through World War I. The manufacturers absorbed their first major assault upon the product in 1905 with the articles by Samuel Hopkins ADAMS in *Collier's* magazine in 1905. Several years later, the Connecticut Medical Society analyzed nine over-the-counter sarsaparillas and found that each contained very little plant extract, a large helping of alcohol, and a variety of other substances, most of no medicinal value. The society found that a few ingredients, such as potassium iodide, did affect users and in fact could be harmful in large doses.

The American Medical Association attacked sarsaparilla as a substance of no known medicinal value, but one which in itself would do no harm. That opinion has not changed. Sarsaparilla products thus passed the minimal standards of the early drug laws and remained on the market through the 1950s. They gradually disappeared only in the face of newer, more potent medicines.

SCHIAPARELLI, GIOVANNI VIRGINIO (1835–1910) Italian astronomer whose first notable discovery was the asteroid Hesperia in 1861. He then went on to demonstrate that meteor swarms have orbits similar to certain comets and speculated that these swarms were probably the remains of spent comets. He also did some work on double stars and conducted extensive observations on Mercury, Venus, and Mars.

It was in 1877, while involved in this latter work at the Milan Observatory, when he was using only a 22.5-centimeter (9-inch) telescope, that he made the observation for which he is best known: He reported seeing groups of straight lines on Mars and called them *canali*, meaning "channels" to Schiaparelli. These observations were followed up using a more powerful telescope that had a 47-centimeter (18.5-inch) aperture. He then observed the channels begin to take on a more linear formation and also appear to alter from one day to the next, sometimes being single lines, sometimes two parallel lines separated by an estimated 80 kilometers (50 miles). This separation was called the gemination, meaning doubling. The word *canali* that Schiaparelli used to describe these formations was unfortunately translated into English in its first, more common meaning, as canals, rather than "channels," thus giving the public and even some astronomers the idea that they were of human construction. There followed much speculation on the possibility of life on Mars.

From his observations of two other planets, Mercury and Venus, Schiaparelli came to the conclusion that they rotated on their axes at the same rate that they revolved around the sun, thus always presenting the same face to the sun. This view, generally accepted till 1960, is now known to be untrue. After Schiaparelli's retirement, he wrote books about astronomy in ancient cultures.

For further reading: Michael J. Crowe, *The Extraterrestrial Life Debate, 1750–1900* (Cambridge University Press, 1986); and Gerard de Vaucouleurs, *The Planet Mars* (Faber and Faber, 1951).

SCHLIEMANN, HEINRICH (1822–1890) German archaeologist and excavator of Mycenae and Tiryns, but best known for his discovery of Troy. He began his career as an agent for an Amsterdam firm and, through the connections he built up, started his own business, very soon amassing a large fortune by selling arms during the Crimea War. Retiring at the age of 36, he then devoted himself to the study of prehistoric archaeology. Notwithstanding, he never took his eyes off the main chance and was not averse to doing sharp deals to make money for himself whenever the opportunity offered. Much is known and more is suspected about him smuggling artifacts illegally out of Turkey. But the recent biography *Schliemann of Troy—Treasure and Deceit* (1995) by David A. Traill goes much further than any previous account, suggesting that Schliemann was not only a pathological liar but that his famous finds at Troy and Mycenae were to varying degrees faked.

The main documents that Traill uses to substantiate his allegations are those written by Schliemann himself, and it is the discrepancies within and between his diaries, his correspondence, and his published accounts, including his books, that Traill draws upon. He revealed that Schliemann claimed credit for ideas that were not his own and that he published reports of discoveries that were very different from his diary records; for example, the so-called Priam's treasure (now in Berlin) was not found inside the walls of Troy, but outside, and the published accounts of the hoard contains articles that were not found at the time but brought in, by an unknown source, to enhance the importance of the find. Also in doubt is the famous mask of Agamemnon (now in the National Museum in Athens) that is said to have been found with other gold masks in shaft graves IV and V at Mycenae. Was it really from the Bronze Age or faked by a 19th-century goldsmith?

For further reading: David Traill, *Schliemann of Troy—Treasure and Deceit* (John Murray, 1995).

SCIENCE FICTION AND SCIENCE The relationship between science fiction and science, which dates back to the 19th century. Works such as Mary Shelley's *Frankenstein* (1818), with its horrific picture of life created through the application of (then current) scientific principles running amuck, paved the way for later writers. Science fiction in the late 19th and early 20th centuries was dominated by two writers: Frenchman Jules Verne and Englishman H. G. Wells. These two illustrate two different schools of science fiction: science fiction based on invention, and science fiction based on speculation.

Verne's work, including *Five Weeks in a Balloon* (1863), *From the Earth to the Moon* (1865), *Twenty Thousand Leagues Under the Sea* (1870), and *The Mysterious Island* (1874), provides examples of science fiction based on invention. In these works and others, Verne predicted the development of air and space travel, submarine warfare, the aqualung, and television. In several ways, however, Verne was simply extrapolating from existing technology. Submarines, for instance, had served in the U.S. Civil War, although with a notable lack of success. Verne's inventions were refinements of tools that already existed; he confirmed this pattern in an interview with the journalist Gordon Jones. "I have always made a point in my romances," he said, "of basing my so-called inventions upon a groundwork of actual fact, and of using in their construction methods and materials which are not entirely without the pale of contemporary engineering skill and knowledge."

Verne labeled his contemporary H. G. Wells "a purely imaginative writer, to be deserving of very high praise." Wells was not interested in telling stories based on current technology: He was concerned with social problems, and he used science fiction to highlight these issues. His science fiction is based on speculation rather than extrapolation. Works such as *The Time Machine* (1895), *The Island of Doctor Moreau* (1896), *The Invisible Man* (1897), and *The War of the Worlds* (1898) present some technologies that were impossible to replicate in Wells's day but have since become practical and many others that are impossible according to our knowledge of the universe's physics. Elements of Wells's stories that defy laws of physics or known scientific data, like his time traveler and his Martian invaders, are placed in a subcategory of science fiction called science fantasy.

Science fiction first became a distinct form of literature in the first two decades of the 20th century under publishers and editors such as Hugo Gernsback (founder and editor of *Amazing Stories*) and John W. Campbell (editor of *Astounding Science Fiction* during its heyday of the 1940s and 1950s). The science fiction of the 1920s drew on the legacy of H. G. Wells: It tended to be heavily adventurous "space opera," with standard plots taken from other forms of fiction and enlivened with bits and pieces of technology. Edgar Rice Burroughs's *A Princess of Mars* (1913) and the "Lensmen" series of E. E. "Doc" Smith are perhaps the best-known representatives of the genre during this period. After he became editor of *Astounding Science Fiction* in 1927, John W. Campbell steered science fiction away from this space-opera form. He mentored many of the best-known science fiction writers of the 20th century, including Isaac Asimov, Arthur C. Clarke, and Robert A. Heinlein. These writers and others like them reintroduced the realism of Verne into the genre, yet kept the vision of Wells. The result was a literary form that dealt with the questions raised by technology.

Modern science fantasy is perhaps best represented through the adventures of Superman. Although a semi-scientific rationalization is given for each of Superman's powers, he continually performs deeds that defy the laws of physics. According to the laws of Earthly physics, Superman should not be able to fly. Animals on Earth that fly, such as birds and bats, do so by using aerodynamic principles—the same principles that keep airplanes in the air. Other creatures that fly, such as insects, are too small for these principles to apply to them.

Superman is not aerodynamic: He should not be able to keep himself in the air. Another example comes from the famous scene where Superman catches Lois Lane in midair after she falls from a great height. According to Newton's law of gravitation, Lois continually accelerates through her fall. She should hit Superman's arms with the same force as if she hit the ground, with similar (fatal) consequences. The fact that she doesn't improves the story but defies science.

The famous television series *Star Trek* also offers examples of Verne-style science fiction, as well as Wells-style science fantasy. The science-fiction writers who worked on the original series actually anticipated scientists in the naming of stellar phenomena such as BLACK HOLES. The matter–antimatter generators that power the starship's warp drive are based on physics currently practiced at science research facilities such as Fermilab in Illinois and the European Laboratory for Particle Physics (CERN) in Geneva, Switzerland. Even the famous transporter beam is theoretically possible, based on current thinking in quantum and post-Einsteinian physics.

Yet, although scientists acknowledge that many of the techniques used on the USS *Enterprise* are theoretically possible, most of them are far outside the reaches of our current technology. Although Fermilab can produce about 50 billion antiparticles in one hour, the energy these release when destroyed by collision with regular particles is far less than the energy required to make them in the first place. The dilithium crystals that make the matter–antimatter particles are unlike any other form we know about. The transporter could theoretically reproduce an entire human being, atom by atom, in another place, but the disks providing the storage capacity needed to save all that information would outweigh the mass of the Earth itself.

There are also elements of *Star Trek* that simply defy the laws of physics as we know them. The crew of the *Enterprise* routinely survive accelerations and decelerations in a period of time that would kill any unprotected human beings. The starship itself accelerates to significant fractions of light speed without any great increase in mass—something that violates Einstein's special theory of relativity.

Science fiction continues to be an important part of society's reaction to science. It warns about some of the dangers inherent in the rapid proliferation of technology; it encourages potential new scientists to dream, to picture new ways of taking the universe apart and putting it together again. It provides us with a means of understanding ourselves as we face the future.

For further reading: Lawrence M. Krauss, *The Physics of Star Trek* (Basic Books, 1995); and H. G. Wells, *The Time Machine/The War of the Worlds: A Critical Edition,* ed. by Frank McConnell (Oxford University Press, 1977).

SCIENTIFIC CREATIONISM See CREATION SCIENCE.

SCIENTISM The use of or appeal to the authority of science to legitimate some particular claim or policy. *The Fontana Dictionary of Modern Thought* defines scientism as "The view that the characteristic induction methods of the natural sciences are the only source of genuine factual knowledge and, in particular, that they alone can yield true knowledge about man and society. This stands in contrast with the explanatory version of DUALISM which insists that the human and social subject-matter of history and the social sciences can be fruitfully investigated only by a method, involving sympathetic intuition of human states of mind, that is proprietary to those disciplines."

In the modern Western science, science and scientists have come to be regarded with great respect and authority. To be scientific is to be accepted; to be unscientific is seen as unacceptable, of dubious merit, if not wrong or wrong-headed. "Scientific" is a badge of merit, since science has great authority in the modern world. Much of the discourse of our society—conversation, newspapers, radio, and television—makes frequent use of appeals to science and its terms; there are constant references to such ideas as evolution, the environment, DNA, genetics, and so on. In their proper place they play important roles, but they are often applied and appealed to in quite different contexts to those from which they originate. This misplaced application of science, scientific, and the terminology of science is what is described by the term scientism.

Such misapplications occur when people draw on widely shared images and notions about the scientific community and its beliefs and practices in order to add weight to arguments that they are advancing, or to values and policies whose adoption they are advocating. Scientism implies an attitude to science: Those who use scientistic language acknowledge and respect the authority of the scientific community, and wish to capitalize on that authority in order to make their discourse more persuasive. In so doing, they reinforce and consolidate that authority. Nonscientists use scientism to promote products for which there is little or no scientific support, by employing scientific appearances, equipment, and scientific or quasi-scientific terminology (for example, BEAUTY TREATMENTS). And perhaps at a more serious level (though all these matters are serious enough) science as authority makes its way into sociology, politics, economics, ethics, and elsewhere. Some obvious examples are SOCIAL DARWINISM, NAZI RACISM, and Marxism—only a few among many.

All this is not to say that cases for which the authority of science is invoked are therefore without merit. The ideas, products, or policies advocated may well be bad, indifferent, or excellent. It is the method of promotion which the use of scientism throws into question. Here are two further examples of the use of scientism. Some scientists claim that direct contact with Nature and inquiry into her laws produces habits of mind that cannot be acquired by other means. Science is not only able to increase the comforts of life and add to material welfare but also to inspire the highest ethical thought and action. While, at least in theory, success in science depends strictly on the discovery of facts or of relationships, in other fields, in politics and in public life, for example, it may be gained by fluent speech and a facile pen. Scientists must concentrate on what is there, on the facts of Nature. By contrast, politicians and writers are more concerned with argument, and through the use of persuasive words and phrases or rhetoric, influence public affairs without necessarily using objective criteria. When scientists resort to this sort of argument they are asserting a special ethical position for scientists and thereby employing a particular type of scientism. And of course, while many scientists have high ethical standards, many do not.

In "Scientific Approach to Ethics" (*Science*, 1957), Anatol Rapoport argues a special relationship between science and ethics, saying: "There is no sharp distinction between scientific outlook and scientific ethics. Both eschew authority—that is coercion in any form—and probably for this reason are irresistibly attractive as means of liberating man from the bonds which, in his ignorance, fear, and ethnocentrism, he has imposed on himself." Rapoport is using the authority of science to validate his ethics—scientism.

For further reading: Iain Cameron and David Edge, *Scientific Images and their Social Uses: An Introduction to the Concept of Scientism* (Butterworths, 1979).

SCIENTOLOGY The religious teachings developed by the well-known science fiction writer L. Ron Hubbard (1911–86), founder of the church of Scientology. In the 1940s Hubbard developed an alternative form of psychotherapy, which he termed DIANETICS. According to Hubbard, dianetics proposed to solve the problems of the individual's true self, or *Thetan* (analogous to the soul in Greek philosophy) attempting to operate in the human body. Problems emerged with the aberrant activity of the mind. Hubbard proposed a method of counseling called auditing to counter the negative activity of the mind. Once such problems were largely erased, when the individual had reached the state of "clear," the Thetan could then begin to operate as a free being.

As dianetics spread in the early 1950s, Hubbard was also expanding his speculation concerning the operation of the "post-clear" individual. Some material came from the observation of auditing sessions and other from his broad reading and thinking. His speculations moved into the metaphysical realm and he began to view the Thetan as an eternal being that repeatedly sought embodiment in the physical. He also saw that among the abilities of the Thetan was its conscious exteriorization from the body. His speculations culminated in teachings on God and the Thetan's ultimate relationship with the divine.

In general, members of the Church of Scientology move through a program of dianetic auditing to the point of clear. They then move progressively through the Scientology program, which is divided into what are termed OT (Operating Thetan) levels. The writings of L. Ron Hubbard used throughout these levels are considered sacred scripture by the church and access to them is limited to church members as they go through the OT levels. In general, the program is seen as providing a final cleansing from the aberrations of the mind, offering techniques for the Operating Thetan, and outlining a more complete view of the universe and the divine. Books by former members who had access to Hubbard's OT writings have presented details of the higher level teachings, but church officials have complained that their reflections have presented a partial and distorted view of these inner teachings. The church has taken significant steps to prevent the spread of the confidential Hubbard writings taken by former members when they separated from the church.

Almost since its founding, the Church of Scientology has been embroiled in a series of public controversies, which have kept it in courts around the world. Hubbard and the church have attacked PSYCHIATRY for its use of what it considered barbaric techniques (such as ELECTRIC SHOCK TREATMENTS and lobotomies), conduct of experiments with dangerous mind-altering drugs, and its cooperation with government agencies in the suppression of political dissidents and religious minorities. At the same time, authorities have opposed the church for what was regarded as practicing medicine without a license, its unorthodox organization that appeared more businesslike than churchly to many observers, and its internal policing system. Early opposition in the United States from the Internal Revenue Service, the American Medical Association, and the Food and Drug Administration had long-term implications as government documents were passed to countries around the world.

Whatever scientific claims were made during the early days of dianetics have been replaced by an understanding of the benefits offered by Scientology as purely religious. Scientology views itself as quite compatible with modern medical science and actively recruits physicians and dentists into its ranks. It does, however, reject psychiatry and actively seeks to destroy psychiatry's influence. In this regard, it has also opposed the development and use of consciousness-altering drugs, such as Prozac, as well as the pharmaceutical companies that produce them.

For further reading, R. G. A. Dolby, *Uncertain Knowledge* (Cambridge University Press, 1996); J. G. Foster,

Enquiry into the Practice and Effects of Scientology (HMSO, 1971); and L. Ron Hubbard, *Dianetics: The Modern Science of Mental Health* (The American Saint Hill Organization, 1950).

SEA SERPENTS Creatures of unknown species, usually having a serpentine or reptilian resemblance, that are said to inhabit the world's seas, often threatening ships and humans. Sea serpents have populated myths for as long as people have known the seas. Additionally, there are documented reports of sailors seeing or being threatened by such creatures in nearly all of the world's oceans. Sea serpents have been blamed for prodigious storms and for destroying ships. Although most are surely mythical, some have been found to have some scientific basis.

The KRAKEN, a gigantic squid long reported in the northern seas, has been shown to be no myth. Although none of these monsters has probably ever grasped a ship in its tentacles and pulled it under the sea, as stories and legends tell, specimens of giant squid of more than 20 meters (60 feet) in length have been documented near New Zealand. Eyewitnesses report even larger ones. Tentacles of immense length have been found in whale stomachs, and the crews of Norwegian fishing boats of the 19th century often reported capturing squid of 8 meters (25 feet) or more long and cutting them up for bait.

Another type of sea serpent was seen in 1966 by two sailors rowing across the Atlantic. One night, one of them, Captain John Ridgway, saw a gigantic "writhing, twisting shape," illuminated by the ocean's natural phosphorescence. Unable to think of another explanation, he concluded that he had seen a sea serpent. He describes the brief and startling encounter in his book about the adventurous voyage, *A Fighting Chance* (J. B. Lippincott, 1966).

As with LAKE MONSTERS, there is a variety of reasons people might think they see sea serpents even when they do not. Misperception and misidentification are perhaps the most common reasons; self-fulfilling expectations are another.

Among the sea serpents, or, more generally, sea monsters, reported in relatively recent times, are Nessielike creatures that are single or double humped and bear some resemblance to the plesiosaur that became extinct during the time of the dinosaurs; giant octopi and jellyfish; mercreatures, described as half-fish and half-human or equine; and gigantic eels and whales. Of these, strong anecdotal reports as well as some physical evidence supports the existence of some form of all but the mercreatures—and even they might be misperceived creatures such as sea cows.

Noted cryptozoologist Bernard HEUVELMANS categorized sea-serpent sightings by type in his 1968 book *In the Wake of the Sea-Serpents.* He studied nearly 600 reports, concluding that 358 were authentic sightings. Among the most common types he found were the long-necked Nessie-type [5–15 meters (15–50 feet) long or more], seen in various locations around the world, and the only amphibious type; the merhorse [9–27 meters (30–90 feet) long], again widespread and bearing some resemblance to Nessie, with the addition of a mane and sometimes whiskers or other hair on the face; the many-humped creature [20 to more than 30 meters (60–100 feet) long] resembling a caterpillar with fins and seen most often in the North Atlantic Gulf Stream; the many-finned creature [6–18 meters (20–60 feet) long] seen in tropical waters and with visible breath or mist coming from its nostrils; the superotter [20–30 meters (60–90 feet) long], seen in northern seas; and the super-eel (most often seen in one of two lengths: 9 meters or 27 meters (30 feet or 90 feet), smooth, undulating, and inhabiting deep, cold water. These classifications are based primarily on sightings, not on physical evidence, such as actual specimens or even photographs.

Sea Serpents

Nevertheless, as scientists delve ever deeper into the world's oceans, they continue to find species that support or help explain the myths of old. Among other things, scientists have discovered amazing, gigantic tube worms; living examples of the COELACANTH, long thought to have become extinct with the dinosaurs; and jellyfish of incredible size.

See also CHAMP; CRYPTOZOOLOGY; LOCH NESS MONSTER; OGOPOGO.

SECOND-RACE THEORY The theory that human beings are of two different races, a male race and a female race. It was an idea put forward by British engineer, William H. Smyth in 1927 in *Did Man and Woman Descend from Different Animals?* It was picked up by an eccentric English antifeminist, Arabella Kenealy who developed it in *The Human Gyroscope* (1934). She explained the separate sexes and the essential maleness and femaleness of all the universe from atoms to galaxies as caused in some way by the gyroscopic effects of rotational motion. So, on Earth, northern peoples are more masculine than southern, the difference resulting from the greater gyroscopic effect nearer the equator, the difference in motion that makes Earth bulge a little at the equator, flatten at the poles—an oblate spheroid. It is not a theory that has gained much support.

SEXUALITY In common usage, the term describing the capacity of humans for sexual behavior, but also used by sociologists in two other ways: *either* as all those attributes connected with sexual activity—appearance, level, desires, roles—*or* as a preference for different forms of sexual activity. This second sociological usage is now becoming common. A version of the NATURE-NURTURE DEBATE centers on sexuality: Is a person's sexuality biologically or socially determined? Sociologists stress that society's rules for sexual behavior are just that—social rules—and they point out that what is acceptable in one country, or at one period in history, is unacceptable in another. What was regarded as sexually attractive in Regency or Victorian England is not so today. What are sexually attractive features in a tribe in central Africa or New Guinea are not in Hollywood. Sexual attractiveness in both men and women varies enormously in different cultures and at different times within the same culture.

The current concerns in the West about sexuality are largely focused on homosexuality. Should gays and lesbians be regarded as a normal sexual subculture, tolerated and not penalized in any way? Should gays and lesbians be allowed in the military? Or in the church? Are their practices forbidden by holy writ? Should there be legislation authorizing the legal marriage of homosexual couples, male or female? Although Western countries have become much more tolerant during the 20th century—homosexuality is no longer a crime—these matters are far from settled.

Attitudes towards sexuality—hetero-, bi-, homosexuality—depend greatly on personal and social prejudices. In Victorian times, for example, there was a great deal of hypocrisy about sex. Respectable women were not supposed to initiate or to enjoy sex; at the same time, prostitution was widespread. In ancient Greece it was regarded as normal for adult men to have sexual relations with young boys before marriage. Some tribes in Sumatra today approve male sexual relationships before marriage. Feminists argue that attitudes towards sexuality, particularly female sexuality, have been and still are male-defined and they seek a feminist definition.

In the light of these various positions, it seems clear that sexuality is largely, if not entirely, socially constructed. But it is possible that individuals have biologically determined *tendencies*. Setting aside all social and biological influences in the nature/nurture dispute proves difficult, if not impossible, especially in this case. Do hormone levels, for example, affect a person's level of sexual activity? Or are social constraints the governing factor? If both, what is the mix? Can biological interference—drugs—alter the level? In the case of homosexuality and bisexuality, the problem is yet more difficult. The homosexual community claims that its sexual orientation is biologically determined. If it is beyond individual control, then social disapproval or regarding it as criminal or immoral are wrong. There has been recent evidence of a gene abnormality in homosexuals, which supports this view, but the research (and the argument) continue.

Against this background, it is not surprising that many social groups—religious movements, sects, and cults—as well as individuals have held unusual views on sexuality. The Mormons advocated and practiced polygamy. Fundamentalist groups, the flat earthers in Zion under their overseer Wilbur Glenn VOLIVA for example, imposed very strict controls on female sexuality. Various writers have insisted that sexual inversion, homosexuality, is a perversion, an evil way of life. Others have maintained that it is a superior form of life, pointing to many great men and women who were inverts. Many eccentric theories and attitudes have been advocated, perhaps the strangest being that of Sir Richard Burton, who believed that there was a strip around the equator where inversion was strongest, which he called the "Sotadic Zone." And, as it were, on the other front there have been and are cults that advocate abstinence and others that practice free and unfettered intercourse (and a mixture of both—see COITUS RESERVATUS), sometimes between any pair of consenting adults at any time, sometimes with the reigning guru having privileged access to the females in his community.

The whole subject of sexuality has always attracted a great many theories. Many claim to have a scientific basis but, with very few exceptions, these gain no support from orthodox science and are pseudoscientific. [See also ELLIS, Henry Havelock.]

For further reading: Kenneth McLeish, ed., *Bloomsbury Guide to Human Thought* (Bloomsbury Reference, 1993).

SHEEP-GOAT A concept in PARAPSYCHOLOGY that explained some significant scoring effects on ESP tests. In 1942, when parapsychology was still less than a decade old, Gertrude Schmeidler suggested that scoring could be affected by the subjects' belief system. Schmeidler concluded that those who believed in the possibility of ESP (whom she labeled "sheep") would score higher than those who did not believe in it (whom she labeled "goats"). The tests she gave bore out her hypothesis. Though the differences between the two groups were small, they were nevertheless statistically significant, and the concept has become an accepted truth in parapsychology.

For further reading: Gertrude Schmeidler and Robert A. McConnell, *ESP and Personality Patterns* (Yale University Press, 1958).

SHELDRAKE, RUPERT (1942–) Biochemist who proposed the hypothesis of FORMATIVE CAUSATION. Sheldrake was educated at Harvard and Cambridge and then became a research fellow of the Royal Society, working on the development of plants and the aging of cells. From 1974 to 1978, he studied the physiology of legume crops at the International Crops Research Institute in Hyderabad, India.

Sheldrake first doubted but soon became convinced that the dominant mechanistic worldview was seriously deficient and helpless before a large range of observations without any satisfactory explanation. He proposed the hypothesis of formative causation, which, he says, applies to all nature, inorganic and organic. All things key into what he calls morphic fields, a sort of collective memory that connects us with others of our kind, past and present, and influences our shape and behavior, by a process of morphic resonance.

Sheldrake's theory states that all systems that in any sense organize themselves—molecules in crystals, plants, animals, human societies—do so in response to a morphic field (a shaping field), a hitherto unknown field akin to the electromagnetic or gravitational fields, a field that is created and builds up as the shaping habit is established and evolves. This means that the universe is not a set organization but keeps evolving as new morphic fields form (presumably crystals respond to a different field from that which governs political systems). Sheldrake's theory suggests that a developing organism resonates with the morphic fields of its species—morphic resonance—and thus draws upon a pooled or collective memory. As one reviewer, David Fideler, puts it, "Sheldrake's thesis challenges the reductionism still inherent in much scientific theory and would help explain many phenomena—and types of human experience—that are otherwise beyond the grasp of scientific understanding."

Sheldrake's hypothesis has obvious implications for genetics. Orthodox Mendelian genetics, explaining inheritance via DNA, is fine as far as it goes, but it leaves much unexplained. In his treatment Sheldrake goes beyond Jean Baptiste LAMARCK. Lamarck posited inheritance of beneficial characteristics acquired during one individual's or generation's lifetime by the next and succeeding generations. Sheldrake posits that the characteristic goes not from an individual only to his or her offspring, but is passed on to, and is transmitted within, the species. Whereas Lamarck only considered living organisms as possible transmitters, parent to offspring, Sheldrake includes all the natural world, living or not, *past or present.*

Sheldrake

In *The Presence of the Past,* Sheldrake discusses biological inheritance and addresses explicitly the dispute between orthodox and Lamarckian theories. He suggests that his process of morphic resonance would resolve the issue, citing experiments with fruit flies which demonstrate that there is an effect to be explained; he writes:

> *For decades, the debate about Lamarckian inheritance has centered not so much on empirical evidence for or against such inheritance as on the question of whether or not such inheritance is theoretically possible. According to the genetic theory of inheritance, it is impossible because the genes cannot be specifically modified as a result of characteristics that organisms acquire in response to their environ-*

ment or through the development of new habits of behavior. Lamarckians have generally assumed that some such genetic modification must take place, but have been unable to suggest how this could happen.

The hypothesis of formative causation provides a new approach, which does not fit into either of these standard positions. Acquired characteristics can be inherited not because the genes are modified, but because this inheritance depends on morphic resonance.

Sheldrake's ideas represent a big shift from the established paradigm and met, not surprisingly, with widespread opposition. The beauty of the established mechanistic theory is that it has been tested by fairly straightforward observation and experiment over a long period. It still has its problems, its controversies, its disputes, and its sub-theories, which are not universally accepted—punctuated equilibrium, selfish genes, and the like. But they are seen to be open to experimental tests within the established paradigm; no revolution is required. Sheldrake's formative causation/morphic field is entirely different.

In a series of papers and books—*A New Science of Life* (1981), *The Presence of the Past* (1988), and *The Rebirth of Nature* (1997), Sheldrake supports his thesis with a wealth of data and of argument. He makes it clear that his theory is falsifiable, based on thorough experimental evidence and proposes ways in which it could be tested. In his work, he also argues that, having excluded the fraud and error with which the subject is plagued, there remains a residue of evidence of real parapsychological phenomena that orthodox science cannot explain. Formative causation, he insists, may open a route to their explanation. He also suggests that formative causation may improve our understanding of the power of prayer, mystical experiences, and similar phenomena. In addition, he posits that creativity is best explained as coming from God (or some superhuman intelligence) but if that is not acceptable, it is better explained by morphic resonance than by the current mechanistic paradigm.

In 1983, the *New Scientist* held a competition to devise an experiment that would test Sheldrake's theory. He adapted the winning entry slightly and conducted it with the aid of a Japanese poet, Shuntaro Tamikawa, and was convinced that the results were positive. (More recently, he reported another experiment on whether or not people were aware that someone out of their line of sight was looking at them. Again he claimed a positive result, a result inexplicable by conventional mechanistic science but possibly providing evidence favoring morphic resonance.)

Needless to say, Sheldrake's claims are hotly contested, not least because he is, in part, motivated by his religious convictions. He is demanding a major paradigm shift, a return to a form of VITALISM from which orthodox science has spent the last 300 years escaping.

For further reading: Rupert Sheldrake, *A New Science of Life: The Hypothesis of Formative Causation* (Blond and Briggs, 1981); Rupert Sheldrake, *The Presence of the Past* (Fontana/Collins, 1988); and Rupert Sheldrake, *The Rebirth of Nature: The Greening of Science and God* (Century, 1997).

SHIATSU Literally meaning "finger pressure" in Japanese. A firm sequence of rhythmic pressure is applied on specific points for three to seven seconds, designed to awaken the ACUPUNCTURE meridians. The meridians are points on the body, sometimes far removed from the area where symptoms or pain are experienced, which, when stimulated, bring about beneficial results. Similar to MASSAGE, the goal of Shiatsu or "ACUPRESSURE" is to deeply relax tensed or exhausted muscles. It differs from massage in that pressure is applied more vigorously, usually with the ball of the thumb, or sometimes even with the thumbnail.

Shiatsu was developed in Japan over the past fifty years by Tokujiro Namikoshi, who used the technique to treat over 100,000 patients suffering from a wide variety of illnesses and conditions. A Shiatsu treatment by a professional may take thirty minutes or more. Self-treatment is generally a matter of several minutes at a time, a few times a day. Shiatsu is said to be a good alternative to aspirin, and to the negative affects of any chemical in the body. It can also be administered to relieve conditions such as headaches, migraine, sore throat, sinus colds, neck fatigue, eyestrain, and toothache, among others. Whether and to what extent Shiatsu is scientifically proven remains a matter of some debate.

SHROUD OF TURIN The linen cloth that reputedly enshrouded the body of Jesus Christ when it was taken down from the Cross. First brought to ecclesiastical attention in the 14th century, the cloth was discovered in a church in Lirey, France. More than 14 feet long and about four-and-a-half feet wide, the cloth is heavily stained with blood, scorch and water marks, and the full-body image, front and back, of a naked man that resembles popular images of Christ. No one knows for sure where the cloth came from, but tradition of the time suggested that a knight, perhaps a Knight Templar, had brought it back from the Holy Land during one of the Crusades.

While it was not officially acknowledged as a Holy Relic by the Church of that time, the shroud was considered valuable and sacred by many people of the area. Over the next several decades, it was passed from one noble family to another and was finally taken to Torino (Turin), Italy, where a special chapel was built for it. It has remained there until today.

In the 20th century the shroud, by now considered by many Roman Catholics to be the most important of all Christian relics, became the subject of much publicity and speculation. Was it actually Jesus's shroud, or was it a fraud? The availability of new, highly reliable scientific tests led to demands that it be studied and judged once and for all. The Catholic Church, owner of the object, long rejected any kind of examination that could damage the shroud. But finally, by the 1980s tests were available that could be performed on minuscule fragments—less than the size of a postage stamp. The Vatican allowed tests to commence.

Up to this time, the arguments (besides faith) for the shroud's authenticity had included the sophisticated rendering of the image; if it were an artistic forgery, the artist would have been far ahead of other artists of his or her time in depicting the human body. In addition, until recently some of the details of the crucifixion shown on the figure were not widely believed to be true. For example, Jesus traditionally was shown with nails piercing the palms of his hands; the shroud shows the nails piercing his wrists; recent historic studies have shown the latter method the most likely to have been used at the time. No analysis has been able to clearly determine whether the pigment is dried blood or another substance, such as iron oxide. Noninvasive microscopic examination showed the weave of the fabric and certain pollens embedded in it to be consistent with those found in Palestine in the first century A.D.

In 1988, the Vatican allowed tiny scraps of the shroud to be independently examined by three laboratories, which used highly sophisticated computerized dating analysis. The results showed that the shroud came from the 13th or 14th century. One might have thought that this would be the end of the shroud as an item of veneration. But as with so many mysteries that depend on faith, while the dating tests were the final confirmation for those who thought the shroud was a clever forgery, to true believers they were simply a minor inconsistency overridden by faith and other evidence.

SIGHT WITHOUT GLASSES The concept that visual defects could be cured by throwing away one's glasses and following a prescribed regimen of eye exercises. The idea was one of the most persuasive pieces of medical pseudoscience of the early part of the 20th century, and the treatment was followed by many thousands of people in Europe and the United States.

The first exponent of this theory was Dr. William Horatio Bates (1860–1931), a New York ophthalmologist who in 1920 published a book called *Cure of Imperfect Eyesight by Treatment without Glasses.*

The core of Bates's book was his theory of accommodation. "Accommodation" is a term for the focusing process that takes place within the eye when a person shifts attention to objects at various distances. Adjustment involves an alteration in the shape of the lens; a tiny muscle called the ciliary muscle causes the lens to become more convex as the eye is focused on closer objects. This change of the lens has been photographed and measured accurately many times and provides a simple explanation of what is happening during accommodation. But Bates always denied that the lens was a factor in accommodation. His belief, so necessary to explain the value of his exercises, was that focusing is accomplished by an alteration in the whole length of the eyeball and that this was brought about by two muscles on the outside of the eye. In support of his theory, Bates produced his own photographs of experiments done on the lens of a fish's eye.

Bates believed that the cause of most, if not all, refractive problems such as nearsightedness, farsightedness, astigmatism, and even presbyopia (the inability to accommodate, the normal result of growing old), was simple strain, brought about by the inability to relax. This inability he saw as "an abnormal condition of the mind, a wrong thought." It was Bates's opinion that glasses were eye crutches that could never cure strain but actually made a cure more impossible to achieve because the eyes kept adjusting to them and consequently the patient needed stronger and stronger lenses. He believed that, if people were taught how to relax their eyes and their complex negative thoughts, their visual problems would disappear in seconds. "If the relaxation is momentary the correction will be momentary; if the relaxation is permanent the correction will be permanent."

To relieve strain, Bates proposed what he called central fixation—learning to see what is in the center of vision without the strain of staring. A few exercises were recommended: (1) *Palming*—Patients were to put their palms over their eyes and think of perfect black; (2) *Shifting and Swinging*—Moving the eyes back and forth so that an illusion is built up of an object swinging in front of them from side to side; and (3) *Reading under adverse conditions*—for example, while lying down or traveling in a bus or train; also reading in dim light or in bright sunlight, and actually looking directly at the Sun—a practice that ophthalmologists should warn their patients *never* to follow, as it causes permanent retinal damage.

Bates died in 1931, but his unconventional methods lived on for the next decade, although his eccentric theory of accommodation was no longer regarded as credible. Before World War II in England and Germany, schools opened teaching the Bates method. Aldous Huxley, the English novelist and essayist (1894–1963), who was afflicted with chronically bad sight, benefited from the Bates system and lent it credence by writing a book about it called *The Art of Seeing*. In Hitler's Germany, the system became quite a cult.

Since many patients believed their sight was improved by following Bates's exercises, one must ask how this came about. It is easily understood how such exercises might improve sight for someone for whom an ophthalmologist had prescribed unnecessary lenses or whose correction was very weak. (Some of Bates's patients had glasses made of virtually plain glass.) Secondly, some eye conditions do improve with age and a few disorders, such as partially crossed or walleyes, respond to orthodox exercises—thought not to those originally prescribed by Bates. The one valuable idea Bates promulgated was the notion that mental factors do affect vision.

For further reading: Martin Gardner, "Throw Away Your Glasses!," *Fads and Fallacies in the Name of Science* (Dover, 1957); and Peter Mansfield, *The Bates Method* (Macdonald Optima, 1992).

SIMPSON, EDWARD (1815–?) Perpetrator of scientific hoaxes. Crediting with the development of a number of archaeological hoaxes, Edward Simpson was born in Yorkshire in the north of England. As a teenager, he worked for several scholars who passed to him a love for the past and some knowledge of paleontology. Later, he became an avid fossil collector and made his living selling specimens to dealers, collectors, and colleges. His career changed in 1843 when one of his clients asked him to manufacture a copy of an arrowhead. Simpson's copy proved to be indistinguishable from the original. He soon discovered that most ancient artifacts could be faked relatively easily. Although Simpson's counterfeits would have been easily spotted today, in the 19th century they were confused with genuine artifacts. They were shipped across the country, and some, especially his arrowheads, were purchased by the British Museum. He became well known in paleontological circles under several nicknames, including Fossil Willie.

The primary problem in Simpson's business appears to have been its growth. He was producing material faster than the small number of excavations could have produced. Investigations were launched in the early 1860s. Then, in 1861, Simpson confided what he had been doing to a dealer in London. The dealer forced Simpson before a group of geologists in January 1862. Simpson demonstrated his skill by producing one of his arrowheads for the group.

Simpson destroyed his credibility and his business together. The exposure made him famous as a forger of artifacts but supplied him with no position from which to make a living. Several years later he was caught and convicted of theft, after which he faded into obscurity.

SKEPTICS' MAGAZINES Magazines addressed to readers who are essentially skeptical about claims for scientific validity. *The Skeptic*, *The Skeptical Inquirer*, and *Free Inquiry* are just three of the several English language skeptical magazines.

The Skeptic, published in Altadena, California, is the quarterly magazine of the SKEPTICS SOCIETY, published since 1993 under the editorship of Michael Shermer, who is also director and CEO of the society. To quote its own explanatory subheading: ". . . devoted to the investigation of extraordinary claims and revolutionary ideas and the promotion of science and critical thinking." The contents of a recent issue included such articles as "Teller about ghosts" and "The question all skeptics are asking: Can a cat be simultaneously alive and dead?" and contained letters such as "Pseudoscience kills" and "Loch Ness photo proof." The purpose of the magazine is primarily to expose and attack what it sees as fraud and irrationality, but, in fairness, it also opens its columns to those it is attacking. Some issues contain an extended debate with articles pro and con some topic in science/pseudoscience.

The *Skeptical Inquirer*, established in the late 1970s, is published bimonthly by the COMMITTEE FOR THE SCIENTIFIC INVESTIGATION OF CLAIMS OF THE PARANORMAL (CSICOP) in Amherst, New York. The committee "encourages the critical investigation of paranormal and fringe-science claims from a responsible, scientific point of view and disseminates factual information about the results of such inquiries to the scientific community and the public." It also promotes science and scientific inquiry into critical thinking, science education, and the use of reason in examining important issues.

Free Inquiry is published quarterly by the Council for Democratic and Secular Humanism from Buffalo, New York. Organizations associated with the council are the Academy of Humanism and the Committee for Scientific Examination of Religion. The journal concerns itself very largely with what it sees as conflicts between secular/humanist and both eastern and western religious views of science, attacking beliefs such as REINCARNATION and faith healing.

See also *FORTEAN TIMES*.

SKEPTICS SOCIETY, THE An organization of scholars, scientists, historians, magicians, and the intellectually curious that sponsors a lecture series at California Institute of Technology and publishes the quarterly magazine *The Skeptic*. Its purpose is to promote science and critical thinking and to disseminate information on pseudoscience, pseudohistory, the paranormal, MAGIC, superstition, fringe claims and groups, revolutionary science, protoscience, and the history of science and pseudoscience. The editor of the magazine is Michael Shermer, and the society's members include Jared Diamond, Stephen Jay GOULD, and John Gribbin.

The Skeptics Society and *The Skeptic* magazine investigate a wide variety of theories and conjectures, looking in particular at how well they stand up under scientific scrutiny. Subjects include, for example, EVOLUTION and CREATION SCIENCE; cult and religion; Holocaust revisionism and extreme Afrocentrism; conspiracy theories; NEAR-DEATH EXPERIENCES and OUT-OF-BODY EXPERIENCES; CRYONICS, life after death, and the quest for immortality.

The organization takes the approach of the 17th-century Dutch philosopher Baruch Spinoza: "I have made a ceaseless effort not to ridicule, not to bewail, not to scorn human actions, but to understand them." It recognizes the limitations and sociocultural influences on science, yet adopts the philosophy of Albert Einstein: "All our science, measured against reality, is primitive and childlike—and yet it is the most precious thing we have."

See also SKEPTICS' MAGAZINES.

SKOTOGRAPHY See THOUGHTOGRAPHY.

SNAPPING A medical theory of conversion advanced by journalists Flo Conway and Jim Siegalman in their 1978 anticult book, *Snapping: America's Epidemic of Sudden Personality Change.* In their volume, they claimed that certain religious groups, popularly called cults, had discovered a new technique of MIND CONTROL. The cults, Conway and Siegalman suggested, used this technique on their members to produce a new form of illness that they termed "information disease." This disease, they said, was characterized by severe disturbances in perception, memory, and informational processing capacities and was caused by an alteration of the neurological pathways in the brain. The genesis of the disease came from the practice of various mind-altering techniques by group members, followed by hours of group indoctrination.

Conway and Siegalman presented no medical data to support their claims of physiological changes in the members studied. But they did present statistical findings that, they claimed, indicated a relationship between participation in a group and long-term negative mental and emotional effects (including NIGHTMARES, amnesia, hallucinations, etc.). However, careful examination of their data failed to support the snapping hypothesis.

For further reading: Flo Conway and Jim Siegelman, *Snapping: America's Epidemic of Sudden Personality Change* (Lippincott, 1978).

SOAL, SAMUEL G. (1889–1975) British parapsychologist whose career ended in charges of fraud. Soal, who was trained as a mathematician, became interested in PSYCHICAL RESEARCH as a result of reading *Raymond* (1916) by Sir Oliver Lodge (1851–1940). In the book, Lodge described communications he had received through several MEDIUMS from his son who had been killed in World War I. Intrigued, Soal visited mediums throughout the 1920s and in particular attended a number of séances with Blanche Cooper. He was present for the famous "Gordon Davis" communications, in which Cooper supposedly transmitted material from a spirit. Later, it turned out that much of the information was correct. However, it also turned out that Davis was still alive.

In the 1930s, Soal became an early convert to PARAPSYCHOLOGY. Following the publication of Joseph Banks RHINE's initial study of *Extrasensory Perception* (1934), Soal tried repeatedly to duplicate Rhine's positive results. He was disappointed when all of his experiments ended in failure and was known in England for several years as a critic of Rhine's work.

Then in 1941 Soal began his own parapsychology experiments with Basil Shackleton. The reported results were phenomenal, and news of them swept through the parapsychological community. Soal's work was hailed as a great step forward. In 1957 Rhine and Pratt, leading figures in the world of parapsychology, listed four experiments that, they asserted, provided conclusive evidence for the reality of ESP. Numbers three and four on their list were Soal's experiments with Shackleton. In the post-World War II period, much of the credibility of parapsychology rested on the work of Soal, a respected academic and scientist who had originally been a skeptic.

Later examination of Soal's finding led to charges that Soal had manipulated his data, a change proved conclusively in 1978.

For further reading: C. E. M. Hansel, *ESP and Parapsychology: A Critical Reevaluation* (Prometheus Books, 1980).

SOCIAL DARWINISM A 19th-century theory proposed by British sociologist Sir Francis GALTON and loosely based on Darwinism, by which the social order is said to be a product of natural selection of those individuals who are best suited to existing living conditions. A "struggle for existence" and "survival of the fittest," terms coined by U.S. sociologist William Sumner, were seen as the ultimate motives for human action and social organization. Although concepts of biological determinism predated Charles DARWIN's *Origin of Species* (1859), such notions became popular in the late 19th and early 20th century, particularly when applied to the rivalries of great powers or to justify or condemn social and political theories. Social Darwinists in both Europe and the United States applied the principles of EVOLUTION as a means of explaining the origin and propriety of human characteristics and institutions, though opinions varied widely over just how Darwin's theories ought to be applied.

This highly interpretive approach to the science of evolution was essentially political; the particular version of Social Darwinism favored by various professional and academic circles tended to accord with the interests and agendas of those circles. In the United States and England, Social Darwinism was seen to reinforce capitalism and laissez-faire as natural and inevitable. German Social Darwinists, on the other hand, advocated state intervention to halt the supposed "degeneration" of the human species. Social Darwinist theories were advanced to describe and address in biological terms a broad range of human conditions including delinquency and criminality, intelligence, mental illness, poverty, susceptibility to disease, and other traits thought to separate the "fit" from the "unfit."

The political influence of Social Darwinism reached its height in the rise of the Nazi regime. By the turn of the century, popular Social Darwinism in Germany had evolved into a new EUGENICS, or "racial hygiene" movement, calling for programs to "improve" the biology of the human species in the face of "counterselective" forces such as war, revolution, and welfare for the sick and the inferior—forces that the capitalistic Social Darwinists of the United States and England also decried. Although initially a progressive movement, promoting genetics and eugenics as tools that might solve social problems for the human race in general, by the mid-1920s the racial hygiene movement in Germany formally promoted Nordic or Aryan supremacy and an anti-Semitic stance. In the next decade, Hitler would declare his National Socialist Party to be the final step in the overcoming of historicism and in the recognition of purely biological values.

SOCIETY FOR PSYCHICAL RESEARCH (SPR) Society founded in London, England, in 1882, branching out from the British National Association of Spiritualists. It was the model for the AMERICAN SOCIETY FOR PSYCHICAL RESEARCH (ASPR), founded soon afterward. Its purpose was to investigate the scientific basis of a variety of apparently paranormal phenomena and report on its findings. Among its founding members were many eminent scientists, including Sir Oliver Lodge, J. J. Thomson, Sir William CROOKES, Sir William Fletcher BARRETT, and Lord Rayleigh. It publishes its *Journal* and its *Proceedings,* in which are recorded the results of many investigations of MEDIUMS, spiritualists, TELEPATHY, ESP—the whole range of paranormal phenomena.

The society, which included both believers in and skeptics of psychical phenomena, set out with the declared intention of approaching the subject scientifically. Oliver Lodge, for example, its president from 1901 to 1904, found the subjective and anecdotal approach (which had characterized almost all work on the subject) unsatisfactory. He sought a basis of testing psychical phenomena that could be statistically analyzed. He was the first to use cards with simple symbols on their faces—the square, the circle, and the triangle—to test telepathy. The idea was taken up much later by Joseph Banks RHINE (1895–1980) and his colleague K. E. Zener, who added two additional symbols to the pack, now known as ZENER CARDS.

Although that tradition has continued, the society's publications, as might be expected from both its membership and its purposes, contain a mixture of the scientific and the pseudoscientific.

SOCIETY FOR RESEARCH ON RAPPORT AND TELEKINESIS See SORRAT.

SOCIOBIOLOGY First used in 1946, a term to describe the effect of biological factors on the behavior of all social animals from humans to ants. The landmark publication was *Sociobiology: The New Synthesis* (1975) by Edward O. Wilson. Its publication met with a mixed reception. Wilson was pleased that it received an overwhelmingly favorable response but was puzzled that there was also a furious reaction by some biologists and sociologists.

Most of sociobiology, over 90 percent, is concerned with nonhuman species: insects, fish, birds, and social animals, especially primates. This, the sociobiology of nonhumans, is not the controversial matter, although there is and always will be some disagreement over interpretation of evidence; it is by the resolution of such disagreements that science advances. So, for example, that individual ants will sacrifice themselves in the interest of the community is biologically determined is not disputed, or that some troops of monkeys collaborate in finding food seems understandably a case of sociobiology. On the other hand, to attribute the behavior of a specific monkey troop that learned to wash root vegetables in the sea to its genes would be stretching things too far. In other words, some social behavior of some animal groups is undoubtedly influenced and perhaps entirely prescribed by their biological makeup. Some is not. One has only to watch a squirrel finding its way to food past a series of obstacles to realize that biological determinism is limited, and it may be that the nearer we get to the human species, the less is biologically determined, the more is intelligently adapted.

Wilson lists the main topics of sociobiology as group size, age composition, mode of organization, including the means of communication, division of labor, and time budgets of both the group and its members. The discipline examines such behaviors as aggression, altruism, and courtship to see what common patterns there are across different species.

It is, of course, when the theory is applied to human social behavior that the furor arises. Who among us is going readily to believe that in humans there is no free will, that it is a delusion, that all or even much of our social behavior is laid down in our genes? The great diversity that anthropologists have discovered in human societies persuades us that our biology has little to do with social orga-

Sociobiology

nization. But if we look again at the anthropologists' findings, there is a residual core, a common content, to all societies, and there are some similarities between humans' behavior and that of other species. The possible biological basis to that shared pattern of behavior is at least worth exploring; that is, the province of human sociobiology.

But even allowing that much, there are still problems. There is a danger of anthropomorphism, taking aspects of human behavior, such as slavery or altruism, using the same terms to describe superficially similar behavior in other species and then assuming similar motivation and therefore genetic origins. These may be, and often are, parallel or analogous behaviors, not derived from some common genetic source but originating in quite different ways. There is also a failure to distinguish between *social behavior*—a group response to physical environment (an ant colony moving en masse to a new nest site when the old one becomes uninhabitable), the evolution of which can be explained by natural selection—and *cultural behavior*—learned social patterns handed down from generation to generation, for example, music, architecture, and plumbing, which are not genetically transmitted. There is no gene for plumbing—it is hard to imagine natural selection operating on plumbing!—and even here the distinction is not total: There may be a genetic predisposition—an aptitude—for music, architecture, and the practical skills essential for plumbing, though the separation of social and genetic influences is extremely difficult. But it is quite clear that the transmission of cultural behaviors, for example, a technological development, from generation to generation has advanced so quickly that it cannot be explained by natural selection. We do not somehow put that technological development into our genes and hand it on to the next generation.

If similar behavior is observed in two very distinct species, there are two possible modes of explanation:

1. The behavior pattern is built in, in the genes, in each species and will inevitably appear—the sociobiology view. The technical term is that such patterns are *homologous.*
2. The common pattern has quite different sources, genetic perhaps in one species, learnt in the other—the skeptics view. These are parallel modes of behavior, *analogous.*

The debate has continued. Wilson published a second book, *On Human Nature* (1978) in which he expanded, and to some extent softened, the thesis of the last chapter of the first book. Wilson and Charles L. Lumsden published a further book *Promethean Fire* (1984) which posited a genetic/cultural positive feedback mechanism to counter the argument that similar patterns of behavior may be parallel, analogous, not genetically determined. Although the argument has moved on, the skeptics on human sociobiology have not gone away. Some observers see sociobiology becoming a subdiscipline of ethology and the dust settling down, but if anything the rift has widened.

The dispute can be summarized as between biological determinism on the one hand and biological potentiality on the other. There is no doubt that human biology affects our behavior; the question is *whether* it opens up a range of possibilities that our exceptionally large brain enables us to explore *or* it genetically determines our behavior, limiting our options rather than expanding them. The critics do not see Wilson and Lumsden's attempt to fuse the two, with their positive feedback loop of gene-culture coevolution, as an acceptable answer but as a fudge, based on some dubious calculation and outdated Darwinism. To return to our earlier example, the skeptics insist that ants are automata, humans are not and that sociobiology is not a science but, based on a preconception, is a pseudoscience.

For further reading: Stephen Jay Gould, *Ever Since Darwin* (Deutsch, 1978/Penguin, 1980); Stephen Jay Gould, *An Urchin in the Storm* (Collins, 1988/Penguin, 1990); Charles L. Lumsden and Edward O. Wilson, *Promethean Fire: Reflections on the Origin of Mind* (Harvard University Press, 1984); Edward O. Wilson, *Sociobiology: The New Synthesis* (Harvard University Press, 1975); and Edward O. Wilson, *On Human Nature* (Harvard University Press, 1978).

SORRAT (SOCIETY FOR RESEARCH ON RAPPORT AND TELEKINESIS) U.S. organization investigating PSYCHOKINESIS and paranormal phenomena. The Society for Research on Rapport and Telekinesis, better known by its acronym SORRAT, was founded in 1960s by poet and author John G. Neihardt (1881–1973). Neihardt had spent six years with the Omaha tribe as a young man and continued his interaction with Native Americans throughout his life. He is best known as the author of *Black Elk Speaks* (1932), a classic of Native American religious literature. His early experience with Native Americans had given him a mystical outlook on life, and in 1908 he had married a spiritualist. He conducted his first research on the paranormal in the 1920s.

From its headquarters at Niehardt's farm near Columbia, Missouri, SORRAT emphasized research on psychokinesis. Neihardt contacted veteran parapsychologist Joseph Banks RHINE to assist him in setting up proper methodological controls. Their experiments began with simple attempts at table tipping. Eventually LEVITATION and even APPORTS were reported. Members of the group experienced periods of trance during which they channeled various entities.

In 1966 Rhine suggested that more controlled experiments in psychokinesis be conducted with the use of an insulated but transparent box in which a target could be securely locked. It was not until 1977, however, when W. E. Cox, who had worked with Rhine, moved to Missouri that the more systematic experiments began. Cox helped construct a 20-liter (5.5-gallon) aquarium tank secured to a wooden platform. A variety of experiments were filmed, but the results have been ambiguous at best. When shown at parapsychology conferences, the films of the experiments never showed the beginning of any reported "paranormal movement" and most who saw the films failed to discern any paranormal events in them.

Various efforts to reproduce the results claimed by SORRAT have failed, and in 1982, at a gathering of parapsychologists at Cambridge, England, a film was shown showing how the effects in the original SORRAT films could have been faked. Thus although Neihardt's group reported a variety of spectacular paranormal results, none have been produced under the exacting conditions demanded by PARAPSYCHOLOGY.

For further reading: John Thomas Richards, *SORRAT: A History of the Neihardt Psychokinesis Experiments, 1961–1981* (Scarecrow Press, 1982).

SOURCEBOOK PROJECT Massive collection of primary source material and rare secondary sources on all kinds of unusual phenomena. The project's originator, William R. CORLISS, a physicist by profession, published his first volume in 1974 and has published nearly a volume a year since that time.

Some of the general subject areas covered by the Sourcebook Project are strange artifacts, weather and geophysical phenomena, biological anomalies, and space mysteries. Some of the specific subject matter includes spook lights, BALL LIGHTNING, waterspouts, wolf children, Tiahuanaco (Bolivia) ruins, plate tectonics, concretions and geodes, kinks in Saturn's rings, pendulum

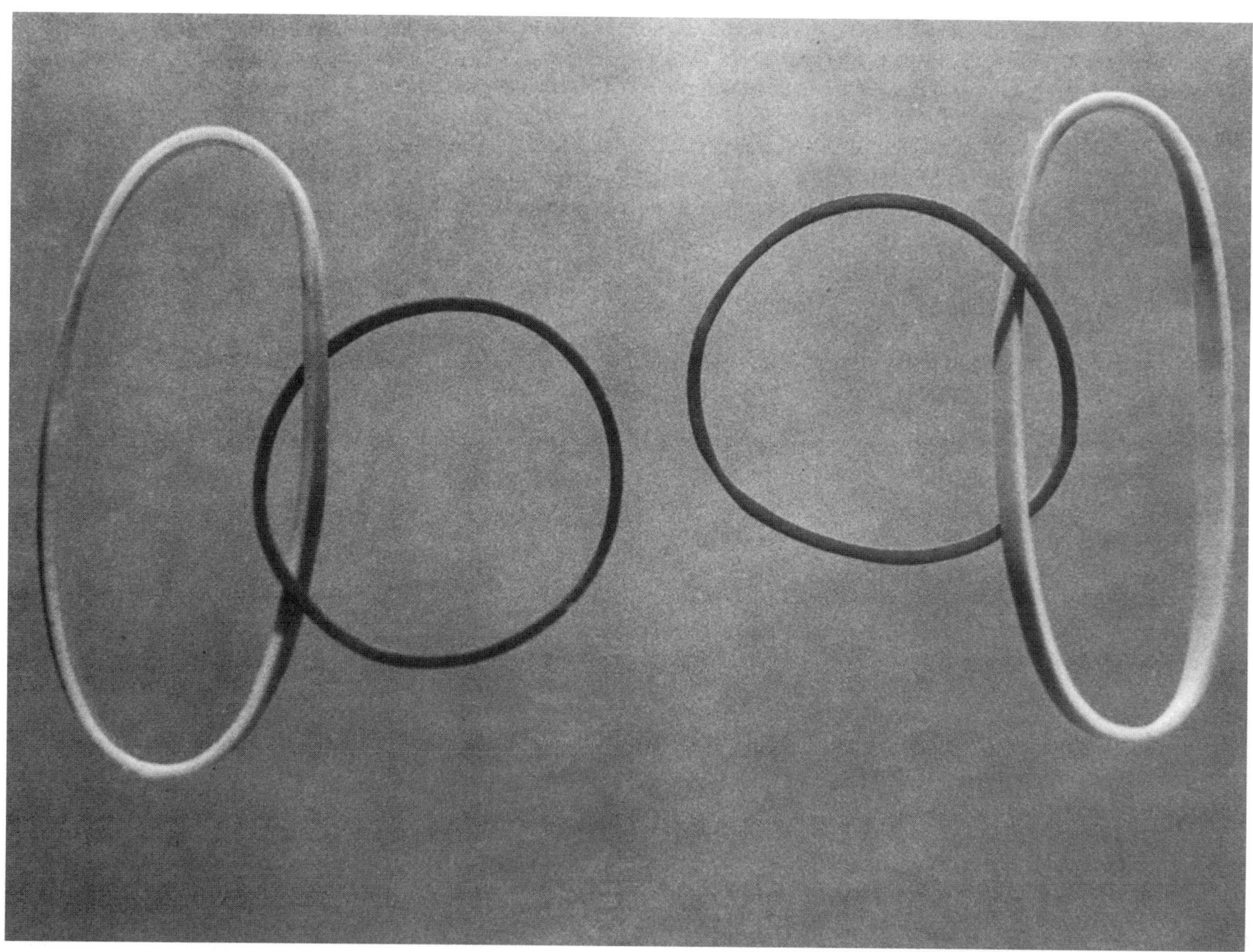

SORRAT

phenomena, infrared cirrus clouds, and water-breathing in mammals.

Corliss has culled articles from some 14,000 volumes of scientific and popular literature from 1820 to the present day. His aim, he has said, is "the collection and consolidation of the unknown and poorly explained to facilitate future research and explanation." In a 1993 article in the Baltimore *Sun,* he was quoted as saying that "science progresses by taking account of anomalies." Indeed, his sourcebooks are used for information and entertainment by both scientists and laypersons. The books contain scientifically illuminating articles, as well as personal accounts and less-well-documented discussions of strange events and artifacts. They contain fact, theory, and opinion, and they are unique in their purpose and contents. While some scientists dismiss them as useless esoterica, others view them as fascinating and inspiring looks at past errors and future promise of their fields.

SOVIET SCIENCE The former Soviet Union's origins and outlook were purportedly science-based. Marxist ideology claimed to be scientific in its version of history. Friedrich Engels in *The Dialectics of Nature* claimed to show that science and dialectical materialism were allied. Vladimir Lenin in his writings and speeches and in exchanges with visitors from the West emphasized the importance of science and technology to the development of the Soviet Union. As the Soviets established their nation, this outlook was built into their industrial and their educational systems. The school curriculum, standard across the country, laid down substantial syllabi in science at every stage. Nigel Grant in *Soviet Education* (1964) wrote: "Scientific subjects are introduced gradually, until by the eighth year they take up nearly half the week's teaching time . . . over the whole of the eight-year course, scientific subjects account for thirty-five percent of study time." Soviet universities also established a considerable base in science. A special town, Akademgorodok (Science City), was built near Novosibirsk, providing opportunities and privileges for those following a career in science and their children.

The Soviet state provided education and opportunities in science. The Soviet population received a good grounding in science, and there was provision for continued adult education throughout life, but there was also a built-in contradiction: The political establishment imposed an ideological straitjacket on what constituted correct and incorrect science. This was seen most clearly in the case of Trofim D. Lysenko, an agricultural geneticist who promoted an essentially Lamarckian theory of plant genetics (see LYSENKOISM).

Lysenko claimed that by treating plant material to germinate and flourish in harsh climatic conditions, resulting resistant characteristics were handed down to the next and subsequent generations; in other words, the genetic makeup of plants could be altered not only by accident (mutation) or by breeding, but by design—the Lamarckian inheritance of acquired characters. This was what the political masters wanted to hear: If true, it opened up vast tracts of the northern Soviet Union to food production to feed the country's growing millions. So Lysenko and his school received official blessing, funding, facilities, and promotion. Those of the traditional school, the great geneticist N. I. Vavilov and others, were dismissed as being wedded to capitalist science—"bourgeois idealism"—demoted and victimized.

The need for scientists and technologists to be looking over their shoulders to see whether their work met with official approval was a terrible restriction on the essential freedom of science to follow up leads, however heretical they might seem. Soviet science undoubtedly suffered as a consequence. The facilities were there—universities and institutes, cheap and plentiful journals and books, the educational backing, status, rewards, and honors—but unfettered freedom was not.

Relations between Western science and Soviet science have waxed and waned. In between the two world wars, many Western scientists, motivated in part by science's view of itself as an international activity, expressed admiration for the Soviet's intentions and achievements.

In 1945 the Australian Government sent Eric Ashby (later Lord Ashby), then a botany professor at the University of Sydney, to the Soviet Union for a year to report on the situation in science. On his return he wrote in *Scientist in Russia* (1947):

> *Russia has endowed science with the authority of religion. Science has a privileged place in the school curriculum; it is the main subject of study in hundreds of institutes of higher education; its plans are woven into the plans for national development; it is admitted to the highest councils of the country; it is generously endowed with money and men. Science carried out with an eye to its practical application is specially encouraged. It follows that the solution of short-term problems arising out of social needs is for many scientists a short cut to preferment; but the scientific worker who is strong minded enough to stick to academic problems is allowed to work as he wishes, and does not have to sacrifice the quality of his work for quick results."*

But Ashby goes on to qualify his admiration: some degree of isolation from the rest of the world, and the indifferent quality of many of the support workers (hack workers) for their first class scientists.

State domination of scientific thinking did not stop at plant genetics; similar orthodoxies were imposed on other branches of science. Engineering and technology, despite their obvious importance to the state, especially in the rebuilding after World War II, also had problems. Loren Graham of the Massachusetts Institute of Technology spent much time over several years in the former Soviet Union investigating the achievements of Soviet engineering and technology. His researchers, recorded in *The Ghost of the Executed Engineer* (1955), show that professionals were severely hampered in their work by the state orthodoxy that influenced everything they did. Graham's executed engineer was Peter Palchinsky, executed in 1928. His sin was to criticize what he saw as the shortcomings of the former USSR's industrialization plan and to suggest ways in which it could be improved. The political chiefs, not themselves engineers, represented this as an underhand way of sneaking capitalism back into the Soviet Union. Palchinsky was tried, found guilty, and shot. Thereafter engineers and technologists had to watch their step. In the end, the Soviets paid a high price.

It would be naive to suppose that the West is entirely free of such behavior—the Scopes trial for example, the echoes of which still rumble on today—but both the Nazi and the Soviet cases constitute extreme and awful warnings.

For further reading: Martin Gardner, *Fads and Fallacies in the Name of Science* (Dover, 1957); and Loren Graham, *The Ghost of the Executed Engineer* (Harvard University Press, 1995).

SPECULATIONS IN SCIENCE AND TECHNOLOGY A speculative journal published four times a year by Science and Technology Letters in Northwood, Middlesex, England. Its editor is Professor Alan Mackay, a crystallographer of Birkbeck College, and it has an international editorial board. It incorporates *Developments in Chemical Engineering*. There are about twelve articles in each issue, a mixture of normal chemical engineering topics and speculative—sometimes highly speculative—articles from across the board of science, technology, and mathematics. The 1992 issues included, for example, articles titled

"Present trends in process safety," "Is the Cosmological Principle inconsistent with the Second Law of Thermodynamics?" and "Greenhouse Warming: a clear and present benefit?" and a book review of *Gas Fluoridation.* There is definite need for a place to publish speculations—off-the-wall thinking, letting the imagination loose—about science, technology, and mathematics, but it is very difficult, if not impossible, to get such unorthodox views published in science and technology journals, which tend to be disciplinary and conservative. This journal provides such a place in the spaces between fairly orthodox—occasionally speculative—chemical engineering contributions, containing the technical stuff of chemical manufacturing along with such stuff as dreams are made of.

SPENCE, LEWIS (1874–1955) A firm advocate of the reality of Atlantis. In the seven books he wrote on the subject, including *The Problem of Atlantis* (1924), he drew on evidence from geology, biology, and archaeology to substantiate the ancient myth. He drew a parallel between the decadence of the ancient Atlanteans and the present-day Europeans, particularly the German Nazis, and predicted a similar fate for Europe unless its inhabitants mended their ways and returned to the true faith. He developed his argument in *Will Europe Follow Atlantis?* (1942). He was particularly incensed by the Nazi claim that Atlantis was the original home of the Nordic race, the Atlanteans in his opinion having been Anglo-Saxons, and focused his attack on the Germans and their Satanic practices in *The Occult Causes of the Present War* (1944).

Spence wrote more than 40 books, including *The Magic and Mysteries of Mexico; The Problem of Lemuria* (1932), *An Encyclopedia of Occultism* (1920), and *The Occult Sciences of Atlantis* (1943). He was vice-president of the Scottish Anthropological and Folklore Society, and in 1951 he was awarded a Royal Pension for Services to Literature.

For further reading: Martin Gardner, *Fads and Fallacies in the Name of Science* (Dover, 1957).

SPIRICOM An apparatus designed in the 1980s to facilitate contact with the dead. The Spiricom appeared in the wake of the interest in ELECTRONIC VOICE PHENOMENA, in which very faint voices were heard on tape recordings, recorded at higher than normal speeds. It was developed by engineer George W. Meek of Metascience Foundation, Inc. At the time of his first announcement of the apparatus, Meek claimed that through the Spiricom he and his colleagues had obtained many hours of conversation recorded at a normal speed from different spirits. These included tapes from a well-known U.S. scientist who had died some 14 years earlier.

The primary component of the apparatus was a transceiver operating on the 30–130 MHz range. It was activated, however, by the input of energy from an operator (a MEDIUM) of highly developed psychic abilities. The medium supposedly fed an unknown energy, which Meek termed bioplasmic energy, into the device. In this regard, the Spiricom was similar to the BLACK BOX developed by Dr. Albert ABRAMS.

The basic information on the Spiricom was released in 1982 with the idea that others would construct copies and improve upon the basic design so that communication with the dead could proceed on a regular basis. How many might have done so is unknown, but apparently the Spiricom failed to live up to the expectations of its developer, and little has been heard about it in the intervening years.

SPIRIT PHOTOGRAPHY The photographing of GHOSTS or other spirits, popularized during the SPIRITUALISM era of the middle and late 19th century. In 1862, a Boston photographer, William H. Mumler, discovered on one of his photographs a human image besides that of his sitter. Although his subject, a Dr. Gardner, had been posing alone, the photographic plate showed an additional figure that Gardner identified as a long-deceased cousin. Gardner's astonishment and subsequent publicizing of his experience led to more ghost photos by Mumler, some of

Spirit Photo

which were clearly shown to be frauds perpetrated by double exposure of the photographic plate. Some, however, were not so easily explained.

During the next several decades, spirit photography increased. Extant photos of the era show amazing images of ghosted figures, both cardboardlike and more realistic, and of spiritualists oozing ectoplasm from every visible orifice. Many of these were later shown to be the result of double exposures using portraits of deceased people and even using images of still-living persons, images projected onto the plates through the use of mechanical projectors or mirrors, or other tricks sometimes involving the collaboration of the sitter and sometimes not.

Today some paranormal researchers attempt to photograph ghosts in an effort to prove their existence. While investigating a haunted house, for example, the researcher may use elaborate photographic setups, including infrared film, multiple cameras and video recorders, Geiger counters, finely tuned thermometers, and other sophisticated tools. They rarely expect to photograph a humanlike ghost but instead look for anomalous elements on the film that suggest unusual energies present in the environment. They often succeed in finding such elements that seem to have no accidental explanation. But more skeptical analysts generally assume these to be the result of some kind of physical—not psychic or spiritual—energy run amok during the course of the photography, sometimes even energy generated by the equipment itself.

See also KIRLIAN PHOTOGRAPHY; PSYCHIC PHOTOGRAPHY.

SPIRITUALISM In philosophy, a way of thinking that believes in immaterial reality, that is, knowledge perceived to be extrasensory, that is by some means other than through the normal five senses. Spiritualism, as opposed to materialism, is a very broad category and could apply to any acceptance of an infinite personal God, the immortality of the soul, or the immateriality of the intellect and will. Thus Plato's and René Descartes's dualistic views of the soul as distinct from but operating from within the body, would mark them both as spiritualists.

There is however a more specific meaning attached to the word, and that is to designate a belief that there is a continuity of life after death and that the dead can communicate with the living through the use of MEDIUMS. There have always been holy men and shamans who have claimed to be able to contact the spirit world, and in the middle ages, necromancy—predicting events by allegedly consulting the dead—was practiced. It was however considered a black art and was frowned upon by the church of Rome, and even today spiritualism's more modern form has still not been accepted by mainstream churches. There are two aspects to the phenomena of spiritualism—the physical side and the mental side. The physical includes LEVITATION, PSYCHOKINESIS, TABLE RAPPING, and in fact any unusual physical manifestation or supernatural happenings indicating that spirits, GHOSTS, or POLTERGEISTS are present. The mental side of spiritualism is the conveying of information from one world to another and includes CLAIRVOYANCE, clairaudience, and TELEPATHY.

Modern spiritualism started in Hydeville, New York, in 1848 when two sisters of the Fox family demonstrated their mediumistic ability to communicate with the dead. The newspapers reported in detail their every move, and amid much publicity many more mediums decided that they too had unusual powers. It is estimated that within five years there were 30 thousand mediums in the United States alone. Fraud became a part of the spiritualist movement from the start. The elder of the Fox sisters, Margaret, confessed at the end of her life in 1880 that she had, from the very first, been fooling people by making the spirit raps by cracking the first joint of one of her big toes.

Most people seem to agree that spiritualism is a form of religion, although there is no fixed convention for the service or "séance," and although it could take place anywhere in a spiritualist church or a private home. There are usually prayers and hymns before the lights are dimmed and the medium, the mediator between the dead and the living, goes into a "trance." After some time has elapsed, greetings and personal messages from the dead are directed to relatives or friends in the assembly. The séance is often, but not always, accompanied by table rapping and turning, AUTOMATIC WRITING, and other demonstrations of PSYCHOKINESIS.

The great upsurge in the 19th century of belief in spiritualism has been put down to the need of people to combine their faith in the new materialism of science with their spirituality. People wanted to believe the findings of science and at the same time needed demonstrations to prove that the soul survived death. Spiritualists actively sought the backing of science, and many scientists were fascinated with spiritualism. During the 20th century Sir Oliver Lodge, the famous British scientist, became involved in devising experiments to test claims of the paranormal. In this he was not alone. At the time, several other eminent scientists in Britain, including Sir William CROOKES, Lord Rayleigh, and J. J. Thomson were also interested. A story is told of Lodge that at one time when he was very aware that his scientific experiments into wireless telegraphy were about to produce useful results, he failed to follow them up, choosing instead to go off to the south of France where he was involved in a series of lengthy experiments in testing the powers of a famous medium—who was later exposed as a fraud.

Although the possibility of having contact with dead friends and relatives still interests people, the spiritualist movement in the United States and Britain is now very reduced in numbers. The decline was, in no short measure, brought about by fraudulent mediums seeking to make money out of the unhappiness of bereaved people. The investigating and exposing of these mediums and spiritualist movement in general, was, to a large extent, done by the famous stage magician Harry HOUDINI who first found his way into spiritualism to try to contact his dead mother. The trickery that he found disgusted him so much that he devoted the last years of his life to exposing fraudulent mediums, believing, rightly, that his training as a magician made him the ideal investigator.

For further reading: Susan Blackmore, *Adventures of a Parapsychologist,* (Prometheus Books 1986); Susan Blackmore and Adam Hart-Davis, *Test Your Psychic Powers* (Thorson, 1995); W. P. Jolly, *Sir Oliver Lodge* (Constable London, 1974); and G. K. Nelson, *Spiritualism and Society* (Routledge, 1969).

SPONTANEOUS GENERATION The belief that under certain conditions living creatures can be produced directly from matter. The concept is well attested in classical sources, notably Aristotle, and is probably based on even more ancient observations of the emergence of organisms such as fly larvae from dead animal tissue or worms form mud. General considerations have distinguished between ABIOGENESIS—generation from inorganic matter—and heterogenesis—generation from living or formerly living tissue. At various times and places, potential candidates for spontaneous generation have included algae, fungi, insects, worms, bacteria, and even small mammals.

Owing to the high regard for Aristotelian science and the ease with which the process could be confirmed by rough observation, belief in spontaneous generation was little doubted by Europeans for centuries. In the 16th and 17th centuries, however, the increasing prestige of experimental science and the development of better instrumentation (particularly microscopes) initiated a complex debate among scientists over the status of spontaneous generation that would last until the 20th century. Notably, the concept continued to elicit support late into the 19th century, even after a great deal of experimental data had been arrayed against it.

Ideological, nationalistic, and philosophical considerations often colored the spontaneous generation debates; for example, in the early 19th century, both French materialists and German *Naturphilosophes* championed the concept, although for very different reasons. The materialists felt that spontaneous generation supported notions of a self-sufficient nature, which could be analyzed solely as the interaction of particles. The *Naturphilosophes* liked the idea because it fit well with their key concepts of the unity of nature and continuous organic development. Resistance came from British naturalists and key French scientists such as Baron Georges CUVIER, who viewed even the most simple living organisms as too complex to have been produced quickly from elemental matter, and who insisted that all life must be produced from existing life forms. Ostensibly based in careful empirical and comparative study, religious and social considerations no doubt factored into the perspectives of these scientists. Many were steeped in the tradition of NATURAL THEOLOGY and therefore wary of the atheism of the materialists.

The spontaneous generation debate was also characterized by unique difficulties surrounding the interpretation of experimental results. Most histories of spontaneous generation cite three key figures—Francesco Redi (1626–97), Lazzaro Spallanzani (1822–95), and Louis PASTEUR—as progressively discrediting the theory via careful experiments. However, all three were predisposed against the idea and did not fully convince all their scientific peers. Proponents of the concept could always argue that the experiments did not meet all the conditions found in nature, and thus spontaneous generation might still be possible. As late as the 1860s for example, German zoologist Ernst HAECKEL believed that abiogenesis occurred in the mud at the bottom of the oceans.

The development of cell biology and biochemistry in the late 19th century reoriented most scientists' views on the fundamental nature of life and in a sense ended the spontaneous generation debate. The focus shifted away from the development of fully formed organisms toward more basic building blocks of life—the cell, protoplasm, and later, genetic material and basic proteins. Conceptual difficulties surrounding the transition from inorganic matter to organic life, while never fully resolved, increasingly were cast into evolutionary schemes that stressed the slow development of increasingly complex molecules and life forms. Today, most biologists and biochemists adhere to an evolutionary model developed by Soviet scientist Aleksandr Oparin in the 1930s, although dissenting opinions, particularly various strains of creationism continue to elicit popular support. Neither support the idea of spontaneous generation as it has been traditionally conceived.

For further reading: John Farley, *The Spontaneous Generation Controversy from Descartes to Oparin* (Johns Hopkins University Press, 1977).

SPONTANEOUS HUMAN COMBUSTION The process of a body catching fire as a result of heat generated by internal chemical action. Many hundreds of cases of spontaneous human combustion have been reported, generally regarding corpses, though a few tales of living people bursting into flame have circulated. One 17th-century tale claims that a German self-ignited after having drunk excessive amounts of brandy. Most reports have been related by police investigators who were puzzled by a partially ignited body near unburnt carpets or furniture. Often these reports involve the corpses of elderly people

Spontaneous Combustion

who may have been the victims of foul play or who may have ignited themselves accidentally.

The physical possibilities of spontaneous human combustion are remote. Fire burns at more than 92°C (200°F). The living human body is generally less than 38°C (100°F). Not only is the human body almost entirely made of water, but aside from fat tissue and methane gas, nothing within the body readily burns. Once a fire starts, it would continue to burn only if the combustion of the burning substance were as high or higher than the point of ignition. Possibility of spontaneous human combustion after death is just as remote; a corpse would tend to cool to room temperature. Cremation requires an enormous amount of heat over a long period of time. If a corpse self-ignited, it would not continue to burn unless the room containing it were so hot as to be on fire itself; a cool body in a cool room would have little chance of bursting into flame.

SPURZHEIM, JOHANN CHRISTOPHE GASPAR (1776–1832) Co-creator and popularizer of the basic principles of PHRENOLOGY. Born into a Lutheran farming family in Germany, Spurzheim initially intended to enter the ministry and studied Latin, Greek, Hebrew, philosophy, and divinity in his hometown and at the University of Trèves. Although many of his later endeavors retained an evangelical sense of mission and optimism, the French invasion of 1799 effectively ended his theological career, and he relocated to Vienna. Here he studied medicine and became the assistant to Franz Joseph GALL, who was beginning to formulate his theories of the cerebral localization of mental and behavioral traits. For the next 13 years, Spurzheim's career closely paralleled Gall's. The two collaborated on the systematic mapping of the brain and the nervous system, incurred the wrath of civil and religious authorities who saw their work as promoting materialism and immorality, and lectured to appreciative audiences throughout Europe. Eventually, they settled in Paris where they published their ideas in a work entitled *Anatomie et Physiologie du système nerveux en général et du cerveau en particular,* four volumes (1810–19). This series formed the basis of the pseudoscience of phrenology (a term Gall disliked but Spurzheim embraced) and helped make psychology a more biologically oriented discipline. It also fostered a widespread cultural fascination with the brain and its function.

Spurzheim worked on the first two volumes only, splitting with Gall around 1813 over a perceived lack of recognition and larger philosophical differences. Most of the latter stemmed from divergent visions regarding the social bearing of their mental science. Gall was something of a cynic, recognizing limits to human goodness, intellectual capacity, and moral vigor as an extension of innate mental constitution. Spurzheim, however, envisioned phrenology as the foundation of an aggressive program of reform that championed individual and collective improvement. In doing so, he backed away from deterministic interpretations of phrenology and asserted that the size and quality of the brain could be enhanced. For the remainder of his career he would spread the gospel of phrenology, achieving his greatest success in England and the United States. He proved to be particularly adept at facing down the many critics of his ideas and published many books touting the advantages of phrenology for reforms in education, penology, medicine, and the treatment of the insane.

Spurzheim's untimely death in Boston in 1832 in the middle of a triumphal lecture tour occasioned a dramatic outpouring of public grief rarely matched in the city's history. It included elaborate funeral services and public processions that were planned and attended by the leading citizens of the city. In death, he became a martyr for the cause of phrenology and perhaps a symbol for the growing sense of anxiety over the social problems that his science purported to address. Phrenological societies formed, literally, at his wake, and his brain, skull, and heart were preserved to inspire the phrenological spirit of later generations (only the skull survives, now in the Harvard collection).

SQUARING THE CIRCLE The idea that there could be a method of taking a circle and constructing from it a square that had exactly the same area and the same perimeter. Although the idea has a long history and although it has long been known that it is impossible, it nevertheless persisted for centuries, so much so that the expression "squaring the circle" has come to represent similar impossible hopes and dreams. One of those claiming to have succeeded was Ramon Lull (1232–1315). He circumscribed and inscribed squares about a circle and then drew the square halfway between the two. But this final square is a very poor approximation to either the area or the perimeter of the circle, and Lull was only repeating a method that had been in use for many years. There are various ways of improving on this method, getting closer to either the area or the perimeter, but obtaining a perfect solution to either can be shown to be a hopeless task.

STEINACH REJUVENATION OPERATION Procedure intended to restore youthfulness to men and women. Throughout the 19th century, many physiologists and anatomists believed that secondary sexual characteristics—the deepening of the voice and the appearance of facial hair in males and the development of breasts in females—were caused by an internal secretion poured into the blood by the sex glands. In 1889, French physiologist Charles-Édouard Brown-Séquard announced the discovery of the potent substance, an extract from the male sex glands. Having inoculated himself with the extract, he claimed that he had renewed his youthfulness. However, colleagues could not reproduce his results, and Brown-Séquard soon passed from the scene permanently.

Other doctors followed Brown-Séquard's lead. In 1903 two French biologists announced that they had discovered the secretion responsible for male secondary sexual characteristics and located it in a part of the male sex gland. Tying off the tubes leading from the gland, they said, would stimulate this portion into producing more of the secretion. This idea was picked up by Austrian physician Eugen Steinach. Through the first few decades of the 20th century doctors on both sides of the Atlantic offered the Steinach rejuvenation operation for aging men. Meanwhile the basic idea underlying it, that secondary sex characteristics are related to secretions of the sex glands, was destroyed by further research, including the observation of changes in men born without sex organs.

The only evidence in favor of the Steinach procedure were the testimonies of people who had the operation and later claimed some rejuvenation. After World War I, in the face of publicity denying the validity of the operation, those who continued to perform it began to emphasize psychological preparation. Patients were selected based upon their complete faith that the procedure would give them renewed youth, a lesson many doctors who specialize in sexual disorders have learned. They understand that many of these disorders are caused simply by strongly held negative attitudes about the self or the body. The Steinach procedure seems to have largely disappeared by World War II.

At the same time that the Steinach procedure was being offered, other nonsurgical rejuvenation remedies appeared on the market that offered gland material that could be taken orally. In the early 1920s, the Lewis Laboratories of Illinois and the Vital-O-Gland Company of Denver were among a handful of concerns around the United States who widely advertised their glandular preparations, until they were driven out of business by the U.S. government.

For further reading: Morris Fishbein, *Fads and Quackery in Healing* (Blue Ribbon Books, 1932).

STEINER, RUDOLF (1861–1925) Scientist, artist, editor, and founder of anthroposophy—a spiritual movement that is the basis for much of the New Age thinking in Europe and North America today. Adherents believe that there is a spiritual world of pure thought that exists detached from the human brain, yet is accessible to those who have been trained to use their very highest faculties of mental awareness. Current New Age writers stress the "supersensible," "psychic," or otherwise mystical realms that are available to the initiated and, they believe, superior to the material world of our five senses.

As a young man, Steiner was attracted to the works of J. W. von Goethe and edited both his scientific work and the standard edition of his complete works. During this period, Steiner wrote *The Philosophy of Freedom* (1894), a book about the necessity to raise philosophy that is free from the clamors of everyday existence into the realm of pure thought. He then moved to Berlin to edit a literary journal and in 1902 became acquainted with the Theosophical Society, a mystical religious group concerned with esoteric doctrine, occult phenomena, monism, and the affinity of Western thought with Asian religions. Gradually, he came to the belief that spiritual perception was independent of the senses, and he felt the need to break away from THEOSOPHY. Consequently in 1912 he founded his own organization—The ANTHROPOSOPHICAL SOCIETY.

Steiner

Steiner had the belief that there was a time in our past when we could connect fully with the spiritual processes through a dreamlike consciousness, and that we had become detached from this facility because of our love of material things. The ability to connect with spiritual things was, he believed, innate, so he reasoned that all that was needed to renew this lost power was an intense course of spiritual training.

In 1913 at Dornach, near Basel, Switzerland, Steiner built his first Goetheanum, a school described by him as giving an education in "spiritual science." The school was burned down in 1922 and immediately rebuilt in molded concrete to Steiner's own design. Steiner died in 1925 but his educational program continued to grow, and by 1969 there were 80 more Steiner schools throughout Europe and the United States.

Other types of institutions followed, such as those for defective children and for drama, speech, art, and eurythmy (movement to speech and music). There is also a general therapeutic clinical center at Arlesheim in Switzerland. In ALTERNATIVE MEDICINE practices, anthroposophical medicine is a set of practices based on Steiner's occult philosophy, which claims to relate humankind to the natural environment with emphasis on color, rhythm, and spirituality. Homeopathic remedies such as herbs are prescribed, and the study of musical instruments along with prayer and MEDITATION are thought to restore the balance of the body and mind, either by strengthening the "etheric body" or by moderating the animalistic "astral body."

The idea that there is such a thing as "pure" thought existing free from a thinking brain is a piece of pseudoscience that has yet to be proved. Antony G. N. Flew, Gilbert Ryle, and other analytic philosophers have said that the "ghost-in-the-machine" doctrine of a "discarnate self" in which either a mind, pure thought, or a soul exists separate from a body, capable of independent existence and causality, is unintelligible and must be questioned on conceptual grounds. Most skeptics reject the idea of a disembodied soul separable from the body, not because of preconceived metaphysical presuppositions but because these claims lack sufficient evidential bases, and the burden of proof is with the believer, not with the skeptic.

For further reading: Gilbert Childs, *Steiner Education in Theory and Practice* (Floris, 1991).

STEPHENS, JOANNA (??–1774) Known for her treatment of kidney stones in the 1730s. She claimed to have infallible treatments and in 1738 advertised in *The Gentleman's Magazine* that she would reveal her secret if she were to receive £5000 in recompense. Five thousand pounds (roughly the equivalent of 7–8 million dollars today) was a sum beyond the means of almost any individual and could only be raised by setting up a fund for public subscription. At first, the fund did not go too well, but after a parliamentary inquiry had declared that Stephens' claims were sufficiently credible to justify public support, the sum was raised and paid to her.

She then published three prescriptions to be used in succession, should the first and then the second not effect a cure. The first was produced by boiling a mixture of herbs, soap, and honey; the second, a powder of snails and burnt eggshells; the third, a pill made from burnt snails, wild seeds, hips, and haws. It was soon realized that none of these was effective, but Joanna Stephens had disappeared—with the proceeds! Nonetheless for many years thereafter, sufferers, including Sir Robert Walpole, continued to use her remedies.

For further reading: Carl Sifakis, *Hoaxes and Scams* (Michael O'Mara, 1994).

STEVENSON, IAN (1918–) Professor of PSYCHIATRY and parapsychologist known primarily for his research on the evidence of survival after death through the examination of memories of past lives. Stevenson was born on October 31, 1918, in Montreal, Quebec, and received his medical degree from McGill University in 1944. He held various teaching positions prior to his becoming chairman of the Department of Psychiatry at the University of Virginia School of Medicine in 1957. He published a number of papers on parapsychological topics, developed a belief in REINCARNATION, and concluded that memories of past lives provided some of the best evidence of that opinion. In 1961, he wrote a prize-winning essay in honor of William JAMES on *The Evidence for Survival from Claimed Memories of Former Incarnations* (1961).

Through the 1960s, Stevenson compiled and correlated the rather detailed accounts of reincarnation. Many of these he had investigated personally. They largely featured young people who claimed to be someone else, usually someone in a nearby town, and backed their claims by showing extraordinary knowledge of details of that prior existence. Stevenson's approach resembled the one used by Tibetan monks to discover the child whom they believe to be the incarnation of a deceased lama: They rely on the child's ability to recognize some articles formerly owned by the lama.

Stevenson's primary argument was made in 1966 with the publication of *Twenty Cases Suggestive of Reincarnation*. He followed that volume with several books that examined additional cases including *Cases of the Reincarnation Type, Vol. 1, Ten Cases in India* (1975); *Cases of the Reincarnation Type, Vol. 2, Ten Cases in Sri Lanka* (1978); and *Cases of the Reincarnation Type, Vol. 3, Twelve Cases in Lebanon & Turkey* (1980).

Stevenson's research has gone against the trend of main parapsychological thought, which has largely given up on the possibility of survival research or has limited the little that has been done to investigation of the mere fact of survival. A few parapsychologists have challenged Stevenson's research, noting that his cases come from countries and subcultures where reincarnation is a dominant belief. They also suggest that, because of his own religious beliefs, he has failed to consider alternative hypotheses for his data.

For further reading: Ian Stevenson, *Twenty Cases Suggestive of Reincarnation* (University of Virginia Press, 1966, rev. ed., 1974).

STHENOMETER An early apparatus developed to detect the existence of an external psychokinetic energy generated by animals, primarily humans. The sthenometer was created by 19th-century psychical researcher Paul Joire, a professor at the Psycho-Physiological Institute in Paris, France. Joire believed that the energy he was attempting to measure was produced by the human nervous system.

The simple device consisted of a dial marked out in 360 degrees. A light needle or pointer (usually a piece of straw) was balanced on a pivot in a glass support. The dial and the needle were then enclosed in a glass cover. Joire discovered that, if the extended fingers of the hand was brought close to the glass cover at a right angle to the needle, the needle would be attracted by the hand and would move toward it. Movement would also take place if certain objects previously held in the hand were brought close to the cover. Cardboard, wood, linen, and water caused the needle to move, but tinfoil, iron, and cotton did not.

Joire received a cold response from colleagues both in France and with the SOCIETY FOR PSYCHICAL RESEARCH in England when he tried to show them his device. They concluded that the demonstrated movements were due to radiating body heat.

STIGMATA Spontaneous development of bruises and wounds, usually bleeding, in places corresponding to the wounds of the crucified Christ, and usually viewed as a sign of saintliness. Some stigmata are, of course, fraudulent. The first officially recognized stigmatic was the 13th-century saint, Francis of Assisi, who was noted for his simple life of material deprivation and devotion to God. One day in 1224, while meditating outside the cave that

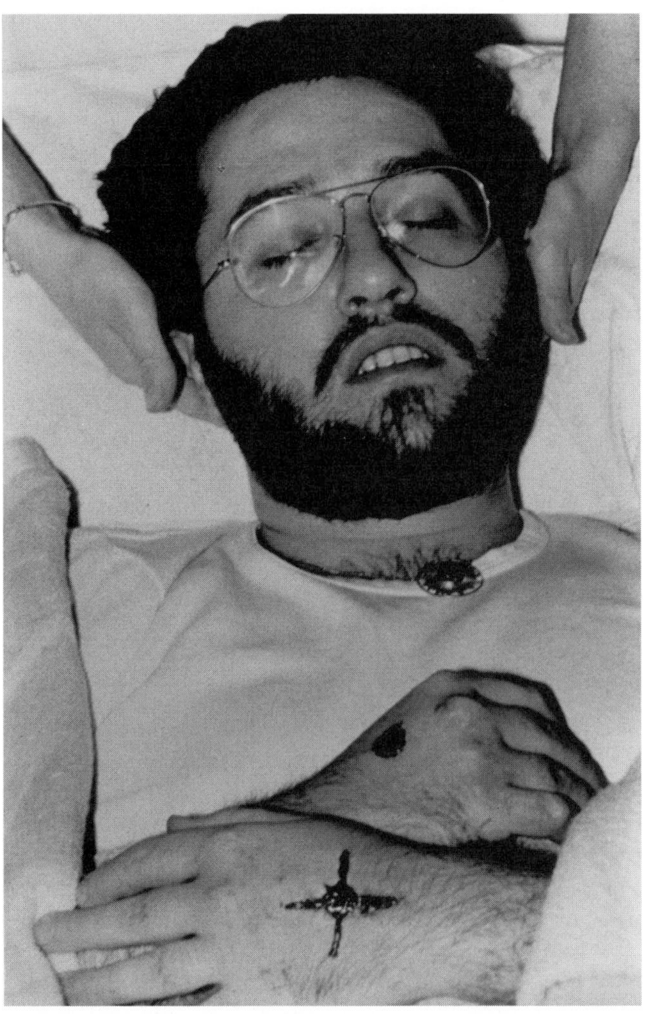

Stigmata

was his home, he saw a vision of Christ on the cross. At the same time, blood began flowing from deep fissures in Francis's hands, feet, and side. Nail-like pieces of skin or bone also appeared, thrusting out through the wounds. Francis's stigmata continued to his death. This is the case with most religious stigmata: Once the phenomena is visited upon a person, it usually lasts for the remainder of his or her life, although it does not necessarily remain continuous; some stigmata have appeared annually, for example, perhaps on Good Friday; some have appeared weekly or monthly—often with the stigmatic showing no signs of wounds or scars in the intervening times.

Doctors and other attendants of various stigmatics have actually measured quantities of blood flowing from the wounds or, in some cases, from the eyes as tears. Padre Pio, for example, was an Italian Capuchin monk who developed stigmata early in the 20th century; he is said to have bled a glassful or more each day.

Over the past several centuries some three hundred stigmatics have been documented. These include several in the 20th century, Padre Pio and Therese Neumann (a Bavarian peasant who was also said to be clairvoyant and an astral traveler, and who died in 1968) probably being the best known.

Many experts have speculated on possible non-miraculous explanations for stigmata. The most popular theory seems to be that stigmata are caused by hysteria or self-HYPNOSIS/autosuggestion. Experiments with hypnosis have shown that the mind is able to cause the body to do quite extraordinary things; a number of case histories document the ability of the mind to create wounds, bleeding, bruises, and so forth, as well as to diminish or stop bleeding, dramatically slow breathing, and inhibit the perception of pain. Such effects can be taught (medical personnel, for example, work with hemophiliacs to help them learn to control their body's tendency to bleed, and yogis spend their lives learning to do some of these things), or they can be spontaneous, brought on by stress or by ALTERED STATES OF CONSCIOUSNESS. Altered states during religious ecstasy or prolonged MEDITATION are not uncommon with stigmatics. Some stigmatics acknowledge that they have urgently prayed for signs from God before their stigmata appeared. Did they perhaps cause the wounds themselves through their ardent desire and autosuggestion?

For further reading: Herbert Thurston, *The Physical Phenomena of Mysticism* (Henry Regnery Company, 1952) and Ian Wilson, *The Bleeding Mind* (Weidenfeld & Nicolson, Ltd., 1988).

STONEHENGE The most famous of Europe's prehistoric monuments, after which a whole class of MEGALITHS is named. Stonehenge lies on Salisbury Plain in southern England. Although the site itself is fairly small—not more than a few hundred square yards—it has won respect not only for its relative completeness but for the complexity of its construction as well. Archaeologists have divided Stonehenge's construction into four distinct phases, beginning with Stonehenge I around 2750 B.C.E. and continuing through Stonehenge IV around 1,200 years later. The site also increased in complexity during that period, changing from a circular trough ringed with a series of small holes to a series of circles and semicircles of *trilithons* (two standing stones capped by a third). Some of these stones may have been imported from Wales, more than 100 miles away. Other holes, posts, and lintels were added to and subtracted from the site over the centuries.

The function and construction of the Stonehenge site has been a subject of much debate over the centuries. In the 12th century, the romance writer Geoffrey of Monmouth suggested that the stones were carried across the Irish Sea by the enchanter Merlin. Geoffrey explained that in 483 the British king Aurelius Ambrosius had defeated an invading army of Saxons and wished to commemorate the victory with a suitable monument. He asked Merlin to transport the Giant's Dance—a ring of stones, supposedly the remains of giants who had ossified while dancing—from a mountain in Ireland to England and had them erected at Salisbury in honor of the British soldiers who had fallen in the fight. In the 17th century, architect Inigo Jones declared that Stonehenge was built by the Romans because they were the only ones with the technical skill to erect such a structure. Dr. Walter Charleton, physician to Charles II of England, believed that the Danish kings of Britain were responsible for Stonehenge and that it was meant as a polling place for the election and inauguration of kings. A century later, antiquarian William Stukeley declared that Stonehenge was erected by Druids, who used it as a place of worship.

Modern opinion on Stonehenge recognizes that Stonehenge is much more ancient than their predecessors believed. However, the opinion on its function is almost as divided as that of the ancients. Most archaeologists agree that the Stonehenge site functions as an astronomical calendar. The monument is carefully placed so that a person looking along the long chalk avenue toward the single menhir called the Heel Stone on the first morning of summer would see the sun rise directly over the Heel Stone. Furthermore, at this particular spot, the sun rises at the winter solstice—halfway around the year from the first day of summer—exactly at right angles to the summer solstice line.

Other modern theorists believe that Stonehenge is one of the major sites that tap Earth's magnetic currents. They believe that ancient people could use the energy collected at Stonehenge and related sites to send messages over long distances and move heavy objects through levitation. They cite as evidence for this belief the ancient Irish legend of the Druid Mog Ruith who, according to the story, was flying a megalith over St. George's Channel when it crashed and he was killed.

For more reading Robert Wernick, *The Monument Builders* (Time Life Books, 1973).

Stonehenge

STONIER, TOM 1927–1999) Scientist who posited that information is a fundamental component of the universe. A university professor, biologist, researcher, teacher, and from 1975 until his retirement in 1990, the first head of the Science and Society Department in the University of Bradford, United Kingdom. His family having emigrated from Germany to the United States in the 1930s, he was educated at Drew, Yale, and Rockefeller Universities. He has published several books including *The Morphology and Physiology of Plant Tumors* (with A. C. Braun, 1958), *Nuclear Disaster* (1963), *The Wealth of Information* (1983), *The Three Cs: Children, Computers and Communication* (with C. Conlin, 1985), *The Communicative Society: A New Era in Human History* (1985), *Information and the Internal Structure of the Universe* (1990), *Beyond Information: The Natural History of Intelligence,* (1992), and *Information and Meaning: An Evolutionary Perspective* (1997).

His publishing and appointments record illustrate his intellectual development: from a technical biologist, through concern about the horrific consequences of nuclear war; from an appointment to the biology faculty of Manhattan College in 1962, to his joining Adam Curle in 1973 in Peace Studies at Bradford; to his appointment two years later to head Bradford's newly created Science and Society School.

While in that post, he became increasingly convinced of the importance of information technology: that we were moving from an industry-based society to an information-based society. He promoted this view and explored the idea that information is a basic component of the universe, that information should be seen as on a par with mass and energy, and that there may be exchanges between these three fundamental components comparable with Einstein's well-known equation for the exchange of energy and mass. His last work was on a General Theory of Information. This is a bold speculation about a direction in which science may develop, that may turn out to be, in Thomas S. KUHN's terms, a paradigm shift. Whether or not it will be accepted by the scientific community remains to be seen.

For further reading: Tom Stonier, *Information and the Internal Structure of the Universe: An Exploration into Information Physics* (Springer-Verlag, 1990); and Tom Stonier, *The Wealth of Information: A Profile of the Post-Industrial Economy* (Methuen/Thames, 1983).

STREICHER, JULIUS (1885–1946) An ardent exponent of NAZI RACISM, claiming scientific grounds for his beliefs and actions: "The blood particles of Jews are completely different from those of a Nordic man. Hitherto one has prevented this fact being proved by microscopic investigation" (from an address in 1935). Although similar xenophobic ideas resurface from time to time, the overwhelming scientific evidence shows that Streicher's theories and similar ideas have no basis. Streicher was the managing editor of a Nazi propagandist journal *Der Stürmer* (1923–45), was area commander of Franconia, and was hanged as a war criminal.

See also RACIAL THEORIES.

STRUCTURAL PATTERNING A form of bodywork/massage developed by Judith Aston out of her critique of ROLFING. Rolfing was a popular MASSAGE technique developed by Ida Rolf that emphasized the manipulation of deep connective tissues (fascia) between the bone joints. Aston had been introduced to rolfing during her recovery from the effects of an automobile accident. When first experienced, rolfing could be quite painful, but many who underwent the treatment felt that the pain was worth the long-term benefits.

Aston, who went on to become a rolfer, began to disagree with her teacher at two primary points. First, she felt that rolfing applied too much force to the body, and that the level of treatment should be adjusted to each individual. Second, from her perspective as a former dancer, she questioned Rolf's emphasis upon the symmetry of the body. She believed that most body movements were, in fact, asymmetrical. Aston introduced a set of movements based upon her dancing experience that were designed to encourage the various parts of the body to work together. Her new synthesis of massage and movement was termed structural patterning.

Structural patterning supports the emphasis within the larger perspective of holistic healing of maintaining continued health by keeping the body fit and toned. Sessions use massage to relieve chronic stress and employ movements to modify the body's structure. Aston founded the Aston Training Center in Mill Valley, California, to train structural patterning practitioners.

STUKELEY, WILLIAM (1687–1765) English physician and archaeologist who made careful surveys of the STONEHENGE and AVEBURY stone circles in Wiltshire. Having mapped these huge prehistoric monuments, he then speculated as to their purpose. He suggested that, by analogy with the nearest equivalents of medieval times, the European cathedrals, they must have had a religious function. They must have been places where the Druids held religious rites and offerings—their temples in effect. Antiquarian John Aubrey had made drawings of Stonehenge for King Charles I in 1663, and it was Aubrey who had first proposed that it was a Druidic temple and a place of sacrifice. Stukeley enthusiastically adopted and embroidered on this explanation. He concluded that Stonehenge and other stone circles had been places for the worship of the serpent and that the Druids were of Phoenician origin.

Stukeley's measurements were meticulous, unlike Aubrey's comparatively sloppy approach, and he made a serious attempt to analyze what he found. He was the first, in 1723, to identify the Avenue of stones running off to the northeast of Stonehenge, and the Cursus, a low mound to the north. He thought that the Stonehenge builders had used a unit of a druid cubit of 52.8 centimeters (20.8 inches), and he tried, using the science of the time, to date the monument. He arrived at a figure of about 460 B.C.E., not accepted today. He recorded his observations and deductions in *Stonehenge, a Temple Restored to the British Druids* (1740). Probably his most important observation was that ". . . . the principal line of the whole work, (points to) the northeast, where the sun rises, when the days are longest." It is on that fact that much of the research on Stonehenge has focused during the last century. In 1771, John Smith, continuing Stukeley's astronomical observations, speculated that Stonehenge was a kind of astronomical calendar, calling it "the Grand Orrery of the Ancient Druids." In 1796, Henry Wansey suggested it was a predictor of eclipses.

Stukeley's Druidic theories remain popular. Every year on Midsummer's Day, the British police have a problem controlling the crowds who assemble to claim their right to hold ceremonies.

For further reading: Gerald S. Hawkins, *Stonehenge Decoded* (Fontana/Collins, 1970).

STUDIEVERENIGING VOOR PSYCHICAL RESEARCH (SVPR) Dutch association of parapsychologists. The Studievereniging voor Psychical Research (Society for Psychical Research) was founded in 1920 and launched its formal program with systematic research on TELEPATHY and CLAIRVOYANCE. Its journal, *Tijdschrift voor Parapsychologie,* was founded in 1928 by a young college student, Wilhelm Heinrich Carl TENHAELF, who would go on to write the first dissertation on PSYCHICAL RESEARCH in the Netherlands in 1933. The work of the growing society was thoroughly disrupted by the German invasion early in World War II. It recovered following the war largely through Tenhaelf's effort. In 1953 Tenhaelf, a professor of psychology at Utrecht, was able to gain the university's support for a chair in parapsychology and a research institute. His position as chair and director of the institute, coupled with his increasing success in research using psychic Gérard Croiset, allowed him to dominate the field in the Netherlands. That same year, the society hosted the First International Conference of Parapsychological Studies in cooperation with the Parapsychological Foundation.

Beginning in 1945, Tenhaelf edited the society's journal and served as its president. He developed a dictatorial leadership style that came to be resented by many of his colleagues, including the equally eminent George Zorab. In 1960, Zorab and other members of SVPR founded the NEDERLANDSE VERENIGING VOOR PARAPSYCHOLOGIE as a rival organization. Shortly thereafter, it was revealed that Tenhaelf had altered his research data to produce a string of positive results. Tenhaelf's last years were spent in disgrace, and the society suffered accordingly. In the years since his death in 1981, the society has moved to recover its role in the parapsychological community.

SUMMERLIN, WILLIAM (1938–) Fraudulent medical researcher. Dr. William Summerlin was a respected immunologist who specialized in research on the problem of rejection of transplanted tissue. Through the 1960s, he had worked at the University of Minnesota and at Stanford. By the early 1970s, he had accepted a position at the Sloan-Kettering Institute of Cancer Research. While there, he published papers that claimed that he had performed skin grafts between two incompatible strains of mice. In 1974, he published another paper that claimed to have repeated the results of his earlier experiment.

In spite of Summerlin's claims, colleagues following his notes were unable to duplicate his results. Nobel Prize winner Sir Peter Medawar called Summerlin's research into question. Pressed to explain himself, Summerlin eventually admitted that he had used mice that were artificially colored to make them appear to be of another strain. His work was reviewed and denounced, and Summerlin was given a permanent leave of absence. In reviewing the case, the press censured Summerlin but also criticized the system of funding in such important areas of research, which tends to reward results rather than research.

SUN STOOD STILL, THE DAY THE Refers to an event recounted in the Old Testament when Joshua commanded the sun to delay going down for a full day and the moon to halt in its movements.

During the 1970s, a story began to circulate in conservative Evangelical Christian circles that some scientists at the National Aeronautics and Space Administration (NASA) had been calculating the positions of the planets in centuries past. Their calculations had apparently run into a problem because they could not account for 24 hours. Seemingly, when refiguring their calculations, taking into account the Joshua story, all of the figures came out correctly. This became a popular anecdotal example, often repeated, of modern science reversing itself and confirming an extraordinary incident recorded in the Bible, which scientists previously claimed to be impossible. In the 1980s, writer Tom McIver attempted to track down the source of the story. He eventually traced it to a conservative Christian engineer who had at one time been a consultant to NASA. In speaking to groups of people, he had told this story, though it has no basis in fact, as a means of convincing his hearers that science really supports the truth of the miraculous occurrences recorded in the Bible. When confronted with his falsehood, he defended the story because it converted some of his audience to his particular version of Christianity—"the end justifies the means."

There are many scientists who believe that their subject need not clash with religious doctrine. Consequently, some have sought to find a scientific explanation to account for unbelievable occurrences and for other miracles that have been recounted in holy books and other ancient texts. Stormy seas and giant waves that swallow whole armies can be understood; the earth could crack asunder and swallow up human beings; Jericho's walls could collapse, not by the blast of a trumpet, but by a coincidental earthquake. But that the sun and the moon should halt in their movements must be beyond belief, a fancy, a poetic image, or a metaphor. Nevertheless, these Old Testament incidents are just two of many unbelievable accounts of cosmic disturbances that engaged the mind of scientist Immanuel VELIKOVSKY in *Worlds in Collision*, first published in 1950. In his book, he propounds the theory that a major series of cosmic catastrophes took place in the second millennium B.C.E. that profoundly influenced the course of civilization in ancient times. The cosmic happening that caused the two events related above was, according to Velikovsky, a giant comet that erupted from the planet Jupiter, passed close to Earth on two occasions, and then settled down as Venus. The first close visit to Earth was precisely at the time that Moses stretched his hand out to cause the Red Sea to divide. The manna that fell from the sky afterward, and by good chance was found to be edible, was a precipitate of elements suspended in the comet's tail. The second visit of the comet was precisely at the time Joshua successfully made the sun and the moon stand still. Both miracles were the result of a temporary arresting of Earth's spin.

Today, with hindsight, we appreciate that subsequent scientific knowledge has shown Velikovsky's book, and his later books in similar vein, to be full of absurdities, a piece of pseudoscientific wishful thinking, but at the time of publication in 1950, many were enthusiastic about persuasive arguments that appeared to be scientific verification of biblical history. The *New York Times* science editor, John J. O'Neill, described the book as "a magnificent piece of scholarly historical research," and the *New York Compass* ranked Velikovsky with Galileo, Kepler, Sir Isaac NEWTON, Charles DARWIN, and Albert Einstein.

Galileo also examined the story of the sun standing still. In his *Letter to the Grand Duchess Christina*, the final section addresses Joshua's miracle. Galileo argues that in the Ptolemaic system, which the church supports, the sun standing still would shorten the day, not lengthen it, just the opposite of what Joshua required. Whereas, if Galileo's premise (that the sun is the source of *all* planetary motion) is accepted, then, the Joshua story would be logical. Galileo insisted, therefore, that he was promoting a theory that is not in conflict with the church and its teachings but is actually supporting them.

For further reading: Martin Gardner, *Science: Good, Bad and Bogus* (Prometheus Books, 1989).

SUNSPOT CYCLES The cyclic occurrence of spots that travel across the surface of the sun as it rotates. The number and area of these spots waxes and wanes. They were first observed by Galileo and much later by Heinrich Samuel Schwabe in 1826. By 1843, Schwabe had detected an average 10-year cycle, which he later corrected to 11 years. In 1908, G. E. Hale showed that the magnetic polarities of the sunspots in one cycle were reversed in the next. The average total cycle is thus 22 years.

Several terrestrial effects have been attributed to this particular cycle. Throughout the 1920s and 1930s, it was proclaimed to be the cause of economic boom and bust and of climatic variations. That the number and extent of sunspots do affect Earth indirectly is very likely. The dark areas on the Sun's surface turned toward Earth reduces the intensity of solar cosmic rays that reach our atmosphere, and this variation in the bombardment of the upper atmosphere undoubtedly has effects near and at ground level. But belief in them as being the cause of so many of our ills has diminished, and it is seen as pseudoscientific rather than scientific.

SUPERORGANISMS Huge entities that behave in a way similar to living organisms—plants or animals—and form self-regulating systems. In response to some disturbance of a superorganism's normal state, it reacts in such a way that it readjusts to a new equilibrium state. An example of such a system, which has been much discussed in recent years, is the GAIA HYPOTHESIS of Professor James Lovelock (1919–).

Traditional ideas held that all living things—plants and animals—evolved on but were distinct from the inanimate planet Earth. Lovelock postulated that the whole Earth—the rocks, the oceans and rivers, the atmosphere and climate, plus all living things—are part of one great organism functioning as a whole, a superorganism. He also proposed the idea that living things and physical conditions evolved together with the biota, the totality of all living organisms, regulating everything. The name he gave to his theory of a self-regulating Earth was "Gaia." The theory has a mathematical basis in the model "Daisyworld."

Lovelock began to develop his idea when he was working for NASA as part of a team that designed equipment to look for life on Mars. He realized that the composition of the atmosphere was the most obvious sign of life on our planet and concluded that the closeness of the Martian atmosphere to chemical equilibrium proved that life was absent there. His conclusion was later confirmed by the two *Viking* space probes.

The Gaia hypothesis is essentially an ecological concept and was first proposed by Lovelock in 1972. The term "Gaia" (*ai* is pronounced *I*) was taken from ancient Greek mythology, chosen because it was the name of the goddess Mother Earth, sometimes known as Ge, as in geology. The earliest text to mention the mythological goddess Gaia was written in the eighth century B.C.E. by Greek writer Hesiod (see his *Theogony*). According to Hesiod, Gaia was the mother and/or grandmother of Zeus, and his rule depended on her consent. Some feminist historians studying the ancient world think that the idea of a mother goddess lies buried in the prehistory of many ancient religions.

Although some biologists have produced data to support the Gaia hypothesis and agree that the biosphere, that part of Earth where living things normally exist, does have a feedback mechanism, the problem is to find out what that self-regulating system is. Some say, quite simply, that it is life itself. The difficulty is that we are always looking at the system from the inside, with no possibility of studying it from the outside. Other biologists have criticized Lovelock for being teleological—that is, assuming a design or divine purpose to life or the universe. Teleological questions ask Why?—Why are we here? Scientific questions ask How?—How did we get here? At the other end of the spectrum of argument, theologians have criticized the Gaia theory for not being teleological enough and for leaving no room for God, the loving creator who cares about his creation. They say that the existence of a watch implies the existence of a concerned watchmaker. Our world, they say, has been made to appear in the Gaia hypothesis as only a neutral living organism, thus forcing us to consider the health of the planet as primary, not the health of human beings, who are a specially chosen species living on it. This is a contentious issue for those who believe that our span of years on Earth has a divine purpose, and Lovelock devotes a whole chapter to it in his second book *The Ages of Gaia*. He writes: "For the present, my belief in God rests at the stage of a positive agnosticism. I am too deeply committed to science for undiluted faith; equally unacceptable to me spiritually is the materialist world of undiluted fact. . . . That Gaia can be both spiritual and scientific is, for me, deeply satisfying."

The Gaia hypothesis is a new idea; many scientists judge it to be both serious and sensible, not a pseudoscience but a protoscience, because as yet it does not fit into the prevailing paradigmatic scheme. To make it more acceptable to the scientific community we must ask: Are its concepts clearly defined and noncontradictory? Are there any tests we could do to verify it? Finally, we must ask the question that has ruled out much crackpot pseudoscience: Is it falsifiable? Care must be taken in its evaluation because in the past many new fields of inquiry have had an uphill struggle against a somewhat rigid orthodox scientific establishment. It is now more than two decades since Lovelock launched his first papers on the scientific community and had them dismissed as New Age mysticism. Nevertheless, his ideas are now beginning to be taken seriously, although reinterpreted in a slightly different way. Today's researchers are looking at superorganisms in the light of modern complexity theory and suggesting that Gaia can be classified as a superorganism. "The serious Gaia theory of today concerns the interactions of a large complex system from which emerge certain properties that we can observe, such as the maintenance of global temperature and cycling of gases," says Lovelock.

In Lovelock's book *Gaia: The Practical Science of Planetary Medicine,* the question is posed: If the planetary biosphere is a living system, is it well? To answer this question, Lovelock looks at some of the serious ecological upsets of this century and shows how the planet has recovered because there has been sufficient space and variety of organisms, like those we find in the rain forests, so that the biosystem may absorb the pollution and cure itself. But Gaia is sick again, and humanity is now recognized to be the major part of the problem. We have a plague of people swarming all over the planet reproducing exponentially and all demanding a lifestyle that is not sustainable without seriously upsetting the "greenhouse" balance. Is Thomas Robert MALTHUS's prediction at last coming true?

For further reading: James Lovelock, *The Ages of Gaia: A Biography of Our Living Earth* (Oxford University Press, 1988); James Lovelock, *Gaia: A New Look at Life on Earth* (Oxford University Press, 1979); and James Lovelock, *Gaia: The Practical Science of Planetary Medicine* (Gaia Books Ltd., 1991).

SYMMES, JOHN CLEVES (1780?–1829) Hollow Earth theorist. Symmes saw Earth as made up of five concentric spheres with openings at each of the poles, several thousand miles in diameter. Water from the oceans flowed through these openings and plants and animal life lived on both the convex and the concave sides of the spheres. He thought this construction, like that of the human body with all its organs inside, would be a very economical way for the Creator to have used materials.

Symmes became obsessed with his ideas, and from 1818 onwards he traveled throughout the various states giving lectures, trying to raise funds, and petitioning Congress to finance a polar expedition. At one point, he actually called for a hundred brave men to accompany him to explore the northern opening, which by then had been dubbed "Symmes's Hole." Despite his fervent lecturing and fund-raising activities, his expedition was never mounted. He fell ill during the winter of 1828 and returned to his home in Hamilton, Ohio, where he died on May 29, 1829.

Two books give very full accounts of the theory—*Symmes' Theory of Concentric Spheres* (1826), written by his friend James McBride, and *The Symmes' Theory of Concentric Spheres* (1878) by his son Americus Symmes. Symmes' theory inspired some 19th-century fiction; in 1820, an anonymous writer published *Symzonia: A Voyage of Discovery.* The story tells of a journey to the southern polar opening and how a strong current sweeps the ship over the rim of the world and into its hollow interior, in which Captain Seaborn and his crew find a continent that they name Symzonia; the continent is inhabited by friendly people dressed in white clothes. On returning home, the explorers make their fortunes by trading Symzonian goods for copra and cacao. An unfinished story by Edgar Allan Poe called "The Narrative of Arthur Gordon Pym" tells of a similar voyage. Other science fiction writers, notably Jules Verne and Edgar Rice Burroughs, exploited the same idea.

See also HOLLOW EARTH DOCTRINE (HOHLWELTLEHRE).

For further reading: Martin Gardner "Flat and Hollow," *Fads and Fallacies in the Name of Science* (Dover 1957).

SYNCHRONICITY A Jungian term for a connecting principle that he thought would give meaning to a series of causal coincidences, as for example the frequent recurrence of a particular number over a short span of time. Carl Gustav JUNG thought that these coincidences were meaningful and would not accept that they could happen in accordance with the normal laws of probability—that is, by pure chance—and that in a very much larger series of observations, these anomalous clusterings would disappear. Instead, he put forward a metaphysical notion where all events of temporal coincidence, like DÉJÀ VU and the act of PRECOGNITION, had spiritual significance within a purposeful universe. Therefore, he thought, their meaning should be sought in their structural relationship as well as in their causal antecedents within the meaningful superstructure. Jung's structuralism entails a form of experiential harmony, a harmony among events, and a harmony between the structure of our understanding and the event structure. He believed that his new notion of synchronicity supplemented the classical and medieval principle of correspondence (see CORRESPONDENCE, THEORIES OF), which, he said, had existed rather naively and unreflectingly up to the time of Leibniz and then had been superseded by scientific thinking.

Since Jung, there have been many studies done on coincidence from a psychological point of view, notably those recently by Ruma Falk and her collaborators. The findings seem very clear and not surprising. First, more of us are impressed when coincidences happen to ourselves than when they happen to others. Second, the rate at which coincidences are experienced varies very much from person to person: Some connect occurrences that are only vaguely alike, and thus overestimation occurs; others see very few connections. Third, reactions to coincidence stories depend on how the story is told; if told in a matter-of-fact way, it is not taken so seriously as when the story is elaborated upon, giving emphasis to details.

For further reading: Martin Gardner, "The Roots of Coincidence," *Science Good, Bad, and Bogus* (Prometheus Books, 1981); and C. G. Jung, *Synchronicity: An Acausal Connecting Principle,* trans. by R. F. C. Hull (Princeton University Press, 1973).

SYNESTHESIA A strange condition in which the senses become intermingled. People see numbers in color or experience words as tastes. Almost any two senses can be involved. The most common fusion is of the visual and the auditory senses; for example, people see colors and shapes while listening to music.

Scientists have recently become very interested in synesthesia. In the past, the condition was considered to be a fanciful notion, in the same way as J. W. von Goethe's *Theory of Color* (1810) was deemed to be a fanciful theory because it maintained that color, among other things, affected our morals. The Theosophical Society drew on many of Goethe's ideas. It was the metaphysical side of Goethe's ideas of color that the theosophists developed, stressing its clairvoyant and telepathic origins. A follower of the theosophists, artist Wassily Kandinsky, wrote a book called *On the Spiritual in Art* where he asserted that "colors had their own internal meaning" that could be "felt only by the more highly developed and sensitive observers." Some "particularly sensitive observers," he wrote, "can stimulate the response of another sense organ, an experience known as synesthesia." By way of evidence, Kandinsky cited the case of a Dresden doctor who reported that one of his patients, another exceptionally sensitive person, could not eat a certain sauce without tasting blue. Scientists of the time thought this was all extremely dubious, but might have been less skeptical if the information had been reported in a less mystical way, and had come from a different quarter.

Recently neuroscientists have been looking at the problem and have agreed that the condition is genuine and diagnosable. A team of psychologists and neurologists in London, headed by Simon Baron-Cohen of the Institute of Psychiatry, are investigating the phenomenon. They accept the currently held view that the various functions of the brain and of the senses are in separate modules (the division of labor model). Individuals who experience synesthesia have brains that are cross-wired in a way that most of ours are not. Baron-Cohen's findings have been published in the journal *Perception.*

Others, notably neurologist Richard Cytowic of Capitol Neurology, a private clinic in Washington, D.C., who has been studying the condition since 1980, believe that the brain of synesthetics may not be different at all and that we do not need to invent special pathways to explain the condition. In his theory, it is our model of the brain that is at fault. Cytowic proposes in his book *The Man Who Tasted Shapes* (1994) a revised model in which the brain is not a passive receiver but, in his words, "goes out and grabs information." He calls this the multiplex brain and believes that it is this concept of the brain that might explain why only some people experience synesthesia.

Both teams confirm that the majority of synesthetics are female (about 6 female:1 male) and that a larger number than expected are lesbian and gay people. There is also a preponderance of right-handedness, and most with the condition seem to have good memories. Both teams suspect that synesthesia has a genetic base. This has been confirmed by a recent study at Cambridge University, England, where a third of those studied had close relatives with the trait. They suggest that it could possibly be X-linked and dominant, that is carried on the X chromosome and evident in everyone carrying it. In three of the families studied, a carrier mother has passed the trait to at least half of her daughters, and in one family an affected father has passed the condition on to all three of his daughters. But all is not straightforward; one of the researchers came upon a synesthetic man who had an identical twin who did not have the condition. The suspicion must be that these twins were not identical; 10 percent of twins who claim to be identical turn out, on closer examination, to be not "identical" but very alike "unidentical" twins.

Synesthesia is an example of a phenomenon that doctors and scientists once deemed pseudoscientific but in the light of further knowledge and research has been found to be scientifically proven.

For further reading: Richard E. Cytowic, *Synesthesia: A Union of the Senses* (Springer, 1989); Richard E. Cytowic, *The Man Who Tasted Shapes* (Abacus, 1994); and Wassily Kandinsky, *Concerning the Spiritual in Art,* trans. by Michael T. H. Sadler (Dover, 1977).

T

TABLE RAPPING A phenomenon of SPIRITUALISM in which POLTERGEISTS or noisy spirits are heard to rap on tables to attract attention, or more frequently a method by which spirits of the dead answer questions posed to them. Table rapping or table tilting is a very laborious way of consulting the spirit world and works rather like the OUIJA BOARD. Several people sit at a small round table with their fingertips upon it, their thumbs touching, and their little fingers stretched out toward their neighbor on either side. Traditionally, the questions posed to the spirit were spelled out with one rap or tilt for letter A, two for B, and so on, but today it is usually speeded up by stating the problem verbally. Questioning usually begins with "Is there anybody there?" and then goes on to more specific queries. The answers come with one rap or one tilt indicating yes, two indicating maybe or don't know, and three meaning no.

Critics of spiritualism do not believe that there is any way of communicating with the dead and point to the confession of Margaret Fox, the elder of the two Fox sisters from Hydesville, New York, who first started the 19th-century spiritualist craze. Her confession came at the end of her long career as a MEDIUM, at the age of 81, when she explained how she had fooled people, making spirit raps by cracking the first joint of a big toe.

Modern parapsychologists do not accept the old 19th-century spiritualist connection to explain table rapping, turning, and the phenomena of noisy GHOSTS. It is now believed that the disturbances do not emanate from ghosts or spirits but come from a certain kind of living individual, often a troubled adolescent. Parapsychologists refer to this type of person as the "focus" or the "agent," and the activity as "recurrent spontaneous psychokinesis (RSPK)." Skeptics say that this kind of rephrasing of and redirecting of the cause does not prove the existence of the phenomena. The least that can be said is that now psychical researchers are admitting that what went on in the last century was indeed trickery.

For further reading: C. E. M. Hansel, *The Search for Psychic Power: ESP and Parapsychology Revisited* (Prometheus Books, 1989).

TALKING APES Communication between apes and humans. During the last hundred years, there have been several attempts to engage in some form of intelligible speech with apes. Richard L. Garner spent many years analyzing the sounds made by apes and eventually claimed to be able to talk to monkeys in their language. He published several books, *The Speech of Monkeys* (1892), *Gorillas and Chimpanzees* (1896), and *Apes and Monkeys* (1900), claiming considerable success. However, nobody since that time has been able to reproduce his results.

Primate studies in recent years have taken a different tack, taking very young chimpanzees and developing a shared sign language. Allen and Beatrice Gardner at the University of Nevada taught American sign language to the chimpanzee Washoe, who became famous. Washoe developed a large vocabulary of signs, which it appeared to use with impressive fluency, not merely responding to and replicating but apparently developing a limited syntax, the use of a basic grammar. Other researchers entered this apparently fruitful field. The gorilla Koko described a zebra as a white tiger. Herbert Terrace trained a chimp named Nim Chimpsky (a variation on the name of the famous linguist Noam Chomsky) to use sign language, which it did with extraordinary facility. But on closer examination, Terrace decided that Nim was neither employing a language nor had any understanding of the meaning of the signs, much less of the grammar, but had learned to imitate his teacher to obtain rewards; that is he was behaving like

Clever Hans (see COUNTING HORSES). Terrace's work has cast doubt on the other apes' sign language achievements. Terrace's conclusions are contested, and primate researchers will continue to explore this subject.

Using the apes' own language is much more difficult, but primate researchers, spending months and years studying small groups in the wild, can recognize the significance of certain calls and mimic them to obtain the appropriate response. Nobody has trained an ape to utter recognizable and meaningful parts of human speech. Dr. Mortimer J. Adler of the University of Chicago tried to used this fact to discredit the theory of EVOLUTION.

For further reading: E. Sue Savage-Rumbaugh and Roger Lewin, *Kanzi: The Ape at the Brink of the Human Mind* (Wiley, 1994); Gordon Stein, ed., *The Encyclopedia of the Paranormal* (Prometheus, 1996); and Joel Wallman, *Aping Language* (Cambridge University Press, 1992).

TAO OF PHYSICS, THE A book by Fritjof Capra, a physicist of some standing and a professor at the University of California, first published in 1975, revised in 1983, and subtitled "An exploration of the parallels between modern physics and Eastern mysticism." The book has three parts: The Way of Physics; The Way of Eastern Mysticism; and The Parallels. It has been enormously popular, selling in large numbers, clearly striking a chord with the public and with many scientists, and being translated into several languages.

Capra explains in his preface that, while sitting on a beach ruminating on the constitution of matter and on cosmic rays, "I 'saw' cascades of energy coming down from outer space, in which particles were created and destroyed in rhythmic pulses; I 'saw' atoms of the elements and those of my body participating in this cosmic dance of energy; I felt its rhythm and I 'heard' its sound, and at that moment I *knew* that this was the Dance of Shiva, The Lord of Dances, worshiped by the Hindus". And later: "It was followed by many similar experiences which helped me gradually to realize that a consistent view of the world is beginning to emerge from modern physics which is harmonious with ancient Eastern wisdom."

The revelatory nature of Capra's beach experience is not that surprising, nor does it necessarily give weight to his claims. Scientists (and others) have often had similar visionary experiences—for example, Kepler, Sir Isaac NEWTON, Kekulé, and Planck. Working on a problem for months or years, the pieces often come together to reveal a possible solution subconsciously; this has been well documented and analyzed by writers on creativity. Nor does such a revelation necessarily imply validity or credibility because a flash of inspiration must be subjected to the same rigorous analysis and experimental confirmation as any other hypothesis or guess before acceptance. Charles DARWIN conceived of evolution by natural selection early in his career and then spent many years in the most painstaking research before publishing *The Origin of Species* (1859).

Looked at in this light, how does *The Tao of Physics* stand up to critical inspection? First, it must be said that Capra is very selective in his choice of modern physics. His modern physics is limited to the quantum theory of subatomic particles. Even that, as he himself admits, "is based on the so-called Copenhagen interpretation of quantum theory which was developed by Bohr and Heisenberg in the late 1920s and is still the most widely accepted model." If some other less mystical model had become, or may yet become "most widely accepted," the parallels with Eastern mysticism might not be so readily drawn. And in the Afterword to the second edition, Capra paints a picture of the current state of quantum theory that best resembles a "dog's breakfast." Where you draw parallels with Eastern mysticism becomes very arbitrary; Capra shows his preference for those bits of the dog's breakfast that best suit his thesis.

The Tao of Physics is an interesting book and certainly worth a read for its exposition of the current state of subatomic physics and of quantum theories. It also offers accounts of Hinduism, Buddhism, Chinese Thought, Taoism and Zen in Part II, though these may also be selective. Part III, The Parallels, must, like all grand claims, be looked at very skeptically, bearing in mind Carl SAGAN's advice: "All science asks is to employ the same levels of skepticism we use in buying a used car or in judging the quality of analgesics or beer from their television commercials."

A similar idea to Capra's appears in a more recent publication *Blackfoot Physics* by F. David Peat, which explores the parallels between Native American science and modern physics.

For further reading: Frijhof Capra, *The Tao of Physics* (Shambhala 1991); Frijhof Capra, *The Turning Point: Science, Society and the Rising Culture* (Wildwood, 1982); F. David Peat, *Blackfoot Physics (Fourth Estate and Penguin/ Arkana, 1996)* and Michael Toms, *Fritjof Capra in Conversation with Michael Toms* (Aslan, 1993).

TAROT CARDS A pack of 78 cards, first used in Northern Italy in the early 14th century, originally as a trick-taking game, but now used by fortune-tellers for DIVINATION. The deck of 78 cards comprises four suits of 14 cards in each, making 56; the other 22 cards are unsuited trumps, decorated with symbolic designs like The Fool (precursor of the modern Joker), the Wheel of Fortune, The Tower, The Lovers, Strength, and other virtues and vices. The cards were first called triumphs (*triumphi* in Latin), and later *tarocchi* in Italian and *tarots* in French.

The tarot shows pictures of conventionalized personages and objects to which meanings are attached for the purpose of fortune-telling. The reader and the client sit at

Tarot Cards

opposite sides of the table; the cards are shuffled and then taken, one at a time, from the top of the pack and laid out on the table in what is called a spread. Many different systems of meanings have been attached to the suit signs, going back to the Ancient Egyptian Book of Thoth, the KABBALAH of Jewish mysticism, the Neoplatonic virtues of the Italian Renaissance, and many others. The cards are interpreted according to the meanings assigned to them within the specific system used, and that their meaning is qualified by the position in which they are laid down—that is, which card is next to which on the table.

Critics believe that the colorful and intricate symbolism that has been attached to the cards has nothing to do with how they are used today. They claim that, as in all types of divination, the information from tarot cards is always modified by the fortune-teller's feelings about, and assessment of, the client's background and concerns. The technique used is a system of specifying from generalizations, starting very broadly to discover the concerns of the client, and then homing in on the specifics.

For further reading: Francis X. King, *The Encyclopedia of Fortune Telling* (Hamlyn, 1988); and Emily Peach, *Discover Tarot: Understanding and Using Tarot Symbolism* (Aquarian, 1990).

TAR WATER Popular medical remedy of the 18th century recommended by philosopher, George Berkeley.

Tar water was made by putting a quart of cold water and a quart of tar together, mixing them, and then allowing the tar to settle. A glass of the clear water that remained behind was then drawn off and mixed with an equal amount of fresh water. Berkeley developed his own procedure of producing tar water by mixing a gallon of water with a quart of tar and allowing the mixture to stand for 48 hours. In 1744 he authored a book on the subject, *Siris, a Chain of philosophical reflections and Enquiries on the Virtue of Tar Water.* Berkeley first used tar water during a local outbreak of smallpox. He found that those who took tar water either did not develop smallpox at all or developed only a mild case of the disease. He also recommended the preparation to sailors for the prevention of scurvy, to students cramped up in unhealthy study quarters, and to the general public for the prevention and treatment of a variety of ailments form hysteria to the plague.

Berkeley responded to his critics by suggesting a controlled experiment. Patients in two hospitals, he recommended, should be given the same diet and bed care, but one group should be treated with contemporary drugs and medicines as seen fit, and the other should be treated only with tar water administered by an "old woman." His suggestion was never followed up.

There was never a decisive confrontation over tar water between physicians and proponents of Berkeley's cure. It continued to be used in both England and the United States for many years but eventually fell out of use. Today, no medicinal properties are known to exist in the remedy, although, like may similar medicines, it did people no harm and may have kept them away from other, more damaging preparations.

TASADAY TRIBE Isolated Philippine natives discovered by Manuel Elizalde in 1971. The Tasaday were a small tribe of 24 people living on the Philippine island of Mindanao. They were so isolated from the 20th century that they believed that they were the only people living in the world. They practiced a Stone Age culture, knew nothing of agriculture, and wore only primitive loincloths made from local plants. The Tasaday were first discovered in June 1971 by a hunter from a nearby tribe, who persuaded them to meet Elizalde. Elizalde worked for the Philippine government as an advisor on national minorities under Ferdinand Marcos. He brought the Tasaday to public notice through skillful use of the media, including the influential magazine *National Geographic.* The organization filmed the Tasaday in a *National Geographic* special, "The Lost Tribes of Mindanao," in December, 1971.

Elizalde cut off contact with the Tasaday in the mid-1970s, claiming that the Tasaday were being corrupted by the media deluge. Once the Marcos regime was overthrown in the 1980s, however, journalists began to try to reestablish contact with the tribe. A Swiss writer named Oswald Iten, who had written extensively about small cultural groups, visited the sites the Tasaday had frequented. He claimed that he found them deserted and that the local people—including members of the original Tasaday, whom he photographed in modern dress—told him that the Tasaday were really members of the Manubo-Blit tribe whom Elizalde had persuaded to masquerade as a Stone Age tribe for the cameras. Iten concluded that Elizalde engineered the discovery of the Tasaday for the sake of publicity and that the Tasaday themselves were a great hoax.

Elizalde fled the Philippines in 1983. Since that time, he has been charged with exploiting the native minorities placed under his care. Other journalists, following Iten's example, have also found differences between the way Elizalde originally presented the Tasaday and the current state of the tribespeople. Photographers from the German magazine *Stern* who visited Mindanao in 1986 brought back pictures showing the same tribespeople whom Iten had photographed in modern dress at the Tasaday caves wearing the Tasaday's leafy loincloths. The German photographers reached the same conclusion as Iten had: that the Tasaday were a hoax masterminded by Elizalde for his own benefit. Most anthropologists hold similar opinions.

TATZELWURM A legendary animal reported to exist in the Swiss, Bavarian, and Austrian Alps that is the subject of cryptozoologic research. The modern record of sightings of the Tatzelwurm (literally "worm with claws") began in 1779. A man named Hans Fuchs encountered two tatzelwurms and was so scared that he suffered a heart attack. He only had time before he died to tell his family what had occurred. From his dying remarks, a relative painted a picture that showed the two large lizard-like animals that Fuchs had described.

Over the next several decades, other sightings occurred, with witnesses reporting having seen a large creature, varying between 60 and 180 centimeters (2–6 feet) long, with a cylindrical body, four legs each of which had three toes, a large mouth with sharp teeth, a short blunt tail, and a light-colored skin. When spotted, the creature generally fled, but on occasion it turned on individuals and tried to bite them. Sketches and descriptions of the tatzelwurm were published in the 1841 *Alpenrosen,* a Swiss almanac, and in *Das Thierleben der Alpenwelt* (1861) by Swiss naturalist Friedrich von Tschudi.

In spite of reports by farmers of having killed a tatzelwurm and the discovery of a 1.5-meter (5-feet)-long skeleton of one, no specimen seems to have found its way into any biological laboratory. In 1934, a photographer named Balkin produced a picture of a tatzelwurm, but it is generally believed to be a simple hoax. Cryptozoologists hold out little hope that such an animal exists, but if it does, they speculate that it could possibly be a new species of otter, a giant salamander, or a lizard similar to the Gila monster of the American Soutwest.

For further reading: Bernard Heuvelmans, *On the Track of Unknown Animals* (Hill & Wang, 1958).

TAYLOR, JOHN G. (1931–) A mathematical physicist at King's College, London, and an ardent believer in Uri GELLER's psychic abilities, particularly his METAL BENDING by mental power. Taylor went further and believed that many children shared this power. So convinced was he that he appeared on television in Britain in support of Geller and this psychic phenomenon, and wrote books and articles on the subject. In *Superminds* (1975), he discusses possible, but highly improbable, physical explanations, eventually deciding that the process must be electromagnetic. In his book *Black Holes* (1973), Taylor engages in flights of fancy about spaceships, extraterrestrials, multiple universes, and immortality. Taylor became an advocate of his fellow London physicist John B. HASTED's experiments on paranormal metal bending. But eventually he became disenchanted, concluding that the demonstrated effects were either fraudulent or inadequately controlled, publishing his account of his disillusionment in *Science and the Supernatural* (1979).

Although Taylor had been convinced that the various paranormal phenomena could be genuine, he was confident that, if so, there had to be some physical process by which they were effected—that they were not, in the strict sense of the word, magical. He considered the many possible force fields or particles of physics, both known and hypothesized, that might be involved: gravity, quarks, the weak force, bosons, neutrinos, magnetic monopoles, tachyons, and so on. He concluded that the only possible way that these effects could be explained was through ELECTROMAGNETIC FIELDS and eventually decided that it could not be done; electromagnetic fields could not be involved. By that time, he had been shown that many of the effects that he had thought genuine were achieved by cheating or trickery. He, with one of his colleagues, published papers in *Nature* in 1978 and 1979. In the former, they explained that their attempts to validate and to detect an electromagnetic link in what they termed "extrasensory phenomena" had failed.

Taylor and Balanovski went to extraordinary lengths to detect electromagnetic signals over the complete range of the spectrum and with very sensitive instruments. They looked at needle rotation, straw rotation, compass needle rotation, metal bending both by contact and at a distance, psychic healing, distant viewing, and TELEPATHY. They also investigated the subjects' sensitivity to electromagnetic radiation. With one exception, no extrasensory effects were observed that could not be caused by electrostatic charges or convection currents; when these were eliminated, the effects disappeared. The exception was metal bending by contact—stroking—and then only when the strict controls to prevent cuing, conscious or otherwise, or any form of trickery were relaxed. They concluded that they questioned the paranormal nature of all the cases they had examined "in which all the subjects always claimed to be in a good psychic state."

In the second paper, Taylor and Balanovski were more emphatic, concluding that "In particular there is no reason to support the common claim that there still may be some scientific explanation which has as yet been undiscovered. The successful reductionist approach of science rules out such a possibility except by utilization of energies impossible to be available to the human body by a factor of billions. We can only conclude that the existence of any of these psychic phenomena we have considered is very doubtful."

For further reading: John G. Taylor, *Science and the Supernatural* (Dutton, 1979); John G. Taylor, *Superminds* (Viking, 1975); John G. Taylor and E. Balanovski, "Can electromagnetism account for extra-sensory phenomena?" *Nature,* November 2, 1978; and John G. Taylor and E. Balanovski, "Is there any scientific explanation of the paranormal?" *Nature,* June 14, 1979.

TEED, CYRUS REED (1839–1908) A propounder of the fantastic theory positing that the whole universe was contained in the hollow of Earth. Humans, along with all the plants and animals, were thought to be living on the inner surface of the shell. When we are observing the cosmos, we are not looking upward and outward but inwards.

Teed was born on a farm in Delaware County, New York, and in his youth was a devout Baptist. Following service in the Union Army during the Civil War, he attended the New York Eclectic Medical College, and established a practice in Utica, New York.

One night in 1869, Teed was seated in his laboratory pondering on the nature of Earth when, he claimed, a beautiful woman came to him in a vision. She told him of the true structure of the universe; of his, Teed's, past incarnations, and of the role he was destined to play as a new messiah. His reading of the scriptures had already convinced him that God had measured the waters in the "hollow" of his hand (Isaiah 40:12), and he knew this meant that Earth was hollow and everything was on the inside. The scientists were wrong; they had everything inside out. He immediately published a pamphlet detailing the whole vision and the basis of his hollow Earth revelation; *The Illumination of Koresh: Marvelous Experience of the Great Alchemist at Utica N.Y.* (1869).

Writing under the pseudonym Koresh, the Hebrew equivalent of Cyrus, he developed his theory in more detail in *The Cellular Cosmogony* (1870). His theory stipulates that the entire cosmos is like an egg and we live on the inner surface of the shell. Inside the hollow is everything else: the Sun, the Moon, stars, planets, and comets. Nothing is on the outside. The shell is 160 kilometers (100 miles) thick and made up of 17 layers, of which the inner five are geological strata. Outside these are five mineral layers followed by seven metallic ones. The atmosphere is dense, and the Sun at the center of the open space is invisible to us. What we see and know as our Sun is a reflection of the real one that rotates from dark to light, making it appear to rise and set. The Moon is a reflection of Earth, and the planets are reflections of "mercurial discs floating between the laminae of the metallic planes." The rest of the heavenly bodies are only points of light explained by Teed by a complex of optical verbiage. Optics are used again to explains why Earth only appears to be convex. To prove this, an experiment was carried out in 1897 by the Koreshan Geodetic Staff on the Gulf Coast of Florida. In later editions of the book, there are photographs showing researchers using a set of three double T-squares, called by Teed a "rectilineator." His assistants extended a straight line for six kilometers (four miles) until it finally plunged into the sea. For Teed, his theory was an article of faith: To believe in Earth's concavity is to know God and not to believe in Earth's concavity is to deny him.

Teed attacked opponents, declaiming: "All that is opposed to Koreshanity is antichrist." He hated all orthodox scientists, whom he termed "humbugs" and "quacks" and likened himself to great thinkers of the past who had a struggle to get themselves accepted. As time went on and he became more and more obsessed with his ideas, Teed eventually abandoned medicine to become an evangelist. He settled in Chicago in 1886, where he founded a "College of Life" and published a magazine *The Guiding Star,* soon renamed *The Flaming Sword.* Twelve years later *The Chicago Herald* said he had 4,000 followers and that he had collected $60,000 from preaching engagements on the West Coast.

In the 1890s, Koresh bought some land in Florida, founded the town of Estero, Florida. He called it "The New Jerusalem," and told the 200 followers who came to live there with him that it would one day be the capital of the world. He taught that after his physical death he would arise and take up to heaven all those who had been faithful to him. He died in 1908. Taking cognizance of his words and waiting for something miraculous to occur, his followers kept vigil over his body. Finally, when the body started to decompose, the health officer stepped in and ordered a burial; he was entombed on Estero Island off the Gulf Coast.

See also HOLLOW EARTH DOCTRINE (HOHLWELTLEHRE); SYMMES, JOHN CLEVES.

For further reading: Martin Gardner, "Flat and Hollow," *Fads and Fallacies in the Name of Science* (Dover, 1957).

TEILHARD DE CHARDIN, PIERRE (1881–1955) Paleontologist and Roman Catholic priest. Born in Sarcenat, France, he entered the Jesuit order in 1899 and was ordained in 1912. He served as a stretcher bearer in World War I, during which was decorated for gallantry, and received his doctorate from the Sorbonne. Deeply interested in EVOLUTION theory, Teilhard sought to convince the church that it ought to embrace the implications of the revolution begun by Charles DARWIN, which, he argued, did not entail a rejection of Christianity. He met with consistent opposition, however, from ecclesiastical superiors. Teilhard was expelled from his teaching position at the Catholic Institute in Paris in 1926 and was "exiled" to China. In China, he participated in research in geology and paleontology from 1926 to 1946 and was a discoverer of the fossilized "Peking man."

The evolution advocated by Teilhard described the basic stuff of the cosmos, living and nonliving, as continually undergoing irreversible changes in the direction of greater complexity of organization. To Teilhard, this "law of complexification" was as significant as the law of gravity and was illustrated by the vast array of organic forms that have appeared in the history of evolution, the most recent of which being humanity.

Teilhard's theory saw the human being, viewed "from without," as a material system in the midst of other material systems. However, each individual experiences him or herself as a conscious being "from within." Thus consciousness is directly identifiable as "spiritual energy." For Teilhard, not only humans but all constituents of the cosmos possess a "conscious inner face." Physical evolution, then, occurs simultaneously to the evolution of consciousness. The more highly integrated a material system, the more advanced is that consciousness. In human beings, from an "involution" or intense concentration of the brain's cells emerges self-conscious thought, the highest stage reached thus far by evolution.

The human capacity for self-conscious thought and the production of cultures has created what Teilhard called the "noösphere," a layer distinct from, yet superimposed onto, the biosphere. This "thinking layer" provides a unique environment for human beings, thus separating them from all other animals. Making an evolutionary convergence possible, the noösphere will manifest itself as a unification of all human cultures into a single world culture, with a parallel movement toward psychic convergence. Evolution will then reach a terminal phase of convergent integration in a Hyperpersonal Consciousness, "Point Omega." This integration of all personal consciousness will be achieved through love, Teilhard suggested, or, more precisely, through the spirit of Christ at work in nature.

On completion of the manuscript of his major work, *Le Phénomène Humain* (*The Phenomenon of Man*), Teilhard petitioned Rome numerous times but was forbidden to publish. It was finally published after his death in 1959, giving rise to controversies both inside and outside the church. Teilhard's views are not easily reconciled with orthodox Christianity or with scientific theory but have nonetheless stirred the imaginations of theologians, scientists, and philosophers alike.

TELEKINESIS See PSYCHOKINESIS.

TELEPATHY Sending thoughts from one mind to another with no physical mechanism involved; thought transference. Telepathy may be either consciously willed or involuntarily experienced. It involves the transmittal of thoughts, feelings, and images from mind to mind; it does not include CLAIRVOYANCE or true PRECOGNITION. Telepathy comes from the Greek words *tele* ("distant") and *pathe* ("occurrence" or "feeling") and was coined in 1882 by Frederick William Henry Myers (1843–1901), a British psychical researcher and a founder of the SOCIETY FOR PSYCHICAL RESEARCH (SPR). Although telepathy has been described in many cultures around the world and through time, the SPR was the first organization to attempt to study psychic phenomena in a methodical manner.

The SPR, established in 1882 and still operating today, included many prominent persons in its membership, from both humanistic and scientific fields. One of their chief goals was to prove to a doubting scientific establishment that parapsychology is real; to do this, they determined to use scientific methods to test such things as telepathy. Their initial experiments involved hypnotizing subjects and asking them to select a card containing a number or image that the experimenter chose. In the 1930s, U.S. parapsychologist Joseph Banks RHINE performed similar laboratory experiments (with nonhypnotized subjects) using a card set he devised, ZENER CARDS. The 25 Zener cards consisted of 5 sets of 5 different symbols. A sender in one location would select the shuffled, face-down cards one by one and concentrate on sending images of the symbols to a recipient in another location. The recipient would attempt to receive the images in the same order the sender sent them. A success rate over a predetermined good guessing rate would be indicative of telepathy occurring.

By the 1980s, telepathy experiments focused on remote viewing, in which the subjects would be placed in an environment in which their normal senses were blocked; a person at a remote location would attempt to send the subject a mental image of a painting or a place or an object. Parapsychological experiments also used computers to generate random numbers that a sender would attempt to send to a recipient in another room. Very promising results were obtained in these and other experiments.

Still, the scientific establishment generally refuses to accept the possibility of telepathy. Their most common criticisms relate to inadequately stringent care on the part of experimenters. An allegation against the Rhine labs of falsifying data fueled this criticism and has never been overcome in the minds of skeptics. Critics also say that it is too easy for astute recipients to receive clues through their normal senses that allow them to guess accurately what they are supposed to perceive. Finally, although parapsychologists claim to be able to replicate results, a basic criteria of the scientific method, skeptical scientists claim they have not been able to do so.

Interestingly, it has long been rumored that the governments of the former Soviet Union and of the United States have done extensive experiments with telepathy, hoping, it has been said, to acquire secret information psychically and to learn how to exert mind control over enemies. U.S. astronaut Edgar Mitchell did perform an unauthorized telepathy experiment aboard *Apollo 14* in 1971. He attempted to send series of random numbers to four recipients on Earth. He said that two of his recipients significantly exceeded normal expectations, suggesting that telepathy was indeed at work.

See also EXTRASENSORY PERCEPTION; HYPNOSIS; REMOTE VIEWING.

TELEPORTATION The movement of something, often a human, through solid objects or from one place to another through paranormal means. Teleportation was a popular feature of many 19th-century séances. The MEDIUM conducting the séance might disappear from the premises only to reappear later, or someone not initially involved in the séance would suddenly be present, his or her arrival in the darkened room usually accompanied by uncomfortable bumps and sounds. One memorable incident allegedly occurred in London in 1871. Medium and spirit photographer Frederick A. Hudson was conducting a séance when one of his guests asked that a Mrs. Guppy, an acquaintance who lived in a different part of London, join them in the dark, sealed room. In a moment or two, one of the guests felt something brush his head; there was a bump on the table, a couple of people screamed, and, when a match was struck to make a light, a dazed Mrs. Guppy, a very heavy woman, was found to be sitting on the table in her dressing gown, pen and still-damp ledger in hand.

While teleportation probably reached its popular height during the spiritualist craze of the 19th century, cases have been reported in the 20th century as well. One occurred during an extended POLTERGEIST incident in India in 1928. A young boy, the older brother of Damodar Ketkar (the focus person in the poltergeist case), suddenly appeared at a home in a town several miles from where he lived. A witness reported that when she looked up, the boy appeared to be doubled over, suspended a couple of inches above the ground, with arms hanging limply, as though someone had grabbed him by the waist and carried him to her door. Saints have also been subjects of teleportation, as have Hindu and other mystics.

Teleportation is accomplished by means of highly advanced physics in television and movie stories such as *Star Trek*. How is it accomplished in real life? Those with a mystical inclination suggest that spiritual or psychic energy can be manipulated to cause teleportation to take place. Those with a religious inclination call it miraculous. Rationalists say it is accomplished through stage magician tricks. Certainly no one has been able to demonstrate teleportation in a laboratory setting, nor has a carefully investigated incident occurred in recent times.

See also APPORTS.

TENHAEFF, WILHELM (1894–1981) Dutch psychical researcher whose career ended amid charges of large-scale fraud. Tenhaeff joined the STUDIEVERENIGING VOOR PSYCHICAL RESEARCH (SVPR) soon after its founding and in

Teleportation

1928 became a cofounder of the *Tijdschrift voor Parapsychologie.* All parapsychological research stopped during the Nazi occupation, but PARAPSYCHOLOGY was reestablished in the Netherlands soon after the war ended. Tenhaeff emerged as the dominant force in the SVPR. About the same time, he discovered Gérard Croiset, who after a brief array of tests was pronounced a remarkable psychic. Tenhaeff not only tested Croiset but worked with him on sharpening and improving his ability. He also encouraged the use of Croiset's abilities in a variety of practical ways, most importantly in assisting police in solving crimes and finding lost children. Tenhaeff publicized Croiset's successes nationally and internationally.

In 1953, Tenhaeff was appointed to the newly established chair in parapsychology at the University of Utrecht, one of the few full-time academic positions in the field. He was also named director of the University's Parapsychology Institute. Croiset joined him near Utrecht in 1956. Among the more notable of the demonstrations of Croiset's abilities conducted by Tenhaeff were the "chair tests." Croiset was asked to describe the person who would sit in a specific chair at a future meeting. Croiset's predictions were sealed and opened at the meeting and his description compared with the person in the chair.

The last years of Tenhaeff's life were punctuated by a series of controversies. In the 1960s, a number of fellow parapsychologists challenged what they felt was his dictatorial leadership style. Unable to have their complaints acted on, many left the Studievereniging voor Psychical Research and founded the rival NEDERLANDSE VERENIGING VOOR PARAPSYCHOLOGIE. Tenhaeff survived their defection; the following year, he issued his book on PRECOGNITION, describing Croiset's chair tests, and in 1964 saw the publication of journalist Jackson Pollock's very popular volume on his protégé, *Croiset the Clairvoyant.* Tenhaeff was viewed by the media as a herald of a more spiritual world practicing a nonmaterialistic science. He was also known to warn audiences of skeptics whom he associated with Bolsheviks.

Through the 1970s, several critics, most notably Piet Hein Hoebans, began to double-check Tenhaeff's books and papers. It became clear that he had been doctoring data for many years. Data were changed and successes were exaggerated prior to publication. Hoebens also demonstrated that Tenhaeff had misrepresented and even fabricated incidents concerning Croiset. Tenhaeff lived the last years of his life struggling against the accusations, which were picked up by his alienated colleagues. He died in 1981, cut off from his fellow scientists. Not a single parapsychological journal reported his demise. Croiset's reputation also suffered, but most have not accused him of participating in Tenhaeff's fraud.

For further reading: Kendrick Frazier, *Science Confronts the Paranormal* (Prometheus Books, 1986).

THEOSOPHY The doctrine and teachings of the Theosophical Society. The literal meaning of the word is "sacred science" or "divine wisdom."

The society was founded in 1875 by a small group with a shared interest in SPIRITUALISM and OCCULTISM, centered around Madame BLAVATSKY and Col. Henry Steel Olcott. Olcott proposed to the group that they should form a society to study a range of phenomena that intrigued them: spiritualism, PYRAMIDOLOGY, APPARITIONS—in short, anything that was considered outside the range of orthodox science and religion. Olcott was elected its first president and William Judge, a lawyer's clerk, its secretary. Although the society purported to be studying occult and psychic phenomena objectively, many of its members were already committed to a belief in them. This was particularly so in the case of Blavatsky, who was writing *Isis Unveiled* (1877). "Was writing" is misleading; Blavatsky claimed that some of the text was produced by her unconsciously while under the influence of her guiding spirits or Masters, and that much was actually written by these beings overnight while she was asleep. The book was a best-seller.

Blavatsky was very much the society's driving force in its early days, and its aims and objectives were strongly influenced by her.

Theosophy proclaimed the existence of a secret (secret only in the sense of being as yet undiscovered) doctrine. The religions and shared beliefs and values held by different cultures, religions, and societies are just outcrops and distortions of a fundamental universal set. With this improved understanding would come worldwide harmony and brotherhood.

Under the influence of Olcott and Blavatsky, the society was attracted to Hinduism and Buddhism. Olcott and Blavatsky, disappointed by their lack of success in the United States, instructed so they claimed by the Masters, moved to India where they established a base, first in Bombay and later in Adyar, near Madras. The U.S. movement was left in the care of Gen. Abner Doubleday and it secretary, Judge.

The subsequent history of the society, and consequently of theosophy, is one of rivalries and disputes. Blavatsky and Olcott were often at odds, she having an edge because she could always summon the Masters to support her. Blavatsky died in 1891, setting off a battle for supremacy between Olcott and Judge. During the early 20th century, Annie BESANT and Rudolph STEINER had a considerable impact on theosophy in Britain, Europe, and India. In 1907, Olcott died and was succeeded as international president by Besant. Five years later, following a prolonged dispute with Besant, Steiner resigned and founded the Anthroposophical Society.

Steiner's break with Besant was not merely over power, though that was a factor. He disagreed basically with the orthodox religion—for religion it had become—of the Theosophical Society. In his view, it was not enough to emphasize the divergence in the West between religion and materialistic science and return to a more oriental attitude that united the two; restoring such a unity required sound and thorough scholarship. Furthermore, it was necessary to establish a spiritually based science teaching that was acceptable to western scholars and was based on Western traditions. The Anthroposophical Society's schools, clinics, bank, and various activities continue to this day with considerable success.

Mainstream theosophy under Besant, based at Adyar, gave its support to the movement for Indian Home Rule and in that respect could claim to have been successful. In Britain in the 1930s, it was associated with pacifism and the Peace Pledge Union. Nevertheless, although theosophy still exists, and still subscribes to the same basic philosophy, it is only a shadow of its former self. The British and European branches have declined to virtual invisibility, having split into numerous splinter groups. The U.S. movement has also suffered several schisms and no longer looms large in a New Age era but still survives, although largely overtaken and absorbed into New Age thinking.

For further reading: Helen Blavatsky, *Isis Unveiled* (Boulton, 1877); Helen Blavatsky, *The Secret Doctrine* (Theosophical Publishing, 1888); Joscelyn Godwin, *The Theosophical Enlightenment* (SUNY Press, 1994); Gordon Stein, ed., *The Encyclopedia of the Paranormal* (Prometheus, 1996); and Peter Washington, *Madame Blavatsky's Baboon: Theosophy and the Emergence of the Western Guru* (Secker & Warburg, 1993).

THOM, ALEXANDER (1894–1985) A professor of engineering science at Oxford University who used his professional descriptive and analytical skills to examine the hundreds of prehistoric stone circles in England and

Scotland. He later extended his surveys to include the many ancient megalithic sites in Brittany. Thom carefully mapped the positions of the stones in more than a hundred of the circles and then tried to explain how and why they came to be the way they are. A common pattern was a combination of a precise semicircle with a distorted semicircle; these inaccurate semicircles formed six groups that could be reproduced by simple geometrical methods. Two groups appeared to be attempts at constructing a circle, using a value of 3 for pi, instead of the more accurate value we use today: 3.14159 . . . . . . Thom also showed that the distorted semicircles in one group were ellipses and concluded that the Britons of 4,000 years ago must have had a better knowledge of geometry than we had hitherto assumed. He also found that they had had a standard measure of length, a unit he called the MEGALITHIC YARD, equal to 82.9 centimeters (2.72 feet).

Thom made careful measurements of the alignment of many of these ancient monuments, in particular of STONEHENGE, and checked them against calculations of astronomical data at the time of their construction. The remarkable correlations he was able to obtain led him to suggest that some, at least of the circles, had been early forms of observatories. This idea has been taken up and elaborated by Sir Fred HOYLE in his book *Stonehenge* (Heinemann, 1977).

For further reading, see Alexander Thom, *Megalithic Sites in Britain* (Oxford University Press, 1967); Alexander Thom, *Megalithic Lunar Observatories*, (Oxford University Press, 1971); and Alexander Thom and A. S. Thom, *Megalithic Remains in Britain and Brittany* (Oxford University Press, 1978).

THOMPSONISM A school of natural medicine popular in 19th-century United States. Samuel Thompson was born in rural New Hampshire. As a youth, he was intrigued by the herbs and medicines used by an elderly woman, the village herbalist. He eventually married, settled down as a farmer, and raised a family. When one of his sons developed scarlet fever, Thompson resorted to steam inhalations and lobelia. His son recovered, and as a result Thompson gave up farming and became an itinerant herb doctor and salesman of the herbal remedies that he had studied.

Thompson had his remedies patented in Washington, and by 1806 he was marketing them through the Friendly Botanic Company. He eventually settled in Massachusetts, where he was opposed by other, more traditional doctors. The death of one of his patients led to his being tried for murder. By the time he was found not guilty, he had become a celebrity. He relocated to Boston and continued to develop his herbal company. Throughout the 1820s, he had salesmen across the country distributing kits of his products. Thompsonism flourished until his death in 1843, after which his practice was absorbed into ECLECTICISM, the continuing school of natural medicine.

See also HERBAL MEDICINE.

For further reading: Morris Fishbein, *Fads and Quackery in Healing* (Blue Ribbon Books, 1932).

THOUGHTOGRAPHY Sometimes called skotography, a form of PSYCHIC PHOTOGRAPHY where the practitioners claim the ability to project, by paranormal means, the image of their thoughts onto a film inside a camera.

The first recorded thoughtographs, then called psychographs, were made by the psychic and spiritualist Duguid in 1878. Between 1920 and 1940, Madge Donohue produced thousands of thoughtographs by putting a plate under her pillow at night and developing it the next morning to reveal her dreams of ancient Egyptian and Greek historical images. Another MEDIUM, Margaret E. W. Fleming, between 1956 and 1976, claimed to make pictures by holding a sheet of photographic paper to her forehead, exposing it briefly to a lightbulb, and then developing it in the normal way. In Japan, it was claimed that certain Buddhist monks could create mental images on film and even erase other images on conventionally exposed film. A Japanese

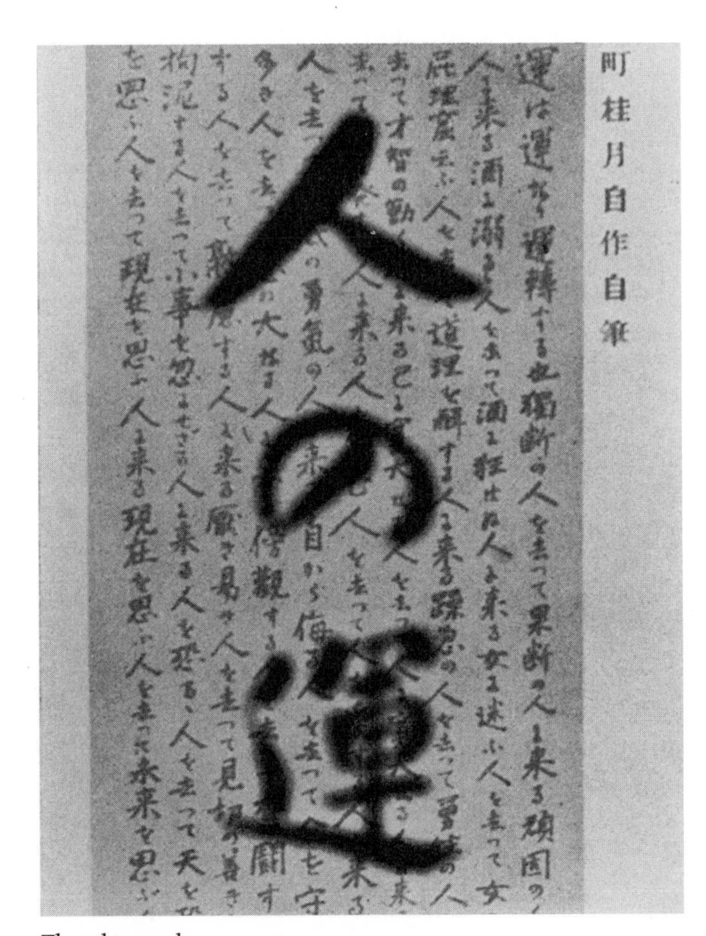

Thoughtography

medium, Masuaki Kiyota, was said to be able to create exposures in sealed Polaroid cameras.

The best-known thoughtographer of recent times was Chicago hotel bellhop Ted Serios who had mastered a technique of working with a Polaroid camera and a small 2.5-centimeter (1-inch)-long paper tube he called a "gizmo," which he placed near the lens of the camera. This, he explained, was just to help him concentrate. Serios became so famous and well thought of that he was called in to help search for the lost submarine *Thresher.* Instead of an image of the submarine, Serios produced a picture from the coronation of Queen Elizabeth, but he managed to talk his way out of it by explaining that the fuzzy image was a symbol of the submarine. Nevertheless, in 1967 *Scientific American* was interested enough to send a team of magicians and photographers to investigate Serios and his psychic claims. One of those present was David Eisendrath of *Popular Photography* who later wrote about the encounter. Unfortunately for Serios, he was caught inserting something into his gizmo but refused to let anyone examine it. The team concluded that sleight of hand was used to put something into the gizmo, probably a low-power lens and a small transparency bearing the image he wished to impress on the film. Placed before a Polaroid camera lens that was focused at infinity, this would have produced a distorted image on the film.

The famous spoon-bender Uri GELLER produced thoughtographs by holding a capped 35-millimeter Nikon camera either against his forehead or at arm's length. Critics of Geller said the whole incident was a publicity exercise.

Thoughtography is trick photography. If it were not so, all the rules of physics and optics developed by science over the past several centuries would have to be rewritten.

TIME Its measurement and the effect it has on many aspects of human activity, the days, the seasons, the years, is important to everyone. Some aspect of this subject has engaged human attention throughout history and probably throughout prehistory. We experience it in our daily existence, through the year, in the cycle of years, and throughout our whole lives. While sometimes we feel that time hangs heavily on our hands, that it passes slowly, and that yet it sometimes passes in a flash, we know rationally that it passes uniformly, at a steady pace. Our clocks and watches tell us so.

For many years, the government of the United Kingdom offered a handsome reward to whoever could build a reliable ship's chronometer. To have such an instrument on board ship made navigation on the open seas possible, more reliably than by dead reckoning. The combination of chronometer and sextant enabled mariners to position themselves accurately in the middle of the ocean, while out of sight of land. That assumed both that the chronometer recorded the passage of time uniformly—our shipboard measure of time—and that the sun passed over the heavens uniformly—our celestial measure of time.

We know intellectually that time proceeds steadily and *in one direction;* it never goes in reverse, nor does it jump ahead or jump back. There are no hiccups in time. Those are judgments that are based on our experience and on which we rely for the safe and dependable conduct of our lives. But the human imagination being what it is, we cannot help asking "What if . . .?" We do that at all sorts of levels, from the idle speculation, daydreaming, through science fiction, to the quasi-academic time societies, to the theoretical physicists exploring time reversal. Somewhere in that spectrum come the paranormal experiments with retrocognition and PRECOGNITION: the ability to recall things long gone—happenings, people, and events from well before our time—and the ability to foresee events yet to come.

There are both national and international societies for the study of time with members from across the arts, the social sciences, and science. They hold annual national conferences and triennial international conferences. The societies publish journals and maintain an on-line annotated bibliography of time-related works. The sort of questions that these societies address are:

- Did time have a beginning and will it have an end?
- Are all life-forms coordinated systems of biological clocks?
- Will globalization of communication influence our individual and collective time horizons?

Some of the questions that concern the societies are of especial interest to theoretical physicists. Equations that they devise allow the theoretician to put in negative time, time running the other way to our experience. In one sense, this has been done for many years; for example, we have substituted time going backward into our equations for the solar system to locate the position of the planets hundreds or thousands of years ago on a particular date. But the current exploration is something more than that; it asks what the universe would be like if time ran in the other direction and whether such a universe could exist. A similar question can be asked at both the fundamental particle level—the very minute—and at the cosmological level—the enormous. Whether this has any significance for us at any practical level is a moot point. Mathematical equations are models, models that help us to understand reality, not rigid formulas that nature must obey.

The special theory of relativity places time in a four-dimensional space–time, space and time being inextricably linked, and shows that time will slow down as the observer's speed approaches the speed of light (time dilation) and also raises some paradoxical questions. Theoreticians have also asked whether time is continuous or whether, like matter and energy, it is discrete—coming in small packets or time quanta—which they estimate might be about 5 billionth billionth millionths of a second (5×10^{-24} secs.)

At the level of the cosmos the big-bang theory, which supposes a beginning to the universe in a colossal explosion, raises the question: If time began at that explosion, what happened before? Or does the verb "happen" have any meaning in that context?

All the foregoing assumes a physical and technological view of time, but the time societies take a much broader view. Time concerns musicians, economists, geographers, psychologists, writers, biologists, historians, sociologists, philosophers, and anthropologists, each discipline bringing a different outlook and asking different questions. A major concern of the time societies is to find links between these different concerns. The Association for the Social Studies of Time in Britain, for example, states its primary function as bringing together those who might not otherwise meet despite their shared interest in time as a subject of inquiry. It has two main foci: first, temporal patterns of human behavior and the assessment of their sociological, economic and political significance; second, the variety of conceptions and perceptions of time that can be found among different people in any one society and in different cultures.

For further reading: J. T. Fraser, *Time, the Familiar Stranger* (Tempus, 1987); *The Study of Time,* Vol. I–IV (Springer-Verlag, 1972–81), Vol. V (University of Massachusetts Press, 1986), and Vol. VI–VIII (International Universities Press, 1989–97); J. Thewlis, ed., *Encyclopaedic Dictionary of Physics,* Vol. 7 (Pergamon, 1962); and Michael Young, *The Metronomic Society* (Thames and Hudson, 1988).

TIME TRAVEL The idea that humans might be able to travel backward or forward in time, an idea that has been of occasional interest, particularly during this last hundred years. H. G. Wells (1866–1946) in *The Time Machine* (1892) imagined a machine that could transport its occupant forward or backward through time and deposit him or her in what would inevitably be an alien culture. Wells puts this thought into the mind of the time traveler as he is about to alight in the year 802,701 C.E.:

> *What might not have happened to men? What if in the interval the race had . . . developed into something inhuman, unsympathetic, and overwhelmingly powerful? I might seem some old-world savage animal, only the more dreadful and disgusting for our common likeness—a foul creature to be incontinently slain.*

At the time of its publication, Wells's fiction was greeted both with interest and ridicule. *The Pall Mall Gazette* invited him to say how he thought humankind might develop in the next million years, and *Punch* satirized his ideas in an anonymous poem "1,000,000 A.D." But in the world of science fiction, the idea of time travel undoubtedly caught on, and many writers have used it during this century.

Is there any practical possibility of travel either backward or forward through time? Apart from the problem of how to go about constructing a time machine, the idea raises difficult questions. If you traveled to an earlier date, would you then be able to change things to alter your subsequent existence favorably? Or might you, by killing your grandfather for example, make your subsequent existence impossible? Suppose instead that you traveled forward, this time for just a few days—sufficient time to see the outcome of an event. You might then change your behavior. But might what would be beneficial to you hurt others?

Time travel can only take place in fiction. It can be used, as Wells used it, as an instrument of social comment to affect social policy; in that role, it *will* be used by the doomsayers to predict disaster unless we change our ways and by optimistic futurologists to paint a golden picture of the future, generally based on some technological revolution.

Other science-fiction writers have explored time travel in different ways. In early stories, it was generally achieved by unexplained means, as in Washington Irving's *Rip Van Winkle* (1820), Charles Dickens's *A Christmas Carol* (1843), Edgar Allan Poe's *A Tale of the Ragged Mountain* (1844), and Mark Twain's *A Connecticut Yankee in King Arthur's Court* (1889). More recently, there has been science fiction in which time travelers accidentally or deliberately alter the course of history, as in Ray Bradbury's *A Sound of Thunder* (1952), in which the transforming event is the premature death of a butterfly in the Mesozoic Era (65 million to 225 million years ago). An interesting variation on this idea is Alfred Bester's *The Men Who Murdered Mohammed* (1958), whose protagonist attempts to effect changes in what he assumes to be *the* time line, the sequence of future events. He succeeds only in annihilating himself, there being as many individual time lines in existence, "like millions of strands of spaghetti," as there are individuals. Robert A. Henlein's *By His Bootstraps* (1941) is an example of the chicken-or-egg type of time-travel story, also known as the closed loop. The hero, having been propelled forward in time, initiates the succession of events by which he was propelled forward in time.

There is, in a sense, another sort of science fiction, that engaged in by theoretical physicists who speculate "what if." They regularly produce equations that describe the experimentalist's world and then ask: What if we substitute this in the equations instead of that, a negative quantity instead of a positive one, an imaginary quantity (an imaginary quantity in the physicists' and mathematicians' world is a quantity containing the square root of minus one, for which there is no real solution) instead of a real one? Or if we make time run backward? The results of these "what ifs" are many varieties of fantasy worlds and universes, worlds in which all objects travel faster than light, for example; in the real world, no object can reach, let alone exceed, the velocity of light in a vacuum. In such worlds, time travels backward. Universes are imagined to be composed entirely or primarily of ANTIMATTER, and time again flows backward.

For the time being, however, we are stuck with the universe in which we live; in this real world, time travel, forward or backward, whether as imagined by science-fiction writers or by theoretical physicists, remains just that: a figment of the imagination.

See also SCIENCE FICTION AND SCIENCE.

For further reading: H. G. Wells, *The Time Machine* (Pan/Heinemann, 1892).

TITANIC British steamship that sank on the night of April 14–15, 1912, while on its maiden voyage. Launched on May 31, 1911, *Titanic* was the largest ship of its time, 882.5 feet long, 104 feet high, with a displacement of 66,000 tons, and a top speed of 25 knots; it could carry up to 3,000 passengers but had lifeboats for only 1,178. *Titanic*'s owners, the White Star Line, made much of both the ship's luxurious first-class accommodations and its compartments, which were supposed to make the ship virtually unsinkable.

The *Titanic* left Southampton bound for New York on April 10, 1912, with more than 2,200 people aboard. At 11:40 P.M. on Sunday, April 14, while steaming at high speed about 350 miles southeast of Newfoundland, in casual defiance of several radio reports of pack ice and icebergs, the ship suffered a collision with a gigantic iceberg that buckled plates along a third of the length of the hull. Thirty minutes later, Captain E. J. Smith ordered the lifeboats launched, but in the general confusion—and as a direct consequence of the preferential treatment accorded first-class passengers—the already inadequate lifeboats were not filled to capacity. Thus, when *Titanic* began its bowfirst plunge at 2:20 A.M., more than 1,500 people were still aboard. The steam ship *Carpathia* arrived on the scene two hours later and rescued 705 persons who were in lifeboats and in the water. The wreck was discovered on September 1, 1985, by a joint U.S.–French expedition.

The sinking of the *Titanic* is said to have been forseen by many psychics. The "evidence" for these paranormal predictions consists of several works of fiction and poetry written between 1890 and 1912, each concerning a great passenger liner that sinks after a collision with an iceberg. In Morgan Robertson's novel, *The Wreck of the Titan,* published in 1898, the ship sinks after striking an iceberg in the North Atlantic in April, and an inadequate complement of lifeboats results in great loss of life. The coincidence is less remarkable when it is considered that, before World War I demonstrated the efficacy of the self-propelled torpedo, icebergs were regarded as practically the only things capable of sinking larger passenger liners. Stories and poems in which such vessels collide with icebergs necessarily had to be set in places and at time where collisions with icebergs could reasonably be expected to occur.

See also PRECOGNITION.

TOFT, MARY (1701?–1763) A woman who claimed to have given birth to rabbits. Housewife Mary Toft emerged out of obscurity in April 1726 in Godalming, Surrey, England. She told a curious story of having been working in the field on St. George's Day when a very large rabbit suddenly appeared and sexually assaulted her. The bizarre nature of the tale was offset by her reputation in the village as a level-headed woman who was not prone to telling such stories. Villagers speculated that the rabbit was some kind of DEMON released for a few hours on St. George's Day.

The story apparently ended at that point, but five months later Toft became ill. The physician who examined her reported feeling what appeared to be something alive in her. He saw her regularly during the next month until, he later reported, she gave birth to five rabbits. His account was published far and wide. A short time later, the physician announced that seven more baby rabbits had been born. Two London physicians who had denounced the whole affair then went to Godalming and observed (or thought they observed) two further rabbits born. Another prominent physician came to Godalming soon afterward and was there for the birth of a 15th rabbit.

At this point, the King stepped in and appointed Sir Richard Manningham to make a final determination on the nature of the affair. Manningham brought Toft to London and kept her under constant observation. No more rabbits were born. Manningham concluded that Toft was a fraud. The rabbits had been brought to Toft by her husband, and she hid them on her person. She was then able to place them so that they appeared to have emerged from her vagina.

Mary Toft's case became the inspiration for the standard conjurer's trick of pulling a rabbit out of a hat.

For further reading: R. Nathaniel St. Andre, *A Short Narrative of an Extraordinary Delivery of Rabbits* (n.d.); and Richard Alfred Davenport, *Sketches of Imposture, Deception and Credulity: An Exact Diary* (T. Tegg & Son, 1837).

TRANSCENDENTAL MEDITATION Usually shortened to TM, a technique to help MEDITATION. "Transcendental" means extending beyond the bounds of any worldly category and is used in the philosophy of religion to describe

Toft

a state of being that is beyond the reach, or apprehension of, the experience of our senses. Doctors have come to believe that certain forms of meditation can be useful in controlling stress and that the TM movement markets a very effective and simple technique. TM claims that their meditation state is a physiologically and psychologically distinct form of consciousness, and as part of its claim for acceptance, it encourages scientific investigation. The technique involves the meditator sitting comfortably, with eyes closed, and mentally repeating a Sanskrit word or sound, a mantra, for 15 to 20 minutes twice a day.

The founder and spiritual leader who brought the Hindu doctrine of TM to the West in 1958 was the Maharishi Mahesh Yogi. But it was not until the British pop group the Beatles became followers of the maharishi at the end of the 1960s that the TM movement took off, with many of the young fans of the Beatles following their example. The maharishi appeared on television and was seen throughout the Western world sitting in the lotus position, surrounded by flowers, his long white hair falling around his face, looking like a depiction of God from a children's Bible.

Followers were initiated by being given a secret mantra (a sacred word or phrase). They were then allowed to proceed through a series of courses, which cost them a considerable amount of money. In exchange, they were promised to be able to make themselves invisible, to walk through solid objects, and to do what TM called "flying"—moving forward while in a cross-legged lotus position, over a mattress spread on the floor, in a series of high bunny jumps. These claims go far beyond what could be considered sensible expectations and would actually break the laws of the physical universe.

The TM movement seemed more sensible when it entered the field of ALTERNATIVE MEDICINE by promoting AYURVEDIC MEDICINE, a system that is based on a traditional Indian approach that includes, as well as meditation, certain purification procedures, rejuvenation therapies, herbal and mineral preparations, exercises, and diet. Ayurvedic medicine purports to be a science of life; it teaches that concentration and control of the thought processes by meditation can calm the mind and slow down harmful bodily processes, and so help to control blood pressure and lower heart and pulse rates. There is no doubt that susceptibility to stress-related disorders and mental problems can be reduced by this practice. Many traditional doctors agree with TM's emphasis on looking after general health rather than curing the disease after it has developed. *Ayurveda* means life knowledge. Many traditional doctors also believe that meditation may temporarily relieve stress but warn that not all illnesses are symptoms of stress and that those following the regime must never fail to consult their doctors if they suspect that their symptoms are physiological and potentially life threatening.

The movement is more like a new religion than a technique for meditation, and it promises not only to improve the well-being of the individual practitioner but also the state of society in general. On the social side, proponents claim that thinking in unison by their followers has diminished the rate of crime and civil disorder in certain Western cities and brought about the fall of the Soviet Union. Both claims are impossible to prove, but, nevertheless, the movement has been accused of meddling in politics.

For further reading: P. Russell, *The TM Technique: An Introduction to Transcendental Meditation as the Teachings of Maharishi Mahesh Yogi* (Routledge & Kegan Paul, 1976); and Anthony Storr, *Feet of Clay: A Study of Gurus* (HarperCollins, 1996).

TRANSCUTANEOUS ELECTRICAL NERVE STIMULATION See ELECTROPHYSIOLOGY.

TRANSMUTATION OF ELEMENTS Changing one element into another.

Plato defined four elements—the earth, air, fire, and water—basic constituents of the universe. He identified each with a regular solid figure—the earth with the cube, air with the octahedron, fire with the tetrahedron, and water with the icosahedron. Any of the last three could be broken into its equilateral triangle faces and reassembled as one of the others—transmutation of the elements. Aristotle had a similar theory but dispensed with the solids, replacing them with hot and cold, wet and dry. Earth was composed of cold and dry, water of cold and wet, air of hot and wet, fire of hot and dry. Again the constituents of one could be reassembled to form another. Evaporating water, for example, was explained by adding heat to cold and wet (water) and turning it into hot and wet (air).

This theory was adopted by the Arabs, who brought it into Europe and was accepted by medieval alchemists, who wished to change other metals into gold. A number of scientists, including Philippus PARACELSUS (1493–1541) and Sir Isaac NEWTON, spent much time searching for the philosopher's stone, which was necessary for the transmutation.

Since then our definition of what constitutes a chemical element has changed radically to the now familiar list of hydrogen, helium, lithium, and so on to the top of the periodic table of 60 years ago, gold (79), mercury (80), and on to uranium (92). From the beginning of the century physicists and chemists had known that transmutation of elements took place spontaneously in radioactive decay, several naturally occurring radioactive heavy elements changing by a succession of decays to lead. By the 1930s they had established that the atom of an element consisted of a positively charged nucleus surrounded by a cloud of negative electrons. The nucleus consisted of electrically positive protons and uncharged neutrons. To transmute an element meant changing the number of protons in the nuclei of its atoms. That happened spontaneously in the naturally radioactive elements and could be made to happen artificially by bombarding a substance with a beam of protons or neutrons, neither an easy nor an economical operation. In 1938 and 1939 the fission of uranium was discovered: Uranium atoms were induced to split into two much smaller atoms with the release of a lot of energy. Atomic bombs and atomic energy were on their way. With them and the accompanying explosion of research, the transmutation of elements became a daily reality.

In 1941 Kenneth T. Bainbridge transmuted mercury, with 80 protons in its nucleus, into gold, with 79. It is ironic that it should have been mercury that was first changed to gold. Aristotle 2500 years earlier had seen all metals as being some mixture of mercury and sulfur; the alchemists had identified mercury as the route to gold, probably because the process of amalgamation with mercury had been used to separate gold from crushed ores. Mercury had been changed to gold (without a philosopher's stone) but not at a practical or commercial level.

The equipment now available made possible a whole range of transmutations, some of which do have practical applications. In nuclear reactors, or by using particle accelerators, elements heavier than uranium can be synthesized: neptunium (93), plutonium (94), up to an element produced as recently as February 1996 and not yet named, with an atomic number of 110. Many of the fission products, the smaller atoms that fission produces, are radioactive versions of the naturally stable elements. Stable elements can be exposed to huge numbers of neutrons in nuclear reactors and made into radioactive versions. These and the fission products have medical, scientific and technological applications.

TRENCH, BRINSLEY LE POER (1911–) A British author of several books, beginning in the early 1960s, centered around a theme of visiting aliens (both ancient and modern), spaceships, interplanetary travel, and so on, using biblical references to back up these speculations. So, for example, in *The Sky People* (1960), he placed the Garden of Eden on another planet and identified Noah's Ark as a spaceship manned by aliens who were rescuing humans (the result of one of their biological experiments—hybridizing apes) from possible extinction. Such flights of fancy were really science fiction, but, together with many similar extravaganzas by Erich VON DÄNIKEN, Immanuel VELIKOVSKY, and others, they struck a chord in the popular imagination and were treated far more seriously than the evidence warranted.

See also SCIENCE FICTION AND SCIENCE.

For further reading: Brinsley Le Poer Trench, *The Sky People* (Neville Spearman, 1960).

TROY, ANCIENT Site of the most renowned battle in ancient literature. To classical readers of Homer, Ilios, Ilium, or Troy—the focus of the poet's *Iliad*—was a real place, still part of the ancient world 7,000 years after its fall. Herodotus the historian asked Egyptian priests for sources that might supplement Homer's account. Thucydides, historian of the Pelopponesian War, writing in the 5th century B.C.E., offered a plausible account of how the Trojan War might have happened. Alexander the Great, before beginning his campaign to conquer Asia, sacrificed to the spirits of Homer's dead warriors. In early Roman times, a town named New Ilion was founded on what was popularly believed to be the former site of Troy. Even the Roman Emperor Julian the Apostate made a special pilgrimage there in 354 to discuss the history of the site with the local Christian bishop.

By the 18th century, however, many believed that the story of Troy was the product of Homer's imagination. Scholars like Jacob Bryant argued that the Trojan War had never occurred and that Troy itself never existed. Others, including the French researcher Jean-Baptiste Lechevalier and the English poet Byron, argued just as vehemently that both the city and the war were historical. Their arguments fired the imaginations of later thinkers, including a German-born American businessman named Heinrich SCHLIEMANN.

During the summer of 1868, Schliemann and Frank Calvert, the British consul in the Dardanelles, explored the northwest coast of Turkey, looking for signs of the site of ancient Ilios or Ilium. Schliemann became convinced—as Charles Maclaren believed 50 years before—that the ruins of Troy lay beneath a hill called Hissarlik on the Plain of Troas near the Menderes River, close by the Hellespont. Beginning in 1871, over a 22-year period, Schliemann and his associates excavated Hissarlik, uncovering nine layers of strata dating back to the fourth millennium B.C.E. In so doing, Schliemann set to rest the question of whether Troy was a real place.

Indeed, Schliemann would become the discoverer not just of ancient Troy but also of Mycenae. (And he might have rediscovered the ancient Minoan civilization on Crete, but for the fact that he was put off digging there by the price the Cretans demanded for their land.) In Schliemann's day, archaeology was still in its infancy. Schliemann undertook four separate excavations: in 1873, 1876, 1882–83, and 1888. Despite the fact that he sought advice from seasoned professionals like Wilhelm Dörpfeld (who had excavated Olympia, joined Schliemann in his final excavations, and continued to work on Troy after Schliemann's death), Schliemann sometimes allowed his enthusiasms to warp his conclusions. He left two lasting impressions of amateurism: one was the great trench he dug across the mound of Hissarlik, destroying in the process some remains of the very period he was looking for; the other was his handling of the golden artifacts he dubbed the "Treasure of Priam," recovered from the site of Troy II (much more ancient than Homer's Troy) and which he illegally transported out of Turkey, thus risking his whole project. Schliemann mistakenly believed that the layer called Troy II contained the Homeric city, which turned out to be located in Troy VIIa. And certain aspects of Schliemann's account of the recovery of the treasure were, in fact, fabrications; his wife Sophie (who, according to his report, helped him recover the treasure, was in Athens on the night it was found). Despite his many faults, however, modern archaeologists and historians of ancient Greece recognize their debt to Schliemann for the depth of his imagination and the strength of his commitment to the dream of rediscovering Troy.

For further reading: Heinrich Schliemann, *Ilios: The City and Country of the Trojans* (1880; repr. Ayer, 1989); William Calder & David A. Traill, eds., *Myth, Scandal & History: The Schliemann Controversy* (Wayne State University Press, 1986); C. W. Ceram, *Gods, Graves & Scholars: The Story of Archaeology,* trans. E. B. Garside & S. Wilkins (repr. Vintage Books, 1986); Leo Deuel, *Memoirs of Heinrich Schliemann* (Harper & Row, 1977); David A. Traill, *Schliemann of Troy: Treasure and Deceit* (St. Martin's Press, 1995).

TUNGUSKA EVENT The 1908 devastation of hundreds of square miles of dense forest near the remote Tunguska River valley in Siberia. Faint aftershocks were felt around the world, and people 80 kilometers (50 miles) and more away saw a gigantic "pillar of fire" from the explosion that caused the devastation. Its cause remains a mystery. It occurred in an inaccessible region at a time when Russia was in a state of prerevolutionary chaos. The government could not mount a scientific expedition to the site until nearly two decades later. Then Leonard Kulik, a Soviet mineralogist, took a crew to the valley and searched for the evidence that would support his theory that a large meteorite had crashed into Earth. He was unable to find sufficient evidence, although he did find a small amount of meteoritic dust and a few small craters. Most puzzling was that he found no large crater, as a meteorite that caused such massive damage would certainly have created. Also, the destruction was less at the center of the explosion than in the surrounding area. The trees seemed to be burned from above and lay like matchsticks, facing outward from the epicenter of the explosion. This does not fit the pattern of a large extraterrestrial stone striking Earth.

Since that time, many theories have been suggested, including a black hole striking Earth and a nuclear explosion from an extraterrestrial spaceship. Most plausible is that it was an extraterrestrial body—most likely a comet or an asteroid—that exploded in the air before it could strike Earth. Most of these space bodies do burn up when they enter Earth's atmosphere, with only a few maintaining enough mass to strike and make a crater. (Meteor Crater in Arizona is one of the rare craters caused by a large body striking Earth; most craters are quite small.)

TWINS The study of which, particularly using identical twins, affords a way of separating the effects of nature and nurture (see NATURE-NURTURE DEBATE).

The use of identical twin studies by scientists in several disciplines has proved an effective way of distinguishing between possible causes. The most well known of these is the attempt to separate nature and nurture in intelligence. Do identical twins, separated at birth and brought up in different environments and cultures, nevertheless score the same IQ when tested years later? Or have their intelligences developed differently because of their different circumstances and the cultural influences on them? The jury is still out and advocates of both sides dispute the significance of the evidence, sometimes bitterly.

Twins

However, twin studies are not limited to the battlefield of intelligence test disputes. Identical twins form the ideal control for any experiment. For example, a recent study on osteoporosis used twins to check whether the condition was inherent or was caused by the patient's history—work, accident, health, etcetera.

Twin studies are a useful way to differentiate the effects of nature and nurture. But, as the case of Sir Cyril BURT shows—incidentally still disputed—they have to be conducted (and interpreted) with caution and rigor, since results can be misinterpreted or manipulated.

See also INTELLIGENCE TESTS.

For further reading: Ronald Fletcher, *Science, Ideology and the Media: The Cyril Burt Scandal* (Transaction, 1991); L. S. Hearnshaw, *Cyril Burt, Psychologist* (Hodder and Stoughton, 1979); and Leon Kamin, *The Science and Politics of I.Q.* (Lawrence Erlbaum, 1974).

TYCHONIAN SOCIETY The most conservative wing of modern creationism. The Tychonian Society supports GEOCENTRISM, the belief that Earth is stationary and that the sun revolves around it. Such a position is a logical extension of a very literal reading of such biblical stories as that of Joshua making the sun stand still (see Joshua 10:12 and Psalms 93:1). Members of the society argue that astronomical data, properly interpreted, support their position. The society was founded in the 1980s by Gerardus Bouw, an instructor at Baldwin-Wallace College, and named for Tycho Brache, the 16th-century astronomer who insisted that the sun revolved around Earth. It is a relatively small organization whose opinions are felt by many to hold the creationist movement up to ridicule.

See also CREATION SCIENCE; SUN STOOD STILL, THE DAY THE.

For further reading: Gerardus D. Bouw, *With Every Wind of Doctrine: Biblical, Historical, and Scientific Perspectives of Geocentricity* (Tychonian Society, 1984); and Raymond A. Eve and Francis B. Harrold, *The Creationist Movement in Modern America* (Twayne Publishers, 1991).

TYCHONIC SYSTEM A model of the planetary system named after Tycho Brahe (1546–1601). Brahe was a Danish nobleman and an energetic and meticulous astronomer, who gathered much accurate data on the planets which, on his death, passed to Kepler. Brahe lived at a time when the Catholic Church was debating the admissibility of Copernicus's theory, which put the sun at the center of the planetary system, the Earth being relegated to being just one of the planets. To fly in the face of the Church's authority was not wise, as the case of Giordano Bruno showed: he was executed after seven years of impenitent imprisonment and torture. Tycho Brahe decided that discretion was the better part of valor and chose a middle way. His system placed the Earth at the center, with the sun and the moon traveling in orbits around it; the remainder of the planets orbited around the sun. This accepted the focal position of the Earth; in Brahe's peculiar subterfuge it did remain the center of the whole system. At the same time his model gave Brahe the convenience of the Copernican system, the planets orbiting around the sun, for observation and calculation.

Although this system (the Tychonic System) is named after Tycho Brahe, he was not its first or only advocate. The system appeared in Brahe's *Progymnasmata* in 1588. Two Greek astronomers, Heracleides in the fourth century B.C. and Seleucus two centuries later, had advanced a heliocentric theory. Christopher Rothmann had arrived at the same compromise a few years earlier than Brahe, and Nicholas Reimer in 1597 published *De Astronomicis Hypothesibus* attacking Brahe roundly for stealing his idea.

As so often is the case, important ideas in science seldom derive from one person in isolation. That this system is attributed to Tycho Brahe is probably because he is seen as having provided an essential step on the road to a synthesis of celestial and terrestrial physics. Copernicus advanced, somewhat timidly, a heliocentric system. Brahe went half way towards accepting it but, at the same time made a huge number of accurate astronomical observations. Building on Brahe's legacy, Kepler showed the planets' orbits to be elliptical and formulated his three laws. Sir Isaac NEWTON put three and three together and showed how a law of universal gravitation made sense of both earthly and heavenly motion. So Brahe is honored by giving his name to the halfway house planetary system—a bit like placing a plaque on a house saying, "Such and such a person lived here." So did a lot of other people.

See also HELIOCENTRISM; TYCHONIAN SOCIETY.

U

UFOLOGY The study of the group of phenomena classed as unidentified flying objects (UFOs). The idea that strange objects seen in the sky might be a significant phenomenon for study arose out of the flying-saucer craze that began in 1947. Most cases could be explained away as misperceptions of natural objects or frauds, but those who took the phenomena seriously maintained that a small proportion of cases could not be dismissed so easily. The study raises questions about how best to use non-expert testimony about transient phenomena. In a popular culture that seeks to link the phenomena to extraterrestrial visitors, such a study can readily be made to look disreputable.

The study of the evidence for, and of explanations of, these observations—ufology—has generated a huge mass of literature, and organizations have been founded in several countries to foster such studies, for example the United States's J. ALLEN HYNEK CENTER FOR UFO STUDIES and the British UFO Society.

The continuing controversy over UFOs can be seen as having three parties: the party represented by the popular press, latching onto any claim, however absurd, and blowing it up to unwarrantable proportions; the serious investigators, such as those in the Hynek Center, who dismiss 95 percent of the claims of sightings as identifiable flying objects, hoaxes, and frauds; and the skeptics who remain unconvinced that there is anything to explain.

The position of the skeptics was put well by Hudson Hoagland writing in *Science* in 1969: "The basic difficulty inherent in any investigation of phenomena such as those of UFOs is that it is impossible for science ever to prove a universal negative. There will always be cases which remain unexplained because of lack of data, lack of repeatability, false reporting, wishful thinking, deluded observers, rumors, lies, and fraud. A residue of unex-

Ufology

plained cases is not a justification for continuing an investigation after overwhelming evidence has disposed of hypotheses of supernormality, such as beings from outer space. . . . Unexplained cases are simply unexplained. They can never constitute evidence for any hypothesis." Despite the lapse of nearly 30 years that remains substantially the position of the skeptics today.

The serious, committed investigators contend that in amongst the mass of reports, most of which serve only to confuse the issue, there are some that are sufficiently credible to warrant continuing scientific examination. The several organizations referred to above continue their investigations and researches. After 50 years and despite efforts by most orthodox scientists to disprove the existence of UFOs, these groups maintain that among all the worthless claims, there remain some that constitute evidence that there are UFOs—genuinely *unidentified* flying objects—that may bring us fresh knowledge of the universe.

The popular attitude to UFOs continues unabated. Claims of sightings continue to come in at the rate of several hundred each year. Most of these are explicable, but there are always a significant proportion in the "unexplained" file. The press never misses a chance of a good story, and the subject attracts so much popular interest that television and the film industry have made UFOs the focus of several popular shows and movies. In December 1996, a policyholder with a British insurance company was paid one million pounds sterling, the first such award, having convinced the firm that he had been abducted by aliens alighting from a UFO.

See also ABDUCTIONS, ALIEN; FLYING SAUCERS, CRASHED; MUTUAL UFO NETWORK; and X-FILES.

For further reading: Jerome Clark, *UFO Encyclopedia* (Omnigraphics, 1990); Jenny Randle, *UFO Retrievals—The Recovery of Alien Spacecraft* (Blandford, 1995); Jenny Randle, *UFOs and How to See Them* (Brockhampton, 1996); and M. Sachs, *The UFO Encyclopedia* (Corgi, 1980).

UNIFICATION CHURCH Founded in Korea in 1954 by the Reverend Sun Myung Moon, espousing millennial and messianic doctrines (see MILLENARIANISM) plus some revelations that Moon claims to be received direct from God. It is also known as the Holy Spirit Association for the Unification of World Christianity, and its members as "Moonies." The movement quickly started to spread and found a foothold in Japan. The church had little success in the Western world until Moon moved to the United States in the early 1970s; now the full-time membership in all Western countries is estimated to be 10,000, although some critics believe this is an underestimation. A favorite recruitment area for the Unification Church is California; recruits are usually young, middle-class, well educated, and celibate.

Moon advances theories of the origin of humanity that are so far from scientific truth that they ought to be contested. His whole story is a fabrication without any evidence to support it; thus it is pseudoscience. Not only biologically unscientific, it is sexist, proclaiming that all the ills of mankind are attributable to the first woman, Eve. Christians and Jews find Moon's interpretation of Judeo-Christian teachings distasteful, distorted, and degrading. Moon's ideology is spread through "workshops" and the publication of a daily newspaper *News-World,* as well as through proselytization.

Moon's theology is contained in his *The Divine Principle.* He believes the Fall of mankind came about through a spiritual/sexual relationship between Eve and the Archangel Lucifer that was followed by a sexual relationship (outside marriage) with Adam. As a result of Eve's sins, God's original intentions for mankind were thwarted and the rest of history converted into a struggle to restore the world's people to their original state of purity, this great mission failing when Christ was killed, so that God can now offer only spiritual, not physical, salvation. According to Moon, the millennium in all senses is now upon us and the second coming of the Messiah imminent. It is his Unification Church that will play the leading role in restoring humanity to its original purity. Though Moon has never proclaimed so himself, many of his followers believe him to be the Messiah.

The organization of the Unification Church, a strictly hierarchical structure with Moon at its head in total command, is run for profit, using sharp business practices. From July 1984 through August 1985 Moon was actually imprisoned in the United States for tax evasion, but he claimed persecution by the authorities. Nevertheless, his Church still owns several valuable businesses and properties—one in Barrytown, New York, is used as a seminary.

Along with newspapers and television, many churches have spoken out against the doctrines and methods of the Unification Church. The Church's tactics have been questioned by the European Parliament; the British government has questioned its status as a charity. The media and "anticult" organizations have accused the Unification Church, not without foundation, of deception, brainwashing, splitting up families, exploiting its followers to amass wealth for Moon, and also of manufacturing armaments—even of being connected with the secret police of the South Korean military dictatorship. The Moonies counter by saying that anything new, however true, is attacked by the authorities.

For further reading: Peter Bishop and Michael Darton, eds., *The Encyclopedia of World Faiths* (Macdonald & Co. Ltd., 1987); D. G. Bromley and A. D. Shupe, *Moonies in America: Cult, Church and Crusade* (Sage, 1979); and F. Sontag, *Sun Myung Moon and the Unification Church* (Abingdon, 1977).

V

VACCINATION Originally the practice of inoculating with a vaccine to prevent smallpox. This was done by introducing into the bloodstream a preparation of cowpox virus (*vacca* is Latin for cow). In 1798 a British doctor, Edward Jenner, became the first to establish the practice when he showed how people inoculated with cowpox not only became immune themselves to the dreaded smallpox but were also prevented from spreading the disease to others. The idea behind inoculation is that introduction of a dilute preparation of a harmful organism encourages the system to produce antibodies that fight off the infection. Prior to inoculation, plagues and deadly diseases killed millions all over the world, especially young children with less immunity than adults. Now we know that even animals can be inoculated against their own specific ailments.

Interestingly, this widely practiced, beneficial scientific procedure derives from another that is still considered pseudoscientific by some, namely HOMEOPATHY. About the time that Jenner was developing the smallpox vaccine a German physician, Samuel Hahnemann (1755–1843), proposed a seemingly similar system of medicine called homeopathy, meaning "like disease." This goes back to the fifth-century B.C.E. Ancient Greek principle, Hippocrates' "law of similars." The law states that disease should be treated by administering an exceedingly small, dilute dose of a substance that, used full-strength, would induce disease symptoms in a healthy person similar to those being treated. For example, quinine was used to cure malaria because the effects upon a healthy person of administering quinine resemble malarial symptoms. Although homeopathy superficially resembles the practices of vaccination and immunization, orthodox doctors find it difficult to accept the homeopathic notion that the greater the remedy's dilution, the greater its effectiveness. This appears inconsistent with medical observations that only a substantial or strong dose of a harmful organism induces the immune system to produce antibodies. Although there have been a few scientific trials, the results have been disputed because of the placebo effect—the patient's belief in the cure rather than the efficacy of the cure itself.

Its early ties to homeopathy may be the reason why parents have been slow to take up offers of vaccination for their children. Another reason is that some vaccination is perceived as risky or dangerous—for instance, DPT vaccinations. Such vaccinations combat three diseases: diphtheria, whooping cough (pertussis), and tetanus. The pertussis component has come under fire, some physicians claiming that previously health children died following DPT injections and that the health of others was impaired through vaccine-related reactions.

Despite this reservation about pertussis, the work of Jenner, and many others subsequently, has demonstrated that vaccination is scientifically sound and that it works, and the suspicious have gradually been won over. Homeopathy has still to convince some doubters—medical professionals and members of the public alike.

For further reading: Harris L. Coulter and Barbara Loe Fisher, *A Shot in the Dark: Why the P in the DPT Vaccination May Be Hazardous to Your Child's Health* (Avery Publishing, 1991), and Patrick C. Pietroni, ed., *The Reader's Digest Family Guide to Alternative Medicine* (Reader's Digest, 1991).

VACUUM A longstanding concept of natural philosophy that provides a useful *tabla rasa* on which to construct our models of reality: a completely empty space devoid of all matter, a void. In the fifth century B.C.E., Leucippus of Miletus and Democritus of Abdera maintained that the world consisted of an infinite number of tiny atoms moving randomly in an infinite void. Aristotle in the fourth century B.C.E. vigorously opposed the concept; a vacuum, total emptiness, in his physics would result in particles moving with infinite velocity, a nonsensical idea. But in the next century Epicurus again favored the void hypothesis. The Stoics, materialists like the Epicureans but otherwise opposed to them, allowed the existence of an infinite void outside the cosmos, that is, outside our universe. Much later, in the 13th and 14th centuries, theologians came to an uneasy compromise with Aristotle's teachings in the course of which they also favored the concept of an infinite void, if only so that God had space to create other universes than our own if he so wished.

In the 17th century, Galileo encouraged his student Evangelista Torricelli to experiment with producing a vacuum in a vessel. Soon after Galileo's death, Torricelli showed that the space above the mercury in a glass barometer tube containing no air was an effective vacuum. But it would be a mistake to think of a Torricellian vacuum as empty. There is mercury vapor in the space, about a million times fewer molecules than in the same volume of air. That, however, is still a lot of molecules—about 500 million million million in each cubic inch.

Since that time, there has been considerable progress in attaining an improved vacuum. Von Guericke and Robert Boyle built the first mechanical vacuum pumps later in the 17th century. The ability to produce a vacuum was then improved, step by step, through Christian Sprengel's pump in the 1870s, the rotary pump, the diffusion pump, to modern pumps producing a very high vacuum at great speed and evacuating huge volumes. Today, pumps are commercially available that will produce a vacuum with only about 50 million molecules per cubic inch, 10 million million times better than Torricelli's. Even the ancients' "infinite void" outside the cosmos is not entirely empty; astronomers estimate that there is about one atom per cubic inch in outer space.

Although we now have a practical appreciation of vacuum, the concept still has its theoretical uses. Throughout the 19th century, the problem of how electromagnetic waves, especially light waves, were propagated through space occupied physicists. Ripples traveled on water and sound waves through air. On what then did light waves travel? On nothing? That hardly seemed feasible, so a substance, the ether, first supposed by Sir Isaac NEWTON, was imagined to occupy all space and to carry the waves. The ether that occupied the vacuum was invisible and intangible, weightless, and, in every way but one, undetectable. The one remaining way had to be some behavior of light waves that would reveal and confirm the existence of the ether. None of the very ingenious experiments to confirm the ether's existence, culminating in the Michelson–Morley experiment near the end of the century, was successful. When Albert Einstein published the Special Theory of Relativity in 1905 he showed that there was no need for the ether. Despite this setback, physicists have not abandoned the vacuum. In the quantum theory of fields, the vacuum is required to have several properties: the electromagnetic vacuum fluctuation, polarizability (which can be thought of as the vacuum comprising an infinite sea of negative-energy electrons), and charge fluctuation. If and when this theory gives way to some new improved hypothesis, it seems probable that the vacuum will oblige and provide a whole new set of properties.

VAMPIRES Primarily a Slavic legend of a blood-sucking creature, supposedly the restless soul of a heretic, criminal, or suicide. Folklore claims that the vampire leaves its burial place or coffin (sometimes in the form of a bat to fly by night) seeking to drink blood from the living and then returns to its dark grave before first light. Its victims become vampires at death. In Rumanian folklore as opposed to Slavic, the draining power of vampires is directed toward their victim's psychic energy rather than their blood.

Slavic folklore describes vampires as being dressed in shrouds. Their whole appearance is uncorrupted by death, although the hair on their head and face is longer than at death. Throughout European history, vampires have been described as having a ruddy lifelike complexion, longish hair that has grown after death, and plump bodies bloated with the blood of their victims. When a stake has been driven into a vampire's chest, the corpse utters a deep moan and blood is seen to come from its mouth.

Forensic pathologists tell us that there is a rational scientific explanation for all of these reported conditions. In many cases of sudden death, the lungs become congealed with a heavily blood-stained frothy fluid. In the agonized struggle to breathe, this may reach the mouth and nose and can appear on the lips and face when the body is being transported. Contrary to common belief, the face of a corpse is not always white. If death occurred when the body was lying face down, then the face may well be ruddy because the blood will pool in the face and chest regions. Contrary to common belief, hair on the head and beard cannot grow after death, but it may appear longer because the skin dries out and retracts slightly, leaving a pronounced stubble on the male chin. During decomposition methane gas often builds up in the thorax and abdominal cavities, giving the corpse a bloated even-fatter-than-in-life appearance. When a vampire hunter drives a stake into the corpse's chest, heavily blood-stained fluid may be forced out of the mouth, and gas may be expelled in a rush past the vocal cords, causing noises

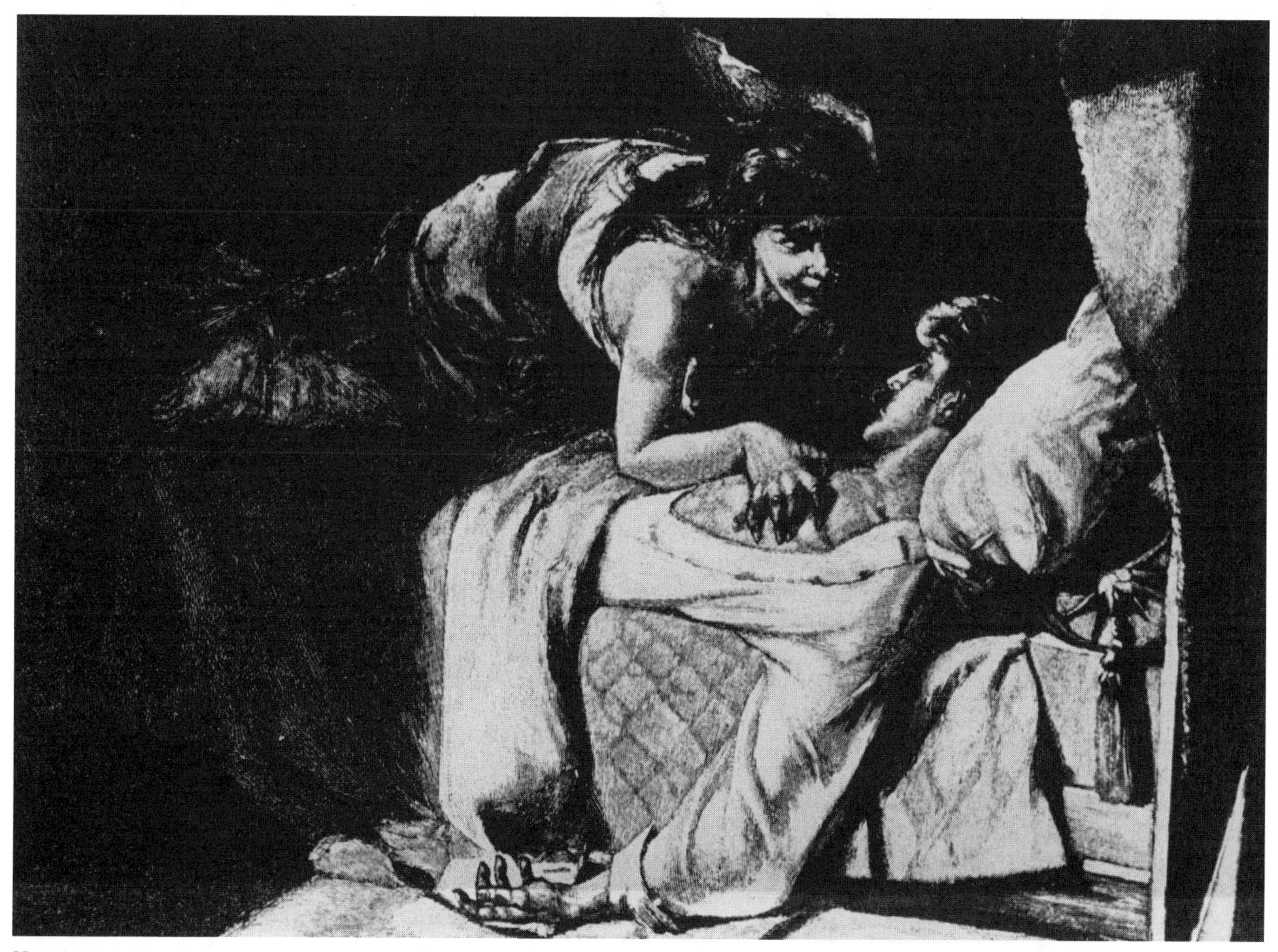

Vampires

that could be interpreted as a groan. Vampires are often depicted with long talonlike nails, but nails fall off as the body decomposes.

Most of the misinformation we have today about vampires comes from romantic literary and cinematic imagination, which is produced for fun and is not meant to be taken seriously.

For further reading: Paul Barber, *Vampires, Burial and Death: Folklore and Reality* (Yale University Press, 1988).

VANDERMEULEN SPIRIT INDICATOR An instrument, also known as Rutot's Spirit Indicator, intended to facilitate contact with spirit entities. The spirit indicator was invented by a young man named Vandermeulen, who died in 1930 while it was still in the testing stage. It consisted of two prisms and a fine wire triangle. The two prisms, one plain and one resinous, were mounted face to face on a board. The wire triangle was connected to the negative pole of a dry cell battery before being suspended between the two prisms. It was believed that spirit entities would generate electricity in the prisms. As the electricity was generated, it would drive the triangle away from the negative prism toward the positive prism where it would contact the wire connected with the positive pole of the battery. At that point a bell would ring. Researchers would then note the spirit's presence and could engage in communication.

Following Vandermeulen's death, his instrument passed to a Belgian scientist named Rutot, who worked with the invention and claimed that he had been able to establish contact with the spirit of the inventor. He published his report of his initial contact and subsequent experiments in the *Bulletin du Conseil de Recherches Métaphysiques de Belgique* (1930). U.S. psychical researcher Hereward Carrington attempted to reproduce Rutot's results but was unable to confirm his findings. The instrument passed into oblivion as another failed attempt at developing a mechanical means to communicate with the dead.

For further reading: Hereward Carrington, *Laboratory Investigations into Psychic Phenomena* (Arno Press, 1975).

VEGETARIANISM A belief in and practice of a diet that avoids meat, fish, fowl, or in some cases, any product derived from animals, such as eggs, cheese, gelatin, et cetera. The purest form of vegetarianism, called the vegan diet, consists of grains, legumes, fruits, vegetables, and fungi. Lacto-vegetarians add dairy products, while lacto-ovo-vegetarians include eggs and dairy products. Although uncommon, some more radical forms of vegetarianism, such as the fruitarian (fruit only) diet and the raw foods diet, have attracted followers. Evidence suggests that any diet, vegetarian or otherwise, ought to include a wide variety of foods.

Some within the conventional medical establishment have questioned the adequacy of a vegetarian diet in providing the individual's complete nutritional needs, particularly in regard to satisfactory levels of protein. However, in the right combinations and proportions, a strictly vegetarian diet devoid of all animal products can meet the body's protein needs, as well as most or perhaps all other nutrient requirements. Scientists now contend that the only essential nutrients not available from plant sources are vitamin B_{12}, or cobalamin, and vitamin D. Current research in edible fungi may soon solve the problem of the vegetarian's B_{12} deficiency, while some 20 minutes daily of direct sunlight stimulates the body's production of enough vitamin D. Many vegetarians make use of dietary supplements to ensure adequate consumption of essential vitamins and minerals.

Those who practice vegetarianism may do so for a variety of reasons. Many religious traditions advise vegetarianism or have been interpreted as such. Mohandas Gandhi (1869–1948), India's inspired nationalist and spiritual leader, was an ardent proponent of vegetarianism. In her book *Diet for a Small Planet*, first published in 1971, Frances Moore Lappé advances a political and ethical rationale for vegetarianism. Others pursue the diet for health reasons, as evidence mounts suggesting a vegetarian diet reduces the risk of several chronic degenerative diseases and conditions such as diabetes, hypertension, coronary artery disease, cardiopathy, obesity, and some types of cancer. In all cases, careful attention to suitable amounts and combinations of nutrients is required. Although arguments persist regarding the health claims of vegetarianism, the American Dietetic Association published a report in 1993 pronouncing vegetarian diets, when properly planned, to be healthful and nutritionally sound. [See also FAD DIETS.]

VELIKOVSKY AFFAIR, THE The controversy that arose from the works of Immanuel VELIKOVSKY beginning with his *Worlds in Collision*, published in 1950. Velikovsky argued in long and scholarly works that Earth was subject to violent catastrophes in historical times and that evidence of these can be found in the myths and legends of all cultures, as well as in Earth's recent geological history. According to Velikovsky, the key events took place between the fifth and eighth centuries B.C.E. when Venus, which had been a comet with a tail, came from the vicinity of Jupiter and nearly collided with Earth. Later, its contact with Mars caused that planet to pass very close to the Earth.

Velikovsky's ideas attracted large popular following, but were also debunked by established scientists. The hostility of the scientific establishment in the United States, threatening a boycott of the publisher's scientific textbook division, caused the original U. S. publisher to transfer *World in Collision* to another publisher that did not publish scientific texts.

Debate continued for two decades. The issues raised by the affair are important, involving the right of those outside the scientific establishment to question scientific orthodoxy.

For further reading: H. H. Bauer, *Beyond Velikovsky: The History of a Public Controversy* (University of Illinois Press, 1984); and R. G. A. Dolby, *Uncertain Knowledge* (Cambridge University Press, 1996).

VELIKOVSKY, IMMANUEL (1895–1979) Best known as the author of *Worlds in Collision* (1950). He was born in Russia, emigrated to the United States in 1939, and emigrated again to Byelorus while it was still part of the USSR. He studied medicine at Edinburgh and Moscow universities, qualifying in 1921, then practicing in Palestine. In the 1930s he studied psychology, first in Zurich and later in Vienna. He was both a psychoanalyst and an astronomer. Between 1921 and 1924, he worked with Einstein, editing the *Scripta Universitatis* at the Bibliothecae Hierosolymitarum, the institution that later became the Hebrew University of Jerusalem. In addition to his best known work, Velikovsky published, on similar lines, *Earth in Upheaval* (1955). He also wrote *Ages in Chaos* (1952), a chronology of the Middle East in the pre-Christian era, and *Oedipus and Akhnaton* (1960), in which he argued that the Pharaoh Akhnaton was the prototype of the mythological Greek character.

See also VELIKOVSKY AFFAIR, THE.

VESTIGES OF THE NATURAL HISTORY OF CREATION A phenomenally successful and controversial book, issued anonymously in October 1844, about the "development hypothesis," as evolutionary theory was sometimes called. *Vestiges*' author was its publisher, Robert Chambers of Edinburgh, who specialized in popular periodicals. He did not admit authorship until shortly before his death in 1871, even though the book had been an immediate sensation, going into four editions in the first seven months after its publication (and eventually into another eight editions).

Chambers was one of several writers who advanced theories of evolution during the 1840s and 1850s. His approach was strongly Lamarckian, in that he held the view that inheritance of acquired characteristics caused evolutionary change; his knowledge of science was thin in the extreme. What distinguished *Vestiges* from other works were its journalistic style, calculated to appeal to the antielitist "ordinary reader," and its bold overview of cosmic evolution from the formation of the planets through the origin of life (ascribed to a "chemico-electrical process") to the fossil record. It suggested that the human being was descended from a large frog.

Many scientists deemed it an outrageous book, and it became one of the most talked-about books of its day. Charles DARWIN deplored its "want of scientific caution" and declared that the unknown author's "geology strikes me as bad & his zoology far worse." Yet he credited it, in a preface to the third and subsequent editions of *Origin of Species* (1859) with "calling attention to the subject [of evolution], in removing prejudice, and in thus preparing the ground for the reception of analogous views."

See also EVOLUTION, PROGRESSIVE; LAMARCK, JEAN BAPTISTE; LAMARCKISM.

VIKINGS IN AMERICA Precolumbian contacts with the New World by Norse explorers. When the scholar Thormod Torfason published his *Historia Vinlandae Antiquae* in 1705, the news of the Scandinavian discovery of America spread throughout northern Europe. Torfason cited two Icelandic sagas to support his views: *Graenlandinga Saga* and *Eirik's Saga*. Both told of the voyage of explorers from the Norse settlements in Iceland and Greenland to a western land. According to the sagas, around 1001 C.E., Leif Eriksson led an expedition to North America. For many years, an ancient Norse presence in the New World went unquestioned. During the 1940s, however, the Harvard historian Samuel Eliot Morrison rejected the stories of the sagas and declared that Columbus had no predecessor. But in the 1950s, archaeologists uncovered a Norse community at L'Anse aux Meadows in Newfoundland, Canada, decisively overturning Morrison's views. They now believe that the L'Anse aux Meadows settlement may be Eriksson's settlement in the land he called Vinland.

Controversy continues about how far into the North American continent the Norse explorers penetrated. The sagas record that Eriksson named the territory of Vinland after the many grape vines that grew wild there. In that case, L'Anse aux Meadows could not be Vinland because grapes do not grow wild that far north. Some historians argue that Eriksson's "grapes" were really wild cranberries, which can still be found in the bogs of Newfoundland. Others—including the Norse scholar Hjalmar R. Holand—claim that Ericksson's grapes were really true grapes and that Vinland was therefore further south. Holand places it in or near modern New York City. Other locations that have been suggested for Vinland include Massachusetts Bay, the coast of Maine, and even tidewater Virginia.

The Vikings may have made other exploratory voyages besides those recorded in the sagas. Barry Fell argues in his *Saga America* (1980) that Vikings sailed as far south as Florida and penetrated the American continent via the Mississippi and Arkansas Rivers as far inland as modern Oklahoma and Colorado. He cites as evidence for this a variety of Native American petroglyphs and rock carvings in what looks to be the runic alphabet used by the Norsemen. Most U.S. archaeologists and historians do not accept Fell's views.

Some Viking finds further north have received a slightly wider acceptance. A tower outside Newport, Rhode Island, has been attributed to Christianized Viking settlers (see MYSTERY HILL). Recently, the VINLAND MAP—believed to be a forgery for 30 years previously—has been reevaluated as a true medieval artifact. One of the most famous of the supposed Viking-era artifacts is the Minnesota KENSINGTON STONE, which tells the story of an expedition that sailed westward from Vinland in the late 14th century. Many scholars believe that the Kensington Stone is a modern forgery, but others, including Holand, argue that this is not proven. In 1930, an iron sword, a shield, and a battle-axe were uncovered at Beardmore, Ontario, just north of Lake Superior. Experts dated them to about 1025—the period when, according to the sagas, Eriksson and his followers were first settling in Vinland. Historians who refused to accept a Viking presence in the New World rejected the artifacts, claiming that they were forgeries. In 1961, after the discoveries at L'Anse aux Meadows definitely established a Viking presence in North America, the Royal Ontario Museum labeled the Beardmore discoveries "not proven." Although a Norse presence in Canada has been established beyond doubt, the extent of their penetration is still in question.

For further reading: Barry Fell, *Saga America* (Times Books, 1980); Hjalmar R. Holand, *Explorations in America before Columbus* (Twayne Publishers, 1956); Magnus Magnusson and Hermann Palsson, trans., *The Vinland Sagas: The Norse Discovery of America* (Penguin, 1965); and R.A. Skelton, Thomas E. Marston, and George D. Painter, *The Vinland Map and the Tatar Relation,* new edition (Yale University Press, 1995).

VINLAND MAP Medieval document that is supposed to be the oldest known European map of North America. The Vinland map is believed to be a European copy of a Norse map that shows the location of Viking colonies in North America around the year 1000 C.E. It is drawn in pen and ink on two parchment leaves measuring about 23 by 25 centimeters (9 by 10 inches) each, and it shows a map of the world as it was known in medieval times. In the upper left-hand corner of the map is an island, labeled in Latin *Vinlandia insula,* or the Isle of Vines. There is also a Latin text giving the story of how Bjarni Herjolfson and Leif Eriksson discovered Vinland by sailing west and south from Greenland.

For about 30 years, scholars regarded the Vinland Map as a modern forgery. There were several good reasons for doing this. The map appeared suddenly in 1957 in the hands of a Swiss bookseller. No references to such a map appeared anywhere in the historical record. Despite the doubts of many major historians, a Connecticut book dealer bought the map for about $4,000 and brought it to Yale University. The university set up a team of scholars, led by George Painter, assistant keeper emeritus of printed books at the British Museum, to determine if the map was authentic. The team worked on the map for eight years and, in 1965, declared that the map was indeed authentic. When the material of the map was submitted for analysis, however, scientists found traces of a modern chemical compound—titanium dioxide—in the ink. Titanium dioxide was not used in inks before 1920. Yale University then declared the Vinland Map a modern forgery and locked it away in its vaults.

In 1985, however, the Vinland Map was reexamined by scientists using new techniques. They found that, although the map did contain traces of titanium dioxide, these traces were not much more than were found in a genuine 15th-century Bible. The scientists concluded that the titanium dioxide was a product of the natural decay of the ink used on the original manuscript. In 1996, Yale University reversed its position on the map's authenticity.

VIRGIL (70–19 B.C.E.) Ancient Roman poet whose *Aeneid* is still used today as an oracular book, purportedly to predict, interpret, and so it is hoped, control the future. In the *Aeneid* Virgil set out a mythological history for Rome to show his people retrospectively where their destiny lay. The admonitory stand and the giving of moral instructions to the reader can be seen to act at the time of writing as a moral injunction to the people of Rome to remind them of their responsibilities. And throughout history it has become a book that many believe to have great oracular powers, much like the I CHING, that helps and advises readers, whatever their problems.

Virgil lived during the time of Caesar and saw the end of the Republic and the start of Imperial Rome. He did not try to write strictly realistic history, but took for his model Homer, who had lived a thousand years prior to him and who wrote legendary narratives, the *Iliad* and the *Odyssey*, about the early Greeks. What Homer did for Greece, Virgil thought he could do for Rome. So Virgil wrote about he imagined origin of Roman nation starting with the sack of Troy in times long before the foundation of Rome itself.

According to Virgil's story, the Trojan prince Aeneas escaped after the fall of Troy and sailed with other Trojan warriors to the west. After much suffering and many adventures they arrived at the coast of Italy, where they settled and founded the city which became Rome. The Trojans succeeded not so much through their own strength as through divine help and encouragement. Virgil depicts the gods communicating with mortals through dreams, visions, omens and also through the words of prophets and clairvoyants (Aeneas himself was supposed to be the son of Venus and a mortal father.) Virgil had no doubt that the affairs of the world are subject to higher powers, and he relates how Rome was destined to rise from weakness and despair to achieve greatness. Aeneas is more than once ready to abandon hope, but every time a voice from heaven shows him the way. Whatever his faults, and Aeneas had many, he never disregarded heavenly advice. Time and again, two important rules of conduct are driven home: one, principally Greek, "avoid excess," and the other principally Roman, "be true," meaning be loyal to the gods, the homeland, family, friends and dependents. An example of this occurs when, after a storm at sea, the Trojans come ashore at Carthage. Aeneas meets and falls in love with Dido, the Queen of Carthage, and, disregarding his destiny, he settles down to live with her. After some months of divine reproof (in the book, Virgil stops calling the "the true") Aeneas obediently sails away. Dido subsequently curses him, and the result, Virgil tells us, was the terrible enmity that arose between Carthage and Rome.

The *Aeneid* was left unfinished at Virgil's death but his reputation stood high. Through the period of the Empire and throughout the Middle Ages and beyond, he was worshiped, first as a divinity and then as a magician, with the *Aeneid* used as an oracle book. Tales of his wonder-working occult powers circulated all over Europe, and his tomb in Naples was visited by people seeking miracles. Those wishing to avail themselves of his didactic wisdom would open the book at random three times, letting a finger on each occasion fall blindly on one line. The three lines indicated would then either be taken literally or interpreted symbolically; either way, answers to questions would be found. A measure of the stature of Virgil in the early 14th century is shown in Dante's choice of him as guide and mentor in his allegory *Divine Comedy*. Throughout the first two books, *The Inferno* and *Purgatory*, Virgil leads Dante literally through fire and water, exhorting him to be brave, and reassuring him by answering all him moral and theological doubts. Nevertheless, as the two friends look over the flowery meadows that surround Paradise, Virgil is not allowed to go further into paradise. Why? Because Virgil was not a Christian.

It is believed that King Charles I of England consulted Virgil's book before the battle of Naseby (1645), which, incidentally, he lost.

For further reading: Virgil *The Aeneid*. Trans. Fitzgerald. New York: Random House, 1983. Dante, *The Divine Comedy*. Trans. Mandelbaum. New York: Knopf, 1994.

VITALISM The concept that bodily functions are due to a "vital principle" or "life force" that is distinct from the physical forces explainable by the laws of chemistry and physics. Many alternative approaches to modern medicine are rooted in vitalism.

The idea of a vital force existing in all living creatures has ancient roots going back at least as far as Aristotle (384–322 B.C.E.). In his general philosophy of nature, an attempt at classification postulated a continuous chain or series in which similar organisms were set next to each other; he had inorganic stones at one end of the chain and human beings at the other. Somewhere early on in the scale, Aristotle had to explain the difference between the nonliving and the living; he did this by theorizing that the heat produced from living organisms was part of their vital heat, which had its source at the center of the body, probably in the heart, and spread around the body before being expelled by the breath.

The exact nature of the vital force was debated by many early philosophers, but vitalism in one form or another remained the preferred thinking behind most science and medicine until 1828. That year, German scientist Friedrich Wöhler (1800–82) synthesized an organic compound from an inorganic substance, a process that the vitalists considered to be impossible.

As the new science of biochemistry developed theories about the life process drew more and more on biochemistry and less and less on vitalist precepts. Outside science, vitalism remained important in the thinking of such teleologists as Hans Driesch (1867–1941). Driesch proposed the existence of a soullike force to which he gave the Aristotelian term *entelechy,* meaning movement from potentiality to actuality. At the same time, French philosopher Henri-Louis BERGSON was arguing for the existence of a single, unique vital impulse that is continually developing in a creative manner rather than a mechanistic one; in this way, he nodded towards evolution. Vitalists claim to be scientific, but in fact they reject the scientific method with its basic postulates of cause and effect and of provability. They often regard subjective experience to be more valid than objective material reality.

Today, vitalism is one of the ideas that form the basis for many pseudoscientific health systems that claim that illnesses are caused by a disturbance or imbalance of the body's vital force.

See also ALTERNATIVE MEDICINE.

For further reading: J. Raso, *"Alternative" Healthcare: A Comprehensive Guide* (Prometheus, 1994).

VITAMIN C The idea that large doses of vitamin C can reduce or eliminate the symptoms of colds and influenza. Vitamin C, ascorbic acid, is an essential ingredient of the human diet. It was established early in the 20th century that small quantities are necessary to maintain human health.

The idea of taking massive doses, as much as 2 grams a day, more than 50 times the normal daily requirement, was given great publicity by the eminent chemist Linus PAULING in the 1970s. Subsequently, Pauling developed plausible circumstantial arguments in favor of the idea and criticized the statistical basis of evidence that was used to discredit it. He was attacked vigorously by members of the medical establishment, who regard megavitamin theories as dangerous fads. However, sales of large dose Vitamin C preparations rose. Many people seem to judge that because Vitamin C preparations are inexpensive and present virtually no risk, it is worth taking them, in case the subjective benefits are not merely due to PLACEBO EFFECTS.

See also MEGAVITAMIN THERAPY.

For further reading: R. G. A. Dolby, *Uncertain Knowledge* (Cambridge University Press, 1996); Linus Pauling, *Vitamin C the Common Cold and the Flu* (Freeman, 1976); Irwin Stone, *The Healing Factor: 'Vitamin C' Against Disease* (Grossett & Dunlap, 1972).

VOLIVA, WILBURGLEN (c. 1880–1942) The leader of the Christian Apostolic Church, a community of about 6,000 members in Zion, Illinois, who believed that the earth is flat. Voliva was more than a "member"; he was, in fact, the community's ruler, its General Overseer, for more than 30 years. He established strict rules for the community's behavior and articulated its beliefs, including that of a flat Earth with the North Pole at the center of the disc and the South Pole around its edges. An around-the-world traveler followed a circular path back to the start: Earth was the center of the universe; the sun was a comparatively small object, 51 kilometers (32 miles) in diameter and a mere 5,000 kilometers (3,000 miles) away; and the stars were distant small objects. Since Voliva's death, Zion's social regulations have become considerably more relaxed, but there are still many flat Earthers in the population, especially among the elderly.

See also FLAT EARTH SOCIETY.

VON DÄNIKEN, ERICH (1935–) The immensely successful author of a series of books about visitation by extraterrestrials. The most widely known is *Chariots of the Gods?* (1969). In his works, Von Däniken claimed that alien beings had visited Earth 10,000 years ago and created humans by altering the genes of apes in their own image. According to his theory, the extraterrestrials did not leave immediately and were subsequently worshiped as gods; archaeological study, he asserted reveals traces of their technology. Von Däniken's ideas were backed up with copious evidence, most of which is more plausibly interpretated in more mundane ways and some of which has been found to be completely spurious. Von Däniken's success may be attributable to his marketing of a product well suited to popular yearnings of the age.

For further reading: R. G. A. Dolby, *Uncertain Knowledge,* (Cambridge University Press, 1996); Erich Von Däniken, *Chariots of the Gods?* (Souvenir, 1969); and Erich Von Däniken, *Erich Von Däniken* (Outlet, 1989).

VOODOO The name given to the religious beliefs and the practice of MAGIC of certain sub-Saharan African peoples, also called *vodou* and *vodun.* (All three terms come from a West African word meaning "god" or "spirit.") During the days of the slave trade, Voodoo was carried to the southern United States and the West Indies, especially Haiti. There are many local forms throughout the Caribbean, such as Santería in Cuba, Pocomania in Jamaica, and Shango (after the Yoruban thunder god) in Trinidad.

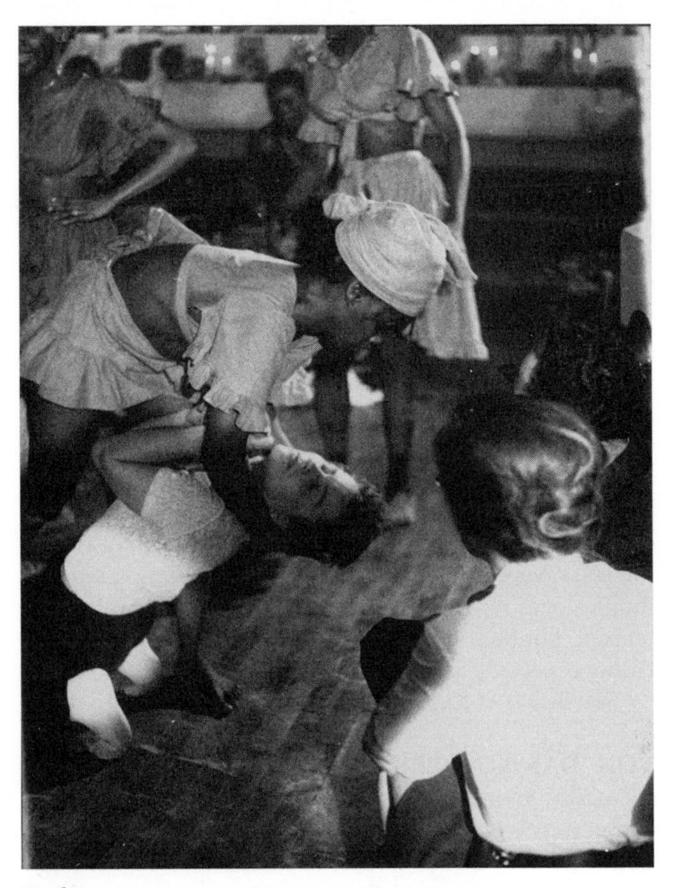

Voodoo

It is in Haiti, however, that Voodoo truly thrives as a religion. Voodoo provided an emotional outlet for black slaves during the colonial period and was a factor in their war of independence from France (1791–1804). Following the success of that uprising and the expulsion of the French, Voodoo consolidated its status. A period of rapprochement began with the Vatican after 1860, however, and Voodoo has coexisted with, and borrowed freely from, Catholicism.

Although lacking organized dogma or theology, Voodoo has spawned elaborate ceremonies, priests (called *houngans*) and priestesses (*mambos*), and a hierarchy of gods (*loas*), who are a mix of African and New World spirits and appropriated Roman Catholic saints. (The Christian God is recognized as a powerful, albeit remote, member of this pantheon.) Others include the eternal female Erzulie, the serpent spirit Damballa, Baron Samedi, the top-hatted, frock-coated, bespectacled keeper of graveyards, and Legba, guardian of gateways and crossroads. Legba is the opener of the way for the other gods when they cross over from the spirit realm into the world of the living.

Central to Voodoo is the belief that the *loas* possess, or "mount," believers and speak through them. The essential function of the Voodoo priest is to bring the living into touch with the inhabitants of the spirit world, who can bless or curse them and whose wishes, therefore, it is important to know. Voodoo is attended by much ritual, some of which, such as the lighting of candles and the sprinkling of water, strongly suggest the Catholic Mass. Drumming, chanting, tumultuous dancing, animal sacrifice, and food offerings play important roles in voodoo ceremonies, which may take place in a peristyle with a central post. Upon being summoned, the *loa* descends the post into the midst of the celebrants, one of whom it invades, displacing the human personality. It then speaks through its host to make its desires known. Possession by a *loa* may last minutes or days; afterward, the displaced human personality returns to its body with no memory of anything that occurred during the period of possession.

Other beliefs include the spiritual causation of diseases and misfortune and the efficacy of charms and spells. A wax doll, baptized in someone's name and incorporating a bit of that person's hair, fingernail parings, or clothing, is often used as weapon: mistreating the doll supposedly causes its living model to suffer.

Voodoo also teaches the grim necessity of ensuring, through a variety of more or less complicated procedures, that the souls of departed loved ones do not pass into the possession of witches or evil spirits. The witch, or *bocor,* has the fearsome power of holding people in a state of undead servitude. These mindless thralls, called ZOMBIES in Haiti and jumbies in the British West Indies, may be put to work in the cane fields and sugar mills or dispatched on errands of mischief by their masters.

Infamous François "Papa Doc" Duvalier, president of Haiti from 1957 to 1971, stage-managed his appearance to resemble Baron Samedi and used Voodoo in tandem with an army of killers, the Tontons Macoutes, to keep the populace in line. It is said that on November 22, 1963, he toasted the spirits to whom he had pleaded for the death of President John F. Kennedy, his archenemy.

VULCAN A hypothetical planet in the solar system that was thought to orbit closer to the Sun than Mercury. In the early 1800s, there were many reports of black objects passing across the face of the Sun, presumed to be the profiles of small planets. Early 19th-century observations were sufficiently positive to encourage leading French astronomer Urbain-Jean-Joseph Le Verrier to investigate the phenomenon. Le Verrier had attempted to explain the high and unique eccentricity of Mercury's orbit and concluded that it could be caused by the gravitational attraction of a belt of asteroids between Mercury and the Sun; a small planet was another possible explanation. He looked at over 50 observations, dismissing all but six of them as inadequate. The remaining six observations, made over the 60 years to 1861, could be accounted for in several ways, of which one, a small planet, seemed to fit best. He calculated the orbit of the postulated planet, which he named Vulcan, as having a period of 33 days and as being at a relatively large angle to the plane of Earth's orbit, the ecliptic. He could now predict the new planet's behavior and did so: Vulcan would next transit the face of the Sun on March 22, 1877. It did not, nor could Vulcan be seen during solar eclipses. There has been no confirmed sighting of Vulcan since Le Verrier's announcement.

If a planet did exist in the orbit calculated by Le Verrier or anywhere near, it would be roughly at half Mercury's distance from the Sun. The temperature on the sunny side of Mercury is about 350°C (about 76°F), above the melting point of lead. The side away from the Sun is near absolute zero. Neither side could support life. The conditions on a planet in the Vulcan position would be much worse. So there is no Vulcan; even if there were, there could be no aliens living on it.

WALLACE, ALFRED RUSSEL (1823–1913) British naturalist who developed a theory of EVOLUTION by means of natural selection at the same time as Charles DARWIN. Wallace, who did most of his work with insects in the Amazon and the Malay Archipelago, concluded that biological evolution must be governed by a mechanism that closely resembled the theory proposed by the economist Thomas Robert MALTHUS (1766–1834): Variations that helped a species to survive would be perpetuated in future generations. Wallace published a book on the subject, *On the Law Which Has Regulated the Introduction of New Species* (1855).

Wallace soon came into contact with Darwin through a mutual friend. Realizing that Darwin's work predated his own, Wallace agreed to combine his work with that of the older man, and they presented their ideas together in July 1858. Wallace continued to present his ideas in books such as *Malay Archipelago* (1869), *Contributions to the Theory of Natural Selection* (1870), and *Island Life* (1880). Later in life, he took a strong stand as an antiracist, declaring that all people, regardless of race, had similar inherent intellectual capacities.

Wallace rejected natural selection late in his career, at least as it related to the human brain. Instead, he argued that the mind was the product of an act of God because it was too complex to be the result of a series of random mutations. People with less-advanced cultures, he believed, were like human ancestors—which meant that those same human ancestors had well-developed, advanced brains before they were ready to use them. Later in life Wallace became a believer in SPIRITUALISM and PHRENOLOGY.

WATER CURES Also called hydrotherapy; a cure for various ailments, water is used to enhance beauty, alleviate stress, or for the derivation of pleasure. Overall, its uses go far beyond quenching thirst or keeping clean; it has always been so.

Water, whether fresh, sea water, or containing naturally occurring substances such as sulfur, is bathed in publicly and privately at different temperatures, taken internally as medicine, or used ritualistically to ward off evil spirits. The body may be immersed in water, float upon it, or be wrapped in dampened towels. Water may be inhaled in the form of steam, applied to wounds in the form of icepacks or applied to the body to prevent swelling.

In religion and folklore water has always been associated with purification and seen as essential for life on Earth. Science, too, has always recognized its importance; three quarters of the Earth's surface is covered by it and life began there.

The curative properties of mineral springs were known to the ancient Romans. First-century naturalist PLINY THE ELDER in his *Natural History* mentioned the popularity of the town of Spa in what is now Belgium. Since then, the name spa has been given to health resorts that provide mineral springs for bathing and drinking. Interest in visiting spa towns waned in Europe during the Middle Ages, and it was not until the 16th century that the old Roman springs were rediscovered. Gradually, the idea of using water to restore health regained popularity, and once more spas became fashionable. In Britain, the ancient town of Bath regained its popularity thus. But the the first hydrotherapy center in Britain to offer a wide range of water cures was founded in the 19th century by Father Sebastian Kneipp (1821–97), a Dominican monk from Bavaria. The "Kneipp method" was an extension of the work of an early pioneer, Vincent Preissnitz, a well-known lay practitioner in Silesia. Hydrotherapy generated another great health movement in the 20th century—that of NATUROPATHY.

Most cultures have recognized the benefit of baths and made water cures part of their medical regimen. It should be remembered, though, that a fair number of so-called "cures" are not so much medical as lifestyle options. There is no doubt people feel better after a course of treatment, but it is now believed that many of the beneficial effects of such therapy are indirect, stemming from being in a beautiful environment and being away from the stress of work. There is also no doubt that drinking water that is free of harmful impurities is good for human beings, and today spa and spring water from many sources is sold in bottles. Whether this type of bottled water is superior to tap water, however, is often debated.

There are many types of baths—whirlpools and aerated baths said to ease stress and strain; sitz baths used for illnesses of the lower abdomen; sweating baths, of the Russian or Turkish kind, used to make the body perspire and so, it is thought, sweat out impurities; and douches, strong streams, or sprays of water are directed onto the body, being thought to stimulate circulation and ease muscular pain. Thalassotherapy uses the supposedly healing powers of seawater and/or seaweed.

Most of these therapies are both harmless and helpful; however, one—colonic irrigation—may be harmful. This "cure" entails injecting water at body temperature into the rectum, sometimes three or four times a week. This treatment is believed to clear the colon of poisons, gases, accumulated fecal matter, and mucous deposits, but the bacteria found in the bowels form part of the ecosystem of the gut and perform a natural and necessary function in the body. Habitual resort to enemas should not be considered a natural therapy and be resorted to only under medical recommendation and supervision.

For further reading: Patrick C. Pietroni, ed., *The Reader's Digest Family Guide to Alternative Medicine* (Reader's Digest, 1991).

WATER DIVINING The specific practice of using a divining rod, usually a forked twig, to look for water. The stick's forked ends are held, one in each hand, the stem pointing upwards. When the dowser walks over the area to be surveyed, the stick is suddenly seen to twist violently and point downward, indicating where to dig for water. Traditional woods for dowsing are hazel, peach or willow, but all types have been used. Other materials such as ivory, metal or even bare hands have been employed.

Some say that water divining is an ancient practice that goes back to Moses striking the rock with his rod, but there is no evidence that ancient cultures utilized it. We do know that in 15th-century Germany, divining with a forked twig was used for the first time to find underground minerals, but, just as today, the practice was controversial. The forked-twig system was later applied in France and England to finding water, and eventually the idea was spread by European settlers to other parts of the world.

The literature on modern dowsing is very extensive. One of the leading references on the subject, *The Divining Rod* (1926) by Sir William Fletcher BARRETT, concluded that the twisting of the rod was due to unconscious muscular actions of the dowser, who, it was thought, must possess a clairvoyant ability to sense the presence of water. Others who wrote at the same time believed that electromagnetic or radio waves controlled this rod. Later, in 1949, Dr. Solcon W. Tromp of Fouad I University, Cairo, published *Psychical Physics* (currently not in print) in which he disclosed his conviction that electromagnetic fields affected what he assumed to be similar fields in the dowser's brain. For Tromp, the dowser and the hands, not the rod, were the important implement in finding water. Many dowsers failed to fulfill Tromp's expectation; he explained that the operator's hands sometimes had reverse polarities.

The scientific community is split over how to assess the frequent success of dowsers. One theory admits that many dowsers are good at assessing the potential locality of underground water by looking for surface indications and that they translate their thoughts into movements of their hands. Psychologists who study body language tell us that thoughts and muscular action are very much tied together: A dowsing rod is delicately balanced, and it only takes a slight muscular movement to point it downwards. However, when this theory was tested under strictly controlled, neutral conditions where the operator could not supplement his dowsing with observational ground knowledge, the results are no better than chance.

Still another theory is based on geologic principles. Water can be found fairly close to the surface in aquifers that can stretch over large areas. Therefore, it is difficult to *avoid* finding small amounts of underground water.

See also DIVINATION; DOWSING.

For further reading: William Barrett and Theodore Besterman, *The Divining Rod* (University Books Carol Group, 1967); and Martin Gardner, *Fads and Fallacies in the Name of Science* (Dover, 1957).

WEEPING STATUES A phenomenon whereby a statue made of stone, terra cotta, or plaster suddenly appears to weep real tears or even blood from its eyes. Many weeping statues of the Virgin Mary have, on closer inspection, turned out to be fakes. The only recent weeping Madonna to be recognized by the Roman Catholic Church is that in Syracuse, Sicily, validated in 1953. There were many eyewitnesses and a couple of amateur films were made showing tears appearing on the face.

Dr. Luigi Garlaschelli, a chemistry researcher at the University of Pavia, described in *Chemistry Today* one way to make a "bleeding" or "weeping" statue. "What is needed is a hollow statue made of porous material, such as plaster or ceramic," he wrote. "The icon must be glazed or painted with some sort of impermeable coating. If the statue is then filled with a liquid (for instance through a tiny hole in the head), the porous material will absorb it by capillary action, but the glazing will stop it from flowing out. If the glazing, however, is imperceptibly scratched away on or around the eyes, tear-like drops will leak out, as if materializing from thin air. If the cavity behind the eyes is small enough, once all the liquid has dripped out there are virtually no traces left in the icon. When I put it to the test, this trick proved to be very satisfactory, baffling all onlookers."

Garlaschelli was not allowed to inspect the glazing of the Syracuse Madonna, which is kept behind glass. But he reported that a careful examination of an exact copy, from the same manufacturer, proved to be made of glazed plaster with a cavity behind the face and then therefore the potential for fraud was present.

Not all such cases are deliberate fraud. For example, a case of a statue of the Virgin that wept blood was reported in the town of Brunssum in southern Holland in the summer of 1968. The statue was five years old and had never been observed weeping before. Consequently hundreds of people came to see it. A specimen of the "blood" was examined at the hospital laboratory in Heerlen and the mystery was soon cleared up. It turned out to be resin which had been used to fix the eyes in place; in the very hot summer of that year, the resin had melted, bringing with it the red/brown pigment from the eyeball itself.

A similar but reverse happening has been reported. Quite recently, throughout India and Asia and also in Britain and the United States where there are Hindu communities, statues of their deities have suddenly appeared to be sipping spoonfuls of milk. Hoaxes were proved in a few cases, but what of the others? Skeptics have pointed out that many of the statues were made of baked clay and were absorbing liquid by capillary action. If the statue is glazed, it needs only a small unglazed area for absorption to take place. In nonabsorbent material like marble, and in metals like brass, small pools of milk have been noticed at the foot of the statue, but no one could explain how they got there. Scientists from the Federal Department of Science and Technology in New Delhi experimented by offering a statue milk mixed with a red dye. While the offering appeared at first to be taken by the idol, it was later observed, because of its red color, to have been deposited over its surface. The explanation emerged after scientists found out that statues were being ritually washed before being fed, and the milk from the tipped spoon was draining over the wet statues, mixing with the water. Without the red dye it would have been imperceptible, especially if the statue was washed regularly.

In the United States and Canada many more weeping statues have been thoroughly investigated; they have all been found to be brought about by natural causes or occasioned by the handiwork of hoaxers, not to be miraculous manifestations.

For further reading: Joe Nickell, *Looking for a Miracle: Weeping Icons, Relics, Stigmata, Visions, and Healing* (Prometheus Books, 1993).

WEGENER, ALFRED LOTHAR (1880–1930) A German glacial meteorologist who was an early proponent of the theory of CONTINENTAL DRIFT. In 1912, he first proposed the theory that the continents had once been united into one single land mass—the super continent Pangaea. In 1915 he published *Die Entstehung der Kontinente und Ozeane (The Origin of Continents and Oceans)* in which he assembled much evidence in support of the theory. He continued to advance the theory until his death, although the great geologists of the time considered it pseudoscientific.

If, as Wegener supposed, the continents were once united, then a number of strange facts about geology and animal distribution are explained. The coal fields of South Wales and Pennsylvania fit together, as do the diamond-bearing beds of Brazil and southwest Africa. Marsupials, such as the kangaroo and opossum are found in Australia and New Guinea but also in South America. Despite the apparent common sense of these arguments, it was not until the 1950s and 1960s that the experimental evidence for the mechanism of continental drift showed Wegener to have been essentially correct.

For further reading: J. Georgi, "Memories of Alfred Wegener," *Continental Drift,* ed. by S. K. Runcorn (Academic Press, 1962); Alfred Wegener, *The Origin of Continents and Oceans* (Methuen, 1978); and Martin Schunzbach, *Alfred Wegener: The Father of Continental Drift* (Sci Tech Pubs., 1986).

WHISTON, WILLIAM (1667–1752) Anglican clergyman and professor of mathematics. In 1701 he became an assistant to Sir Isaac NEWTON in Cambridge, succeeding him as professor of mathematics in 1703. In 1696 he published *The New Theory of the Earth,* an attempt to reconcile the new Newtonian understanding of the world with the biblical accounts. Whiston's theory was that the planetary system was formed by a giant comet in a perfect state: exactly circular orbits and a year of 360 days. The misbehavior of Adam and Eve started Earth's rotation. On Friday, November 28, 2349 B.C.E., another giant comet produced the flood, and slowly Earth settled into its present imperfect state—divine retribution for human sinfulness. Whiston's thesis was a serious scholarly attempt to produce a unified theory for his time, using the best scientific and mathematical knowledge at his disposal and combining it with the indisputable word of God, the Bible.

Whiston's theories extended beyond Earth. He believed that other planets and planetary systems were inhabited with beings with similar problems to ours, and he also proposed that the interior of Earth, the sun, the planets, and comets were inhabited and that some planets had very different occupants: beings that were invisible, having no physical substance. He was deprived of his professorship in 1710 because of the unpopularity of his ideas.

For further reading: James E. Force, *William Whiston* (Cambridge University Press, 1985); Martin Gardner, *Fads and Fallacies in the Name of Science* (Dover, 1957); and Stephen Jay Gould, *Time's Arrow, Time's Cycle* (Penguin, 1990).

WILD MAN OF BORNEO Anthropological hoax. The general image in Western society of tribal cultures as primitive and uncultured allowed the development of the Wild Man of Borneo as a popular attraction in San Francisco in the late 19th century. The original wild man appeared in the 1880s at a freak show on Market Street in the Barbary Coast district. However, the person posing as the wild man was, in fact, an unknown actor who allowed himself to be covered with tar and horsehair. He was exhibited in a cage and made mock attacks on his audience while devouring pieces of raw meat thrown into the cage to him. The sounds he made at the gawkers sounded like "Oofty Goofty," so those words soon became his name.

The Wild Man was advertised as an attraction that both educated and entertained the public. Very successful as an attraction, his career was cut short by a basic health problem: the tar coating his skin did not allow him to perspire; poisons built up in his body and threatened his life. Unable to peel the tar coating off, he was finally doused in tar solvent, which saved his life but ended his career as a wild man. However, he had become a public personality, and his career in entertainment did not end. He developed an act in which he sang a song in a local pub, and then allowed himself to be kicked out of the establishment. In the process he discovered he was somewhat impervious to pain. Soon the former wild man developed a more lucrative act: he allowed people from the audience to hit him with a large stick. This act came to an end in 1890 when he mistakenly allowed boxing champion John L. Sullian to hit him. He died six years later, without ever revealing his true identity, and was buried under his public name, Oofty Goofty.

WITCHCRAFT The practices of a witch in the exercising of supernatural powers supposedly to conjure up, or become possessed by, spirits. The nature of these spirits and how witchcraft is presumed to be practiced has been perceived in many ways in different cultures throughout the world.

The legacy of the medieval christian church made witchcraft satanic and synonymous with the devil, sorcery, the occult, and black MAGIC, connecting it to the many negative cultural associations of the word "black." This way of thinking rose from the church's firm belief in the dualistic character of the universe: God and Satan, good and evil, heaven and hell, benediction and malediction. Given this basic assumption, it is not a big jump to suppose that, just as some people are in God's league, there are others in league with the devil. During the Middle Ages and also into the early modern period, church courts and inquisitions used the notion of "evil powers" to intimidate people whose opinions disagreed with those of the church of Rome. Later, christian churches of all denominations protected their own specific belief systems by persecuting those who thought differently. So witchcraft became a catch-all category for charges against dissidents, the specific accusations being the conjuring up of natural disasters (storms, floods, hail, drought and the like) and the bringing about of diseases and even plagues.

Given the climate of opinion of the times, allegations of witchcraft were much easier to advance than to defend because no motive other than pure demonic malice was needed. In defense, empirical reality was not an issue; for example, an alibi was not accepted because witches were thought capable of bilocation. The punishment when found guilty was to be burned at the stake. In the 16th and 17th centuries, during the fierce conflict between Catholics and Protestants, the rooting out of witches became more and more misogynist. The town of Toulouse, France, was reported to have burned 400 witches in 1577, mainly middle-aged and older women; between 1587 and 1593 two villages in the Trier region survived the ordeal with a female population of one each.

The wicked witch of European myth and fairy tale follows a similar profile to the witches of medieval and early modern times. These literary characters, directed toward young children, were always, without exception, female and were usually portrayed as malicious figures dressed in black with a tall pointed black hat, owning a black cat, flying on a broomstick, cackling, and cursing all and sundry, and often given to cannibalism (as in the story of Hansel and Gretel). The witch was never a mother, always a stepmother or a lonely malcontent. Feminist historians, putting to one side the many psychoanalytical explanations, say that at a time when women's equality was just starting to become an issue, this form of early nursery indoctrination had the effect of keeping the demands of women's equality in check, while the pure image of the mother remained revered.

Neopagans follow a good white witch "wicca" (the word *wicca* is Old English and means "witch," *wiccan* means the practice of this form of white magic sorcery). Some modern pagans have construed the persecution of

Witchcraft

early modern witchcraft, not as attacks on heretics, but as the christian church's attempt to exterminate the vestiges of ancient religions such as Druidism. The first to put forward this theory was Margaret Murray in the 1920s and 1930s and, not surprisingly, since then the theory has been subjected to many attacks. Undaunted followers of modern Wicca continue their mission to revive paganism, and in the United States, Wicca membership rose from 40,000 in 1979 to 100,000 in 1991. Wicca followers believe in sympathetic magic and other New Age magical and psychic phenomena, but do not believe in an evil power like Satan, and will not practice any form of sorcery or evil magic that will harm others. Many modern witches wrongly assume that in the past there was one uniform Pagan religion, and others, although better informed, have helped to created a syncretic paganism by squeezing diverse ancient traditions into their own religious mold. Early forms of witchcraft have left no written history to put forward their own point of view; all our information comes down through Roman or Christian writers. Nevertheless, deficiency of information must not be used to disparage a point of view; most world religions have dim and uncertain pasts, employ wishful thinking, and are very selective in what they claim as their history.

Anthropological literature offers numerous research reports that describe witchcraft practices in tribal culture. Some tribal rituals are of a healing nature performed by a shaman, medicine man, or witch doctor. Other rituals are designed to bring harm, perhaps on the enemies of the tribe, by spreading disease in their adversaries' cattle or drying up their wells. Other negative actions could be to stop neighbors' superior magic from doing harm by using preemptive strikes against them. Even so, destructive practices committed by the witch doctor are never attributed to collusion with the devil; they are just seen as combating one magic with another. One of the social functions of a belief in tribal witchcraft is a form of justice–magic resulting in a revenge hex. Individuals were kept in order by fear of punishment through the hex; the Zulus of South Africa, for example, explained any misfortunes that came their way as a consequence of hexes brought about by community quarrels. In many tribes, the shaman fulfills the role of the tribal psychiatrist, casting out devils or retrieving the lost souls of mentally ill people. The magician's effectiveness in healing or cursing depends on the credulity of the community.

Adherents to any form of witchcraft, malevolent or benevolent, believe in the power of paranormal forces to alter reality or to change the course of nature. But rain cannot be brought on by a dance nor victory won by a chant. Because of the unscientific nature of the belief in witchcraft, it must be put into the category of pseudoscience.

For further reading: Julia O'Faolain and Lauro Martines, eds., "Witches," *Not in God's Image: Women in History from the Greeks to the Victorians* (Harper Torchbooks, 1973); Burton Russell Jeffrey, *A History of Witchcraft: Sorcerers, Heretics and Pagans* (Thames and Hudson, 1980); and Hans Sebald, *Witchcraft—The Heritage of a Heresy* (Elsevier, 1978).

WURZBURG STONES See BERINGER, JOHANN.

XENOGLOSSIA The ability to speak in a foreign language unknown to the speaker. *Xenoglossia* refers to "speaking in tongues" through psychic rather than religious connections. It is considered by students of the paranormal to be one of a number of mental phenomena, including TELEPATHY, PRECOGNITION, CLAIRVOYANCE, EXTRASENSORY PERCEPTION, OUT-OF-BODY EXPERIENCES, and NEAR-DEATH EXPERIENCES. Xenoglossia has also been connected by psychologists with supposed past-life memories and suppressed memories.

Several great MEDIUMS of the past used xenoglossia as proof of their ability to communicate with the spirit world. Nineteenth-century medium Laura Edmonds, for instance, was supposed to be able to speak in Greek, Spanish, and Chippewa while she was in a trance. In the early 20th century Brazilian psychic Carlos Mirabelli demonstrated an ability to write Arabic—a language he did not know—using his psychic powers, a combination of AUTOMATIC WRITING and xenoglossia. In general, xenoglossia was only one, and not necessarily the most important one, of the powers that such psychics demonstrated.

Xenoglossia is important in psychology because of its connection with supposed past-life memories and REINCARNATION. Ian STEVENSON, a professor at the University of Virginia during the 1960s and 1970s, published a series of studies about spontaneous recollections of past lives in 1974. His statistics showed that some young children, usually between the ages of four and nine, occasionally recall details of past lives in other countries. These details can include the languages they spoke—languages that they had had no opportunity to learn. Stevenson also showed that some adults, regressing through HYPNOSIS, claim to be able to speak languages they spoke in earlier lives and speak local dialects matching those of the places where they spent those lives.

Xenoglossia is documented only through anecdotal or statistical evidence, and most scientists do not accept it as scientifically verifiable phenomenon. Some psychologists suggest that xenoglossia may be the results of unresolved personal feelings; others look at it as a way of releasing emotions. Skeptics classify reported cases of xenoglossia as a statistical error or a "measurement error"—a concept that takes into account the errors that creep into any human activity. Skeptics explain away xenoglossia by saying the reported cases are statistically insignificant: that there are not enough reported cases to study the phenomenon seriously.

YETI Hairy hominid, also called the ABOMINABLE SNOWMAN, reputed to live in the Himalaya Mountains. Called by Tibetans *Metoh-Kangmi* ("manlike thing that is not a man;" misinterpreted by a translator as "wild man of the snows"), the creature became better known in the West by its nickname, Yeti. Although the Yeti has been reported in Western literature for at least two centuries (and in Tibetan folklore for much longer), it found its permanent place in the Western popular imagination through the well-publicized expeditions of Eric Shipton and Sir Edmund Hillary in the 1950s. Shipton, a mountaineer and explorer, in 1951 took photographs of footprints that his local Sherpa guides believed were made by a Yeti. In 1960, Hillary and his crew, the first westerners to climb to the top of Mount Everest (in 1953), went back to the Himalayas hoping to capture a Yeti or at least to find scientific evidence of one. They did find footprints, and they were given an alleged yeti scalp by local Sherpas. Later analysis showed the scalp, which was covered with red and black hair, to be from an animal, not a hominid. Hillary denounced the Yeti as a fraud or a delusion.

The Yeti is reported to range between about 1.5 and 2.5 meters (5–8 feet) tall (a few reports say 4.6 meters [15 feet]) and to be covered with red and black fur every place except its face, palms, knees, and footsoles. Its footprints have been described as up to 38 centimeters long and 15 centimeters wide (15 inches long and 6 inches wide), and it carries with it an obnoxious odor.

Possible explanations for the Yeti include misperceptions of langur monkeys, Himalayan blue bears, serow goats, and even wandering ascetic monks. The footprints photographed by Shipton and others are generally considered the strongest evidence for the Yeti's existence; yet they are often explained by skeptics as prints made by ordinary animals, such as bears and snow leopards or even humans, that have become distorted by wind and by the melting and freezing of the snow in which they are embedded.

See also ALMAS; BIGFOOT; CRYTOZOOLOGY; YOWIE.

YOGA The name commonly used to refer both to a system of Hindu philosophy and to the method prescribed within that system for attaining *samādhi,* the state of inner peace and equilibrium that Hindu philosophers customarily describe as a mystical union with the universal consciousness. Following conventional usage, *Yoga* (capitalized) here denotes the former sense of the term, and *yoga* (lowercase) the latter sense.

As a fully articulated metaphysical system in its own right, Yoga ranks among the six orthodox schools of Hindu philosophy. The guiding philosophy of the Yoga school is elaborated in the *Yoga Sūtra,* which was compiled by Patañjali (second century, B.C.E.). Scholars have long noted that the Yoga system differs very little from the older, more established Samkhya system, especially with respect to the underlying metaphysical principles shared by the two systems. A noteworthy addition to the Yoga system is the introduction of a deity, Iśvara, who serves primarily as an object on which yogis can focus their attention while performing their spiritual exercises. Because Iśvara plays no active role in the metaphysical system of the Yoga philosophy, however, some scholars speculate that he was introduced for largely pragmatic reasons, perhaps to attract theistically inclined apostates from the Samkhya system.

The underlying metaphysics of the Yoga system derives from a basic distinction between self (*purusa*) and Nature (*prakrti*). This distinction corresponds roughly to familiar Western metaphysical distinctions between subject/object and mind/body. According to the teachings of the *Yoga*

Sūtra, the *purusa* comprises the enduring, eternal element of human existence, while the *prakrti* comprises the afflicted, temporal element of human existence. From this basic metaphysical distinction issues the guiding ethical/religious imperative of the Yoga system: to free one's *purusa* from the constraints of *prakrti* and thereby accede to a state of *samādhi.* To isolate one's own most *purusa* and purge it of all bodily taint, one must undertake a rigorous discipline of spiritual exercises designed to free the mind from the afflictions of the body. In the Yoga Sūtra, Patañjali elaborates an eight-stage regimen of mental and physical concentration that, on completion, will deliver the yogi to a purified state of release and tranquility. In the process of attaining liberation, the yogi gradually gains control over all sensory influences and adventitious desires so that, by dint of an act of will, he can suspend the conventional mental states and processes that are typically associated with human consciousness. At this point he is prepared, though not necessarily destined, to enter into a state of mystical union with the universal consciousness. Those yogis who achieve *samādhi,* known as *jīvanmuktas,* live in total freedom from affliction and consequently accrue no additional karma (a person's karma is a sort of debit balance acquired as a result of his or her actions in the present life or in previous incarnations). On exhausting the karma acquired in their past incarnations, these *jīvanmuktas* will exit the cycle of rebirth and enter into permanent union with the World Soul, or Brahman.

As a method for attaining liberation and release within the Yoga system, yoga comprises an eight-stage discipline of spiritual and physical exercises, which some scholars compare to the "Noble Eightfold Path" prescribed by the Buddha. The eight stages described by Patañjali are: (1) abstinence and restraint; (2) discipline; (3) posture; (4) breath-control; (5) retraction of the senses; (6) fixed attention; (7) contemplation; and (8) concentration (*samādhi*), at which point the soul reclaims its essence and freedom within the monistic unity of the universal consciousness. Broadly speaking, the yogi's progress through these eight stages describes a gradual transition from physical discipline to mental discipline, which culminates in the yogi's power to refuse all external distractions and even to perform certain feats of telekinesis. The *Yoga Sūtra* warns that the physical discipline attained by yogis on the path to liberation must never be treated simply as an end in itself, but always as a means to the advanced mental discipline required to achieve *samādhi.* This admonition is particularly appropriate in light of the widespread legends surrounding the magical and supernatural powers displayed by some yogis. These alleged powers include the ability to become quite small or even invisible; to grow to an enormous size; to transport oneself to another place by dint of an act of will; to gain knowledge of the past and future; to speak with the dead and make the deceased appear; and to read the thoughts of others. These ancillary gifts and powers, though indeed impressive, tend to distract aspiring yogis from the genuine aim of the Yoga philosophy—namely, to secure the release of one's purusa from the afflictions of body and nature.

A related distraction from the metaphysical truth of the Yoga system is largely responsible for the relatively recent popularity of yoga as a pseudoscience. Individuals who possess little or no knowledge of Yoga philosophy now flock to schools devoted to the instruction and practice of Hatha yoga. Especially (though not exclusively) in Western cultures and societies, yoga is prized for its alleged benefits to dancers and athletes who seek the competitive advantage that lies in an expanded range of bodily flexibility; to convalescents who seek freedom from chronic muscle and joint pain; to overworked, high-strung, and troubled individuals who seek temporary release from their nagging obsessions and anxieties; and to underachieving individuals who seek to focus their energies and concentrate their attention. Although the value of these benefits is indisputable, they tend to arise as a result of treating the physical discipline involved in yoga as an end unto itself. To reap the "magical" benefits of yogic discipline with no concern for spiritual purification may be extremely popular, and even salutary, but it also involves a corruption of the guiding philosophy of the Yoga system.

For further reading: Sarvepalli Radhakrishnan and Charles A. Moore eds., *A Sourcebook in Indian Philosophy* (Princeton University Press, 1957).

YOWIE Fabled hairy hominid, also called yahoo, native to Australia. Like the American BIGFOOT, the Himalayan YETI, and the Chinese ALMA, the yowie is an elusive creature said to live in remote regions, primarily high forest areas. The yowie has been reported as being a longtime inhabitant of Aboriginal lore, and accounts of sightings by European-Australians have been reported since at least the mid-19th century and up to today. The yowie is usually described as taller than a human male, black or dark skinned, and covered with gray hair. Its footprint is larger than a human's and, at least at one site, has only four toes. A handprint found after one sighting show that its little finger is somewhat splayed, so as to suggest a second thumb. Some witnesses suggest that it may be related to orangutans or apes, while others suggest that it could be the remnants of an earlier form of humankind. However, because there is no fossil evidence of any primates other than humans on the Australian continent, the yowie is not generally acknowledged by Australian scientists and remains a fringe creature even for modern cryptozoologists, but anecdotal lore remains prolific.

See also CRYPTOZOOLOGY.

Z

ZEN Refers popularly to a tradition of Buddhism and, more literally, to the MEDITATION practiced in this tradition as the primary means of attaining enlightenment, which is the objective shared by all traditions of Buddhism. Although all schools and traditions of Buddhism value introspection and spirituality, the Zen school is unique in the unusual emphasis that it places on meditation. Zen Buddhists generally believe that a Buddha-nature resides within each person. There is no need, consequently, to seek enlightenment outside oneself. To disclose one's own Buddha-nature, one need only focus one's attention inward, through the meditation techniques prescribed and/or perfected by a Zen master. Meditation enables the Zen practitioner to escape the cage of discursive thinking and to explode the conventional boundaries of egoconsciousness.

The meditation practiced by Zen Buddhists comprises a rigorous regimen of mind-body discipline, which is intended to concentrate and focus the mind. As a means of helping their disciples to refine their meditation techniques, Zen Masters often prescribe kōans, which are short, epigrammatic parables in which seemingly important questions about the meaning of life are typically answered with illogical, nonsensical replies. The paradoxical "teachings" dispensed by these kōans are intended to assist the Zen practitioner in transcending the limitations of binary logic and discursive thinking. Many Zen Masters believe that intense meditation upon a single *kōan* can precipitate a sudden flash of intuition, thereby delivering the Zen practitioner to an abrupt awakening or enlightenment.

Although Zen meditation typically commences by attaching itself to a single idea or object (such as a particular kōans), the ultimate goal of meditation is to dispense with all objects and ideas of consciousness and to throw off the discursive, teleological frame of ego consciousness itself. All striving, even the striving for self-realization that leads the Zen initiate to take up meditation in the first place, must be set aside in a posture of anticipatory releasement. When one's mind is finally free of all conceptions, of all objectives and subjective desires, one may experience the embodied truth of enlightenment. In the course of its own progress, Zen meditation must undergo a radical transformation from a technique deployed by initiates to facilitate enlightenment into enlightenment itself.

Although Zen is popularly associated with Japan and Japanese sects of Buddhism, its roots can be traced proximately to the Ch'an school in China and ultimately to the Māhāyana (or "Greater Vehicle") tradition in India. In fact, both the Japanese word *Zen* (from *zenna*) and the Chinese word Ch'an are transliterations of the Sanskrit word dhyāna (meditation). The tradition of Zen Buddhism is popularly understood to have migrated first to China in the sixth century with the journey of Bodhidharma from India to China, and then it migrated to Japan in the 12th century.

Like all schools and traditions of Buddhism, Zen takes as its primary objective the direct, intuitive experience of the Buddha-nature that resides within each individual. Buddhism originated in India as a religion or way of life based on the teachings and exemplary conduct of Siddhartha Gautama, who, upon attaining Enlightenment in 528 B.C.E., took the title *Buddha,* which means "The Enlightened (or Awakened) One." For the most part, Buddhists acknowledge no official dogma, no theology, and no tradition of divine revelation. Although not, strictly speaking, a species of humanism, Buddhism appeals for its wisdom and teachings to no founding divinity. Its basic concern is with human existence and the Buddha-nature that resides within each individual. The vital core of Buddhism is perhaps best conveyed by the "Four Noble Truths" professed by the Buddha in his famous Sermon in the Deer Park: (1) All existence is suffering; (2) All suffering is caused by craving; (3) All suffering can be ended; (4) The way to end suffering is to follow the "Noble Eightfold Path" to self-abnegation. The culmination of the practice of Buddhism is known as nirvāna (or *satori* in Japanese), which denotes both the successful transcendence of ego consciousness and a quasi-mystical embodiment of the pure, incorporeal Buddha-nature.

Buddhists generally believe that the generative power of the human mind must be turned against the ego consciousness that ordinarily guides and contours human experience. The suffering that is endemic to the human condition can be negated, but only through a systematic dismantling of the ego (or self) that invests this suffering with meaning. Appealing to a karmic principle of rebirth and transmigration, the Buddha teaches that this process of self-abnegation may require several (or even many) lifetimes to complete.

The Ch'an-Zen tradition of Buddhism was introduced into Japan on several occasions, but it finally took root through the efforts of two charismatic teachers in the 12th and 13th centuries: Eisai and Dōgen. Dōgen in particular taught that Buddhahood cannot be pursued as a goal for one's own personal aggrandizement but only for its own sake. He consequently objected to what he perceived to be an overreliance on methods and techniques that presupposed a teleological frame of reference for Zen meditation. Even the pursuit of sudden enlightenment (through, for example, sustained meditation on a particular kōan) might hinder one's progress toward self-realization, especially if this pursuit focuses the mind's attention on a single task or goal. Dōgen thus emphasizes the importance of becoming the type of person for whom enlightenment and self-realization are valued in themselves and not as the valedictory achievements of one's individual ego consciousness.

The teachings of Eisai and Dōgen have been extremely influential in Japan. The rigorous discipline required of Zen practitioners is valued by soldiers and warriors, while the elegant simplicity with which the teachings of Zen Buddhism express the truth of human existence has inspired artists and poets.

The relatively recent appropriation of Zen Buddhism as a pseudoscience is partially attributable to the burgeoning interest among Western cultures in alternative forms of spirituality. Zen meditation and relaxation techniques are regularly marketed in the West for their purported benefits in the areas of stress reduction, performance enhancement, productivity development, attention deficit reduction, and crisis management. Although these selective applications of Zen meditation techniques often tend to trivialize the rigorous mind-body discipline required of devoted Zen practitioners, the attraction to Zen Buddhism as a pseudoscience remains strong in the West.

For further reading Robert M. Pirsig, *Zen and the Art of Motorcycle Maintenance* (Corgi, 1976).

ZENER CARDS Deck of cards developed for research in EXTRASENSORY PERCEPTION (ESP). The cards were designed by Karl E. Zener, a colleague of Joseph Banks RHINE at Duke University's Department of Psychology in the 1930s. Rhine had developed ESP tests using cards but ran into a number of methodological problems. He asked Zener, who had done special research in the psychology of perception, to design a set of cards whose symbols were more easily distinguished than the already existing alternatives.

The resultant Zener cards pictured five symbols: a circle, a square, wavy lines, a triangle, and a cross. The complete deck had 25 cards, each symbol appearing on 5 cards. There were no color differences as in ordinary playing cards. The number of cards and symbols made statistical analysis easier. They have been widely used in ESP tests ever since their introduction.

As Rhine's work proceeded, Zener turned against him and emerged as a hostile critic of PARAPSYCHOLOGY.

For further reading: Arthur S. Berger and Joyce Berger, *The Encyclopedia of Parapsychology and Psychical Research* (Paragon House, 1991).

ZIMBABWE Remains of a medieval native civilization in southeast Africa. In 1871, German explorer Karl Mauch (1837–75) set out on an expedition from South Africa. He was heading north, looking for the fabled city of Ophir, perhaps the home of the Queen of Sheba or the site of KING SOLOMON'S MINES. However, Mauch was less interested in archaeological exploration than he was in expanding German power and influence in the aftermath of the Franco-Prussian War. On September 5, 1871, he came across a series of stone ruins north of the Limpopo River in what was then known as Rhodesia. The ruins—built of individual granite stones without the use of mortar to hold them together—fascinated Mauch. Some of them were on top of a stony hill, later named the Acropolis; others were in the plain below. The site, known to the local natives as Great Zimbabwe, meaning "stone enclosure," covered an area of more than 24 hectares (60 acres).

Mauch was extremely impressed by the grandeur of the site. He compared it to the great cathedrals and castles of Europe, and he could not conceive of it being built by Africans, whom he despised. He reported that local natives, when questioned, declared that the great wall had been built in the distant past by a foreign race of white people. In his report on the site, Mauch dated the site to before the birth of Christ. He also believed that he had found evidence that the ancient

Zimbabwe

inhabitants of the lost city practiced Jewish rites of sacrifice. Mauch concluded, "The Queen of Sheba is the Queen of Zimbabwe."

The ruins became a matter of contention in the British colony's history. Prominent British and colonial authorities, including archaeologist James Theodore Bent (1852–97) and journalist R. N. Hall both declared that Great Zimbabwe had been built by a non-African race, perhaps Phoenicians, Arabs, or other Middle Easterners. Hall, who was a colonial official as well as a journalist, worked on this assumption while examining the site. He was so convinced of the truth of this theory that he routinely discarded artifacts of African manufacture, declaring that they were later intrusions. When archaeologist David Randall-MacIver excavated Great Zimbabwe in 1905, however, he proved that Zimbabwe was built by native Africans and that it dated only from the medieval period, between about 1000 and 1450—far too late to be Solomon's Ophir.

Later investigations confirmed Randall-MacIver's conclusions. The great walls that Mauch had compared to European fortresses were representations in stone of native African huts still in use in his day. Archaeologists such as Gertrude Caton-Thompson showed that Great Zimbabwe was an important trading post between the copper and gold mines of the African interior and the traders of the East African coast. Discovery of glass beads from Arabia and glazed pottery from as far away as China helped confirm the site's date and its importance to international trade. The site was first occupied by the Karanga people—ancestors of the tribes who still live in the area—around 950. It began to decline in the 15th century when the trade routes shifted away from the city. As the trade failed, the citizens left the city, finally leaving it deserted by the 18th or early 19th century.

Despite the evidence produced by such archaeologists as Randall-MacIver and Caton-Thompson, white Rhodesians were unwilling to accept Great Zimbabwe as a native

African city. They continued to believe that it had been built by a group of Middle-Eastern colonists in ancient times. It was not until the 1960s that radio-carbon dating proved the medieval date of the center and not until the 1970s that Great Zimbabwe was shown conclusively to be of African origin. Part of the reason for the resistance among white Rhodesians was because of the significance of Zimbabwe to anticolonialist black Rhodesians: To them, Great Zimbabwe was a symbol of the power of black Africa, and they adopted it as the symbol of their independence movement. Although Britain offered Rhodesia independence in 1965 if it would accept a black majority rule, white Rhodesians rebelled and declared independence unilaterally. From 1965 until 1978, the country was embroiled in civil war. In 1979, Great Britain negotiated a truce between the two sides and agreed to take over administration of the country again. When the country finally gained its dependence from Great Britain in 1980, the newly elected black African politicians quickly renamed their nation Zimbabwe.

For further reading: Gertrude Caton-Thompson, *The Zimbabwe Culture: Ruins and Reactions* (Negro Universities Press, 1970); David Chanaiwa, *The Zimbabwe Controversy: A Case of Colonial Historiography* (Syracuse University, 1973); Brian M. Fagan, *In the Beginning: An Introduction to Archaeology* (Little, Brown & Co., 1972); Peter Garlake, *Great Zimbabwe* (Stein and Day, 1973); and D. W. Phillipson, *African Archaeology*, 2d ed. (Cambridge University Press, 1993).

ZOMBIES "The living dead," or *zombie cadavre;* people supposedly raised from the dead by Voodoo (vodun) *houngans* (priests) or *bokors* (evil sorcerers) and used as slaves. Thanks to horror novels and movies, zombies have become a part of Western folklore. They originated, however, in Haitian Voodoo belief and are not a part of Voodoo belief in other parts of the world where Voodoo and Voodoolike religions are practiced.

Haitians have long lived with the fear that evil sorcerers could capture the bodies of their dead, expel their souls, and make them into slaves for as long as the bokor wanted them. Zombies are thought to have no mind or will of their own and to be entirely subject to the commands of their owner. Several well-documented 20th-century cases exist of people who claim to have survived zombification. One is that of Clairvius Narcisse, who died in 1962 in a hospital. His family buried him. Eighteen years later, Narcisse walked up to his sister on the street and announced that he was back. He told her that his slave master had been killed and he had escaped, gradually regaining much of his memory and his physical functioning. Narcisse was thoroughly tested and investigated, and no one could find holes in his story.

Although most people do not believe in zombies, the publicity gained by Narcisse and a few other "restored" zombies led Western scientists to be curious about the phenomenon. Could there be a scientific explanation? The answer seemed to lie in some kind of drug. It was long known that Voodoo priests and bokors used herbs and powders to effect some of their "magic." Could they have some secret knowledge of drugs that could make someone into a zombie, and could this knowledge be used in Western medicine, for example, to improve anesthesiology?

In 1982, ethnobotanist Wade Davis traveled to Haiti to seek the answers. His investigations led him to conclude that a concoction of certain substances, including poisonous plants, toxins from poisonous toads, and, most important, a nerve poison (tetrodotoxin) from a puffer fish could indeed make someone appear to be dead. The antidote, according to a *houngan* Davis worked with, included the potent plant *Datura stramonium,* or "zombie cucumber," which, carefully used, could keep a person in a drugged but controllable state.

Davis's findings were controversial, but they seem to be the best explanation to date for the mystery of the zombie.

For further reading: Wade Davis, *The Serpent and the Rainbow* (Warner Books, 1985).

ZONE THERAPY See REFLEXOLOGY.

ZWAAN RAYS An energy field supposedly capable of stimulating the activity of the psychic senses. In 1948, Dutch researcher N. Zwaan demonstrated an apparatus, which supposedly created the Zwaan rays, at a gathering of the International Spiritualist Federation Congress in London. The following year, the Spirit Communication Society was founded in Manchester, England, at which time an improved Zwaan-ray machine was demonstrated, and in 1954 an optimistic report was issued. However, the Zwaan rays were never shown as having more than subjective effects upon people who already believed in them, and the Zwaan apparatus in subsequent years was abandoned.

For further reading: Mark Dyne, *Electronic Communication for the Spiritual Emancipation of the People* (Spirit Electronic Communication Society, 1954).

SELECTED BIBLIOGRAPHY

Abbott, David P. *Behind the Scenes with the Medium.* Chicago: Open Court, 1907.

Abell, George O., and Barry Singer, eds. *Science and the Paranormal: Probing the Existence of the Supernatural.* New York: Scribner's, 1981.

Ackerknecht, Erwin H., and Henri V. Vallois. *Franz Joseph Gall, Inventor of Phrenology and His Collection.* Madison: University of Wisconsin Medical School, 1956.

Adler, Margot. *Drawing Down the Moon.* Boston: Beacon Press, 1986.

Alcock, James E. *Parapsychology: Science or Magic?* Elmsford, N.Y.: Pergamon, 1981.

———. *Science and Supernature: A Critical Appraisal of Parapsychology.* Buffalo, N.Y.: Prometheus Books, 1990.

Alexander, F. M. *The Use of the Self.* Los Angeles, Calif.: Centerline Press, 1985.

Anderson, George. *It Must Be Mind-Reading.* Chicago: Ireland Magic, 1963.

———. *You, Too, Can Read Minds.* Chicago: Magic, 1968.

Annemann, Theodore. *Practical Mental Effects.* New York: Holden's Magic Shops, 1944.

Appleby, Joyce, Lynn Hunt, and Margaret Jacob. *Telling the Truth About History.* New York: Norton, 1994.

Archer, Fred. *Crime and the Psychic World.* New York: William Morrow, 1969.

Armitage, A. *Edmond Halley.* Port Chester, N.Y.: Nelson, 1966.

Asimov, Isaac. *The Collapsing Universe.* New York: Walker, 1977.

———. *Extraterrestrial Civilizations.* New York: Crown, 1979.

Au, Whitlow. *The Sonar of Dolphins.* New York: Springer-Verlag, 1993.

Aveni, Anthony. *Behind the Crystal Ball: Magic, Science, and the Occult from Antiquity through the New Age.* New York: Random House, 1996.

Bach, Edward. *Heal Thyself.* London: Fowler, 1931.

Bach, Marcus. *The Will to Believe.* Englewood Cliffs, N.J.: Prentice-Hall, 1955.

Baggally, W. W. *Telepathy: Genuine and Fraudulent.* London: Methuen, 1922.

Bagnall, Oscar. *The Origin and Properties of the Human Aura.* 1937. rev. ed. New Hyde Park, N.Y.: University Books, 1970.

Baker, Robert A. *Hidden Memories: Voices and Visions from Within.* Amherst, N.Y.: Prometheus Books, 1996.

Bander, Peter. *Carry on Talking: How Dead Are the Voices?* London: Colin Smythe Ltd., 1972.

Banville, John. *Doctor Copernicus.* New York: Random House, 1993.

———. *Kepler.* New York: Random House, 1993.

Barbanell, Maurice. *This Is Spiritualism.* London: Herbert Jenkins, 1959.

Barber, Paul. *Vampires, Burial and Death: Folklore and Reality.* New Haven, Conn.: Yale University Press, 1988.

Barnes, Barry. *T. S. Kuhn and Social Science.* New York: Macmillan, 1982.

Barnum, P. T. *The Humbugs of the World.* New York: G. W. Carleton & Co., 1865. Reprint. Detroit, Mich.: Singing Tree Press, 1970.

Barrett, S. and W. T. Jarvis, *The Health Robbers: A Close Look at Quackery in America.* Buffalo, N.Y.: Prometheus Books, 1993.

Barrett, William. *Death-Bed Visions.* London: Methuen, 1926.

———. *The Divining Rod.* New York: University Books, 1967.

Barron, Randall F. *Cryogenic Systems.* 2d ed. New York: Oxford University Press, 1985.

Barthelemy-Madaule, Madeleine. *Lamarck the Mythical Precursor.* Translated by M. H. Shank. Cambridge, Mass.: MIT Press, 1982.

Basil, Robert, ed. *Not Necessarily the New Age: Critical Essays.* Buffalo, N.Y.: Prometheus Books, 1988.

Bauer, Henry H. *Beyond Velikovsky: The History of a Public Controversy.* Champaign: University of Illinois Press, 1984.

———. *The Enigma of Loch Ness: Making Sense of a Mystery.* Champaign: University of Illinois Press, 1986.

Bayless, Raymond. *Experiences of a Psychical Researcher.* New Hyde Park, N.Y.: University Books, 1972.

Becker, Carl B. *Paranormal Experience and Survival of Death.* Albany, N.Y.: SUNY Press, 1993.

Beloff, John. *Parapsychology: A Concise History.* New York: St. Martin's, 1993.
———. *The Relentless Question: Reflections on the Paranormal.* Jefferson, N.C.: McFarland & Co., 1990.
Benjamin, Elsie. *Man at Home in the Universe: A Study of the Great Evolutionary Cycle: the "Globes," the "Rounds," "Races," "Root-Races" and "Sub-Races."* San Diego, Calif.: Point Loma, 1981.
Benson, Herbert, and Miriam Z. Klipper. *The Relaxation Response.* New York: Avon, 1976.
Benson, Herbert, and William Proctor. *Beyond the Relaxation Response.* New York: Berkeley Publishing Co., 1985.
Benwell, Gwen, and Arthur Waugh. *Sea Enchantress: The Tale of the Mermaid and Her Kin.* New York: Citadel Press, 1965.
Berger, Arthur S., and Joyce Berger. *The Encyclopedia of Parapsychology and Psychical Research.* New York: Paragon House, 1991.
Berlitz, Charles. *The Bermuda Triangle.* Garden City, N.Y.: Doubleday, 1974.
Bernard, Raymond W. *The Hollow Earth: The Greatest Geographical Discovery in History Made by Admiral Richard E. Byrd in the Mysterious Land beyond the Poles—the True Origin of the Flying Saucers.* New York: Bell Publishing Co., 1969.
Berra, Tim M. *Evolution and the Myth of Creationism.* Stanford, Calif.: Stanford University Press, 1990.
Berstein, Morey. *The Search for Bridey Murphy.* Garden City, N.Y.: Doubleday, 1956.
Binet, A., and T. Simon. *The Development of Intelligence in Children.* New York: Arno Press, 1973.
Bird, Christopher. *The Divining Hand.* New York: Dutton, 1979.
Bishop, Peter, and Michael Darton, eds. *The Encyclopedia of World Faiths.* London: Macdonald and Co. Ltd, 1987.
Blackmore, Simon Augustine. *Spiritism Facts and Frauds.* New York: Benziger Brothers, 1924.
Blackmore, Susan J. *The Adventures of a Parapsychologist: In Search of the Light.* Amherst, N.Y.: Prometheus Books, 1996.
———. *Beyond the Body: An Investigation of Out-of-the-Body Experiences.* Chicago: Academy Chicago, 1992.
———. *Dying To Live: Near-Death Experiences.* Reprint. Buffalo, N.Y.: Prometheus Books, 1993.
Blackmore, Susan J., and Adam Hart-Davis. *Test Your Psychic Powers.* New York: Sterling, 1997.
Blavatsky, Helena P. *Isis Unveiled.* 2 vols. Reprint. Pasadena, Calif.: Theosophical University Press, 1988.
———. *The Secret Doctrine.* 2 vols. Reprint. Pasadena, Calif.: Theosophical University Press, 1988.
Blunsdon, Norman. *A Popular Dictionary of Spiritualism.* London: Arco Publications, 1961.
Boning, Richard A. *The Cardiff Giant.* Baldwin, N.Y.: Dexter & Westbrook, 1972.
Booth, John. *Psychic Paradoxes.* Buffalo, N.Y.: Prometheus Books, 1984.
Bord, Janet. *Fairies: Real Encounters with Little People.* New York: Carroll & Graf, 1997.
Bord, Janet, and Colin Bord. *Unexplained Mysteries of the Twentieth Century.* Chicago: Contemporary Books, 1990.
———. *The Enchanted Land: Myths & Legends of Britain's Landscape.* San Francisco: HarperSanFrancisco, 1995.
Bouw, Geradus D. *With Every Wind of Doctrine: Biblical, Historical, and Scientific Perspectives of Geocentricity.* Cleveland, Ohio: Tychonian Society, 1984.
Bowers, Edwin F. *Spiritualism's Challenge.* New York: National Library Press, 1936.
Bowler, Peter J. *Evolution: The History of an Idea.* rev. ed. Berkeley: University of California Press, 1989.
Boylan, Richard J., and Lee K. Boylan. *Close Extraterrestrial Encounters: Positive Experiences with Mysterious Visitors.* Tigard, Oreg.: Wild Flower Press, 1994.
Braden, Charles S. *Spirits in Rebellion.* Dallas: Southern Methodist University Press, 1963.
Brandon, Ruth. *The Spiritualists.* Buffalo, N.Y.: Prometheus Books, 1983.
Brennan, Richard P. *Levitating Trains and Kamikaze Genes—Technological Literacy for the 1990s.* New York: Wiley, 1990.
Briggs, Katherine Mary. *An Encyclopedia of Fairies, Hobgoblins, Brownies, Bogies, and Other Supernatural Creatures.* New York: Pantheon Books, 1976.
Broad, C. D. *The Mind and its Place in Nature.* London: Paul, Trench, Trubner, 1925.
Broad, W., and N. Wade. *Betrayers of the Truth.* New York: Oxford University Press, 1982.
Bromley, David G., and Anson D. Shupe. *Moonies in America: Cult, Church, and Crusade.* Beverly Hills: Sage, 1979.
Broughton, Richard S. *Parapsychology: The Controversial Science.* New York: Ballantine, 1991.
Brown, Slater. *The Heyday of Spiritualism.* New York: Hawthorn Books, 1970.
Brush, Stephen G. *History of Modern Planetary Physics.* Vol. 1. Cambridge: Cambridge University Press, 1996.
Bryan, C. D. B. *Close Encounters of the Fourth Kind: Alien Abductions, UFOs and the Conference at MIT.* New York: Alfred A. Knopf, 1995.
Butler, Kurt A. *A Consumer's Guide to Alternative Medicine: A Close Look at Homeopathy, Acupuncture, Faith-Healing, and Other Unconventional Treatments.* Buffalo, N.Y.: Prometheus Books, 1992.
Buzan, Tony, and Barry Buzan. *The Mind Map Book: How to Use Radiant Thinking to Maximize Your Brain's Untapped Potential.* New York: NAL-Dutton, 1996.
Byrd, Richard Evelyn. *Alone.* New York: St. Martin's, 1986.
Cabot, Laurie. *Power of the Witch.* New York: Bantam, 1989.
Calder, Nigel. *The Key to the Universe.* New York: Viking, 1977.
Cameron, Iain, and David Edge. *Scientific Images and Their Social Uses: An Introduction to the Concept of Scientism.* London: Butterworths, 1979.
Camp, L. Sprague de. *Lost Continents: The Atlantis Theme in History, Science, and Literature.* New York: Dover, 1970.
Campbell, Steuart. *The Loch Ness Monster: The Evidence.* Reprint. New York: Macmillan, 1991.
Candland, Douglas Keith. *Feral Children and Clever Animals: Reflections on Human Nature.* New York: Oxford University Press, 1993.
Cannell, J. C. *The Secrets of Houdini.* New York: Dover Publications, 1973.

Caponigro, Paul. *Megaliths*. Boston: Little, Brown & Co., 1986.
Capra, Fritjof. *The Tao of Physics: An Exploration of the Parallels Between Modern Physics and Eastern Mysticism*. Boston: Shambhala, 1991.
———. *The Turning Point: Science, Society and the Rising Culture*. New York: Bantam, 1984.
Carr, John Dickson. *Life of Sir Arthur Conan Doyle*. New York: Carroll & Graf, 1987.
Carrington, Hereward. *Laboratory Investigations into Psychic Phenomena*. Philadelphia: David McKay Co., 1939. Reprint. New York: Arno Press, 1975.
———. *Personal Experiences in Spiritualism*. London: T. Wetner Lurie, n.d.
———. *The Physical Phenomena of Spiritualism*. New York: American Universities, 1920.
———. *Sideshow and Animal Tricks*. Atlanta, Ga.: Pinchpenny Press, 1973.
Carrington, Richard. *Mermaids and Mastodons: A Book of Natural and Unnatural History*. New York: Rinehart & Co., 1957.
Carson, Gerald. *Cornflake Crusade*. New York: Rinehart & Co., 1957.
Carter, Howard. *The Tomb of Tutankhamen*. Philadelphia: David McKay Co., 1972.
Caton-Thompson, Gertrude. *The Zimbabwe Culture: Ruins and Reactions*. New York: Negro Universities Press, 1970.
Chanaiwa, David. *The Zimbabwe Controversy: A Case of Colonial Historiography*. Syracuse, N.Y.: Syracuse University, 1973.
Chandler, Russell. *Understanding the New Age*. Dallas: Word, 1988.
Chandrasekhar, S. *Eddington, the Most Distinguished Astrophysicist of His Time*. Cambridge: Cambridge University Press, 1983.
Childs, Gilbert. *Steiner Education in Theory and Practice*. London: Floris Books, 1991.
Christian, William A., Jr. *Apparitions in Late Medieval and Renaissance Spain*. Princeton, N.J.: Princeton University Press, 1981.
Christopher, Milbourne. *ESP, Seers and Psychics*. New York: Thomas Y. Crowell, 1970.
———. *Houdini: The Untold Story*. New York: Thomas Y. Crowell, 1970.
———. *The Illustrated History of Magic*. New York: Thomas Y. Crowell, 1973.
———. *Mediums, Mystics and the Occult*. New York: Thomas Y. Crowell, 1975.
———. *One Man's Mental Magic*. New York: Tannen Publications, 1952.
———. *Panorama of Magic*. New York: Dover Publications, 1962.
Churchland, P. *Matter and Consciousness*. Cambridge, Mass.: MIT Press, 1988.
Churchward, James. *Children of Mu*. Reprint. Albuquerque, N. Mex.: Brotherhood of Life, Inc., 1988.
———. *Cosmic Forces of Mu*. 2 vols. Reprint. Albuquerque, N. Mex.: Brotherhood of Life, Inc., 1992.
———. *The Lost Continent of Mu*. Reprint. Albuquerque, N. Mex.: Brotherhood of Life, Inc., 1994.
———. *The Sacred Symbols of Mu*. Reprint. Albuquerque, N. Mex.: Brotherhood of Life, Inc., 1988.
Claiborne, Robert. *The First Americans*. New York: Time-Life Books, 1973.
Clare, Anthony. *Psychiatry in Dissent*. New York: Tavistock, 1980.
Clark, Jerome. *Encyclopedia of Strange and Unexplained Physical Phenomena*. Detroit, Mich.: Gale Research Company, 1993.
———. *UFO Directory: A Guide to Organizations, Research Collections, Museums, Publications, Periodicals, Web Sites and Other Resources Concerning UFO Phenomena*. Detroit, Mich.: Omnigraphics, 1997.
———. *The UFO Encyclopedia*. 3 vols. *Vol. 1, UFOs in the 1980s*. Detroit, Mich: Omnigraphics, 1990. *Vol. 2, The Emergence of a Phenomenon: UFOs from the Beginning through 1959*. Detroit, Mich.: Omnigraphics, 1992. *Vol. 3, High Strangeness: UFOs from 1960 through 1979*. Detroit, Mich.: Omnigraphics, 1996.
———. *Unexplained!* Detroit, Mich.: Visible Ink, 1993.
Clarke, Arthur C. *Astounding Days: A Science Fictional Autobiography*. New York: Bantam, 1989.
———. *The View from Serendip*. New York: Random House, 1977.
Close, Frank. *Too Hot to Handle: The Race for Cold Fusion*. Princeton, N.J.: Princeton University Press, 1991.
Cloyd, E. L. *James Burnett, Lord Monboddo*. New York: Clarendon Press, 1972.
Cohen, Bernard. *Science and the Founding Fathers*. Cambridge, Mass.: Harvard University Press, 1995.
Cohen, Morris R. *Reason and Nature: An Essay on the Meaning of Scientific Method*. New York: Dover, 1978.
Cohn, Norman. *Europe's Inner Demons*. New York: Basic Books, 1975.
———. *The Pursuit of the Millennium*. London: Secker & Warburg, 1957.
Conant, J. B. *Harvard Case Histories in Experimental Science*. Cases 6 & 7. Cambridge, Mass.: Harvard University Press, 1957.
Condon, E. U. *Scientific Study of Unidentified Flying Objects*. New York: Bantam Books, 1969.
Constable, George. *The Neanderthals*. New York: Time-Life Books, 1973.
Conway, Flo, and Jim Siegelman. *Snapping: America's Epidemic of Sudden Personality Change*. Philadelphia: Lippincott, 1978.
Cooter, Roger. *The Cultural Meaning of Popular Science: Phrenology and the Organization of Consent in Nineteenth-Century Britain*. New York: Oxford University Press, 1984.
Corballis, Michael C. *The Lopsided Ape: Evolution of the Generative Mind*. New York: Oxford University Press, 1991.
Coren, Michael. *The Life of Sir Arthur Conan Doyle*. London: Bloomsbury, 1996.
Corinda. *Thirteen Steps to Mentalism*. New York: Louis Tannen, 1968.
Corinda, and Ralph W. Read, eds. *The Complete Guide to Billet-Switching*. New York: Louis Tannen, 1976.
Corliss, William R. *The Moon and the Planets*. Glen Arm, Md.: The Sourcebook Project, 1985.

———. *Neglected Geological Anomalies*. Glen Arm, Md.: The Sourcebook Project, 1990.

———. *Unknown Earth: A Handbook of Geological Enigmas*. Glen Arm, Md.: The Sourcebook Project, 1980.

Corliss, William R., ed. *Ancient Man*. Glen Arm, Md.: The Sourcebook Project, 1978.

———. *Strange Phenomena: A Sourcebook of Unusual Natural Phenomena*. Vols. G 1–2. Glen Arm, Md.: The Author, 1974.

Cornforth, Maurice. *The Open Philosophy and the Open Society: A Reply to Dr. Popper's Refutations of Marxism*. London: Lawrence & Wishart, 1968.

Corydon, Bent, and L. Ron Hubbard Jr. *L. Ron Hubbard: Messiah or Madman?* Secaucus, N.J.: Lyle Stuart, Inc., 1987.

Costello, Peter. *In Search of Lake Monsters*. New York: Coward, McCann & Geoghegan, 1974.

Coulter, Harris L., and Barbara Loe Fisher. *A Shot in the Dark: Why the P in the DPT Vaccination May Be Hazardous to Your Child's Health*. Garden City Park, N.Y.: Avery Publishing, 1991.

Couttie, Bob. *Forbidden Knowledge: The Paranormal Paradox*. Lutterworth Press, 1988.

Covina, Gina. *The Ouija Board*. New York: Simon & Schuster, 1979.

Crabtree, Adam. *Animal Magnetism, Early Hypnotism, and Psychical Research, 1766-1935: An Annoted Bibliography*. Hackensack, N.J.: Kraus International, 1988.

Crail, Ted. *Apetalk and Whalespeak the Quest for Interspecies Communication*. Boston: J. P. Tarcher/Houghton Mifflin, 1981.

Cranston, Ruth. *The Miracle of Lourdes*. New York: Doubleday, 1988.

Cranston, Sylvia. *H.P.B.: The Extraordinary Life and Influence of Helena Blavatsky*. New York: Putnam, 1993.

Cranston, Sylvia, and Carey Williams. *Reincarnation: A New Horizon in Science, Religion, and Society*. Pasadena, Calif.: Theosophical University Press, 1984.

Crawford, Charles, and Graham T. T. Molitor. *The Evolution of Electro-magnetic Fields as a Public Policy Issue: Analysis and Response Options*. Hill and Knowlton, 1991.

Cromer, Alan. *Uncommon Sense: The Heretical Nature of Science*. New York: Oxford University Press, 1993.

Crossley-Holland, Kevin. *The Stones Remain: Megalithic Sites of Britain*. London: Rider, 1989.

Crowe, M. J. *The Extraterrestrial Life Debate 1750–1900*. New York: Cambridge University Press, 1986.

Culver, R. B., and P. A. Ianna. *Astrology: True or False? A Scientific Evaluation*. Buffalo, N.Y.: Prometheus Books, 1988.

———. *The Gemini Syndrome: A Scientific Explanation of Astrology*. Buffalo, N.Y.: Prometheus Books, 1984.

Curry, Paul. *Magician's Magic*. New York: Franklin Watts, 1965.

Cytowic, Richard E. *The Man Who Tasted Shapes*. Grand Rapids, Mich.: Abacus, 1994.

———. *Synesthesia: A Union of the Senses*. New York: Springer, 1989.

Damer, T. Edward. *Attacking Faulty Reasoning*. 2d ed. Belmont, Calif.: Wadsworth, 1987.

Daniel, Glyn. *Megaliths in History*. London: Thames & Hudson, 1972.

Darnton, Robert. *Mesmerism and the End of the Enlightenment in France*. Cambridge, Mass.: Harvard University Press, 1968.

Dash, Mike. *Borderlands*. London: Heinemann, 1997.

Davenport, Reuben Briggs. *The Death Blow to Spiritualism*. New York: G. W. Dillingham, 1888.

Davenport, Richard Alfred. *Sketches of Imposture, Deception and Credulity: An Exact Diary*. London: T. Tegg & Son, 1837.

Davies, John D. *Phrenology, Fad and Science: A 19th-Century American Crusade*. New York: Archon Books, 1971.

Davies, Paul. *God and the New Physics*. New York: Touchstone Books, 1984.

Davis, Wade. *The Serpent and the Rainbow*. New York: Warner Books, 1985.

Dawkins, Richard. *The Blind Watchmaker*. New York: Norton, 1986.

Day, Michael H. *Guide to Fossil Man*. 4th ed. Chicago: University of Chicago Press, 1986.

de Mille, Richard. *Castaneda's Journey: The Power and the Allegory*. Santa Barbara, Calif.: Capra Press, 1976.

Delgado, Pat. *Crop Circles: Conclusive Evidence?* London: Bloomsbury, 1992.

Delgado, Pat, and Colin Andrews. *Crop Circles: The Last Evidence*. London: Bloomsbury, 1990.

Desmond, Adrian. *Darwin: The Life of a Tormented Evolutionist*. New York: Norton, 1994.

Devereaux, Paul. *Earth Lights Revelation*. London: Blandford Press, 1990.

———. *Shamanism and the Mystery Lines: Ley Lines, Spirit Paths, Shape-Shifting, and Out-of-Body Travel*. St. Paul, Minn.: Llewellyn, 1993.

Dexter, Will. *This is Magic: Secrets of Conjurers' Craft*. New York: Bell, 1948.

Dinsdale, Tim. *Loch Ness Monster*. 4th ed. New York: Routledge, 1982.

Dinshah, H. Jay. *Health Can Be Harmless*. Malaga, N.J.: American Vegan Society, 1974.

Dolby, R. G. A. *Uncertain Knowledge: An Image of Science for a Changing World*. Cambridge: Cambridge University Press, 1996.

Dossey, Larry. *Healing Words*. New York: Harper, 1995.

Doyle, Arthur Conan. *The History of Spiritualism*. New York: George H. Doran Co., 1926.

Duncan, John Charles. *Astronomy*. New York: Harper, 1955.

Dunninger, Joseph. *Houdini's Spirit Exposé: From Houdini's Own Manuscripts, Records, and Photographs, and Dunninger's Psychical Investigations*. New York: Experimenter, 1928.

———. *Inside the Medium's Cabinet*. New York: David Kemp, 1935.

Dunninger, Joseph, as told to Walter B. Gibson. *Dunninger's Secrets*. Secaucus, N.J.: Lyle Stuart, Inc., 1974.

Dyne, Mark. *Electronic Communication for the Spiritual Emancipation of the People*. Manchester, U.K.: Spirit Electronic Communication Society, 1954.

Easton, S. C. *Rudolf Steiner: Herald of a New Epoch.* Hudson, N.Y.: Anthroposophic Press, 1980.

Eberhart, George M., ed. *Monsters: A Guide to Information on Unaccounted-for Creatures, Including Bigfoot, Many Water Monsters, and Other Irregular Animals.* New York: Garland, 1983.

Eddington, Arthur. *The Expanding Universe.* New York: Penguin, 1940.

Eddy, Mary Baker. *Science and Health with Key to the Scripture.* Boston: Trustees Under the Will of Mary Baker G. Eddy, 1906.

Edey, Maitland A. *Lost World of the Aegean.* New York: Time-Life Books, 1975.

———. *The Missing Link.* New York: Time-Life Books, 1972.

Edey, Maitland A., and Donald Johanson. *Lucy: The Beginnings of Humankind.* New York: Simon & Schuster, 1981.

Edge, Hoyt. L., Robert L. Morris, et al. *Foundations of Parapsychology: Exploring the Boundaries of Human Capability.* New York: Routledge, 1986.

Edmunds, Simeon. *Spiritualism: A Critical Survey.* London: Aquarian Press, 1966.

Edwards, B. *Drawing on the Right Side of the Brain.* Boston: J.P. Tarcher, 1979.

Edwards, Paul. *Reincarnation: A Critical Examination.* Buffalo, N.Y.: Prometheus Books, 1996.

Ehrlich, Paul, and S. Shirley Feldman. *The Race Bomb: Skin Color, Prejudice, and Intelligence.* New York: New York Times Book Co., 1977.

Ehrlich, Paul R. *The Population Bomb.* London: Pan/Vallantine, 1972.

Eisenbud, Jule. *Parapsychology and the Unconscious.* Berkeley, Calif.: North Atlantic Books, 1983.

Ellenberger, Henri F. *The Discovery of the Unconscious: The History and Evolution of Dynamic Psychiatry.* New York: Basic Books, 1981.

Ellis, Ida. *Planchette and Automatic Writing.* London: Blackpool, 1904.

Ellis, Havelock. *Sex in Relation to Society.* London: Heinemann, 1945.

Ernst, Bernard M. L., and Hereward Carrington. *Houdini and Conan Doyle: The Story of a Strange Friendship.* New York: Albert and Charles Boni, Inc., 1932.

Evans, Bergen. *The Natural History of Nonsense.* New York: Alfred A. Knopf, 1953.

Evans, Henry Ridgely. *Hours with the Ghosts.* Chicago: Laird and Lee, 1897.

———. *The Spirit World Unmasked.* Chicago: Laird and Lee, 1897.

Eve, Raymond A., and Francis B. Harrold. *The Creationist Movement in Modern America.* Boston: Twayne Publishers, 1991.

Eysenck, Hans J. *The Decline and Fall of the Freudian Empire.* New York: Penguin, 1985.

Eysenck, Hans J., and Carl Sargent. *Explaining the Unexplained: Mysteries of the Paranormal.* London: Prion, 1993.

Fagan, Brian M. *In the Beginning: An Introduction to Archaeology.* Boston: Little, Brown & Co., 1972.

———. *Snapshots of the Past.* Walnut Creek, Calif.: Alta Mira Press, 1995.

Fancher, Raymond E. *The Intelligence Men: Makers of the IQ Controversy.* New York: Norton, 1987.

Farley, John. *The Spontaneous Generation Controversy from Descartes to Oparin.* Baltimore, Md.: Johns Hopkins University Press, 1977.

Farrar, Janet, Stewart Farrar, and Gavin Bone. *The Pagan Path.* West Kennebunk, Maine: Phoenix Publishing Inc., 1995.

Fay, Charles Eden. *Mary Celeste: The Odyssey of an Abandoned Ship.* Custer, Wash.: Peabody Museum, 1942.

Feder, Kenneth L. *Frauds, Myths, and Mysteries: Science and Pseudoscience in Archaeology.* 2d ed. Mountain View, Calif.: Mayfield Publishing Co., 1995.

Fell, Barry. *America BC: Ancient Settlers in the New World.* New York: Times Books, 1976.

———. *Saga America: A Startling New Theory on the Old World Settlement of America Before Columbus.* New York: Times Books, 1980.

Fernald, Dodge. *The Hans Legacy: A Story of Science.* Hillsdale, N.J.: Lawrence Erlbaum, 1983.

Ferreri, C. A., and R. B. Wainwright. *Breakthrough for Dyslexia and Learning Disabilities.* Pompano Beach, Fla.: Exposition Press, 1985.

Ferry, Georgina, ed. *The Understanding of Animals.* London: Basil Blackwell Ltd., 1984.

Field, Geoffrey G. *Evangelist of Race: The Germanic Vision of Houston Stewart Chamberlain.* New York: Columbia University Press, 1981.

Finnis, J. M. *Natural Law and Natural Rights.* New York: Oxford University Press, 1980.

Finucane, R. C. *Appearances of the Dead: A Cultural History of Ghosts.* Buffalo, N.Y.: Prometheus Books, 1984.

Fishbein, Morris. *Fads and Quackery in Healing.* New York: Blue Ribbon Books, 1932.

Fisher, John. *Body Magic.* New York: Stein and Day, 1979.

Fletcher, Horace. *Fletcherism, What It Is.* New York: Frederick A. Stokes Co., 1913.

Fletcher, Ronald. *Science, Ideology and the Media: The Cyril Burt Scandal.* New Brunswick, N.J.: Transaction, 1991.

Flew, Antony. *The Logic of Immortality.* Cambridge, Mass.: Blackwell, 1987.

———. *The Logic of Mortality.* Cambridge, Mass.: Blackwell, 1987.

———. *A New Approach to Psychical Research.* Watts, 1953.

Flew, Antony, ed. *Readings in the Philosophical Problems of Parapsychology.* Buffalo, N.Y.: Prometheus Books, 1986.

Fodor, Nandor. *The Haunted Mind.* New York: Helix Press-Garrett Publications, 1959.

———. *Encyclopedia of Psychic Science.* New Hyde Park, N.Y.: University Books, 1966.

Force, James E. *William Whiston: Honest Newtonian.* Cambridge: Cambridge University Press, 1985.

Ford, Arthur A., and Marguerite Harmon Bro. *Nothing So Strange.* New York: Harper, 1958.

Forrest, D. W. *Francis Galton: The Life and Work of a Victorian Genius.* New York: Taplinger Publishing Co., 1974.

Fortey, Richard. *Life: An Unauthorized Biography.* New York: HarperCollins, 1997.

Foster, J. G. *Enquiry into the Practice and Effects of Scientology.* London: HMSO, 1971.

Franklin, Richard L. *Overcoming the Myth of Self-Worth: Reason and Fallacy in What You Say to Yourself.* Appleton, Wis.: R.L. Franklin, 1994.

Frazer, James G. *The Golden Bough: A Study in Magic and Religion.* New York: Random House, 1987.

Frazer, J. T., et al., eds. *The Study of Time Series.* Volumes I to IV. New York: Springer-Verlag, 1972–1981. Volume V. Cambridge, Mass.: University of Massachusetts Press, 1986. Volumes VI to VIII. Madison, Conn.: International University Press, 1989–1997.

Frazier, Kendrick. *Science Confronts the Paranormal.* Buffalo, N.Y.: Prometheus Books, 1986.

Frazier, Kendrick, ed. *Paranormal Borderlands of Science.* Buffalo, N.Y.: Prometheus Books, 1981.

Freedman, Rita. *Beauty Bound.* New York: Lexington/Columbus, 1988.

Freeman, Derek. *Margaret Mead and Samoa: The Making and Unmaking of an Anthropological Myth.* Cambridge, Mass.: Harvard University Press, 1989.

Freemantle, F., and C. Trungpa, trans. *Tibetan Book of the Dead: The Great Liberation through Hearing in the Bardo.* Boston: Shambhala, 1992.

Freud, Sigmund. *Psychopathology of Everyday Life.* New York: Penguin, 1938.

Frikell, Samri. *Spirit Mediums Exposed.* New York: New York Metropolitan Fiction, 1930.

Fuller, John G. *Arigo: Surgeon With the Rusty Knife.* Santa Cruz, Calif: Devin, 1975.

Fuller, Robert C. *Mesmerism and the American Cure of Souls.* Philadelphia: University of Pennsylvania Press, 1982.

Fuller, Uriah [pseud.]. *Confessions of a Psychic: A Factual Account of How Fake Psychics Perform Seemingly Incredible Paranormal Feats.* Teaneck, N.J.: Karl Fulves, 1975.

———. *Further Confessions of a Psychic: Inside Secrets of Seemingly Incredible Psychic Feats.* Teaneck, N.J.: Karl Fulves, 1980.

Gabbard, Glen O., and Stuart W. Twemlow. *With the Eyes of the Mind: An Empirical Analysis of Out-of-Body States.* New York: Praeger, 1984.

Gaddis, Vincent. *Invisible Horizons: True Mysteries of the Sea.* Philadelphia: Chilton Books, 1965.

———. *Mysterious Fires and Lights.* New York: David McKay Co., 1967.

Gaffron, Norma. *Bigfoot.* San Diego, Calif.: Greenhaven Books, 1989.

———. *Great Mysteries: The Bermuda Triangle.* San Diego, Calif.: Greenhaven Press, 1994.

Gaines, Steven S. *Marjoe: The Life of Marjoe Gortner.* New York: Harper and Row, 1973.

Gardner, Edward L. *Fairies: The Cottingley Photographs and Their Sequel.* Wheaton, Ill.: Theosophical Society Publishing House, 1966.

Gardner, Gerald. *Witchcraft Today.* New York: Magickal Childe, 1991.

Gardner, Martin. *Fads and Fallacies in the Name of Science.* New York: Dover, 1957.

———. *How Not to Test a Psychic: A Study of the Remarkable Experiments with Renowned Clairvoyant Pavel Stepanek.* Buffalo, N.Y.: Prometheus Books, 1989.

———. *The New Age: Notes of a Fringe Watcher.* Buffalo, N.Y.: Prometheus Books, 1991.

———. *Science: Good, Bad and Bogus.* Buffalo, N.Y.: Prometheus Books, 1989.

Gardner, Martin, ed. *The Wreck of the Titanic Foretold?* Buffalo, N.Y.: Prometheus Books, 1986.

Garinger, Alan. *Water Monsters.* San Diego, Calif.: Greenhaven Press, 1991.

Garlake, Peter. *Great Zimbabwe.* New York: Stein and Day, 1973.

Gasman, Daniel. *The Scientific Origins of National Socialism.* Macdonald, 1971.

Gauquelin, Michael. *The Truth about Astrology.* Cambridge, Mass.: Blackwell, 1983.

Georgi, J. "Memories of Alfred Wegener," *Continental Drift.* Edited by S. K. Runcorn. San Diego, Calif.: Academic Press, 1962.

Gettings, F. *The Book of the Hand: An Illustrated History of Palmistry.* London: Hamlyn, 1965.

———. *Ghosts in Photographs: The Extraordinary Story of Spirit Photography.* New York: Harmony, 1978.

Gevitz, Norman, ed. *Other Healers:Unorthodox Medicine in America.* Baltimore, Md.: Johns Hopkins Press, 1988.

Gherman, Beverly. *The Mysterious Rays of Dr. Röntgen.* New York: Atheneum, 1994.

Gibbon, Charles. *The Life of George Combe.* London: Macmillan and Co., 1878.

Gilkey, Langdon. *Maker of Heaven and Earth: Christian Doctrine of Creation in the Light of Modern Knowledge.* Lanham, Md.: University Press of America, 1986.

———. *Nature, Reality and the Sacred: The Nexus of Science and Religion.* Minneapolis, Minn.: Augsburg Fortress, 1993.

Gillispie, C. C. *Genesis and Geology.* New York: Harper, 1951.

Gilovich, Tom. *How We Know What Isn't So: The Fallibility of Human Reason in Everyday Life.* New York: Free Press, 1991.

Gish, D. T. *Evolution? The Fossils Say No!* San Diego, Calif.: Creation-Life Pubs., 1979.

Godfrey, L. R. *Scientists Confront Creationism.* New York: Norton, 1983.

Godwin, Joscelyn. *The Theosophical Enlightenment.* Albany, N.Y.: SUNY Press, 1994.

Goethe, J. W. von. *Theory of Colours.* Translated by C. L. Eastlake. Cambridge, Mass.: MIT Press, 1987.

Gordon, Henry. *Channeling into the New Age.* Buffalo, N.Y.: Prometheus Books, 1988.

Gordon, Stuart. *The Book of Hoaxes: An A–Z of Famous Fakes, Frauds and Cons.* London: Headline, 1995.

Gosden, Roger. *Cheating Time: Science, Sex and Aging.* New York: Macmillan, 1996.

Gould, Stephen Jay. *Dinosaur in a Haystack.* New York: Norton, 1995.

———. *Ever Since Darwin.* New York: Norton, 1992.

———.*The Flamingo's Smile: Reflections in Natural History.* New York: Norton, 1987.

———. *Hen's Teeth and Horse's Toes: Further Reflections in Natural History.* New York: Norton, 1994.
———. *The Mismeasure of Man.* New York: Norton, 1996.
———. *The Panda's Thumb: More Reflections in Natural History.* New York: Norton, 1992.
———. *Time's Arrow—Time's Cycle: Myth and Metaphor in the Discovery of Geological Time.* Reprint. Cambridge, Mass.: Harvard University Press, 1988.
———. *An Urchin in the Storm.* San Francisco, Calif: Collins, 1988; New York: Penguin, 1990.
Graham, Loren. *The Ghost of the Executed Engineer.* Cambridge, Mass.: Harvard University Press, 1995.
Green, Elmer, and Alyce Green. *Beyond Biofeedback.* New York: Delacorte, 1977.
Gregory, Richard L., ed. *The Oxford Companion to the Mind.* New York: Oxford University Press, 1987.
Gresham, William Lindsay. *Nightmare Alley.* New York: Rinehart, 1946.
Grim, Patrick, ed. *Philosophy of Science and the Occult,* 2d ed. Albany: SUNY Press, 1990.
Grosskurth, Phyllis. *Havelock Ellis: A Biography.* New York: Allen Lane/Alfred A. Knopf, 1980.
Guiley, Rosemary E. *Harper's Encyclopedia of Mystical and Paranormal Experience.* New York: HarperCollins, 1991.
Habermas, Gary R., and J. P. Moreland. *Immortality: The Other Side of Death.* Nashville, Tenn.: Thomas Nelson, 1992.
Hadingham, Evan. *Lines to the Mountain Gods: Nazca and the Mysteries of Peru.* New York: Random House, 1987.
Haldane, J. B. S. *Fact and Faith.* London: Watts & Co., 1934.
Hall, Trevor H. *The Enigma of Daniel Home: Medium or Fraud?* Buffalo, N.Y.: Prometheus Books, 1984.
———. *The Medium and the Scientists: The Story of Florence Cook and William Crookes.* Buffalo, N.Y.: Prometheus Books, 1985.
———. *The Spiritualists.* New York: Helix Press, Garrett Publications, 1962.
Hallam, A. *A Revolution in the Earth Sciences: From Continental Drift to Plate Tectonics.* Oxford: Clarendon Press, 1973.
Haller, Mark. *Eugenics: Hereditarian Attitudes in American Thought.* New Brunswick, N.J.: Rutgers University Press, 1963.
Hamilton, Iain. *Koestler: A Biography.* London: Secker & Warburg, 1982.
Hamilton-Patterson, James, and Carol Andrews. *Mummies: Death and Life in Ancient Egypt.* New York: Viking, 1979.
Hansel, C. E. M. *ESP: A Scientific Evaluation.* New York: Scribner, 1966.
———. *ESP and Parapsychology: A Critical Re-evaluation.* Buffalo, N.Y.: Prometheus Books, 1980.
———. *The Search for Psychic Power: ESP and Parapsychology Revisited.* Buffalo, N.Y.: Prometheus Books, 1989.
Haraway, Donna. *Primate Visions: Gender, Race, and Nature in the World of Modern Science.* New York: Routledge, 1989.
Hardinge, Emma. *Modern American Spiritualism.* New Hyde Park, N.Y.: University Books, 1970.
Hardy, Alister. *The Biology of God: A Scientist's Study of Man the Religious Animal.* New York: Taplinger, 1975.
Harris, Melvin. *Investigating the Unexplained.* Buffalo, N.Y.: Prometheus Books, 1986.
Hasted, John. *The Metal-Benders.* New York: Routledge & Kegan Paul, 1981.
Hawking, Stephen J. *Black Holes and Baby Universes and Other Essays.* New York: Bantam Books, 1993.
———. *A Brief History of Time.* New York: Bantam Books, 1988.
Hawkins, Gerald S. *Stonehenge Decoded.* Garden City, N.Y.: Doubleday, 1965.
Hearney, John J. *The Sacred and the Psychic: Parapsychology and Christian Theology.* Mahwah, N.J.: Paulist Press, 1984.
Hearnshaw, L. S. *Cyril Burt Psychologist.* London: Hodder & Stoughton Ltd., 1979.
Herrnstein, R. J., and C. Murray. *The Bell Curve.* New York: Free Press, 1994.
Hess, David J. *Science in the New Age: The Paranormal, Its Defenders and Debunkers, and American Culture.* Madison: University of Wisconsin Press, 1993.
Heuvelmans, Bernard. *In the Wake of the Sea-Serpents.* New York: Hill and Wang, 1968.
———. *On the Track of Unknown Animals.* New York: Hill and Wang, 1958.
Heyerdahl, Thor. *American Indians in the Pacific: The Theory behind the Kon-Tiki Expeditions.* Chicago, Ill.: Rand McNally, 1953.
———. *Easter Island: The Mystery Solved.* New York: Random House, 1989.
———. *Kon-Tiki: Across the Pacific by Raft.* Chicago, Ill.: Rand McNally, 1950.
Hibbard, Whitney S., and Raymond S. Worring. *Psychic Criminology: An Operations Manual for Using Psychics in Criminal Investigations.* Springfield, Ill.: Charles C. Thomas, 1981.
Hilgard, Ernest R. *Divided Consciousness: Multiple Controls in Thought and Action.* New York: John Wiley and Sons, 1986.
Hill, J. Arthur. *Spiritualism: Its History, Phenomena, and Doctrine.* New York: George Doran, 1919.
Hoagland, Richard C. *Monuments of Mars: A City on the Edge of Forever.* Revised and enlarged. Berkeley, Calif.: North Atlantic Books, 1992.
Hobson, J. Allan, M.D. *The Chemistry of Conscious States.* Boston: Little, Brown & Company, 1994.
———. *The Dreaming Brain.* Basic Books, 1988.
Holand, Hjalmar R. *Explorations in America before Columbus.* New York: Twayne Publishers, 1956.
Holbrook, Stewart H. *The Golden Age of Quackery.* New York: Macmillan Company, 1959.
Holton, Gerald. *Science and Anti-Science.* Cambridge, Mass.: Harvard University Press, 1993.
———. *Thematic Origins of Scientific Thought: Kepler to Einstein.* Cambridge, Mass.: Harvard University Press, 1974.
Houdini, Harry. *Houdini Exposes the Tricks Used by the Boston Medium "Margery."* New York: Adams Press, circa 1924.
———. *A Magician Among the Spirits.* New York: Harper, 1924; New York: Arno Press, 1972.
———. *Miracle Mongers and their Methods: A Complete Exposé.* Buffalo, N.Y.: Prometheus Books, 1981.

———. *The Right Way to Do Wrong*. Boston: Harry Houdini, 1906.

Houdini, Harry, and Joseph Dunninger. *Magic and Mystery: The Incredible Psychic Investigations of Houdini and Dunninger.* New York: Tower Publications, 1968.

Howard, Jane. *Margaret Mead: A Life*. New York: Simon & Schuster, 1984.

Hoy, David. *The Bold and Subtle Miracles of Dr. Faust.* Chicago: Ireland Magic, 1963.

Hoyle, Fred. *Cosmic Life Force*. London: Dent, 1988.

———. *Home is Where the Wind Blows*. New York: Oxford University Press, 1997.

———. *The Small World of Fred Hoyle*. London: M. Joseph, 1986.

———. *Ten Faces of the Universe*. San Francisco: W. H. Freeman, 1977.

Hoyle, Fred, and Chandra Wickramsinghe. *Lifecloud: The Origin of Life in the Universe*. London: Dent, 1978.

Hubbard, L. Ron. *Dianetics: The Modern Science of Mental Healing*. Los Angeles, Calif.: The American Saint Hill Organization, 1950.

———. *Scientology: The Fundamentals of Thought*. Los Angeles, Calif.: Bridge Publications, 1988.

———. *What Is Scientology?* Los Angeles, Calif.: Bridge Publications, 1992.

Hubbard, L. Ron., Jr. *L. Ron Hubbard: Messiah or Madman?* Secaucus, N.J.: Lyle Stuart, Inc., 1987.

Hull, Burling. *Encyclopedic Dictionary of Mentalism*. Vol. 2. Calgary, Alberta, Canada: Micky Hades Enterprises, 1973.

———. *The Last Word in Blindfold Methods: 12 Sensational Blindfolds*. Woodside, N.Y.: Burling Hull, 1946.

———. *Thirty-Three Rope Ties and Chain Releases*. New York: Stage Magic, 1947.

Hyman, Ray. *The Elusive Quarry: A Scientific Appraisal of Psychical Research*. Buffalo, N.Y.: Prometheus Books, 1989.

Hynek, J. Allen. *The Hynek UFO Report*. New York: Dell, 1977.

Inglis, Brian. *Natural and Supernatural*. London: Abacus (Sphere Books), 1979.

———. *Science and Parascience: A History of the Paranormal, 1914–1939*. London: Hodder & Stoughton, 1984.

Irwin, H. J. *Flight of Mind: A Psychological Study of the Out-of-Body Experience*. Lanham, Md.: Scarecrow Press, 1985.

Iverson, Jeffrey. *More Lives Than One?* London: Pan Books, 1977.

Jackson, Herbert G., Jr. *The Spirit Rappers*. New York: Doubleday, 1972.

Jacobs, David M. *Secret Life: First-Hand Accounts of UFO Abductions*. New York: Simon & Schuster, 1992.

Jacobs, J. *Reversals of the Earth's Magnetic Field*. New York: Cambridge University Press, 1994.

Jacoby, Arnold. *Señor Kon-Tiki: The Biography of Thor Heyerdahl*. Chicago: Rand McNally, 1967.

James, William. *Varieties of Religious Experience*. New York: Penguin, 1982.

Jameson, Eric. *The Natural History of Quackery*. London: Michael Joseph, 1961.

Janelle, Pierre. *The Catholic Reformation*. New York: Collier-Macmillan, 1975.

Jeffrey, Burton Russell. *A History of Witchcraft: Sorcerers, Heretics and Pagans*. London: Thames and Hudson, 1980.

Jensen, Arthur R. *Straight Talk About Mental Tests*. New York: Free Press, 1981.

Johnson, Paul K. *The Masters Revealed*. Albany, N.Y.: SUNY Press, 1994.

Jolly, W. P. *Sir Oliver Lodge*. London: Constable, 1974.

Jung, C. G. *Synchronicity: An Acausal Connecting Principle*. Translated by R. F. C. Hull. Princeton, N.J.: Princeton University Press, 1973.

Kabat-Zinn, Jon. *Full Catastrophe Living*. London: Piatkus Books, 1996.

Kagan, Daniel, and Ian Summers. *Mute Evidences*. New York: Bantam Books, 1984.

Kahane, Howard. *Logic and Contemporary Rhetoric: The Use of Reason in Everyday Life*. 7th ed. Belmont, Calif.: Wadsworth, 1992.

Kamin, Leon. *The Science and Politics of I.Q.* Hillsdale, N.J.: Lawrence Erlbaum, 1974.

Kandinsky, Wassily. *Concerning the Spiritual in Art*. Translated by Michael T. H. Sadler. New York: Dover, 1977.

Kaufman, Martin. *Homeopathy in America*. Baltimore, Md.: Johns Hopkins Press, 1971.

Kaye, Marvin. *The Handbook of Mental Magic*. New York: Stein and Day, 1975.

Keane, John. *Tom Paine: A Political Life*. Boston: Little, Brown & Co., 1995.

Keene, Lamar. *The Psychic Mafia*. Amherst, N.Y.: Prometheus Books, 1997.

Kellogg, John Harvey. *The Natural Diet of Man*. Battle Creek, Mich.: Modern Medicine Publishing Co., 1923.

Kenawell, William W. *The Quest at Glastonbury: A Biographical Study of Frederick Bligh Bond*. London: Helix, 1965.

Kerr, Howard. *Mediums and Spirit Rappers, and Roaring Radicals*. Urbana: University of Illinois Press, 1972.

Kevles, Daniel. *In the Name of Eugenics: Genetics and the Uses of Human Heredity*. New York: Alfred A. Knopf, 1985.

Kevles, Daniel, and Leroy Hood. *The Code of Codes: Scientific and Social Issues in the Human Genome Project*. Cambridge, Mass.: Harvard University Press, 1992.

Kilminster, C. W. *Eddington's Search for a Fundamental Theory: A Key to the Universe*. Cambridge: Cambridge University Press, 1995.

Kilner, Walter J. *The Human Atmosphere*. 1911. Reprinted as *The Human Aura*. New Hyde Park, N.Y.: University Books, 1965.

King, Francis X. *The Encyclopedia of Fortune Telling*. London: Hamlyn, 1988.

Kinsbourne, Marcel, ed. *Asymmetrical Function of the Brain*. Cambridge: Cambridge University Press, 1978.

Kinsey, A. C., et al. *Sexual Behavior in the Human Female*. Philadelphia: W.B. Saunders, 1953.

———. *Sexual Behavior in the Human Male*. Philadelphia: W.B. Saunders, 1948.

Kitcher, Philip. *Abusing Science: The Case Against Creationism*. Cambridge, Mass.: MIT Press, 1982.

Klass, Philip J. *UFO Abductions: A Dangerous Game*. Buffalo, N.Y.: Prometheus Books, 1989.

———. *UFOs: The Public Deceived*. Buffalo, N.Y.: Prometheus Books, 1983.

Klein, Alexander. *Grand Deception: The World's Most Spectacular and Successful Hoaxes, Imposters, Ruses, and Frauds.* Philadelphia: J. B. Lippincott, 1955.

Kline, M. V., ed. *The Scientific Report on "The Search for Bridey Murphy."* New York: Viking Press, 1956.

Knight, Damon. *Charles Fort: Prophet of the Unexplained.* New York: Doubleday, 1970.

Knight, Marcus. *Spiritulism, Reincarnation, and Immortaility.* London: Gerald Duckworth, 1950.

Koestler, Arthur. *The Case of the Midwife Toad.* New York: Random House, 1971.

———. *The Sleepwalkers: A History of Man's Changing Vision of the Universe.* New York: Penguin, 1990.

Krantz, Grover S. *Big Footprints: A Scientific Inquiry into the Reality of Bigfoot.* Boulder, Colo.: Johnson Books, 1992.

Krause, Michael. *Relativism: Interpretation and Confrontation.* South Bend, Ind.: University of Notre Dame, 1989.

Krauss, Lawrence M. *The Physics of Star Trek.* New York: Basic Books, 1995.

Kreskin, A. *The Amazing World of Kreskin.* New York: Random House, 1973.

———. *Secrets of the Amazing Kreskin: The World's Foremost Mentalist Reveals How You Can Expand Your Powers.* Amherst, N.Y.: Prometheus Books, 1991.

Kumar, Krishan, et al. *Utopias and the Millennium.* London: Reaktion Books, 1993.

Kurian, George T., and Graham T. Molitor, eds. *Encyclopedia of the Future.* Vol. 2. New York: Macmillan, 1996.

Kurtz, Paul. *The New Skepticism: Inquiry and Reliable Knowledge.* Amherst, N.Y.: Prometheus Books, 1992.

———. *The Transcendental Temptation: A Critique of Religion and the Paranormal.* Amherst, N.Y.: Prometheus Books, 1991.

Kurtz, Paul, ed. *A Skeptic's Handbook of Parapsychology.* Buffalo, N.Y.: Prometheus Books, 1985.

Kusche, Lawrence. *The Bermuda Triangle Mystery—Solved.* Buffalo, N.Y.: Prometheus Books, 1986.

LaBarre, Weston. *Seeing Castaneda: Reactions to the "Don Juan" Writings of Carlos Castaneda.* Edited by Daniel C. Noel. New York: Putnam, 1976.

Lamarck, Jean Baptiste. *Zoological Philosophy.* 1809. Translated by Hugh Elliot. Reprint. New York: Hafner, 1963.

Lamont, Corliss. *The Illusion of Immortality.* 1935. 5th ed. New York: Continuum, 1990.

Lawton, George. *The Drama of Life after Death.* New York: Henry Holt, 1932.

Lee, Richard Borshay. *The Kung San: Men, Women, and Work in a Foraging Society.* Cambridge: Cambridge University Press, 1979.

Lehmann, A., and J. Myers. *Magic, Witchcraft and Religion.* 2d ed. Mountain View, Calif.: Mayfield, 1989.

Leibowitz, Judith, and Connington Leibowitz. *The Alexander Technique.* New York: Harper & Row, 1990.

Lemesurier, P. *The Great Pyramid Decoded.* rev. ed. Boston: Element Books, 1996.

Lévi-Strauss, Claude. *The Savage Mind.* Chicago, Ill.: University of Chicago Press, 1968.

Lewontin, Richard C. *Biology and Ideology: The Doctrine of DNA.* New York: HarperCollins, 1993.

Ley, Willy. *The Lungfish, the Dodo, and the Unicorn: An Excursion into Romantic Zoology.* New York: Viking Press, 1948.

———. *Salamanders and Other Wonders: Still More Adventures of a Romantic Naturalist.* New York: Viking Press, 1955.

Lim, Robin. *After the Baby's Birth . . . A Woman's Way to Wellness.* Berkeley, Calif.: Celestial Arts, 1991.

Lindner, Robert. *The Fifty-Minute Hour: A Collection of True Psychoanalytic Tales.* New York and Toronto: Rinehart, 1954.

Lindzey, Gardner, Calvin S. Hall, and Richard F. Thompson. *Psychology.* 3d ed. New York: Worth, 1988.

Lockley, Ronald M. *Whales, Dolphins and Porpoises.* David and Charles, 1979.

Lodge, Oliver J. *Raymond, or Life and Death, with Examples of the Evidence for the Survival of Memory and Affection after Death.* London: Methuen, 1916.

Loftus, Elizabeth, and Katherine Ketcham. *The Myth of Repressed Memory.* New York: St. Martin's Press, 1994.

Longridge, George. *Spiritualism and Christianity.* London: A. R. Mowbray, 1926.

Lovejoy, Arthur O. *The Great Chain of Being.* Cambridge, Mass.: Harvard University Press, 1936.

Ludwig, Jan, ed. *Philosophy and Parapsychology.* Buffalo, N.Y.: Prometheus Books, 1978.

Lumsden, Charles L., and Edward O. Wilson, *Promethean Fire: Reflections on the Origin of Mind.*Cambridge, Mass.: Harvard University Press, 1984.

Lurie, Edward. *Louis Agassiz: A Life in Science.* Chicago: University of Chicago Press, 1960.

———. *Nature and the American Mind: Louis Agassiz and the Culture of Science.* Chicago: University of Chicago Press, 1974.

Lyons, Arthur, and Marcello Truzzi. *The Blue Sense: Psychic Detectives and Crimes.* New York: Mysterious Press, 1991.

MacCormack, Carol, and Marilyn Strathern, eds. *Nature, Culture, Gender*, Cambridge: Cambridge University Press, 1980.

MacDougall, Curtis D. *Hoaxes.* New York: Dover Publications, 1958.

Mack, John E. *Abduction: Human Encounters with Aliens.* New York: Scribner, 1994.

———. *Nightmares and Human Conflict.* Boston: Little, Brown & Co., 1970.

Mackal, Roy P. *A Living Dinosaur? In Search of Mokele-Mbembe.* Kinderhook, N.Y.: E.J. Brill, 1987.

———. *The Monsters of Loch Ness.* Chicago: Swallow Press, 1976.

———. *Searching for Hidden Animals.* New York: Doubleday, 1980.

MacKay, Charles. *Extraordinary Popular Delusions and the Madness of Crowds.* New York: Crown Publishing Group, 1980.

Magnusson, Magnus, and Hermann Palsson, trans. *The Vinland Sagas: The Norse Discovery of America.* New York: Penguin, 1965.

Majno, Guido. *The Healing Hand: Man and Wound in the Ancient World.* Cambridge, Mass.: Harvard University Press, 1975.

Manaka, Yoshio. *The Layman's Guide to Acupuncture*. New York: John Weatherhill, 1972.
Mann, W. Edward. *Orgone, Reich and Eros*. New York: Simon & Schuster, 1973.
Mansfield, Peter. *The Bates Method*. Boston: Charles E. Tuttle, 1994.
Marks, David, and Richard Kammann. *The Psychology of the Psychic*. Buffalo, N.Y.: Prometheus Books, 1980.
Masson, Jeffrey Moussaieff. *Against Therapy: Emotional Tyranny and the Myth of Psychological Healing*. New York: Atheneum, 1988.
Mavin, Max. *Max Mavin's Book of Fortunetelling: The Complete Guide to Augury, Soothsaying, and Divination*. Englewood Cliffs, N.J.: Prentice Hall, 1992.
McClenon, James. *Deviant Science: The Case of Parapsychology*. Philadelphia: University of Pennsylvania Press, 1984.
———. *Wondrous Events: Foundations of Religious Belief*. Philadelphia: University of Pennsylvania Press, 1994.
McGervey, John D. *Probabilities in Everyday Life*. Chicago: Nelson-Hall, 1986.
McHargue, Georgess. *Facts, Frauds, and Phantasms: A Survey of the Spiritualist Movement*. New York: Doubleday, 1972.
McLeish, Kenneth, ed. *Bloomsbury Guide to Human Thought*. London: Bloomsbury Reference, 1993.
McLellan, David. *Marxism: Essential Writings*. New York: Oxford University Press, 1988.
McLynn, Frank. *Carl Gustav Jung*. New York: Bantam, 1996.
McNaughten, Hugh. *Emile Coué: The Man and His Works*. Society of Metaphysicians, 1995.
McPhee, John. *Basin and Range*. New York: Farrar, Straus & Giroux, 1981.
Meade, Marion. *Madame Blavatsky: The Woman Behind the Myth*. New York: Putnam, 1980.
Medhurst, R. G., in association with K. M. Goldney and M. R. Barrington. *Crookes and the Spirit World*. New York: Taplinger Publishing Company, 1972.
Mellars, Paul, ed. *The Emergence of Modern Humans: An Archaeological Perspective*. Ithaca, N.Y.: Cornell University Press, 1990.
Melton, J. Gordon. *New Thought: A Reader*. Santa Barbara, Calif.: Institute for the Study of American Religion, 1990.
Melton, J. Gordon, Jerome Clark, and Aidan Kelley. *New Age Encyclopedia*. Detroit, Mich.: Gale Research, 1990.
Menotti, Giancarlo. *The Medium*. New York: G. Schirmer, 1947.
Michell, John. *Natural Likeness: Faces and Figures in Nature*. New York: Dutton, 1979.
———. *The New View Over Atlantis*. 1969. London: Thames and Hudson, 1983.
Miller, Richard Alan. *The Magical and Ritual Use of Aphrodisiacs*. New York: Destiny Books, 1985.
Minter, William. *King Solomon's Mines Revisited: Western Interests and the Burdened History of Southern Africa*. New York: Basic Books, 1986.
Mitchell, Edgar D., ed. *Psychic Explorations: A Challenge for Science*. New York: Putnam, 1974.
Montagu, Ashley, ed. *Race and IQ*. expanded ed. New York: Oxford University Press, 1996.
Montgomery, Ruth. *Companions Along the Way*. New York: Coward, McCann & Geoghegan, Inc., 1974.
———. *A Gift of Prophecy: The Phenomenal Jeane Dixon*. New York: William Morrow & Company, 1965.
———. *Here and Hereafter*. New York: Coward, McCann & Geoghegan, Inc., 1968.
———. *A World Beyond*. New York: Coward, McCann & Geoghegan, Inc. 1971.
Montgomery, Ruth, and Joanne Garland. *Ruth Montgomery: Herald of the New Age*. Garden City, N.Y.: Doubleday & Company, Inc., 1986.
Moon, Mary. *Ogopogo, The Okanagan Mystery*. Vancouver: J. J. Douglas, 1977.
Moore, Noel Brooke, and Richard Parker. *Critical Thinking*. Palo Alto, Calif.: Mayfield, 1991.
Moore, R. Laurence. *In Search of White Crows*. New York: Oxford University Press, 1977.
Morehouse, David. *Psychic Warrior*. New York: Penguin, 1997.
Morgan, Elaine. *The Aquatic Ape*. Lanham, Md.: Madison Books, 1982.
———. *The Descent of the Child: Human Evolution from a New Perspective*. New York: Oxford University Press, 1995.
———. *The Descent of Woman: The Classic Study of Evolution*. 4th ed. London: Souvenir Press, 1977.
———. *The Scars of Evolution—What Our Bodies Tell Us about Human Origins*. Reprint. New York: Oxford University Press, 1994.
Morris, Henry M. *A History of Modern Creationism*. Green Forest, Ark.: Master Books, 1984.
———. *What Is Creation Science?* rev. ed. Green Forest, Ark.: Master Books, 1987.
Mulholland, John. *Beware Familiar Spirits*. 1938. Reprint. New York: Scribner, 1979.
Mumford, Michael D., Andrew M. Rose, and David A. Goslin. *An Evaluation of Remote Viewing: Research and Applications*. American Institute for Research, 1995.
Murphy, Charles John Vincent. *Struggle: The Life and Exploits of Commander Richard E. Byrd*. New York: Frederick A. Stokes Co., 1928.
Murphy, Gardner, and R. O. Ballou, eds. *William James on Psychical Research*. London: Chatto and Windus, 1961.
Murray, Elaine. *A Layman's Guide to New Age & Spiritual Terms*. Nevada City, Calif.: Blue Dolphin Publishing, 1993.
Musés, Charles Arthur, and Arthur M. Young, eds. *Consciousness and Reality: The Human Pivot Point*. New York: Outerbridge & Lazard/Dutton, 1972.
Napier, John. *Bigfoot*. New York: Dutton, 1973.
Neher, Andrew. *The Psychology of Transcendence*. 2d ed. New York: Dover, 1990.
Nelkin, Dorothy. *Science Textbook Controversies and the Politics of Equal Time*. Cambridge, Mass.: MIT Press, 1977.
Nelson, G. K. *Spiritualism and Society*. New York: Routledge, 1969.
Nelson, Robert. *How to Read Sealed Messages*. Columbus, Ohio: Nelson Enterprises, 1961.
———. *Secret Methods of Private Readers!* Columbus, Ohio: Nelson Enterprises, 1964.

Nelson, Robert, and E. J. Moore. *Super Prediction Tricks*. Columbus, Ohio: Nelson Enterprises, n.d.

Nicholas, Elizabeth. *The Devil's Sea*. New York: Award Books, 1975.

Nickell, Joe. *Entities: Angels, Spirits, Demons, and Other Alien Beings*. Amherst, N.Y.: Prometheus Books, 1995.

———. *Looking for a Miracle: Weeping Icons, Relics, Stigmata, Visions, and Healing*. Buffalo, N.Y.: Prometheus Books, 1993.

Nickell, Joe, ed. *Psychic Sleuths: ESP and Sensational Cases*. Buffalo, N.Y.: Prometheus Books, 1994.

Nickell, Joe, and John F. Fischer. *Mysterious Realms: Probing Paranormal, Historical, and Forensic Enigmas*. Buffalo, N.Y.: Prometheus Books, 1992.

Nolen, William A., M.D., *Healing: A Doctor in Search of a Miracle*. New York: Random House, 1974.

Noll, Richard. *The Jung Cult: Origins of a Charismatic Movement*. Princeton, N.J.: Princeton University Press, 1994.

Noll, Richard, ed. *Vampires, Werewolves, and Demons: Twentieth-Century Reports in the Psychiatric Literature*. Springfield, Ill.: Charles C. Thomas, 1983.

North, Anthony. *The Paranormal: A Guide to the Unexplained*. London: Blandford, 1996.

Numbers, Ronald L. *The Creationists: The Evolution of Scientific Creationism*. Berkeley: University of California Press, 1993.

O'Donnell, Elliot. *The Menace of Spiritualism*. New York: Frederick A. Stokes Co., 1920.

O'Faolain, Julia, and Lauro Martines, eds. *Not in God's Image: Women in History from the Greeks to the Victorians*. New York: Harper, 1973.

Ofshe, Richard, and Ethan Watters. *Making Monsters: False Memories, Psychotherapy, and Sexual Hysteria*. New York: Scribner, 1994.

Olby, R. C., ed. *Late Eighteenth Century European Scientists*. Elmsford, N.Y.: Pergamon Press, 1966.

Olsen, Kristin Gottschalk. *The Encyclopedia of Alternative Health Care*. New York: Pocket Books, 1989.

Oppenheim, Janet. *The Other World: Spiritualism and Psychical Research in England, 1850–1914*. Cambridge: Cambridge University Press, 1985.

Ord-Hume, Arthur W. J. G. *Perpetual Motion: The History of An Obsession*. London: Allen & Unwin Ltd., 1977.

Orliac, Catherine. *Easter Island: Mystery of the Stone Giants*. New York: Abrams, 1995.

Osis, K., and E. Haraldsson. *At the Hour of Death*. New York: Avon, 1977.

Ostrander, Sheila, and Lynn Schroeder. *Psychic Discoveries Behind the Iron Curtain*. Englewood Cliffs, N.J.: Prentice Hall, 1970.

Otten, Charlotte F., ed. *A Lycanthropy Reader: Werewolves in Western Culture*. New York: Dorset Press, 1986.

Owen, A. *The Darkened Room: Women, Power, and Spiritualism in Late Nineteenth-Century England*. Virago, 1989.

Owen, Iris M., and Margaret Sparrow. *Conjuring Up Philip*. New York: Harper & Row, 1976.

Park, James. *Icons: An A–Z Guide to the People Who Shaped Our Time*. New York: Collier Books, 1991.

Pasachoff, Jay M. *Astronomy: Earth to Science*. 5th ed. Orlando, Fla.: Harcourt Brace, 1997.

Pauling, Linus. *Vitamin C: The Common Cold and the Flu*. New York: W.H. Freeman, 1976.

Peach, Emily. *Discover Tarot: Understanding and Using Tarot Symbolism*. London: Aquarian Press, 1990.

Peale, Norman Vincent. *The Power of Positive Thinking*. 3d ed. New York: Simon & Schuster, 1987.

Pearsall, Ronald. *The Table-Rappers*. New York: St. Martin's Press, 1972.

Pearson, R. B. *The Dreams and Lies of Louis Pasteur*. Sumeria, 1994.

Peat, David F. *Blackfoot Physics*. New York: Fourth Estate and Penguin/Arkana, 1996.

———. *Synchronicity: The Bridge Between Matter and Mind*. New York: Bantam, 1987.

Peebles, Curtis. *Watch the Skies! A Chronicle of the Flying Saucer Myth*. Washington and London: Smithsonian Institution Press, 1994.

Peel, Robert. *Christian Science, Its Encounter with American Cultures*. Garden City, N.Y.: Doubleday & Company, 1965.

Percival, F., and Alexander H. Johnstone. *Polywater: A Library Exercise for Chemistry Degree Students*. London: The Chemical Society London, 1978.

Perutz, Max. *Is Science Necessary? Essays on Science and Scientists*. Oxford: Oxford University Press, 1991.

Pfungst, Oskar. *Clever Hans*. 1907. Edited by Robert Rosenthal. Holt Rinehart and Winston, 1965.

Phillipson, D. W. *African Archaeology*. 2d ed. New York: Cambridge University Press, 1993.

Piattelli-Palmarini, Massimo. *Inevitable Illusions*. New York: John Wiley, 1994.

Pidgeon, Charles [pseud.]. *Revelations of a Spirit Medium*. Edited by Harry Price and Eric J. Dingwall. London: Kegan Paul, Trench, Trubner, 1922.

Pietroni, Patrick C., ed. *Reader's Digest Family Guide to Alternative Medicine*. Pleasantville, N.Y.: Reader's Digest, 1991.

Pirsig, Robert M. *Zen and the Art of Motorcycle Maintenance: An Inquiry into Values*. New York: Bantam, 1984.

Pledge, H. T. *Science Since Fifteen-Hundred: A Short History of Mathematics, Physics, Chemistry, Biology*. Magnolia, Mass.: Peter Smith, 1990.

Pliny the Elder. *Natural History: A Selection*. New York: Penguin, 1991.

Podmore, Frank. *Mediums of the Nineteenth Century*. Vols. 1 and 2. New Hyde Park, N.Y.: University Books, 1963.

Popper, Karl. *Conjectures and Refutations*. New York: Harper & Row, 1963.

———. *The Open Society and Its Enemies*. New York: Routledge & Kegan Paul, 1966.

Prendergast, Mark. *Victims of Memory*. New York: HarperCollins, 1996.

Price, Harry. *Confessions of a Ghost Hunter*. New York: Causeway Books, 1974.

———. *The End of Borley Rectory*. London: Harrap, 1946.

———. *The Most Haunted House in England*. London: Longman Green, 1940.

Prideaux, Tom. *Cro-Magnon Man*. New York: Time-Life Books, 1973.

Prince, Walter Franklin. *Noted Witnesses for Psychical Occurrences*. Boston: Boston Society for Psychical Research, 1928; New Hyde Park, N.Y.: University Books, 1963.

Pritchard, Andrea, et al., eds. *Alien Discussions: Proceedings of the Abduction Study Conference Held at MIT.* Cambridge, Mass.: North Cambridge Press, 1994.

Proskauer, Julien J. *Spook Crooks!* New York: A. L. Burton, 1932.

Rachman, Stanley J. *Aversion Therapy and Behavior Disorders*. New York: Routledge & Kegan Paul, 1996.

Radner, H., and Morris N. Young. *Houdini on Magic*. New York: Dover Publications, 1963.

Rago, D. Scott. *Parapscychology: A Century of Inquiry.* New York: Dell, 1975.

Rama, Swami, Rudolf Ballentine, and Allan Weinstock. *Yoga and Psychotherapy.* Honesdale, Pa.: Himalaya Institute, 1976.

Rampa, T. Lobsang. *The Third Eye*. Reprint. New York: Ballantine, 1986.

———. *The Rampa Story.* New York: Ballantine, n.d.

Randi, James. *Conjuring*. New York: St. Martin, 1993.

———. *An Encyclopedia of Claims, Frauds, and Hoaxes of the Occult and Supernatural*. New York: St. Martin's Press, 1995.

———. *The Faith Healers*. Buffalo, N.Y.: Prometheus Books, 1989.

———. *Flim-Flam! The Truth About Unicorns, Parapsychology & Other Delusions*. Reprint. Buffalo, N.Y.: Prometheus Books, 1982.

———. *The Magic of Uri Geller.* New York: Ballantine Books, 1975.

———.*The Mask of Nostradamus: The Prophecies of the World's Most Famous Seer.* Reprint. Buffalo, N.Y.: Prometheus Books, 1993.

———. *The Truth About Uri Geller.* rev. ed. Buffalo, N.Y.: Prometheus Books, 1982.

Randle, Kevin D. *A History of UFO Crashes*. New York: Avon, 1995.

Randle, Kevin D., and Donald R. Schmitt. *The Truth About the UFO Crash at Roswell*. New York: Avon, 1994.

Randles, Jenny. *Alien Contacts and Abductions: The Real Story from the Other Side*. New York: Sterling, 1994.

———. *Sixth Sense: Psychic Powers and Your Five Senses*. New York: State Mutual Book & Periodical Service, 1987.

———. *UFO Retrievals—The Recovery of Alien Spacecraft*. London: Blandford, 1995.

———. *UFOs and How to See Them*. New York: Sterling, 1993.

Raso, J. *"Alternative" Healthcare: A Comprehensive Guide*. Buffalo, N.Y.: Prometheus Books, 1994.

———. *Mystical Diets*. Buffalo, N.Y.: Prometheus Books, 1993.

Raudive, Konstantin. *Breakthrough: An Amazing Experiment in Electronic Communication with the Dead*. New York: Taplinger, 1971.

Rauscher, William V., and Allen Spraggett. *The Spiritual Frontier.* New York: Doubleday, 1975.

Rawlins, Dennis. *Peary at the North Pole: Fact or Fiction?* Fairfield, Conn.: Luce, 1973.

Read, J. *Prelude to Chemistry: An Outline of Alchemy*. Cambridge, Mass.: MIT Press, 1966.

Reed, Graham. *The Psychology of Anomalous Experience*. rev. ed. Buffalo, N.Y.: Prometheus Books, 1988.

Reich, Wilhelm. *The Mass Psychology of Fascism*. Translated by Vincent R. Carfagno. New York: Souvenir Press, 1997.

———. *Selected Writings*. New York: Farrar, Straus & Giroux, 1979.

Reichenbach, Karl von. *The Odic Force: Letters on Od and Magnetism*. New Hyde Park, N.Y.: University Books, 1968.

———. *Researches on Magnetism, Electricity, Heat, Light, Crystallization and Chemical Attraction in Relation to the Vital Force*. New Hyde Park, N.Y. University Books, 1974.

Reilly, S. W. *Table-Lifting Methods Used by Fake Mediums*. Chicago: Ireland Magic, 1957.

Reystak, R. M. *The Mind*. New York: Bantam, 1988.

Rhyner, Hans H. *Ayurveda: The Gentle Health System*. New York: Sterling, 1994.

Richards, John Thomas. *SORRAT: A History of the Neihardt Psychokinesis Experiments, 1961-1981*. Metuchen, N.J.: Scarecrow Press, 1982.

Ring, Kenneth. *The Omega Project: Near-Death Experiences, UFO Encounters, and Mind at Large*. New York: William Morrow, 1992.

Rinn, Joseph F. *Searchlight on Psychical Research*. London: Rider, 1954.

Roach, Ted. *The Physics of a Flying Saucer*. Roach Industries, 1997.

Robbins, Rossell Hope. *The Encyclopedia of Witchcraft and Demonology.* New York: Crown, 1960.

Roberts, Bechofer, C. E. *The Truth about Spiritualism*. London: Eyre and Spottiswood, 1932.

Robinson, John A. T. *Honest to God*. Valley Forge, Pa.: SCM Press, 1963.

Robinson, Paul. *Modernization of Sex: Havelock Ellis, Arthur Kinsey, William Masters and Virginia Johnson*. Ithaca, N.Y.: Cornell University Press, 1988.

Robinson, William E. *Spirit Slate Writing and Kindred Phenomena*. New York: Munn and Company Scientific American Office, 1898.

Roede, Machteld, Jan Wind, John Patrick, and Vernon Reynolds, eds. *The Aquatic Ape: Fact or Fiction*. London: Souvenir, 1991.

Rogo, D. Scott. *Parapsychology: A Century of Inquiry.* New York: Taplinger, 1975.

Roll, William G. *Theory and Experiment in Psychical Research*. Stratford, N.H.: Ayer, 1975.

Romains, Jules. *Eyeless Sight*. New York: Citadel Press, 1924.

Roman, Sanaya, and Duane Packer. *Opening to Channel—How to Connect with Your Guide*. Fort Collins, Colo.: Kramer, 1987.

Ronan, C. A. *Edmond Halley—Genius in Eclipse*. New York: Doubleday, 1969.

The Roper Organization. *Unusual Personal Experiences: An Analysis of the Data from Three National Surveys*. Roper, 1992.

Rose, Steven, Richard C. Lewontin, and Leon J. Kamin. *Not in our Genes*. New York: Penguin, 1984.

Rothstein, William G. *American Physicians in the Nineteenth Century.* Baltimore, Md.: Johns Hopkins University Press, 1972.

Routledge, Kathrine Pease. *The Mystery of Easter Island: The Story of an Expedition.* New York: AMS Press, 1978.

Rudaux, Lucien, and Gérard de Vaucouleurs, eds. *Larousse Encyclopedia of Astronomy.* London: Hamlyn, 1966.

Rushkoff, Douglas. *The Ecstasy Club.* London: Hodder and Stoughton Ltd., 1997.

Russell, Jeffrey B. *A History of Witchcraft.* London: Thames and Hudson, 1980.

Russell, P. *The TM Technique: An Introduction to Transcendental Meditation as the Teachings of Maharishi Mahesh Yogi.* New York: Routledge & Kegan Paul, 1976.

Sachs, Margaret. *The UFO Encyclopedia.* New York: Putnam, 1980.

Sagan, Carl. *Cosmos.* New York: Ballantine, 1985.

———. *Communication with Extraterrestrial Intelligence.* Cambridge, Mass.: MIT Press, 1973.

———. *The Demon-Haunted World: Science as a Candle in the Dark.* New York: Ballantine, 1997.

———. *The Dragons of Eden: Speculations on the Evolution of Human Intelligence.* New York: Random House, 1977.

Sagan, Carl, and Ann Druyan. *Shadows of Forgotten Ancestors: A Search for Who We Are.* New York: Ballantine, 1993.

Sagan, Carl, and I. S. Shklovskii. *Intelligent Life in the Universe.* New York: Dell, 1967.

Sagan, Carl, and Richard Turco. *A Path Where No Man Thought: Nuclear Winter and the End of the Arms Race.* New York: Random House, 1990.

Sagan, Carl, and Thornton Page, eds. *UFOs: A Scientific Debate.* Ithaca, N.Y.: Cornell University Press, 1972.

Samarin, William J. *Tongues of Men and Angels.* New York: Macmillan, 1972.

Sanderson, Stuart, ed. *The Secret Commonwealth & A Short Treatise of Charms and Spells by Robert Kirk.* Lanham, Md.: Rowman & Littlefield, 1976.

Savage-Rumbaugh, E. Sue, and Roger Lewin. *Kanzi: The Ape at the Brink of the Human Mind.* New York: Wiley, 1994.

Schick, Theodore, Jr., and Lewis Vaughn. *How to Think About Weird Things: Critical Thinking for a New Age.* Mountain View, Calif.: Mayfield, 1995.

Schmeidler, Gertrude R. *Parapsychology and Psychology: Matches and Mismatches.* Jefferson, N.C.: McFarland & Co., 1988.

Schmeidler, Gertrude, and Robert A. McConnell. *ESP and Personality Patterns.* New Haven, Conn.: Yale University Press, 1958.

Schnabel, Jim. *Aliens, Abductions, and the UFO Obsession.* London: Hamish Hamilton, 1994.

———. *Round in Circles: Physicists, Poltergeists, Pranksters and the Secret History of the Cropwatchers.* London: Hamish Hamilton, 1993; London: Penguin Books, 1994.

Schultz, Ted, ed. *The Fringes of Reason: A Whole Earth Catalog: A Field Guide to New Age Frontiers, Unusual Beliefs & Eccentric Sciences.* New York: Harmony, 1989.

Schwarz, Richard W. *John Harvey Kellogg: Father of the Health Food Industry.* Reprint. Berrien Springs, Mich.: Andrews University Press, 1996.

Seagraves, Kelly. *Sons of God Return.* New York: Pyramid Books, 1975.

Sebald, Hans. *Witchcraft—The Heritage of a Heresy.* New York: Elsevier, 1978.

Sebeok, Thomas, and Jean Umiker-Sebeok, eds. *Speaking of Apes: A Critical Anthology of Two-Way Communication with Man.* New York: Plenum Press, 1980.

Seybert Commission. *Preliminary Report of the Commission by University of Pennsylvania to Investigate Modern Spiritualism.* Philadelphia: J. B. Lippincott, 1887.

Seymour, Percy. *Astrology: The Evidence of Science.* London: Lennard Publishing, 1988.

Shebar, Sharon Sigmond. *The Cardiff Giant.* New York: J. Messner, 1983.

Sheldrake, Rupert. *A New Science of Life: The Hypothesis of Formative Causation.* Los Angeles and New York: J. P. Tarcher, 1981.

———. *The Presence of the Past: Morphic Resonance and the Habits of Nature.* Rochester, Vt.: Park Street Press, 1995.

———. *The Rebirth of Nature: The Greening of Science and God.* Rochester, Vt.: Park Street Press, 1994.

Shepard, Leslie, ed. *Encyclopedia of Occultism & Parapsychology.* 3d ed. Detroit, Mich.: Gale Research, 1991.

Shuker, Karl. *In Search of Prehistoric Survivors: Do Giant "Extinct" Creatures Still Exist?* London: Blandford, 1995.

Siegel, Harvey. *Relativism Refuted.* Dordrecht, Netherlands: D. Reidel, 1987.

Siegel, R. K., and L. J. West, eds. *Hallucinations: Behavior, Experience and Theory.* New York: Wiley, 1975.

Sifakis, Carl. *Hoaxes and Scams: A Compendium of Deceptions, Ruses and Swindles.* New York: Facts On File, 1994.

Skelton, R. A., Thomas E. Marston, and George D. Painter. *The Vinland Map and the Tartar Relation.* New ed. New Haven, Conn.: Yale University Press, 1995.

Skorupski, J. *Symbol and Theory.* Cambridge: Cambridge University Press, 1976.

Sladek, John. *The New Apocrypha.* New York: Stein and Day, 1973.

Smart, Ninian. *The Religious Experience of Mankind.* 3d ed. Paramus, N.J.: Prentice Hall, 1984.

Sommerlott, Robert. *"Here, Mr. Splitfoot!"* New York: Viking Press, 1971.

Sontag, F. *Sun Myung Moon and the Unification Church.* Nashville, Tenn.: Abingdon, 1977.

Spence, Lewis. *Encyclopedia of the Occult.* New Hyde Park, N.Y.: University Books, 1968.

Spraggett, Allen. *The Unexplained.* New York: New American Library, 1967.

Spraggett, Allen, and William Rauscher. *Arthur Ford: The Man who Spoke with the Dead.* New York: New American Library, 1973.

Stark, Raymond. *The Book of Aphrodisiacs.* New York: Stein & Day, 1981.

States, Bert O. *Seeing in the Dark: Reflections on Dreaming.* New Haven, Conn.: Yale University Press, 1996.

Steele, David, ed. *The History of Scientific Ideas.* Hutchinson Educational, 1970.

Stein, Gordon. *Encyclopedia of Hoaxes*. Detroit, Mich.: Gale Research, 1993.

———. *The Sorcerer of Kings: The Case of Daniel Dunglass Home and William Crookes*. Buffalo, N.Y.: Prometheus Books, 1993.

Stein, Gordon, ed. *The Encyclopedia of the Paranormal*. Buffalo, N.Y.: Prometheus Books, 1996.

Steiner, Lee R. *Where Do People Take Their Troubles?* Boston: Houghton Mifflin Co., 1945.

Steiner, Robert A. *Don't Get Taken! Bunco and Buncum Exposed—How to Protect Yourself*. El Cerrito, Calif.: Wide-Awake Books, 1989.

Stemman, Roy. *Spirits and Spirit Worlds*. New York: Doubleday and Co., 1976.

Stenger, Victor J. *Physics and Psychics: The Search for a World Beyond the Senses*. Buffalo, N.Y.: Prometheus Books, 1990.

Stern, Madeline. *Heads and Headlines: The Phrenological Fowlers*. Norman: University of Oklahoma Press, 1971.

Stevenson, Ian. *Twenty Cases Suggestive of Reincarnation*. 2d enl. rev. ed. Charlottesville: University Press of Virginia, 1980.

Stone, I. *The Healing Factor: "Vitamin C" Against Disease*. New York: Grossett & Dunlap, 1972.

Stonier, Tom. *Information and the Internal Structure of the Universe: An Exploration into Information Physics*. New York: Springer-Verlag, 1990.

———. *The Wealth of Information: A Profile of the Post-Industrial Economy*. London: Methuen/Thames, 1983.

Stormer, Chris. *Reflexology—The Definitive Guide*. London: Hodder and Stoughton, 1995.

Storr, Anthony. *Feet of Clay: A Study of Gurus*. New York: HarperCollins, 1996.

Story, Ronald. *The Space Gods Revealed: A Close Look at the Theories of Erich Von Däniken*. New York: Putnam, 1969.

Szasz, Thomas. *The Myth of Mental Illness*. Hoeber-Harper, 1961.

Taberner, Peter V. *Aphrodisiacs: The Science and the Myth*. London: Croom Helm, 1985.

Tabori, Paul. *Crime and the Occult*. New York: Taplinger, 1974.

Tanner, Amy. *Studies in Spiritism*. Buffalo, N.Y.: Prometheus Books, 1993.

Tanner, Don. *How To Do Headline Predictions*. Chicago: Ireland Magic, 1957.

Tansley, David. *Radionics—Interface with the Ether Field*. Health Science Press, 1975.

Taylor, Ann. *Annie Besant*. New York: Oxford University Press, 1992.

Taylor, John G. *Science and the Supernatural*. New York: Dutton, 1979.

———. *Superminds*. New York: Viking, 1975.

Thom, Alexander. *Megalithic Lunar Observatories*. New York: Oxford University Press, 1971.

———. *Megalithic Sites in Britain*. New York: Oxford University Press, 1967.

Thom, A., and A. S. Thom. *Megalithic Remains in Britain and Brittany*. New York: Oxford University Press, 1978.

Thurston, Herbert. *The Church and Spiritualism*. Milwaukee, Wis.: Bruce, 1933.

———. *The Psychical Phenomena of Mysticism*. London: Nurns, Oates, 1955.

Tietze, Thomas R. *Margery*. New York: Harper & Row, 1973.

Tipler, Frank J. *The Physics of Immortality: Modern Cosmology, God, and the Resurrection of the Dead*. New York: Doubleday, 1994.

Toms, Michael. *Fritjof Capra in Conversation with Michael Toms*. Santa Rosa, Calif.: Aslan Publishing, 1993.

Toth, M., and G. Nielson. *Pyramid Power: The Secret Energy of the Ancients Revealed*. Rochester, Vt.: Destiny Books, 1976.

Toumey, Christopher P. *God's Own Scientists*. New Brunswick, N.J.: Rutgers University Press, 1994.

Traill, David. *Schliemann of Troy: Treasure and Deceit*. New York: St. Martin's Press, 1995.

Tyler, V. E. *The Honest Herbal*. Binghamton, N.Y.: Haworth, 1993.

Ullman, Montague, and Stanley Krippner. *Dream Telepathy: Experiments in ESP.* Jefferson, N.C.: McFarland & Co., 1989.

Underhill, Evelyn. *The Essentials of Mysticism and Other Essays*. Rockport, Mass.: Oneworld, 1995.

Vallee, Jacques. *Passport to Magonia*. Chicago: Henry Regnery Co., 1969.

Vandenberg, Philip. *The Curse of the Pharaohs*. Philadelphia: J.B. Lippincott, 1975.

Vaucouleurs, Gerard de. *The Planet Mars*. Winchester, Mass.: Faber and Faber, 1951.

Vogt, Evon, and Ray Hyman. *Water Witching U.S.A.* 2d ed. Chicago: University of Chicago Press, 1979.

Von Däniken, Erich. *Chariots of the Gods?* London: Souvenir, 1969.

———. *Erich Von Däniken*. New York: Outlet, 1989.

Wallman, Joel. *Aping Language*. Cambridge: Cambridge University Press, 1992.

Warner, John Harley. *The Therapeutic Perspective: Medical Practice, Knowledge, and Identity in America, 1820-1885*. Cambridge, Mass.: Harvard University Press, 1986.

Washington, Peter. *Madame Blavatsky's Baboon: Theosophy and the Emergence of the Western Guru*. London: Secker & Warburg, 1993.

Webb, James. *The Occult Establishment*. Glasgow: Richard Drew, 1976.

———. *The Occult Underground*. La Salle, Ill.: Library Press, 1974.

Wegener, Alfred. *The Origin of Continents and Oceans*. London: Methuen, 1978.

Weiner, Joseph S. *The Piltdown Forgery*. New York: Oxford University Press, 1955.

Wells, H. G. *The Time Machine/The War of the Worlds: A Critical Edition*. Edited by Frank McConnell. New York: Oxford University Press, 1977.

Wernick, Robert. *The Monument Builders*. New York: Time-Life Books, 1973.

West, John Anthony. *The Case for Astrology*. New York: Viking Arkana, 1991.

———. *Serpent in the Sky: The High Wisdom of Ancient Egypt*. Wheaton, Ill.: Quest Books, 1993.

Westwood, Jennifer, ed. *The Atlas of Mysterious Places*. New York: Weidenfeld & Nicolson, 1987.

Whalen, William J. *Minority Religions in America*. New York: Alba House, 1972.
White, Rhea A. *Parapsychology: New Sources of Information, 1973–1989*. Scarecrow Press, 1990.
Whitehead, Harriet. *Renunciation and Reformulation*. Ithaca, N.Y.: Cornell University Press, 1987.
Whitfeld, Charles. *Healing the Child Within*. Deerfield Beach, Fla.: Health Communications, Inc., 1987.
Whorton, James C. *Crusaders for Fitness*. Princeton, N.J.: Princeton University Press, 1982.
Wilhelm, Richard. *The I Ching*. New York: Routledge & Kegan Paul, 1968.
Wills, Pauline Wills. *The Reflexology Manual*. London: Headline, 1995.
Wilson, Colin. *The Psychic Detectives*. London: Pan Books, 1984.
———. *The Quest for Wilhelm Reich*. New York: Doubleday, 1981.
Wilson, Colin, and Damon Wilson. *The Encyclopedia of Unsolved Mysteries*. Chicago: Contemporary Books, 1988.
Wilson, Edward O. *On Human Nature*. Cambridge, Mass.: Harvard University Press, 1978.
———. *Sociobiology: The New Synthesis*. Cambridge, Mass.: Harvard University Press, 1975.
Wilson, Ian. *The Bleeding Mind*. London: Weidenfeld & Nicolson, Ltd., 1988.
———. *Mind Out of Time: Reincarnation Claims Investigated*. London: Gollancz, 1981.
Wilson, Robert Anton. *The New Inquisition: Irrational Rationalism and the Citadel of Science*. Phoenix: Falcon Press, 1986.
Winter, J. A. *Dianetics: A Doctor's Report*. New York: Julian Press, 1987.
Wiseman, Richard, and Robert L. Morris. *Guidelines for Testing Psychic Claimants*. Buffalo, N.Y.: Prometheus Books, 1995.
Witchell, Nicholas. *The Loch Ness Story*. London: Transaction, 1993.
Wolf, Theta A. *Alfred Binet*. Chicago: Chicago University Press, 1973.
Wolman, Benjamin B., ed. *Handbook of Parapsychology*. New York: Van Nostrand Reinhold, 1977.
Wood, Michael. *In Search of the Trojan War*. New York: Facts On File, 1985.
Woodman, Jim. *Nazca: Journey to the Sun*. New York: Pocket Books, 1977.
Woodward, Ian. *The Werewolf Delusion*. New York: Paddington Press, 1979.
Wright, Lawrence. *Remembering Satan: A Case of Recovered Memory and the Shattering of an American Family*. New York: Knopf, 1994.
Yapko, Michael D. *True and False Memories of Childhood Sexual Trauma: Suggestions of Abuse*. New York: Simon & Schuster, 1994.
Yapp, Nick. *Hoaxers and Their Victims*. London: Parkwest, 1993.
Young, Michael. *The Metronomic Society*. London: Thames and Hudson, 1988.
Young, Robert M. *Mind, Brain, and Adaptation in the Nineteenth Century: Cerebral Localization and Its Biological Context from Gall to Ferrier*. New York: Oxford University Press, 1990.
Zaleski, Carol. *Otherworld Journeys: Accounts of Near-Death Experiences in Medieval and Modern Times*. Reprint. New York: Oxford University Press, 1987.
Zarzynski, Joseph W. *Champ: Beyond the Legend*. Port Henry, N.Y.: Bannister Publications, 1984.
Zimdars-Swartz, Sandra L. *Encountering Mary*. Princeton, N.J.: Princeton University Press, 1991.
Zollschan, G. K., J. F. Schumaker, and G. F. Walsh, eds. *Exploring the Paranormal: Perspectives on Belief and Experience*. Garden City Park, N.Y.: Avery, 1989.
Zolotow, Maurice. *It Takes All Kinds*. New York: Random House, 1952.
Zusne, Leonard, and Warren H. Jones. *Anomalistic Psychology: A Study of Magical Thinking*. Hillsdale, N.J.: Lawrence Erlbaum Associates, 1989.
Zwicky, J. F., A. W. Hafner, S. Barrett, and W. T. Jarvis. *Reader's Guide to "Alternative" Health Methods*. American Medical Association, 1993.

INDEX

PHOTO CREDITS

Page 3. The anaesthetist applies acupuncture during surgery in a Beijing hospital. (By permission of Archive Photos/Camera Press.)

Page 6. Two alchemists at work from Ulstadius; an early 16th-century woodcut. (By permission of the Fortean Picture Library.)

Page 8. Woman undergoing vibrasound therapy at the Festival of Mind, Body, and Spirit, 1991. (By permission of Guy Lyon Playfair/Fortean Picture Library.)

Page 14. Joan of Arc, who claimed to have seen apparitions of religious figures and to have heard what she believed were the voices of saints. (By permission of Archive Photos.)

Page 15. From J. Buschius, 1702. (Facts On File.)

Page 17. An early depiction of the proposed Asian origins of Native American tribes on the North American continent. (By permission of the Debry Collection.)

Page 18. Title page from *Astrolgaster of the Figure-Caster*, John Melton's critical appraisal of astrology and other mantic sciences, 1620. (By permission of the Fortean Picture Library.)

Page 22. Automatic writing sample in Old French by a psychic in Florida, 1988. (By permission of the Fortean Picture Library.)

Page 23. Avebury henge, Wiltshire. (By permission of Janet and Colin Bord/Fortean Picture Library.)

Page 28. Reverend Sabine Baring-Gould. (By permission of the Fortean Picture Library.)

Page 28. Sir William F. Barrett. (By permission of the Fortean Picture Library.)

Page 33. A frame from a film on Bigfoot, taken by Roger Patterson at Bluff Creek, northern California, 1967. (By permission of René Dahinden/Fortean Picture Library.)

Page 38. Madame Blavatsky. (By permission of the Fortean Picture Library.)

Page 46. Illustration of a man-eating plant from "The Purple Terror," *Strand Magazine*, 1899, by Fred M. White. (By permission of the Fortean Picture Library.)

Page 47. Cattle mutilation at Morrill Farm in Piermost, New Hampshire, 1978. (By permission of Loren Coleman/Fortean Picture Library.)

Page 55. Photograph of first specimen of modern-day *Latimeria* coelacanth, 1939. (By permission of the Fortean Picture Library.)

Page 57. Physical scientist cautiously works with various gases in laboratory. (By permission of Bayard Brattstrom/Visuals Unlimited.)

Page 63. Bank's horse. (By permission of the Fortean Picture Library.)

Page 65. Clarence Darrow, who defended the right of John T. Scopes to teach the theory of evolution, at one time considered to be pseudoscientific, in a public school. (By permission of New York Times Co./Archive Photos.)

Page 66. Sir William Crookes. (By permission of the Fortean Picture Library.)

Page 67. Photograph of crop circles found in southern England, 1995. There are 96 circles and three rings, which are roughly 400 feet from the top to bottom circles. (By permission of the Fortean Picture Library.)

Page 69. Aleister Crowley, dressed as a magician, wearing a serpent crown and shown with a wand, cup, sword, bell, vial of holy oil, and The Book of the Law. (By permission of the Fortean Picture Library.)

Page 71. Valentine Sapunov, a cryptozoologist and doctor of biological sciences from St. Petersburg, Russia, holds a plaster cast of a "snowman's" footprint from the early 1990s. (By permission of the Fortean Picture Library.)

Page 78. Ape photographed by geologist François de Loys, 1929, in Venezuela, South America. (By permission of the Fortean Picture Library.)

Page 80. Demonic guardian on the western façade of the Cathedral of Notre Dame, Paris. (By permission of the Fortean Picture Library.)

Page 85. An illustration from 16th-century mathematician, physician, and astrologer Jerome Cardan's *On the Variety of Things*. It depicts an attempt to divine the future. The past is seen in an earthenware vessel filled with oil, the present in a bronze vessel filled with oil, and the future in a glass vessel filled with water. "The instruments" at the top are a poplar wand, a curved cutting tool, and a cucumber root. Latin inscription indicates that the shape is a piece of wood with 19 holes in it, designed to "receive the influences of the stars." (By permission of the Fortean Picture Library.)

Page 87. Hamish Miller dowsing with a Y-shaped stick or a divining rod. (By permission of Paul Broadhurst/Fortean Picture Library.)

Page 88. During dream and sleep analysis, psychoanalysts seek to decipher the mind's ability to unfold its illusions of reality. (By permission of Stuart Bratesman/Visuals Unlimited.)

Page 94. Restored stone statues on Easter Island. (By permission of Klaus Aarsleff/Fortean Picture Library.)

Page 99. A microelectron photograph of a dividing human T lymphocyte, enlarged 6,000 times. (By permission of David M. Phillips/Visuals Unlimited.)

Page 101. Photograph of living species arranged in ascending order from shark to man by Thane L. Bierwert, 1942. As species progressed up the evolutionary scale, survival capabilities increased. (Courtesy of the Department of Library Services, American Museum of Natural History, neg. no. 318686.)

Page 103. A woodcut of an exorcism by a bishop of Notre Dame of a person possessed by a devil, France, 1566. (By permission of the Fortean Picture Library.)

Page 109. Healer performing a karma diagnosis, Philippines, 1984. (By permission of Andreas Trottman/Fortean Picture Library.)

Page 111. Cover of *Fate* magazine, May 1958, showing a rain of frogs. (By permission of Llewellyn Publications/Fortean Picture Library.)

Page 112. Following visions of the Blessed Virgin Mary at Fatima, Portugal, 1917, the crowd in Cova da Iria witnessed a solar phenomenon on October 13, 1917. (By permission of the Fortean Picture Library.)

Page 113. An experimental firewalk staged by Harry Price for the University of London Council for Psychical Investigation, 1935. (By permission of the Fortean Picture Library.)

Page 118. Charles Fort at his checkers board. (By permission of the Charles Fort Institute/Fortean Picture Library.)

Page 120. Cover of the *Fortean Times*. (By permission of the Fortean Picture Library.)

Page 122. Sigmund Freud, a pioneer in the development of psychiatry and psychoanalysis. (By permission of Archive Photos.)

Page 129. Woman undergoing Ganzfeld experiment in Cambridge, 1981, while the experimenter, Carl Sargent, adjusts sound and lighting. (By permission of Guy Lyon Playfair/Fortean Picture Library.)

Page 121. Uri Gellar, with Dr. Elmar R. Gruber, performing experiments on a Luxembourg TV show, 1988. (By permission of Dr. Elmar R. Gruber/Fortean Picture Library.)

Page 144. Some of the herbal medicines relied upon in the 19th century. (By permission of K. Gregory Catalog No. 44.)

Page 150. Hanging from his ankles from the corner of a building, Harry Houdini strips off his straitjacket to a cheering crowd below. (By permission of Archive Photos/Authenticated News International.)

Page 155. Dr. J. Allen Hynek at the "swamp gas" press conference, Detroit, 1966. (By permission of the Fortean Picture Library.)

Page 156. Women in unconscious state induced by hypnosis; from G. Gessmann's *Magnetismus and Hypnotismus*, Leipzig, 1887. (By permission of the Fortean Picture Library.)

Page 160. *I Ching*, the ancient Chinese text, was used by the Chinese, with the aid of sticks (yarrow stalks), as a method of predicting the future. (By permission of the Fortean Picture Library.)

Page 170. Levitation of St. Joseph of Copertino. (By permission of the Fortean Picture Library.)

Page 171. Carl G. Jung. (By permission of the Fortean Picture Library.)

Page 173. Engraving of the Kabbalah, 1677. (By permission of the Fortean Picture Library.)

Page 174. John A. Keel. (By permission of the Fortean Picture Library.)

Page 178. Kirlian picture of fingers and toes, taken for diagnostic purposes by naturopathy researcher Peter Mandel, Germany. (By permission of Dr. Elamar R. Gruber/Fortean Picture Library.)

Page 180. Philip Klass. (By permission of Dennis Stacy/Fortean Picture Library).

Page 189. Photograph of a man levitating a young girl. (By permission of the Fortean Picture Library.)

Page 191. Aerial view of Saintbury ley, Gloucestershire, England, which runs for about 3 1/2 miles, with the significant features encircled. (By permission of Paul Revereux/Fortean Picture Library.)

Page 193. A photograph thought to be of Loch Morar, Inverness, Scotland, 1988. The dark spot in the center disappeared a few minutes after the picture was taken. (By permission of Lars Thomas/Fortean Picture Library.)

Page 194. Cesare Lombroso, psychic researcher. (By permission of the Fortean Picture Library.)

Page 198. Woodcut depicting werewolf with a baby in its jaws. (By permission of the Fortean Picture Library.)

Page 204. Marfa mystery lights, as photographed by James Crocker in Texas, 1986. (By permission of James Crocker/Fortean Picture Library.)

Page 208. Photograph of the materialization of the dead poet Giuseppe Parini sitting between South American medium Carlo Mirabelli (left, in trance) and Dr. Carlos de Castro (right) at the Cesare Lombroso Academy of Psychic Studies, Brazil, 1920s. (By permission of the Fortean Picture Library.)

Page 209. Polish medium Franek Kluski at work in Warsaw, 1919. (By permission of the Fortean Picture Library.)

Page 213. Dr. Donald H. Menzel. (By permission of the Fortean Picture Library.)

Page 219. Ivan T. Sanderson's outline drawing of the "Minnesota Iceman," with measurements, from his 1968 examination of a hair-covered corpse that was frozen in a block of ice. (By permission of the Fortean Picture Library.)

Page 221. Dr. Karl Shuker holding model of Congolese Mobele-mbembe. (By permission of Dr. Karl Shuker/Fortean Picture Library.)

Page 230. Photograph of strange markings, also referred to as Nazca lines, on the Nazca Desert, Peru. (By permission of Klaus Aarsleff/Fortean Picture Library.)

Page 247. Cover of *The Onza*, 1961. (By permission of the Fortean Picture Library.)

Page 251. Photograph representing an out-of-body experience. (By permission of Philip Panton/Fortean Picture Library.)

Page 256. Plate from the earliest printed book on palmistry, *Die Kunst Ciromantia,* by Johann Hartief, which deals with special signs and was printed about 1475. (By permission of the Fortean Picture Library.)

Page 257. Portrait of Paracelsus from Conrad Horlacher's *Eröffnete Geheimnisse Des Steins Der Weisen*, 1718. (By permission of the Fortean Picture Library.)

Page 258. An experiment at Cambridge University, where an observer on a moving swing watches a circle, which is varied in size to appear constant. (By permission of Archive Photos/Express Newspapers.)

Page 260. An illustration of Louis Pasteur, the French chemist who contributed some of the most valuable work in the history of science. (By permission of Archive Photos/American Stock.)

Page 266. Phrenological diagram. (By permission of the Fortean Picture Library.)

Page 272. Photograph depicting poltergeist activity (furniture upended and small items thrown around) in Chester, England, May 1985. This is said to have occurred during the time when computer messages were being received from a man living in the 16th century. (By permission of Ken Webster/Fortean Picture Library.)

Page 274. Nineteenth-century engraving of a woman possessed by a demon. (By permission of the Fortean Picture Library.)

Page 288. In the 1920s, Baron von Schrenck-Notzing experimented with the psychokinetic influence on sales. (By permission of the Fortean Picture Library.)

Page 298. A demonstration in reflexology at the Festival of Mind, Body, and Spirit, 1991. (By permission of Guy Lyon Playfair/Fortean Picture Library.)

Page 308. Ivan T. Sanderson. (By permission of August C. Roberts/Fortean Picture Library.)

Page 313. An engraving from *The World of Wonders*, 1886, of a sea serpent as seen from the ship *City of Baltimore* in 1879. (By permission of the Fortean Picture Library.)

Page 315. Dr. Rupert Sheldrake in 1983. (By permission of Dennis Stacy/Fortean Picture Library.)

Page 321. Behavior patterns and social interaction, both inherited and learned, can be observed in this flock of snow geese in flight. (By permission of Leonard Lee Rue/Visuals Unlimited.)

Page 323. A SORRAT psychokinesis experiment of paranormally linked rubber bands, 1993. (By permission of Dr. J. T. Richards/Fortean Picture Library.)

Page 325. A spirit photograph showing W. T. Stead with an unidentified spirit form, taken by clairvoyant photographer Robert Boursnell in London, c. 1896. (By permission of the Fortean Picture Library.)

Page 328. Officials inspecting the debris at the house where Mrs. M. H. Reeser is thought to have died of spontaneous human combustion, in St. Petersburg, Florida, 1951. (By permission of the Fortean Picture Library.)

Page 330. Rudolf Steiner. (By permission of the Fortean Picture Library.)

Page 332. 1993 photograph of Giorgio Bongiovanni, who received a stigmata during a visit to Fatima. A wound to his hand formed the shape of a cross, and the five passion marks are said to bleed almost daily. (By permission of Dr. Elmao R. Gruber/Fortean Picture Library.)

Page 333. Stonehenge, Wiltshire, England, at sunset. (By permission of Janet and Colin Bord/Fortean Picture Library.)

Page 343. The three major arcana of the 19th-century Italian tarot deck: the Stars, the Pope, and the Moon. (By permission of the Fortean Picture Library.)

Page 348. Illustration depicting the materialization of a boy who had apparently been teleported into the room. (By permission of the Fortean Picture Library.)

Page 350. An example of thoughtography by Tenshin Takeushi, 1915, from T. Fukurai's *Clairvoyance and Thoughtography*, 1931. (By permission of the Fortean Picture Library.)

Page 354. Eighteenth-century engraving of Englishwoman Mary Toft, who claimed to have given birth to rabbits and later confessed to hoaxing. (By permission of the Fortean Picture Library.)

Page 357. Identical twins, the ideal control for many scientific experiments. (By permission of Archive Photos/Lamert.)

Page 359. UFOs photographed in 1966 by 15-year-old Stephen Pratt in Conisbrough, South Yorkshire, England. (By permission of S. C. Pratt/Fortean Picture Library.)

Page 363. An illustration of a female vampire. (By permission of the Fortean Picture Library.)

Page 368. In a Paris nightclub in 1969, a Voodoo priestess puts a woman in a trance as dancers perform to drum sounds said to be summoning the voices of ancestors. (By permission of Archive Photos/Archive France.)

Page 375. Present-day witches engaging in a ceremony in London. (By permission of Klaus Aarsleff/Fortean Picture Library.)

Page 383. A scene on the hilltop Citadel of Zimbabwe. (By permission of Hamish Brown/Fortean Picture Library.)